Techniques for Finding the Inverse (If It Exists) of an $n \times n$ Matrix A

- **2 × 2 case**: The inverse of $\begin{bmatrix} a & b \\ c & d \end{bmatrix}$ exists if and only if $ad - bc \neq 0$. In that case, the inverse is given by $\frac{1}{ad - bc}\begin{bmatrix} d & -b \\ -c & a \end{bmatrix}$. (Section 2.4)
- **Row reduction**: Row reduce $[\mathbf{A} \mid \mathbf{I}_n]$ to $[\mathbf{I}_n \mid \mathbf{A}^{-1}]$ (where $\mathbf{A}^{-1}$ does not exist if the process stops prematurely). Advantages: easily computerized; relatively efficient. (Section 2.4)
- **Adjoint matrix**: $\mathbf{A}^{-1} = (1/|\mathbf{A}|)\mathcal{A}$, where $\mathcal{A}$ is the adjoint matrix of **A**. Advantage: gives an algebraic formula for $\mathbf{A}^{-1}$. Disadvantage: not very efficient, because $|\mathbf{A}|$ and all n^2 cofactors of **A** must be calculated first. (Section 3.3)

Techniques for Finding the Determinant of an $n \times n$ Matrix A

- **2 × 2 case**: $|\mathbf{A}| = ad - bc$ if $\mathbf{A} = \begin{bmatrix} a & b \\ c & d \end{bmatrix}$. (Sections 2.4 and 3.1)
- **3 × 3 case**: Use the "basketweaving" technique. (Section 3.1)
- **Row reduction**: Row reduce to an upper triangular form matrix, keeping track of the effect of each row operation on the determinant. Advantages: easily computerized; relatively efficient. (Section 3.2)
- **Cofactor expansion**: Multiply each element along any row or column of **A** by its cofactor and sum the results. Advantage: useful for matrices with many zero entries. Disadvantage: not as fast as row reduction. (Section 3.3)

  Also remember: $|\mathbf{A}| = 0$ if **A** is row equivalent to a matrix with a row or column of zeroes or with two equal rows or with two equal columns.

Techniques for Finding the Eigenvalues of an $n \times n$ Matrix A

- **Characteristic polynomial**: Find the roots of $p_{\mathbf{A}}(x) = |x\mathbf{I}_n - \mathbf{A}|$. Disadvantages: tedious to calculate $p_{\mathbf{A}}(x)$; polynomial may not factor easily; becomes more difficult to factor as degree increases. (Section 6.1)
- **Power method:** Use for dominant eigenvalues. Choose an initial approximation to a unit eigenvector for the eigenvalue, and repeatedly multiply **A** by the approximation and normalize the result. The eigenvalue is ± length of the final resultant vector. Advantage: iterative method, easy to computerize. Disadvantage: may fail to converge. (Section 9.3)

Elementary Linear Algebra

Elementary Linear Algebra

Stephen Andrilli
La Salle University

David Hecker
St. Joseph's University

PWS-KENT Publishing Company
Boston

PWS-KENT
Publishing Company

20 Park Plaza
Boston, Massachusetts 02116

PWS-KENT Publishing Company is a division of Wadsworth, Inc.

Library of Congress Cataloging-in-Publication Data

Andrilli, Stephen Francis
Elementary linear algebra / by Stephen Andrilli, David Hecker.
p. cm.
Includes index.
ISBN 0-534-17964-9
1. Algebras, Linear. I. Hecker, David II. Title.
QA184.A54 1993
512′.5– dc20 92-29674
CIP

Printed in the United States of America.
92 93 94 95 96 — 10 9 8 7 6 5 4 3 2 1

Development Editor: *Alan Venable*
Editorial Assistant: *Leslie With*
Production: *Del Mar Associates*
Print Buyer: *Diana Spence*
Designer: *Janet Ashford*
Copy Editor: *Rebecca Smith*
Technical Illustrator: *Kristi Mendola*
Cover: *John Odam*
Compositor: *Publication Services, Inc.*
Printer: *R. R. Donnelley & Sons, Harrisonburg, VA*

To our wives, Ene and Lyn, for all
their help and encouragement

Contents

Figures

Prerequisites for Chapters 8 and 9

SECTION	PREREQUISITE
Section 8.1	Section 1.5
Section 8.2	Section 2.1
Section 8.3	Section 2.1
Section 8.4	Section 2.1
Section 8.5	Section 2.4
Section 8.6	Section 4.7
Section 8.7	Section 4.7
Section 8.8	Section 6.2
Section 8.9	Section 6.3
Section 9.1	Section 2.2
Section 9.2	Section 2.5
Section 9.3	Section 6.1

Preface

This textbook is intended for a sophomore- or junior-level introductory course in linear algebra. We assume that the student has had a first course in calculus.

Philosophy of the Textbook

Why another linear algebra textbook? In teaching elementary linear algebra, we encountered three major problems:

(1) Students had difficulty reading linear algebra textbooks. Frequently they were too terse, especially where proofs of important results were concerned.
(2) Students invariably ran into trouble as the largely computational first half of the course gave way to the more theoretical second half. Students were asked to work on a much higher level of abstraction and had difficulty writing proofs involving such nontrivial concepts as vector spaces, spanning, and linear independence.
(3) Most textbooks contained few or no guidelines about reading and writing simple mathematical proofs. Yet many instructors use the linear algebra course as a vehicle for introducing students to proofs. Most textbooks do not help students overcome their initial anxiety in this area.

This textbook addresses these problems. Above all, we have striven for clarity and used straightforward language throughout the book. When forced to choose between brevity and clarity, we usually chose the latter. We strongly encourage students to take advantage of the book's presentation by reading it thoroughly.

Linear algebra is an unusual subject in that instructors must gloss over many "details" in order to cover all the standard course material. This textbook includes many of these details so students can become familiar with the underlying concepts that instructors do not have time to cover in class.

To facilitate a smooth transition to the second half of the course, we get students to work on proofs right away. In particular, we begin with a general discussion of vectors and matrices. These are the basic algebraic tools necessary for elementary linear algebra, and their fundamental properties provide a natural springboard for reading and writing simple algebraic proofs.

We have also included a special section (Section 1.3) on general proof techniques, with various examples (using vectors) of the standard methods of proof and some caveats about the methods. The early placement of this section gives students a strong foundation and helps to build their confidence in reading and writing proofs.

To make this textbook as complete as possible, we have included clear, readable proofs of virtually all *nontrivial* theorems in Chapters 1 through 6. The only exceptions are Theorem 2.3 (uniqueness of reduced row echelon form), Lemma 3.3 (parity of permutations), and Theorem 3.11 (cofactor expansion). Thus, the textbook can also serve as a useful reference for proofs of the fundamental results in linear algebra.

We have found that "clever" or "sneaky" proofs, in which the last line suddenly produces "a rabbit out of a hat," invariably frustrate students. They are given no clear insight into the deductive process and often come away with less understanding than before. In fact, such proofs tend to reinforce students' mistaken belief that they will never become competent in the art of writing proofs. Worse yet, students do not understand the general strategy used in such "clever" proofs. These are no help as models when students are called on to write similar proofs on their own.

In this textbook, proofs longer than one paragraph are usually written in a "top-down" manner, a concept borrowed from the technique of structured programming. A complex theorem is broken down into a secondary series of results, which together are sufficient to prove the original theorem. The same process is used again on each of the secondary results, until each remaining result is simple enough to be proved directly. Hence, the student has a clear outline of the logical argument at all times and can more easily reproduce the proof if called on to do so. This "goal-oriented" method of writing proofs is much closer to the actual deductive process originally used to create the proof.

Features

In addition to an emphasis on clarity and proof techniques, this textbook has the following features:

Numerous examples and exercises: There are over 200 examples in the text, at least one for each new concept or application, to ensure that students fully understand the material before proceeding. Almost every theorem has a corresponding example to illustrate its usefulness.

The text also contains an unusually large number of exercises. There are almost 700 numbered exercises, and many of these have multiple parts, for a total of 1600. Some are purely computational, to ensure that students know how to perform various calculations. There are also exercises in almost every section that ask students to write short proofs. Finally, some exercises encourage students to explore further consequences of the material.

Exercises within a section are generally ordered by increasing difficulty, usually starting with basic computations and moving to more theoretical problems and proofs. Answers are provided at the end of the book for approximately half the computational exercises; these problems are marked with a star ($\star$).

Solid algebraic foundations: We begin with the building blocks of linear algebra: vectors and matrices. Systems of linear equations are then discussed in the context of matrices, and the relationship between matrix algebra and systems of linear equations is more fully understood. Students are then gradually introduced to more advanced algebraic concepts involving vector spaces and linear transformations. Where applicable, geometry is invoked to motivate the algebra.

Careful coverage of vector space topics: Many students have difficulties when the abstract topics of vector space, subspace, span, linear independence, and basis are introduced. They represent a sharp transition from concrete problem solving to theoretical conceptualizing. Student understanding of these topics is critical. We have introduced each of these vector space concepts in a separate section in order to give students sufficient time to become comfortable with each one before proceeding. We have also included a special section (Section 4.6) on constructing bases that illustrates how to reduce a finite spanning set to a basis or how to enlarge a linearly independent set to a finite basis, since these procedures are troublesome for many students. Finally, complex vector spaces are introduced in Chapter 7, only after students have fully mastered the fundamental material on real vector spaces, linear transformations, and eigenvalues in Chapters 4 through 6.

Applications: Linear algebra is a subject with a multitude of practical applications. Although there is never enough time in the course to cover all the desired applications, we have included many standard ones so instructors can choose their favorites. A few short applications are included in Chapters 1 through 6. However, Chapter 8 is devoted entirely to applications, including least squares, Markov chains, differential equations, and quadratic forms. Each of these topics can be covered as soon as students have completed the prerequisites, as indicated on page xii. Instructors may choose to assign some of these applications as reading assignments outside of class.

Algorithms and computational methods: We have included detailed step-by-step methods for carrying out many of the fundamental processes in elementary linear algebra: performing Gauss-Jordan row reduction (Section 2.1), finding the inverse of a matrix (Section 2.4), finding the determinant via row reduction (Section 3.2), finding a basis using row reduction (Section 4.5), shrinking a finite spanning set to a basis (Section 4.6), enlarging a linearly independent set to a basis (Section 4.6), performing the Gram-Schmidt Process (Sections 5.5 and 7.2), diagonalizing a linear operator (Section 6.2), orthogonally diagonalizing a symmetric operator (Section 6.3), and finding the dominant eigenvalue using the power method (Section 9.3).

Subsections: Almost every section is divided into several subsections to enhance clarity and readability. These subsections are individually titled to highlight the main themes of the section.

Summary charts: The book contains several useful charts that summarize important results. Appropriate versions of these charts have been printed on the inside front and back covers for easy reference.

Answer Book with Sample Tests: This separate publication includes all the answers to computational problems in the text as well as a number of alternate forms of chapter tests.

Use of Computers

Many computer software packages and programmable calculators are available to reduce the amount of tedious computations involved in a typical elementary linear algebra course. Our philosophy is that, once a student has mastered the concepts of row reduction and matrix multiplication, there is no need to waste precious time in or out of class with these rote computations. Therefore, we encourage the use of a computer or calculator to speed up the problem-solving process, especially after Chapter 3. However, a computer or calculator is not a requirement, except for certain topics in Chapter 9 (Numerical Methods).

This textbook is not "tied" to any particular package or calculator. We expect students to use whatever is available at their school or is affordable to them. Any standard software package or calculator that can perform row reduction and standard matrix operations can be used (such as DERIVE, Mathematica, MAX, LINEAR-KIT, HP-28S, HP-48S, TI-85).

Chapter-by-Chapter Summary

The first six chapters constitute the fundamental material covered in most elementary linear algebra courses:

- **Chapter 1** introduces vectors and matrices and their fundamental operations and properties. This chapter includes a special section on proof techniques, illustrating some of the most important methods of proof and pointing out some of the pitfalls.
- **Chapter 2** begins with the solution of systems of linear equations using the Gauss-Jordan row reduction method. This topic is followed by a discussion of the uniqueness of reduced row echelon form, rank, row space, inverses of matrices, and elementary matrices.
- **Chapter 3** introduces the determinant (using permutations) and shows its usefulness in working with systems of linear equations. The chapter ends with a discussion of cofactor expansion and Cramer's Rule.
- **Chaper 4** begins a treatment of the abstract concepts of vector spaces and subspaces. Span, linear independence, basis and dimension, and coordinatization are covered. Several methods for finding bases are illustrated.
- **Chapter 5** introduces linear transformations. The matrix of a linear transformation and its kernel and range are covered. One-to-one and onto linear transformations and isomorphisms are discussed thoroughly. The chapter ends with orthogonality, the Gram-Schmidt Process, orthogonal matrices, and orthogonal complements.

- **Chapter 6** covers eigenvalues and eigenvectors, diagonalization, orthogonal diagonalization, and some applications.

The remaining three chapters contain ancillary material:

- **Chapter 7** introduces complex vector spaces and general inner product spaces.
- **Chapter 8** is devoted to applications of linear algebra, including elementary graph theory, Ohm's Law, least squares, Markov chains, function spaces, rotation of axes, differential equations, and quadratic forms.
- **Chapter 9** dicusses important considerations when using a computer or calculator to perform linear algebra computations. Numerical methods such as Gaussian elimination, **LDU** decomposition, and the power method for calculating eigenvalues are covered.

There are three appendices:

- **Appendix A** contains proofs of several theorems that were omitted from the main text because of length or complexity.
- **Appendix B** contains a review of basic function terminology and properties, as well as a treatment of one-to-one and onto functions.
- **Appendix C** contains a review of the basic properties of complex numbers.

Guide for the Instructor

Chapters 1 through 7 have been written in a sequential fashion. Each section is generally needed as a prerequisite for what follows. Therefore, we recommend that the sections of these chapters be covered in order. Exceptions:

- **Section 1.3** (An Introduction to Proofs) can be covered any time after Section 1.2).
- **Section 2.5** (Elementary Matrices) and **Section 3.3** (Cofactor Expansion and the Adjoint) can be omitted. Only a handful of places in the remainder of the text use the material from these sections.
- **Section 5.6** (Orthogonal Complements) is needed only for several examples in Section 6.2 (Diagonalization), for Section 6.3 (Orthogonal Diagonalization), and for parts of Chapter 7.

This textbook has been classroom tested several times for both three- and four-credit courses in elementary linear algebra, with classes consisting mostly of mathematics and computer science majors. The sample three-credit course below (Course A) covers Chapters 1 through 6 (excluding Sections 2.5, 3.3, 5.6, and 6.3) and some of Chapter 8. The sample four-credit course below (Course B) covers all of Chapters 1 through 6 and some of Chapters 8 and 9. In the sample timetables, R means that the section was assigned to students for reading outside of class. In both courses, students were required to submit elementary proofs from the exercises in the textbook for critique and grading.

	Course A (3 credits)	Course B (4 credits)
Chapter 1	6 classes	6 classes
Section 8.1		R
Chapter 2	5 classes	7 classes
Section 9.1		1 class
Section 8.3	R	
Section 8.4		R
Chapter 3	3 classes	4 classes
Chapter 4	10 classes	12 classes
Section 8.6		R
Section 8.7	R	1 class
Chapter 5	9 classes	10 classes
Chapter 6	4 classes	6 classes
Section 9.3	R	R
Section 8.8		1 class
Subtotal	37 classes	48 classes
Review sessions	3 classes	4 classes
Tests	3 classes	4 classes
Total	42 classes	56 classes

Acknowledgments

We thank all those who have helped in the publication of this book, especially the many editors who have helped us through the eight-year-long process of creation. Special thanks go to Kevin Howat, our original editor at Wadsworth, who had faith in us and encouraged us very much, and to our current editors Steve Quigley and Alexander Kugushev, who have had a very keen interest in our work. We also thank Becky Smith for her fine editing of the manuscript, Nancy Sjoberg for managing the project throughout the production stages, and the production staff at Wadsworth for all their help.

Special accolades should be given to our department chairs: Samuel Wiley, Charles Hofmann, Agnes Rash, and Jonathan Hodgson, who strongly supported our (eternal, all-consuming) writing effort. We also thank Douglas Riddle for his sage advice to first-time authors.

We appreciate the support of the Academic Computing staff at St. Joseph's University, who helped us print the manuscript using LATEX to make such beautiful copies for class testing and final submission. In particular, we are especially grateful to Joseph Petragnani, Jeff Bachovchin, Sally Milliken, and Wayne Lyle. We also thank the Academic Computing and Technology staff at La Salle University who also helped with the printing of the manuscript, especially Ralph Romano and Thomas Pasquale.

We sincerely thank those students who have helped to classroom-test earlier versions of this manuscript over the last decade, both at La Salle University and St. Joseph's University. Their comments and suggestions have guided us in shaping this text in many ways. Agnes Rash also deserves thanks for testing the manuscript in an independent study course.

We also want to thank our wives Ene and Lyn for taking on extra burdens at home so that we could work on this text. We also thank Ene, who conveniently works at St. Joseph's, for agreeing to be our courier and for carrying revisions of the manuscript back and forth between us.

We are also grateful to everyone mentioned above for putting up with our sense of humor (or lack thereof) during the long preparation of this manuscript.

Finally, we acknowledge those reviewers who have given us many worthwhile suggestions, especially C. S. Ballantine, Oregon State University; Yuh-ching Chen, Fordham University; Susan Jane Colley, Oberlin College; Roland di Franco, University of the Pacific; Colin Graham, Northwestern University; K. G. Jinadasa, Illinois State University; Ralph Kelsey, Denison University; Masood Otarod; University of Scranton; J. Bryan Sperry, Pittsburg State University; and Robert Tyler, Susquehanna University.

A Final Note

Expanding our original kernel of ideas into a full-fledged linear algebra textbook encompassed a much longer range of time than we expected. But now, at last, we have the basis to declare our linear independence from this work, which has spanned most of the last decade. As a result, we hope our lives will undergo a transformation and become more normalized. We are now free to explore some new dimensions—at least until it is time to prepare a second (even more homogeneous) edition.

Stephen Andrilli
David Hecker

1

Vectors and Matrices

In linear algebra, the most fundamental object of study is the vector. We formally define vectors in Sections 1.1 and 1.2 and describe many of their algebraic and geometric properties. The interplay between algebraic manipulation and geometric intuition is a recurring theme in linear algebra, and we will use this link in later chapters to establish many important results.

In Section 1.3 we examine techniques that are useful for reading and writing proofs. In Sections 1.4 and 1.5 we introduce the matrix, another fundamental object in linear algebra, whose most basic properties parallel those of the vector. However, for several reasons, including the nature of matrix multiplication, we will eventually find many differences between the applications, interpretation, and more advanced properties of vectors and matrices.

1.1 Fundamental Operations with Vectors

In this section, we introduce vectors and consider two operations on vectors: scalar multiplication and addition. We use $\mathbb{R}$ to denote the set of all **real numbers** (that is, all coordinate values on the real number line).

Definition of a Vector

DEFINITION

A **real n-vector** is an ordered sequence of n real numbers (sometimes referred to as an **ordered n-tuple** of real numbers). The set of all n-vectors is denoted $\mathbb{R}^n$.

For example, $\mathbb{R}^2$ is the set of all 2-vectors (ordered 2-tuples = ordered pairs) of real numbers; it includes $[2, -4]$ and $[-6.2, 3.14]$ and $[0, 0]$. $\mathbb{R}^3$ is the set of all

3-vectors (ordered 3-tuples = ordered triples) of real numbers; it includes $[2, -3, 0]$ and $[-\sqrt{2}, 42.7, \pi]$.†

The vector in $\mathbb{R}^n$, which has all n entries equal to zero, is called the **zero n-vector**. In $\mathbb{R}^2$ and $\mathbb{R}^3$, the zero vectors are $[0, 0]$ and $[0, 0, 0]$, respectively.

Two vectors in $\mathbb{R}^n$ are **equal** if and only if all corresponding entries (called **coordinates**) in their n-tuples agree. That is, $[x_1, x_2, \ldots, x_n] = [y_1, y_2, \ldots, y_n]$ if and only if $x_1 = y_1$, $x_2 = y_2, \ldots,$ and $x_n = y_n$. Thus, the order in which the numbers appear in an n-tuple is as important as which numbers appear.

Geometric Interpretation of Vectors

Vectors in $\mathbb{R}^2$ frequently represent movement from one point in a coordinate plane to another. From the initial point $(3, 2)$ to a terminal point $(1, 5)$, there is a net decrease of 2 units along the x-axis and a net increase of 3 units along the y-axis. A vector representing this change would thus be $[-2, 3]$, as indicated by the arrow in Figure 1.1.

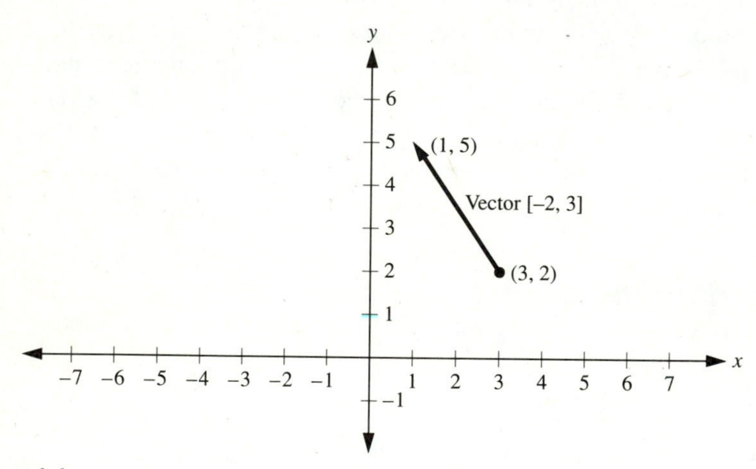

Figure 1.1
Movement represented by the vector $[-2, 3]$

Vectors can be positioned at any desired starting point. For example, the vector $[-2, 3]$ could also represent a movement from the initial point $(9, -6)$ to the terminal point $(7, -3)$.‡

---

†Many textbooks distinguish between *row* vectors, such as $[2, -3]$, and *column* vectors:

$$\begin{bmatrix} 2 \\ -3 \end{bmatrix}$$

However, in this textbook, we express vectors as row or column vectors as the situation warrants.

‡We use italicized capital letters and parentheses for the points of a coordinate system, such as $A = (3, 2)$, and boldface lowercase letters and brackets for vectors, such as $\mathbf{x} = [3, 2]$.

Vectors in $\mathbb{R}^3$ have a geometric interpretation analogous to that of vectors in $\mathbb{R}^2$. But a 3-vector is used to represent movement between points in three-dimensional space rather than on a two-dimensional plane. For example, the vector $[2, -2, 6]$ can be used to represent movement from the initial point $(2, 3, -1)$ to the terminal point $(4, 1, 5)$, as shown in Figure 1.2.

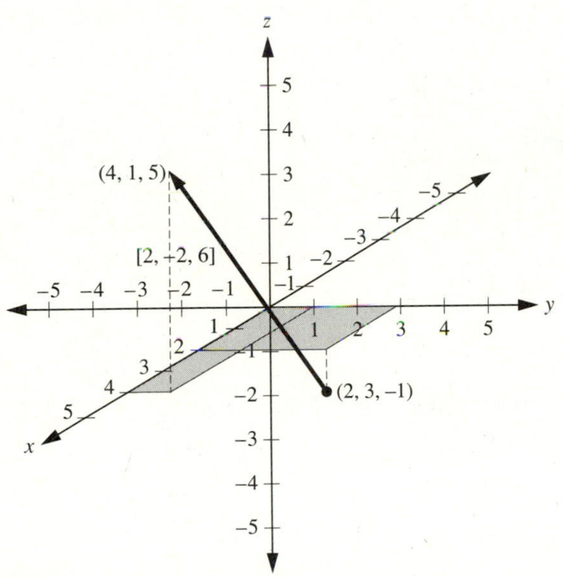

Figure 1.2
The vector $[2, -2, 6]$ with initial point $(2, 3, -1)$

Three-dimensional movements are usually graphed on a two-dimensional page by slanting the x-axis at an angle to create the optical illusion of three mutually perpendicular axes. Positions and movements are determined on such a graph by breaking them down into their components parallel to each of the coordinate axes. The appropriate distances are traveled in each of these three directions to arrive at the desired point.

Visualizing vectors in $\mathbb{R}^4$ and higher dimensions as geometric objects is more difficult. However, the same algebraic principles are involved. For example, the vector $\mathbf{x} = [2, 7, -3, 10]$ could represent a movement between the points $(5, -6, 2, -1)$ and $(7, 1, -1, 9)$ in a four-dimensional coordinate system.

Note that a single number (for example, -10 or 2.6) usually indicates only a magnitude, not a corresponding direction. Such a number is called a **scalar** to distinguish it from a vector.

Length of a Vector

The distance between the initial and terminal points of a vector is the **length** of the vector. Recall the **distance formula** in the plane: the distance between two points (x_1, y_1) and (x_2, y_2) is $d = \sqrt{(x_2 - x_1)^2 + (y_2 - y_1)^2}$ (see Figure 1.3). This formula arises from the Pythagorean Theorem for right triangles. The 2-vector between the points is $[a_1, a_2]$, where $a_1 = x_2 - x_1$ and $a_2 = y_2 - y_1$, so $d = \sqrt{a_1^2 + a_2^2}$. This formula motivates the following definition:

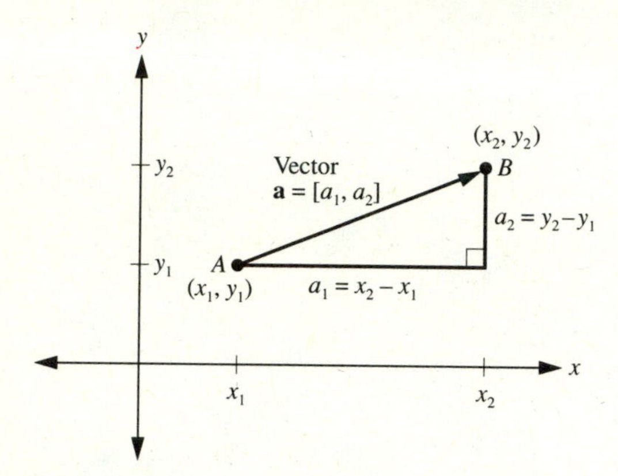

Figure 1.3
The line segment (and vector) connecting points A and B, with length $\sqrt{(x_2 - x_1)^2 + (y_2 - y_1)^2} = \sqrt{a_1{}^2 + a_2{}^2}$

DEFINITION

The **length** (also known as the **norm** or **magnitude**) of a vector $\mathbf{a} = [a_1, a_2, \ldots, a_n]$ in $\mathbb{R}^n$ is $\|\mathbf{a}\| = \sqrt{a_1^2 + a_2^2 + \cdots + a_n^2}$.

Example 1

The length of the vector $\mathbf{a} = [4, -3, 0, 2]$ is given by $\|\mathbf{a}\| = \sqrt{4^2 + (-3)^2 + 0^2 + 2^2}$ $= \sqrt{16 + 9 + 4} = \sqrt{29}$. ■

Note that the length of any vector in $\mathbb{R}^n$ is always nonnegative (that is, ≥ 0). (Do you know why this statement is true?) Also, the only vector with length 0 in $\mathbb{R}^n$ is the zero vector $[0, 0, \ldots, 0]$. (Why?)

Vectors of length 1 play such an important role in linear algebra that they have a special name.

DEFINITION

Any vector of length 1 is called a **unit vector**.

In $\mathbb{R}^2$, the vector $\left[\frac{3}{5}, -\frac{4}{5}\right]$ is a unit vector, because $\sqrt{(\frac{3}{5})^2 + (-\frac{4}{5})^2} = 1$. Similarly, $\left[0, \frac{3}{5}, 0, -\frac{4}{5}\right]$ is a unit vector in $\mathbb{R}^4$. Certain unit vectors are particularly useful:

those with a single coordinate equal to 1 and all other coordinates equal to 0. In $\mathbb{R}^2$ these vectors are denoted $\mathbf{i} = [1, 0]$ and $\mathbf{j} = [0, 1]$; in $\mathbb{R}^3$ they are denoted $\mathbf{i} = [1, 0, 0]$, $\mathbf{j} = [0, 1, 0]$, and $\mathbf{k} = [0, 0, 1]$. In $\mathbb{R}^n$, these vectors, often called the **standard unit vectors**, are denoted $\mathbf{e}_1 = [1, 0, 0, \ldots, 0]$, $\mathbf{e}_2 = [0, 1, 0, \ldots, 0]$, ..., $\mathbf{e}_n = [0, 0, 0, \ldots, 1]$.

Scalar Multiplication and Parallel Vectors

DEFINITION

Let $\mathbf{x} = [x_1, x_2, \ldots, x_n]$ be a vector in $\mathbb{R}^n$, and let c be any scalar (real number). Then $c\mathbf{x}$, the **scalar multiple of x by** c, is the vector $[cx_1, cx_2, \ldots, cx_n]$, obtained by multiplying each coordinate of $\mathbf{x}$ by c.

For example, if $\mathbf{x} = [4, -5]$, then $2\mathbf{x} = [8, -10]$, $-3\mathbf{x} = [-12, 15]$, and $-\frac{1}{2}\mathbf{x} = [-2, \frac{5}{2}]$. These vectors are graphed in Figure 1.4. From the graph, you can see that the vector $2\mathbf{x}$ points in the same direction as $\mathbf{x}$ but is twice as long. The vectors $-3\mathbf{x}$ and $-\frac{1}{2}\mathbf{x}$ indicate movements in the direction opposite to $\mathbf{x}$, with $-3\mathbf{x}$ being three times as long as $\mathbf{x}$ and $-\frac{1}{2}\mathbf{x}$ being half as long.

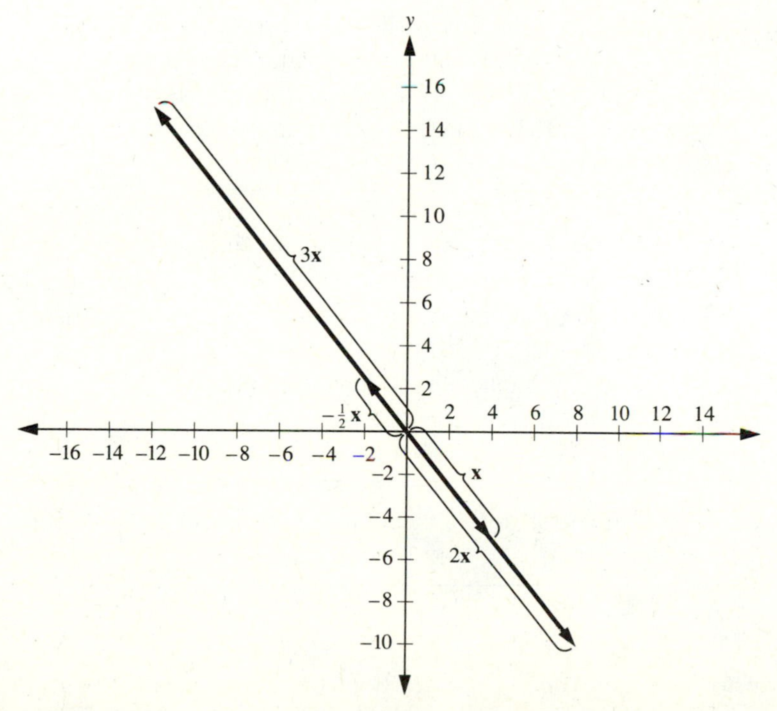

Figure 1.4
Scalar multiples of $\mathbf{x} = [4, -5]$ (all vectors drawn with initial point at origin)

Let us generalize these observations to $\mathbb{R}^n$. First, multiplication by c **expands (dilates)** the length of the vector when $|c| > 1$ and **shrinks (contracts)** the length when $|c| < 1$. Scalar multiplication by 1 or -1 does not affect the length of the vector. Scalar multiplication by 0 always yields the zero vector. These properties are all special cases of the following theorem:

THEOREM 1.1

Let $\mathbf{x} \in \mathbb{R}^n$, and let c be any real number (scalar). Then $\|c\mathbf{x}\| = |c|\,\|\mathbf{x}\|$. That is, the length of $c\mathbf{x}$ is the absolute value of c times the length of $\mathbf{x}$.

The proof of Theorem 1.1 is left for you to do in Exercise 22 at the end of this section.

We have noted intuitively that in $\mathbb{R}^2$ the scalar multiple of a vector $c\mathbf{x}$ is in the same direction as $\mathbf{x}$ when c is positive and in the direction opposite to $\mathbf{x}$ when c is negative, but we have not yet discussed how to measure "direction" in higher-dimensional coordinate systems. Therefore, we will use scalar multiplication to give a precise mathematical definition for the concept of vectors having the same or opposite directions:

DEFINITION

Two vectors $\mathbf{x}$ and $\mathbf{y}$ in $\mathbb{R}^n$ are **in the same direction** if and only if there is a positive real number c such that $\mathbf{y} = c\mathbf{x}$. The vectors $\mathbf{x}$ and $\mathbf{y}$ are **in opposite directions** if and only if there is a negative real number c such that $\mathbf{y} = c\mathbf{x}$. Two vectors are **parallel** if and only if they are either in the same direction as each other or in the opposite direction.

Hence, the vectors $[1, -3, 2]$ and $[3, -9, 6]$ are in the same direction, because $[3, -9, 6] = 3[1, -3, 2]$ (or because $[1, -3, 2] = \frac{1}{3}[3, -9, 6]$), as shown in Figure 1.5. Similarly, the vectors $[-3, 6, 0, 15]$ and $[4, -8, 0, -20]$ are in opposite directions, because $[4, -8, 0, -20] = -\frac{4}{3}[-3, 6, 0, 15]$.

Theorem 1.1 clearly implies the following result:

COROLLARY 1.2

If $\mathbf{x}$ is a nonzero vector in $\mathbb{R}^n$, then a unit vector $\mathbf{u}$ in the same direction as $\mathbf{x}$ is given by $\mathbf{u} = \left(1/\|\mathbf{x}\|\right)\mathbf{x}$.

Proof of Corollary 1.2

The vector $\mathbf{u}$ in Corollary 1.2 is clearly in the same direction as $\mathbf{x}$, because it is a positive scalar multiple of $\mathbf{x}$ (the scalar is $1/\|\mathbf{x}\|$). Also, Theorem 1.1 tells us that $\|\mathbf{u}\| = \|\left(1/\|\mathbf{x}\|\right)\mathbf{x}\| = \left(1/\|\mathbf{x}\|\right)\|\mathbf{x}\| = 1$, so $\mathbf{u}$ is a unit vector. ■

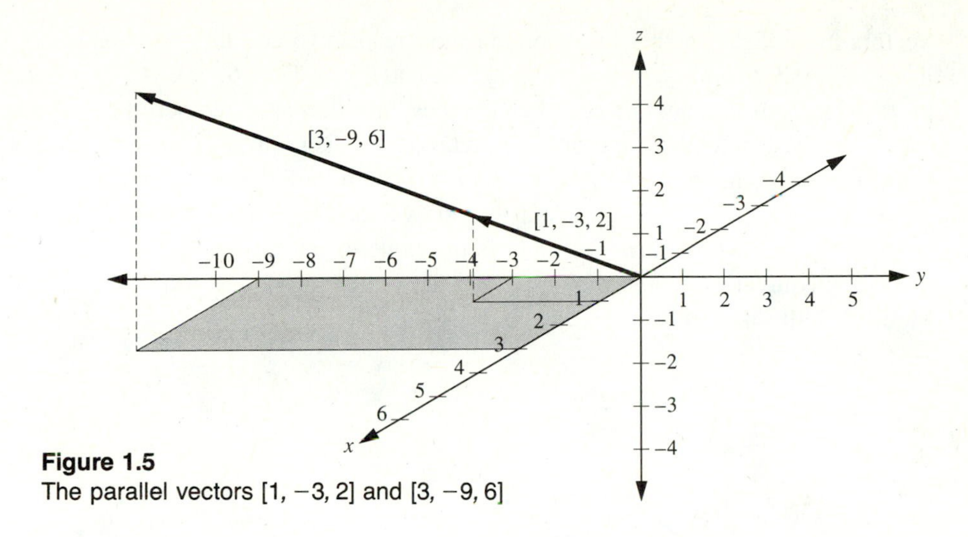

Figure 1.5
The parallel vectors [1, −3, 2] and [3, −9, 6]

Example 2

Consider the vector [2, 3, −1, 1] in $\mathbb{R}^4$. Because $\|[2, 3, -1, 1]\| = \sqrt{15}$, a unit vector in the same direction as [2, 3, −1, 1] is $(1/\sqrt{15})[2, 3, -1, 1] = [2/\sqrt{15}, 3/\sqrt{15}, -1/\sqrt{15}, 1/\sqrt{15}]$. ■

The process of "dividing" a vector by its length to obtain a unit vector in the same direction is called **normalizing** the vector (see Figure 1.6).

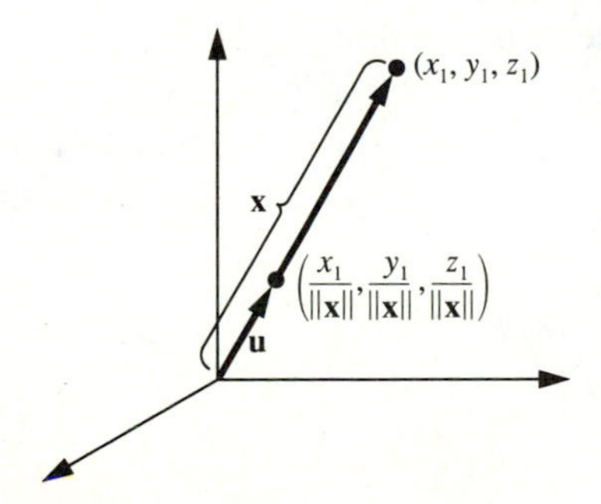

Figure 1.6
Normalizing a vector **x** to obtain a unit vector **u** in the same direction (with $\|\mathbf{x}\| > 1$)

Addition and Subtraction with Vectors

DEFINITION

Let $\mathbf{x} = [x_1, x_2, \ldots, x_n]$ and $\mathbf{y} = [y_1, y_2, \ldots, y_n]$ be vectors in $\mathbb{R}^n$. Then $\mathbf{x}+\mathbf{y}$, the **sum** of **x** and **y**, is the vector $[x_1 + y_1, x_2 + y_2, \ldots, x_n + y_n]$ in $\mathbb{R}^n$.

Vectors are added together by summing their respective coordinates. For example, if $\mathbf{x} = [2, -3, 5]$ and $\mathbf{y} = [-6, 4, -2]$, then $\mathbf{x} + \mathbf{y} = [2 - 6, -3 + 4, 5 - 2] = [-4, 1, 3]$. Vectors cannot be added unless they have the same number of coordinates.

There is a natural geometric interpretation for the sum of two vectors, **x** and **y**, in a plane or in space. Draw the vector **x**. Then draw the vector **y** from the terminal point of **x**. The sum of **x** and **y** is the vector whose *initial* point is the same as that of **x** and whose *terminal* point is the same as that of **y**. The total movement $(\mathbf{x} + \mathbf{y})$ is equivalent to first moving along the vector **x** and then along the vector **y**. Figure 1.7 illustrates this procedure in $\mathbb{R}^2$.

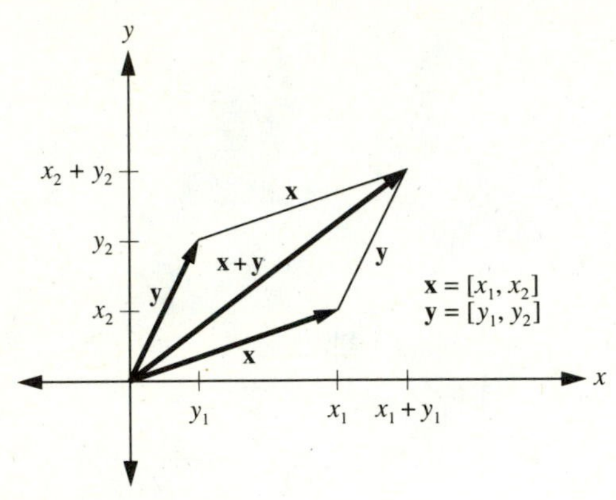

Figure 1.7
Addition of vectors in $\mathbb{R}^2$

We use the notation $-\mathbf{y}$ to denote the scalar multiple $-1\mathbf{y}$. Thus, we can now define **subtraction** of vectors in a natural way: if **x** and **y** are both vectors in $\mathbb{R}^n$, let $\mathbf{x} - \mathbf{y}$ be the vector $\mathbf{x} + (-\mathbf{y})$. A geometric interpretation of subtraction is shown in Figure 1.8: movement **x** followed by movement $-\mathbf{y}$. An alternate interpretation is described in Exercise 11.

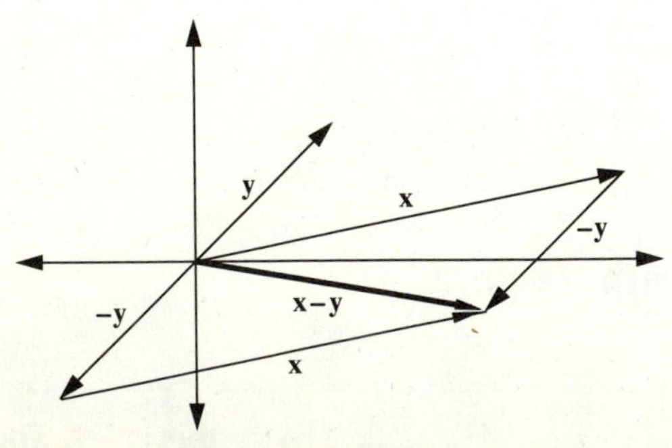

Figure 1.8
Subtraction of vectors in $\mathbb{R}^2$: $\mathbf{x} - \mathbf{y} = \mathbf{x} + (-\mathbf{y})$

Fundamental Properties of Addition and Scalar Multiplication

Theorem 1.3 contains the basic properties of addition and scalar multiplication of vectors. Later, we will explain why these properties are so important.

THEOREM 1.3

Let $\mathbf{x} = [x_1, x_2, \ldots, x_n]$ and $\mathbf{y} = [y_1, y_2, \ldots, y_n]$ and $\mathbf{z} = [z_1, z_2, \ldots, z_n]$ be any vectors in $\mathbb{R}^n$, and let c and d be any real numbers (scalars). Let $\mathbf{0}$ represent the zero vector in $\mathbb{R}^n$. Then

(1)	$\mathbf{x} + \mathbf{y} = \mathbf{y} + \mathbf{x}$	Commutative Law of Addition
(2)	$\mathbf{x} + (\mathbf{y} + \mathbf{z}) = (\mathbf{x} + \mathbf{y}) + \mathbf{z}$	Associative Law of Addition
(3)	$\mathbf{0} + \mathbf{x} = \mathbf{x} + \mathbf{0} = \mathbf{x}$	Existence of Identity Element for Addition
(4)	$\mathbf{x} + (-\mathbf{x}) = (-\mathbf{x}) + \mathbf{x} = \mathbf{0}$	Existence of Inverse Elements for Addition
(5)	$c(\mathbf{x} + \mathbf{y}) = c\mathbf{x} + c\mathbf{y}$	Distributive Laws of Scalar Multiplication over Addition
(6)	$(c + d)\mathbf{x} = c\mathbf{x} + d\mathbf{x}$	
(7)	$(cd)\mathbf{x} = c(d\mathbf{x})$	Associativity of Scalar Multiplication
(8)	$1\mathbf{x} = \mathbf{x}$	Existence of Identity Element for Scalar Multiplication

These **commutative**, **associative**, and **distributive laws** are so named because they resemble the corresponding laws for real numbers.

In part (3) of the theorem, the vector $\mathbf{0}$ is called an **identity element** for addition because it preserves (doesn't change) the identity of any vector to which it is added. A similar statement is true for the scalar 1 with the operation of scalar multiplication, as you can see in part (8). In part (4), the vector $-\mathbf{x}$ is called the **additive inverse element of x** because it "cancels out $\mathbf{x}$"; that is, $\mathbf{x} + (-\mathbf{x}) = \mathbf{0}$, the zero vector.

Each part of the theorem is proved by calculating the entries in each coordinate of the vectors and applying a corresponding law for real-number arithmetic. We illustrate this *coordinate-wise* proof technique by giving the proof of part (6) of Theorem 1.3. You are asked to prove other parts of the theorem in Exercise 23.

Proof of Theorem 1.3, Part (6)

$(c + d)\mathbf{x} = (c + d)[x_1, x_2, \ldots, x_n]$	
$= [(c + d)x_1, (c + d)x_2, \ldots, (c + d)x_n]$	definition of scalar multiplication
$= [cx_1 + dx_1, cx_2 + dx_2, \ldots, cx_n + dx_n]$	coordinate-wise use of distributive law in $\mathbb{R}$
$= [cx_1, cx_2, \ldots, cx_n] + [dx_1, dx_2, \ldots, dx_n]$	definition of vector addition

$$= c[x_1, x_2, \ldots, x_n] + d[x_1, x_2, \ldots, x_n]$$ definition of scalar multiplication

$$= c\mathbf{x} + d\mathbf{x}$$ ■

Linear Combinations of Vectors

DEFINITION

Let $\mathbf{v}_1, \mathbf{v}_2, \ldots, \mathbf{v}_k$ be vectors in $\mathbb{R}^n$. Then the vector $\mathbf{v}$ is said to be a **linear combination** of the vectors $\mathbf{v}_1, \mathbf{v}_2, \ldots, \mathbf{v}_k$ if and only if there are scalars $c_1, c_2, \ldots, c_k$ such that $\mathbf{v} = c_1\mathbf{v}_1 + c_2\mathbf{v}_2 + \cdots + c_k\mathbf{v}_k$. That is, a linear combination of vectors is a sum of scalar multiples of those vectors.

For example, the vector $[-2, 8, 5, 0]$ is a linear combination of the vectors $[3, 1, -2, 2]$, $[1, 0, 3, -1]$, and $[4, -2, 1, 0]$, because $2[3, 1, -2, 2] + 4[1, 0, 3, -1] - 3[4, -2, 1, 0] = [-2, 8, 5, 0]$.

You are probably familiar with linear combinations of the vectors **i**, **j**, and **k** in $\mathbb{R}^3$ from calculus. Recall that any vector in $\mathbb{R}^3$ can be expressed in a unique way as a linear combination of these three vectors. For example, $[3, -2, 5] = 3[1, 0, 0] - 2[0, 1, 0] + 5[0, 0, 1] = 3\mathbf{i} - 2\mathbf{j} + 5\mathbf{k}$. In general, $[a, b, c] = a\mathbf{i} + b\mathbf{j} + c\mathbf{k}$. Also, every vector in $\mathbb{R}^n$ can be expressed as a linear combination of the standard unit vectors $\mathbf{e}_1 = [1, 0, 0, \ldots, 0]$, $\mathbf{e}_2 = [0, 1, 0, \ldots, 0], \ldots, \mathbf{e}_n = [0, 0, \ldots, 0, 1]$.

Physical Applications of Addition and Scalar Multiplication

Frequently, addition and scalar multiplication of vectors can be used to solve real-world problems. Here we present two examples from elementary physics. In the first, we use the following trigonometric fact: if $\mathbf{v}$ is a vector in $\mathbb{R}^2$ forming an angle of θ with the positive x-axis, then $\mathbf{v} = [\|\mathbf{v}\|\cos\theta, \|\mathbf{v}\|\sin\theta]$. This fact is illustrated in Figure 1.9.

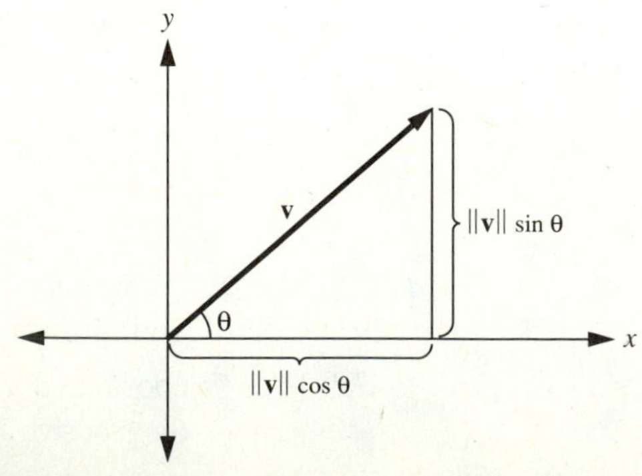

Figure 1.9
The vector $\mathbf{v} = [\|\mathbf{v}\|\cos\theta, \|\mathbf{v}\|\sin\theta]$ forming an angle of θ with the positive x-axis

Example 3: Resultant Velocity

Suppose a man swims 5 km/hr (kilometers per hour) in calm water. If he is swimming toward the east in a wide stream having a northwest current of 3 km/hr, what is his **resultant velocity** (net speed and direction)?

The velocities of the swimmer and current are shown as vectors in Figure 1.10, where we have, for convenience, placed the swimmer at the origin of a coordinate system. Now, $\mathbf{v}_1 = [5, 0]$ and $\mathbf{v}_2 = [3\cos 135°, 3\sin 135°] = [-3\sqrt{2}/2, 3\sqrt{2}/2]$. Thus, the total (resultant) velocity of the swimmer is the vector sum of these velocities, $\mathbf{v}_1 + \mathbf{v}_2$, which is $[5 - 3\sqrt{2}/2, 3\sqrt{2}/2]$, which is approximately $[2.88, 2.12]$. Hence, each hour the swimmer is actually traveling about 2.9 km (kilometers) east and 2.1 km north. The resultant speed of the swimmer is $\|[5 - 3\sqrt{2}/2, 3\sqrt{2}/2]\|$, which is about 3.58 km/hr. ■

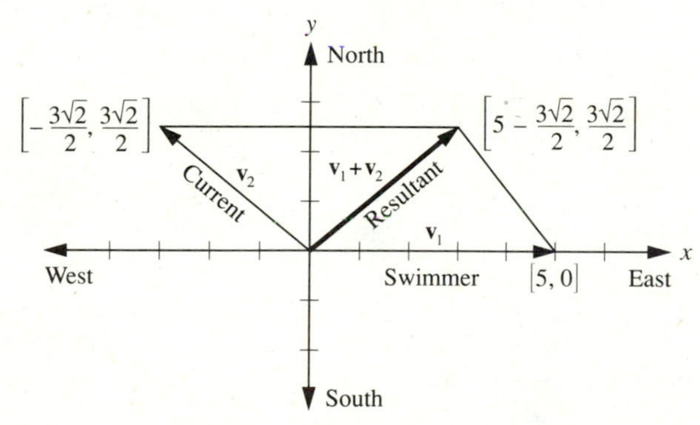

Figure 1.10
Velocity $\mathbf{v}_1$ of swimmer, velocity $\mathbf{v}_2$ of current, and resultant velocity $\mathbf{v}_1 + \mathbf{v}_2$

Example 4: Newton's Second Law

Newton's famous **Second Law of Motion** asserts that the sum of the vector forces on an object, $\mathbf{f}$, is equal to the scalar multiple of the mass of the object, m, times the vector acceleration of the object, $\mathbf{a}$; that is, $\mathbf{f} = m\mathbf{a}$. For example, suppose a mass of 5 kg (kilograms) in a three-dimensional coordinate system has two forces acting on it: a force $\mathbf{f}_1$ of 10 newtons[†] in the direction of the vector $[-2, 1, 2]$ and a force $\mathbf{f}_2$ of 20 newtons in the direction of the vector $[6, 3, -2]$. What is the acceleration of the object?

To get a vector representation for $\mathbf{f}_1$ and $\mathbf{f}_2$, we use the technique of normalization of vectors. Because $[-2, 1, 2]/\|[-2, 1, 2]\|$ is a unit vector in the di-

[†] 1 newton = 1 kg-m/sec^2 (kilogram-meter/second^2), or the force needed to push 1 kg at a speed 1 m/sec (meter per second) faster every second; 1 m (meter) = 3.281 ft (feet), and 1 kg = 2.205 lb (pounds).

rection of $[-2, 1, 2]$, force $\mathbf{f}_1 = 10([-2, 1, 2]/\|[-2, 1, 2]\|)$. Similarly, force $\mathbf{f}_2 = 20([6, 3, -2]/\|[6, 3, -2]\|)$. The net force on the object is $\mathbf{f} = \mathbf{f}_1 + \mathbf{f}_2$. Thus, the net acceleration on the object, $\mathbf{a}$, is

$$\frac{1}{m}\mathbf{f} = \frac{1}{m}(\mathbf{f}_1 + \mathbf{f}_2) = \frac{1}{5}\left(10\left(\frac{[-2, 1, 2]}{\|[-2, 1, 2]\|}\right) + 20\left(\frac{[6, 3, -2]}{\|[6, 3, -2]\|}\right)\right)$$

which equals $\frac{2}{3}[-2, 1, 2] + \frac{4}{7}[6, 3, -2] = \left[\frac{44}{21}, \frac{50}{21}, \frac{4}{21}\right]$. The length of $\mathbf{a}$ is approximately 3.18, so pulling out a factor of 3.18 from each coordinate, we can express the vector $\mathbf{a}$ as approximately 3.18 [0.66, 0.75, 0.06], where the vector [0.66, 0.75, 0.06] is a *unit* vector. Hence, the acceleration is about 3.18 m/sec^2 in the direction of the vector [0.66, 0.75, 0.06]. ■

If the sum of the forces on an object is the zero vector, then the object is said to be in **equilibrium**: there is no acceleration in any direction. Exercise 20 deals with equilibrium of forces.

Exercises—Section 1.1

Note: A ★ next to an exercise indicates that the answer for that exercise appears in the back of the book.

1. In each of the following cases, find a vector that represents a movement from the first (initial) point to the second (terminal) point. Then use this vector to find the distance between the given points.

★(a) $(-4, 3), (5, -1)$ (b) $(2, -1, 4), (-3, 0, 2)$
★(c) $(1, -2, 0, 2, 3), (0, -3, 2, -1, -1)$

2. In each of the following cases, draw a directed line segment in space that represents the movement associated with each of the vectors if the initial point is $(1, 1, 1)$. What is the terminal point in each case?

★(a) $[2, 3, 1]$ (b) $[-1, 4, 2]$
★(c) $[0, -3, -1]$ (d) $[2, -1, -1]$

3. In each of the following cases, find the initial point given the vector and the terminal point.

★(a) $[-1, 4], (6, -9)$ (b) $[2, -2, 5], (-4, 1, 7)$
★(c) $[3, -4, 0, 1, -2], (2, -1, -1, 5, 4)$

4. In each of the following cases, find a vector describing a movement from the first (initial) point to a terminal point that is two-thirds of the distance to the second point.

★(a) $(-4, 7, 2), (10, -10, 11)$ (b) $(2, -1, 0, -7), (-11, -1, -9, 2)$

5. In each of the following cases, find a unit vector in the same direction as the given vector. Is the resulting (normalized) vector longer or shorter than the original? Why?

★(a) $[3, -5, 6]$ (b) $[4, 1, 0, -2]$

★(c) $[0.6, -0.8]$ (d) $\left[\frac{1}{5}, -\frac{2}{5}, -\frac{1}{5}, \frac{1}{5}, \frac{2}{5}\right]$

6. Which of the following pairs of vectors are parallel?

★(a) $[12, -16]$, $[9, -12]$ (b) $[4, -14]$, $[-2, 7]$

★(c) $[-2, 3, 1]$, $[6, -4, -3]$ (d) $[10, -8, 3, 0, 27]$, $\left[\frac{5}{6}, -\frac{2}{3}, \frac{3}{4}, 0, -\frac{5}{2}\right]$

7. If $\mathbf{x} = [-2, 4, 5]$, $\mathbf{y} = [-1, 0, 3]$, and $\mathbf{z} = [4, -1, 2]$, find the following:

★(a) $3\mathbf{x}$ (b) $-2\mathbf{y}$ ★(c) $\mathbf{x} + \mathbf{y}$

(d) $\mathbf{y} - \mathbf{z}$ ★(e) $4\mathbf{y} - 5\mathbf{x}$ (f) $2\mathbf{x} + 3\mathbf{y} - 4\mathbf{z}$

8. Given $\mathbf{x}$ and $\mathbf{y}$ as follows, calculate $\mathbf{x} + \mathbf{y}$, $\mathbf{x} - \mathbf{y}$, and $\mathbf{y} - \mathbf{x}$, and sketch $\mathbf{x}, \mathbf{y}$, $\mathbf{x} + \mathbf{y}$, $\mathbf{x} - \mathbf{y}$, and $\mathbf{y} - \mathbf{x}$ in the same coordinate system.

★(a) $\mathbf{x} = [-1, 5]$, $\mathbf{y} = [2, -4]$

(b) $\mathbf{x} = [10, -2]$, $\mathbf{y} = [-7, -3]$

★(c) $\mathbf{x} = [2, 5, -3]$, $\mathbf{y} = [-1, 3, -2]$

(d) $\mathbf{x} = [1, -2, 5]$, $\mathbf{y} = [-3, -2, -1]$

9. Show that the points $(7, -3, 6)$, $(11, -5, 3)$, and $(10, -7, 8)$ are the vertices of an isosceles triangle. Is this an equilateral triangle?

10. A certain clock has a minute hand that is 10 cm (centimeters) long. Find the vector representing the displacement of the tip of the minute hand of the clock:

★(a) From 12 PM to 12:15 PM ★(b) From 12 PM to 12:40 PM (Hint: use trigonometry)

(c) From 12 PM to 1 PM

11. Show that if $\mathbf{x}$ and $\mathbf{y}$ are vectors in $\mathbb{R}^2$, then $\mathbf{x}+\mathbf{y}$ and $\mathbf{x}-\mathbf{y}$ are the two diagonals of the parallelogram whose sides are $\mathbf{x}$ and $\mathbf{y}$.

12. Consider the picture in $\mathbb{R}^3$ in Figure 1.11. Verify that $\mathbf{x} + (\mathbf{y} + \mathbf{z})$ is a diagonal of the parallelepiped with sides $\mathbf{x}, \mathbf{y}, \mathbf{z}$. Does $(\mathbf{x}+\mathbf{y})+\mathbf{z}$ represent the same diagonal vector? Why or why not?

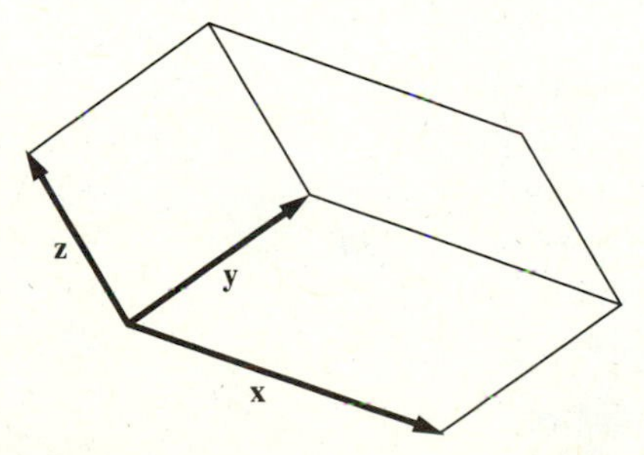

Figure 1.11
Parallelepiped with sides **x, y, z**

★**13.** At a certain green on a golf course, a golfer takes three putts to sink the ball. If the first putt moved the ball 1 m (meter) southwest, the second putt moved the ball 0.5 m east, and the third putt moved the ball 0.2 m northwest, what single putt (expressed as a vector) would have had the same final result?

14. (a) Show that every unit vector in $\mathbb{R}^2$ is of the form $[\cos(\theta_1), \cos(\theta_2)]$, where θ_1 is the angle the vector makes with the positive x-axis and θ_2 is the angle the vector makes with the positive y-axis.

(b) Show that every unit vector in $\mathbb{R}^3$ is of the form $[\cos(\alpha_1), \cos(\alpha_2), \cos(\alpha_3)]$, where α_1, α_2, and α_3 are the angles the vector makes with the x-, y-, and z-axes, respectively. (Note: The coordinates of this unit vector are often called the **direction cosines** of the vector.)

★**15.** A rower can propel a boat 4 km/hr on a calm river. If the rower rows northwestward against a current of 3 km/hr southward, what is the net velocity of the boat? What is its resultant speed?

16. A singer is walking 3 km/hr southwestward on a moving parade float that is being pulled northward at 4 km/hr. What is the net velocity of the singer? What is the singer's resultant speed?

★**17.** A woman rowing on a wide river wants the resultant (net) velocity of her boat to be 8 km/hr westward. If the current is moving 2 km/hr northeastward, what velocity vector should the rower maintain?

★**18.** Using Newton's Second Law of Motion, find the acceleration vector on a 20 kg object in a three-dimensional coordinate system when the following three forces are simultaneously applied:

(i) A force of 4 newtons in the direction of the vector $[3, -12, 4]$
(ii) A force of 2 newtons in the direction of the vector $[0, -4, -3]$
(iii) A force of 6 newtons in the direction of the unit vector $\mathbf{k}$

19. Using Newton's Second Law of Motion, find the resultant sum of the forces on a 30 kg object in a three-dimensional coordinate system undergoing an acceleration of 6 m/sec^2 in the direction of the vector $[-2, 3, 1]$.

★**20.** Two forces, $\mathbf{a}$ and $\mathbf{b}$, are simultaneously applied along cables attached to a weight, as in Figure 1.12, to keep the weight in equilibrium by balancing the force of gravity (which is $m\mathbf{g}$, where m is the mass of the weight and $\mathbf{g} = [0, -g]$ is the downward acceleration due to gravity). Solve for the coordinates of forces $\mathbf{a}$ and $\mathbf{b}$ in terms of m and g.

21. (a) Prove that the length of each vector in $\mathbb{R}^n$ is nonnegative (that is, ≥ 0).

(b) Prove that the only vector in $\mathbb{R}^n$ of length 0 is the zero vector.

22. Prove Theorem 1.1.

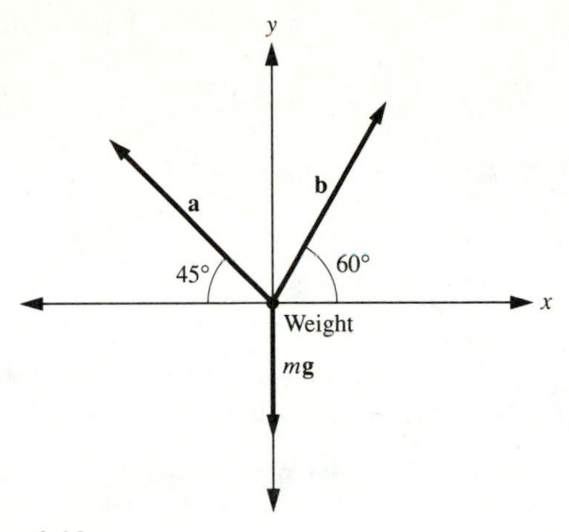

Figure 1.12
Forces in equilibrium

23. Prove parts (2), (4), (5), and (7) of Theorem 1.3.

24. If $\mathbf{x}$ is a vector in $\mathbb{R}^n$ and $c_1 \neq c_2$, show that $c_1\mathbf{x} = c_2\mathbf{x}$ implies that $\mathbf{x} = \mathbf{0}$ (zero vector).

1.2 The Dot Product

We have already examined scalar multiplication and addition of vectors. We now discuss another important operation: the dot product. After explaining several properties of the dot product, we show how it can be used to calculate the angle between vectors and to "project" one vector onto another.

Definition and Properties of the Dot Product

DEFINITION

Let $\mathbf{x} = [x_1, x_2, \ldots, x_n]$ and $\mathbf{y} = [y_1, y_2, \ldots, y_n]$ be two vectors in $\mathbb{R}^n$. The **dot (inner) product** of $\mathbf{x}$ and $\mathbf{y}$ is given by

$$\mathbf{x} \cdot \mathbf{y} = x_1y_1 + x_2y_2 + \cdots + x_ny_n = \sum_{k=1}^{n} x_ky_k.$$

For example, if $\mathbf{x} = [2, -4, 3]$ and $\mathbf{y} = [1, 5, -2]$, then $\mathbf{x} \cdot \mathbf{y} = (2)(1) + (-4)(5) + (3)(-2) = -24$. Notice that this type of multiplication involves two vectors and the result is a *scalar*, whereas scalar multiplication involves a scalar and a vector and the result is a *vector*. Also, the dot product is not defined for vectors having different numbers of coordinates.

The next theorem states some elementary results involving the dot product:

THEOREM 1.4

If $\mathbf{x}$, $\mathbf{y}$, and $\mathbf{z}$ are any vectors in $\mathbb{R}^n$ and if c is any scalar, then

(1) $\mathbf{x} \cdot \mathbf{y} = \mathbf{y} \cdot \mathbf{x}$	Commutativity of dot product
(2) $\mathbf{x} \cdot \mathbf{x} = \|\mathbf{x}\|^2 \geq 0$	Relationship between
(3) $\mathbf{x} \cdot \mathbf{x} = 0$ if and only if $\mathbf{x} = \mathbf{0}$	dot product and length
(4) $c(\mathbf{x} \cdot \mathbf{y}) = (c\mathbf{x}) \cdot \mathbf{y} = \mathbf{x} \cdot (c\mathbf{y})$	Scalar multiplication property of dot product
(5) $\mathbf{x} \cdot (\mathbf{y} + \mathbf{z}) = (\mathbf{x} \cdot \mathbf{y}) + (\mathbf{x} \cdot \mathbf{z})$	Distributive properties of dot
(6) $(\mathbf{x} + \mathbf{y}) \cdot \mathbf{z} = (\mathbf{x} \cdot \mathbf{z}) + (\mathbf{y} \cdot \mathbf{z})$	product over addition

The proofs of parts (1), (2), (4), (5), and (6) are done by expanding the expressions on each side of the equation to be proved and then showing that they are equal. We illustrate this with the proof of part (5). The remaining parts are left for you to prove in Exercise 6.

Proof of Theorem 1.4, Part (5)

Let $\mathbf{x} = [x_1, x_2, \ldots, x_n]$, $\mathbf{y} = [y_1, y_2, \ldots, y_n]$, and $\mathbf{z} = [z_1, z_2, \ldots, z_n]$. Then,

$$\begin{aligned}
\mathbf{x} \cdot (\mathbf{y} + \mathbf{z}) &= [x_1, x_2, \ldots, x_n] \cdot ([y_1, y_2, \ldots, y_n] + [z_1, z_2, \ldots, z_n]) \\
&= [x_1, x_2, \ldots, x_n] \cdot [y_1 + z_1, y_2 + z_2, \ldots, y_n + z_n] \\
&= x_1(y_1 + z_1) + x_2(y_2 + z_2) + \cdots + x_n(y_n + z_n) \\
&= (x_1y_1 + x_2y_2 + \cdots + x_ny_n) + (x_1z_1 + x_2z_2 + \cdots + x_nz_n).
\end{aligned}$$

Also,

$$\begin{aligned}
(\mathbf{x} \cdot \mathbf{y}) + (\mathbf{x} \cdot \mathbf{z}) &= ([x_1, x_2, \ldots, x_n] \cdot [y_1, y_2, \ldots, y_n]) \\
&\quad + ([x_1, x_2, \ldots, x_n] \cdot [z_1, z_2, \ldots, z_n]) \\
&= (x_1y_1 + x_2y_2 + \cdots + x_ny_n) + (x_1z_1 + x_2z_2 + \cdots + x_nz_n).
\end{aligned}$$

Because $\mathbf{x} \cdot (\mathbf{y} + \mathbf{z})$ and $(\mathbf{x} \cdot \mathbf{y}) + (\mathbf{x} \cdot \mathbf{z})$ simplify to the same expression, they are equal. ■

The properties described in Theorem 1.4 allow us to simplify expressions involving the dot product in ways similar to those used in elementary algebra. For example,

$$
\begin{aligned}
(5\mathbf{x} - 4\mathbf{y}) \cdot (-2\mathbf{x} + 3\mathbf{y}) &= [(5\mathbf{x} - 4\mathbf{y}) \cdot (-2\mathbf{x})] + [(5\mathbf{x} - 4\mathbf{y}) \cdot (3\mathbf{y})] \\
&= [(5\mathbf{x}) \cdot (-2\mathbf{x})] + [(-4\mathbf{y}) \cdot (-2\mathbf{x})] + [(5\mathbf{x}) \cdot (3\mathbf{y})] \\
&\quad + [(-4\mathbf{y}) \cdot (3\mathbf{y})] \\
&= -10(\mathbf{x} \cdot \mathbf{x}) + 8(\mathbf{y} \cdot \mathbf{x}) + 15(\mathbf{x} \cdot \mathbf{y}) - 12(\mathbf{y} \cdot \mathbf{y}) \\
&= -10\,\|\mathbf{x}\|^2 + 23(\mathbf{x} \cdot \mathbf{y}) - 12\,\|\mathbf{y}\|^2.
\end{aligned}
$$

Work: An Application of the Dot Product

In elementary physics, if a vector force $\mathbf{f}$ is exerted on an object that undergoes a vector movement ("displacement") $\mathbf{d}$, then the **work** done by the force is defined to be $\mathbf{f} \cdot \mathbf{d}$. Work is measured in *joules*, where 1 joule is the work done when a force of 1 newton moves an object 1 m (meter).

Example 1

Suppose that a force of 8 newtons is exerted on an object in the direction of the vector $[1, -2, 1]$ and that the object travels 5 m in the direction of the vector $[2, -1, 0]$. Then, setting $\mathbf{f}$ equal to 8 times a unit vector in the direction of $[1, -2, 1]$ and setting $\mathbf{d}$ equal to 5 times a unit vector in the direction of $[2, -1, 0]$, the total work performed on the object is

$$\mathbf{f} \cdot \mathbf{d} = 8\left(\frac{[1, -2, 1]}{\|[1, -2, 1]\|}\right) \cdot 5\left(\frac{[2, -1, 0]}{\|[2, -1, 0]\|}\right) = \frac{40(2 + 2 + 0)}{\sqrt{6}\sqrt{5}} \qquad \text{joules,}$$

which is approximately 29.2 joules. ■

Inequalities Involving the Dot Product

The next theorem is a very famous and important result that gives an upper and lower bound on the value of the dot product. Later in this section, we use it to calculate the angle between two nonzero vectors in $\mathbb{R}^n$.

THEOREM 1.5: Cauchy-Schwarz Inequality

If $\mathbf{x}$ and $\mathbf{y}$ are vectors in $\mathbb{R}^n$, then $|\mathbf{x} \cdot \mathbf{y}| \leq (\|\mathbf{x}\|)(\|\mathbf{y}\|)$.

Proof of Theorem 1.5

If either $\|\mathbf{x}\| = \mathbf{0}$ or $\|\mathbf{y}\| = \mathbf{0}$, the theorem is obviously true. Hence, we need to examine only the case when both $\|\mathbf{x}\|$ and $\|\mathbf{y}\|$ are nonzero. We need to prove that $-(\|\mathbf{x}\|)(\|\mathbf{y}\|) \leq \mathbf{x} \cdot \mathbf{y} \leq (\|\mathbf{x}\|)(\|\mathbf{y}\|)$. This statement is true if and only if

$$-1 \leq \frac{\mathbf{x} \cdot \mathbf{y}}{(\|\mathbf{x}\|)(\|\mathbf{y}\|)} \leq 1.$$

Now, the term $(\mathbf{x} \cdot \mathbf{y})/(\|\mathbf{x}\|)(\|\mathbf{y}\|)$ is equal to $(\mathbf{x}/\|\mathbf{x}\|) \cdot (\mathbf{y}/\|\mathbf{y}\|)$. Note that $\mathbf{x}/\|\mathbf{x}\|$ and $\mathbf{y}/\|\mathbf{y}\|$ are both *unit* vectors. Thus, it is enough to show that $-1 \leq \mathbf{a} \cdot \mathbf{b} \leq 1$ for any unit vectors $\mathbf{a}$ and $\mathbf{b}$.

The term $\mathbf{a} \cdot \mathbf{b}$ occurs as part of the expansion of $(\mathbf{a} + \mathbf{b}) \cdot (\mathbf{a} + \mathbf{b})$, as well as part of $(\mathbf{a} - \mathbf{b}) \cdot (\mathbf{a} - \mathbf{b})$. The first expansion gives

$$
\begin{aligned}
(\mathbf{a} + \mathbf{b}) \cdot (\mathbf{a} + \mathbf{b}) &= \|\mathbf{a} + \mathbf{b}\|^2 \geq 0, && \text{using part (2) of Theorem 1.4} \\
(\mathbf{a} \cdot \mathbf{a}) + (\mathbf{b} \cdot \mathbf{a}) + (\mathbf{a} \cdot \mathbf{b}) + (\mathbf{b} \cdot \mathbf{b}) &\geq 0, && \\
\|\mathbf{a}\|^2 + 2(\mathbf{a} \cdot \mathbf{b}) + \|\mathbf{b}\|^2 &\geq 0, && \text{by parts (1) and (2) of Theorem 1.4} \\
1 + 2(\mathbf{a} \cdot \mathbf{b}) + 1 &\geq 0, && \text{because } \mathbf{a} \text{ and } \mathbf{b} \text{ are unit vectors} \\
\mathbf{a} \cdot \mathbf{b} &\geq -1. &&
\end{aligned}
$$

A similar argument beginning with $(\mathbf{a} - \mathbf{b}) \cdot (\mathbf{a} - \mathbf{b}) = \|\mathbf{a} - \mathbf{b}\|^2 \geq 0$ can be used to show that $\mathbf{a} \cdot \mathbf{b} \leq 1$. (You are asked to supply this argument in Exercise 8.) Hence, $-1 \leq \mathbf{a} \cdot \mathbf{b} \leq 1$. ■

Example 2

Let $\mathbf{x} = [-1, 4, 2, 0, -3]$ and let $\mathbf{y} = [2, 1, -4, -1, 0]$. We verify the Cauchy-Schwarz Inequality in this specific case. Now, $\mathbf{x} \cdot \mathbf{y} = -2 + 4 - 8 + 0 + 0 = -6$. Also, $\|\mathbf{x}\| = \sqrt{1 + 16 + 4 + 0 + 9} = \sqrt{30}$, and $\|\mathbf{y}\| = \sqrt{4 + 1 + 16 + 1 + 0} = \sqrt{22}$. Then, $|\mathbf{x} \cdot \mathbf{y}| \leq (\|\mathbf{x}\|)(\|\mathbf{y}\|)$, because $|-6| = 6 \leq \sqrt{(30)(22)} = 2\sqrt{165}$ (which is about 25). ■

Another useful result, sometimes known as Minkowski's Inequality, is stated in the next theorem:

THEOREM 1.6: Triangle Inequality

If $\mathbf{x}$ and $\mathbf{y}$ are vectors in $\mathbb{R}^n$, then $\|\mathbf{x} + \mathbf{y}\| \leq \|\mathbf{x}\| + \|\mathbf{y}\|$.

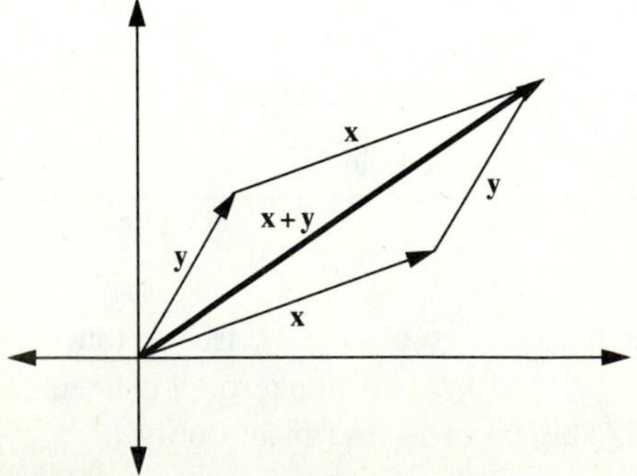

Figure 1.13
Triangle inequality in $\mathbb{R}^2$: $\|\mathbf{x} + \mathbf{y}\| \leq \|\mathbf{x}\| + \|\mathbf{y}\|$

It is easy to give a geometric proof of this theorem in $\mathbb{R}^2$ and $\mathbb{R}^3$, because the theorem simply asserts that the length of $\mathbf{x} + \mathbf{y}$, one side of the triangles in Figure 1.13, is never larger than the sum of the lengths of the other two sides, $\mathbf{x}$ and $\mathbf{y}$. The following algebraic proof extends this result to $\mathbb{R}^n$ for $n > 3$:

Proof of Theorem 1.6

It is enough to show that $\|\mathbf{x} + \mathbf{y}\|^2 \le (\|\mathbf{x}\| + \|\mathbf{y}\|)^2$. (Why?) But

$$\begin{aligned}
\|\mathbf{x} + \mathbf{y}\|^2 &= (\mathbf{x} + \mathbf{y}) \cdot (\mathbf{x} + \mathbf{y}) \\
&= (\mathbf{x} \cdot \mathbf{x}) + 2(\mathbf{x} \cdot \mathbf{y}) + (\mathbf{y} \cdot \mathbf{y}) \\
&= \|\mathbf{x}\|^2 + 2(\mathbf{x} \cdot \mathbf{y}) + \|\mathbf{y}\|^2 \\
&\le \|\mathbf{x}\|^2 + 2|\mathbf{x} \cdot \mathbf{y}| + \|\mathbf{y}\|^2 \\
&\le \|\mathbf{x}\|^2 + 2(\|\mathbf{x}\|)(\|\mathbf{y}\|) + \|\mathbf{y}\|^2 \qquad \text{by the Cauchy-Schwarz Inequality} \\
&= (\|\mathbf{x}\| + \|\mathbf{y}\|)^2.
\end{aligned}$$

■

The Angle between Two Vectors

The dot product enables us to find the angle θ between two nonzero vectors $\mathbf{x}$ and $\mathbf{y}$ in $\mathbb{R}^2$ or $\mathbb{R}^3$ that begin at the same initial point. There are actually *two* angles formed by the vectors $\mathbf{x}$ and $\mathbf{y}$. In the discussion that follows, we always choose the angle θ between two vectors to be the "smaller" of the two choices—that is, the angle measuring between 0 and π radians.

Consider the vector $\mathbf{x} - \mathbf{y}$ in Figure 1.14, which begins at the terminal point of $\mathbf{y}$ and ends at the terminal point of $\mathbf{x}$. Because $0 \le \theta \le \pi$, it follows from the Law of Cosines that $\|\mathbf{x} - \mathbf{y}\|^2 = \|\mathbf{x}\|^2 + \|\mathbf{y}\|^2 - 2(\|\mathbf{x}\|)(\|\mathbf{y}\|)\cos\theta$. But we also have

$$\begin{aligned}
\|\mathbf{x} - \mathbf{y}\|^2 &= (\mathbf{x} - \mathbf{y}) \cdot (\mathbf{x} - \mathbf{y}) \\
&= (\mathbf{x} \cdot \mathbf{x}) - 2(\mathbf{x} \cdot \mathbf{y}) + (\mathbf{y} \cdot \mathbf{y}) \\
&= \|\mathbf{x}\|^2 - 2(\mathbf{x} \cdot \mathbf{y}) + \|\mathbf{y}\|^2.
\end{aligned}$$

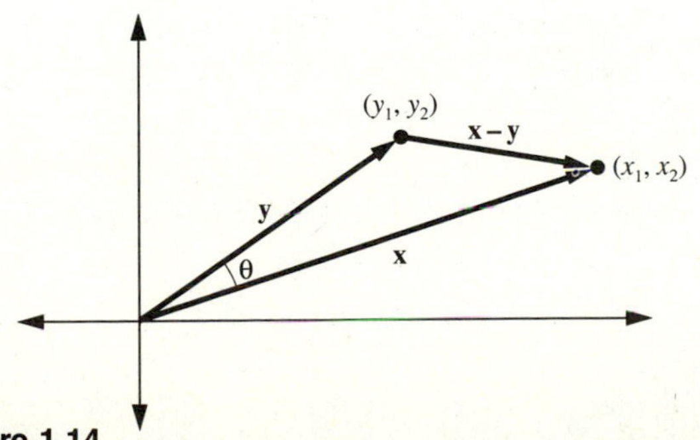

Figure 1.14
The angle θ between two nonzero vectors $\mathbf{x}$ and $\mathbf{y}$ in $\mathbb{R}^2$

Hence, $-2\|\mathbf{x}\|\,\|\mathbf{y}\|\cos\theta = -2(\mathbf{x}\cdot\mathbf{y})$, which implies $\|\mathbf{x}\|\,\|\mathbf{y}\|\cos\theta = \mathbf{x}\cdot\mathbf{y}$, and so

$$\cos\theta = \frac{\mathbf{x}\cdot\mathbf{y}}{(\|\mathbf{x}\|)(\|\mathbf{y}\|)}.$$

Example 3

Suppose $\mathbf{x} = [6, -4]$ and $\mathbf{y} = [-2, 3]$ and θ is the angle between $\mathbf{x}$ and $\mathbf{y}$. Then

$$\cos\theta = \frac{\mathbf{x}\cdot\mathbf{y}}{(\|\mathbf{x}\|)(\|\mathbf{y}\|)} = \frac{(6)(-2)+(-4)(3)}{\sqrt{52}\sqrt{13}} = -\frac{12}{13},$$

which is approximately -0.9231. Using a calculator, we find that θ is about 2.74 radians, or 157°. (Remember that θ is between 0 and π radians.) ■

The previous derivation for the formula for $\cos\theta$ works well in $\mathbb{R}^2$ and $\mathbb{R}^3$. But in higher-dimensional spaces, we are outside the geometry of everyday experience, and we have not yet defined the angle between two vectors in such cases. However, by the Cauchy-Schwarz Inequality, the term $(\mathbf{x}\cdot\mathbf{y})/\|\mathbf{x}\|\,\|\mathbf{y}\|$ always has a value between -1 and 1 for any nonzero vectors $\mathbf{x}$ and $\mathbf{y}$ in $\mathbb{R}^n$. Thus, there will always be a unique θ between 0 and π radians such that $\cos\theta$ equals this value. Therefore, we can define the angle between two vectors in $\mathbb{R}^n$ algebraically in a manner consistent with the situation in $\mathbb{R}^2$ and $\mathbb{R}^3$:

DEFINITION

Let $\mathbf{x}$ and $\mathbf{y}$ be two nonzero vectors in $\mathbb{R}^n$, for $n \geq 2$. Then the **angle between x and y** is the unique angle between 0 and π radians whose cosine is $(\mathbf{x}\cdot\mathbf{y})/(\|\mathbf{x}\|)(\|\mathbf{y}\|)$.

Example 4

For the vectors $\mathbf{x} = [-1, 4, 2, 0, -3]$ and $\mathbf{y} = [2, 1, -4, -1, 0]$, we can calculate that $(\mathbf{x}\cdot\mathbf{y})/(\|\mathbf{x}\|)(\|\mathbf{y}\|) = -6/(2\sqrt{165})$, which is approximately -0.234. Using a calculator, we find that the angle θ between $\mathbf{x}$ and $\mathbf{y}$ is approximately 1.8 radians, or 103.5°. ■

The next theorem is an immediate consequence of the last definition:

THEOREM 1.7

Let $\mathbf{x}$ and $\mathbf{y}$ be nonzero vectors in $\mathbb{R}^n$, and let θ be the angle between $\mathbf{x}$ and $\mathbf{y}$. Then

(1)	$\mathbf{x} \cdot \mathbf{y} > 0$	if and only if	$0 \leq \theta < \frac{\pi}{2}$	radians (0° or *acute*)
(2)	$\mathbf{x} \cdot \mathbf{y} = 0$	if and only if	$\theta = \frac{\pi}{2}$	radians (90°)
(3)	$\mathbf{x} \cdot \mathbf{y} < 0$	if and only if	$\frac{\pi}{2} < \theta \leq \pi$	radians (180° or *obtuse*)

Special Cases: Orthogonal and Parallel Vectors

You have just seen that the angle θ between two nonzero vectors $\mathbf{x}$ and $\mathbf{y}$ equals $\frac{\pi}{2}$ radians if and only if $\mathbf{x} \cdot \mathbf{y} = 0$. (Why?) Such vectors play an important role in linear algebra:

DEFINITION

Two vectors $\mathbf{x}$ and $\mathbf{y}$ in $\mathbb{R}^n$ are **orthogonal** (**perpendicular**) if and only if $\mathbf{x} \cdot \mathbf{y} = 0$.

Example 5

The vectors $\mathbf{x} = [2, -5]$ and $\mathbf{y} = [-10, -4]$ are orthogonal in $\mathbb{R}^2$ because $\mathbf{x} \cdot \mathbf{y} = 0$. Geometrically, this fact means that the vectors $\mathbf{x}$ and $\mathbf{y}$ form a right angle when given the same initial point, as illustrated in Figure 1.15.

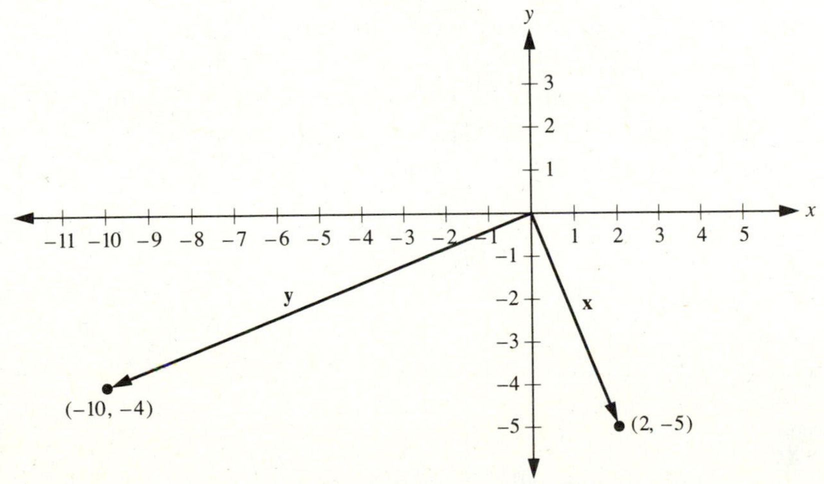

Figure 1.15
The orthogonal vectors $\mathbf{x} = [2, -5]$ and $\mathbf{y} = [-10, -4]$ ■

Notice that in $\mathbb{R}^2$ the special unit vectors $\mathbf{i}$ and $\mathbf{j}$ are orthogonal, because $\mathbf{i} \cdot \mathbf{j} = 0$. (Why?) In $\mathbb{R}^3$, the vectors $\mathbf{i}$, $\mathbf{j}$, and $\mathbf{k}$ are **mutually orthogonal**; that is, the dot product of any pair of these vectors equals zero. In general, in $\mathbb{R}^n$ the standard unit vectors $\mathbf{e}_1 = [1, 0, 0, \ldots, 0], \mathbf{e}_2 = [0, 1, 0, \ldots, 0], \ldots, \mathbf{e}_n = [0, 0, 0, \ldots, 1]$ form a mutually orthogonal set of vectors.

The next theorem gives an alternate way of describing parallel vectors in terms of the angle between them. A proof for the case $\mathbf{x} \cdot \mathbf{y} = +\|\mathbf{x}\| \, \|\mathbf{y}\|$ appears in Section 1.3 (see Result 4), and the proof of the other case is similar.

THEOREM 1.8

Let $\mathbf{x}$ and $\mathbf{y}$ be nonzero vectors in $\mathbb{R}^n$. Then $\mathbf{x}$ and $\mathbf{y}$ are parallel if and only if $\mathbf{x} \cdot \mathbf{y} = \pm\|\mathbf{x}\| \, \|\mathbf{y}\|$ (that is, $\cos\theta = \pm 1$, where θ is the angle between $\mathbf{x}$ and $\mathbf{y}$).

Example 6

Consider the vectors $\mathbf{x} = [8, -20, 4]$ and $\mathbf{y} = [6, -15, 3]$, which are parallel (from the definition of parallel vectors in Section 1.1). Now,

$$\frac{\mathbf{x} \cdot \mathbf{y}}{(\|\mathbf{x}\|)(\|\mathbf{y}\|)} = \frac{48 + 300 + 12}{\sqrt{480}\,\sqrt{270}} = \frac{360}{\sqrt{129600}} = 1.$$

Hence, if θ is the angle between $\mathbf{x}$ and $\mathbf{y}$, then $\cos\theta = 1$, showing that $\mathbf{x}$ and $\mathbf{y}$ are parallel according to Theorem 1.8 as well. ■

Projection Vectors

To conclude this section, we consider the projection of one vector onto another. This concept is especially useful in physics, engineering, computer graphics, and statistics. Suppose $\mathbf{a}$ and $\mathbf{b}$ are two nonzero vectors, both in $\mathbb{R}^2$ or both in $\mathbb{R}^3$, drawn at the same initial point. Let θ represent the angle between $\mathbf{a}$ and $\mathbf{b}$. Drop a perpendicular line segment from the terminal point of $\mathbf{b}$ to the straight line l containing the vector $\mathbf{a}$, as in Figure 1.16.

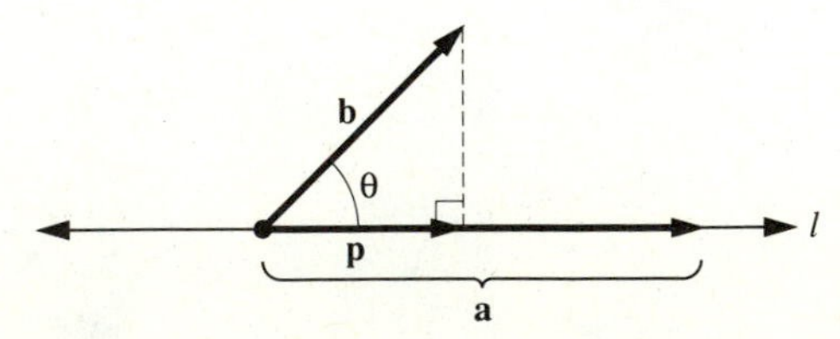

Figure 1.16
The projection **p** of the vector **b** onto **a** (when θ is acute)

By the projection $\mathbf{p}$ of $\mathbf{b}$ onto $\mathbf{a}$, we mean the vector from the initial point of $\mathbf{a}$ (or $\mathbf{b}$) to the point where the dropped perpendicular meets the line l. Note that $\mathbf{p}$ is in the same direction as $\mathbf{a}$ when $0 \le \theta < \pi/2$ radians (see Figure 1.16) and in the opposite direction to $\mathbf{a}$ when $\pi/2 < \theta \le \pi$ radians, as in Figure 1.17.

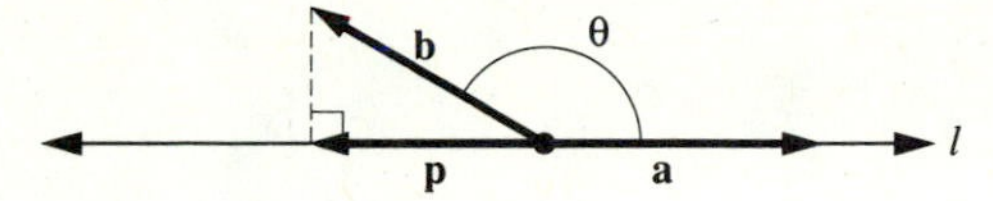

Figure 1.17
The projection **p** of **b** onto **a** (when θ is obtuse)

Using elementary trigonometry, we see that when $0 \leq \theta \leq \pi/2$, the vector **p** has length $\|\mathbf{b}\| \cos\theta$ and is in the direction of the unit vector $\mathbf{a}/\|\mathbf{a}\|$. Also, when $\pi/2 < \theta \leq \pi$, **p** has length $-\|\mathbf{b}\|\cos\theta$ and is in the direction of the unit vector $-\mathbf{a}/\|\mathbf{a}\|$. Therefore, we can express **p** in all cases as

$$\mathbf{p} = \left(\|\mathbf{b}\| \cos\theta\right)\left(\frac{\mathbf{a}}{\|\mathbf{a}\|}\right).$$

But we know that $\cos\theta = (\mathbf{a} \cdot \mathbf{b}) / \left(\|\mathbf{a}\|\right)\left(\|\mathbf{b}\|\right)$, and hence

$$\mathbf{p} = \left(\frac{\mathbf{a} \cdot \mathbf{b}}{\|\mathbf{a}\|^2}\right)\mathbf{a}.$$

The vector **p** is often denoted by $\mathbf{proj_a b}$, the projection of vector **b** onto **a**.

Example 7

Let $\mathbf{a} = [4, 0, -3]$ and $\mathbf{b} = [3, 1, -7]$. Then

$$\mathbf{proj_a b} = \mathbf{p} = \left(\frac{\mathbf{a} \cdot \mathbf{b}}{\|\mathbf{a}\|^2}\right)\mathbf{a} = \frac{(4)(3) + (0)(1) + (-3)(-7)}{\left(\sqrt{16+0+9}\right)^2}\mathbf{a} = \frac{33}{25}\mathbf{a}$$

$$= \frac{33}{25}[4, 0, -3] = \left[\frac{132}{25}, 0, -\frac{99}{25}\right]. \qquad \blacksquare$$

Next, we generalize the concept of projection vectors to $\mathbb{R}^n$ so that the definition in $\mathbb{R}^n$ will be consistent with the results we obtained in $\mathbb{R}^2$ and $\mathbb{R}^3$:

DEFINITION

If **a** and **b** are nonzero vectors in $\mathbb{R}^n$, then $\mathbf{proj_a b}$, the **projection vector of b onto a**, is the vector

$$\mathbf{p} = \left(\frac{\mathbf{a} \cdot \mathbf{b}}{\|\mathbf{a}\|^2}\right)\mathbf{a}.$$

One use of the projection vector is to separate ("decompose") a given vector **b** into the sum of two **component vectors**. Let **a** be a nonzero vector. Notice that if $\mathbf{proj_a b} \neq 0$, then by definition it is parallel to **a**, because it is a scalar multiple of **a**

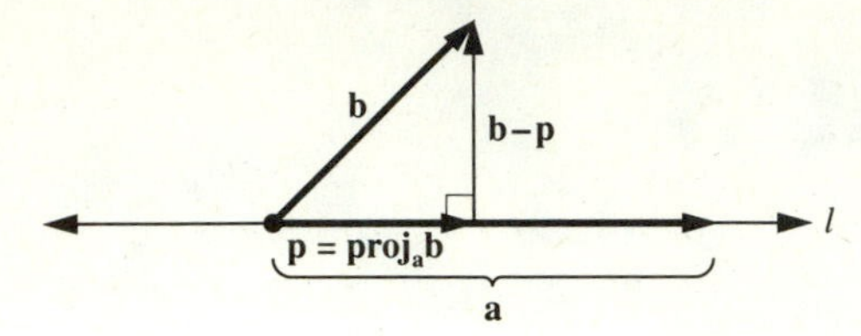

Figure 1.18
Decomposition of a vector **b** into two components, one parallel to **a** and the other orthogonal to **a**

(see Figure 1.18). Also, $\mathbf{b} - \mathbf{proj_a b}$ is orthogonal to **a**, because

$$\begin{aligned}(\mathbf{b} - \mathbf{proj_a b}) \cdot \mathbf{a} &= \mathbf{b} \cdot \mathbf{a} - (\mathbf{proj_a b}) \cdot \mathbf{a} \\ &= \mathbf{b} \cdot \mathbf{a} - \left(\frac{\mathbf{a} \cdot \mathbf{b}}{\|\mathbf{a}\|^2}\right)(\mathbf{a} \cdot \mathbf{a}) \\ &= \mathbf{b} \cdot \mathbf{a} - \left(\frac{\mathbf{a} \cdot \mathbf{b}}{\|\mathbf{a}\|^2}\right)\|\mathbf{a}\|^2 \\ &= 0.\end{aligned}$$

Because the sum of $\mathbf{proj_a b}$ and $\mathbf{b} - \mathbf{proj_a b}$ is obviously **b** itself, we have proved the following theorem:

THEOREM 1.9

> Let **a** be a nonzero vector in $\mathbb{R}^n$, and let **b** be any vector in $\mathbb{R}^n$. Then **b** can be decomposed as the sum of two component vectors, $\mathbf{proj_a b}$ and $\mathbf{b} - \mathbf{proj_a b}$, where the first (if nonzero) is parallel to **a** and the second is orthogonal to **a**.

Example 8

Consider the vectors $\mathbf{a} = [4, 0, -3]$ and $\mathbf{b} = [3, 1, -7]$ in Example 7. From that example, the component of **b** in the direction of the vector **a** is $\mathbf{proj_a b} = \mathbf{p} = [132/25, 0, -99/25]$. But, then, the component of **b** that is orthogonal to **a** (and **p** as well) is $\mathbf{b} - \mathbf{p} = [-57/25, 1, -76/25]$. We can easily check that $\mathbf{b} - \mathbf{p}$ is orthogonal to **a** as follows:

$$(\mathbf{b} - \mathbf{p}) \cdot \mathbf{a} = \left(-\frac{57}{25}\right)(4) + (1)(0) + \left(-\frac{76}{25}\right)(-3) = -\frac{228}{25} + \frac{228}{25} = 0. \quad \blacksquare$$

Exercises—Section 1.2

Note: Some of the exercises ask for proofs. If you have difficulty with these, you may find it helpful to try them again after working through Section 1.3, where proof techniques are discussed.

1. Use a calculator to find the angle θ (to the nearest degree) between the given vectors $\mathbf{x}$ and $\mathbf{y}$:

★(a) $\mathbf{x} = [-4, 3]$, $\mathbf{y} = [6, -1]$

(b) $\mathbf{x} = [0, -3, 2]$, $\mathbf{y} = [1, -7, -4]$

★(c) $\mathbf{x} = [7, -4, 2]$, $\mathbf{y} = [-6, -10, 1]$

(d) $\mathbf{x} = [-18, -4, -10, 2, -6]$, $\mathbf{y} = [9, 2, 5, -1, 3]$

2. Show that the points $A_1(9, 19, 16), A_2(11, 12, 13)$, and $A_3(14, 23, 10)$ are the vertices of a right triangle. (Hint: Construct the vectors between the points and check for an orthogonal pair.)

3. (a) Show that $[a, b]$ and $[-b, a]$ are orthogonal. Show that $[a, -b]$ and $[b, a]$ are orthogonal.

(b) Show that the lines given by the equations $ax + by + c = 0$ and $bx - ay + d = 0$ (where $a, b, c, d \in \mathbb{R}$) are perpendicular by finding a vector in the direction of each line and showing that these vectors are orthogonal. (Hint: Watch out for the case when a or b equals zero.)

★ **4.** Calculate (in joules) the total work done by a force of 26 newtons acting in the direction of the vector $-2\mathbf{i} + 4\mathbf{j} + 5\mathbf{k}$ on an object displaced a total of 10 m in the direction of the vector $-\mathbf{i} + 2\mathbf{j} + 2\mathbf{k}$.

5. Why isn't it true that if $\mathbf{x}, \mathbf{y}, \mathbf{z} \in \mathbb{R}^n$, then $\mathbf{x} \cdot (\mathbf{y} \cdot \mathbf{z}) = (\mathbf{x} \cdot \mathbf{y}) \cdot \mathbf{z}$?

6. Prove parts (1), (2), (3), (4), and (6) of Theorem 1.4.

★ **7.** Does the Cancellation Law of algebra hold for the dot product; that is, assuming that $\mathbf{z} \neq \mathbf{0}$, does $\mathbf{x} \cdot \mathbf{z} = \mathbf{y} \cdot \mathbf{z}$ always imply that $\mathbf{x} = \mathbf{y}$?

8. Finish the proof of Theorem 1.5 by showing that for unit vectors $\mathbf{a}$ and $\mathbf{b}$, $(\mathbf{a} - \mathbf{b}) \cdot (\mathbf{a} - \mathbf{b}) \geq 0$ implies $\mathbf{a} \cdot \mathbf{b} \leq 1$.

9. Prove that if $(\mathbf{x} + \mathbf{y}) \cdot (\mathbf{x} - \mathbf{y}) = 0$, then $\|\mathbf{x}\| = \|\mathbf{y}\|$. This proof shows that, if the diagonals of a parallelogram (in $\mathbb{R}^2$ or $\mathbb{R}^3$) are perpendicular, then the parallelogram is a rhombus.

10. Prove that, for any vectors $\mathbf{x}, \mathbf{y}$ in $\mathbb{R}^n$, $\frac{1}{2}\left(\|\mathbf{x} + \mathbf{y}\|^2 + \|\mathbf{x} - \mathbf{y}\|^2\right) = \|\mathbf{x}\|^2 + \|\mathbf{y}\|^2$. This equation is known as the **Parallelogram Identity**, because it asserts that the sum of the squares (of the lengths) of all four sides of a parallelogram equals the sum of the squares of the diagonals.

11. (a) Prove that, for vectors $\mathbf{x}$, $\mathbf{y}$ in $\mathbb{R}^n$, $\|\mathbf{x} + \mathbf{y}\|^2 = \|\mathbf{x}\|^2 + \|\mathbf{y}\|^2$ if and only if $\mathbf{x} \cdot \mathbf{y} = 0$.

(b) Prove that, if $\mathbf{x}, \mathbf{y}, \mathbf{z}$ are a mutually orthogonal set of vectors in $\mathbb{R}^n$, then $\|\mathbf{x} + \mathbf{y} + \mathbf{z}\|^2 = \|\mathbf{x}\|^2 + \|\mathbf{y}\|^2 + \|\mathbf{z}\|^2$.

(c) Prove that, if $\mathbf{x}$ and $\mathbf{y}$ are vectors in $\mathbb{R}^n$, then $\mathbf{x} \cdot \mathbf{y} = \frac{1}{4}\left(\|\mathbf{x} + \mathbf{y}\|^2 - \|\mathbf{x} - \mathbf{y}\|^2\right)$. This result, a form of the **Polarization Identity**, shows a way of defining the dot product of vectors using the norms of vectors.

12. Given **x**, **y**, **z** in $\mathbb{R}^n$, with **x** orthogonal to both **y** and **z**, prove that **x** is orthogonal to $c_1\mathbf{y} + c_2\mathbf{z}$, where $c_1, c_2 \in \mathbb{R}$.

★**13.** Let $\mathbf{x} = [a, b, c]$ be a vector in $\mathbb{R}^3$. If θ_1, θ_2, and θ_3 are the angles that **x** forms with the x-, y-, and z-axes, respectively, find formulas for $\cos\theta_1, \cos\theta_2$, and $\cos\theta_3$ in terms of a, b, c, and show that $\cos^2\theta_1 + \cos^2\theta_2 + \cos^2\theta_3 = 1$. (Note: $\cos\theta_1$, $\cos\theta_2$, and $\cos\theta_3$ are commonly known as the **direction cosines** of the vector **x**. See Exercise 14(b) in Section 1.1.)

★**14.** What is the length of the diagonal of a cube that has a side of length s? Using vectors, find the angle that the diagonal of a cube makes with one of the edges of the cube.

15. Calculate $\mathbf{proj_a b}$, the projection vector of **b** onto **a**, for each of the following, and verify that $\mathbf{b} - \mathbf{proj_a b}$ is orthogonal to **a**:

★(a) $\mathbf{a} = [2, 1, 5]$, $\mathbf{b} = [1, 4, -3]$
(b) $\mathbf{a} = [-5, 3, 0]$, $\mathbf{b} = [3, -7, 1]$
★(c) $\mathbf{a} = [1, 0, -1, 2]$, $\mathbf{b} = [3, -1, 0, -1]$

16. (a) Suppose that **a** is orthogonal to **b** in $\mathbb{R}^n$. What is $\mathbf{proj_a b}$? (Why?) Give a geometric interpretation in $\mathbb{R}^2$ or $\mathbb{R}^3$.
(b) Suppose **a** and **b** are parallel vectors in $\mathbb{R}^n$. What is $\mathbf{proj_a b}$? (Why?) Give a geometric interpretation in $\mathbb{R}^2$ or $\mathbb{R}^3$.

★**17.** What are the projections of the general vector $[a, b, c]$ onto each of the vectors **i**, **j**, and **k** in turn?

18. Let $\mathbf{x} = [-6, 2, 7]$ represent the force on an object in a three-dimensional coordinate system. Decompose **x** into two component forces in directions parallel and orthogonal to each vector given.

★(a) $[2, -3, 4]$ (b) $[-1, 2, -1]$ ★(c) $[3, -2, 6]$

19. Show that, if l is any line through the origin in $\mathbb{R}^3$ and **x** is any vector with its initial point at the origin, then the **reflection** of **x** through the line l (acting as a mirror) is equal to $2(\mathbf{proj_r x}) - \mathbf{x}$, where **r** is any nonzero vector parallel to the line l (see Figure 1.19).

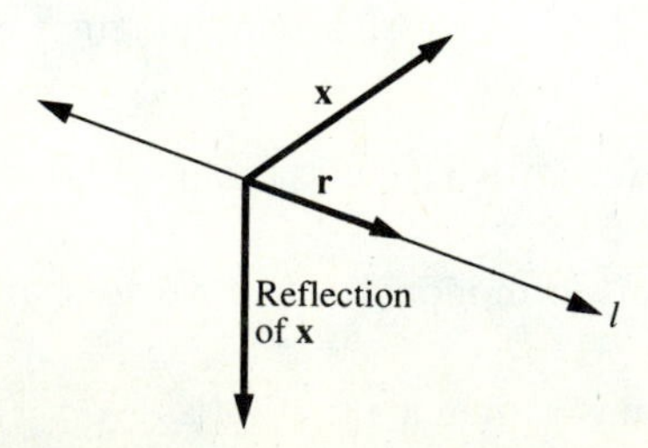

Figure 1.19
Reflection of **x** through the line l

20. Prove the **Reverse Triangle Inequality**; that is, for any vectors $\mathbf{x}$ and $\mathbf{y}$ in $\mathbb{R}^n$, we have $\big|\|\mathbf{x}\| - \|\mathbf{y}\|\big| \leq \|\mathbf{x}+\mathbf{y}\|$. (Hint: Consider the cases $\|\mathbf{x}\| \leq \|\mathbf{y}\|$ and $\|\mathbf{x}\| \geq \|\mathbf{y}\|$ separately.)

21. Let $\mathbf{x}$ and $\mathbf{y}$ be nonzero vectors in $\mathbb{R}^n$:
(a) Prove that $\mathbf{y} = c\mathbf{x} + \mathbf{w}$ for some scalar c and some vector $\mathbf{w}$ such that $\mathbf{w}$ is orthogonal to $\mathbf{x}$.
(b) Show that the vector $\mathbf{w}$ and the scalar c in part (a) are unique; that is, show that if $\mathbf{y} = c\mathbf{x} + \mathbf{w}$ and $\mathbf{y} = d\mathbf{x} + \mathbf{v}$, where $\mathbf{w}$ and $\mathbf{v}$ are both orthogonal to $\mathbf{x}$, then $c = d$ and $\mathbf{w} = \mathbf{v}$.

1.3 An Introduction to Proofs

Throughout this book, you will spend a great deal of time studying the statements and proofs of theorems, and in the exercises, you will frequently be asked to write proofs. Hence, in this section we discuss several methods of proving theorems in order to sharpen your skills in the reading and writing of proofs.

The "results" (not all new) proved in this section are intended only to illustrate various proof techniques. Therefore, they are not labeled as "theorems" and are not needed in the remainder of the book.

Proof Technique: Direct Proof

The most familiar method for proving theorems is known as **direct proof**, which is simply a straightforward argument of logical steps in support of the statement to be proved. As an example, we give a direct proof for the following familiar result:

RESULT 1

Let $\mathbf{x}$ be a vector in $\mathbb{R}^n$. Then $\mathbf{x} \cdot \mathbf{x} = \|\mathbf{x}\|^2$.

You may recognize this result as part of Theorem 1.4.

Proof of Result 1

Step 1:	Let $\mathbf{x} = [x_1, \ldots, x_n]$	because $\mathbf{x} \in \mathbb{R}^n$
Step 2:	$\mathbf{x} \cdot \mathbf{x} = x_1^2 + \cdots + x_n^2$	definition of dot product
Step 3:	$\|\mathbf{x}\| = \sqrt{x_1^2 + \cdots + x_n^2}$	definition of $\|\mathbf{x}\|$
Step 4:	$\|\mathbf{x}\|^2 = x_1^2 + \cdots + x_n^2$	squaring both sides of step 3
Step 5:	$\mathbf{x} \cdot \mathbf{x} = \|\mathbf{x}\|^2$	from steps 2 and 4 ■

Each step in a direct proof should follow immediately from a definition, a previous step, a known fact, or some combination of these. Also, the reasons behind each step should be clearly stated when the intended reader might not follow the argument. The preceding format clarifies the logical steps involved in a proof, but this type of presentation is infrequently used by mathematicians. A paragraph version of the same argument, which is more typical, would look something like this:

Proof of Result 1

If $\mathbf{x}$ is a vector in $\mathbb{R}^n$, then we can express $\mathbf{x}$ as $[x_1, \ldots, x_n]$ for some real numbers $x_1, \ldots, x_n$. Now, $\mathbf{x} \cdot \mathbf{x} = x_1^2 + \cdots + x_n^2$, by definition of the dot product. However, $\|\mathbf{x}\| = \sqrt{x_1^2 + \cdots + x_n^2}$, by definition of the length of a vector. Therefore, $\|\mathbf{x}\|^2 = \mathbf{x} \cdot \mathbf{x}$, because both sides are equal to $x_1^2 + \cdots + x_n^2$. ■

When the paragraph form is used, it should contain the same information as the step-by-step form and should be presented in such a way that a corresponding step-by-step formulation occurs naturally in the reader's mind. We present most proofs in this book in paragraph style. But you may want to begin writing your proofs in the step-by-step format and then change to paragraph style once you have more confidence and experience with proofs.

The preceding proof also illustrates that stating the definitions of the terms involved is usually a good beginning when tackling any proof. For example, the first four of the five steps in the step-by-step proof of Result 1 merely involve writing down what each side of the equation $\mathbf{x} \cdot \mathbf{x} = \|\mathbf{x}\|^2$ means. The final result is then obvious. Not all results can be proved this easily, but stating the definitions clarifies what you must prove. Then proving difficult statements is often easier.

Working "Backward" to Discover a Proof

A method often used in cases where there is no obvious direct proof is to work "backward"—that is, to start with the desired conclusion and to work in reverse toward the given facts. Although these "reversed" steps do not themselves constitute a proof, they may provide sufficient insight into the problem to make construction of a "forward" proof easier. We illustrate this technique with the following result:

RESULT 2

Let $\mathbf{x}$ and $\mathbf{y}$ be nonzero vectors in $\mathbb{R}^n$ with $\mathbf{x} \cdot \mathbf{y} \geq 0$. Then $\|\mathbf{x} + \mathbf{y}\| > \|\mathbf{y}\|$.

We begin with the desired conclusion $\|\mathbf{x} + \mathbf{y}\| > \|\mathbf{y}\|$ and try to work "backward" toward the given fact $\mathbf{x} \cdot \mathbf{y} \geq 0$, as follows:

$$\|\mathbf{x}+\mathbf{y}\| > \|\mathbf{y}\|$$
$$\|\mathbf{x}+\mathbf{y}\|^2 > \|\mathbf{y}\|^2$$
$$(\mathbf{x}+\mathbf{y})\cdot(\mathbf{x}+\mathbf{y}) > \|\mathbf{y}\|^2$$
$$\mathbf{x}\cdot\mathbf{x} + 2\mathbf{x}\cdot\mathbf{y} + \mathbf{y}\cdot\mathbf{y} > \|\mathbf{y}\|^2$$
$$\|\mathbf{x}\|^2 + 2\mathbf{x}\cdot\mathbf{y} + \|\mathbf{y}\|^2 > \|\mathbf{y}\|^2$$
$$\|\mathbf{x}\|^2 + 2\mathbf{x}\cdot\mathbf{y} > 0$$

At this point, we cannot easily go any further "backward." However, the last inequality can easily be proved in a "forward" manner, starting with $\mathbf{x}\cdot\mathbf{y} \geq 0$. Therefore, we can incorporate the above steps *in reverse order* in the following "forward" proof of Result 2:

Proof of Result 2

$\|\mathbf{x}\|^2 > 0$	$\mathbf{x}$ is nonzero
$2(\mathbf{x}\cdot\mathbf{y}) \geq 0$	because $\mathbf{x}\cdot\mathbf{y} \geq 0$
$\|\mathbf{x}\|^2 + 2\mathbf{x}\cdot\mathbf{y} > 0$	from above
$\|\mathbf{x}\|^2 + 2\mathbf{x}\cdot\mathbf{y} + \|\mathbf{y}\|^2 > \|\mathbf{y}\|^2$	
$\mathbf{x}\cdot\mathbf{x} + 2\mathbf{x}\cdot\mathbf{y} + \mathbf{y}\cdot\mathbf{y} > \|\mathbf{y}\|^2$	from Theorem 1.4
$(\mathbf{x}+\mathbf{y})\cdot(\mathbf{x}+\mathbf{y}) > \|\mathbf{y}\|^2$	from Theorem 1.4
$\|\mathbf{x}+\mathbf{y}\|^2 > \|\mathbf{y}\|^2$	from Theorem 1.4
$\|\mathbf{x}+\mathbf{y}\| > \|\mathbf{y}\|$	by taking the square root of both sides and using the fact that length is always nonnegative ■

When "working backward," keep in mind that your steps must be reversed when writing out the final, polished proof. Therefore, each step must be carefully examined to determine if it is "reversible." For example, if t is a real number, the following step is valid:

$$t > 5$$
$$\Longrightarrow t^2 > 25 \qquad \text{squaring both sides}$$

However, reversing this step yields

$$t^2 > 25$$
$$\Longrightarrow t > 5$$

which is clearly invalid if $t < -5$. Notice that we were very careful in the proof of Result 2 when we took the square root of both sides to make sure that the step was indeed valid.

"If *A* then *B*" Proofs

Frequently, a theorem will be given in the form "If A then B," where A and B represent statements. An example of this form is "If $\|\mathbf{x}\| = 0$, then $\mathbf{x} = \mathbf{0}$" for vectors $\mathbf{x}$ in $\mathbb{R}^n$, where A is the statement "$\|\mathbf{x}\| = 0$" and B is the statement "$\mathbf{x} = \mathbf{0}$." The entire "If A then B" statement is called an **implication**; A alone is referred to as the **premise**, and B is called the **conclusion**. The meaning of "If A then B" is that, whenever A is found to be true, B will be true as well. Returning to our example, the implication "If $\|\mathbf{x}\| = 0$, then $\mathbf{x} = \mathbf{0}$" means that, if we know that $\|\mathbf{x}\| = 0$ for some particular vector $\mathbf{x}$ in $\mathbb{R}^n$, then we can conclude that $\mathbf{x}$ must be the zero vector.

Note that the implication "If A then B" asserts nothing at all about the truth or falsity of B unless A is true. The case where A is false is irrelevant.† Therefore, to prove "If A then B" true, we assume that A is true and try to prove that B is also true.

The next result illustrates the proof of an implication:

RESULT 3

If $\mathbf{x}$ and $\mathbf{y}$ are nonzero vectors in $\mathbb{R}^n$ such that $\mathbf{x} \cdot \mathbf{y} > 0$, then the angle between $\mathbf{x}$ and $\mathbf{y}$ is acute.

You may recognize this result as part of Theorem 1.7.

Proof of Result 3

The premise in this result is "$\mathbf{x}$ and $\mathbf{y}$ are nonzero vectors and $\mathbf{x} \cdot \mathbf{y} > 0$." The conclusion is "the angle between $\mathbf{x}$ and $\mathbf{y}$ is acute." We begin by assuming that both parts of the premise are true:

Step 1:	$\mathbf{x}$ and $\mathbf{y}$ are nonzero	first part of premise
Step 2:	$\|\mathbf{x}\| > 0$ and $\|\mathbf{y}\| > 0$	Theorem 1.4, parts (2) and (3)
Step 3:	$\mathbf{x} \cdot \mathbf{y} > 0$	second part of premise
Step 4:	$\cos\theta = \dfrac{\mathbf{x} \cdot \mathbf{y}}{\|\mathbf{x}\|\,\|\mathbf{y}\|}$, where θ is the angle between $\mathbf{x}$ and $\mathbf{y}$	definition of the angle between two vectors

†In formal logic, when A is false, the implication "If A then B" is considered true but worthless, because it tells us absolutely nothing about B. For example, the implication "If every vector in $\mathbb{R}^3$ is a unit vector, then the inflation rate will be 8% next year" is considered true, because the premise "every vector in $\mathbb{R}^3$ is a unit vector" is clearly false. However, even though this implication is true, it is useless. It tells us nothing about next year's inflation rate, because an implication can only assert the conclusion when the premise is true. Next year's rate of inflation is free to take any value, including 8%.

Step 5: $\cos\theta > 0$	any quotient of positive real numbers is positive
Step 6: θ is acute	$0 \le \theta \le \pi$ by definition, and in this range, $\cos\theta$ is positive only if $0 < \theta < \frac{\pi}{2}$ ■

An implication is not always written in the standard form. Several other ways of expressing "If A then B" are summarized here:

Some Equivalent Forms for "If A then B"

A implies B	B if A
$A \Rightarrow B$	A is a sufficient condition for B
A only if B	B is a necessary condition for A

Another common practice is to place some of the conditions of the premise before the "If . . . then" structure. Usually this is done to create a context in which the implication is to be understood. For example, Result 3 might be rewritten as

> Let $\mathbf{x}$ and $\mathbf{y}$ be nonzero vectors in $\mathbb{R}^n$. If $\mathbf{x} \cdot \mathbf{y} > 0$, then the angle between $\mathbf{x}$ and $\mathbf{y}$ is acute.

The condition that "$\mathbf{x}$ and $\mathbf{y}$ are nonzero vectors in $\mathbb{R}^n$" sets the stage for the implication to come. Hence, this condition is really part of the premise and may be used as given information when proving the implication.

"*A* If and Only If *B*" Proofs

Some theorems have the form "A if and only if B." This is really a combination of two statements: "If A then B" and "If B then A." Both of these statements must be shown true to fully complete the proof of the original statement. In essence, you are showing that A and B are logically equivalent: the "If A then B" half means that whenever A is true, B must follow; the "If B then A" half means that whenever B is true, A must follow. Therefore, A is true exactly when B is true, and vice versa. Here is an example of an "If and only if" argument:

RESULT 4

> Let $\mathbf{x}$ and $\mathbf{y}$ be nonzero vectors in $\mathbb{R}^n$. Then $\mathbf{x} \cdot \mathbf{y} = \|\mathbf{x}\|\,\|\mathbf{y}\|$ if and only if $\mathbf{y}$ is a positive scalar multiple of $\mathbf{x}$.

This result is a special case of Theorem 1.8.

In an "If and only if" proof, it is usually a good idea to begin by stating the two halves of the "If and only if" statement, so that you have a clearer idea of what is given and what must be proved in each half. In the case of Result 4, the two things that must be proved are

(1) Suppose that $\mathbf{y} = c\mathbf{x}$ for some positive $c \in \mathbb{R}$. Prove that $\mathbf{x} \cdot \mathbf{y} = \|\mathbf{x}\|\,\|\mathbf{y}\|$.
(2) Suppose that $\mathbf{x} \cdot \mathbf{y} = \|\mathbf{x}\|\,\|\mathbf{y}\|$. Prove that there is a positive real number c such that $\mathbf{y} = c\mathbf{x}$.

Proof of Result 4

Part (1): We suppose that $\mathbf{y} = c\mathbf{x}$ for some positive real number c. Then,

$$\begin{aligned}
\mathbf{x} \cdot \mathbf{y} &= \mathbf{x} \cdot (c\mathbf{x}) && \text{because } \mathbf{y} = c\mathbf{x} \\
&= c(\mathbf{x} \cdot \mathbf{x}) && \text{Theorem 1.4, part (4)} \\
&= c\|\mathbf{x}\|^2 && \text{Theorem 1.4, part (2)} \\
&= \|\mathbf{x}\|(c\|\mathbf{x}\|) && \text{associative law of multiplication of real numbers} \\
&= \|\mathbf{x}\|(|c|\,\|\mathbf{x}\|) && \text{because } c > 0 \\
&= \|\mathbf{x}\|\,\|c\mathbf{x}\| && \text{Theorem 1.1} \\
&= \|\mathbf{x}\|\,\|\mathbf{y}\|. && \text{because } \mathbf{y} = c\mathbf{x}
\end{aligned}$$

Part (2): Now we assume that $\mathbf{x} \cdot \mathbf{y} = \|\mathbf{x}\|\,\|\mathbf{y}\|$ and show that there is a positive real number c such that $\mathbf{y} = c\mathbf{x}$. We know by Theorem 1.9 that $\mathbf{y}$ can be expressed as $\mathbf{proj}_{\mathbf{x}}\mathbf{y} + \mathbf{w}$, where $\mathbf{w}$ is orthogonal to $\mathbf{x}$. Our strategy is to show first that $\mathbf{proj}_{\mathbf{x}}\mathbf{y}$ is a scalar multiple, $c\mathbf{x}$, of $\mathbf{x}$ with $c > 0$ and then to show that $\mathbf{w} = \mathbf{0}$. Then we will have $\mathbf{y} = c\mathbf{x}$ with $c > 0$, and the proof is done.

First, we show that $c > 0$. Now,

$$\begin{aligned}
\mathbf{proj}_{\mathbf{x}}\mathbf{y} &= \left(\frac{\mathbf{x} \cdot \mathbf{y}}{\|\mathbf{x}\|^2}\right)\mathbf{x} && \text{formula for } \mathbf{proj}_{\mathbf{x}}\mathbf{y} \\
&= \left(\frac{\|\mathbf{x}\|\,\|\mathbf{y}\|}{\|\mathbf{x}\|^2}\right)\mathbf{x} && \text{because } \mathbf{x} \cdot \mathbf{y} = \|\mathbf{x}\|\,\|\mathbf{y}\| \\
&= \left(\frac{\|\mathbf{y}\|}{\|\mathbf{x}\|}\right)\mathbf{x}.
\end{aligned}$$

Letting $c = \|\mathbf{y}\|/\|\mathbf{x}\|$, we see that c is clearly positive.

Finally, we finish the proof by showing that $\mathbf{w} = \mathbf{0}$. Now,

$$\begin{aligned}
\|\mathbf{w}\|^2 &= \mathbf{w} \cdot \mathbf{w} && \text{Theorem 1.4, part (2)} \\
&= (\mathbf{y} - c\mathbf{x}) \cdot (\mathbf{y} - c\mathbf{x}) && \text{because } \mathbf{y} = c\mathbf{x} + \mathbf{w} \\
&= (\mathbf{y} \cdot \mathbf{y}) - 2c(\mathbf{x} \cdot \mathbf{y}) + c^2(\mathbf{x} \cdot \mathbf{x}) && \text{distributive law of dot product over addition}
\end{aligned}$$

$$= \|\mathbf{y}\|^2 - 2c\|\mathbf{x}\|\,\|\mathbf{y}\| + c^2\|\mathbf{x}\|^2 \qquad \text{Theorem 1.4, part (2)}$$

$$= \|\mathbf{y}\|^2 - 2\|\mathbf{y}\|^2 + \|\mathbf{y}\|^2, \qquad \text{because } c = \frac{\|\mathbf{y}\|}{\|\mathbf{x}\|}$$

which equals zero, and so $\mathbf{w} = \mathbf{0}$. The proof is complete. ■

Note that two proofs are required to prove an "If and only if" type of statement: one for each of the implications involved. Also, each half is not necessarily just a reversal of the steps in the other half. Sometimes the two parts must be proved very differently, as for Result 4.

There are several other frequently used alternate forms for "If and only if":

Some Equivalent Forms for "*A* If and Only If *B*"

A iff *B*

$A \Longleftrightarrow B$

If *A* then *B*, and if *B* then *A*

A is a necessary and sufficient condition for *B*

B is a necessary and sufficient condition for *A*

"If *A*, then *B* or *C*" Proofs

Sometimes we are asked to prove statements of the form "If *A*, then *B* or *C*."† This is an implication whose conclusion has two parts. Note that there are only two possibilities for the first part of the conclusion, *B*: it is either true or false. In the case where *B* is true, the proof is finished, because we were asked to check that either *B* or *C* holds, and *B* does hold. Therefore, we are allowed to make the extra assumption that *B* is false to assist in the proof that *C* is true. That is, proving "If *A*, then *B* or *C*" is equivalent to proving "If *A* is true and *B* is false, then *C*."

For example, suppose we are asked to prove the following result:

RESULT 5

If $\mathbf{x}$ is a nonzero vector in $\mathbb{R}^2$, then $\mathbf{x} \cdot [1, 0] \neq 0$ or $\mathbf{x} \cdot [0, 1] \neq 0$.

†When we use *or* in this text, it is in the **inclusive** sense. That is, a statement such as "*n* is even or prime" means that *n* could be an even number or *n* could be a prime number or it could be both. Therefore, the statement "*n* is even or prime" is true for $n = 2$, which is both even and prime, as well as for $n = 6$ (even but not prime) and $n = 7$ (prime but not even). However, in English the word *or* is frequently used in the **exclusive** sense, as in "You may have the prize behind the curtain *or* the cash in my hand." Clearly, in this case you are not meant to have both prizes. The "exclusive or" is rarely used in mathematics, and whenever a phrase of the form "*A* or *B*" appears in this text, it means "*A* or *B* or both."

In this case, A = "$\mathbf{x}$ is a nonzero vector in $\mathbb{R}^2$," B = "$\mathbf{x} \cdot [1, 0] \neq 0$," and C = "$\mathbf{x} \cdot [0, 1] \neq 0$." Therefore, we can prove Result 5 by proving the equivalent statement instead:

If $\mathbf{x}$ is a nonzero vector in $\mathbb{R}^n$ and $\mathbf{x} \cdot [1, 0] = \mathbf{0}$, then $\mathbf{x} \cdot [0, 1] \neq \mathbf{0}$.

This is easy to do with a direct proof. (Try it!)

Finally, note that an alternate way of attacking the proof of "If A, then B or C" is to assume instead that C is false and use this assumption to prove that B is true.

Proof Technique: Proof by Contrapositive

Related to the implication "If A then B" is its **contrapositive**: "If not B, then not A." For example, for an integer n, the statement "If n^2 is even, then n is even" has the contrapositive "If n is odd (that is, not even), then n^2 is odd." A statement and its contrapositive are always logically equivalent; that is, they are either both true or both false together. Therefore, proving the contrapositive of any statement (known as **proof by contrapositive**) has the effect of proving the original statement as well. In many cases, the contrapositive is easier to prove. The next result illustrates this method:

RESULT 6

Let $\mathbf{x}$ be a vector in $\mathbb{R}^n$. If $\|\mathbf{x}\| = 0$, then $\mathbf{x} = \mathbf{0}$.

Proof of Result 6

To prove this result, we give a direct proof of its contrapositive: if $\mathbf{x} \neq \mathbf{0}$, then $\|\mathbf{x}\| \neq \mathbf{0}$.

Step 1:	Let $\mathbf{x} = [x_1, \ldots, x_n]$	because $\mathbf{x} \in \mathbb{R}^n$
Step 2:	For some i, $1 \le i \le n$, we have $x_i \neq 0$	because $\mathbf{x} \neq \mathbf{0}$ from premise of contrapositive
Step 3:	$\|\mathbf{x}\| = \sqrt{x_1^2 + \cdots + x_i^2 + \cdots + x_n^2}$	
Step 4:	$\|\mathbf{x}\|^2 = x_1^2 + \cdots + x_i^2 + \cdots + x_n^2$	
Step 5:	$x_i^2 > 0$ and $x_j^2 \ge 0$ for $1 \le j \le n$	
Step 6:	$\|\mathbf{x}\|^2 > 0$	
Step 7:	$\|\mathbf{x}\| \neq 0$	

Fill in the missing reasons for steps 3 through 7 to complete the proof of the contrapositive and hence the proof of the result itself. ■

Converse and Inverse

You have just seen that the contrapositive of "If A then B" is "If not B, then not A." Two other related statements of interest are "If B then A," which is the **converse** of "If A then B," and "If not A, then not B," which is the **inverse** of "If A then B."

Notice that, when a statement "If A then B" and its converse "If B then A" are taken together, they form the familiar "If and only if" type of statement.

Although the converse and inverse may resemble the contrapositive, take care: neither the converse nor the inverse is logically equivalent to the original statement. However, the converse and inverse of a statement are equivalent to each other, and these two are both true or both false together.

For example, consider the implication "If $\mathbf{x} = \mathbf{y}$, then $\mathbf{x} \cdot \mathbf{y} = \|\mathbf{x}\|^2$," regarding vectors in $\mathbb{R}^n$. We then have

If $\mathbf{x} = \mathbf{y}$, then $\mathbf{x} \cdot \mathbf{y} = \|\mathbf{x}\|^2$	Original statement	equivalent to each other
If $\mathbf{x} \cdot \mathbf{y} \neq \|\mathbf{x}\|^2$, then $\mathbf{x} \neq \mathbf{y}$	Contrapositive	
If $\mathbf{x} \cdot \mathbf{y} = \|\mathbf{x}\|^2$, then $\mathbf{x} = \mathbf{y}$	Converse	equivalent to each other
If $\mathbf{x} \neq \mathbf{y}$, then $\mathbf{x} \cdot \mathbf{y} \neq \|\mathbf{x}\|^2$	Inverse	

Notice that in this case the original statement and its contrapositive are both true; the converse and the inverse are both false (see Exercise 5).

Beware! It is also possible for a statement and its converse to have the same truth value: they may be both true together or both false together. For example, the converse of Result 6 is "If $\mathbf{x} = \mathbf{0}$, then $\|\mathbf{x}\| = 0$," and this is also a true statement. The moral here is that, because a statement and its converse are logically independent of each other, proving the converse (or inverse) is never acceptable as a valid proof of the original statement.

Proof Technique: Proof by Contradiction

Another common method of proof is **proof by contradiction**. In this technique, we assume that the statement to be proved is false, and we use this assumption to contradict a known fact. This proof technique relies on the consistency of mathematics. In effect, we prove a result by showing that if it were false, it would be inconsistent with some other true statement. We prove the next result using this technique. (Recall that a set $\{\mathbf{x}_1, \ldots, \mathbf{x}_k\}$ of nonzero vectors is mutually orthogonal if and only if $\mathbf{x}_i \cdot \mathbf{x}_j = 0$ whenever $i \neq j$.)

RESULT 7

Let $S = \{\mathbf{x}_1, \ldots, \mathbf{x}_k\}$ be a set of mutually orthogonal nonzero vectors in $\mathbb{R}^n$. Then no vector in S can be expressed as a linear combination of the other vectors in S.

Proof of Result 7

To prove this result by contradiction, we assume that it is false; that is, *some* vector in S, say $\mathbf{x}_i$, *can* be expressed as a linear combination of the other vectors in S. We then show that this assumption leads to a contradiction.

Our assumption is $\mathbf{x}_i = a_1\mathbf{x}_1 + \cdots + a_{i-1}\mathbf{x}_{i-1} + a_{i+1}\mathbf{x}_{i+1} + \cdots + a_k\mathbf{x}_k$, for some $a_1, \ldots, a_{i-1}, a_{i+1}, \ldots, a_k \in \mathbb{R}$. Then

$$\begin{aligned}\mathbf{x}_i \bullet \mathbf{x}_i &= \mathbf{x}_i \bullet (a_1\mathbf{x}_1 + \cdots + a_{i-1}\mathbf{x}_{i-1} + a_{i+1}\mathbf{x}_{i+1} + \cdots + a_k\mathbf{x}_k)\\ &= a_1(\mathbf{x}_i \bullet \mathbf{x}_1) + \cdots + a_{i-1}(\mathbf{x}_i \bullet \mathbf{x}_{i-1}) + a_{i+1}(\mathbf{x}_i \bullet \mathbf{x}_{i+1}) + \cdots + a_k(\mathbf{x}_i \bullet \mathbf{x}_k)\\ &= a_1(0) + \cdots + a_{i-1}(0) + a_{i+1}(0) + \cdots + a_k(0)\\ &= 0.\end{aligned}$$

Hence, $\mathbf{x}_i = \mathbf{0}$, by part (3) of Theorem 1.4. This equation contradicts the given fact that $\mathbf{x}_1, \ldots, \mathbf{x}_k$ are all nonzero vectors, thus completing the proof. ■

A mathematician generally constructs a proof by contradiction by assuming that the given statement is false and then experimenting—seeing where this assumption leads—until some absurdity creeps up. Of course, any "blind alleys" encountered in the experimenting process should not appear in the final draft of the proof.

Proof Technique: Proof by Induction

The method of **proof by induction** is used to show that a statement involving some integer variable is true for all values of that variable greater than or equal to some initial value i. For example, A = "For every integer $n \geq 1$, $1^2 + 2^2 + \cdots + n^2 = n(n+1)(2n+1)/6$" is a statement of this type. It can be proved by induction for each integer n greater than or equal to the initial value $i = 1$. You may have seen this proof in your calculus course.

There are two steps in any induction proof, the **base step** and the **inductive step**:

> **(1) Base step:** Prove that the desired statement is true for the initial value i of the (integer) variable.
>
> **(2) Inductive step:** Prove that *if* the statement is true for an integer value k of the variable (with $k \geq i$), *then* the statement must be true for the next integer value $k + 1$ as well.

These two steps together show that the statement to be proved is true for every integer greater than or equal to the initial value i, because the inductive step sets up a "chain of implications," as illustrated in Figure 1.20. First, the base step implies that the initial statement, A_i, is true. But A_i is the premise for the first implication in the chain. Because this premise is true, the inductive step tells us that the conclusion of this implication, A_{i+1}, must also be true. However, A_{i+1} is the premise of the second implication; hence the inductive step tells us that A_{i+2}, the conclusion of that implication, must be true. Following the chain, we see that the statement is true for each integer value $\geq i$.

A_i	$\Rightarrow$	A_{i+1}	$\Rightarrow$	A_{i+2}	$\Rightarrow$	A_{i+3}	$\Rightarrow$	$\cdots$
Statement at initial value i		Statement when variable equals $i+1$		Statement when variable equals $i+2$		Statement when variable equals $i+3$		

Figure 1.20
Chain of implications set up by the inductive step

The process of induction can be likened to knocking down a line of dominoes—one domino for each integer greater than or equal to the initial value. Keep in mind that the base step is needed to knock over the first domino and thus start the entire process. Without the base step, you cannot be sure that the given statement is true for any integer value whatsoever.

We illustrate the induction technique by proving the following result:

RESULT 8

Let $r \neq 1$ be a given real number, and let n be an integer with $n \geq 0$. Then,

$$1 + r + r^2 + \cdots + r^n = \frac{1 - r^{n+1}}{1 - r}.$$

Proof of Result 8

The integer induction variable is n, with initial value $n = 0$.

Base step: The base step of an induction proof is typically proved by plugging in the initial value for the integer variable and verifying that the result is true in that case. When $n = 0$, the left-hand side of the equation in the result has only one term, namely, 1. The right-hand side yields

$$\frac{1 - r^{0+1}}{1 - r} = \frac{1 - r}{1 - r} = 1.$$

Because both sides of the equation are equal when $n = 0$, we have completed the base step.

Inductive step: The inductive step requires us to prove the following:

Let $k \geq 0$. If $1 + r + r^2 + \cdots + r^k = (1 - r^{k+1})/(1 - r)$, then $1 + r + r^2 + \cdots + r^k + r^{k+1} = (1 - r^{(k+1)+1})/(1 - r)$.

Therefore, we may assume that the premise

$$1 + r + r^2 + \cdots + r^k = \frac{1 - r^{k+1}}{1 - r}$$

is true and use it to prove the conclusion

$$1 + r + r^2 + \cdots + r^k + r^{k+1} = \frac{1 - r^{k+2}}{1 - r}.$$

We proceed as follows:

$$\begin{aligned}
1 + r + r^2 + \cdots + r^k + r^{k+1} &= \left(1 + r + \cdots + r^k\right) + r^{k+1} \\
&= \frac{1 - r^{k+1}}{1 - r} + r^{k+1} \qquad \text{by the premise} \\
&= \frac{1 - r^{k+1}}{1 - r} + r^{k+1}\left(\frac{1 - r}{1 - r}\right) \\
&= \frac{1 - r^{k+1}}{1 - r} + \frac{r^{k+1} - r^{k+2}}{1 - r} \\
&= \frac{1 - r^{k+2}}{1 - r}.
\end{aligned}$$

Thus, we prove the conclusion and complete the inductive step. Because we have completed both parts of the induction proof, the proof is finished. ■

The proof of Result 8 illustrates the role of both parts of a proof by induction. Note that in the inductive step we are proving an implication, and so we get the powerful advantage of being allowed to assume true the premise of that implication. This premise is called the **induction hypothesis**. In Result 8, the induction hypothesis is

$$1 + r + r^2 + \cdots + r^k = \frac{1 - r^{k+1}}{1 - r}.$$

It allows us to make the crucial substitution for $1 + r + r^2 + \cdots + r^k$ in the inductive step. A successful proof by induction ultimately depends on assuming the induction hypothesis to reach the final conclusion.

Negating Statements with Quantifiers and Connectives

When considering some statement "A," we are frequently interested in its **negation**, or logical opposite, "not A." For example, negating a statement is used in constructing the contrapositive of an implication, as well as in a proof by contradiction. Of course, "not A" should be true precisely when A is false, and "not A" should be false exactly when A is true. That is, A and "not A" must always have opposite truth values. Finding the negation of a simple statement is usually easy. However, when the statement involves **quantifiers** (such as *all*, *every*, *some*, *no*, or *none*) or involves **connectives** (such as *and* or *or*), the negation process can be tricky.

We first discuss negating statements with quantifiers. As an example, suppose S represents some set of vectors in $\mathbb{R}^3$ and A = "All vectors in S are unit vectors." The correct negation of A is "not A" = "Some vector in S is not a unit vector." These statements have opposite truth values in all cases. Students frequently make the error of giving B = "No vector in S is a unit vector" as the negation of A. This answer is incorrect, because if S happened to contain some unit vectors as well as some nonunit vectors, then both A and B would be false. Hence, A and B do not have opposite truth values in all cases.

Next consider C = "There is a real number c such that $\mathbf{y} = c\mathbf{x}$," referring to specific vectors $\mathbf{x}$ and $\mathbf{y}$. Then "not C" = "No real number c exists such that $\mathbf{y} = c\mathbf{x}$." Alternately, "not C" = "For every real number c, $\mathbf{y} \neq c\mathbf{x}$."

There are two types of quantifiers: **universal quantifiers** (such as *every*, *all*, *no*, and *none*), which say that a statement is true or false in every instance, and **existential quantifiers** (such as *some* and *there exists*), which claim that there is at least one instance in which the condition in the statement is satisfied. The statements "A" and "not C" in the preceding examples involve universal quantifiers; "not A" and "C" use existential quantifiers. These examples follow a general pattern:

Rules for Negating Statements with Quantifiers

The negation of a statement involving a universal quantifier uses an existential quantifier.

The negation of a statement involving an existential quantifier uses a universal quantifier.

Hence, negating a statement changes the type of quantifier used.

Next, consider the negation of statements with connectives. Sometimes it is necessary to negate a statement containing two clauses that are connected by the word *and* or *or*. The formal rules for negating such statements are known as **DeMorgan's Laws**:

Rules for Negating Statements with Connectives (DeMorgan's Laws)

The negation of "A or B" is "(not A) and (not B)."

The negation of "A and B" is "(not A) or (not B)."

Note that when negating, *or* is converted to *and* and vice versa.

Figure 1.21 illustrates the laws for quantifiers and connectives. In the figure, S refers to a set of vectors in $\mathbb{R}^3$, and n represents a positive integer. Some of the

Original Statement	Negation of the Statement
n is an even number or a prime.	n is odd and not prime.
$\mathbf{x}$ is a unit vector and $\mathbf{x} \in S$.	$\|\mathbf{x}\| \neq 1$ or $\mathbf{x} \notin S$.
Some prime numbers are odd.	Every prime number is even.
There is a unit vector in S.	No elements of S are unit vectors.
There is a vector $\mathbf{x}$ in S with $\mathbf{x} \cdot [1, 1, -1] = 0$.	For every vector $\mathbf{x}$ in S, $\mathbf{x} \cdot [1, 1, -1] \neq 0$.
All numbers divisible by 4 are even.	Some number divisible by 4 is odd.
Every vector in S is either a unit vector or is parallel to $[1, -2, 1]$.	There is a non-unit vector in S that is not parallel to $[1, -2, 1]$.
For every nonzero vector $\mathbf{x}$ in $\mathbb{R}^3$, there is a vector in S that is parallel to $\mathbf{x}$.	There is a nonzero vector $\mathbf{x}$ in $\mathbb{R}^3$ that is not parallel to any vector in S.
There is a real number K such that for every $\mathbf{x} \in S$, $\|\mathbf{x}\| \leq K$.	For every real number K there is a vector $\mathbf{x} \in S$ such that $\|\mathbf{x}\| > K$.

Figure 1.21
Several statements and their negations

statements are true and others are false, but each statement has the opposite truth value of its negation.

Disproving Statements

Frequently we are required to prove that a given statement is false rather than true. Essentially, to disprove a statement A, we must instead prove the statement "not A." There are two important cases.

Case 1, statements involving universal quantifiers: A statement A with a universal quantifier can be disproved by finding a single **counterexample** that makes A false. For example, consider B = "For all $\mathbf{x}$ and $\mathbf{y}$ in $\mathbb{R}^3$, $\|\mathbf{x}+\mathbf{y}\| = \|\mathbf{x}\|+\|\mathbf{y}\|$," which involves a universal quantifier. We disprove B by finding a counterexample—that is, by finding a specific case where B is false. Letting $\mathbf{x} = [3, 0, 0]$ and $\mathbf{y} = [0, 0, 4]$, we get $\|\mathbf{x} + \mathbf{y}\| = \|[3, 0, 4]\| = \sqrt{9 + 16} = \sqrt{25} = 5$. However, $\|\mathbf{x}\| = 3$ and $\|\mathbf{y}\| = 4$. Clearly, in this case $\|\mathbf{x} + \mathbf{y}\| \neq \|\mathbf{x}\| + \|\mathbf{y}\|$, thus giving us a counterexample and disproving B.

Sometimes we are asked to disprove an implication "If A then B." This implication is equivalent to a statement involving a universal quantifier, because it asserts "In all cases in which A is true, B is also true." Therefore,

> Disproving "If A then B" entails the construction of a counterexample for which A is true but B is false.

To illustrate, consider C = "If $\mathbf{x}$ and $\mathbf{y}$ are unit vectors in $\mathbb{R}^4$, then $\mathbf{x} \cdot \mathbf{y} = 1$." To disprove C, we must find a counterexample in which the premise "$\mathbf{x}$ and $\mathbf{y}$ are unit vectors in $\mathbb{R}^4$" is true and the conclusion "$\mathbf{x} \cdot \mathbf{y} = 1$" is false. Hence, we look for two unit vectors $\mathbf{x}$ and $\mathbf{y}$ in $\mathbb{R}^4$ such that $\mathbf{x} \cdot \mathbf{y} \neq 1$. Consider $\mathbf{x} = [1, 0, 0, 0]$ and $\mathbf{y} = [0, 1, 0, 0]$, which are clearly unit vectors in $\mathbb{R}^4$. Then $\mathbf{x} \cdot \mathbf{y} = 0$, which does not equal 1. This counterexample disproves C.

Case 2, statements involving existential quantifiers: Recall that the negation of a statement involving an existential quantifier uses a universal quantifier. For example, consider D = "There is a nonzero vector $\mathbf{x}$ in $\mathbb{R}^2$ such that $\mathbf{x} \cdot [1, 0] = 0$ and $\mathbf{x} \cdot [0, 1] = 0$." To disprove D, we must prove "not D" = "For every nonzero vector $\mathbf{x}$ in $\mathbb{R}^2$, either $\mathbf{x} \cdot [1, 0] \neq 0$ or $\mathbf{x} \cdot [0, 1] \neq 0$."† We cannot prove this statement by giving a single example, because we must show that "not D" is true for every nonzero vector in $\mathbb{R}^2$. A formal proof of "not D" is needed. ("Not D" is the same statement as Result 5. You were asked to supply this proof earlier in this section.)

The moral here is that we cannot disprove a statement that involves an existential quantifier by supplying a counterexample. Instead, a detailed proof of the negation must be given.

Exercises—Section 1.3

1. (a) Give a direct proof that, if $\mathbf{x}$ and $\mathbf{y}$ are vectors in $\mathbb{R}^n$, then $\|4\mathbf{x} + 7\mathbf{y}\| \leq 7(\|\mathbf{x}\| + \|\mathbf{y}\|)$.

★(b) Can you generalize your proof in part (a) to draw any conclusions about $\|c\mathbf{x} + d\mathbf{y}\|$, where $c, d \in \mathbb{R}$? What about $\|c\mathbf{x} - d\mathbf{y}\|$?

2. (a) Give a direct proof that every integer of the form $6j - 5$ is also of the form $3k + 1$, where j and k are integers.

★(b) Find a counterexample to show that the converse of part (a) is not true—that is, that not every integer of the form $3k + 1$ is of the form $6j - 5$.

3. Let $\mathbf{x}$ and $\mathbf{y}$ be nonzero vectors in $\mathbb{R}^n$. Prove that $\mathbf{proj_x y} = \mathbf{0}$ if and only if $\mathbf{proj_y x} = \mathbf{0}$.

4. Let $\mathbf{x}$ and $\mathbf{y}$ be nonzero vectors in $\mathbb{R}^n$. Prove that $\|\mathbf{x} + \mathbf{y}\| = \|\mathbf{x}\| + \|\mathbf{y}\|$ if and only if $\mathbf{y} = c\mathbf{x}$ for some $c > 0$. (Hint: Be sure to prove both parts of this statement. Using Result 4 may make one half of the proof easier.)

★ **5.** Consider the statement A = "If $\mathbf{x} \cdot \mathbf{y} = \|\mathbf{x}\|^2$, then $\mathbf{x} = \mathbf{y}$."

(a) Show that A is false by exhibiting a counterexample.

(b) State the contrapositive of A.

(c) Can your counterexample from part (a) also serve to show that the contrapositive from part (b) is false?

†Notice that, in addition to the quantifier being changed, the *and* connective changes to *or*.

6. Prove the following statements. Each has the form "If A, then B or C."
(a) If $\|\mathbf{x} + \mathbf{y}\| = \|\mathbf{x}\|$, then $\mathbf{y} = \mathbf{0}$ or $\mathbf{x}$ is not orthogonal to $\mathbf{y}$.
(b) If $\mathbf{proj_x y} = \mathbf{x}$, then either $\mathbf{x}$ is a unit vector or $\mathbf{x} \cdot \mathbf{y} \neq 1$.

7. Prove the following by contrapositive: assume that $\mathbf{x}$ and $\mathbf{y}$ are vectors in $\mathbb{R}^n$. If $\mathbf{x} \cdot \mathbf{y} \neq 0$, then $\|\mathbf{x} + \mathbf{y}\|^2 \neq \|\mathbf{x}\|^2 + \|\mathbf{y}\|^2$.

8. State the contrapositive, converse, and inverse of each of the following statements regarding vectors in $\mathbb{R}^n$:
★(a) If $\mathbf{x}$ is a unit vector, then $\mathbf{x}$ is nonzero.
(b) Let $\mathbf{x}$ and $\mathbf{y}$ be nonzero vectors. If $\mathbf{x} \parallel \mathbf{y}$, then $\mathbf{y} = \mathbf{proj_x y}$.
★(c) Let $\mathbf{x}$ and $\mathbf{y}$ be nonzero vectors. If $\mathbf{proj_x y} = \mathbf{0}$, then $\mathbf{proj_y x} = \mathbf{0}$.

9. (a) State the converse of Result 2.
(b) Show that this converse is false by finding a counterexample.

10. Each of the following statements has the opposite truth value as its converse; that is, one of them is true and the other is false. In each case,

(i) State the converse of the given statement.
(ii) Determine which is true—the statement or its converse.
(iii) Prove the one from part (ii) that is true.
(iv) Show that the other one is false by finding a counterexample.

(a) Let $\mathbf{x}$, $\mathbf{y}$, and $\mathbf{z}$ be vectors in $\mathbb{R}^n$. If $\mathbf{x} \cdot \mathbf{y} = \mathbf{x} \cdot \mathbf{z}$, then $\mathbf{y} = \mathbf{z}$.
★(b) Let $\mathbf{x}$ and $\mathbf{y}$ be vectors in $\mathbb{R}^n$. If $\mathbf{x} \cdot \mathbf{y} = 0$, then $\|\mathbf{x} + \mathbf{y}\| \geq \|\mathbf{y}\|$.
(c) Assume that $\mathbf{x}$ and $\mathbf{y}$ are vectors in $\mathbb{R}^n$ with $n > 1$. If $\mathbf{x} \cdot \mathbf{y} = 0$, then $\mathbf{x} = \mathbf{0}$ or $\mathbf{y} = \mathbf{0}$.

11. Prove the following by contradiction: there does not exist a set of three mutually orthogonal nonzero vectors in $\mathbb{R}^2$. (Hint: Assume that such a set of vectors $[x_1, x_2]$, $[y_1, y_2]$, and $[z_1, z_2]$ does exist. First, show that at least one of x_1, y_1, or z_1 is nonzero. After suitable relabeling of the vectors, you may assume that x_1 is nonzero. Next, show that you may also assume that $y_1 \neq 0$. Let $a = x_2/x_1$ and $b = y_2/y_1$. Then, prove that $[1, a]$, $[1, b]$, and $[z_1, z_2]$ are also mutually orthogonal. You have now simplified the forms of the vectors to make your work easier. Finally, get a contradiction by showing that not all of the three possible dot products yield zero as an answer.)

12. Recall that a number is **irrational** if it cannot be expressed as the quotient of two integers. Use a proof by contradiction to show that $\sqrt{2}$ is an irrational number. (Hint: The following is an outline of the proof. Fill in the details.

(i) Assume, instead, that $\sqrt{2}$ is a rational number. Thus, there are integers p and q such that $p/q = \sqrt{2}$, with p/q in lowest terms.
(ii) Prove that $p^2 = 2q^2$.
(iii) Prove that p is even.

(iv) Prove that q is even.

(v) Find the contradiction.

13. Prove the following by induction: for each integer $m \geq 1$, let $\mathbf{x}_1, \ldots, \mathbf{x}_m$ be vectors in $\mathbb{R}^n$. Then, $\|\mathbf{x}_1 + \mathbf{x}_2 + \cdots + \mathbf{x}_m\| \leq \|\mathbf{x}_1\| + \|\mathbf{x}_2\| + \cdots + \|\mathbf{x}_m\|$.

14. Let $\mathbf{x}_1, \ldots, \mathbf{x}_k$ be a mutually orthogonal set of nonzero vectors in $\mathbb{R}^n$. Use induction to show that

$$\left\| \sum_{i=1}^{k} \mathbf{x}_i \right\|^2 = \sum_{i=1}^{k} \|\mathbf{x}_i\|^2.$$

15. Prove by induction: let $\mathbf{x}_1, \ldots, \mathbf{x}_k$ be unit vectors in $\mathbb{R}^n$, and let $a_1, \ldots, a_k$ be real numbers. Then, for every $\mathbf{y}$ in $\mathbb{R}^n$,

$$\left(\sum_{i=1}^{k} a_i \mathbf{x}_i \right) \cdot \mathbf{y} \leq \left(\sum_{i=1}^{k} |a_i| \right) \|\mathbf{y}\|.$$

16. Let $\mathbf{x} = [x_1, \ldots, x_n]$ be a vector in $\mathbb{R}^n$. Prove that $\|\mathbf{x}\| \leq \sum_{i=1}^{n} |\mathbf{x}_i|$. (Hint: Use a proof by induction on n to prove that $\sqrt{\sum_{i=1}^{n} x_i^2} \leq \sum_{i=1}^{n} |x_i|$. For the inductive step, let $\mathbf{y} = [x_1, \ldots, x_k, x_{k+1}]$, $\mathbf{z} = [x_1, \ldots, x_k, 0]$, and $\mathbf{w} = [0, 0, \ldots, 0, x_{k+1}]$. Note that $\mathbf{y} = \mathbf{z} + \mathbf{w}$. Then apply the Triangle Inequality.)

★17. Which steps in the following argument cannot be "reversed"? Why? Assume that $y = f(x)$ is a nonzero function and that d^2y/dx^2 exists for all x.

Step 1: $\quad y = x^2 + 2 \implies y^2 = x^4 + 4x^2 + 4$

Step 2: $\quad y^2 = x^4 + 4x^2 + 4 \implies 2y\dfrac{dy}{dx} = 4x^3 + 8x$

Step 3: $\quad 2y\dfrac{dy}{dx} = 4x^3 + 8x \implies \dfrac{dy}{dx} = \dfrac{4x^3 + 8x}{2y}$

Step 4: $\quad \dfrac{dy}{dx} = \dfrac{4x^3 + 8x}{2y} \implies \dfrac{dy}{dx} = \dfrac{4x^3 + 8x}{2(x^2 + 2)}$

Step 5: $\quad \dfrac{dy}{dx} = \dfrac{4x^3 + 8x}{2(x^2 + 2)} \implies \dfrac{dy}{dx} = 2x$

Step 6: $\quad \dfrac{dy}{dx} = 2x \implies \dfrac{d^2y}{dx^2} = 2$

18. State the negation of each of the following statements involving quantifiers and connectives. (No claim is made about their truth or falsity.)

★(a) There is a unit vector in $\mathbb{R}^3$ perpendicular to $[1, -2, 3]$.

(b) $\mathbf{x} = \mathbf{0}$ or $\mathbf{x} \cdot \mathbf{y} > 0$, where $\mathbf{x}$ and $\mathbf{y}$ are vectors in $\mathbb{R}^n$.

★(c) $\mathbf{x} \neq \mathbf{0}$ and $\|\mathbf{x} + \mathbf{y}\| = \|\mathbf{y}\|$, where $\mathbf{x}$ and $\mathbf{y}$ are vectors in $\mathbb{R}^n$.

(d) For every vector $\mathbf{x}$ in $\mathbb{R}^n$, $\mathbf{x} \cdot \mathbf{x} > 0$.

★(e) For every $\mathbf{x} \in \mathbb{R}^3$, there is a nonzero $\mathbf{y} \in \mathbb{R}^3$ such that $\mathbf{x} \cdot \mathbf{y} = 0$.

(f) There is an $\mathbf{x} \in \mathbb{R}^4$ such that for every $\mathbf{y} \in \mathbb{R}^4$, $\mathbf{x} \cdot \mathbf{y} = 0$.

19. State the contrapositive, converse, and inverse of the following statements involving connectives. (No claim is made about their truth or falsity.)

★(a) If $\mathbf{x} \cdot \mathbf{y} = 0$, then either $\mathbf{x} = \mathbf{0}$ or $\|\mathbf{x} - \mathbf{y}\| > \|\mathbf{y}\|$.

(b) If $\mathbf{x} \neq \mathbf{0}$ and $\mathbf{x} \cdot \mathbf{y} = 0$, then $\|\mathbf{x} - \mathbf{y}\| > \|\mathbf{y}\|$.

20. Prove the following by contrapositive: Let $\mathbf{x}$ be a vector in $\mathbb{R}^n$. If $\mathbf{x} \cdot \mathbf{y} = 0$ for every vector $\mathbf{y}$ in $\mathbb{R}^n$, then $\mathbf{x} = \mathbf{0}$.

21. Disprove: if $\mathbf{x}$ and $\mathbf{y}$ are vectors in $\mathbb{R}^n$, then $\|\mathbf{x} - \mathbf{y}\| \leq \|\mathbf{x}\| - \|\mathbf{y}\|$.

22. Use Result 2 to disprove the following: there is a vector $\mathbf{x}$ in $\mathbb{R}^3$ such that $\mathbf{x} \cdot [1, -2, 2] = 0$ and $\|\mathbf{x} + [1, -2, 2]\| < 3$.

1.4 Fundamental Operations with Matrices

We now introduce a new algebraic structure: the matrix. Matrices are two-dimensional arrays created by combining vectors into rows and columns. In this section, we discuss several elementary types of matrices. We also examine three operations on matrices—addition, scalar multiplication, and the transpose operation—and give their basic properties. Finally, we introduce symmetric and skew-symmetric matrices.

Definition of a Matrix

DEFINITION

An **m × n matrix** is a rectangular array of real numbers, arranged in m rows and n columns. The elements of a matrix are called the **entries**. The expression $m \times n$ denotes the **size** of the matrix.

For example, each of the following is a matrix, with its correct size indicated below it:

$$\mathbf{A} = \underbrace{\begin{bmatrix} 2 & 3 & -1 \\ 4 & 0 & -5 \end{bmatrix}}_{2 \times 3 \text{ matrix}} \qquad \mathbf{B} = \underbrace{\begin{bmatrix} 4 & -2 \\ 1 & 7 \\ -5 & 3 \end{bmatrix}}_{3 \times 2 \text{ matrix}} \qquad \mathbf{C} = \underbrace{\begin{bmatrix} 1 & 2 & 3 \\ 4 & 5 & 6 \\ 7 & 8 & 9 \end{bmatrix}}_{3 \times 3 \text{ matrix}}$$

$$\mathbf{D} = \underbrace{\begin{bmatrix} 7 \\ 1 \\ -2 \end{bmatrix}}_{3 \times 1 \text{ matrix}} \qquad \mathbf{E} = \underbrace{\begin{bmatrix} 4 & -3 & 0 \end{bmatrix}}_{1 \times 3 \text{ matrix}} \qquad \mathbf{F} = \underbrace{\begin{bmatrix} 4 \end{bmatrix}}_{1 \times 1 \text{ matrix}}$$

Here are some conventions to keep in mind when working with matrices:

- We use a single (or subscripted) bold capital letter to denote a matrix (such as $\mathbf{A}$, $\mathbf{B}$, $\mathbf{C}_1$, $\mathbf{C}_2$), in contrast to the lowercase bold letters used to represent vectors. The capital letters $\mathbf{I}$ and $\mathbf{O}$ are usually reserved for special types of matrices discussed later.
- The size of a matrix is always specified by stating the number of rows first and then the number of columns. For example, a 3×4 matrix always has three rows and four columns, never four rows and three columns.
- An $m \times n$ matrix can be thought of either as a collection of m row vectors, each having n coordinates, or as a collection of n column vectors, each having m coordinates. A matrix with just one row (or column) is essentially equivalent to a vector whose coordinates are written in row (or column) form.
- A specific matrix entry is often symbolized by the lowercase letter corresponding to the matrix name, together with a pair of subscripts representing the particular row and column numbers. For example, in the previous matrix $\mathbf{A}$, a_{23} is the entry -5 in the second row and third column. Often we write a_{ij} to represent the entry in the ith row and j^{th} column of $\mathbf{A}$. A typical 3×4 matrix $\mathbf{C}$ has entries symbolized by

$$\mathbf{C} = \begin{bmatrix} c_{11} & c_{12} & c_{13} & c_{14} \\ c_{21} & c_{22} & c_{23} & c_{24} \\ c_{31} & c_{32} & c_{33} & c_{34} \end{bmatrix}.$$

- The notation $\mathcal{M}_{mn}$ refers to the set of all matrices with real-number entries having m rows and n columns. For example, the set $\mathcal{M}_{34}$ is the set of all matrices having three rows and four columns. A typical matrix in the set $\mathcal{M}_{34}$ has the form of the preceding matrix $\mathbf{C}$.
- The **main diagonal** entries of a matrix $\mathbf{A}$ are $a_{11}, a_{22}, a_{33}, \ldots$. The main diagonal entries lie on a diagonal line drawn down and to the right, beginning from the upper-left corner of the matrix.

Matrices occur naturally in many contexts. For example, two-dimensional tables of numbers (that is, tables having rows and columns) are matrices. The following table represents a 50×3 matrix with integer entries:

State	Population (1990)	Area (sq. mi.)	Year admitted to union
Alabama	4040587	51609	1819
Alaska	550043	589757	1959
Arizona	3665228	113909	1912
⋮	⋮	⋮	⋮
Wyoming	453588	97914	1890

Special Types of Matrices

We now describe a few important types of matrices.

A **square matrix** is a matrix having the same number of rows as columns. That is, a matrix is square if and only if it is an $n \times n$ matrix, for some integer n. For example, the following matrices are square:

$$\mathbf{A} = \begin{bmatrix} 5 & 0 \\ 9 & -2 \end{bmatrix} \quad \text{and} \quad \mathbf{B} = \begin{bmatrix} 1 & 2 & 3 \\ 4 & 5 & 6 \\ 7 & 8 & 9 \end{bmatrix}.$$

A **diagonal matrix** is a square matrix in which all of the entries that are not on the main diagonal are zero. That is, $\mathbf{A}$ is a diagonal matrix if and only if $a_{ij} = 0$ for $i \neq j$. For example, the following matrices are diagonal matrices:

$$\mathbf{E} = \begin{bmatrix} 6 & 0 & 0 \\ 0 & 7 & 0 \\ 0 & 0 & -2 \end{bmatrix}, \ \mathbf{F} = \begin{bmatrix} 4 & 0 & 0 & 0 \\ 0 & 0 & 0 & 0 \\ 0 & 0 & -2 & 0 \\ 0 & 0 & 0 & 0 \end{bmatrix}, \text{ and } \mathbf{G} = \begin{bmatrix} -4 & 0 \\ 0 & 5 \end{bmatrix}.$$

The matrices

$$\mathbf{H} = \begin{bmatrix} 4 & 3 \\ 0 & 1 \end{bmatrix}, \text{ and } \mathbf{J} = \begin{bmatrix} 0 & 4 & 3 \\ -7 & 0 & 6 \\ 5 & -2 & 0 \end{bmatrix},$$

however, are *not* diagonal. (The main diagonal elements have been circled in each case.) We use the notation $\mathcal{D}_n$ to represent the **set of all $n \times n$ diagonal matrices**.

An **identity matrix** is a diagonal matrix with all main diagonal entries equal to 1. That is, an $n \times n$ matrix $\mathbf{A}$ is an identity matrix if and only if $a_{ij} = 0$ for $i \neq j$ and $a_{ii} = 1$ for $1 \leq i \leq n$. The identity matrix of size $n \times n$ is denoted by $\mathbf{I}_n$. For example, the following are identity matrices:

$$\mathbf{I}_2 = \begin{bmatrix} 1 & 0 \\ 0 & 1 \end{bmatrix} \quad \text{and} \quad \mathbf{I}_4 = \begin{bmatrix} 1 & 0 & 0 & 0 \\ 0 & 1 & 0 & 0 \\ 0 & 0 & 1 & 0 \\ 0 & 0 & 0 & 1 \end{bmatrix}.$$

None of the other matrices shown so far are identity matrices. If the size of the identity matrix is clear from the context, $\mathbf{I}$ alone may be used.

An **upper triangular matrix** is a square matrix with all entries below the main diagonal equal to zero. That is, an $n \times n$ matrix $\mathbf{A}$ is an upper triangular matrix if and only if $a_{ij} = 0$ for $i > j$. For example, these matrices are upper triangular:

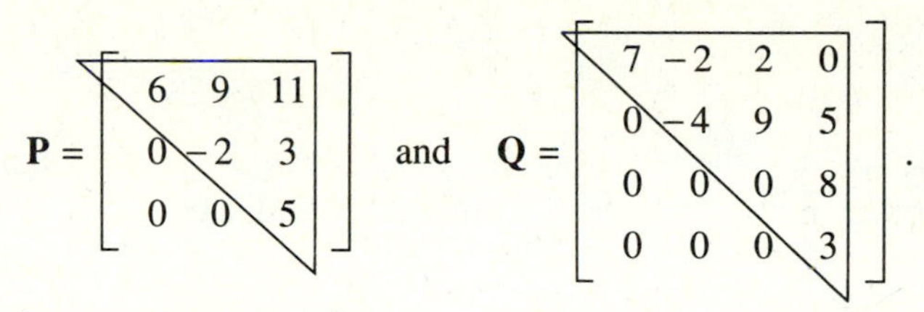

Similarly, a **lower triangular matrix** is one in which all entries above the main diagonal equal zero; for example:

$$\mathbf{R} = \begin{bmatrix} 3 & 0 & 0 \\ 9 & -2 & 0 \\ 14 & -6 & 1 \end{bmatrix}.$$

We use the notation $\mathcal{U}_n$ to represent the **set of all $n \times n$ upper triangular matrices** and $\mathcal{L}_n$ to represent the **set of all $n \times n$ lower triangular matrices**.

A **zero matrix** is any matrix all of whose entries are zero. $\mathbf{O}_{mn}$ denotes the $m \times n$ zero matrix, and $\mathbf{O}_n$ denotes the $n \times n$ zero matrix. For example,

$$\mathbf{O}_{23} = \begin{bmatrix} 0 & 0 & 0 \\ 0 & 0 & 0 \end{bmatrix} \quad \text{and} \quad \mathbf{O}_2 = \begin{bmatrix} 0 & 0 \\ 0 & 0 \end{bmatrix}.$$

If the size of the zero matrix is clear from the context, **O** alone may be used.

As with vectors, we say that two $m \times n$ matrices **A** and **B** are **equal** if and only if all their corresponding entries are equal. That is, **A** and **B** are equal if $a_{ij} = b_{ij}$ for all i, $1 \le i \le m$, and for all j, $1 \le j \le n$.

One point that should be noted is that, as vectors, the following expressions may be considered equal:

$$[3, -2, \ \ 4] \quad \text{and} \quad \begin{bmatrix} 3 \\ -2 \\ 4 \end{bmatrix}.$$

But as matrices, they are not equal; the former is a 1×3 matrix, and the latter is a 3×1 matrix.

Addition and Scalar Multiplication with Matrices

DEFINITION

Let **A** and **B** both be $m \times n$ matrices. The **sum** of **A** and **B** is the $m \times n$ matrix $(\mathbf{A} + \mathbf{B})$ whose $(i, j)^{th}$ entry is equal to $a_{ij} + b_{ij}$.

As with vectors, matrices are summed simply by adding their corresponding entries together. For example,

$$\begin{bmatrix} 6 & -3 & 2 \\ -7 & 0 & 4 \end{bmatrix} + \begin{bmatrix} 5 & -6 & -3 \\ -4 & -2 & -4 \end{bmatrix} = \begin{bmatrix} 11 & -9 & -1 \\ -11 & -2 & 0 \end{bmatrix}.$$

Notice that the definition does not allow addition of matrices with different sizes. For example, the following matrices cannot be added:

$$\mathbf{A} = \begin{bmatrix} -2 & 3 & 0 \\ 1 & 4 & -5 \end{bmatrix} \quad \text{and} \quad \mathbf{B} = \begin{bmatrix} 6 & 7 \\ -2 & 5 \\ 4 & -1 \end{bmatrix}.$$

A is a 2×3 matrix, and **B** is a 3×2 matrix.

DEFINITION

Let **A** be an $m \times n$ matrix, and let c be a scalar. Then the matrix $c\mathbf{A}$, the **scalar multiplication** of c and **A**, is the $m \times n$ matrix whose $(i, j)^{th}$ entry is equal to ca_{ij}.

Scalar multiplication with matrices is done just as with vectors, by multiplying every entry by the given scalar. For example, if $c = -2$ and

$$\mathbf{A} = \begin{bmatrix} 4 & -1 & 6 & 7 \\ 2 & 4 & 9 & -5 \end{bmatrix}, \quad \text{then} \quad -2\mathbf{A} = \begin{bmatrix} -8 & 2 & -12 & -14 \\ -4 & -8 & -18 & 10 \end{bmatrix}.$$

Note that if **A** is any $m \times n$ matrix, then $0\mathbf{A} = \mathbf{O}_{mn}$.

For a matrix **A**, let $-\mathbf{A}$ denote the matrix $-1\mathbf{A}$, the scalar product of -1 and **A**. For example, if

$$\mathbf{A} = \begin{bmatrix} 3 & -2 \\ 10 & 6 \end{bmatrix}, \quad \text{then} \quad -1\mathbf{A} = -\mathbf{A} = \begin{bmatrix} -3 & 2 \\ -10 & -6 \end{bmatrix}.$$

Also, we define **subtraction** of matrices as $\mathbf{A} - \mathbf{B} = \mathbf{A} + (-\mathbf{B})$.

As with vectors, sums of scalar multiples of matrices are called **linear combinations** of the matrices. For example, $-2\mathbf{A} + 6\mathbf{B} - 3\mathbf{C}$ is a linear combination of the matrices **A**, **B**, and **C**.

Fundamental Properties of Addition and Scalar Multiplication

The properties in the next theorem should look familiar, because they are similar to the vector properties of Theorem 1.3. These properties allow us to simplify and rearrange linear combinations of matrices in ways analogous to those used in elementary algebra.

THEOREM 1.10

Let **A**, **B**, and **C** be $m \times n$ matrices (elements of the set $\mathcal{M}_{mn}$), and let c and d be scalars. Then

(1)	$\mathbf{A} + \mathbf{B} = \mathbf{B} + \mathbf{A}$	Commutative Law of Addition
(2)	$\mathbf{A} + (\mathbf{B} + \mathbf{C}) = (\mathbf{A} + \mathbf{B}) + \mathbf{C}$	Associative Law of Addition
(3)	$\mathbf{O}_{mn} + \mathbf{A} = \mathbf{A} + \mathbf{O}_{mn} = \mathbf{A}$	Existence of Identity Element for Addition
(4)	$\mathbf{A} + (-\mathbf{A}) = (-\mathbf{A}) + \mathbf{A} = \mathbf{O}_{mn}$	Existence of Inverse Elements for Addition
(5)	$c(\mathbf{A} + \mathbf{B}) = c\mathbf{A} + c\mathbf{B}$	Distributive Laws of Scalar
(6)	$(c + d)\mathbf{A} = c\mathbf{A} + d\mathbf{A}$	Multiplication over Addition
(7)	$(cd)\mathbf{A} = c(d\mathbf{A})$	Associativity of Scalar Multiplication
(8)	$1(\mathbf{A}) = \mathbf{A}$	Existence of Identity Element for Scalar Multiplication

To prove each property, calculate corresponding entries on both sides and apply an appropriate law of real numbers to show that these entries agree. We prove part (1) as an example and leave some of the remaining parts for you to prove in Exercise 10.

Proof of Theorem 1.10, Part (1)

For any i, j, where $1 \le i \le m$ and $1 \le j \le n$, the $(i, j)^{th}$ entry of $(\mathbf{A} + \mathbf{B})$ is the sum of the entries a_{ij} and b_{ij} from **A** and **B**, respectively. Similarly, the $(i, j)^{th}$ entry of $\mathbf{B} + \mathbf{A}$ is the sum of b_{ij} and a_{ij}. But $a_{ij} + b_{ij} = b_{ij} + a_{ij}$, by the commutative property of addition for real numbers. Because the corresponding entries of $\mathbf{A} + \mathbf{B}$ and $\mathbf{B} + \mathbf{A}$ agree, we must have $\mathbf{A} + \mathbf{B} = \mathbf{B} + \mathbf{A}$. ■

The Transpose of a Matrix and Its Properties

DEFINITION

If **A** is an $m \times n$ matrix, then the **transpose** matrix $\mathbf{A}^T$ is the $n \times m$ matrix whose $(i, j)^{\text{th}}$ entry is the same as the $(j, i)^{\text{th}}$ entry of the original matrix **A**.

Thus, the transpose operation moves the $(i, j)^{\text{th}}$ entry of **A** to the $(j, i)^{\text{th}}$ entry of $\mathbf{A}^T$. Notice that the entries on the main diagonal do not move as you convert **A** to $\mathbf{A}^T$. However, all entries above the main diagonal are moved below it, and vice versa. For example,

$$\text{if} \quad \mathbf{A} = \begin{bmatrix} 6 & 10 \\ -2 & 4 \\ 3 & 0 \\ 1 & 8 \end{bmatrix} \quad \text{and} \quad \mathbf{B} = \begin{bmatrix} 1 & 5 & -3 \\ 0 & -4 & 6 \\ 0 & 0 & -5 \end{bmatrix},$$

$$\text{then} \quad \mathbf{A}^T = \begin{bmatrix} 6 & -2 & 3 & 1 \\ 10 & 4 & 0 & 8 \end{bmatrix} \quad \text{and} \quad \mathbf{B}^T = \begin{bmatrix} 1 & 0 & 0 \\ 5 & -4 & 0 \\ -3 & 6 & -5 \end{bmatrix}.$$

Notice that the transpose of an upper triangular matrix (such as **B**) is lower triangular, and vice versa. The examples above also illustrate that the transpose operation takes the rows of **A** and makes them the columns of $\mathbf{A}^T$. Similarly, the columns of **A** become the rows of $\mathbf{A}^T$.

Three useful properties of the transpose are given in the next theorem. We prove one of these properties and leave the others for you to prove in Exercise 11.

THEOREM 1.11

Let **A** and **B** both be $m \times n$ matrices, and let c be a scalar. Then
(1) $(\mathbf{A}^T)^T = \mathbf{A}$
(2) $(\mathbf{A} + \mathbf{B})^T = \mathbf{A}^T + \mathbf{B}^T$
(3) $(c\mathbf{A})^T = c(\mathbf{A}^T)$

Proof of Theorem 1.11, Part (2)

Notice that both $(\mathbf{A} + \mathbf{B})^T$ and $(\mathbf{A}^T) + (\mathbf{B}^T)$ are $n \times m$ matrices. (Why?) We need to show that the $(i, j)^{th}$ entries of both are equal, for $1 \leq i \leq n$ and $1 \leq j \leq m$. Now, by definition of the transpose, the $(i, j)^{th}$ entry of $(\mathbf{A} + \mathbf{B})^T$ equals the $(j, i)^{th}$ entry of $\mathbf{A} + \mathbf{B}$, which is $a_{ji} + b_{ji}$. But the $(i, j)^{th}$ entry of $\mathbf{A}^T + \mathbf{B}^T$ equals the $(i, j)^{th}$ entry of $\mathbf{A}^T$ plus the $(i, j)^{th}$ entry of $\mathbf{B}^T$, which is also $a_{ji} + b_{ji}$. ■

Symmetric and Skew-Symmetric Matrices

DEFINITION

A matrix **A** is **symmetric** if and only if $\mathbf{A} = \mathbf{A}^T$. A matrix **A** is **skew-symmetric** if and only if $\mathbf{A} = -\mathbf{A}^T$.

In Exercise 5, you are asked to show that any symmetric or skew-symmetric matrix is a square matrix.

Example 1

Consider the following matrices:

$$\mathbf{A} = \begin{bmatrix} 2 & 6 & 4 \\ 6 & -1 & 0 \\ 4 & 0 & -3 \end{bmatrix} \quad \text{and} \quad \mathbf{B} = \begin{bmatrix} 0 & -1 & 3 & 6 \\ 1 & 0 & 2 & -5 \\ -3 & -2 & 0 & 4 \\ -6 & 5 & -4 & 0 \end{bmatrix}.$$

A is symmetric and **B** is skew-symmetric, because their respective transposes are

$$\mathbf{A}^T = \begin{bmatrix} 2 & 6 & 4 \\ 6 & -1 & 0 \\ 4 & 0 & -3 \end{bmatrix} \quad \text{and} \quad \mathbf{B}^T = \begin{bmatrix} 0 & 1 & -3 & -6 \\ -1 & 0 & -2 & 5 \\ 3 & 2 & 0 & -4 \\ 6 & -5 & 4 & 0 \end{bmatrix}.$$

These transposes are equal to **A** and $-\mathbf{B}$, respectively. However, neither of the following matrices is symmetric or skew-symmetric (why?):

$$\mathbf{C} = \begin{bmatrix} 3 & -2 & 1 \\ 2 & 4 & 0 \\ -1 & 0 & -2 \end{bmatrix} \quad \text{and} \quad \mathbf{D} = \begin{bmatrix} 1 & -2 \\ 3 & 4 \\ 5 & -6 \end{bmatrix}.$$

■

Notice that an $n \times n$ matrix **A** is symmetric [skew-symmetric] if and only if $a_{ij} = a_{ji}$ $[a_{ij} = -a_{ji}]$ for all i, j such that $1 \leq i, j \leq n$. In other words, the main diagonal of a symmetric or skew-symmetric matrix acts like a mirror in which the entries above the diagonal are reflected into equal (for symmetric) or opposite (for skew-symmetric) entries below the diagonal. The main diagonal elements are, of course, reflected into themselves. In particular, *all of the main diagonal elements of a skew-symmetric matrix must be zeroes*, because $a_{ii} = -a_{ii}$ only if $a_{ii} = 0$.

Notice that any diagonal matrix is equal to its transpose, and so such matrices are automatically symmetric.

The next theorem shows that every square matrix can be decomposed in a unique way as the sum of a symmetric and a skew-symmetric matrix:

THEOREM 1.12

> Every square matrix **A** can be expressed uniquely as the sum of two matrices **S** and **V**, where **S** is symmetric and **V** is skew-symmetric.

An outline of the proof of Theorem 1.12 is given in Exercise 13. In addition to asking you to prove the theorem, the exercise asks you to show that the symmetric matrix **S** in the theorem equals $\frac{1}{2}(\mathbf{A} + \mathbf{A}^T)$ and that $\mathbf{V} = \frac{1}{2}(\mathbf{A} - \mathbf{A}^T)$.

Example 2

The matrix

$$\mathbf{A} = \begin{bmatrix} -4 & 2 & 5 \\ 6 & 3 & 7 \\ -1 & 0 & 2 \end{bmatrix}$$

can be decomposed into the sum of a symmetric matrix **S** and a skew-symmetric matrix **V** as follows:

$$\mathbf{S} = \frac{1}{2}(\mathbf{A} + \mathbf{A}^T) = \frac{1}{2}\left(\begin{bmatrix} -4 & 2 & 5 \\ 6 & 3 & 7 \\ -1 & 0 & 2 \end{bmatrix} + \begin{bmatrix} -4 & 6 & -1 \\ 2 & 3 & 0 \\ 5 & 7 & 2 \end{bmatrix}\right) = \begin{bmatrix} -4 & 4 & 2 \\ 4 & 3 & \frac{7}{2} \\ 2 & \frac{7}{2} & 2 \end{bmatrix},$$

and

$$\mathbf{V} = \frac{1}{2}(\mathbf{A} - \mathbf{A}^T) = \frac{1}{2}\left(\begin{bmatrix} -4 & 2 & 5 \\ 6 & 3 & 7 \\ -1 & 0 & 2 \end{bmatrix} - \begin{bmatrix} -4 & 6 & -1 \\ 2 & 3 & 0 \\ 5 & 7 & 2 \end{bmatrix}\right) = \begin{bmatrix} 0 & -2 & 3 \\ 2 & 0 & \frac{7}{2} \\ -3 & -\frac{7}{2} & 0 \end{bmatrix}.$$

Notice that **S** and **V** really are, respectively, symmetric and skew-symmetric and that **S** + **V** really does equal **A**. ■

Exercises—Section 1.4

1. For the matrices,

$$\mathbf{A} = \begin{bmatrix} -4 & 2 & 3 \\ 0 & 5 & -1 \\ 6 & 1 & -2 \end{bmatrix} \quad \mathbf{B} = \begin{bmatrix} 6 & -1 & 0 \\ 2 & 2 & -4 \\ 3 & -1 & 1 \end{bmatrix} \quad \mathbf{C} = \begin{bmatrix} 5 & -1 \\ -3 & 4 \end{bmatrix}$$

$$\mathbf{D} = \begin{bmatrix} -7 & 1 & -4 \\ 3 & -2 & 8 \end{bmatrix} \quad \mathbf{E} = \begin{bmatrix} 3 & -3 & 5 \\ 1 & 0 & -2 \\ 6 & 7 & -2 \end{bmatrix} \quad \mathbf{F} = \begin{bmatrix} 8 & -1 \\ 2 & 0 \\ 5 & -3 \end{bmatrix}$$

compute the following, if possible:

★ (a) $\mathbf{A} + \mathbf{B}$ (b) $\mathbf{C} + \mathbf{D}$ ★ (c) $4\mathbf{A}$
(d) $2\mathbf{A} - 4\mathbf{B}$ ★ (e) $\mathbf{C} + 3\mathbf{F} - \mathbf{E}$ (f) $\mathbf{A} - \mathbf{B} + \mathbf{E}$
★ (g) $2\mathbf{A} - 3\mathbf{E} - \mathbf{B}$ (h) $2\mathbf{D} - 3\mathbf{F}$ ★ (i) $\mathbf{A}^T + \mathbf{E}^T$
(j) $(\mathbf{A} + \mathbf{E})^T$ (k) $4\mathbf{D} + 2\mathbf{F}^T$ ★ (l) $2\mathbf{C}^T - 3\mathbf{F}$
(m) $5(\mathbf{F}^T - \mathbf{D}^T)$ ★ (n) $((\mathbf{B} - \mathbf{A})^T + \mathbf{E}^T)^T$

★**2.** Calculate the transpose of each of the following matrices. Also, indicate which of these matrices are square, diagonal, upper or lower triangular, symmetric, or skew-symmetric.

$$\mathbf{A} = \begin{bmatrix} -1 & 4 \\ 0 & 1 \\ 6 & 0 \end{bmatrix} \quad \mathbf{B} = \begin{bmatrix} 2 & 0 \\ 0 & -1 \end{bmatrix} \quad \mathbf{C} = \begin{bmatrix} -1 & 1 \\ -1 & 1 \end{bmatrix} \quad \mathbf{D} = \begin{bmatrix} -1 \\ 4 \\ 2 \end{bmatrix}$$

$$\mathbf{E} = \begin{bmatrix} 0 & 0 & 6 \\ 0 & -6 & 0 \\ -6 & 0 & 0 \end{bmatrix} \quad \mathbf{F} = \begin{bmatrix} 1 & 0 & 0 & 1 \\ 0 & 0 & 1 & 1 \\ 0 & 1 & 0 & 0 \\ 1 & 1 & 0 & 1 \end{bmatrix} \quad \mathbf{G} = \begin{bmatrix} 6 & 0 & 0 \\ 0 & 6 & 0 \\ 0 & 0 & 6 \end{bmatrix}$$

$$\mathbf{H} = \begin{bmatrix} 0 & -1 & 6 & 2 \\ 1 & 0 & -7 & 1 \\ -6 & 7 & 0 & -4 \\ -2 & -1 & 4 & 0 \end{bmatrix} \quad \mathbf{J} = \begin{bmatrix} 0 & 1 & 0 & 0 \\ 1 & 0 & 1 & 1 \\ 0 & 1 & 1 & 1 \\ 0 & 1 & 1 & 0 \end{bmatrix} \quad \mathbf{K} = \begin{bmatrix} 1 & 2 & 3 & 4 \\ -2 & 1 & 5 & 6 \\ -3 & -5 & 1 & 7 \\ -4 & -6 & -7 & 1 \end{bmatrix}$$

$$\mathbf{L} = \begin{bmatrix} 1 & 1 & 1 \\ 0 & 1 & 1 \\ 0 & 0 & 1 \end{bmatrix} \quad \mathbf{M} = \begin{bmatrix} 0 & 0 & 0 \\ 1 & 0 & 0 \\ 1 & 1 & 0 \end{bmatrix} \quad \mathbf{N} = \begin{bmatrix} 1 & 0 & 0 \\ 0 & 1 & 0 \\ 0 & 0 & 1 \end{bmatrix}$$

$$\mathbf{P} = \begin{bmatrix} 0 & 1 \\ 1 & 0 \end{bmatrix} \quad \mathbf{Q} = \begin{bmatrix} -2 & 0 & 0 \\ 4 & 0 & 0 \\ -1 & 2 & 3 \end{bmatrix} \quad \mathbf{R} = \begin{bmatrix} 6 & 2 \\ 3 & -2 \\ -1 & 0 \end{bmatrix}$$

3. Express each of the following matrices as the sum of a symmetric and a skew-symmetric matrix:

$\star$(a) $\begin{bmatrix} 3 & -1 & 4 \\ 0 & 2 & 5 \\ 1 & -3 & 2 \end{bmatrix}$ (b) $\begin{bmatrix} 1 & 0 & -4 \\ 3 & 3 & -1 \\ 4 & -1 & 0 \end{bmatrix}$ (c) $\begin{bmatrix} 2 & 3 & 4 & -1 \\ -3 & 5 & -1 & 2 \\ -4 & 1 & -2 & 0 \\ 1 & -2 & 0 & 5 \end{bmatrix}$

4. Prove that if $\mathbf{A}^T = \mathbf{B}^T$, then $\mathbf{A} = \mathbf{B}$.

5. (a) Prove that any symmetric or skew-symmetric matrix is a square matrix.
(b) Prove that every diagonal matrix is a symmetric matrix.
$\star$(c) Describe completely every matrix that is both diagonal and skew-symmetric.

6. Assume that **A** and **B** are square matrices of the same size.
(a) If **A** and **B** are diagonal, prove that $\mathbf{A} + \mathbf{B}$ is diagonal.
(b) If **A** and **B** are upper triangular, prove that $\mathbf{A} + \mathbf{B}$ is upper triangular.
(c) If **A** and **B** are symmetric, prove that $\mathbf{A} + \mathbf{B}$ is symmetric.

7. Use induction to prove that, if $\mathbf{A}_1, \ldots, \mathbf{A}_n$ are upper triangular matrices of the same size, then $\sum_{i=1}^{n} \mathbf{A}_i$ is upper triangular.

8. (a) If **A** is a symmetric matrix, show that $\mathbf{A}^T$ and $c\mathbf{A}$ are also symmetric.
(b) If **A** is a skew-symmetric matrix, show that $\mathbf{A}^T$ and $c\mathbf{A}$ are also skew-symmetric.

9. (a) Show that $(\mathbf{I}_n)^T = \mathbf{I}_n$.
(b) The **Kronecker Delta** δ_{ij} is defined as follows: $\delta_{ij} = \begin{cases} 1 & \text{if } i = j \\ 0 & \text{if } i \neq j \end{cases}$.
If $\mathbf{A} = \mathbf{I}_n$, explain why $a_{ij} = \delta_{ij}$.

10. Prove parts (4), (5), and (7) of Theorem 1.10.

11. Prove parts (1) and (3) of Theorem 1.11.

12. Let $\mathbf{A}$ be an $m \times n$ matrix. Prove that if $c\mathbf{A} = \mathbf{O}_{mn}$, the $m \times n$ zero matrix, then $c = 0$ or $\mathbf{A} = \mathbf{O}_{mn}$.

13. This exercise provides an outline for the proof of Theorem 1.12. Let $\mathbf{A}$ be an $n \times n$ matrix:

(a) Prove that $\frac{1}{2}(\mathbf{A} + \mathbf{A}^T)$ is a symmetric matrix.
(b) Prove that $\frac{1}{2}(\mathbf{A} - \mathbf{A}^T)$ is a skew-symmetric matrix.
(c) Show that $\mathbf{A} = \frac{1}{2}(\mathbf{A} + \mathbf{A}^T) + \frac{1}{2}(\mathbf{A} - \mathbf{A}^T)$.
(d) Suppose that $\mathbf{S}_1$ and $\mathbf{S}_2$ are symmetric matrices and that $\mathbf{V}_1$ and $\mathbf{V}_2$ are skew-symmetric matrices such that $\mathbf{S}_1 + \mathbf{V}_1 = \mathbf{S}_2 + \mathbf{V}_2$. Derive a second equation involving $\mathbf{S}_1$, $\mathbf{S}_2$, $\mathbf{V}_1$, and $\mathbf{V}_2$ by taking the transpose of both sides of the equation and simplifying.
(e) Add the two equations from part (d) together, and use the result to prove that $\mathbf{S}_1 = \mathbf{S}_2$.
(f) Prove that the matrices $\mathbf{V}_1$ and $\mathbf{V}_2$ from part (d) are equal.
(g) Explain how all parts of this exercise together prove Theorem 1.12.

14. The **trace** of a square matrix $\mathbf{A}$ is the sum of the elements along the main diagonal.

★(a) Find the trace of each square matrix in Exercise 2.
(b) If $\mathbf{A}$ and $\mathbf{B}$ are both $n \times n$ matrices, prove that:
(i) trace($\mathbf{A} + \mathbf{B}$) = trace($\mathbf{A}$) + trace($\mathbf{B}$) (ii) trace($c\mathbf{A}$) = c(trace($\mathbf{A}$))
(iii) trace($\mathbf{A}$) = trace($\mathbf{A}^T$)
★(c) Suppose that trace($\mathbf{A}$) = trace($\mathbf{B}$) for two $n \times n$ matrices $\mathbf{A}$ and $\mathbf{B}$. Does $\mathbf{A} = \mathbf{B}$? Prove your answer.

1.5 Matrix Multiplication

We now turn to another operation on matrices: multiplication. However, we do not simply multiply corresponding matrix entries together, as you might expect. Instead, matrix multiplication is defined as a generalization of the dot product of vectors, which is a sum of products of corresponding coordinates. Also, matrix multiplication is not a commutative operation.

Definition of Matrix Multiplication

Two matrices $\mathbf{A}$ and $\mathbf{B}$ can be multiplied (in that order) only if the number of columns of $\mathbf{A}$ is equal to the number of rows of $\mathbf{B}$. In that case,

Size of product $\mathbf{AB}$ = (number of rows of $\mathbf{A}$) $\times$ (number of columns of $\mathbf{B}$).

That is, if $\mathbf{A}$ is an $m \times n$ matrix, then $\mathbf{AB}$ is defined only when the number of rows of $\mathbf{B}$ is n—that is, when $\mathbf{B}$ is an $n \times p$ matrix, for some integer p. In this case, $\mathbf{AB}$

is an $m \times p$ matrix, because **A** has m rows and **B** has p columns. The actual entries of **AB** are given by the following:

DEFINITION

If **A** is an $m \times n$ matrix and **B** is an $n \times p$ matrix, their matrix product $\mathbf{C} = \mathbf{AB}$ is the $m \times p$ matrix whose $(i, j)^{th}$ entry is the dot product of the ith row of **A** with the jth column of **B**. That is,

$$\underbrace{\begin{bmatrix} a_{11} & a_{12} & a_{13} & \cdots & a_{1n} \\ a_{21} & a_{22} & a_{23} & \cdots & a_{2n} \\ \vdots & \vdots & \vdots & \ddots & \vdots \\ a_{i1} & a_{i2} & a_{i3} & \cdots & a_{in} \\ \vdots & \vdots & \vdots & \ddots & \vdots \\ a_{m1} & a_{m2} & a_{m3} & \cdots & a_{mn} \end{bmatrix}}_{m \times n \text{ matrix } \mathbf{A}} \underbrace{\begin{bmatrix} b_{11} & b_{12} & \cdots & b_{1j} & \cdots & b_{1p} \\ b_{21} & b_{22} & \cdots & b_{2j} & \cdots & b_{2p} \\ b_{31} & b_{32} & \cdots & b_{3j} & \cdots & b_{3p} \\ \vdots & \vdots & \ddots & \vdots & \ddots & \vdots \\ b_{n1} & b_{n2} & \cdots & b_{nj} & \cdots & b_{np} \end{bmatrix}}_{n \times p \text{ matrix } \mathbf{B}} =$$

$$\underbrace{\begin{bmatrix} c_{11} & c_{12} & \cdots & c_{1j} & \cdots & c_{1p} \\ c_{21} & c_{22} & \cdots & c_{2j} & \cdots & c_{2p} \\ \vdots & \vdots & \ddots & \vdots & \ddots & \vdots \\ c_{i1} & c_{i2} & \cdots & c_{ij} & \cdots & c_{ip} \\ \vdots & \vdots & \ddots & \vdots & \ddots & \vdots \\ c_{m1} & c_{m2} & \cdots & c_{mj} & \cdots & c_{mp} \end{bmatrix}}_{m \times p \text{ matrix } \mathbf{C}},$$

where $c_{ij} = a_{i1}b_{1j} + a_{i2}b_{2j} + a_{i3}b_{3j} + \cdots + a_{in}b_{nj} = \sum_{k=1}^{n} a_{ik}b_{kj}$.

Because the number of columns in the matrix **A** equals the number of rows in **B**, it follows that each row of **A** contains the same number of entries as each column of **B**. Thus, it is possible to perform the dot products used to calculate the matrix **C**.

We will work through some examples of matrix multiplication:

Example 1

Consider

$$\mathbf{A} = \begin{bmatrix} 5 & -1 & 4 \\ -3 & 6 & 0 \end{bmatrix} \quad \text{and} \quad \mathbf{B} = \begin{bmatrix} 9 & 4 & -8 & 2 \\ 7 & 6 & -1 & 0 \\ -2 & 5 & 3 & -4 \end{bmatrix}.$$

These matrices can be multiplied because **A** is a 2×3 matrix and **B** is a 3×4 matrix, which means that the number of columns of **A** does equal the number of rows of **B** (three in each case). Now, the product matrix $\mathbf{C} = \mathbf{AB}$ is a 2×4 matrix, because **A** has two rows and **B** has four columns. To calculate each of the entries of **C**, we must find the dot product of the appropriate row of **A** with the appropriate row of **B**. For example, to find the entry c_{11}, we take the dot product of the first row of **A** with the first column of **B**:

$$c_{11} = [5, -1, 4] \cdot \begin{bmatrix} 9 \\ 7 \\ -2 \end{bmatrix} = (5)(9) + (-1)(7) + (4)(-2) = 45 - 7 - 8 = 30.$$

To find the entry c_{23}, we must take the dot product of the second row of **A** with the third column of **B**:

$$c_{23} = [-3, 6, 0] \cdot \begin{bmatrix} -8 \\ -1 \\ 3 \end{bmatrix} = (-3)(-8) + (6)(-1) + (0)(3) = 24 - 6 + 0 = 18.$$

The other entries are computed similarly, yielding

$$\mathbf{C} = \begin{bmatrix} 30 & 34 & -27 & -6 \\ 15 & 24 & 18 & -6 \end{bmatrix}.$$

■

Example 2

Consider the following matrices:

$$\mathbf{D} = \underbrace{\begin{bmatrix} -2 & 1 \\ 0 & 5 \\ 4 & -3 \end{bmatrix}}_{3 \times 2 \text{ matrix}}, \quad \mathbf{E} = \underbrace{\begin{bmatrix} 1 & -6 \\ 0 & 2 \end{bmatrix}}_{2 \times 2 \text{ matrix}}, \quad \mathbf{F} = \underbrace{\begin{bmatrix} -4 & 2 & 1 \end{bmatrix}}_{1 \times 3 \text{ matrix}},$$

$$\mathbf{G} = \underbrace{\begin{bmatrix} 7 \\ -1 \\ 5 \end{bmatrix}}_{3 \times 1 \text{ matrix}}, \quad \text{and} \quad \mathbf{H} = \underbrace{\begin{bmatrix} 5 & 0 \\ 1 & -3 \end{bmatrix}}_{2 \times 2 \text{ matrix}}.$$

The only possible products of two of these matrices are

$$\mathbf{DE} = \begin{bmatrix} -2 & 14 \\ 0 & 10 \\ 4 & -30 \end{bmatrix}, \quad \mathbf{DH} = \begin{bmatrix} -9 & -3 \\ 5 & -15 \\ 17 & 9 \end{bmatrix}, \quad \mathbf{GF} = \begin{bmatrix} -28 & 14 & 7 \\ 4 & -2 & -1 \\ -20 & 10 & 5 \end{bmatrix},$$

$$\mathbf{FG} = [-25] \ (1 \times 1 \text{ matrix}), \quad \mathbf{FD} = [12 \quad 3] \ (1 \times 2 \text{ matrix}),$$

$$\mathbf{EH} = \begin{bmatrix} -1 & 18 \\ 2 & -6 \end{bmatrix}, \quad \mathbf{HE} = \begin{bmatrix} 5 & -30 \\ 1 & -12 \end{bmatrix},$$

$$\mathbf{EE} = \begin{bmatrix} 1 & -18 \\ 0 & 4 \end{bmatrix}, \quad \text{and} \quad \mathbf{HH} = \begin{bmatrix} 25 & 0 \\ 2 & 9 \end{bmatrix}.$$

No other products are defined. ■

Example 2 points out the fact that the order in which matrix multiplication is performed is extremely important. In fact, in Example 2 we see each of the following possibilities for the product of two matrices:

- Neither product may be defined (for example, **DG** and **GD**).
- The product may be defined one way but not the other. (For example, **DE** is defined but **ED** is not.)
- The product may be defined both ways, but the resulting sizes may not be the same. (For example, **FG** is a 1×1 matrix, but **GF** is a 3×3 matrix.)
- The product may be defined both ways, and the resulting sizes may agree, but the entries of the products may differ. (For example, **EH** and **HE** are both 2×2 matrices, but they have different entries.)

It is possible for the product of certain matrices to be the same when multiplied in either order. For example, if $\mathbf{A}$ is any 2×2 matrix, then $\mathbf{AI}_2$ will always equal $\mathbf{I}_2\mathbf{A}$, where $\mathbf{I}_2$ is the identity matrix $\begin{bmatrix} 1 & 0 \\ 0 & 1 \end{bmatrix}$. To illustrate, if $\mathbf{A} = \begin{bmatrix} -4 & 2 \\ 5 & 6 \end{bmatrix}$, then $\begin{bmatrix} -4 & 2 \\ 5 & 6 \end{bmatrix}\begin{bmatrix} 1 & 0 \\ 0 & 1 \end{bmatrix} = \begin{bmatrix} 1 & 0 \\ 0 & 1 \end{bmatrix}\begin{bmatrix} -4 & 2 \\ 5 & 6 \end{bmatrix}$, because both products are equal to $\begin{bmatrix} -4 & 2 \\ 5 & 6 \end{bmatrix}$, which is $\mathbf{A}$ itself. In fact, in Exercise 13, you are asked to show that, if $\mathbf{A}$ is any $m \times n$ matrix, then $\mathbf{AI}_n = \mathbf{I}_m\mathbf{A} = \mathbf{A}$. This is why $\mathbf{I}$ is called the **(multiplicative) identity** matrix—because it preserves the "identity" of any matrices multiplied by it.

When two matrices $\mathbf{A}$ and $\mathbf{B}$ have the unusual property that $\mathbf{AB} = \mathbf{BA}$, then we say that $\mathbf{A}$ and $\mathbf{B}$ **commute**, or that "**A commutes** with **B**." But as you have seen, in general there is no commutative law for multiplication of matrices, even though there is a commutative law for addition.

Often we need to find only a particular row or column of a matrix product. The rules for doing so are summarized in the following general principle:

> Assume that the matrix product $\mathbf{AB}$ is defined. Then the kth row of $\mathbf{AB}$ is the product (k^{th} row of $\mathbf{A}$)$\mathbf{B}$ and the lth column of $\mathbf{AB}$ is the product $\mathbf{A}$(l^{th} column of $\mathbf{B}$).

An Application of Matrix Multiplication

Matrix products are vital in modeling various geometric transformations. They are also widely used in graph theory, coding theory, physics, and chemistry. The next example shows a simple application in business:

Example 3

Suppose that three branches of a company are currently selling four types of videotapes—say, Tapes W, X, Y, and Z. The number of each type of tape sold during the past week by each branch is shown in matrix **A** below. The shipping cost and profit collected for each tape sold is shown in matrix **B**.

$$\mathbf{A} = \begin{array}{r} \\ \text{Branch 1} \\ \text{Branch 2} \\ \text{Branch 3} \end{array} \begin{array}{c} \begin{array}{cccc} \text{Tape W} & \text{Tape X} & \text{Tape Y} & \text{Tape Z} \end{array} \\ \begin{bmatrix} 130 & 160 & 240 & 190 \\ 210 & 180 & 320 & 240 \\ 170 & 200 & 340 & 220 \end{bmatrix} \end{array}$$

$$\mathbf{B} = \begin{array}{r} \\ \text{Tape W} \\ \text{Tape X} \\ \text{Tape Y} \\ \text{Tape Z} \end{array} \begin{array}{c} \begin{array}{cc} \text{Shipping cost} & \text{Profit} \end{array} \\ \begin{bmatrix} \$3 & \$3 \\ \$4 & \$2 \\ \$3 & \$4 \\ \$2 & \$2 \end{bmatrix} \end{array}$$

The product **C** = **AB** represents the total shipping costs and profits last week for all types of tapes combined:

$$\mathbf{C} = \begin{array}{r} \\ \text{Branch 1} \\ \text{Branch 2} \\ \text{Branch 3} \end{array} \begin{array}{c} \begin{array}{cc} \text{Total shipping cost} & \text{Total profit} \end{array} \\ \begin{bmatrix} \$2130 & \$2050 \\ \$2790 & \$2750 \\ \$2770 & \$2710 \end{bmatrix} \end{array}.$$

To understand this result, let us consider a particular case. The entry in the second row and second column of **C** is calculated by taking the dot product of the second row of **A** with the second column of **B**:

$$(210)(\$3) + (180)(\$2) + (320)(\$4) + (240)(\$2) = \$2750.$$

We are multiplying the number of each type of tape sold by Branch 2 times the profit per tape, which must equal the total profit for Branch 2. ■

In Example 3, if we wanted the results for just Branch 3, we would need to compute only the third row of **C**:

$$\underbrace{[170 \quad 200 \quad 340 \quad 220]}_{\text{Third row of } \mathbf{A}} \underbrace{\begin{bmatrix} \$3 & \$3 \\ \$4 & \$2 \\ \$3 & \$4 \\ \$2 & \$2 \end{bmatrix}}_{\mathbf{B}} = \underbrace{[\$2770 \quad \$2710]}_{\text{Third row of } \mathbf{C}}.$$

Properties of Matrix Multiplication

If the zero matrix **O** is multiplied times any matrix **A**, or if **A** is multiplied times **O**, the result is **O** (see Exercise 12). The following theorem lists some other important properties of matrix multiplication:

THEOREM 1.13

Suppose that **A**, **B**, and **C** are matrices for which the following sums and products are defined. Let c be a scalar. Then

(1)	$\mathbf{A}(\mathbf{BC}) = (\mathbf{AB})\mathbf{C}$	Associative Law of Multiplication
(2)	$\mathbf{A}(\mathbf{B} + \mathbf{C}) = \mathbf{AB} + \mathbf{AC}$	Distributive Laws of Matrix Multiplication over Addition
(3)	$(\mathbf{A} + \mathbf{B})\mathbf{C} = \mathbf{AC} + \mathbf{BC}$	
(4)	$c(\mathbf{AB}) = (c\mathbf{A})\mathbf{B} = \mathbf{A}(c\mathbf{B})$	Associative Law of Scalar and Matrix Multiplication

The proof of part (1) of Theorem 1.13 is more difficult than the others, and so it is included in Appendix A for the interested reader. You are asked to provide the proofs of parts (2), (3), and (4) in Exercise 11.

Although Theorem 1.13 shows that matrix multiplication has some properties in common with ordinary multiplication of real numbers, other properties besides the commutative law do fail. For example, the **cancellation laws** in algebra do not hold in general. That is, if $\mathbf{AB} = \mathbf{AC}$, with **A** not the zero matrix, we might expect **B** to equal **C**. But this does not always happen. For example, if

$$\mathbf{A} = \begin{bmatrix} 2 & 1 \\ 6 & 3 \end{bmatrix}, \quad \mathbf{B} = \begin{bmatrix} -1 & 0 \\ 5 & 2 \end{bmatrix}, \quad \text{and} \quad \mathbf{C} = \begin{bmatrix} 3 & 1 \\ -3 & 0 \end{bmatrix},$$

then

$$\mathbf{AB} = \begin{bmatrix} 2 & 1 \\ 6 & 3 \end{bmatrix}\begin{bmatrix} -1 & 0 \\ 5 & 2 \end{bmatrix} = \begin{bmatrix} 3 & 2 \\ 9 & 6 \end{bmatrix},$$

and

$$\mathbf{AC} = \begin{bmatrix} 2 & 1 \\ 6 & 3 \end{bmatrix}\begin{bmatrix} 3 & 1 \\ -3 & 0 \end{bmatrix} = \begin{bmatrix} 3 & 2 \\ 9 & 6 \end{bmatrix}.$$

And yet **B** is not equal to **C**. Similarly, just because **AB** equals **CB** for some matrices **A**, **B**, **C**, it does not necessarily follow that **A** equals **C**. (Can you find an example?)

In particular, **AB** can equal the zero matrix without either **A** or **B** being equal to the zero matrix. For example, consider

$$\mathbf{A} = \begin{bmatrix} 2 & 1 \\ 6 & 3 \end{bmatrix} \quad \text{and} \quad \mathbf{B} = \begin{bmatrix} -1 & 2 \\ 2 & -4 \end{bmatrix}.$$

Then

$$\mathbf{AB} = \begin{bmatrix} 2 & 1 \\ 6 & 3 \end{bmatrix}\begin{bmatrix} -1 & 2 \\ 2 & -4 \end{bmatrix} = \begin{bmatrix} 0 & 0 \\ 0 & 0 \end{bmatrix}.$$

And yet neither **A** nor **B** is the zero matrix $\mathbf{O}_2$.

Powers of Square Matrices

Any square matrix can be multiplied by itself, because the number of rows and columns of the matrix is the same. In fact, square matrices are the only matrices that can be multiplied by themselves. (Why?) The various nonnegative powers of a square matrix are defined in a natural way:

DEFINITION

Let **A** be any $n \times n$ matrix. Then the (nonnegative) powers of **A** are given by $\mathbf{A}^0 = \mathbf{I}_n$, $\mathbf{A}^1 = \mathbf{A}$, and for $k \geq 2$, $\mathbf{A}^k = (\mathbf{A}^{k-1})(\mathbf{A})$.

Example 4

Suppose that $\mathbf{A} = \begin{bmatrix} 2 & 1 \\ -4 & 3 \end{bmatrix}$. Then

$$\mathbf{A}^2 = (\mathbf{A})(\mathbf{A}) = \begin{bmatrix} 2 & 1 \\ -4 & 3 \end{bmatrix}\begin{bmatrix} 2 & 1 \\ -4 & 3 \end{bmatrix} = \begin{bmatrix} 0 & 5 \\ -20 & 5 \end{bmatrix},$$

$$\text{and} \quad \mathbf{A}^3 = (\mathbf{A}^2)(\mathbf{A}) = \begin{bmatrix} 0 & 5 \\ -20 & 5 \end{bmatrix}\begin{bmatrix} 2 & 1 \\ -4 & 3 \end{bmatrix} = \begin{bmatrix} -20 & 15 \\ -60 & -5 \end{bmatrix}.$$ ■

Example 5

The identity matrix $\mathbf{I}_n$ is a square matrix, and so we can consider its nonnegative powers. However, because $\mathbf{I}_n\mathbf{A} = \mathbf{A}$, for any $n \times n$ matrix **A**, we see that $\mathbf{I}_n\mathbf{I}_n = \mathbf{I}_n$. It follows easily that $\mathbf{I}_n^k = \mathbf{I}_n$, for all $k \geq 0$. ■

The next theorem asserts that two familiar laws of exponents in algebra are still valid when the exponents represent powers of a square matrix. The proof can be done easily by induction and is left for you to do in Exercise 16.

THEOREM 1.14

If $\mathbf{A}$ is a square matrix, and if s and t are nonnegative integers, then
(1) $\mathbf{A}^{s+t} = (\mathbf{A}^s)(\mathbf{A}^t)$
(2) $(\mathbf{A}^s)^t = \mathbf{A}^{st} = (\mathbf{A}^t)^s$

As an example of part (1) of this theorem, we have $\mathbf{A}^{4+6} = (\mathbf{A}^4)(\mathbf{A}^6) = \mathbf{A}^{10}$. As an example of part (2), we have $(\mathbf{A}^3)^2 = \mathbf{A}^{(3)(2)} = (\mathbf{A}^2)^3 = \mathbf{A}^6$.

One law of exponents in elementary algebra that does not carry over to matrix algebra is $(xy)^q = x^q y^q$. In fact, if $\mathbf{A}$ and $\mathbf{B}$ are square matrices of the same size, it usually happens that $(\mathbf{AB})^q \neq \mathbf{A}^q\mathbf{B}^q$, if q is an integer ≥ 2. This result occurs because the *order* of matrix multiplication is important. Even in the simplest case $q = 2$, we usually have $(\mathbf{AB})(\mathbf{AB}) \neq (\mathbf{AA})(\mathbf{BB})$.

Example 6

Let

$$\mathbf{A} = \begin{bmatrix} 2 & -4 \\ 1 & 3 \end{bmatrix} \quad \text{and} \quad \mathbf{B} = \begin{bmatrix} 3 & 2 \\ -1 & 5 \end{bmatrix}.$$

Then

$$(\mathbf{AB})^2 = \begin{bmatrix} 10 & -16 \\ 0 & 17 \end{bmatrix}^2 = \begin{bmatrix} 100 & -432 \\ 0 & 289 \end{bmatrix}.$$

However,

$$\mathbf{A}^2\mathbf{B}^2 = \begin{bmatrix} 0 & -20 \\ 5 & 5 \end{bmatrix}\begin{bmatrix} 7 & 16 \\ -8 & 23 \end{bmatrix} = \begin{bmatrix} 160 & -460 \\ -5 & 195 \end{bmatrix}.$$

Hence, for these matrices $\mathbf{A}$ and $\mathbf{B}$, we have $(\mathbf{AB})^2 \neq \mathbf{A}^2\mathbf{B}^2$. ■

The Transpose of a Matrix Product

The next theorem presents an interesting result about the transpose of a matrix product:

THEOREM 1.15

If $\mathbf{A}$ is an $m \times n$ matrix and $\mathbf{B}$ is an $n \times p$ matrix, then $(\mathbf{AB})^T = \mathbf{B}^T\mathbf{A}^T$.

This result may seem unusual at first, because you might expect $(\mathbf{AB})^T$ to equal $\mathbf{A}^T\mathbf{B}^T$. But notice that the product $\mathbf{A}^T\mathbf{B}^T$ may not even be defined, because $\mathbf{A}^T$ is an $n \times m$ matrix and $\mathbf{B}^T$ is a $p \times n$ matrix. This theorem asserts instead that the transpose of the product of two matrices is precisely the product of their transposes *in reverse order*.

Proof of Theorem 1.15

Because **AB** is an $m \times p$ matrix and $\mathbf{B}^T$ is a $p \times n$ matrix and $\mathbf{A}^T$ is an $n \times m$ matrix, it follows that $(\mathbf{AB})^T$ and $\mathbf{B}^T\mathbf{A}^T$ are both $p \times m$ matrices. Hence, the proof will be complete if we can show that the $(i, j)^{th}$ entries of $(\mathbf{AB})^T$ and $\mathbf{B}^T\mathbf{A}^T$ are equal, for $1 \leq i \leq p$ and $1 \leq j \leq m$. Now, the $(i, j)^{th}$ entry of $(\mathbf{AB})^T$ is the $(j, i)^{th}$ entry of **AB**, which equals $[j^{th}$ row of $\mathbf{A}] \cdot [i^{th}$ column of $\mathbf{B}]$. However, the $(i, j)^{th}$ entry of $\mathbf{B}^T\mathbf{A}^T$ equals $[i^{th}$ row of $\mathbf{B}^T] \cdot [j^{th}$ column of $\mathbf{A}^T]$, which equals $[i^{th}$ column of $\mathbf{B}] \cdot [j^{th}$ row of $\mathbf{A}]$. Thus, the $(i, j)^{th}$ entries of $(\mathbf{AB})^T$ and $\mathbf{B}^T\mathbf{A}^T$ are equal. ■

Example 7

For the matrices **A** and **B** of Example 6, we have

$$\mathbf{AB} = \begin{bmatrix} 10 & -16 \\ 0 & -17 \end{bmatrix}, \quad \mathbf{B}^T = \begin{bmatrix} 3 & -1 \\ 2 & 5 \end{bmatrix}, \quad \text{and} \quad \mathbf{A}^T = \begin{bmatrix} 2 & 1 \\ -4 & 3 \end{bmatrix}.$$

Hence,

$$\mathbf{B}^T\mathbf{A}^T = \begin{bmatrix} 3 & -1 \\ 2 & 5 \end{bmatrix}\begin{bmatrix} 2 & 1 \\ -4 & 3 \end{bmatrix} = \begin{bmatrix} 10 & 0 \\ -16 & 17 \end{bmatrix} = (\mathbf{AB})^T.$$

Notice, however, that

$$\mathbf{A}^T\mathbf{B}^T = \begin{bmatrix} 2 & 1 \\ -4 & 3 \end{bmatrix}\begin{bmatrix} 3 & -1 \\ 2 & 5 \end{bmatrix} = \begin{bmatrix} 8 & 3 \\ -6 & 19 \end{bmatrix} \neq (\mathbf{AB})^T.$$

■

♦ **Application:** You now have covered the prerequisites for Section 8.1, "Graph Theory."

Exercises—Section 1.5

Note: Exercises 1 through 4 refer to the following matrices:

$$\mathbf{A} = \begin{bmatrix} -2 & 3 \\ 6 & 5 \\ 1 & -4 \end{bmatrix} \quad \mathbf{B} = \begin{bmatrix} -5 & 3 & 6 \\ 3 & 8 & 0 \\ -2 & 0 & 4 \end{bmatrix} \quad \mathbf{C} = \begin{bmatrix} 11 & -2 \\ -4 & -2 \\ 3 & -1 \end{bmatrix}$$

$$\mathbf{D} = \begin{bmatrix} -1 & 4 & 3 & 7 \\ 2 & 1 & 7 & 5 \\ 0 & 5 & 5 & -2 \end{bmatrix} \quad \mathbf{E} = \begin{bmatrix} 1 & 1 & 0 & 1 \\ 1 & 0 & 1 & 0 \\ 0 & 0 & 0 & 1 \\ 1 & 0 & 1 & 0 \end{bmatrix} \quad \mathbf{F} = \begin{bmatrix} 9 & -3 \\ 5 & -4 \\ 2 & 0 \\ 8 & -3 \end{bmatrix}$$

$$\mathbf{G} = \begin{bmatrix} 5 & 1 & 0 \\ 0 & -2 & -1 \\ 1 & 0 & 3 \end{bmatrix} \quad \mathbf{H} = \begin{bmatrix} 6 & 3 & 1 \\ 1 & -15 & -5 \\ -2 & -1 & 10 \end{bmatrix} \quad \mathbf{J} = \begin{bmatrix} 8 \\ -1 \\ 4 \end{bmatrix}$$

$$\mathbf{K} = \begin{bmatrix} 2 & 1 & -5 \\ 0 & 2 & 7 \end{bmatrix} \quad \mathbf{L} = \begin{bmatrix} 10 & 9 \\ 8 & 7 \end{bmatrix} \quad \mathbf{M} = \begin{bmatrix} 7 & -1 \\ 11 & 3 \end{bmatrix}$$

$$\mathbf{N}=\begin{bmatrix}0 & 0\\0 & 0\end{bmatrix}\qquad \mathbf{P}=\begin{bmatrix}3 & -1\\4 & 7\end{bmatrix}\qquad \mathbf{Q}=\begin{bmatrix}1 & 4 & -1 & 6\\8 & 7 & -3 & 3\end{bmatrix}$$

$$\mathbf{R}=[-3\quad 6\quad -2]\quad \mathbf{S}=[6\quad -4\quad 3\quad 2]\quad \mathbf{T}=[4\quad -1\quad 7]$$

1. Indicate which of the following products can be performed. If the multiplication can be performed, calculate the result.

(a) **AB** ★(b) **BA** ★(c) **JM** (d) **DF** ★(e) **RJ**
★(f) **JR** ★(g) **RT** (h) **SF** (i) **KN** ★(j) $\mathbf{F}^2$
(k) $\mathbf{B}^2$ ★(l) $\mathbf{E}^3$ (m) $(\mathbf{TJ})^3$ ★(n) **D(FK)** (o) **(CL)G**

2. Determine whether these pairs of matrices commute:

★(a) **L** and **M** (b) **G** and **H** ★(c) **A** and **K**
★(d) **N** and **P** (e) **F** and **Q**

3. Find only the indicated row or column of the given matrix product:

★(a) The second row of **BG** (b) The third column of **DE**
★(c) The first column of **SE** (d) The third row of **FQ**

★ **4.** Determine whether each of the following equations holds for the given matrices by performing the computations on each side of each equation. Also, if the equation follows from theorems in this section, specify which theorems (and parts, if appropriate) apply.

(a) $(\mathbf{RG})\mathbf{H} = \mathbf{R}(\mathbf{GH})$ (b) $\mathbf{LP} = \mathbf{PL}$
(c) $\mathbf{E}(\mathbf{FK}) = (\mathbf{EF})\mathbf{K}$ (d) $\mathbf{K}(\mathbf{A} + \mathbf{C}) = \mathbf{KA} + \mathbf{KC}$
(e) $(\mathbf{QF})^T = \mathbf{F}^T\mathbf{Q}^T$ (f) $\mathbf{L}(\mathbf{ML}) = \mathbf{L}^2\mathbf{M}$
(g) $\mathbf{GC} + \mathbf{HC} = (\mathbf{G} + \mathbf{H})\mathbf{C}$ (h) $(\mathbf{GH})^T = \mathbf{G}^T\mathbf{H}^T$
(i) $\mathbf{R}(\mathbf{J} + \mathbf{T}^T) = \mathbf{RJ} + \mathbf{RT}^T$ (j) $(\mathbf{AK})^T = \mathbf{A}^T\mathbf{K}^T$
(k) $(\mathbf{Q} + \mathbf{F}^T)\mathbf{E}^T = \mathbf{QE}^T + (\mathbf{EF})^T$

★ **5.** Assume that the following matrices give information about the number of employees and their wages and benefits (per year) at four different retail outlets. Calculate the total amount of salaries and fringe benefits paid out by each outlet per year to its employees.

	Executives	Salespersons	Others
Outlet 1	3	7	8
Outlet 2	2	4	5
Outlet 3	6	14	18
Outlet 4	3	6	9

	Salary	Fringe Benefits
Executives	$30000	$7500
Salespersons	$22500	$4500
Others	$15000	$3000

★ **6.** The matrix **A** that follows represents the percentage of nitrogen, phosphates, and potash in three fertilizers. The matrix **B** represents the amount (in tons) of each type of fertilizer that is spread on three different fields. Use this information and matrix operations to find the total amount of nitrogen, phosphates, and potash spread on each field:

$$\mathbf{A} = \begin{array}{l} \\ \text{Fertilizer 1} \\ \text{Fertilizer 2} \\ \text{Fertilizer 3} \end{array} \begin{array}{c} \begin{array}{ccc} \text{Nitrogen} & \text{Phosphate} & \text{Potash} \end{array} \\ \begin{bmatrix} 10\% & 10\% & 5\% \\ 25\% & 5\% & 5\% \\ 0\% & 10\% & 20\% \end{bmatrix} \end{array}$$

$$\mathbf{B} = \begin{array}{l} \\ \text{Fertilizer 1} \\ \text{Fertilizer 2} \\ \text{Fertilizer 3} \end{array} \begin{array}{c} \begin{array}{ccc} \text{Field 1} & \text{Field 2} & \text{Field 3} \end{array} \\ \begin{bmatrix} 5 & 2 & 4 \\ 2 & 1 & 1 \\ 3 & 1 & 3 \end{bmatrix} \end{array}$$

★ **7.** (a) Find a nondiagonal matrix **A** such that $\mathbf{A}^2 = \mathbf{I}_2$.

(b) Find a nondiagonal matrix **A** such that $\mathbf{A}^2 = \mathbf{I}_3$. (Hint: Modify your answer to part (a).)

(c) Find a nonidentity matrix **A** such that $\mathbf{A}^3 = \mathbf{I}_3$.

8. Let **A** be an $m \times n$ matrix, and let **B** be an $n \times m$ matrix, with $m, n \geq 5$. Each of the following sums represents an entry of either **AB** or **BA**. Determine which matrix product is involved and which entry of that product is represented in each case.

★(a) $\sum_{k=1}^{n} a_{3k} b_{k4}$ (b) $\sum_{q=1}^{n} a_{4q} b_{q1}$

★(c) $\sum_{k=1}^{m} a_{k2} b_{3k}$ (d) $\sum_{q=1}^{m} b_{2q} a_{q5}$

9. Let **A** be an $m \times n$ matrix, and let **B** be an $n \times m$ matrix, where $m, n \geq 4$. Use sigma (Σ) notation to express the following entries symbolically:

★(a) In the product **AB**, the entry in the third row and second column

(b) In the product **BA**, the entry in the fourth row and first column

10. (a) Consider the unit vectors **i**, **j**, and **k** in $\mathbb{R}^3$. Show that, if **A** is an $m \times 3$ matrix, then $\mathbf{Ai}$ = first column of **A**, $\mathbf{Aj}$ = second column of **A**, and $\mathbf{Ak}$ = third column of **A**.

(b) Generalize part (a) to a similar result involving an $m \times n$ matrix **A** and the standard unit vectors $\mathbf{e}_1, \ldots, \mathbf{e}_n$ in $\mathbb{R}^n$.

(c) Let **A** be an $m \times n$ matrix. Use part (b) to show that, if $\mathbf{Ax} = \mathbf{0}$ for all vectors $\mathbf{x} \in \mathbb{R}^n$, then $\mathbf{A} = \mathbf{O}_{mn}$.

11. Prove parts (2), (3), and (4) of Theorem 1.13.

12. Show that the product of any $m \times n$ matrix with $\mathbf{O}_{np}$ is $\mathbf{O}_{mp}$.

13. Let **A** be an $m \times n$ matrix. Prove that $\mathbf{AI}_n = \mathbf{I}_m\mathbf{A} = \mathbf{A}$.

14. (a) Prove that the product of two diagonal matrices is diagonal. (Hint: If $\mathbf{C} = \mathbf{AB}$, where **A** and **B** are diagonal, show that $c_{ij} = 0$ when $i \neq j$.)

(b) Prove that the product of two upper triangular matrices is upper triangular. (Hint: Show that if **A** and **B** are upper triangular and $\mathbf{C} = \mathbf{AB}$, then $c_{ij} = 0$ when $i > j$ by showing that all the terms $a_{ik}b_{kj}$ in the formula for c_{ij} have at least one factor equal to zero. Consider two cases: $i > k$ and $i \leq k$.)

(c) Prove that the product of two lower triangular matrices is lower triangular. (Hint: Use Theorem 1.15 and part (b) of this exercise.)

15. Show that if $c \in \mathbb{R}$ and **A** is a square matrix, then $(c\mathbf{A})^n = c^n\mathbf{A}^n$ for any integer $n \geq 1$. (Hint: Use a proof by induction.)

16. Prove each part of Theorem 1.14 using the method of induction. (Hint: Use induction on t for both parts. Part (1) will be useful in proving part (2).)

17. (a) Show that $\mathbf{AB} = \mathbf{BA}$ only if **A** and **B** are square matrices of the same size.

(b) Prove that two square matrices **A** and **B** of the same size commute if and only if $(\mathbf{A} + \mathbf{B})^2 = \mathbf{A}^2 + 2\mathbf{AB} + \mathbf{B}^2$.

18. If **A**, **B**, and **C** are all square matrices of the same size, show that **AB** commutes with **C** if **A** and **B** both commute with **C**.

19. Show that matrices **A** and **B** commute if and only if $\mathbf{A}^T$ and $\mathbf{B}^T$ commute.

20. Let **A** and **B** both be $n \times n$ matrices.

(a) Show that $\mathbf{AA}^T$ and $\mathbf{A}^T\mathbf{A}$ are both symmetric matrices.

(b) Show that $(\mathbf{AB})^T = \mathbf{BA}$ if **A** and **B** are both symmetric or both skew-symmetric.

(c) If **A** and **B** are both symmetric, show that **AB** is symmetric if and only if **A** and **B** commute.

21. Recall the definition of the **trace** of a matrix given in Exercise 14 of Section 1.4. If **A** and **B** are both $n \times n$ matrices, show that

(a) Trace($\mathbf{AA}^T$) is the sum of the squares of all entries of **A**.

(b) If trace($\mathbf{AA}^T$) = 0, then $\mathbf{A} = \mathbf{O}_n$. (Hint: Use part (a) of this exercise.)

(c) Trace(**AB**) = trace(**BA**). (Hint: Work out the terms for trace(**AB**) and trace(**BA**) in the 3×3 case to get an idea of how to attack the proof in the general $n \times n$ case.)

22. An **idempotent** matrix is a square matrix **A** for which $\mathbf{A}^2 = \mathbf{A}$. (It follows that if **A** is idempotent, then $\mathbf{A}^n = \mathbf{A}$ for every integer $n \geq 1$.)

★(a) Find a 2×2 matrix (besides $\mathbf{I}_n$ and $\mathbf{O}_n$) that is idempotent.

(b) Show that $\begin{bmatrix} -1 & 1 & 1 \\ -1 & 1 & 1 \\ -1 & 1 & 1 \end{bmatrix}$ is idempotent.

(c) If **A** is an $n \times n$ idempotent matrix, show that $\mathbf{I}_n - \mathbf{A}$ is also idempotent.

(d) Use the results of parts (b) and (c) to get another example of an idempotent matrix.

(e) Let **A** and **B** be $n \times n$ matrices. Show that **A** is idempotent if both $\mathbf{AB} = \mathbf{A}$ and $\mathbf{BA} = \mathbf{B}$.

23. (a) Let **A** be an $m \times n$ matrix, and let **B** be an $n \times p$ matrix. Prove that $\mathbf{AB} = \mathbf{O}_{mp}$ if and only if every row of **A**, considered as a vector, is orthogonal to each column of **B**.

★(b) Find a 2×3 matrix **A** and a 3×2 matrix **B** such that neither **A** nor **B** is a zero matrix and $\mathbf{AB} = \mathbf{O}_2$.

(c) Using your answers from part (b), find a matrix **C** such that $\mathbf{AB} = \mathbf{AC}$ but $\mathbf{B} \neq \mathbf{C}$.

★**24.** What form does a 2×2 matrix have if it commutes with every other 2×2 matrix?

25. Let **A** be an $n \times n$ matrix. Consider the $n \times n$ matrix $\mathbf{B}_{ij}$, which has all entries zero except for an entry of 1 in the $(i, j)^{th}$ position.

(a) Show that the jth column of $\mathbf{AB}_{ij}$ equals the ith column of **A** and that all other columns of $\mathbf{AB}_{ij}$ have only zero entries.

(b) Show that the ith row of $\mathbf{B}_{ij}\mathbf{A}$ equals the jth row of **A** and that all other rows of $\mathbf{B}_{ij}\mathbf{A}$ have only zero entries.

(c) Use the results in parts (a) and (b) to prove that an $n \times n$ matrix **A** commutes with all other $n \times n$ matrices if and only if $\mathbf{A} = c\mathbf{I}_n$, for some $c \in \mathbb{R}$. (Hint: Use $\mathbf{AB}_{kk} = \mathbf{B}_{kk}\mathbf{A}$, for $1 \leq k \leq n$, to prove $a_{ij} = 0$ for $i \neq j$. Then use $\mathbf{AB}_{ij} = \mathbf{B}_{ij}\mathbf{A}$ to show $a_{ii} = a_{jj}$.)

2

Systems of Linear Equations

The original attempts to solve systems of linear equations inspired much of the development of linear algebra as we know it today. In Sections 2.1 and 2.2, we present Gauss-Jordan row reduction, the most important method for solving linear systems. The study of linear systems leads to more results concerning matrices, including the rank and the row space of a matrix in Sections 2.2 and 2.3, inverses of matrices in Section 2.4, and elementary matrices in Section 2.5.

2.1 Solving Systems of Linear Equations

In this section we introduce systems of linear equations and the Gauss-Jordan row reduction method for solving such systems. After giving several examples of the method and discussing it informally, we give a formal description of the algorithm.

A **linear equation** is an equation involving one or more variables in which only the operations of multiplication by real numbers and summing of terms are allowed. For example, $6x - 3y = 4$ and $8x_1 + 3x_2 - 4x_3 = -20$ are examples of linear equations in two and three variables, respectively.

When linear equations are collected together, we have a **system of linear equations**. For example, the following is a system with four equations and three variables:

$$\begin{cases} 3x_1 - 2x_2 - 5x_3 = 4 \\ 2x_1 + 4x_2 - x_3 = 2 \\ 6x_1 - 4x_2 - 10x_3 = 8 \\ -4x_1 + 8x_2 + 9x_3 = -6 \end{cases}$$

We often want to know the solutions to a given system. The ordered triple, or 3-tuple, $(x_1, x_2, x_3) = (4, -1, 2)$ is a solution to the preceding system because each equation in the system is satisfied for these values of x_1, x_2, and x_3. Notice that

$(-\frac{3}{2}, \frac{3}{4}, -2)$ is another solution for that same system. These two particular solutions are part of the complete set of all solutions for that system.

Let us define linear systems and their solutions in a more formal manner:

DEFINITION

A **system** of m (simultaneous) linear equations in n variables

$$\begin{cases} a_{11}x_1 & + & a_{12}x_2 & + & a_{13}x_3 & + & \cdots & + & a_{1n}x_n & = & b_1 \\ a_{21}x_1 & + & a_{22}x_2 & + & a_{23}x_3 & + & \cdots & + & a_{2n}x_n & = & b_2 \\ \vdots & & \vdots & & \vdots & & \ddots & & \vdots & & \vdots \\ a_{m1}x_1 & + & a_{m2}x_2 & + & a_{m3}x_3 & + & \cdots & + & a_{mn}x_n & = & b_m \end{cases}$$

is a collection of m equations, each containing a linear combination of the same n variables summing to a scalar. A **particular solution** to a system of linear equations in the variables $x_1, x_2, \ldots, x_n$ is an n-tuple $(s_1, s_2, \ldots, s_n)$ that satisfies each equation in the system when s_1 is substituted for x_1, s_2 for x_2, and so on. The **(complete) solution set** for a system of linear equations in n variables is the collection of all n-tuples that form solutions to the system.

The coefficients of $x_1, x_2, \ldots, x_n$ in the definition of a linear system can be collected together in an $m \times n$ **coefficient matrix**:

$$\mathbf{A} = \begin{bmatrix} a_{11} & a_{12} & \cdots & a_{1n} \\ a_{21} & a_{22} & \cdots & a_{2n} \\ \vdots & \vdots & \ddots & \vdots \\ a_{m1} & a_{m2} & \cdots & a_{mn} \end{bmatrix}.$$

If we also let

$$\mathbf{X} = \begin{bmatrix} x_1 \\ x_2 \\ \vdots \\ x_n \end{bmatrix} \quad \text{and} \quad \mathbf{B} = \begin{bmatrix} b_1 \\ b_2 \\ \vdots \\ b_m \end{bmatrix}$$

then the entire linear system is equivalent to the matrix equation $\mathbf{AX} = \mathbf{B}$. (Verify this.)

An alternate way to express this system is to form the **augmented matrix**:

$$[\mathbf{A} \mid \mathbf{B}] = \left[\begin{array}{cccc|c} a_{11} & a_{12} & \cdots & a_{1n} & b_1 \\ a_{21} & a_{22} & \cdots & a_{2n} & b_2 \\ \vdots & \vdots & \ddots & \vdots & \vdots \\ a_{m1} & a_{m2} & \cdots & a_{mn} & b_m \end{array}\right].$$

This augmented matrix contains all the vital information from the original system. Each row of $[\mathbf{A}|\mathbf{B}]$ represents one equation in the original system, and each column to the left of the vertical bar represents one of the variables in the system.

Example 1

The system

$$\begin{cases} 4w - 2x + y - 3z = 5 \\ 3w + x + 5z = 12 \end{cases}$$

can be represented by the augmented matrix

$$[\mathbf{A}|\mathbf{B}] = \left[\begin{array}{cccc|c} 4 & -2 & 1 & -3 & 5 \\ 3 & 1 & 0 & 5 & 12 \end{array}\right],$$

$$\text{where} \quad \mathbf{A} = \begin{bmatrix} 4 & -2 & 1 & -3 \\ 3 & 1 & 0 & 5 \end{bmatrix} \quad \text{and} \quad \mathbf{B} = \begin{bmatrix} 5 \\ 12 \end{bmatrix}.$$

We can also express this system in $\mathbf{AX} = \mathbf{B}$ form as

$$\begin{bmatrix} 4 & -2 & 1 & -3 \\ 3 & 1 & 0 & 5 \end{bmatrix} \begin{bmatrix} w \\ x \\ y \\ z \end{bmatrix} = \begin{bmatrix} 5 \\ 12 \end{bmatrix}.$$

■

Number of Solutions to a System

There are only three possibilities for the solution set of a linear system: a single solution, an infinite number of solutions, or no solutions. There is no other possibility for the number of solutions because, if at least two solutions exist, we can show that an infinite number of solutions must exist. You are asked to prove this statement in Exercise 11. In the special case of a system of two equations and two variables—say, x and y—it is easy to see why there are only these three possibilities for the number of solutions. The set of ordered pairs satisfying each equation forms a line in the xy-plane. Hence, the solutions to the system are the points where the two lines intersect. But any two given lines in the plane either intersect in exactly one point (unique solution) or are equal (infinite number of solutions, with all points on the common line) or are parallel (no solutions).

For example, the system

$$\begin{cases} 4x_1 - 3x_2 = 0 \\ 2x_1 + 3x_2 = 18 \end{cases}$$

(where x_1 and x_2 are used instead of x and y) has the unique solution $(3, 4)$, because that is the intersection point of the two lines.

On the other hand, the system

$$\begin{cases} 4x - 6y = 10 \\ 6x - 9y = 15 \end{cases}$$

has an infinite number of solutions, because the two given lines are really the same, and so every point on one line is also on the other.

Finally, consider the system

$$\begin{cases} 2x_1 + x_2 = 3 \\ 2x_1 + x_2 = 1 \end{cases}.$$

This system has no solutions at all, because the two lines are parallel but not equal. (Their slopes are both -2.) The solution set for this system is the empty set $\{\} = \emptyset$. All three of these systems are pictured in Figure 2.1.

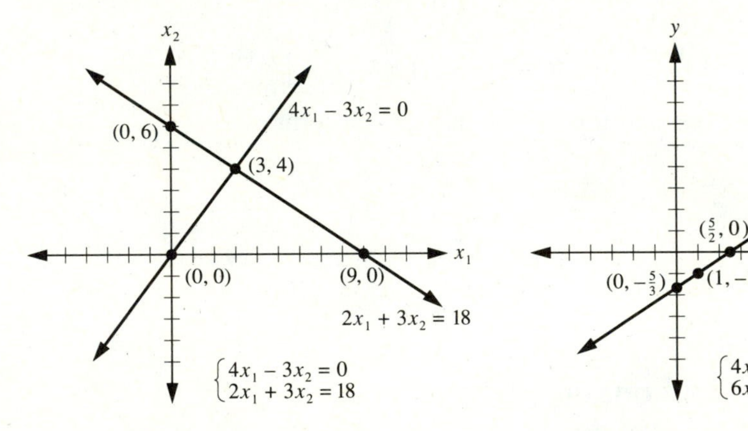

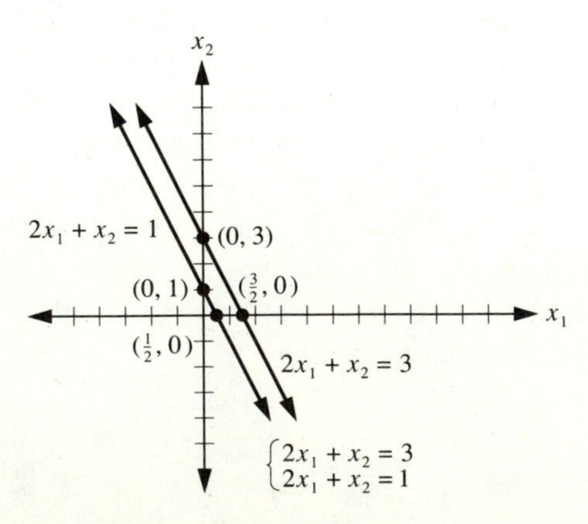

Figure 2.1
Three systems: unique solution, infinite number of solutions, no solution

Any system that has at least one solution (either unique or infinitely many) is said to be **consistent**. A system whose solution set is empty is called **inconsistent**. The first two systems in Figure 2.1 are consistent, and the last one is inconsistent.

Introduction to Gauss-Jordan Row Reduction

Many methods are available for finding the complete solution set for a given linear system. The one we present here, **Gauss-Jordan row reduction**, is highly systematic. For this reason, computers can easily be programmed to perform this technique. At first we work through several examples by hand so you can fully master the strategy involved, but we recommend that later you use computer software to perform row reduction when necessary. A formal algorithm for the Gauss-Jordan row reduction process appears later in this section. The proof that this process gives the correct solution set is presented in Section 2.2.

We begin with the basic steps in the algorithm: row operations. These operations are performed on the augmented matrix for a linear system until it is transformed into a new augmented matrix that reveals the complete solution set. The three allowable operations are

ROW OPERATIONS

(I) Multiplying a row by a *nonzero* scalar
(II) Adding a scalar multiple of one row to another row
(III) Switching the positions of two rows in the matrix

Let us illustrate the use of these operations with an example:

Example 2

Our goal is to solve the following system of linear equations:

$$\begin{cases} 5x - 5y - 15z = 40 \\ 4x - 2y - 6z = 19 \\ 3x - 6y - 17z = 41 \end{cases}.$$

The augmented matrix associated with this system is

$$\left[\begin{array}{ccc|c} 5 & -5 & -15 & 40 \\ 4 & -2 & -6 & 19 \\ 3 & -6 & -17 & 41 \end{array}\right].$$

We perform row operations on this matrix to give it a simpler form.

The first operation we use is of type (I): multiply row 1 by $\frac{1}{5}$. This operation places 1 in the (1, 1) position of the matrix, which aids our simplification process.

The resulting matrix is

$$\left[\begin{array}{rrr|r} 1 & -1 & -3 & 8 \\ 4 & -2 & -6 & 19 \\ 3 & -6 & -17 & 41 \end{array}\right].$$

The next row operation is performed on this resultant matrix. It is of type (II): we add $(-4) \times$ (row 1) to row 2 (a scalar multiple of one row added to another row). To perform this operation, we first do the following side calculation:

$$\begin{array}{rrrr|r} (-4) \times (\text{row } 1) & -4 & 4 & 12 & -32 \\ (\text{row } 2) & 4 & -2 & -6 & 19 \\ \hline (\text{sum}) & 0 & 2 & 6 & -13 \end{array}.$$

The resulting sum is now substituted in place of the old row 2, producing

$$\left[\begin{array}{rrr|r} 1 & -1 & -3 & 8 \\ 0 & 2 & 6 & -13 \\ 3 & -6 & -17 & 41 \end{array}\right].$$

Note that, even though we multiplied row 1 by -4 in the side calculation, row 1 itself was not changed when the type (II) row operation was finished. Only row 2 was altered by the operation.

The next row operation is also of type (II).

Type (II) operation: replace row 3 with $(-3) \times$ (row 1) + (row 3)

Side calculation					Resultant matrix
$(-3) \times$ (row 1)	-3	3	9	-24	$\left[\begin{array}{rrr\|r} 1 & -1 & -3 & 8 \\ 0 & 2 & 6 & -13 \\ 0 & -3 & -8 & 17 \end{array}\right]$
(row 3)	3	-6	-17	41	
(sum)	0	-3	-8	17	

Now each augmented matrix has a corresponding linear system. For example, the last matrix is associated with

$$\begin{cases} x - y - 3z = 8 \\ \quad\;\; 2y + 6z = -13 \\ \;\; -3y - 8z = 17 \end{cases}.$$

Note that x has been eliminated from the second and third equations, which makes this system simpler than the original.

Next we eliminate y from the first and third equations. We first perform a type (I) operation to obtain 1 in the (2, 2) position. This operation changes the coefficient of y in the second equation to 1, which makes it easier to solve for y when the row reduction is complete. Additionally, the 1 makes it easier to obtain the desired zeroes above and below the (2, 2) position.

Type (I) operation: multiply row 2 by $\frac{1}{2}$

$$\text{Resultant matrix} = \left[\begin{array}{ccc|c} 1 & -1 & -3 & 8 \\ 0 & 1 & 3 & -\frac{13}{2} \\ 0 & -3 & -8 & 17 \end{array}\right]$$

We now use two type (II) operations to change the $(1, 2)$ and $(3, 2)$ entries of the matrix to zeroes.

Type (II) operation: replace row 1 with $(1) \times (\text{row } 2) + (\text{row } 1)$

Side calculation

$$\begin{array}{rccc|c} (1) \times (\text{row } 2) & 0 & 1 & 3 & -\frac{13}{2} \\ (\text{row } 1) & 1 & -1 & -3 & 8 \\ \hline (\text{sum}) & 1 & 0 & 0 & \frac{3}{2} \end{array}$$

Resultant matrix

$$\left[\begin{array}{ccc|c} 1 & 0 & 0 & \frac{3}{2} \\ 0 & 1 & 3 & -\frac{13}{2} \\ 0 & -3 & -8 & 17 \end{array}\right]$$

Type (II) operation: replace row 3 with $3 \times (\text{row } 2) + (\text{row } 3)$

Side calculation

$$\begin{array}{rccc|c} (3) \times (\text{row } 2) & 0 & 3 & 9 & -\frac{39}{2} \\ (\text{row } 3) & 0 & -3 & -8 & 17 \\ \hline (\text{sum}) & 0 & 0 & 1 & -\frac{5}{2} \end{array}$$

Resultant matrix

$$\left[\begin{array}{ccc|c} 1 & 0 & 0 & \frac{3}{2} \\ 0 & 1 & 3 & -\frac{13}{2} \\ 0 & 0 & 1 & -\frac{5}{2} \end{array}\right].$$

The last matrix corresponds to

$$\begin{cases} x & & & = & \frac{3}{2} \\ & y & + \; 3z & = & -\frac{13}{2} \\ & & z & = & -\frac{5}{2} \end{cases}.$$

To finish solving the system, we eliminate z from the first and second equations. Luckily, we already have 1 in the $(3, 3)$ position and zero in the $(1, 3)$ position, as desired. We perform the following operation to place a zero in the $(2, 3)$ position.

Type (II) operation: replace row 2 with $(-3) \times$ (row 3) + (row 2)

Side calculation					Resultant matrix
$(-3) \times$ (row 3)	0	0	-3	$\frac{15}{2}$	
(row 2)	0	1	3	$-\frac{13}{2}$	
(sum)	0	1	0	1	

$$\left[\begin{array}{ccc|c} 1 & 0 & 0 & \frac{3}{2} \\ 0 & 1 & 0 & 1 \\ 0 & 0 & 1 & -\frac{5}{2} \end{array}\right]$$

The process is now complete. The final matrix is associated with the system

$$\begin{cases} x & = \frac{3}{2} \\ \quad y & = 1 \\ \quad\quad z & = -\frac{5}{2} \end{cases},$$

whose (unique) solution is the ordered triple $(\frac{3}{2}, 1, -\frac{5}{2})$. You can easily check by substitution that $(\frac{3}{2}, 1, -\frac{5}{2})$ is a solution to the original system. As you will see, the row reduction process always produces the complete solution set for a linear system, and so $(\frac{3}{2}, 1, -\frac{5}{2})$ is the unique solution to the original problem. ■

The Strategy in Its Simplest Form

When simplifying the augmented matrix associated with a system, we work on one column at a time. Initially, we begin with the first nonzero column of the matrix. After a column is completely simplified, we move to the next column to the right. Note that in Example 2 we completely simplified the first column before simplifying the second.

While working on a particular column, one row is singled out as special. This is called the **home row**. We begin with the first row as the home row and move down as the row reduction progresses. The matrix entry that lies in both the current home row and in the current column is called the **pivot element** (or **pivot**, for short). These are the basic steps of the method:

(1) The first step when working with a new column is to convert the pivot element to 1. Usually, you do this by multiplying the home row by the reciprocal of the current pivot element. This was the purpose of the type (I) operations used in Example 2. Occasionally, the pivot element is zero, in which case its reciprocal is undefined. We deal with this situation later in this section.

(2) After converting the pivot element to 1, use type (II) operations to obtain zero for every other entry in the current column. This step eliminates the variable corresponding to that column from each equation in the system, except in the home row. This was the purpose of the type (II) operations used in Example 2. When obtaining zero in a given position of the current column, be careful to add an appropriate multiple of the *home row* to the row containing that position. Any other type (II) operation could destroy work done in previous columns.

(3) Once all entries in the current column are zero, except for the pivot element, advance to the next column to the right, and advance the home row to the next row down.

(4) In solving linear systems, stop either when you run out of rows to use as the home row or when you run out of columns before the augmentation bar (or both).

Using Type (III) Operations

So far, we have used only type (I) and type (II) operations. However, if the pivot element is zero when you begin work on a new column, it is impossible to convert the pivot to 1 using a type (I) operation. Frequently, this problem can be solved by using a type (III) operation to switch the home row with another row *below* it.

Example 3

Let us solve the following system using Gauss-Jordan row reduction:

$$\begin{cases} 3x + y = -5 \\ -6x - 2y = 10 \\ 4x + 5y = 8 \end{cases}.$$

The augmented matrix is

$$\left[\begin{array}{cc|c} 3 & 1 & -5 \\ -6 & -2 & 10 \\ 4 & 5 & 8 \end{array}\right].$$

We begin with row 1 as the home row and start with column 1. The pivot is in the $(1, 1)$ position. Our first task is to convert it to 1.

Type (I) operation: multiply row 1 by $\frac{1}{3}$

$$\text{Resultant matrix} = \left[\begin{array}{cc|c} 1 & \frac{1}{3} & -\frac{5}{3} \\ -6 & -2 & 10 \\ 4 & 5 & 8 \end{array}\right]$$

Next, we use type (II) operations to produce zeros in the rest of the first column by adding some multiple of the home row (the first row) to the other rows.

Type (II) operation: replace row 2 with $6 \times$ (row 1) + (row 2)

Type (II) operation: replace row 3 with $(-4) \times$ (row 1) + (row 3)

$$\text{Resultant matrix} = \left[\begin{array}{cc|c} 1 & \frac{1}{3} & -\frac{5}{3} \\ 0 & 0 & 0 \\ 0 & \frac{11}{3} & \frac{44}{3} \end{array}\right]$$

These steps finish the simplification of the first column. Therefore, we advance to column 2 and advance the home row to row 2. The pivot is now the (2, 2) entry. Our goal is to convert this to 1, but because the pivot is zero, a type (I) operation will not work. Instead, we first perform a type (III) operation to change the pivot to a nonzero number.

Type (III) operation: switch row 2 and row 3

$$\text{Resultant matrix} = \left[\begin{array}{cc|c} 1 & \frac{1}{3} & -\frac{5}{3} \\ 0 & \frac{11}{3} & \frac{44}{3} \\ 0 & 0 & 0 \end{array}\right]$$

Using a type (I) operation, we can now convert the pivot to 1.

Type (I) operation: multiply row 2 by $\frac{3}{11}$

$$\text{Resultant matrix} = \left[\begin{array}{cc|c} 1 & \frac{1}{3} & -\frac{5}{3} \\ 0 & 1 & 4 \\ 0 & 0 & 0 \end{array}\right]$$

Finally, we "zero out" the remaining nonzero entry of the second column.

Type (II) operation: replace row 1 with $-\frac{1}{3} \times$ (row 2) + (row 1)

$$\text{Resultant matrix} = \left[\begin{array}{cc|c} 1 & 0 & -3 \\ 0 & 1 & 4 \\ 0 & 0 & 0 \end{array}\right]$$

The second column is now completely simplified. Because there are no more columns to the left of the augmentation bar, we stop. The final resultant matrix corresponds to the following system:

$$\begin{cases} x \quad\quad = -3 \\ \quad\; y = 4 \\ \quad\; 0 = 0 \end{cases}.$$

The third equation is always satisfied, no matter what values x and y have, and so provides us with no information. The first two equations, however, give us the (unique) solution for our original system: $x = -3$, $y = 4$. ■

The general rule is this:

> When starting a new column, if the pivot element is zero, look for a nonzero number in the pivot column below the home row. If you find one, switch the home row with the row containing this nonzero number.

This method converts the pivot to a nonzero number. Then perform a type (I) operation to turn the pivot element into 1. When using a type (III) operation, make sure you switch the home row with a row below it. If you were to switch the home row with a row above it, you would undo the simplification of the previous columns.

Notation for the Row Operations

To save space, you can use a shorthand notation for the row operations. For example, (I): $<3> \longleftarrow \frac{1}{2} <3>$ represents the row operation of type (I) where each entry of row 3 is being replaced by $\frac{1}{2}$ times that entry. Also, (II): $<2> \longleftarrow -3<4> + <2>$ is the type (II) row operation where row 2 is replaced by $-3 \times$ (row 4) + (row 2). Finally, (III): $<2> \longleftrightarrow <3>$ is the type (III) row operation where the second and third rows are exchanged.

Skipping a Column

Occasionally, when we are starting a new column, the pivot element and all lower entries in this column are zero. Here, a type (III) operation cannot help. In such cases, we move to the next column to the right to start a new column, skipping the current column. However, we maintain the same home row. Hence, the new pivot element is located one column to the right, horizontally.

We illustrate the use of this rule in the next two examples. Example 4 involves an inconsistent system, and Example 5 involves infinitely many solutions. In fact, this problem arises in only these two cases.

Inconsistent Systems

Example 4

Let us solve the following system using Gauss-Jordan row reduction:

$$\begin{cases} 3x_1 - 6x_2 + 3x_4 = 9 \\ -2x_1 + 4x_2 + 2x_3 - x_4 = -11 \\ 4x_1 - 8x_2 + 6x_3 + 7x_4 = -5 \end{cases}.$$

First, we set up the augmented matrix

$$\left[\begin{array}{rrrr|r} 3 & -6 & 0 & 3 & 9 \\ -2 & 4 & 2 & -1 & -11 \\ 4 & -8 & 6 & 7 & -5 \end{array}\right].$$

We begin with the first row as home row and start with the first column. The pivot element is the (1, 1) entry. First, we convert the pivot to 1.

$$\text{(I)}: <1> \longleftarrow \tfrac{1}{3} <1>$$

$$\text{Resultant matrix} = \left[\begin{array}{rrrr|r} 1 & -2 & 0 & 1 & 3 \\ -2 & 4 & 2 & -1 & -11 \\ 4 & -8 & 6 & 7 & -5 \end{array}\right]$$

Next we zero out the rest of the first column using type (II) row operations.

$$\text{(II)}: <2> \longleftarrow 2 <1> + <2>$$

$$\text{(II)}: <3> \longleftarrow -4 <1> + <3>$$

$$\text{Resultant matrix} = \left[\begin{array}{rrrr|r} 1 & -2 & 0 & 1 & 3 \\ 0 & 0 & 2 & 1 & -5 \\ 0 & 0 & 6 & 3 & -17 \end{array}\right]$$

We are finished with the first column, so we advance the home row and the current column. Thus, the pivot is now the (2, 2) entry, which unfortunately is zero. We search for a nonzero entry below the pivot but do not find one; the second column is completely simplified. Hence, we skip this column and advance to column 3. However, we maintain row 2 as the home row.

We now change the new pivot element (the (2, 3) entry) into 1.

$$\text{(I)}: <2> \longleftarrow \tfrac{1}{2} <2>$$

$$\text{Resultant matrix} = \left[\begin{array}{rrrr|r} 1 & -2 & 0 & 1 & 3 \\ 0 & 0 & 1 & \frac{1}{2} & -\frac{5}{2} \\ 0 & 0 & 6 & 3 & -17 \end{array}\right]$$

Zeroing out the rest of the third column, we obtain

$$\text{(II)}: <3> \longleftarrow -6 <2> + <3>$$

$$\text{Resultant matrix} = \left[\begin{array}{rrrr|r} 1 & -2 & 0 & 1 & 3 \\ 0 & 0 & 1 & \frac{1}{2} & -\frac{5}{2} \\ 0 & 0 & 0 & 0 & -2 \end{array}\right]$$

Since we are finished with the third column, the pivot moves diagonally to the (3, 4) entry, which is also zero. There is no row below the home row (now row 3) with which an exchange can be made, and so the fourth column is finished. We have reached the augmentation bar, so we stop. The resultant system is

$$\begin{cases} x_1 - 2x_2 + x_4 = 3 \\ x_3 + \frac{1}{2}x_4 = -\frac{5}{2} \\ 0 = -2 \end{cases}.$$

Regardless of the values of x_1, x_2, x_3, and x_4, the last equation, $0 = -2$, is *never* satisfied. This equation has no solutions. Any solution to the system must satisfy every equation in the system. Therefore, this system is inconsistent, as is the original system with which we started. ■

General inconsistent systems are handled similarly to the system in Example 4. The final augmented matrix always contains a row with all zeroes on the left of the augmentation bar and a nonzero number on the right. This row corresponds to the equation $0 = c$, for some $c \neq 0$, which clearly has no solutions. In fact, if you encounter a row of the form

$$[0 \quad 0 \quad \cdots \quad 0 \mid c],$$

with $c \neq 0$, at any stage of the row reduction, you should stop and declare the system to be inconsistent.

Beware! A row of all zeroes, with zero on the right of the augmentation bar, does not imply that the system is inconsistent. Such a row is ignored because it provides no information regarding the solution set of the system, as in Example 3.

Infinite Solution Sets

Example 5

Let us solve the following system:

$$\begin{cases} 3x_1 + x_2 + 7x_3 + 2x_4 = 13 \\ 2x_1 - 4x_2 + 14x_3 - x_4 = -10 \\ 5x_1 + 11x_2 - 7x_3 + 8x_4 = 59 \\ 2x_1 + 5x_2 - 4x_3 - 3x_4 = 39 \end{cases}.$$

The original augmented matrix for this system is

$$\left[\begin{array}{rrrr|r} 3 & 1 & 7 & 2 & 13 \\ 2 & -4 & 14 & -1 & -10 \\ 5 & 11 & -7 & 8 & 59 \\ 2 & 5 & -4 & -3 & 39 \end{array}\right].$$

After row reducing the first two columns, we obtain

$$\left[\begin{array}{cccc|c} 1 & 0 & 3 & \frac{1}{2} & 3 \\ 0 & 1 & -2 & \frac{1}{2} & 4 \\ 0 & 0 & 0 & 0 & 0 \\ 0 & 0 & 0 & -\frac{13}{2} & 13 \end{array}\right].$$

There is no nonzero pivot in column 3, so we advance to column 4 and use row operation (III): $< 3 > \longleftrightarrow < 4 >$ to put a nonzero number into the $(3, 4)$ position:

$$\left[\begin{array}{cccc|c} 1 & 0 & 3 & \frac{1}{2} & 3 \\ 0 & 1 & -2 & \frac{1}{2} & 4 \\ 0 & 0 & 0 & -\frac{13}{2} & 13 \\ 0 & 0 & 0 & 0 & 0 \end{array}\right].$$

Continuing the row reduction leads to the final augmented matrix

$$\left[\begin{array}{cccc|c} 1 & 0 & 3 & 0 & 4 \\ 0 & 1 & -2 & 0 & 5 \\ 0 & 0 & 0 & 1 & -2 \\ 0 & 0 & 0 & 0 & 0 \end{array}\right].$$

This matrix corresponds to

$$\begin{cases} x_1 \quad\quad + 3x_3 \quad\quad = 4 \\ \quad\quad x_2 - 2x_3 \quad\quad = 5 \\ \quad\quad\quad\quad\quad\quad x_4 = -2 \\ \quad\quad\quad\quad\quad\quad 0 = 0 \end{cases}.$$

We discard the last equation, because it gives no information about the solution set.

The third equation gives the value -2 for x_4, but values for the other three variables are not uniquely determined—there are infinitely many solutions. We can let x_3 take on any value whatsoever, which then determines the values for x_1 and x_2. For example, if we let $x_3 = 5$, then solving the first equation for x_1 yields $x_1 = -11$. The second equation produces $x_2 = 15$. Thus, *one* solution is $x_1 = -11$, $x_2 = 15$, $x_3 = 5$, $x_4 = -2$. Different solutions can be found by choosing alternate values for x_3. For example, letting $x_3 = -4$ gives the solution $x_1 = 16$, $x_2 = -3$, $x_3 = -4$, $x_4 = -2$. All such solutions will satisfy the original system.

How can we express the complete solution set? If we use a variable, say c, to represent x_3, then $x_1 = 4 - 3c$ and $x_2 = 5 + 2c$. Of course, $x_4 = -2$. Thus, the infinite solution set can be expressed as

$$\{(4 - 3c,\ 5 + 2c,\ c,\ -2) \mid c \in \mathbb{R}\}.$$

■

After row reduction, the columns with nonzero pivot elements are often labeled as **pivot columns**, and the others are called **nonpivot columns**. Recall that the columns to the left of the augmentation bar correspond to the variables x_1, x_2, and so on in the system. The variables for nonpivot columns are called **independent variables**, and the others are **dependent variables**. In Example 5, only columns 1, 2, and 4 are pivot columns. Hence, x_3 is an independent variable; x_1, x_2, and x_4 are dependent variables.

If a given system is consistent, solutions are found by letting each independent variable take on any real value whatsoever. The values of the dependent variables are then calculated from these choices.

Example 6

Suppose that the final matrix produced after row reduction is

$$\left[\begin{array}{cccccc|c} 1 & -2 & 0 & 3 & 5 & 0 & 17 \\ 0 & 0 & 1 & 4 & 23 & 0 & -9 \\ 0 & 0 & 0 & 0 & 0 & 1 & 16 \\ 0 & 0 & 0 & 0 & 0 & 0 & 0 \end{array}\right],$$

which corresponds to the system

$$\begin{cases} x_1 - 2x_2 + 3x_4 + 5x_5 = 17 \\ x_3 + 4x_4 + 23x_5 = -9 \\ x_6 = 16 \end{cases}.$$

Note that we have ignored the row of zeroes. Because the pivot columns are columns 1, 3, and 6, we see that x_2, x_4, and x_5 are the independent variables. Therefore, we can let x_2, x_4, and x_5 take on any real values—say, $x_2 = b$, $x_4 = d$, and $x_5 = e$. We now solve the equations in the system for the dependent variables x_1, x_3, and x_6, yielding $x_1 = 17 + 2b - 3d - 5e$, $x_3 = -9 - 4d - 23e$, and $x_6 = 16$. Hence, the solution set is

$$\{(17 + 2b - 3d - 5e,\ b,\ -9 - 4d - 23e,\ d,\ e, 16) \mid b, d, e \in \mathbb{R}\}.$$

Particular solutions can be found by choosing values for b, d, and e. For example, choosing $b = 1$, $d = -1$, and $e = 0$ yields $(22, 1, -5, -1, 0, 16)$. ■

Reduced Row Echelon Form

In the Gauss-Jordan method, each new pivot element must be to the right of its predecessor and immediately below it. The effect is a "staircase" pattern of pivots, as in this matrix:

$$\left[\begin{array}{cccccc|c} 1 & -3 & 0 & -2 & 4 & 0 & -3 \\ 0 & 0 & 1 & -3 & 2 & 0 & -5 \\ 0 & 0 & 0 & 0 & 0 & 1 & 2 \\ 0 & 0 & 0 & 0 & 0 & 0 & 0 \end{array}\right].$$

Each step on the staircase begins with a nonzero pivot, although the steps are not necessarily uniform in width. Finally, notice that all entries below the staircase are zero and that all entries above a nonzero pivot are zero. When a matrix satisfies these conditions, it is said to be in **reduced row echelon form**. The next definition is a more formal statement of this set of conditions:

DEFINITION

An (augmented) matrix is in **reduced row echelon form** if and only if all the following conditions hold:

(1) The first nonzero entry in each row is 1.
(2) Each successive row has its first nonzero entry in a later column.
(3) All entries above and below the first nonzero entry of each row are zero.
(4) All full rows of zeroes are the final rows of the matrix.

Technically speaking, this definition requires us to row reduce all columns in order to put an augmented matrix into reduced row echelon form. However, as you have seen, the solution set of a linear system can actually be determined without simplifying the column to the right of the augmentation bar.

Formal Algorithm for Gauss-Jordan Row Reduction

We now give a formal algorithm for Gauss-Jordan row reduction. The choice of pivots and the use of the row operations in this algorithm lead to a final augmented matrix in reduced row echelon form:

ALGORITHM FOR GAUSS-JORDAN ROW REDUCTION

Step 1: Given a system of linear equations, create its associated augmented matrix.

Step 2: Choose the first row as the "home row."

Step 3: Repeat the following steps for each of the columns in turn:

(a) If all entries from the home row to the last row (inclusive) of the current column are zero, skip steps 3(b) to 3(f) for this column and advance to the next column. Maintain the same home row.

(b) If the home row entry of the current column is zero and if there is a later row whose entry in this column is nonzero, use row operation type (III) to exchange the home row with the nearest such row below it.

(c) Designate the home row entry of the current column as the current pivot element.

(d) Use row operation type (I) on the home row to convert the current pivot entry to 1.

(e) Use row operation type (II) to add multiples of the home row to other rows to convert all nonzero entries except the pivot in the current column to zero.

(f) If the home row of the matrix is the last nonzero row, stop. Otherwise, advance the home row to the next row.

The exact row operations in steps 3(d) and 3(e) can be specified. Suppose that **C** is the augmented matrix and its ith row is the current "home row." If the pivot element of this row is in the jth column, then in step 3(d) perform the type (I) row operation $< i > \leftarrow \frac{1}{c_{ij}} < i >$. Also, in step 3(e), to zero out the (k, j) entry, using the ith row as home row, perform the type (II) operation $< k > \leftarrow -c_{kj} < i > + < k >$.

The Gauss-Jordan algorithm also implies the following:

NUMBER OF SOLUTIONS OF A LINEAR SYSTEM

Let $\mathbf{AX} = \mathbf{B}$ be a system of linear equations. Let **C** be the reduced row echelon form augmented matrix obtained by row reducing $[\mathbf{A} \mid \mathbf{B}]$.

If there is a row of **C** having all zeroes to the left of the augmentation bar but with its last entry nonzero, then $\mathbf{AX} = \mathbf{B}$ has no solution.

Otherwise, if one of the columns of **C** to the left of the augmentation bar has no nonzero pivot element, then **AX** = **B** has an infinite number of solutions. The nonpivot columns correspond to (independent) variables that can take on any value, and the values of the remaining (dependent) variables are determined from those.

Otherwise, **AX** = **B** has a unique solution.

Application: Curve Fitting

Example 7

Let us find a quadratic equation of the form $y = ax^2 + bx + c$ that goes through the points $(-2, 20)$, $(1, 5)$, and $(3, 25)$ in the xy-plane. By substituting each of the (x, y) pairs in turn into the equation, we get

$$\begin{cases} 20 = a(-2)^2 + b(-2) + c \\ 5 = a(1)^2 + b(1) + c \\ 25 = a(3)^2 + b(3) + c \end{cases},$$

which leads to the system

$$\begin{cases} 4a - 2b + c = 20 \\ a + b + c = 5 \\ 9a + 3b + c = 25 \end{cases}.$$

Using the Gauss-Jordan algorithm on the augmented matrix associated with this system leads to the final augmented matrix:

$$\left[\begin{array}{ccc|c} 1 & 0 & 0 & 3 \\ 0 & 1 & 0 & -2 \\ 0 & 0 & 1 & 4 \end{array}\right].$$

Hence, $a = 3$, $b = -2$, and $c = 4$, and so the desired quadratic equation is $y = 3x^2 - 2x + 4$. ■

Application: Balancing Chemical Equations

In chemical reactions, we often know the reactants (initial substances) and products (results of the reaction). For example, it is known that the reactants phosphoric acid and calcium hydroxide produce calcium phosphate and water. This reaction can be symbolized as

$$\underset{\text{Phosphoric acid}}{H_3PO_4} + \underset{\text{Calcium hydroxide}}{Ca(OH)_2} \rightarrow \underset{\text{Calcium phosphate}}{Ca_3(PO_4)_2} + \underset{\text{Water}}{H_2O}$$

It is often desirable to determine the **empirical formula** for the reaction, an equation containing the minimal integer multiples of the reactants and products so that the number of atoms of each element agrees on both sides. This procedure is called **balancing** the equation. In the preceding example, we are looking for minimal positive integer values of a, b, c, and d such that

$$aH_3PO_4 + bCa(OH)_2 \rightarrow cCa_3(PO_4)_2 + dH_2O$$

balances the number of hydrogen (H), phosphorus (P), oxygen (O), and calcium (Ca) atoms on both sides.[†] Considering each element in turn, we get

$$\begin{cases} 3a + 2b = 2d & (H) \\ a = 2c & (P) \\ 4a + 2b = 8c + d & (O) \\ b = 3c & (Ca) \end{cases}.$$

Bringing the c and d terms to the left side of each equation, we get the following augmented matrix for this system:

$$\left[\begin{array}{cccc|c} 3 & 2 & 0 & -2 & 0 \\ 1 & 0 & -2 & 0 & 0 \\ 4 & 2 & -8 & -1 & 0 \\ 0 & 1 & -3 & 0 & 0 \end{array}\right].$$

After applying the Gauss-Jordan algorithm, the final augmented matrix is

$$\left[\begin{array}{cccc|c} 1 & 0 & 0 & -\frac{1}{3} & 0 \\ 0 & 1 & 0 & -\frac{1}{2} & 0 \\ 0 & 0 & 1 & -\frac{1}{6} & 0 \\ 0 & 0 & 0 & 0 & 0 \end{array}\right].$$

The only variable corresponding to a nonpivot column is d. We choose $d = 6$ because this is the minimum positive integer value we can assign to d so that a, b, and c are also integers. (Why?) We then have $a = 2$, $b = 3$, and $c = 1$. Thus, the balanced chemical equation for this reaction is

$$2H_3PO_4 + 3Ca(OH)_2 \rightarrow Ca_3(PO_4)_2 + 6H_2O.$$

♦ **Application:** You now have covered the prerequisites for Section 8.2, "Ohm's Law"; Section 8.3, "Least-Squares Approximations"; and Section 8.4, "Markov Chains."

[†]In expressions like $(OH)_2$ and $(PO_4)_2$, the number immediately following the parentheses indicates that every term in the unit should be considered to appear that many times. Hence, $(PO_4)_2$ is equivalent to PO_4PO_4, for our purposes.

Exercises—Section 2.1

★**1.** Which of these matrices are not in reduced row echelon form? Why?

(a) $\begin{bmatrix} 1 & 0 & 0 & 0 \\ 0 & 0 & 1 & 0 \\ 0 & 1 & 0 & 0 \end{bmatrix}$ (b) $\begin{bmatrix} 1 & -2 & 0 & 0 \\ 0 & 0 & 0 & 0 \\ 0 & 0 & 1 & 0 \\ 0 & 0 & 0 & 1 \end{bmatrix}$ (c) $\begin{bmatrix} 1 & 0 & 0 & 3 \\ 0 & 2 & 0 & -2 \\ 0 & 0 & 3 & 0 \end{bmatrix}$

(d) $\begin{bmatrix} 1 & -4 & 0 & 0 \\ 0 & 0 & 1 & 0 \\ 0 & 0 & 2 & 0 \end{bmatrix}$ (e) $\begin{bmatrix} 1 & 0 & 4 \\ 0 & 1 & -2 \\ 0 & 0 & 0 \end{bmatrix}$ (f) $\begin{bmatrix} 1 & -2 & 0 & -2 & 3 \\ 0 & 0 & 1 & 5 & 4 \\ 0 & 0 & 0 & 0 & 1 \end{bmatrix}$

2. Use the Gauss-Jordan algorithm to convert these matrices to reduced row echelon form, and draw in the correct staircase pattern:

★(a) $\left[\begin{array}{ccc|c} 5 & 20 & -18 & -11 \\ 3 & 12 & -14 & 3 \\ -4 & -16 & 13 & 13 \end{array}\right]$ ★(b) $\begin{bmatrix} -2 & 1 & 1 & 15 \\ 6 & -1 & -2 & -36 \\ 1 & -1 & -1 & -11 \\ -5 & -5 & -5 & -14 \end{bmatrix}$

★(c) $\left[\begin{array}{cccc|c} -5 & 10 & -19 & -17 & 20 \\ -3 & 6 & -11 & -11 & 14 \\ -7 & 14 & -26 & -25 & 31 \\ 9 & -18 & 34 & 31 & -37 \end{array}\right]$ (d) $\begin{bmatrix} 2 & -5 & -20 \\ 0 & 2 & 7 \\ 1 & -5 & -19 \\ -5 & 16 & 64 \\ 3 & -9 & -36 \end{bmatrix}$

★(e) $\left[\begin{array}{ccccc|c} -3 & 6 & -1 & -5 & 0 & -5 \\ -1 & 2 & 3 & -5 & 10 & 5 \end{array}\right]$

(f) $\begin{bmatrix} -2 & 1 & -1 & -1 & 3 \\ 3 & 1 & -4 & -2 & -4 \\ 7 & 1 & -6 & -2 & -3 \\ -8 & -1 & 6 & 2 & 3 \\ -3 & 0 & 2 & 1 & 2 \end{bmatrix}$

3. Use the Gauss-Jordan algorithm to solve each of the following systems of linear equations. Give the full solution set in each case:

★(a) $\begin{cases} -5x_1 - 2x_2 + 2x_3 = 14 \\ 3x_1 + x_2 - x_3 = -8 \\ 2x_1 + 2x_2 - x_3 = -3 \end{cases}$

(b) $\begin{cases} 3x_1 - 3x_2 - 2x_3 = 23 \\ -6x_1 + 4x_2 + 3x_3 = -38 \\ -2x_1 + x_2 + x_3 = -11 \end{cases}$

★(c) $\begin{cases} 3x_1 - 2x_2 + 4x_3 = -54 \\ -x_1 + x_2 - 2x_3 = 20 \\ 5x_1 - 4x_2 + 8x_3 = -83 \end{cases}$

(d) $\begin{cases} -2x_1 + 3x_2 - 4x_3 + x_4 = -17 \\ 8x_1 - 5x_2 + 2x_3 - 4x_4 = 47 \\ -5x_1 + 9x_2 - 13x_3 + 3x_4 = -44 \\ -4x_1 + 3x_2 - 2x_3 + 2x_4 = -25 \end{cases}$

★(e) $\begin{cases} 6x_1 - 12x_2 - 5x_3 + 16x_4 - 2x_5 = -53 \\ -3x_1 + 6x_2 + 3x_3 - 9x_4 + x_5 = 29 \\ -4x_1 + 8x_2 + 3x_3 - 10x_4 + x_5 = 33 \end{cases}$

(f) $\begin{cases} 5x_1 - 5x_2 - 15x_3 - 3x_4 = -34 \\ -2x_1 + 2x_2 + 6x_3 + x_4 = 12 \end{cases}$

★(g) $\begin{cases} 4x_1 - 2x_2 - 7x_3 = 5 \\ -6x_1 + 5x_2 + 10x_3 = -11 \\ -2x_1 + 3x_2 + 4x_3 = -3 \\ -3x_1 + 2x_2 + 5x_3 = -5 \end{cases}$

(h) $\begin{cases} 5x_1 - x_2 - 9x_3 - 2x_4 = 26 \\ 4x_1 - x_2 - 7x_3 - 2x_4 = 21 \\ -2x_1 + 4x_3 + x_4 = -12 \\ -3x_1 + 2x_2 + 4x_3 + 2x_4 = -11 \end{cases}$

★4. Solve the following problem by using a linear system: A certain number of nickels, dimes, and quarters totals \$16.50. There are twice as many dimes as quarters, and the total number of nickels and quarters is twenty more than the number of dimes. Find the correct number of each type of coin.

★5. Find the quadratic equation $y = ax^2 + bx + c$ that goes through the points $(3, 18)$, $(2, 9)$, and $(-2, 13)$.

6. Find the cubic equation $y = ax^3 + bx^2 + cx + d$ that goes through the points $(1, 1)$, $(2, -18)$, $(-2, 46)$, and $(3, -69)$.

★7. The general equation of a circle is $x^2 + y^2 + ax + by = c$. Find the equation of the circle that goes through the points $(6, 8)$, $(8, 4)$, and $(3, 9)$.

8. Find the values of A, B, C (and D in part (b)) in the following partial fractions problems:

★(a) $\dfrac{5x^2 + 23x - 58}{(x-1)(x-3)(x+4)} = \dfrac{A}{x-1} + \dfrac{B}{x-3} + \dfrac{C}{x+4}$

(b) $\dfrac{-3x^3 + 29x^2 - 91x + 94}{(x-2)^2(x-3)^2} = \dfrac{A}{(x-2)^2} + \dfrac{B}{x-2} + \dfrac{C}{(x-3)^2} + \dfrac{D}{x-3}$

9. Find the minimal integer values for the variables that will balance each of the following chemical equations:†

★(a) $aC_6H_6 + bO_2 \rightarrow cCO_2 + dH_2O$

(b) $aC_8H_{18} + bO_2 \rightarrow cCO_2 + dH_2O$

★(c) $aAgNO_3 + bH_2O \rightarrow cAg + dO_2 + eHNO_3$

(d) $aHNO_3 + bHCl + cAu \rightarrow dNOCl + eHAuCl_4 + fH_2O$

10. When performing a type (I) row operation, you must always multiply a row by a nonzero scalar. Explain why the scalar must be nonzero.

11. Prove that, if more than one solution to a system of linear equations exists, then an infinite number of solutions exists. (Hint: Show that if $\mathbf{X}_1 = (s_1, s_2, \ldots, s_n)$ and $\mathbf{X}_2 = (t_1, t_2, \ldots, t_n)$ are different solutions to $\mathbf{AX} = \mathbf{B}$, then $\mathbf{X}_1 + c(\mathbf{X}_2 - \mathbf{X}_1)$ is also a solution, for every real number c. Be sure to show that all these solutions are different.)

2.2 Equivalent Systems and Rank

In this section, we continue our discussion of the solution sets of systems of linear equations. First we prove that the Gauss-Jordan algorithm produces the correct solution set. Then we note that every matrix has a unique corresponding matrix in reduced row echelon form and use this fact to define the rank of the matrix. Next we introduce homogeneous systems and prove some important results about their solution sets. Finally, we exhibit a method for solving multiple systems having the same coefficient matrix.

Equivalent Systems and Row Equivalence of Matrices

The first two definitions below involve related concepts. The connection between them is shown in Theorem 2.2.

DEFINITION

Two systems of m linear equations in n variables are **equivalent** if and only if they have exactly the same solution set.

†The chemical elements used in these equations are silver (Ag), gold (Au), carbon (C), chlorine (Cl), hydrogen (H), nitrogen (N), and oxygen (O). The compounds are water (H_2O), carbon dioxide (CO_2), benzene (C_6H_6), octane (C_8H_{18}), silver nitrate ($AgNO_3$), nitric acid (HNO_3), hydrochloric acid (HCl), nitrous chloride ($NOCl$), and hydrogen tetrachloroaurate (III) ($HAuCl_4$).

For example, the systems

$$\begin{cases} 2x - y = 1 \\ 3x + y = 9 \end{cases} \quad \text{and} \quad \begin{cases} x + 4y = 14 \\ 5x - 2y = 4 \end{cases}$$

are equivalent, because the solution set of both is exactly $\{(2, 3)\}$.

DEFINITION

> An (augmented) matrix **D** is **row equivalent** to a matrix **C** if and only if **D** is obtained from **C** by a finite number of applications of any row operations of types (I), (II), and (III).

For example, given any matrix, the Gauss-Jordan row reduction process produces a matrix in reduced row echelon form that is row equivalent to the original matrix.

It is relatively easy to see that, if **D** is row equivalent to **C**, then **C** is also row equivalent to **D**. The reason is that each row operation is reversible; that is, the effect of any row operation can be undone by performing another row operation. Figure 2.2 shows how to reverse row operations. Notice that a row operation of type (I) is reversed by using the reciprocal $1/c$ and that an operation of type (II) is reversed by using the additive inverse $-c$. (Do you see why?)

Thus, if **D** is obtained from **C** by the sequence

$$\mathbf{C} \overset{\text{Op}_1}{\longrightarrow} \mathbf{A}_1 \overset{\text{Op}_2}{\longrightarrow} \mathbf{A}_2 \overset{\text{Op}_3}{\longrightarrow} \cdots \overset{\text{Op}_n}{\longrightarrow} \mathbf{A}_n \overset{\text{Op}_{n+1}}{\longrightarrow} \mathbf{D},$$

then **C** can be obtained from **D** using the reverse (or **inverse**) operations in reverse order:

$$\mathbf{D} \overset{\text{Op}_{n+1}^{-1}}{\longrightarrow} \mathbf{A}_n \overset{\text{Op}_n^{-1}}{\longrightarrow} \mathbf{A}_{n-1} \overset{\text{Op}_{n-1}^{-1}}{\longrightarrow} \cdots \overset{\text{Op}_2^{-1}}{\longrightarrow} \mathbf{A}_1 \overset{\text{Op}_1^{-1}}{\longrightarrow} \mathbf{C}.$$

(Op_i^{-1} represents the inverse operation to Op_i, as indicated in Figure 2.2.) This argument is a sketch for the proof of the following theorem. You are asked to fill in the details of the proof in Exercise 13.

Type of operation	Operation	Reverse Operation
(I)	$<i> \leftarrow c<i>$	$<i> \leftarrow \frac{1}{c}<i>$
(II)	$<j> \leftarrow c<i> + <j>$	$<j> \leftarrow -c<i> + <j>$
(III)	$<i> \longleftrightarrow <j>$	$<i> \longleftrightarrow <j>$

Figure 2.2
Row operations and their inverses

THEOREM 2.1

If a matrix **D** is row equivalent to a matrix **C**, then **C** is row equivalent to **D**.

The next theorem asserts that, if two augmented matrices are obtained from each other using only row operations, then their corresponding linear systems have the same solution set. This result guarantees that the Gauss-Jordan algorithm provided in Section 2.1 is correct, because the only operations allowed in that algorithm were row operations of types (I), (II), and (III). Therefore, the final augmented matrix produced by a Gauss-Jordan row reduction process represents a system equivalent to the original—that is, a system with precisely the same solution set.

THEOREM 2.2

Let $\mathbf{AX} = \mathbf{B}$ be a system of linear equations. If $[\mathbf{C}|\mathbf{D}]$ is row equivalent to $[\mathbf{A}|\mathbf{B}]$, then the system $\mathbf{CX} = \mathbf{D}$ is equivalent to $\mathbf{AX} = \mathbf{B}$.

Proof of Theorem 2.2 (abridged)

Let S_A represent the complete solution set of the system $\mathbf{AX} = \mathbf{B}$, and let S_C be the solution set of $\mathbf{CX} = \mathbf{D}$. Our goal is to prove that, if $[\mathbf{C}|\mathbf{D}]$ is row equivalent to $[\mathbf{A}|\mathbf{B}]$, then $S_A = S_C$. It will be enough to show that $[\mathbf{C}|\mathbf{D}]$ row equivalent to $[\mathbf{A}|\mathbf{B}]$ implies $S_A \subseteq S_C$. This fact, together with Theorem 2.1, implies the reverse inclusion, $S_C \subseteq S_A$. (Why?)

Also, it is enough to assume that $[\mathbf{C}|\mathbf{D}]$ can be produced from $[\mathbf{A}|\mathbf{B}]$ by a single use of a row operation, because a simple induction argument can then extend the result to the case where any (finite) number of row operations are required to produce $[\mathbf{C}|\mathbf{D}]$ from $[\mathbf{A}|\mathbf{B}]$. Therefore, we need only consider the effect of each row operation in turn. We present the proof for a type (II) row operation and leave the proofs for the other types to you (see Exercise 13).

Type (II) Operation: Suppose that the original system has the form

$$\begin{cases} a_{11}x_1 + a_{12}x_2 + a_{13}x_3 + \cdots + a_{1n}x_n = b_1 \\ a_{21}x_1 + a_{22}x_2 + a_{23}x_3 + \cdots + a_{2n}x_n = b_2 \\ \quad \vdots \qquad\quad \vdots \qquad\quad \vdots \qquad \ddots \qquad \vdots \qquad\quad \vdots \\ a_{m1}x_1 + a_{m2}x_2 + a_{m3}x_3 + \cdots + a_{mn}x_n = b_m \end{cases}$$

and that the row operation used is $< j > \longleftarrow q < i > + < j >$ (where $i \neq j$). When this row operation is applied to the augmented matrix corresponding to the original system, all rows remain unchanged, except for the jth row. The jth equation in the new system corresponding to the new j^{th} row then has the form

$$(qa_{i1} + a_{j1})x_1 + (qa_{i2} + a_{j2})x_2 + \cdots + (qa_{in} + a_{jn})x_n = qb_i + b_j.$$

We must show that any solution $(s_1, s_2, \ldots, s_n)$ of the original system is a solution of the new one. Now, $(s_1, s_2, \ldots, s_n)$ is also a solution of every equation in the new system other than the jth one, since none of these equations have changed. Also, because it is a solution of both the ith and j^{th} equations in the original system, we have

$$a_{i1}s_1 + a_{i2}s_2 + \cdots + a_{in}s_n = b_i \quad \text{and} \quad a_{j1}s_1 + a_{j2}s_2 + \cdots + a_{jn}s_n = b_j,$$

which yields

$$(qa_{i1} + a_{j1})s_1 + (qa_{i2} + a_{j2})s_2 + \cdots + (qa_{in} + a_{jn})s_n = qb_i + b_j.$$

Hence, $(s_1, s_2, \ldots, s_n)$ is also a solution of the jth equation of the new system. ■

Rank of a Matrix

You have probably already realized that when the Gauss-Jordan row reduction process is performed on a matrix, only one final augmented matrix can result. This fact is stated in the next theorem, which we present without proof:

THEOREM 2.3

Every matrix is row equivalent to a unique matrix in reduced row echelon form.

Because each matrix has a unique corresponding matrix in reduced row echelon form, we can make the following definition:

DEFINITION

Let **A** be a matrix. Then the **rank** of **A** is the number of nonzero rows (that is, rows with nonzero pivot elements) in the unique reduced row echelon form matrix that is row equivalent to **A**.

Example 1

Consider the following matrix:

$$\mathbf{A} = \begin{bmatrix} 2 & 1 & 4 \\ 3 & 2 & 5 \\ 0 & -1 & 1 \end{bmatrix}.$$

The rank of **A** is 3, because row reduction reveals that **A** is row equivalent to the following reduced row echelon form matrix:

$$\mathbf{I}_3 = \begin{bmatrix} 1 & 0 & 0 \\ 0 & 1 & 0 \\ 0 & 0 & 1 \end{bmatrix}.$$

This matrix has three nonzero rows, and hence three pivot elements.

Similarly, the matrix

$$\mathbf{B} = \begin{bmatrix} 3 & 1 & 0 & 1 & -9 \\ 0 & -2 & 12 & -8 & -6 \\ 2 & -3 & 22 & -14 & -17 \end{bmatrix}$$

has rank 2 because it is row equivalent to the reduced row echelon form matrix

$$\begin{bmatrix} 1 & 0 & 2 & -1 & -4 \\ 0 & 1 & -6 & 4 & 3 \\ 0 & 0 & 0 & 0 & 0 \end{bmatrix},$$

which has two nonzero rows, and hence two nonzero pivots. ■

Homogeneous Systems

DEFINITION

A system of linear equations having matrix form $\mathbf{AX} = \mathbf{O}$, where $\mathbf{O}$ represents a zero column matrix, is called a **homogeneous system**.

For example, the following are homogeneous systems:

$$\begin{cases} 2x - 3y = 0 \\ -4x + 6y = 0 \end{cases} \quad \text{and} \quad \begin{cases} 5x_1 - 2x_2 + 3x_3 = 0 \\ 6x_1 + x_2 - 7x_3 = 0 \\ -x_1 + 3x_2 + x_3 = 0 \end{cases}.$$

Many important types of linear systems are homogeneous. For example, the systems for Exercise 9 of Section 2.1, obtained from balancing chemical equations, are all homogeneous systems.

Notice that a homogeneous system always has at least one solution and hence is consistent, since all of the variables can be set equal to zero to satisfy all of the equations. This special solution $(0, 0, \ldots, 0)$ is called the **trivial solution**. Any other solution of a homogeneous system is called a **nontrivial solution**. For example, $(0, 0)$ is the trivial solution to the first example given of a homogeneous system, but $(9, 6)$ is a nontrivial solution. Whenever a homogeneous system has a nontrivial solution, it actually has infinitely many solutions. (Why?)

An important result about homogeneous systems is the following:

THEOREM 2.4

Let $\mathbf{AX} = \mathbf{O}$ be a homogeneous system in n variables:

(1) If $\text{rank}(\mathbf{A}) < n$, then the system has a nontrivial solution.

(2) If $\text{rank}(\mathbf{A}) = n$, then the system has only the trivial solution.

Note that the presence of a nontrivial solution when $\text{rank}(\mathbf{A}) < n$ means that the system has an infinite number of solutions.

Proof of Theorem 2.4

After the Gauss-Jordan row reduction process is applied to $\mathbf{AX} = \mathbf{O}$, the number of nonzero pivots equals $\text{rank}(\mathbf{A})$. Suppose $\text{rank}(\mathbf{A}) < n$. Then at least one of the n columns on the left side of the augmented matrix $[\mathbf{A}|\mathbf{O}]$ has no nonzero pivot element. Thus, the system has at least one independent variable. Now, because this system is homogeneous, it is consistent. Therefore, the solution set is nonempty, with particular solutions found by choosing arbitrary values for all independent variables and then solving for the dependent variables. Hence, a nontrivial solution can be found by choosing a nonzero value for one of the independent variables.

On the other hand, suppose $\text{rank}(\mathbf{A}) = n$. Then every column on the left side of $[\mathbf{A}|\mathbf{O}]$ is a pivot column, and each variable must equal zero. Hence, in this case $\mathbf{AX} = \mathbf{O}$ has only the trivial solution. ■

Example 2

Consider the following 3×3 homogeneous systems:

$$\begin{cases} 2x_1 + x_2 + 4x_3 = 0 \\ 3x_1 + 2x_2 + 5x_3 = 0 \\ \quad\; - x_2 + x_3 = 0 \end{cases} \quad \text{and} \quad \begin{cases} 4x_1 - 8x_2 - 2x_3 = 0 \\ 3x_1 - 5x_2 - 2x_3 = 0 \\ 2x_1 - 8x_2 + x_3 = 0 \end{cases}.$$

After the row reduction process, the final augmented matrices for these systems are, respectively,

$$\left[\begin{array}{ccc|c} 1 & 0 & 0 & 0 \\ 0 & 1 & 0 & 0 \\ 0 & 0 & 1 & 0 \end{array}\right] \quad \text{and} \quad \left[\begin{array}{ccc|c} 1 & 0 & -\frac{3}{2} & 0 \\ 0 & 1 & -\frac{1}{2} & 0 \\ 0 & 0 & 0 & 0 \end{array}\right].$$

The first system has only the trivial solution because the rank of its coefficient matrix equals 3. However, the second system has a nontrivial solution because the rank of its coefficient matrix equals 2, which is less than 3. The complete solution set for the second system is

$$\left\{\left(\tfrac{3}{2}c, \tfrac{1}{2}c, c\right) \middle| \, c \in \mathbb{R}\right\}.$$

■

The following corollary follows immediately from Theorem 2.4:

COROLLARY 2.5

A homogeneous system $\mathbf{AX} = \mathbf{O}$ of m linear equations in n variables has a nontrivial solution if $m < n$.

Example 3

Consider the following homogeneous system:

$$\begin{cases} x_1 & - & 3x_2 & + & 2x_3 & - & 4x_4 & + & 8x_5 & + & 17x_6 & = & 0 \\ 3x_1 & - & 9x_2 & + & 6x_3 & - & 12x_4 & + & 24x_5 & + & 49x_6 & = & 0 \\ -2x_1 & + & 6x_2 & - & 5x_3 & + & 11x_4 & - & 18x_5 & - & 40x_6 & = & 0 \end{cases}.$$

Because this system has fewer equations than variables, by Corollary 2.5 it has a nontrivial solution. To find all the solutions, we row reduce to obtain the final augmented matrix:

$$\left[\begin{array}{cccccc|c} 1 & -3 & 0 & 2 & 4 & 0 & 0 \\ 0 & 0 & 1 & -3 & 2 & 0 & 0 \\ 0 & 0 & 0 & 0 & 0 & 1 & 0 \end{array}\right].$$

The second, fourth, and fifth columns are nonpivot columns, so we can let x_2, x_4, and x_5 take on any real values—say, b, d, and e, respectively. The values of the remaining variables are determined from $x_1 - 3b + 2d + 4e = 0$, $x_3 - 3d + 2e = 0$, and $x_6 = 0$. The complete solution set is

$$\{(3b - 2d - 4e, b, 3d - 2e, d, e, 0) \mid b, d, e \in \mathbb{R}\}.$$

■

Solving Several Systems Simultaneously

In many cases, we need to solve two or more systems that have the same coefficient matrix. Suppose we wanted to solve both of these systems:

$$\begin{cases} 3x_1 & + & x_2 & - & 2x_3 & = & 1 \\ 4x_1 & & & - & x_3 & = & 7 \\ 2x_1 & - & 3x_2 & + & 5x_3 & = & 18 \end{cases} \quad \text{and} \quad \begin{cases} 3x_1 & + & x_2 & - & 2x_3 & = & 8 \\ 4x_1 & & & - & x_3 & = & -1 \\ 2x_1 & - & 3x_2 & + & 5x_3 & = & -32 \end{cases}.$$

It is not necessary to do two separate row reductions starting with these augmented matrices:

$$\left[\begin{array}{ccc|c} 3 & 1 & -2 & 1 \\ 4 & 0 & -1 & 7 \\ 2 & -3 & 5 & 18 \end{array}\right] \quad \text{and} \quad \left[\begin{array}{ccc|c} 3 & 1 & -2 & 8 \\ 4 & 0 & -1 & -1 \\ 2 & -3 & 5 & -32 \end{array}\right].$$

Two row reductions would be wasteful, since an identical sequence of row operations is used to transform the first three columns of both matrices into the same "staircase" pattern of pivots. Instead, we can create a "simultaneous" matrix:

$$\left[\begin{array}{ccc|cc} 3 & 1 & -2 & 1 & 8 \\ 4 & 0 & -1 & 7 & -1 \\ 2 & -3 & 5 & 18 & -32 \end{array}\right].$$

Row reducing this matrix completely yields

$$\left[\begin{array}{ccc|cc} 1 & 0 & 0 & 2 & -1 \\ 0 & 1 & 0 & -3 & 5 \\ 0 & 0 & 1 & 1 & -3 \end{array}\right].$$

By considering each of the right-hand columns separately, we discover that the unique solution of the first system is $x_1 = 2$, $x_2 = -3$, and $x_3 = 1$ and that the unique solution of the second system is $x_1 = -1$, $x_2 = 5$, and $x_3 = -3$.

Any number of systems with the same coefficient matrix can be handled in the same manner: there will be one column on the right-hand side of the augmented matrix for each system.

Exercises—Section 2.2

1. For each pair of given matrices **A** and **B**, give a reason why **A** and **B** are row equivalent:

★(a) $\mathbf{A} = \begin{bmatrix} 1 & 0 & 0 \\ 0 & 1 & 0 \\ 0 & 0 & 1 \end{bmatrix}$, $\quad \mathbf{B} = \begin{bmatrix} 1 & 0 & 0 \\ 0 & -5 & 0 \\ 0 & 0 & 1 \end{bmatrix}$

(b) $\mathbf{A} = \begin{bmatrix} 12 & 9 & -5 \\ 4 & 6 & -2 \\ 0 & 1 & 3 \end{bmatrix}$, $\quad \mathbf{B} = \begin{bmatrix} 0 & 1 & 3 \\ 4 & 6 & -2 \\ 12 & 9 & -5 \end{bmatrix}$

★(c) $\mathbf{A} = \begin{bmatrix} 3 & 2 & 7 \\ -4 & 1 & 6 \\ 2 & 5 & 4 \end{bmatrix}$, $\quad \mathbf{B} = \begin{bmatrix} 3 & 2 & 7 \\ -2 & 6 & 10 \\ 2 & 5 & 4 \end{bmatrix}$

2. (a) Find the reduced row echelon form **B** of the matrix **A** below, keeping track of the row operations used:

$$\mathbf{A} = \begin{bmatrix} 4 & 0 & -20 \\ -2 & 0 & 11 \\ 3 & 1 & -15 \end{bmatrix}.$$

★(b) Use your answer to part (a) to give a sequence of row operations that converts **B** back to **A**. Check your answer. (Hint: Use the inverses of the row operations from part (a), but in reverse order.)

★**3.** (a) Verify that the following matrices are row equivalent by showing they have the same reduced row echelon form:

$$\mathbf{A} = \begin{bmatrix} 1 & 0 & 9 \\ 0 & 1 & -3 \\ 0 & -2 & 5 \end{bmatrix} \quad \text{and} \quad \mathbf{B} = \begin{bmatrix} -5 & 3 & 0 \\ -2 & 1 & 0 \\ -3 & 0 & 1 \end{bmatrix}.$$

(b) Find a sequence of row operations that converts **A** into **B**. (Hint: Let **C** be the common matrix in reduced row echelon form corresponding to **A** and **B**. In part (a) you found a sequence of row operations that converts **A** to **C** and another sequence that converts **B** to **C**. Reverse the operations in the second sequence to obtain a sequence that converts **C** to **B**. Combine the first sequence with these "reversed" operations to create a sequence from **A** to **B**.)

4. Verify that the given matrices are not row equivalent by showing that their corresponding matrices in reduced row echelon form are different:

$$\mathbf{A} = \begin{bmatrix} 1 & -2 & 0 & 0 & 3 \\ 2 & -5 & -3 & -2 & 6 \\ 0 & 5 & 15 & 10 & 0 \\ 2 & 6 & 18 & 8 & 6 \end{bmatrix} \quad \text{and} \quad \mathbf{B} = \begin{bmatrix} 0 & 0 & 1 & 1 & 0 \\ 0 & 0 & 0 & 0 & 1 \\ 0 & 1 & 3 & 2 & 0 \\ -1 & 2 & 0 & 0 & -3 \end{bmatrix}.$$

5. Find the rank of each of the following matrices:

$\star$(a) $\begin{bmatrix} 1 & -1 & 3 \\ 2 & 0 & 4 \\ -1 & -3 & 1 \end{bmatrix}$ (b) $\begin{bmatrix} -1 & 3 & 2 \\ 2 & -6 & -4 \end{bmatrix}$ $\star$(c) $\begin{bmatrix} 4 & 0 & 0 \\ 0 & 0 & 0 \\ 0 & 0 & 5 \end{bmatrix}$

(d) $\begin{bmatrix} 3 & 5 & 2 \\ 4 & 2 & 3 \\ -1 & 2 & 4 \end{bmatrix}$ $\star$(e) $\begin{bmatrix} -1 & -1 & 0 & 0 \\ 0 & 0 & 2 & 3 \\ 4 & 0 & -2 & 1 \\ 3 & -1 & 0 & 4 \end{bmatrix}$ (f) $\begin{bmatrix} 1 & 1 & -1 & 0 & 1 \\ 2 & -4 & 3 & 1 & 0 \\ 3 & 15 & -13 & -2 & 7 \end{bmatrix}$

6. Corollary 2.5 implies that each of the following homogeneous systems has a nontrivial solution. Determine the complete solution set for each system, and give one particular nontrivial solution:

$\star$(a) $\begin{cases} -2x_1 - 3x_2 + 2x_3 - 13x_4 = 0 \\ -4x_1 - 7x_2 + 4x_3 - 29x_4 = 0 \\ x_1 + 2x_2 - x_3 + 8x_4 = 0 \end{cases}$

(b) $\begin{cases} 2x_1 + 4x_2 - x_3 + 5x_4 + 2x_5 = 0 \\ 3x_1 + 3x_2 - x_3 + 3x_4 = 0 \\ -5x_1 - 6x_2 + 2x_3 - 6x_4 - x_5 = 0 \end{cases}$

$\star$(c) $\begin{cases} 7x_1 + 28x_2 + 4x_3 - 2x_4 + 10x_5 + 19x_6 = 0 \\ -9x_1 - 36x_2 - 5x_3 + 3x_4 - 15x_5 - 29x_6 = 0 \\ 3x_1 + 12x_2 + 2x_3 + 6x_5 + 11x_6 = 0 \\ 6x_1 + 24x_2 + 3x_3 - 3x_4 + 10x_5 + 20x_6 = 0 \end{cases}$

7. Find the complete solution set for each of the following homogeneous systems:

★(a) $\begin{cases} -2x_1 + x_2 + 8x_3 = 0 \\ 7x_1 - 2x_2 - 22x_3 = 0 \\ 3x_1 - x_2 - 10x_3 = 0 \end{cases}$

(b) $\begin{cases} 5x_1 - 2x_3 = 0 \\ -15x_1 - 16x_2 - 9x_3 = 0 \\ 10x_1 + 12x_2 + 7x_3 = 0 \end{cases}$

★(c) $\begin{cases} 2x_1 + 6x_2 + 13x_3 + x_4 = 0 \\ x_1 + 4x_2 + 10x_3 + x_4 = 0 \\ 2x_1 + 8x_2 + 20x_3 + x_4 = 0 \\ 3x_1 + 10x_2 + 21x_3 + 2x_4 = 0 \end{cases}$

(d) $\begin{cases} 2x_1 - 6x_2 + 3x_3 - 21x_4 = 0 \\ 4x_1 - 5x_2 + 2x_3 - 24x_4 = 0 \\ -x_1 + 3x_2 - x_3 + 10x_4 = 0 \\ -2x_1 + 3x_2 - x_3 + 13x_4 = 0 \end{cases}$

8. Find the complete solution set for each given homogeneous system. Can Theorem 2.4 or Corollary 2.5 be used here? Why or why not?

★(a) $\begin{cases} -2x_1 + 6x_2 + 3x_3 = 0 \\ 5x_1 - 9x_2 - 4x_3 = 0 \\ 4x_1 - 8x_2 - 3x_3 = 0 \\ 6x_1 - 11x_2 - 5x_3 = 0 \end{cases}$

(b) $\begin{cases} -x_1 + 4x_2 + 19x_3 = 0 \\ 5x_1 + x_2 - 11x_3 = 0 \\ 4x_1 - 5x_2 - 32x_3 = 0 \\ 2x_1 + x_2 - 2x_3 = 0 \\ x_1 - 2x_2 - 11x_3 = 0 \end{cases}$

9. Assume for each type of system below that there is at least one variable with a nonzero coefficient. Find the smallest and largest rank possible for the corresponding augmented matrix in each case:

★(a) Four equations, three variables, nonhomogeneous

(b) Three equations, four variables

★(c) Three equations, four variables, inconsistent

(d) Five equations, three variables, nonhomogeneous, consistent

★10. Solve the systems $\mathbf{AX} = \mathbf{B}_1$ and $\mathbf{AX} = \mathbf{B}_2$ simultaneously, as illustrated in this section, where

$$\mathbf{A} = \begin{bmatrix} 9 & 2 & 2 \\ 3 & 2 & 4 \\ 27 & 12 & 22 \end{bmatrix}, \quad \mathbf{B}_1 = \begin{bmatrix} -6 \\ 0 \\ 12 \end{bmatrix}, \quad \text{and} \quad \mathbf{B}_2 = \begin{bmatrix} -12 \\ -3 \\ 8 \end{bmatrix}.$$

11. Solve the systems $\mathbf{AX} = \mathbf{B}_1$ and $\mathbf{AX} = \mathbf{B}_2$ simultaneously, as illustrated in this section, where

$$\mathbf{A} = \begin{bmatrix} 12 & 2 & 0 & 3 \\ -24 & -4 & 1 & -6 \\ -4 & -1 & -1 & 0 \\ -30 & -5 & 0 & -6 \end{bmatrix}, \quad \mathbf{B}_1 = \begin{bmatrix} 3 \\ 8 \\ -4 \\ 6 \end{bmatrix}, \quad \text{and} \quad \mathbf{B}_2 = \begin{bmatrix} 2 \\ 4 \\ -24 \\ 0 \end{bmatrix}.$$

12. Let $\mathbf{A}$ be a diagonal $n \times n$ matrix. Prove that $\mathbf{A}$ is row equivalent to $\mathbf{I}_n$ if and only if $a_{ii} \neq 0$, for all i, $1 \leq i \leq n$.

13. (a) Finish the proof of Theorem 2.1 by showing that the three inverse row operations given in Figure 2.2 correctly reverse their corresponding type (I), (II), and (III) row operations.

(b) Finish the proof of Theorem 2.2 by showing that, when a single row operation of type (I) or type (III) is applied to a linear system $\mathbf{AX} = \mathbf{B}$, every solution of the original system is also a solution of the new system.

★14. Let $\mathbf{A}$ be an $m \times n$ matrix. If $\mathbf{B}$ is a nonzero m-vector, explain why the systems $\mathbf{AX} = \mathbf{B}$ and $\mathbf{AX} = \mathbf{O}$ are not equivalent.

15. (a) Show that, if five distinct points in the plane are given, then they must lie on a conic section: an equation of the form $ax^2 + bxy + cy^2 + dx + ey + f = 0$. (Hint: Create the corresponding homogeneous system of five equations and use Corollary 2.5.)

(b) Is this result also true when fewer than five points are given? Why or why not?

16. Show that the proof of Theorem 2.4 does not necessarily work for a nonhomogeneous system.

17. Consider the homogeneous system $\mathbf{AX} = \mathbf{O}$ having m equations and n variables:

(a) Prove that, if $\mathbf{X}_1 = (s_1, s_2, \ldots, s_n)$ and $\mathbf{X}_2 = (t_1, t_2, \ldots, t_n)$ are both solutions to this system, then $\mathbf{X}_1 + \mathbf{X}_2$ and any scalar multiple $c\mathbf{X}_1$ are also solutions.

★(b) Give a counterexample to show that the results of part (a) do not necessarily hold if the system is nonhomogeneous.

(c) Consider a nonhomogeneous system $\mathbf{AX} = \mathbf{B}$ having the same coefficient matrix as the homogeneous system $\mathbf{AX} = \mathbf{O}$. Prove that, if $\mathbf{X}_1$ is a solution of this nonhomogeneous system and if $\mathbf{X}_2 \neq \mathbf{0}$ is a solution of the homogeneous system $\mathbf{AX} = \mathbf{O}$, then $\mathbf{X}_1 + \mathbf{X}_2$ is another solution of $\mathbf{AX} = \mathbf{B}$.

(d) Show that if $\mathbf{AX} = \mathbf{B}$ has a unique solution, with $\mathbf{B} \neq \mathbf{O}$, then the corresponding homogeneous system $\mathbf{AX} = \mathbf{O}$ can have only the trivial solution. (Hint: Use part (c).)

18. Prove that the following homogeneous system has a nontrivial solution if and only if $ad - bc = 0$:

$$\begin{cases} ax_1 + bx_2 = 0 \\ cx_1 + dx_2 = 0 \end{cases}.$$

(Hint: First suppose that $a \neq 0$, and show that in applying the Gauss-Jordan algorithm, the second column has a nonzero pivot element if and only if $ad - bc \neq 0$. Then consider the case $a = 0$.)

★19. Show that the converse to Theorem 2.2 is not true by exhibiting two inconsistent systems (with the same number of equations and variables) whose corresponding augmented matrices are not row equivalent.

20. Suppose that $\mathbf{AX} = \mathbf{O}$ is a homogeneous system of n equations in n variables:

(a) If the system $\mathbf{A}^2\mathbf{X} = \mathbf{O}$ has a nontrivial solution, show that $\mathbf{AX} = \mathbf{O}$ also has a nontrivial solution. (Hint: Prove the contrapositive.)

(b) Generalize the result of part (a) to show that, if the system $\mathbf{A}^t\mathbf{X} = \mathbf{O}$ has a nontrivial solution for some positive integer t, then $\mathbf{AX} = \mathbf{O}$ has a nontrivial solution. (Hint: Use a proof by induction.)

2.3 Row Space of a Matrix

In this section we introduce an important set of linear combinations of vectors associated with a matrix, called the row space of the matrix. We then show that row equivalent matrices have the same row space.

Linear Combinations of Vectors

In Section 1.1, we introduced linear combinations of vectors. Recall that a linear combination of vectors is a sum of scalar multiples of the vectors.

Example 1

Let $\mathbf{a}_1 = [-4, 1, 2]$, $\mathbf{a}_2 = [2, 1, 0]$, and $\mathbf{a}_3 = [6, -3, -4]$ in $\mathbb{R}^3$. Consider the vector $[-18, 15, 16]$. Because

$$[-18, 15, 16] = 2[-4, 1, 2] + 4[2, 1, 0] - 3[6, -3, -4],$$

the vector $[-18, 15, 16]$ is a linear combination of the vectors $\mathbf{a}_1$, $\mathbf{a}_2$, and $\mathbf{a}_3$.

Now consider the vector $[16, -3, 8]$. This vector is not a linear combination of $\mathbf{a}_1$, $\mathbf{a}_2$, and $\mathbf{a}_3$. If it were, the equation

$$[16, -3, 8] = c_1[-4, 1, 2] + c_2[2, 1, 0] + c_3[6, -3, -4]$$

would have a solution. Equating coordinates, we get the following system:

$$\begin{cases} -4c_1 + 2c_2 + 6c_3 = 16 & \text{first coordinate} \\ c_1 + c_2 - 3c_3 = -3 & \text{second coordinate} \\ 2c_1 \quad\quad - 4c_3 = 8 & \text{third coordinate} \end{cases}$$

After row reducing the augmented matrix for this system, we obtain

$$\left[\begin{array}{ccc|c} 1 & 0 & -2 & -\frac{11}{3} \\ 0 & 1 & -1 & \frac{2}{3} \\ 0 & 0 & 0 & \frac{46}{3} \end{array}\right].$$

The third row of this final matrix indicates there are no values of c_1, c_2, and c_3 that together satisfy the equation

$$[16, -3, 8] = c_1[-4, 1, 2] + c_2[2, 1, 0] + c_3[6, -3, -4].$$

Therefore, $[16, -3, 8]$ is not a linear combination of the vectors $[-4, 1, 2]$, $[2, 1, 0]$, and $[6, -3, -4]$. ■

The next example shows that a vector $\mathbf{x}$ can sometimes be expressed as a linear combination of vectors $\mathbf{a}_1, \mathbf{a}_2, \ldots, \mathbf{a}_k$ in more than one way:

Example 2

To determine whether $[14, -21, 7]$ is a linear combination of $[2, -3, 1]$ and $[-4, 6, -2]$, we need to find scalars c_1 and c_2 such that

$$[14, -21, 7] = c_1[2, -3, 1] + c_2[-4, 6, -2].$$

This is equivalent to solving the system

$$\begin{cases} 2c_1 - 4c_2 = 14 \\ -3c_1 + 6c_2 = -21 \\ c_1 - 2c_2 = 7 \end{cases}.$$

After row reducing the appropriate augmented matrix, we obtain

$$\left[\begin{array}{cc|c} 1 & -2 & 7 \\ 0 & 0 & 0 \\ 0 & 0 & 0 \end{array}\right].$$

Because c_2 is an independent variable, we may take c_2 to be any real value. Then $c_1 = 2c_2 + 7$. Hence, there are an infinite number of solutions to the system.

For example, we could let $c_2 = 1$, which forces $c_1 = 2(1) + 7 = 9$, yielding

$$[14, -21, 7] = 9[2, -3, 1] + 1[-4, 6, -2].$$

On the other hand, we could let $c_2 = 0$, which forces $c_1 = 7$, yielding

$$[14, -21, 7] = 7[2, -3, 1] + 0[-4, 6, -2].$$

Thus, we have expressed $[14, -21, 7]$ as a linear combination of $[2, -3, 1]$ and $[-4, 6, -2]$ in more than one way. ■

It is possible to have a linear combination of a single vector: any scalar multiple of $\mathbf{a}$ is a linear combination of $\mathbf{a}$. For example, if $\mathbf{a} = [3, -1, 5]$, then $-2\mathbf{a} = [-6, 2, -10]$ is a linear combination of $\mathbf{a}$.

The Row Space of a Matrix

Suppose $\mathbf{A}$ is an $m \times n$ matrix. Recall that each of the m rows of $\mathbf{A}$ is a vector with n entries—that is, a vector in $\mathbb{R}^n$.

DEFINITION

Let $\mathbf{A}$ be an $m \times n$ matrix. The subset of $\mathbb{R}^n$ consisting of all vectors that are linear combinations of the rows of $\mathbf{A}$ is called the **row space** of $\mathbf{A}$.

Example 3

Consider this matrix:

$$\mathbf{A} = \begin{bmatrix} 3 & 1 & -2 \\ 4 & 0 & 1 \\ -2 & 4 & -3 \end{bmatrix}.$$

We want to determine whether $[5, 17, -20]$ is in the row space of $\mathbf{A}$. If so, $[5, 17, -20]$ can be expressed as a linear combination of the rows of $\mathbf{A}$, as follows:

$$[5, 17, -20] = c_1[3, 1, -2] + c_2[4, 0, 1] + c_3[-2, 4, -3].$$

Equating the coordinates on each side leads to the following system:

$$\begin{cases} 3c_1 + 4c_2 - 2c_3 = 5 \\ c_1 \qquad\quad + 4c_3 = 17 \\ -2c_1 + c_2 - 3c_3 = -20 \end{cases}.$$

Using row reduction yields this associated augmented matrix:

$$\left[\begin{array}{ccc|c} 1 & 0 & 0 & 5 \\ 0 & 1 & 0 & -1 \\ 0 & 0 & 1 & 3 \end{array}\right].$$

Hence, $c_1 = 5$, $c_2 = -1$, and $c_3 = 3$, and

$$[5, 17, -20] = 5[3, 1, -2] - 1[4, 0, 1] + 3[-2, 4, -3].$$

Therefore $[5, 17, -20]$ is in the row space of $\mathbf{A}$. ■

Example 4

The vector $[3, 5]$ is not in the row space of $\mathbf{B} = \begin{bmatrix} 2 & -4 \\ -1 & 2 \end{bmatrix}$ because there is no way to express $[3, 5]$ as a linear combination of the rows $[2, -4]$ and $[-1, 2]$ of $\mathbf{B}$. If

$$[3, 5] = c_1[2, -4] + c_2[-1, 2],$$

equating the coordinates on each side leads to the following system:

$$\begin{cases} 2c_1 - c_2 = 3 \\ -4c_1 + 2c_2 = 5 \end{cases}.$$

But this system is inconsistent. (Verify.) ■

If $\mathbf{A}$ is any $m \times n$ matrix, then the vector $[0, 0, \ldots, 0]$ in $\mathbb{R}^n$ is always in the row space of $\mathbf{A}$. This fact is true because the zero vector can always be expressed as a linear combination of the rows of $\mathbf{A}$ simply by multiplying each row by zero and adding the results.

Similarly, each individual row of $\mathbf{A}$ is in the row space of $\mathbf{A}$, because any particular row of $\mathbf{A}$ can be expressed as a linear combination of the rows of $\mathbf{A}$ simply by multiplying that row by 1, multiplying all other rows by zero, and summing.

Row Equivalence Determines the Row Space

The following lemma is used in the proof of Theorem 2.7:

LEMMA 2.6

Suppose that $\mathbf{x}$ is a linear combination of $\mathbf{q}_1, \ldots, \mathbf{q}_k$, and suppose also that each of $\mathbf{q}_1, \ldots, \mathbf{q}_k$ is itself a linear combination of $\mathbf{r}_1, \ldots, \mathbf{r}_l$. Then $\mathbf{x}$ is a linear combination of $\mathbf{r}_1, \ldots, \mathbf{r}_l$.

If we create a matrix $\mathbf{Q}$ whose rows are the vectors $\mathbf{q}_1, \ldots, \mathbf{q}_k$ and a matrix $\mathbf{R}$ whose rows are the vectors $\mathbf{r}_1, \ldots, \mathbf{r}_l$, then Lemma 2.6 can be rephrased:

> If $\mathbf{x}$ is in the row space of $\mathbf{Q}$ and each row of $\mathbf{Q}$ is in the row space of $\mathbf{R}$, then $\mathbf{x}$ is in the row space of $\mathbf{R}$.

Proof of Lemma 2.6

Because $\mathbf{x}$ is a linear combination of $\mathbf{q}_1, \ldots, \mathbf{q}_k$, we can write $\mathbf{x} = c_1\mathbf{q}_1 + c_2\mathbf{q}_2 + \cdots + c_k\mathbf{q}_k$ for some scalars $c_1, c_2, \ldots, c_k$. But, since each of $\mathbf{q}_1, \ldots, \mathbf{q}_k$ can be expressed as a linear combination of $\mathbf{r}_1, \ldots, \mathbf{r}_l$, there are scalars $d_{11}, \ldots, d_{kl}$ such that

$$\begin{cases} \mathbf{q}_1 = d_{11}\mathbf{r}_1 + d_{12}\mathbf{r}_2 + \cdots + d_{1l}\mathbf{r}_l \\ \mathbf{q}_2 = d_{21}\mathbf{r}_1 + d_{22}\mathbf{r}_2 + \cdots + d_{2l}\mathbf{r}_l \\ \quad\vdots \qquad\quad \vdots \qquad\qquad \vdots \qquad\quad \ddots \qquad \vdots \\ \mathbf{q}_k = d_{k1}\mathbf{r}_1 + d_{k2}\mathbf{r}_2 + \cdots + d_{kl}\mathbf{r}_l \end{cases}.$$

Substituting these equations into the equation for $\mathbf{x}$, we obtain

$$\begin{aligned} \mathbf{x} = \; & c_1(d_{11}\mathbf{r}_1 + d_{12}\mathbf{r}_2 + \cdots + d_{1l}\mathbf{r}_l) \\ + \; & c_2(d_{21}\mathbf{r}_1 + d_{22}\mathbf{r}_2 + \cdots + d_{2l}\mathbf{r}_l) \\ & \quad \vdots \qquad\qquad \vdots \qquad\quad \ddots \qquad \vdots \\ + \; & c_k(d_{k1}\mathbf{r}_1 + d_{k2}\mathbf{r}_2 + \cdots + d_{kl}\mathbf{r}_l) \end{aligned}.$$

Collecting all $\mathbf{r}_1$ terms, all $\mathbf{r}_2$ terms, and so on, we get

$$\begin{aligned} \mathbf{x} = \; & (c_1 d_{11} + c_2 d_{21} + \cdots + c_k d_{k1})\mathbf{r}_1 \\ + \; & (c_1 d_{12} + c_2 d_{22} + \cdots + c_k d_{k2})\mathbf{r}_2 \\ & \quad \vdots \qquad\qquad \vdots \qquad\quad \ddots \qquad \vdots \\ + \; & (c_1 d_{1l} + c_2 d_{2l} + \cdots + c_k d_{kl})\mathbf{r}_l \end{aligned}.$$

Thus, $\mathbf{x}$ can be expressed as a linear combination of $\mathbf{r}_1, \mathbf{r}_2, \ldots, \mathbf{r}_l$. ■

The next theorem illustrates a connection between row equivalence and row space:

THEOREM 2.7

Suppose that **A** and **B** are row equivalent matrices. Then the row space of **A** equals the row space of **B**.

In other words, if **A** and **B** are row equivalent, then any vector that is a linear combination of the rows of **A** must be a linear combination of the rows of **B**, and vice versa. Theorem 2.7 assures us that we do not gain or lose any linear combinations of the rows when we perform row operations.

Proof of Theorem 2.7 (abridged)

Let **A** and **B** be row equivalent $m \times n$ matrices. We will show that, if $\mathbf{x}$ is a vector in the row space of **A**, then $\mathbf{x}$ is in the row space of **B**. (The same argument can then be used to show that if $\mathbf{x}$ is in the row space of **B**, then $\mathbf{x}$ is in the row space of **A**.)

Suppose $\mathbf{x}$ is in the row space of $\mathbf{A}$. Then $\mathbf{x}$ is a linear combination of the rows of $\mathbf{A}$. If we can show that each row of $\mathbf{A}$ is in the row space of $\mathbf{B}$, then Lemma 2.6 tells us that $\mathbf{x}$ is in the row space of $\mathbf{B}$, and we are done.

Thus, we only need to show that each row of $\mathbf{A}$ is in the row space of $\mathbf{B}$. Now, $\mathbf{A}$ and $\mathbf{B}$ are row equivalent. Hence, there is a finite sequence of row operations and intermediate matrices $\mathbf{D}_1, \ldots, \mathbf{D}_k$ such that

$$\mathbf{B} \longrightarrow \mathbf{D}_1 \longrightarrow \mathbf{D}_2 \longrightarrow \cdots \longrightarrow \mathbf{D}_{k-1} \longrightarrow \mathbf{D}_k \longrightarrow \mathbf{A}.$$

Each step in the chain represents one row operation.

Suppose we can show that, if a matrix $\mathbf{T}$ is created from $\mathbf{S}$ with a single row operation, then each row of $\mathbf{T}$ is in the row space of $\mathbf{S}$. This will complete the proof, because a simple induction argument extends the result to the case where any (finite) number of row operations are performed (see Exercise 4).

Let us consider the case where an operation of type (II) is performed. That is, suppose that $\mathbf{S}$ is an $m \times n$ matrix and that $\mathbf{T}$ is created from $\mathbf{S}$ with a single row operation of type (II). Let $\mathbf{s}_1, \ldots, \mathbf{s}_m$ be the vectors representing the rows of $\mathbf{S}$. If $\mathbf{t}_1, \ldots, \mathbf{t}_m$ are the rows of $\mathbf{T}$, then some row of $\mathbf{T}$, say $\mathbf{t}_j$, is equal to $c\mathbf{s}_i + \mathbf{s}_j$, for some scalar c, and for $i \neq j$. Every other row $\mathbf{t}_k$ of $\mathbf{T}$ equals the corresponding row $\mathbf{s}_k$ of $\mathbf{S}$. Then, for $k \neq j$, we can write

$$\mathbf{t}_k = 0\mathbf{s}_1 + 0\mathbf{s}_2 + \cdots + 1\mathbf{s}_k + \cdots + 0\mathbf{s}_m,$$

which shows that for $k \neq j$, $\mathbf{t}_k$ is a linear combination of the rows of $\mathbf{S}$. We can also write

$$\mathbf{t}_j = 0\mathbf{s}_1 + 0\mathbf{s}_2 + \cdots + c\mathbf{s}_i + \cdots + 1\mathbf{s}_j + \cdots + 0\mathbf{s}_m,$$

which shows that $\mathbf{t}_j$ is also a linear combination of the rows of $\mathbf{S}$. Hence, each row of $\mathbf{T}$ is in the row space of $\mathbf{S}$. A similar argument for row operations of types (I) and (III) concludes the proof. (See Exercise 5.) ■

Because a matrix $\mathbf{A}$ and its reduced row echelon form matrix $\mathbf{B}$ have exactly the same row space, Theorem 2.7 implies that the rank of a matrix indicates precisely how many distinct rows of a matrix are "important." However, $\mathbf{B}$ is usually a "simpler-looking" matrix than $\mathbf{A}$, and it gives us the essential part of $\mathbf{A}$. In particular, in $\mathbf{B}$ we get the minimal number of nonzero rows needed to produce the same row space. (We prove this assertion in Chapter 4.) For example, if a matrix $\mathbf{A}$ has five rows and its reduced row echelon form matrix $\mathbf{B}$ has two nonzero rows, then in a sense, three rows of $\mathbf{A}$ are redundant.

Exercises—Section 2.3

Note: To save time, you should use an appropriate software package to perform nontrivial row reductions.

1. In each case, express the vector $\mathbf{x}$ as a linear combination of the other vectors, if possible:

★(a) $\mathbf{x} = [-3, -6]$, $\mathbf{a}_1 = [1, 4]$, $\mathbf{a}_2 = [-2, 3]$

(b) $\mathbf{x} = [5, 9, 5]$, $\mathbf{a}_1 = [2, 1, 4]$, $\mathbf{a}_2 = [1, -1, 3]$, $\mathbf{a}_3 = [3, 2, 5]$

★(c) $\mathbf{x} = [2, -1, 4]$, $\mathbf{a}_1 = [3, 6, 2]$, $\mathbf{a}_2 = [2, 10, -4]$

(d) $\mathbf{x} = [2, 2, 3]$, $\mathbf{a}_1 = [6, -2, 3]$, $\mathbf{a}_2 = [0, -5, -1]$, $\mathbf{a}_3 = [-2, 1, 2]$

★(e) $\mathbf{x} = [7, 2, 3]$, $\mathbf{a}_1 = [1, -2, 3]$, $\mathbf{a}_2 = [5, -2, 6]$, $\mathbf{a}_3 = [4, 0, 3]$

(f) $\mathbf{x} = [1, 1, 1, 1]$, $\mathbf{a}_1 = [2, 1, 0, 3]$, $\mathbf{a}_2 = [3, -1, 5, 2]$, $\mathbf{a}_3 = [-1, 0, 2, 1]$

★(g) $\mathbf{x} = [2, 3, -7, 3]$, $\mathbf{a}_1 = [3, 2, -2, 4]$, $\mathbf{a}_2 = [-2, 0, 1, -3]$, $\mathbf{a}_3 = [6, 1, 2, 8]$

(h) $\mathbf{x} = [-3, 1, 2, 0, 1]$, $\mathbf{a}_1 = [-6, 2, 4, -1, 7]$

2. In each case, determine whether the given vector is in the row space of the given matrix:

★(a) $[7, 1, 18]$, with $\begin{bmatrix} 3 & 6 & 2 \\ 2 & 10 & -4 \\ 2 & -1 & 4 \end{bmatrix}$

(b) $[4, 0, -3]$, with $\begin{bmatrix} 3 & 1 & 1 \\ 2 & -1 & 5 \\ -4 & -3 & 3 \end{bmatrix}$

★(c) $[2, 2, -3]$, with $\begin{bmatrix} 4 & -1 & 2 \\ -2 & 3 & 5 \\ 6 & 1 & 9 \end{bmatrix}$

(d) $[1, 2, 5, -1]$, with $\begin{bmatrix} 2 & -1 & 0 & 3 \\ 7 & -1 & 5 & 8 \end{bmatrix}$

★(e) $[1, 11, -4, 11]$, with $\begin{bmatrix} 2 & -4 & 1 & -3 \\ 7 & -1 & -1 & 2 \\ 3 & 7 & -3 & 8 \end{bmatrix}$

★**3.** (a) Express the vector $[13, -23, 60]$ as a linear combination of the vectors

$$\mathbf{q}_1 = [-1, -5, 11], \mathbf{q}_2 = [-10, 3, -8], \text{ and } \mathbf{q}_3 = [7, -12, 30].$$

(b) Express each of the vectors $\mathbf{q}_1$, $\mathbf{q}_2$, and $\mathbf{q}_3$ in turn as a linear combination of the vectors $\mathbf{r}_1 = [3, -2, 4]$, $\mathbf{r}_2 = [2, 1, -3]$, and $\mathbf{r}_3 = [4, -1, 2]$.

(c) Use the results of parts (a) and (b) to express the vector $[13, -23, 60]$ as a linear combination of the vectors $\mathbf{r}_1$, $\mathbf{r}_2$, and $\mathbf{r}_3$. (Hint: Use the method given in the proof of Lemma 2.6.)

4. Suppose the following chain of matrices is given:

$$\mathbf{B} \longrightarrow \mathbf{D}_1 \longrightarrow \mathbf{D}_2 \longrightarrow \cdots \longrightarrow \mathbf{D}_{k-1} \longrightarrow \mathbf{D}_k \longrightarrow \mathbf{A}.$$

If each row of $\mathbf{A}$ is in the row space of $\mathbf{D}_k$, each row of $\mathbf{D}_k$ is in the row space of $\mathbf{D}_{k-1}$, and so on, prove that each row of $\mathbf{A}$ is in the row space of $\mathbf{B}$. (Hint: Use Lemma 2.6 and a proof by induction.)

5. Finish the proof of Theorem 2.7 by showing that, if a matrix $\mathbf{T}$ is created from a matrix $\mathbf{S}$ by a single row operation of type (I) or of type (III), then each row of $\mathbf{T}$ is in the row space of $\mathbf{S}$.

6. Let $\mathbf{x}_1, \ldots, \mathbf{x}_{n+1}$ be vectors in $\mathbb{R}^n$.
   (a) Show that there exist real numbers $a_1, \ldots, a_{n+1}$, not all zero, such that the linear combination $a_1\mathbf{x}_1 + \cdots + a_{n+1}\mathbf{x}_{n+1}$ equals the zero vector. (Hint: Solve an appropriate homogeneous system.)
   (b) Using part (a), show that

$$\mathbf{x}_i = b_1\mathbf{x}_1 + \cdots + b_{i-1}\mathbf{x}_{i-1} + b_{i+1}\mathbf{x}_{i+1} + \cdots + b_{n+1}\mathbf{x}_{n+1},$$

   for some i, $1 \le i \le n+1$, and some $b_1, \ldots, b_{i-1}, b_{i+1}, \ldots, b_{n+1} \in \mathbb{R}$.

7. (a) Find two different 3×5 matrices $\mathbf{A}$ and $\mathbf{B}$, each having rank 3, that are in reduced row echelon form. Do these matrices have the same row space? (Is every row of $\mathbf{A}$ in the row space of $\mathbf{B}$, and vice versa?) Prove your answer.
   (b) Repeat part (a) with 3×6 matrices of different rank (both in reduced row echelon form).
   (c) Make a conjecture (a statement that you guess is true) based on the examples you provided in parts (a) and (b).
   (d) Prove your conjecture from part (c). (You may find this difficult.)
   (e) Use part (d) to prove the converse to Theorem 2.7.

2.4 Inverses of Matrices

In this section, we consider whether a given $n \times n$ (square) matrix $\mathbf{A}$ has a corresponding $n \times n$ *multiplicative* inverse matrix. That is, we want a matrix $\mathbf{A}^{-1}$ such that $\mathbf{A}\mathbf{A}^{-1} = \mathbf{I}_n$ ($\mathbf{I}_n$ being the $n \times n$ identity matrix for multiplication). Interestingly, not all square matrices have multiplicative inverses, but most do. In this section, we examine some properties of multiplicative inverses and illustrate some methods for finding these inverses when they exist.

Definition of the Multiplicative Inverse

In linear algebra, when the word *inverse* is used, it usually refers to *multiplicative inverse* rather than the "additive inverse" of Theorem 1.10, part (4).

DEFINITION

Let $\mathbf{A}$ be an $n \times n$ matrix. Then an $n \times n$ matrix $\mathbf{B}$ is a **(multiplicative) inverse** of $\mathbf{A}$ if and only if $\mathbf{AB} = \mathbf{BA} = \mathbf{I}_n$.

Note that, if $\mathbf{B}$ is an inverse of $\mathbf{A}$, then $\mathbf{A}$ is also an inverse of $\mathbf{B}$, as can easily be seen by switching the roles of $\mathbf{A}$ and $\mathbf{B}$ in the definition.

Example 1

The matrices

$$\mathbf{A} = \begin{bmatrix} 1 & -4 & 1 \\ 1 & 1 & -2 \\ -1 & 1 & 1 \end{bmatrix} \quad \text{and} \quad \mathbf{B} = \begin{bmatrix} 3 & 5 & 7 \\ 1 & 2 & 3 \\ 2 & 3 & 5 \end{bmatrix}$$

are inverses of each other because

$$\underbrace{\begin{bmatrix} 1 & -4 & 1 \\ 1 & 1 & -2 \\ -1 & 1 & 1 \end{bmatrix}}_{\mathbf{A}} \underbrace{\begin{bmatrix} 3 & 5 & 7 \\ 1 & 2 & 3 \\ 2 & 3 & 5 \end{bmatrix}}_{\mathbf{B}} = \underbrace{\begin{bmatrix} 1 & 0 & 0 \\ 0 & 1 & 0 \\ 0 & 0 & 1 \end{bmatrix}}_{\mathbf{I}_3} = \underbrace{\begin{bmatrix} 3 & 5 & 7 \\ 1 & 2 & 3 \\ 2 & 3 & 5 \end{bmatrix}}_{\mathbf{B}} \underbrace{\begin{bmatrix} 1 & -4 & 1 \\ 1 & 1 & -2 \\ -1 & 1 & 1 \end{bmatrix}}_{\mathbf{A}}.$$

However, the matrix $\mathbf{C} = \begin{bmatrix} 2 & 1 \\ 6 & 3 \end{bmatrix}$ has no inverse at all, because there is no 2×2 matrix $\begin{bmatrix} a & b \\ c & d \end{bmatrix}$ such that

$$\begin{bmatrix} 2 & 1 \\ 6 & 3 \end{bmatrix} \begin{bmatrix} a & b \\ c & d \end{bmatrix} = \begin{bmatrix} 1 & 0 \\ 0 & 1 \end{bmatrix}.$$

If there were such a matrix, then multiplying out the left side of this equation would give

$$\begin{bmatrix} 2a + c & 2b + d \\ 6a + 3c & 6b + 3d \end{bmatrix} = \begin{bmatrix} 1 & 0 \\ 0 & 1 \end{bmatrix}.$$

This result would force $2a + c = 1$ and $6a + 3c = 0$, but these are contradictory equations, since $6a + 3c = 3(2a + c)$. ■

When checking to see whether two given square matrices $\mathbf{A}$ and $\mathbf{B}$ are inverses, we do not need to actually multiply both products $\mathbf{AB}$ and $\mathbf{BA}$, as the next theorem asserts:

THEOREM 2.8

Let $\mathbf{A}$ and $\mathbf{B}$ be $n \times n$ matrices. If either product $\mathbf{AB}$ or $\mathbf{BA}$ equals $\mathbf{I}_n$, then the other product also equals $\mathbf{I}_n$, and $\mathbf{A}$ and $\mathbf{B}$ are inverses of each other.

The proof of this theorem is tedious, and so it is placed in Appendix A for the interested reader.

DEFINITION

A square matrix is **singular** if and only if it does not have an inverse. A square matrix is **nonsingular** if and only if it has an inverse.

For example, the 2×2 matrix $\mathbf{C}$ from Example 1 is a singular matrix; we proved that it does not have an inverse. Another example of a singular matrix is the zero $n \times n$ matrix $\mathbf{O}_n$. (Why?) On the other hand, the 3×3 matrix $\mathbf{A}$ from Example 1 is nonsingular, because we found a 3×3 inverse $\mathbf{B}$ for $\mathbf{A}$.

Properties of the Matrix Inverse

The next theorem is important because it shows that the inverse of a matrix must be unique (when it exists):

THEOREM 2.9: Uniqueness of Matrix Inverse

If $\mathbf{B}$ and $\mathbf{C}$ are both inverses of an $n \times n$ matrix $\mathbf{A}$, then $\mathbf{B} = \mathbf{C}$.

Proof of Theorem 2.9

$$\mathbf{B} = \mathbf{B}\mathbf{I}_n = \mathbf{B}(\mathbf{A}\mathbf{C}) = (\mathbf{B}\mathbf{A})\mathbf{C} = \mathbf{I}_n\mathbf{C} = \mathbf{C}. \qquad \blacksquare$$

Because Theorem 2.9 assures us that a nonsingular matrix $\mathbf{A}$ can have exactly one inverse, we use the notation $\mathbf{A}^{-1}$ to denote the inverse of $\mathbf{A}$.

For a nonsingular matrix $\mathbf{A}$, we can use the matrix inverse to define negative integral powers of $\mathbf{A}$:

DEFINITION

Let $\mathbf{A}$ be a nonsingular $n \times n$ matrix. Then the negative powers of $\mathbf{A}$ are given as follows: $\mathbf{A}^{-1}$ is the (unique) inverse of $\mathbf{A}$, and for $k \geq 2$, $\mathbf{A}^{-k} = (\mathbf{A}^{-1})^k$.

Example 2

We already know that the matrix

$$\mathbf{A} = \begin{bmatrix} 1 & -4 & 1 \\ 1 & 1 & -2 \\ -1 & 1 & 1 \end{bmatrix}$$

from Example 1 has the matrix

$$\mathbf{A}^{-1} = \begin{bmatrix} 3 & 5 & 7 \\ 1 & 2 & 3 \\ 2 & 3 & 5 \end{bmatrix}$$

as its inverse. Since $\mathbf{A}^{-3} = (\mathbf{A}^{-1})^3$, we have

$$\mathbf{A}^{-3} = \begin{bmatrix} 3 & 5 & 7 \\ 1 & 2 & 3 \\ 2 & 3 & 5 \end{bmatrix}^3 = \begin{bmatrix} 272 & 445 & 689 \\ 107 & 175 & 271 \\ 184 & 301 & 466 \end{bmatrix}.$$

■

THEOREM 2.10

Let $\mathbf{A}$ and $\mathbf{B}$ be nonsingular matrices of the same size. Then
(1) $(\mathbf{A}^{-1})^{-1} = \mathbf{A}$
(2) $(\mathbf{A}^k)^{-1} = (\mathbf{A}^{-1})^k = \mathbf{A}^{-k}$, for any integer k
(3) $(\mathbf{AB})^{-1} = \mathbf{B}^{-1}\mathbf{A}^{-1}$
(4) $(\mathbf{A}^T)^{-1} = (\mathbf{A}^{-1})^T$

Note that part (3) says that the inverse of a product equals the product of the inverses in *reverse* order. Hence, if $\mathbf{A}$ and $\mathbf{B}$ are both nonsingular, then $\mathbf{AB}$ is also nonsingular.

To prove each of the parts of this theorem, show that the right side is the inverse of the term in parentheses on the left side by multiplying them together and observing that their product is the identity matrix. We prove parts (3) and (4) here and leave the others for you in Exercise 15:

Proof of Theorem 2.10 (abridged)

Part (3): We must show that $\mathbf{B}^{-1}\mathbf{A}^{-1}$ (right side) is the inverse of $\mathbf{AB}$ (in parentheses on the left side). Multiplying them together gives $(\mathbf{AB})(\mathbf{B}^{-1}\mathbf{A}^{-1}) = \mathbf{A}(\mathbf{BB}^{-1})\mathbf{A}^{-1} = \mathbf{AI}_n\mathbf{A}^{-1} = \mathbf{AA}^{-1} = \mathbf{I}_n$.

Part (4): We must show that $(\mathbf{A}^{-1})^T$ (right side) is the inverse of $\mathbf{A}^T$ (in parentheses on the left side). Multiplying them together gives $\mathbf{A}^T(\mathbf{A}^{-1})^T = (\mathbf{A}^{-1}\mathbf{A})^T$ (by Theorem 1.15) $= (\mathbf{I}_n)^T = \mathbf{I}_n$, since $\mathbf{I}_n$ is symmetric. ■

Using a proof by induction, part (3) of Theorem 2.10 generalizes as follows: if $\mathbf{A}_1, \mathbf{A}_2, \ldots, \mathbf{A}_k$ are nonsingular matrices of the same size, then

$$(\mathbf{A}_1\mathbf{A}_2\cdots\mathbf{A}_k)^{-1} = \mathbf{A}_k^{-1}\cdots\mathbf{A}_2^{-1}\mathbf{A}_1^{-1}.$$

(See Exercise 15.) Notice that the order of the matrices on the right side is reversed. Theorem 1.14 can also be generalized to show that the laws of exponents hold for negative integer powers, as follows:

THEOREM 2.11: Expanded version of Theorem 1.14

If $\mathbf{A}$ is a nonsingular matrix and if s and t are integers, then
(1) $\mathbf{A}^{s+t} = (\mathbf{A}^s)(\mathbf{A}^t)$
(2) $(\mathbf{A}^s)^t = \mathbf{A}^{st} = (\mathbf{A}^t)^s$

The proof of this theorem is easy but tedious. We ask you to prove some special cases in Exercise 17.

Recall that in Section 1.5 we observed that, if $\mathbf{AB} = \mathbf{AC}$ for three matrices $\mathbf{A}$, $\mathbf{B}$, and $\mathbf{C}$, it does not necessarily follow that $\mathbf{B}$ equals $\mathbf{C}$. However, if $\mathbf{A}$ is a nonsingular matrix, then $\mathbf{B} = \mathbf{C}$, because you can multiply both sides of $\mathbf{AB} = \mathbf{AC}$ by $\mathbf{A}^{-1}$ on the left to effectively cancel out the $\mathbf{A}$'s.

Inverses for 2 × 2 Matrices

Up to this point we have studied many properties of the matrix inverse, but we have not discussed methods for finding inverses. In fact, there is an immediate way to find the inverse (if it exists) of a 2×2 matrix:

THEOREM 2.12

The matrix $\begin{bmatrix} a & b \\ c & d \end{bmatrix}$ has an inverse if and only if $\delta = ad - bc \neq 0$. In that case, the inverse is

$$\begin{bmatrix} a & b \\ c & d \end{bmatrix}^{-1} = \frac{1}{\delta}\begin{bmatrix} d & -b \\ -c & a \end{bmatrix}.$$

For a 2×2 matrix $\mathbf{A} = \begin{bmatrix} a & b \\ c & d \end{bmatrix}$, the condition $\delta = ad - bc \neq 0$ is both a necessary and a sufficient condition for the inverse to exist. The quantity $\delta = ad - bc$ is called the **determinant** of $\mathbf{A}$. We will discuss determinants in more detail in Chapter 3.

Theorem 2.12 is easy to prove once you realize that

$$\begin{bmatrix} a & b \\ c & d \end{bmatrix}\begin{bmatrix} d & -b \\ -c & a \end{bmatrix} = \begin{bmatrix} \delta & 0 \\ 0 & \delta \end{bmatrix},$$

whether $\delta = 0$ or not. Then you can easily show that if $\delta \neq 0$, then $\mathbf{A}$ has an inverse, and vice versa.

Example 3

The matrix $\begin{bmatrix} 12 & -4 \\ 9 & -3 \end{bmatrix}$ has no inverse, since $\delta = (12)(-3) - (-4)(9) = 0$. On the other hand, the matrix $\mathbf{M} = \begin{bmatrix} -5 & 2 \\ 9 & -4 \end{bmatrix}$ does have an inverse, because $\delta =$

$(-5)(-4) - (2)(9) = 2 \neq 0$. This inverse is

$$\mathbf{M}^{-1} = \frac{1}{2}\begin{bmatrix} -4 & -2 \\ -9 & -5 \end{bmatrix} = \begin{bmatrix} -2 & -1 \\ -\frac{9}{2} & -\frac{5}{2} \end{bmatrix}.$$

(Verify.) ■

Inverses of Larger Matrices

Let $\mathbf{A}$ be an $n \times n$ matrix. We now describe a process for calculating $\mathbf{A}^{-1}$, if it exists. We augment $\mathbf{A}$ with the n columns of the identity matrix $\mathbf{I}_n$. That is, we create the $n \times (2n)$ augmented matrix $[\mathbf{A}|\mathbf{I}_n]$. We then use row reduction to transform the first n columns into $\mathbf{I}_n$. If $\mathbf{A}$ is nonsingular, this process transforms the last n columns of $[\mathbf{A}|\mathbf{I}_n]$ into $\mathbf{A}^{-1}$. That is,

$$[\mathbf{A}|\mathbf{I}_n] \quad \text{row reduces to} \quad [\mathbf{I}_n|\mathbf{A}^{-1}].$$

Before proving that this procedure always gives the (correct) inverse for $\mathbf{A}$, let us consider some examples:

Example 4

Let us find the inverse of the matrix

$$\mathbf{A} = \begin{bmatrix} 2 & -6 & 5 \\ -4 & 12 & -9 \\ 2 & -9 & 8 \end{bmatrix}.$$

We first enlarge this to a 3×6 matrix by adjoining the identity matrix $\mathbf{I}_3$:

$$\left[\begin{array}{ccc|ccc} 2 & -6 & 5 & 1 & 0 & 0 \\ -4 & 12 & -9 & 0 & 1 & 0 \\ 2 & -9 & 8 & 0 & 0 & 1 \end{array}\right].$$

Row reduction yields

$$\left[\begin{array}{ccc|ccc} 1 & 0 & 0 & \frac{5}{2} & \frac{1}{2} & -1 \\ 0 & 1 & 0 & \frac{7}{3} & 1 & -\frac{1}{3} \\ 0 & 0 & 1 & 2 & 1 & 0 \end{array}\right].$$

The last three columns give the inverse of the original matrix $\mathbf{A}$:

$$\mathbf{A}^{-1} = \begin{bmatrix} \frac{5}{2} & \frac{1}{2} & -1 \\ \frac{7}{3} & 1 & -\frac{1}{3} \\ 2 & 1 & 0 \end{bmatrix}.$$

You should check that this matrix really is the inverse of **A** by showing that its product with **A** is equal to $\mathbf{I}_3$. ■

Using Row Reduction to Show That a Matrix Is Singular

As you have seen, not every (square) matrix has an inverse. The row reduction process that we have illustrated "breaks down" when we try to compute an inverse for a singular matrix. The only way the process can stop prematurely is if we reach a column whose main diagonal entry and all other entries below it are zero. In that case, there is no way to use an operation of either type (I) or type (III) to place a nonzero element in the main diagonal position for that column. In effect, we cannot transform the leftmost columns into the identity matrix. This situation is illustrated in the following example:

Example 5

Let us attempt to find an inverse for the singular matrix

$$\mathbf{A} = \begin{bmatrix} 4 & 2 & 8 & 1 \\ -2 & 0 & -4 & 1 \\ 1 & 4 & 2 & 0 \\ 3 & -1 & 6 & -2 \end{bmatrix}.$$

Beginning with $[\mathbf{A}|\mathbf{I}_4]$ and simplifying the first two columns, we obtain

$$\left[\begin{array}{cccc|cccc} 1 & 0 & 2 & -\frac{1}{2} & 0 & -\frac{1}{2} & 0 & 0 \\ 0 & 1 & 0 & \frac{3}{2} & \frac{1}{2} & 1 & 0 & 0 \\ 0 & 0 & 0 & -\frac{11}{2} & -2 & -\frac{7}{2} & 1 & 0 \\ 0 & 0 & 0 & 1 & \frac{1}{2} & \frac{5}{2} & 0 & 1 \end{array}\right].$$

Continuing on to the third column, we see that the (3, 3) entry is zero. Thus, a type (I) operation cannot be used to make the pivot 1. Because the (4, 3) entry is also zero, no type (III) operation (switching the home row with a row below it) will make the pivot nonzero. We conclude that there is no way to transform the first four columns into the identity matrix $\mathbf{I}_4$ using the row reduction process. In such situations, the original matrix **A** has no inverse. ■

Formal Algorithm for Finding Inverses

We now state a formal algorithm for determining whether a given (square) matrix has an inverse and, if it does, for calculating that inverse. The algorithm is based on the row reduction process illustrated above:

ALGORITHM FOR FINDING THE INVERSE OF A MATRIX (IF IT EXISTS)

Suppose that $\mathbf{A}$ is a given $n \times n$ matrix:

Step 1. Augment $\mathbf{A}$ to a $n \times 2n$ matrix, whose first n columns constitute $\mathbf{A}$ itself and whose last n columns constitute $\mathbf{I}_n$.

Step 2: Row reduce $[\mathbf{A}|\mathbf{I}_n]$ as far as possible.

Step 3: If the first n columns of $[\mathbf{A}|\mathbf{I}_n]$ cannot be converted into $\mathbf{I}_n$, then $\mathbf{A}$ is singular. Stop.

Step 4: Otherwise, $\mathbf{A}$ is nonsingular, and the last n columns of the augmented matrix in reduced row echelon form constitute $\mathbf{A}^{-1}$.

We now show that this algorithm is valid. We must prove that, for a given square matrix $\mathbf{A}$, the algorithm correctly predicts whether $\mathbf{A}$ has an inverse and, if it does, calculates its (unique) inverse.

Suppose first that $\mathbf{A}$ is a nonsingular $n \times n$ matrix $\mathbf{A}$. We first show that $\text{rank}(\mathbf{A}) = n$. Consider the homogeneous system $\mathbf{AX}=\mathbf{O}$. Since $\mathbf{A}^{-1}$ exists, we can multiply both sides by $\mathbf{A}^{-1}$ to see that this system has the unique solution $\mathbf{X} = \mathbf{O}$. (Verify.) Then, by Theorem 2.4, $\text{rank}(\mathbf{A}) = n$.

Now, because $\mathbf{AA}^{-1} = \mathbf{I}_n$, we see that $\mathbf{A}\begin{bmatrix} i^{th} \\ \text{column} \\ \text{of } \mathbf{A}^{-1} \end{bmatrix} = \begin{bmatrix} i^{th} \\ \text{column} \\ \text{of } \mathbf{I}_n \end{bmatrix}$. Therefore, the solutions of the systems whose augmented matrices are

$$\left[\begin{array}{c|c} \mathbf{A} & \begin{array}{c}\text{1st} \\ \text{column} \\ \text{of } \mathbf{I}_n\end{array} \end{array}\right], \left[\begin{array}{c|c} \mathbf{A} & \begin{array}{c}\text{2nd} \\ \text{column} \\ \text{of } \mathbf{I}_n\end{array} \end{array}\right], \ldots, \left[\begin{array}{c|c} \mathbf{A} & \begin{array}{c}n^{th} \\ \text{column} \\ \text{of } \mathbf{I}_n\end{array} \end{array}\right]$$

must be the columns of $\mathbf{A}^{-1}$. Now, $\text{rank}(\mathbf{A}) = n$, so in solving these systems, all columns of $\mathbf{A}$ become pivot columns during the row reduction process. Hence, each of these systems has a unique solution.

Therefore, using the technique of solving simultaneous systems (from Section 2.2), we find that row reduction of

$$\left[\begin{array}{c|ccccc} \mathbf{A} & \begin{array}{c}\text{1st} \\ \text{column} \\ \text{of } \mathbf{I}_n\end{array} & \begin{array}{c}\text{2nd} \\ \text{column} \\ \text{of } \mathbf{I}_n\end{array} & \begin{array}{c}\text{3rd} \\ \text{column} \\ \text{of } \mathbf{I}_n\end{array} & \cdots & \begin{array}{c}n^{th} \\ \text{column} \\ \text{of } \mathbf{I}_n\end{array} \end{array}\right] = [\mathbf{A}|\mathbf{I}_n]$$

produces

$$\left[\begin{array}{c|ccccc} \mathbf{I}_n & \begin{array}{c}\text{1st} \\ \text{column} \\ \text{of } \mathbf{A}^{-1}\end{array} & \begin{array}{c}\text{2nd} \\ \text{column} \\ \text{of } \mathbf{A}^{-1}\end{array} & \begin{array}{c}\text{3rd} \\ \text{column} \\ \text{of } \mathbf{A}^{-1}\end{array} & \cdots & \begin{array}{c}n^{th} \\ \text{column} \\ \text{of } \mathbf{A}^{-1}\end{array} \end{array}\right] = [\mathbf{I}_n|\mathbf{A}^{-1}].$$

Hence, in this case, step 2 of the algorithm will complete, and then step 4 will identify, the correct inverse of $\mathbf{A}$.

Now consider the case where **A** is singular. Because an inverse for **A** cannot be found, at least one of the original n systems, say,

$$\left[\begin{array}{c|c} \mathbf{A} & \begin{array}{c} k^{th} \\ \text{column} \\ \text{of } \mathbf{I}_n \end{array} \end{array}\right],$$

has no solutions. This result occurs in the Gauss-Jordan row reduction process only if the final augmented matrix contains a row of the form

$$[0 \quad 0 \quad 0 \quad \cdots \quad 0 \mid r],$$

where $r \neq 0$. Then there is a row that contains no pivot element. Thus, we cannot obtain $\mathbf{I}_n$ to the left of the augmentation bar, and so step 3 of the inverse algorithm correctly concludes that **A** is singular.

Because the inverse algorithm is correct, we have the following theorem:

THEOREM 2.13

An $n \times n$ matrix **A** is nonsingular if and only if the rank of **A** is n.

Solving a System Using the Inverse of the Coefficient Matrix

Suppose $\mathbf{AX} = \mathbf{B}$ has the same number of equations as variables. Theorems 2.4 and 2.13 together tell us that $\mathbf{AX} = \mathbf{B}$ has a unique solution if and only if **A** is nonsingular. In that case, we can multiply both sides of $\mathbf{AX} = \mathbf{B}$ on the left by $\mathbf{A}^{-1}$ to get

$$\begin{aligned} \mathbf{A}^{-1}(\mathbf{AX}) &= \mathbf{A}^{-1}\mathbf{B} \\ (\mathbf{A}^{-1}\mathbf{A})\mathbf{X} &= \mathbf{A}^{-1}\mathbf{B} \\ \mathbf{I}_n\mathbf{X} &= \mathbf{A}^{-1}\mathbf{B} \\ \mathbf{X} &= \mathbf{A}^{-1}\mathbf{B}. \end{aligned}$$

Thus, we have proved the following:

THEOREM 2.14

Let $\mathbf{AX} = \mathbf{B}$ represent a system where the coefficient matrix **A** is square. Then the system has a unique solution ($\mathbf{X} = \mathbf{A}^{-1}\mathbf{B}$) if and only if **A** is nonsingular.

Therefore, if $\mathbf{A}^{-1}$ is known, the matrix **X** of variables can be found by a simple matrix multiplication of $\mathbf{A}^{-1}$ and **B**.

Example 6

Let us solve the system

$$\begin{cases} -7x_1 + 5x_2 + 3x_3 = 6 \\ 3x_1 - 2x_2 - 2x_3 = -3 \\ 3x_1 - 2x_2 - x_3 = 2 \end{cases}$$

using the inverse of the coefficient matrix. This system has matrix form

$$\underbrace{\begin{bmatrix} -7 & 5 & 3 \\ 3 & -2 & -2 \\ 3 & -2 & -1 \end{bmatrix}}_{\mathbf{A}} \underbrace{\begin{bmatrix} x_1 \\ x_2 \\ x_3 \end{bmatrix}}_{\mathbf{X}} = \underbrace{\begin{bmatrix} 6 \\ -3 \\ 2 \end{bmatrix}}_{\mathbf{B}}.$$

The inverse matrix of **A** is easily calculated to be

$$\mathbf{A}^{-1} = \begin{bmatrix} 2 & 1 & 4 \\ 3 & 2 & 5 \\ 0 & -1 & 1 \end{bmatrix}.$$

Therefore, by Theorem 2.14, $\mathbf{X} = \mathbf{A}^{-1}\mathbf{B}$, and so

$$\begin{bmatrix} x_1 \\ x_2 \\ x_3 \end{bmatrix} = \begin{bmatrix} 2 & 1 & 4 \\ 3 & 2 & 5 \\ 0 & -1 & 1 \end{bmatrix} \begin{bmatrix} 6 \\ -3 \\ 2 \end{bmatrix} = \begin{bmatrix} 17 \\ 22 \\ 5 \end{bmatrix}.$$

■

This method is not as efficient as Gauss-Jordan row reduction, because it involves finding an inverse and performing a matrix multiplication. It is sometimes used when many systems must be solved, all having the same nonsingular coefficient matrix. In that case, the inverse of the coefficient matrix can be calculated first, and then each system can be solved with a single matrix multiplication. This method is similar to the one in Section 2.2 for solving several systems simultaneously, in that many systems can be solved by row reducing only one matrix.

♦ **Application:** You now have covered the prerequisites for Section 8.5, "Hill Substitution: An Introduction to Coding Theory."

Exercises—Section 2.4

Note: You should be using appropriate computer software to perform nontrivial row reductions.

1. Verify that the following pairs of matrices are inverses:

(a) $\begin{bmatrix} 10 & 41 & -5 \\ -1 & -12 & 1 \\ 3 & 20 & -2 \end{bmatrix}$, $\begin{bmatrix} 4 & -18 & -19 \\ 1 & -5 & -5 \\ 16 & -77 & -79 \end{bmatrix}$

(b) $\begin{bmatrix} 1 & 0 & -1 & 5 \\ -1 & 1 & 0 & -3 \\ 0 & 2 & -3 & 7 \\ 2 & -1 & -2 & 12 \end{bmatrix}$, $\begin{bmatrix} 1 & -4 & 1 & -2 \\ 4 & 6 & -2 & 1 \\ 5 & 11 & -4 & 3 \\ 1 & 3 & -1 & 1 \end{bmatrix}$

2. Determine whether each of the following matrices is nonsingular by calculating its rank:

★(a) $\begin{bmatrix} 4 & -9 \\ -2 & 3 \end{bmatrix}$ (b) $\begin{bmatrix} 3 & -1 & 4 \\ 2 & -2 & 1 \\ -1 & 3 & 2 \end{bmatrix}$ ★(c) $\begin{bmatrix} -6 & -6 & 1 \\ 2 & 3 & -1 \\ 8 & 6 & -1 \end{bmatrix}$

(d) $\begin{bmatrix} -10 & -3 & 1 & -3 \\ 18 & 5 & -2 & 6 \\ 6 & 2 & -1 & 6 \\ 12 & 3 & -1 & 3 \end{bmatrix}$ ★(e) $\begin{bmatrix} 2 & 1 & -7 & 14 \\ -6 & -3 & 19 & -38 \\ 1 & 0 & -3 & 6 \\ 2 & 1 & -6 & 12 \end{bmatrix}$

3. Find the inverse, if it exists, for each of the following 2×2 matrices:

★(a) $\begin{bmatrix} 4 & 2 \\ 9 & -3 \end{bmatrix}$ (b) $\begin{bmatrix} 10 & -5 \\ -4 & 2 \end{bmatrix}$ ★(c) $\begin{bmatrix} -3 & 5 \\ -12 & -8 \end{bmatrix}$

(d) $\begin{bmatrix} 1 & 2 \\ 4 & -3 \end{bmatrix}$ ★(e) $\begin{bmatrix} -6 & 12 \\ 4 & -8 \end{bmatrix}$ (f) $\begin{bmatrix} -\frac{1}{2} & \frac{3}{4} \\ \frac{1}{3} & -\frac{1}{2} \end{bmatrix}$

4. Find the inverse, if it exists, by using row reduction for each of the following:

★(a) $\begin{bmatrix} -4 & 7 & 6 \\ 3 & -5 & -4 \\ -2 & 4 & 3 \end{bmatrix}$ (b) $\begin{bmatrix} 5 & 7 & -6 \\ 3 & 1 & -2 \\ 1 & -5 & 2 \end{bmatrix}$

★c) $\begin{bmatrix} 2 & -2 & 3 \\ 8 & -4 & 9 \\ -4 & 6 & -9 \end{bmatrix}$ (d) $\begin{bmatrix} 0 & 0 & -2 & -1 \\ -2 & 0 & -1 & 0 \\ -1 & -2 & -1 & -5 \\ 0 & 1 & 1 & 3 \end{bmatrix}$

★(e) $\begin{bmatrix} 2 & 0 & -1 & 3 \\ 1 & -2 & 3 & 1 \\ 4 & 1 & 0 & -1 \\ 1 & 3 & -2 & -5 \end{bmatrix}$ (f) $\begin{bmatrix} 3 & 3 & 0 & -2 \\ 14 & 15 & 0 & -11 \\ -3 & 1 & 2 & -5 \\ -2 & 0 & 1 & -2 \end{bmatrix}$

5. Assuming that all main diagonal entries are nonzero, find the inverse of each of the following:

(a) $\begin{bmatrix} a_{11} & 0 \\ 0 & a_{22} \end{bmatrix}$ (b) $\begin{bmatrix} a_{11} & 0 & 0 \\ 0 & a_{22} & 0 \\ 0 & 0 & a_{33} \end{bmatrix}$ ★(c) $\begin{bmatrix} a_{11} & 0 & 0 & \cdots & 0 \\ 0 & a_{22} & 0 & \cdots & 0 \\ \vdots & \vdots & \vdots & \ddots & \vdots \\ 0 & 0 & 0 & \cdots & a_{nn} \end{bmatrix}$

★6. The following matrices are useful in computer graphics for rotating vectors (see Section 5.1). Find the inverse of each matrix, and then state what the matrix and its inverse are when $\theta = \frac{\pi}{6}, \frac{\pi}{4}$, and $\frac{\pi}{2}$:

(a) $\begin{bmatrix} \cos\theta & -\sin\theta \\ \sin\theta & \cos\theta \end{bmatrix}$ (b) $\begin{bmatrix} \cos\theta & -\sin\theta & 0 \\ \sin\theta & \cos\theta & 0 \\ 0 & 0 & 1 \end{bmatrix}$

7. In each case, find the inverse of the coefficient matrix and use it to solve the system by matrix multiplication:

★(a) $\begin{cases} 5x_1 - x_2 = 20 \\ -7x_1 + 2x_2 = -31 \end{cases}$ (b) $\begin{cases} -5x_1 + 3x_2 + 6x_3 = 4 \\ 3x_1 - x_2 - 7x_3 = 11 \\ -2x_1 + x_2 + 2x_3 = 2 \end{cases}$

★(c) $\begin{cases} \quad\quad - 2x_2 + 5x_3 + x_4 = 25 \\ -7x_1 - 4x_2 + 5x_3 + 22x_4 = -15 \\ 5x_1 + 3x_2 - 4x_3 - 16x_4 = 9 \\ -3x_1 - x_2 \quad\quad + 9x_4 = -16 \end{cases}$

★8. A matrix with the property $\mathbf{A}^2 = \mathbf{I}_n$ is called an **involutory** matrix:

(a) Find an example of a 2×2 involutory matrix other than $\mathbf{I}_2$.

(b) Find an example of a 3×3 involutory matrix other than $\mathbf{I}_3$.

(c) What is $\mathbf{A}^{-1}$ if $\mathbf{A}$ is involutory?

9. (a) Give an example to show that $\mathbf{A} + \mathbf{B}$ can be singular if $\mathbf{A}$ and $\mathbf{B}$ are both nonsingular.

(b) Give an example to show that $\mathbf{A} + \mathbf{B}$ can be nonsingular if $\mathbf{A}$ and $\mathbf{B}$ are both singular.

(c) Give an example to show that even when $\mathbf{A}$, $\mathbf{B}$, and $\mathbf{A} + \mathbf{B}$ are all nonsingular, $(\mathbf{A} + \mathbf{B})^{-1}$ is not necessarily equal to $\mathbf{A}^{-1} + \mathbf{B}^{-1}$.

★10. Let $\mathbf{A}$, $\mathbf{B}$, and $\mathbf{C}$ be $n \times n$ matrices:

(a) Suppose that $\mathbf{AB} = \mathbf{O}_n$, and $\mathbf{A}$ is nonsingular. What must $\mathbf{B}$ be?

(b) If $\mathbf{AB} = \mathbf{I}_n$, is it possible for $\mathbf{AC}$ to equal $\mathbf{O}_n$ without $\mathbf{C} = \mathbf{O}_n$? Why or why not?

★11. If $\mathbf{A}^5 = \mathbf{I}_n$, which powers of $\mathbf{A}$ are equal to $\mathbf{A}^{-1}$?

★**12.** If the matrix product $\mathbf{A}^{-1}\mathbf{B}$ is known, how could you calculate $\mathbf{B}^{-1}\mathbf{A}$ without necessarily knowing what $\mathbf{A}$ and $\mathbf{B}$ are?

13. Let $\mathbf{A}$ be a symmetric nonsingular matrix. Prove that $\mathbf{A}^{-1}$ is symmetric.

14.★(a) You have already seen in this section that every square matrix containing a row of zeroes must be singular. Why must every square matrix containing a column of zeroes be singular?

(b) Why must every diagonal matrix with at least one zero main diagonal entry be singular?

(c) Why must every upper triangular matrix with no zero entries on the main diagonal be nonsingular?

(d) Use part (c) and the transpose to show that every lower triangular matrix with no zero entries on the main diagonal must be nonsingular.

15. (a) Prove parts (1) and (2) of Theorem 2.10.

(b) Use the method of induction to prove the following generalization of part (3) of Theorem 2.10: if $\mathbf{A}_1, \mathbf{A}_2, \ldots, \mathbf{A}_n$ are nonsingular matrices of the same size, then $(\mathbf{A}_1\mathbf{A}_2\cdots\mathbf{A}_n)^{-1} = \mathbf{A}_n^{-1}\cdots\mathbf{A}_2^{-1}\mathbf{A}_1^{-1}$.

16. If $\mathbf{A}$ is a nonsingular matrix and $c \in \mathbb{R}$, prove that $(c\mathbf{A})^{-1} = (\frac{1}{c})\mathbf{A}^{-1}$.

17. (a) Prove part (1) of Theorem 2.11 if $s < 0$ and $t < 0$.

(b) Prove part (2) of Theorem 2.11 if $s \geq 0$ and $t < 0$.

18. Assume that $\mathbf{A}$ and $\mathbf{B}$ are nonsingular $n \times n$ matrices. Prove that $\mathbf{A}$ and $\mathbf{B}$ commute (that is, $\mathbf{AB} = \mathbf{BA}$) if and only if $(\mathbf{AB})^2 = \mathbf{A}^2\mathbf{B}^2$.

19. Prove that, if $\mathbf{A}$ and $\mathbf{B}$ are nonsingular matrices of the same size, then $\mathbf{AB} = \mathbf{BA}$ if and only if $(\mathbf{AB})^q = \mathbf{A}^q\mathbf{B}^q$ for every positive integer q. (Hint: First show by induction that if $\mathbf{AB} = \mathbf{BA}$, then $\mathbf{AB}^q = \mathbf{B}^q\mathbf{A}$, for any positive integer q. Finish the proof with a second induction argument to show $(\mathbf{AB})^q = \mathbf{A}^q\mathbf{B}^q$.)

20. Prove that, if $\mathbf{A}$ is an $n \times n$ matrix and $\mathbf{A} - \mathbf{I}_n$ is nonsingular, then for every integer $k \geq 0$, $\mathbf{I}_n + \mathbf{A} + \mathbf{A}^2 + \mathbf{A}^3 + \cdots + \mathbf{A}^k = (\mathbf{A}^{k+1} - \mathbf{I}_n)(\mathbf{A} - \mathbf{I}_n)^{-1}$. (Compare this statement with Result 8 in Section 1.3.)

2.5 Elementary Matrices

In this section, we introduce elementary matrices and show that performing a row operation on a matrix is equivalent to multiplying it by an elementary matrix. We conclude with some useful properties of elementary matrices. Elementary matrices are used only sparingly in this book.†

---

†Outside of this section, elementary matrices are used in the proofs of Theorem 3.4, Lemma 3.8, and Theorem 3.9; in Figure 3.3; in Section 9.2; and in a handful of exercises.

Elementary Matrices

DEFINITION

An $n \times n$ matrix is an **elementary matrix of type (I), (II)**, or **(III)** if and only if it is obtained by performing a single row operation of type (I), (II), or (III), respectively, on the identity matrix $\mathbf{I}_n$.

That is, an elementary matrix is a matrix that is one step away from an identity matrix in terms of row operations.

Example 1

The row operation $< 2 > \longleftarrow -3 < 2 >$ converts the identity matrix

$$\mathbf{I}_3 = \begin{bmatrix} 1 & 0 & 0 \\ 0 & 1 & 0 \\ 0 & 0 & 1 \end{bmatrix} \quad \text{into} \quad \mathbf{A} = \begin{bmatrix} 1 & 0 & 0 \\ 0 & -3 & 0 \\ 0 & 0 & 1 \end{bmatrix}.$$

Hence, $\mathbf{A}$ is an elementary matrix of type (I), because it is the result of a single row operation of type (I) on $\mathbf{I}_3$.

Next, consider

$$\mathbf{B} = \begin{bmatrix} 1 & 0 & -2 \\ 0 & 1 & 0 \\ 0 & 0 & 1 \end{bmatrix}.$$

Since $\mathbf{B}$ is obtained from $\mathbf{I}_3$ by performing the single type (II) row operation $< 1 > \longleftarrow -2 < 3 > + < 1 >$, $\mathbf{B}$ is an elementary matrix of type (II).

Finally,

$$\mathbf{C} = \begin{bmatrix} 0 & 1 \\ 1 & 0 \end{bmatrix}$$

is an elementary matrix of type (III), because it is obtained by performing the single type (III) row operation $< 1 > \longleftrightarrow < 2 >$ on $\mathbf{I}_2$. ■

Representing a Row Operation as Multiplication by an Elementary Matrix

The next theorem shows the connection between row operations and matrix multiplication:

THEOREM 2.15

Let $\mathbf{A}$ and $\mathbf{B}$ be $m \times n$ matrices. If $\mathbf{B}$ is obtained from $\mathbf{A}$ by performing one row operation and if $\mathbf{E}$ is the $m \times m$ elementary matrix obtained by performing the same row operation on $\mathbf{I}_m$, then $\mathbf{B} = \mathbf{EA}$.

In other words, the effect of a single row operation on **A** can be obtained by multiplying **A** on the left by the appropriate elementary matrix. The proof of this theorem is straightforward but tedious (see Exercise 3).

Example 2

Consider the matrices

$$\mathbf{A} = \begin{bmatrix} 2 & -3 & 0 & 1 \\ 1 & 6 & -2 & -2 \\ 0 & 5 & 3 & 4 \end{bmatrix} \quad \text{and} \quad \mathbf{B} = \begin{bmatrix} 2 & -3 & 0 & 1 \\ 1 & 6 & -2 & -2 \\ -3 & -13 & 9 & 10 \end{bmatrix}.$$

Notice that **B** is obtained from **A** by performing the type (II) row operation: $< 3 > \leftarrow -3 < 2 > + < 3 >$. The elementary matrix

$$\mathbf{E} = \begin{bmatrix} 1 & 0 & 0 \\ 0 & 1 & 0 \\ 0 & -3 & 1 \end{bmatrix}$$

is obtained by performing this same row operation on $\mathbf{I}_3$. Notice that

$$\mathbf{EA} = \begin{bmatrix} 1 & 0 & 0 \\ 0 & 1 & 0 \\ 0 & -3 & 1 \end{bmatrix} \begin{bmatrix} 2 & -3 & 0 & 1 \\ 1 & 6 & -2 & -2 \\ 0 & 5 & 3 & 4 \end{bmatrix} = \begin{bmatrix} 2 & -3 & 0 & 1 \\ 1 & 6 & -2 & -2 \\ -3 & -13 & 9 & 10 \end{bmatrix} = \mathbf{B}.$$

That is, **B** can also be obtained from **A** by multiplying **A** on the left by the appropriate elementary matrix. ■

Using Elementary Matrices to Show Row Equivalence

If two matrices **A** and **B** are row equivalent, there is some finite sequence of, say, k row operations that converts **A** into **B**. But according to Theorem 2.15, performing each of these row operations is equivalent to multiplying (on the left) by an appropriate elementary matrix. Hence, there must be a sequence of k elementary matrices $\mathbf{E}_1, \mathbf{E}_2, \ldots, \mathbf{E}_k$ such that $\mathbf{B} = \mathbf{E}_k(\cdots(\mathbf{E}_3(\mathbf{E}_2(\mathbf{E}_1\mathbf{A})))\cdots)$. In fact, the converse is true as well, because if $\mathbf{B} = \mathbf{E}_k(\cdots(\mathbf{E}_3(\mathbf{E}_2(\mathbf{E}_1\mathbf{A})))\cdots)$ for some collection of elementary matrices $\mathbf{E}_1, \mathbf{E}_2, \ldots, \mathbf{E}_k$, then **B** can be obtained from **A** through a sequence of k row operations. Hence:

THEOREM 2.16

Two $m \times n$ matrices **A** and **B** are row equivalent if and only if there is a (finite) sequence $\mathbf{E}_1, \mathbf{E}_2, \ldots, \mathbf{E}_k$ of elementary matrices such that $\mathbf{B} = \mathbf{E}_k \cdots \mathbf{E}_2\mathbf{E}_1\mathbf{A}$.

Example 3

Consider the matrix $\mathbf{A} = \begin{bmatrix} 0 & 1 & -4 \\ 2 & 5 & 9 \end{bmatrix}$. We perform a series of row operations to obtain a row equivalent matrix **B**. Next to each operation we give its corresponding elementary matrix:

$$\mathbf{A} = \begin{bmatrix} 0 & 1 & -4 \\ 2 & 5 & 9 \end{bmatrix} \quad \left(\text{(III)} :< 1 > \longleftrightarrow < 2 >, \quad \mathbf{E}_1 = \begin{bmatrix} 0 & 1 \\ 1 & 0 \end{bmatrix} \right)$$

$$\Longrightarrow \begin{bmatrix} 2 & 5 & 9 \\ 0 & 1 & -4 \end{bmatrix} \quad \left(\text{(I)} :< 1 > \longleftarrow \tfrac{1}{2} < 1 >, \quad \mathbf{E}_2 = \begin{bmatrix} \frac{1}{2} & 0 \\ 0 & 1 \end{bmatrix} \right)$$

$$\Longrightarrow \begin{bmatrix} 1 & \frac{5}{2} & \frac{9}{2} \\ 0 & 1 & -4 \end{bmatrix} \quad \left(\text{(II)} :< 1 > \longleftarrow -\tfrac{5}{2} < 2 > + < 1 >, \quad \mathbf{E}_3 = \begin{bmatrix} 1 & -\frac{5}{2} \\ 0 & 1 \end{bmatrix} \right)$$

$$\Longrightarrow \begin{bmatrix} 1 & 0 & \frac{29}{2} \\ 0 & 1 & -4 \end{bmatrix} = \mathbf{B}.$$

Alternately, the same result **B** is obtained if we multiply **A** on the left by the product of elementary matrices $\mathbf{E}_3\mathbf{E}_2\mathbf{E}_1$:

$$\mathbf{B} = \begin{bmatrix} 1 & 0 & \frac{29}{2} \\ 0 & 1 & -4 \end{bmatrix} = \underbrace{\begin{bmatrix} 1 & -\frac{5}{2} \\ 0 & 1 \end{bmatrix}}_{\mathbf{E}_3} \underbrace{\begin{bmatrix} \frac{1}{2} & 0 \\ 0 & 1 \end{bmatrix}}_{\mathbf{E}_2} \underbrace{\begin{bmatrix} 0 & 1 \\ 1 & 0 \end{bmatrix}}_{\mathbf{E}_1} \underbrace{\begin{bmatrix} 0 & 1 & -4 \\ 2 & 5 & 9 \end{bmatrix}}_{\mathbf{A}}.$$

(Verify that this final product really does equal **B**.) It is important to write the product in the reverse of the order in which the row operations were performed. ■

Inverses of Elementary Matrices

Recall that every row operation has a corresponding inverse row operation. The exact form for the inverse of a row operation of each type is given in Figure 2.2 in Section 2.2. These inverse row operations can be used to find inverses of elementary matrices:

THEOREM 2.17

Every elementary matrix **E** is nonsingular, and its inverse $\mathbf{E}^{-1}$ is an elementary matrix of the same type ((I), (II), or (III)).

Proof of Theorem 2.17

Any $n \times n$ elementary matrix **E** is formed by performing a single row operation of type (I), (II), or (III) on $\mathbf{I}_n$. If we then perform its inverse operation, the result is $\mathbf{I}_n$

again. But the inverse row operation has the same type as the original row operation, and so its elementary matrix **F** has the same type as **E**. Now, by Theorem 2.15, the product **FE** must equal $\mathbf{I}_n$. Hence, **F** and **E** are inverses and have the same type. ■

Example 4

Suppose that we want the inverse of the elementary matrix

$$\mathbf{B} = \begin{bmatrix} 1 & 0 & -2 \\ 0 & 1 & 0 \\ 0 & 0 & 1 \end{bmatrix}.$$

The row operation corresponding to **B** is (II): $<1> \longleftarrow -2<3> + <1>$. Hence, the inverse operation is (II): $<1> \longleftarrow 2<3> + <1>$, whose elementary matrix is

$$\mathbf{B}^{-1} = \begin{bmatrix} 1 & 0 & 2 \\ 0 & 1 & 0 \\ 0 & 0 & 1 \end{bmatrix}.$$

■

Nonsingular Matrices Expressed as a Product of Elementary Matrices

Suppose that we can convert a matrix **A** to a matrix **B** using row operations. Then, by Theorem 2.16, $\mathbf{B} = \mathbf{E}_k \cdots \mathbf{E}_2\mathbf{E}_1\mathbf{A}$, for some elementary matrices $\mathbf{E}_1, \mathbf{E}_2, \ldots, \mathbf{E}_k$. But we can multiply both sides by $\mathbf{E}_k^{-1}$ through $\mathbf{E}_1^{-1}$ (in that order) to obtain $\mathbf{E}_1^{-1}\mathbf{E}_2^{-1} \cdots \mathbf{E}_k^{-1}\mathbf{B} = \mathbf{A}$. Now, by Theorem 2.17, each of the inverses $\mathbf{E}_1^{-1}$, $\mathbf{E}_2^{-1}, \ldots, \mathbf{E}_k^{-1}$ is also an elementary matrix. Therefore, we have found a product of elementary matrices that converts **B** into the original matrix **A**. We can use this fact to express a nonsingular matrix as a product of elementary matrices, as in the next example:

Example 5

Suppose that we want to express the nonsingular matrix $\mathbf{A} = \begin{bmatrix} -5 & -2 \\ 7 & 3 \end{bmatrix}$ as a product of elementary matrices. We begin by row reducing **A**, keeping track of the row operations used:

$$\mathbf{A} = \begin{bmatrix} -5 & -2 \\ 7 & 3 \end{bmatrix} \qquad \text{(I): } <1> \longleftarrow -\tfrac{1}{5}<1>$$

$$\Longrightarrow \begin{bmatrix} 1 & \frac{2}{5} \\ 7 & 3 \end{bmatrix} \qquad \text{(II): } <2> \longleftarrow -7<1> + <2>$$

$$\Longrightarrow \begin{bmatrix} 1 & \frac{2}{5} \\ 0 & \frac{1}{5} \end{bmatrix} \qquad \text{(I): } <2> \longleftarrow 5 <2>$$

$$\Longrightarrow \begin{bmatrix} 1 & \frac{2}{5} \\ 0 & 1 \end{bmatrix} \qquad \text{(II): } <1> \longleftarrow -\tfrac{2}{5} <2> + <1>$$

$$\Longrightarrow \begin{bmatrix} 1 & 0 \\ 0 & 1 \end{bmatrix} = \mathbf{I}_2.$$

Reversing this process, we get a series of row operations that start with $\mathbf{I}_2$ and end with $\mathbf{A}$. The inverse of each row operation, in reverse order, is listed below, together with its corresponding elementary matrix:

$$\text{(II): } <1> \longleftarrow \tfrac{2}{5} <2> + <1> \qquad \mathbf{F}_1 = \begin{bmatrix} 1 & \frac{2}{5} \\ 0 & 1 \end{bmatrix}$$

$$\text{(I): } <2> \longleftarrow \tfrac{1}{5} <2> \qquad \mathbf{F}_2 = \begin{bmatrix} 1 & 0 \\ 0 & \frac{1}{5} \end{bmatrix}$$

$$\text{(II): } <2> \longleftarrow 7 <1> + <2> \qquad \mathbf{F}_3 = \begin{bmatrix} 1 & 0 \\ 7 & 1 \end{bmatrix}$$

$$\text{(I): } <1> \longleftarrow -5 <1> \qquad \mathbf{F}_4 = \begin{bmatrix} -5 & 0 \\ 0 & 1 \end{bmatrix}$$

Therefore, we can express $\mathbf{A}$ as the product

$$\mathbf{A} = \underbrace{\begin{bmatrix} -5 & 0 \\ 0 & 1 \end{bmatrix}}_{\mathbf{F}_4} \underbrace{\begin{bmatrix} 1 & 0 \\ 7 & 1 \end{bmatrix}}_{\mathbf{F}_3} \underbrace{\begin{bmatrix} 1 & 0 \\ 0 & \frac{1}{5} \end{bmatrix}}_{\mathbf{F}_2} \underbrace{\begin{bmatrix} 1 & \frac{2}{5} \\ 0 & 1 \end{bmatrix}}_{\mathbf{F}_1} \underbrace{\begin{bmatrix} 1 & 0 \\ 0 & 1 \end{bmatrix}}_{\mathbf{I}_2}.$$

You should verify that this product is really equal to $\mathbf{A}$. ■

Example 5 motivates the following corollary of Theorem 2.16. We leave the proof for you to do in Exercise 9.

COROLLARY 2.18

> An $n \times n$ matrix $\mathbf{A}$ is nonsingular if and only if $\mathbf{A}$ is the product of a finite collection of $n \times n$ elementary matrices.

Exercises—Section 2.5

1. For each elementary matrix below, determine the corresponding row operation. Also, use the inverse operation to find the inverse of the given matrix:

★(a) $\begin{bmatrix} 1 & 0 & 0 \\ 0 & 0 & 1 \\ 0 & 1 & 0 \end{bmatrix}$ ★(b) $\begin{bmatrix} 1 & 0 & 0 \\ 0 & -2 & 0 \\ 0 & 0 & 1 \end{bmatrix}$ (c) $\begin{bmatrix} 1 & 0 & 0 \\ 0 & 1 & 0 \\ -4 & 0 & 1 \end{bmatrix}$

(d) $\begin{bmatrix} 1 & 0 & 0 & 0 \\ 0 & 6 & 0 & 0 \\ 0 & 0 & 1 & 0 \\ 0 & 0 & 0 & 1 \end{bmatrix}$ ★(e) $\begin{bmatrix} 1 & 0 & 0 & 0 \\ 0 & 1 & 0 & 0 \\ 0 & 0 & 1 & -2 \\ 0 & 0 & 0 & 1 \end{bmatrix}$ (f) $\begin{bmatrix} 0 & 0 & 0 & 1 \\ 0 & 1 & 0 & 0 \\ 0 & 0 & 1 & 0 \\ 1 & 0 & 0 & 0 \end{bmatrix}$

2. Express each of the following as a product of elementary matrices (if possible), in the manner of Example 5:

★(a) $\begin{bmatrix} 4 & 9 \\ 3 & 7 \end{bmatrix}$ ★(b) $\begin{bmatrix} -3 & 2 & 1 \\ 13 & -8 & -9 \\ 1 & -1 & 2 \end{bmatrix}$ (c) $\begin{bmatrix} 0 & 0 & 5 & 0 \\ -3 & 0 & 0 & -2 \\ 0 & 6 & -10 & -1 \\ 3 & 0 & 0 & 3 \end{bmatrix}$

3. Prove Theorem 2.15 by showing that, if a single row operation of any type is performed on a matrix **A**, the result in each case is the same as multiplying **A** on the left by the corresponding elementary matrix.

4. Let **A** and **B** be $m \times n$ matrices. Prove that **A** and **B** are row equivalent if and only if $\mathbf{B} = \mathbf{PA}$, for some nonsingular $m \times m$ matrix **P**.

5. Consider the homogeneous system $\mathbf{AX} = \mathbf{O}$, where **A** is an $n \times n$ matrix. Show that this system has a nontrivial solution if and only if **A** cannot be expressed as the product of elementary $n \times n$ matrices.

6. Prove that, if **U** is an upper triangular matrix with all main diagonal entries nonzero, then $\mathbf{U}^{-1}$ exists and is upper triangular. (Hint: Show that the algorithm for calculating the inverse of a matrix does not produce a row of zeroes on the left side of the augmented matrix. Also, show that for each row reduction step, the corresponding elementary matrix is upper triangular. Conclude that $\mathbf{U}^{-1}$ is the product of upper triangular matrices, which is easily seen to be upper triangular; see Exercise 14 in Section 1.5.)

7. If **E** is an elementary matrix, show that $\mathbf{E}^T$ is also an elementary matrix. What is the relationship between the row operation corresponding to **E** and the row operation corresponding to $\mathbf{E}^T$?

8. Let **F** be an elementary $n \times n$ matrix. Show that the product $\mathbf{AF}^T$ is the matrix obtained by performing a "column" operation on **A** analogous to one of the three types of row operations. (Hint: What is $(\mathbf{AF}^T)^T$?)

9. Prove Corollary 2.18.

10. Let **A** and **B** be $m \times n$ and $n \times p$ matrices, respectively, and let **E** be an $m \times m$ elementary matrix.

(a) Prove that rank(**EA**) = rank(**A**).
(b) Show that, if **A** has k rows of all zeroes, then rank(**A**) $\leq m - k$.
(c) Show that, if **A** is in reduced row echelon form, then rank(**AB**) $\leq$ rank(**A**). (Hint: Use part (b).)
(d) Use parts (a) and (c) to prove that for a general matrix **A**, rank(**AB**) $\leq$ rank(**A**).

3

Determinants

In this chapter, we introduce the determinant: a real number associated with each square matrix that provides much useful information. In Section 3.1, we consider some basic properties of determinants. In Section 3.2, we examine a technique for finding the determinant of a given square matrix using row reduction and consider the relationship between a system of n linear equations in n variables and the determinant of its coefficient matrix. In Section 3.3, we illustrate an alternate technique for finding determinants, known as cofactor expansion. The chapter concludes with an alternate method for finding a matrix inverse using the adjoint matrix and another method for solving certain linear systems, known as Cramer's Rule.

3.1 Introduction to Determinants

Determinants of 2 × 2 and 3 × 3 Matrices

In Section 2.4 we proved that a 2×2 matrix $\begin{bmatrix} a_{11} & a_{12} \\ a_{21} & a_{22} \end{bmatrix}$ has an inverse if and only if $a_{11}a_{22} - a_{12}a_{21} \neq 0$. The value $a_{11}a_{22} - a_{12}a_{21}$ is referred to as the **determinant** of the matrix $\begin{bmatrix} a_{11} & a_{12} \\ a_{21} & a_{22} \end{bmatrix}$. For example, the determinant of $\mathbf{A} = \begin{bmatrix} 4 & -3 \\ 2 & 5 \end{bmatrix}$ is $(4)(5) - (-3)(2) = 26$.

For the 3×3 matrix

$$\begin{bmatrix} a_{11} & a_{12} & a_{13} \\ a_{21} & a_{22} & a_{23} \\ a_{31} & a_{32} & a_{33} \end{bmatrix},$$

we define the determinant to be the following expression, which has six terms:

$$a_{11}a_{22}a_{33} + a_{12}a_{23}a_{31} + a_{13}a_{21}a_{32} - a_{13}a_{22}a_{31} - a_{11}a_{23}a_{32} - a_{12}a_{21}a_{33}.$$

This expression may look complicated, but its terms can easily be obtained by multiplying together the entries linked by arrows below, where the first two columns of the original 3×3 matrix have been repeated. Notice that the arrows pointing right indicate terms with a positive sign, while those pointing left indicate terms with a negative sign:

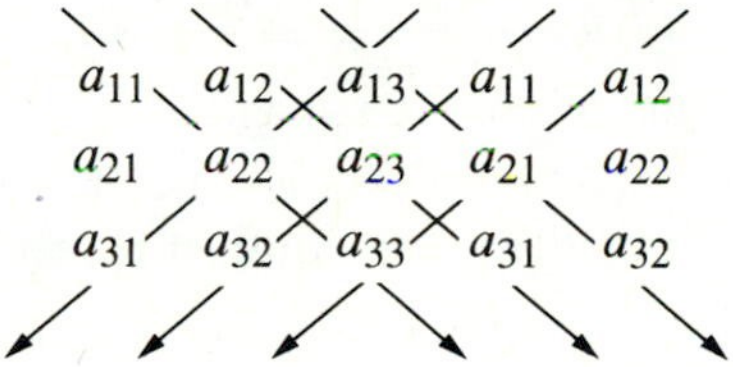

$$\begin{matrix} a_{11} & a_{12} & a_{13} & a_{11} & a_{12} \\ a_{21} & a_{22} & a_{23} & a_{21} & a_{22} \\ a_{31} & a_{32} & a_{33} & a_{31} & a_{32} \end{matrix}$$

This technique is sometimes referred to as the "basketweaving" method for calculating the determinant of a 3×3 matrix.

Example 1

Find the determinant of the following matrix:

$$\begin{bmatrix} 4 & -2 & 3 \\ -1 & 5 & 0 \\ 6 & -1 & -2 \end{bmatrix}.$$

Repeating the first two columns and forming terms using the "basketweaving" method, we have

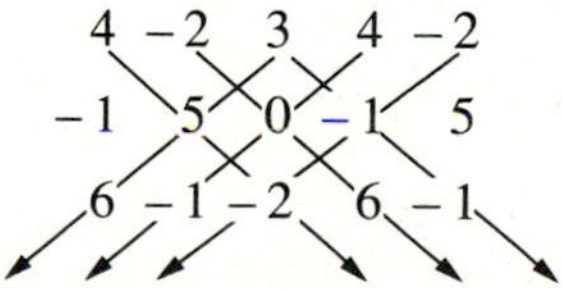

$$\begin{matrix} 4 & -2 & 3 & 4 & -2 \\ -1 & 5 & 0 & -1 & 5 \\ 6 & -1 & -2 & 6 & -1 \end{matrix},$$

which gives $(4)(5)(-2) + (-2)(0)(6) + (3)(-1)(-1) - (3)(5)(6) - (4)(0)(-1) - (-2)(-1)(-2)$, which reduces to $-40 + 0 + 3 - 90 - 0 - (-4)$. Thus, the determinant is -123. ■

We place absolute value signs around a matrix to represent the determinant, even though its value could be negative. For example, the determinant of the matrix in

Example 1 can be expressed as follows:

$$\begin{vmatrix} 4 & -2 & 3 \\ -1 & 5 & 0 \\ 6 & -1 & -2 \end{vmatrix} = -123.$$

Another common notation for the determinant of a matrix $\mathbf{A}$ is $\det(\mathbf{A})$.

Geometric Uses of the Determinant

The next theorem shows two applications of the determinant in geometry and illustrates why a determinant is sometimes interpreted as a volume.

THEOREM 3.1

(1) Suppose that $\mathbf{x} = [x_1, x_2]$ and $\mathbf{y} = [y_1, y_2]$ are two nonparallel vectors in $\mathbb{R}^2$ beginning at a common point (see Figure 3.1(a)). Then the area of the parallelogram determined by $\mathbf{x}$ and $\mathbf{y}$ is the absolute value of

$$\begin{vmatrix} x_1 & x_2 \\ y_1 & y_2 \end{vmatrix}.$$

(2) Suppose that $\mathbf{x} = [x_1, x_2, x_3]$, $\mathbf{y} = [y_1, y_2, y_3]$, and $\mathbf{z} = [z_1, z_2, z_3]$ are three vectors not all in the same plane beginning at a common initial point (see Figure 3.1(b)). Then the volume of the parallelepiped determined by $\mathbf{x}$, $\mathbf{y}$, and $\mathbf{z}$ is the absolute value of

$$\begin{vmatrix} x_1 & x_2 & x_3 \\ y_1 & y_2 & y_3 \\ z_1 & z_2 & z_3 \end{vmatrix}.$$

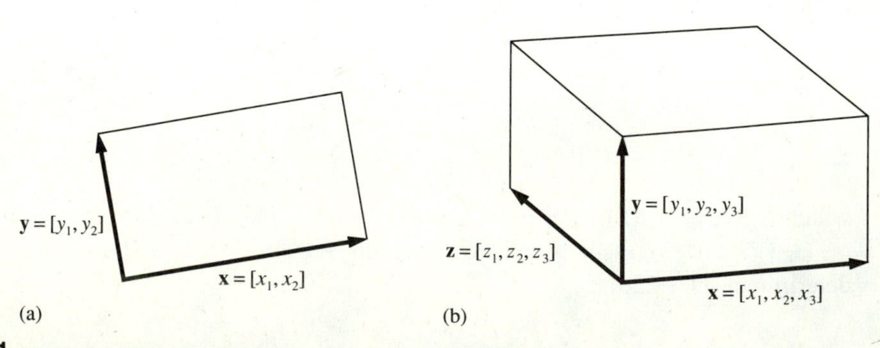

Figure 3.1
(a) The parallelogram determined by **x** and **y** (Theorem 3.1);
(b) The parallelepiped determined by **x**, **y**, and **z** (Theorem 3.1)

The proof of this theorem is straightforward (see Exercises 10 and 12).

Example 2

The volume of the parallelepiped whose sides are $\mathbf{x} = [-2, 1, 3]$, $\mathbf{y} = [3, 0, -2]$, and $\mathbf{z} = [-1, 3, 7]$ is given by the absolute value of the determinant

$$\begin{vmatrix} -2 & 1 & 3 \\ 3 & 0 & -2 \\ -1 & 3 & 7 \end{vmatrix}.$$

By the "basketweaving" method, this determinant is easily seen to equal -4, so the volume is $|-4| = 4$. ■

Permutations

Notice that the six terms in the formula for a 3×3 determinant are really just all possible terms of the form

$$a_{1\square}a_{2\square}a_{3\square}.$$

The boxes (for the second subscripts) are filled in with the digits 1, 2, and 3 so that each digit is used in precisely one box in each term. That is, each term in the determinant formula represents a different arrangement of the digits 1, 2, and 3 in the boxes.

Returning to the 2×2 case, notice that the two terms in the $a_{11}a_{22} - a_{12}a_{21}$ determinant formula arise from the two ways that the boxes in

$$a_{1\square}a_{2\square}$$

can be filled in with the digits 1 and 2, with each digit placed in exactly one box in each term. Again, notice that both terms of the determinant formula represent a different arrangement of the digits 1 and 2 in the boxes.

To generalize this idea to define the determinant of an $n \times n$ matrix, the following definition is needed:

DEFINITION

> An arrangement (ordering) of the elements of a set is a **permutation** of the set. The set of all possible permutations of the numbers $1, 2, \ldots, n$ is referred to as S_n.

There is only one arrangement of the element of the set $\{1\}$ (namely, 1), and so there is only one permutation in S_1. As you have seen, there are only two arrangements of the elements in the set $\{1, 2\}$, so there are two permutations in S_2: 1 2 and 2 1. There are six arrangements of the elements in the set $\{1, 2, 3\}$, so there are six permutations in S_3: 1 2 3, 1 3 2, 2 1 3, 2 3 1, 3 1 2, and 3 2 1. There are twenty-four elements of S_4:

1 2 3 4	1 2 4 3	1 3 2 4	1 3 4 2	1 4 2 3	1 4 3 2
2 1 3 4	2 1 4 3	2 3 1 4	2 3 4 1	2 4 1 3	2 4 3 1
3 1 2 4	3 1 4 2	3 2 1 4	3 2 4 1	3 4 1 2	3 4 2 1
4 1 2 3	4 1 3 2	4 2 1 3	4 2 3 1	4 3 1 2	4 3 2 1

Recall that, for a positive integer n, the expression $n!$ (n factorial) means $n(n-1)(n-2)\cdots(3)(2)(1)$. For example, $5! = 5 \times 4 \times 3 \times 2 \times 1 = 120$. The following result can be easily proved.

THEOREM 3.2

The number of permutations in S_n, for $n \geq 1$, is $n!$.

Even and Odd Permutations

We now turn our attention to the positive and negative signs in the determinant formulas. Notice in the 2×2 and 3×3 determinant expressions that half the terms have positive signs in front and the other half have negative signs. The reason is that there are two types of permutations in S_n, for $n \geq 2$.

DEFINITION

A permutation in S_n is an **even [odd] permutation** if it can be obtained from the special permutation 1 2 3 ... n by interchanging pairs of numbers an even [odd] number of times.

Example 3

The permutation 3 4 1 2 in S_4 is an even permutation because, starting with 1 2 3 4 and interchanging pairs of numbers as follows

$$1\,2\,3\,4 \longrightarrow 3\,2\,1\,4 \longrightarrow 3\,4\,1\,2,$$

the total number of pairs switched is two, an even number. The permutation 2 3 4 1 in S_4 is an odd permutation because it can be created from 1 2 3 4 with three interchanges (an odd number):

$$1\,2\,3\,4 \longrightarrow 2\,1\,3\,4 \longrightarrow 2\,4\,3\,1 \longrightarrow 2\,3\,4\,1.$$

■

How do we know that a permutation cannot be both even and odd? For example, we labeled the permutation 3 4 1 2 as even. Is it possible to find some odd number of interchanges starting with 1 2 3 4 that also produces 3 4 1 2 as a result? Surprisingly, the answer is no. Although we can find many different ways of switching pairs of numbers to produce 3 4 1 2 from 1 2 3 4, it turns out that all involve an even number of interchanges. This point is crystallized in the next lemma. This lemma is usually proved in an abstract algebra course, and so we do not give its proof here:

LEMMA 3.3

If $n \geq 2$, then any permutation in S_n is either even or odd but not both. Exactly half of the $n!$ permutations in S_n are even.

Categorizing the permutations in S_2 and S_3, we have the following:

	Even			Odd		
S_2	1 2			2 1		
S_3	1 2 3	2 3 1	3 1 2	1 3 2	2 1 3	3 2 1

(Verify.) When we list the terms for the determinant of a 2×2 or 3×3 matrix $\mathbf{A} = [a_{ij}]$, we place a positive sign in front of a term when the boxes of $a_{1\square}a_{2\square}$ or $a_{1\square}a_{2\square}a_{3\square}$ contain an even permutation, and a negative sign for an odd permutation. This is the secret to the formal definition of the determinant.

Formal Definition of the Determinant

DEFINITION

Let $\mathbf{A} = [a_{ij}]$ be an $n \times n$ (square) matrix. The **determinant** of $\mathbf{A}$, denoted $|\mathbf{A}|$, is the sum

$$\Sigma \pm a_{1\square}a_{2\square}a_{3\square}\cdots a_{n\square},$$

where the empty boxes for the second subscripts are filled with each permutation in S_n in turn and where $+$ is placed in front of terms involving even permutations and $-$ is placed in front of terms involving odd permutations.

For example, the determinant of a 4×4 matrix $\mathbf{A} = [a_{ij}]$ is a sum of twenty-four terms, each of the form $\pm a_{1\square}a_{2\square}a_{3\square}a_{4\square}$, where a term has a positive or negative sign in front depending on whether its boxes are filled with an even or odd permutation in S_4. The term $+a_{13}a_{24}a_{31}a_{42}$ has a positive sign in front since $\begin{matrix}3 & 4 & 1 & 2\end{matrix}$ is an even permutation; the term $-a_{12}a_{23}a_{34}a_{41}$ has a negative sign in front since $\begin{matrix}2 & 3 & 4 & 1\end{matrix}$ is an odd permutation. In Exercise 5 you are asked to write out the entire sum of twenty-four terms for this 4×4 determinant.[†]

We can theoretically calculate the determinant for any given $n \times n$ matrix by working out all $n!$ terms. But this process is exhausting for matrices of size 4×4, which have twenty-four terms (each with four factors) to sum.[‡] With larger matrices the process is worse. Finding the determinant of a 10×10 matrix using the definition would involve working out $10! = 3628800$ terms of the form $\pm a_{1\square}a_{2\square}a_{3\square}\cdots a_{10\square}$, each involving a product of ten factors. Therefore, our next goal is to find shortcuts for calculating determinants.

Properties of the Determinant

Suppose $\mathbf{A}$ is a 4×4 matrix. Consider the term $-a_{12}a_{24}a_{31}a_{43}$ in the formula for $|\mathbf{A}|$. This term involves the circled elements in the following matrix:

$$\begin{bmatrix} a_{11} & \textcircled{a_{12}} & a_{13} & a_{14} \\ a_{21} & a_{22} & a_{23} & \textcircled{a_{24}} \\ \textcircled{a_{31}} & a_{32} & a_{33} & a_{34} \\ a_{41} & a_{42} & \textcircled{a_{43}} & a_{44} \end{bmatrix}.$$

Note that each of these circled elements is in a different row and column. In fact, after a little thought, you can see that the terms in the determinant formula represent all possible products of matrix entries, with the condition that *for each term, exactly one entry from each row and column is chosen*. This observation is needed for the proof of the next theorem:

THEOREM 3.4

Let $\mathbf{A}$ be an $n \times n$ matrix.

(1) If $\mathbf{A}$ has a row or column of zeroes, then $|\mathbf{A}| = 0$.

(2) If $\mathbf{A}$ has two equal rows, then $|\mathbf{A}| = 0$.

(3) If $\mathbf{A}$ is upper or lower triangular, then $|\mathbf{A}|$ is the product of the elements on the main diagonal; that is, $|\mathbf{A}| = a_{11}a_{22}\cdots a_{nn}$.

[†] Although it is rarely needed, the formal definition of a determinant implies that a 1×1 matrix $\mathbf{A} = [a_{11}]$ has determinant $|\mathbf{A}| = a_{11}$.

[‡] Beware! The "basketweaving" shortcut for finding the determinant of a 3×3 matrix does not work for 4×4 (or larger) matrices.

Part (1) of Theorem 3.4 is easy to prove (see Exercise 16). The proof of part (2) is long and is placed in Appendix A for the interested reader.

Proof of Theorem 3.4, Part (3)

Let **A** be an upper triangular matrix. Then **A** has this form:

$$\begin{bmatrix} a_{11} & a_{12} & a_{13} & \cdots & a_{1n} \\ 0 & a_{22} & a_{23} & \cdots & a_{2n} \\ 0 & 0 & a_{33} & \cdots & a_{3n} \\ \vdots & \vdots & \vdots & \ddots & \vdots \\ 0 & 0 & 0 & \cdots & a_{nn} \end{bmatrix}.$$

Any term in the formula for $|\mathbf{A}|$ has n factors, each from a different row and column of **A**.

For a term in the formula for $|\mathbf{A}|$ to be nonzero, all factors in the term must be nonzero. The only possible nonzero factor in the first column is a_{11}. But if a_{11} is chosen, no other first-row entries can be chosen for the same term. Thus, the only possible nonzero factor we can choose in the second column is a_{22}. Similarly, this choice eliminates the remaining second-row entries from consideration. Proceeding through the remaining columns, we see that the only possible nonzero term in the formula for $|\mathbf{A}|$ is the one with all the main diagonal elements as its factors. That is, $|\mathbf{A}| = +a_{11}a_{22}\cdots a_{nn}$. (This term has a positive sign in front. Why?) A similar argument for lower triangular matrices completes the proof. ■

Example 4

Consider these matrices:

$$\mathbf{A} = \begin{bmatrix} -4 & 0 & 5 \\ 6 & 0 & -2 \\ -1 & 0 & 6 \end{bmatrix}, \quad \mathbf{B} = \begin{bmatrix} 6 & -2 & 1 \\ 0 & -3 & 5 \\ 6 & -2 & 1 \end{bmatrix}, \quad \text{and} \quad \mathbf{C} = \begin{bmatrix} 3 & 1 & 6 \\ 0 & -2 & 3 \\ 0 & 0 & 4 \end{bmatrix}.$$

Now, $|\mathbf{A}| = 0$ by part (1) of Theorem 3.4, because **A** has a column of zeroes. Similarly, if a square matrix has a row of zeroes, its determinant is zero.

Also, $|\mathbf{B}| = 0$, using part (2) of Theorem 3.4, because **B** has two identical rows.

Finally, $|\mathbf{C}| = -24$, using part (3) of Theorem 3.4, because **C** is upper triangular and the product of the main diagonal elements is $(3)(-2)(4) = -24$. (This determinant is amusing to calculate using the "basketweaving" method. Try it.) ■

Notice that, by part (3) of Theorem 3.4, $|\mathbf{I}_n| = 1$ for all $n \geq 1$, because $\mathbf{I}_n$ is an upper triangular matrix with all main diagonal entries equal to 1.

Exercises—Section 3.1

1. Calculate the determinant of each of the following matrices directly:

★(a) $\begin{bmatrix} -2 & 5 \\ 3 & 1 \end{bmatrix}$ (b) $\begin{bmatrix} 5 & -3 \\ 2 & 0 \end{bmatrix}$ ★(c) $\begin{bmatrix} 6 & -12 \\ -4 & 8 \end{bmatrix}$

(d) $\begin{bmatrix} \cos\theta & \sin\theta \\ -\sin\theta & \cos\theta \end{bmatrix}$ ★(e) $\begin{bmatrix} 2 & 0 & 5 \\ -4 & 1 & 7 \\ 0 & 3 & -3 \end{bmatrix}$ (f) $\begin{bmatrix} 3 & -2 & 4 \\ 5 & 1 & -2 \\ -1 & 3 & 6 \end{bmatrix}$

★(g) $\begin{bmatrix} 5 & 0 & 0 \\ 3 & -2 & 0 \\ -1 & 8 & 4 \end{bmatrix}$ (h) $\begin{bmatrix} -6 & 0 & 0 \\ 0 & 2 & 0 \\ 0 & 0 & 5 \end{bmatrix}$ ★(i) $\begin{bmatrix} 3 & 1 & -2 \\ -1 & 4 & 5 \\ 3 & 1 & -2 \end{bmatrix}$

★(j) $[-3]$

2. Use Theorem 3.4 to find the determinant of each of the following, and state the reason for your answer:

★(a) $\begin{bmatrix} 5 & -1 & 4 \\ 5 & -1 & 4 \\ 2 & 8 & 0 \end{bmatrix}$ (b) $\begin{bmatrix} 3 & 0 & 0 \\ 1 & -2 & 0 \\ 5 & 10 & 6 \end{bmatrix}$

★(c) $\begin{bmatrix} 2 & 5 & -2 & 3 \\ 0 & 3 & 4 & -1 \\ 0 & 0 & -1 & 6 \\ 0 & 0 & 0 & 5 \end{bmatrix}$ (d) $\begin{bmatrix} 1 & -3 & 6 & -2 \\ 4 & 0 & 3 & 5 \\ 1 & -3 & 6 & -2 \\ -2 & 4 & 1 & 0 \end{bmatrix}$ ★(e) $\begin{bmatrix} -2 & 3 & 5 \\ 0 & 0 & 0 \\ 1 & 4 & -3 \end{bmatrix}$

(f) $\begin{bmatrix} 5 & 6 & 0 & -2 \\ -2 & -1 & 0 & 8 \\ 5 & 4 & 0 & 7 \\ 4 & -2 & 0 & -3 \end{bmatrix}$ ★(g) $\begin{bmatrix} 2 & -3 & 4 & 5 \\ 0 & -3 & 4 & 5 \\ 0 & 0 & 4 & 5 \\ 0 & 0 & 0 & 5 \end{bmatrix}$

3. Which of the following permutations are odd? Which are even?

★(a) 3 2 1 (b) 2 3 1 4 ★(c) 2 1 4 3 ★(d) 3 5 4 1 2 (e) 5 4 3 2 1

4. Would the following terms of various determinant formulas for $\mathbf{A} = [a_{ij}]$ be preceded by positive or negative signs? Explain why. Also, indicate the size of the matrix $\mathbf{A}$ in each case:

★(a) $a_{13}a_{22}a_{31}$ (b) $a_{12}a_{23}a_{31}$

★(c) $a_{14}a_{21}a_{33}a_{42}$ (d) $a_{12}a_{23}a_{34}a_{41}$

★(e) $a_{11}a_{25}a_{34}a_{42}a_{53}$ (f) $a_{12}a_{23}a_{36}a_{44}a_{51}a_{65}$

★(g) $a_{32}a_{21}a_{13}$ (Think!) (h) $a_{24}a_{31}a_{43}a_{12}$ (Think!)

5. Work out all twenty-four terms for the formula for $|\mathbf{A}|$ if $\mathbf{A} = [a_{ij}]$ is a 4×4 matrix. Be sure to indicate which terms are preceded by a negative sign.

6. Let **A** be an upper (or lower) triangular matrix. Prove that $|\mathbf{A}| \neq 0$ if and only if all the main diagonal elements of **A** are nonzero.

7. (a) Prove that $|c\mathbf{A}| = c^n|\mathbf{A}|$, if **A** is an $n \times n$ matrix. (Hint: Consider the formula $\Sigma \pm a_{1\square}a_{2\square}\cdots a_{n\square}$ for $|\mathbf{A}|$.)

(b) Use part (a) together with part (2) of Theorem 3.1 to explain why, when each side of a parallelepiped is doubled, the volume is multiplied by 8.

8. (a) Show that the cross product $\mathbf{a} \times \mathbf{b} = [a_2b_3 - a_3b_2, a_3b_1 - a_1b_3, a_1b_2 - a_2b_1]$ of $\mathbf{a} = [a_1, a_2, a_3]$ and $\mathbf{b} = [b_1, b_2, b_3]$ can be expressed in "determinant notation" as

$$\begin{vmatrix} \mathbf{i} & \mathbf{j} & \mathbf{k} \\ a_1 & a_2 & a_3 \\ b_1 & b_2 & b_3 \end{vmatrix}.$$

(b) Show that $\mathbf{a} \times \mathbf{b}$ is orthogonal to both **a** and **b**.

9. Calculate the area of the parallelogram in $\mathbb{R}^2$ determined by:

★(a) $\mathbf{x} = [3, 2]$, $\mathbf{y} = [4, 5]$
(b) $\mathbf{x} = [-4, 3]$, $\mathbf{y} = [-2, 6]$
★(c) $\mathbf{x} = [5, -1]$, $\mathbf{y} = [-3, 3]$
(d) $\mathbf{x} = [-2, 3]$, $\mathbf{y} = [6, -9]$

10. Prove part (1) of Theorem 3.1. (Hint: See Figure 3.2. The area of the parallelogram is the length of the base **x** multiplied by the length of the perpendicular height **h**. Note that if $\mathbf{p} = \mathbf{proj_x y}$, then $\mathbf{h} = \mathbf{y} - \mathbf{p}$.)

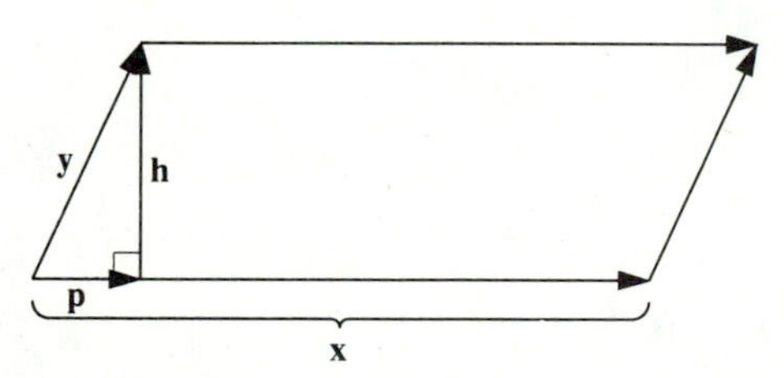

Figure 3.2
Parallelogram determined by **x** and **y**

11. Calculate the volumes of the parallelepipeds in $\mathbb{R}^3$ determined by:

★(a) $\mathbf{x} = [-2, 3, 1]$, $\mathbf{y} = [4, 2, 0]$, $\mathbf{z} = [-1, 3, 2]$
(b) $\mathbf{x} = [1, 2, 3]$, $\mathbf{y} = [0, -1, 0]$, $\mathbf{z} = [4, -1, 5]$
★(c) $\mathbf{x} = [-3, 4, 0]$, $\mathbf{y} = [6, -2, 1]$, $\mathbf{z} = [0, -3, 3]$
(d) $\mathbf{x} = [1, -2, 0]$, $\mathbf{y} = [3, 2, -1]$, $\mathbf{z} = [5, -2, -1]$

12. Prove part (2) of Theorem 3.1. (Hint: See Figure 3.3. Let **h** be the perpendicular dropped from **z** to the plane of the parallelogram. From Exercise 10, you know that $\mathbf{x} \times \mathbf{y}$ is perpendicular to both $\mathbf{x}$ and $\mathbf{y}$, and so $\mathbf{h}$ is actually the projection of $\mathbf{z}$ onto $\mathbf{x} \times \mathbf{y}$. Hence, the volume of the parallelepiped is the area of the parallelogram determined by $\mathbf{x}$ and $\mathbf{y}$ multiplied by the length of $\mathbf{h}$. A calculation similar to that in Exercise 10 shows that the area of the parallelogram is $\sqrt{(x_2y_3 - x_3y_2)^2 + (x_1y_3 - x_3y_1)^2 + (x_1y_2 - x_2y_1)^2}$.)

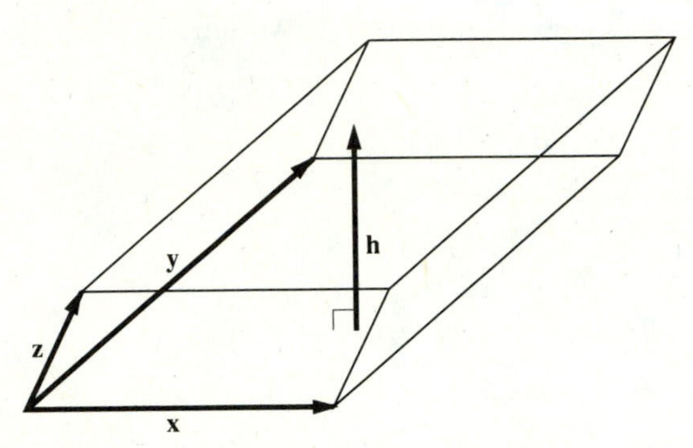

Figure 3.3
Parallepiped determined by **x**, **y**, and **z**

13. Solve the following determinant equations:

★(a) $\begin{vmatrix} x & 2 \\ 5 & x+3 \end{vmatrix} = 0$ (b) $\begin{vmatrix} 15 & x-4 \\ x+7 & -2 \end{vmatrix} = 0$

★(c) $\begin{vmatrix} x-3 & 5 & -19 \\ 0 & x-1 & 6 \\ 0 & 0 & x-2 \end{vmatrix} = 0$

★**14.** Find the determinant of the following matrix:

$$\mathbf{A} = \begin{bmatrix} 0 & 0 & 0 & 0 & 0 & a_{16} \\ 0 & 0 & 0 & 0 & a_{25} & a_{26} \\ 0 & 0 & 0 & a_{34} & a_{35} & a_{36} \\ 0 & 0 & a_{43} & a_{44} & a_{45} & a_{46} \\ 0 & a_{52} & a_{53} & a_{54} & a_{55} & a_{56} \\ a_{61} & a_{62} & a_{63} & a_{64} & a_{65} & a_{66} \end{bmatrix}.$$

(Hint: Consider the proof of part (3) of Theorem 3.4. Think about whether to use a positive or negative sign.)

15. (a) Show that the determinant of the matrix

$$\begin{bmatrix} 1 & 1 & 1 \\ a & b & c \\ a^2 & b^2 & c^2 \end{bmatrix}$$

is equal to $(b-a)(c-a)(c-b)$. (This is called the **Vandermonde determinant**.)

⋆(b) Calculate the determinant of

$$\begin{bmatrix} 1 & 1 & 1 \\ 2 & 3 & -2 \\ 4 & 9 & 4 \end{bmatrix},$$

using part (a).

16. Prove part (1) of Theorem 3.4 by using the comments directly before the statement of the theorem.

3.2 Using Row Reduction to Calculate Determinants

In this section, we give a shortcut for finding the determinant of any matrix by using row reduction to create an upper triangular form, whose determinant is known by part (3) of Theorem 3.4. We can then prove a fundamental result: a matrix is singular if and only if its determinant is zero.

Effect of Row Operations on the Determinant

The following theorem describes explicitly how each type of row operation affects the determinant:

THEOREM 3.5

Let $\mathbf{A}$ be an $n \times n$ matrix, with determinant $|\mathbf{A}|$.

(1) If the row operation $< i > \longleftarrow c < i >$ of type (I) is performed on $\mathbf{A}$, the resulting matrix $\mathbf{B}$ has a determinant equal to c times $|\mathbf{A}|$; that is, $|\mathbf{B}| = c|\mathbf{A}|$.

(2) If the row operation $< j > \longleftarrow c < i > + < j >$ of type (II) is performed on $\mathbf{A}$, the resulting matrix $\mathbf{B}$ has the same determinant as $\mathbf{A}$; that is, $|\mathbf{B}| = |\mathbf{A}|$.

(3) If the row operation $< i > \longleftrightarrow < j >$ of type (III) is performed on $\mathbf{A}$, the resulting matrix $\mathbf{B}$ has the opposite determinant as $\mathbf{A}$; that is, $|\mathbf{B}| = -|\mathbf{A}|$.

The proof of part (1) of Theorem 3.5 is easy (see Exercise 21).

Proof of Theorem 3.5 (abridged)

Part (2): Suppose that $\mathbf{B}$ is the result of the operation $< j > \longleftarrow c < i > + < j >$ of type (II) on $\mathbf{A}$, where $i < j$. (A completely analogous proof can be done for the case where $i > j$.) Then the entries of **B** are the same as those of **A** except in the jth row; that is, for $k \neq j$, we have $b_{kl} = a_{kl}$, but $b_{jl} = c(a_{il}) + a_{jl}$. Then,

$$\begin{aligned}|\mathbf{B}| &= \Sigma \pm b_{1\square}b_{2\square}\cdots b_{i\square}\cdots b_{j\square}\cdots b_{n\square} \\ &= \Sigma \pm a_{1\square}a_{2\square}\cdots a_{i\square}\cdots(ca_{i\square} + a_{j\square})\cdots a_{n\square}.\end{aligned}$$

Splitting this expression into two sums, we have

$$|\mathbf{B}| = c\left(\Sigma \pm a_{1\square}a_{2\square}\cdots a_{i\square}\cdots a_{j\square}\cdots a_{n\square}\right) \quad + \quad \Sigma \pm a_{1\square}a_{2\square}\cdots a_{i\square}\cdots a_{j\square}\cdots a_{n\square}.$$

The first sum is zero, by part (2) of Theorem 3.4, because it is equivalent to c times the determinant of a matrix whose i^{th} and j^{th} rows have the same entries (those of the ith row of the original matrix **A**). But the second sum is equal to $|\mathbf{A}|$ itself. Hence, $|\mathbf{B}| = 0 + |\mathbf{A}| = |\mathbf{A}|$.

Part (3): It is easy to show that the following sequence of row operations has the same effect on a matrix as the one row operation $< i > \longleftrightarrow < j >$:

$$\begin{aligned}< i > &\longleftarrow < i > + < j > \\ < j > &\longleftarrow - < j > \\ < j > &\longleftarrow < i > + < j > \\ < i > &\longleftarrow < i > - < j >\end{aligned}$$

By part (2) of Theorem 3.5, the three type (II) operations in this sequence do not change the determinant of the matrix, and by part (1), the operation $<j> \longleftarrow -<j>$ changes only the sign of the determinant. Thus, the total effect on the determinant of switching two rows of a matrix is to change its sign. ■

Let us now look at some examples involving row operations and the determinant:

Example 1

Let

$$\mathbf{A} = \begin{bmatrix} 5 & -2 & 1 \\ 4 & 3 & -1 \\ 2 & 1 & 0 \end{bmatrix}.$$

You can quickly verify by the "basketweaving" method that $|\mathbf{A}| = 7$. Consider the following matrices:

$$\mathbf{B}_1 = \begin{bmatrix} 5 & -2 & 1 \\ 4 & 3 & -1 \\ -6 & -3 & 0 \end{bmatrix}, \quad \mathbf{B}_2 = \begin{bmatrix} 5 & -2 & 1 \\ 4 & 3 & -1 \\ 12 & -3 & 2 \end{bmatrix}, \quad \text{and} \quad \mathbf{B}_3 = \begin{bmatrix} 4 & 3 & -1 \\ 5 & -2 & 1 \\ 2 & 1 & 0 \end{bmatrix}.$$

Now, $\mathbf{B}_1$ is obtained from **A** by the operation $< 3 > \longleftarrow -3 < 3 >$ of type (I). Hence, part (1) of Theorem 3.5 asserts that $|\mathbf{B}_1| = -3|\mathbf{A}| = (-3)(7) = -21$.

Next, $\mathbf{B}_2$ is obtained from $\mathbf{A}$ by the operation $\langle 3 \rangle \longleftarrow 2\langle 1 \rangle + \langle 3 \rangle$ of type (II). By part (2) of Theorem 3.5, $|\mathbf{B}_2| = |\mathbf{A}| = 7$.

Finally, $\mathbf{B}_3$ is obtained by the operation $\langle 1 \rangle \longleftrightarrow \langle 2 \rangle$ of type (III). Then, by part (3) of Theorem 3.5, $|\mathbf{B}_3| = -|\mathbf{A}| = -7$.

You can use "basketweaving" on $\mathbf{B}_1$, $\mathbf{B}_2$, and $\mathbf{B}_3$ to verify that the values given for their determinants are indeed correct. ■

Calculating the Determinant by Row Reduction

We now illustrate how to use row operations to find the determinant of a given matrix $\mathbf{A}$. The strategy is to calculate an upper triangular matrix $\mathbf{B}$ that is row equivalent to $\mathbf{A}$.

Example 2

Let

$$\mathbf{A} = \begin{bmatrix} 0 & -14 & -8 \\ 1 & 3 & 2 \\ -2 & 0 & 6 \end{bmatrix}.$$

We find $|\mathbf{A}|$ by row reducing $\mathbf{A}$ to produce an upper triangular form, as follows:

$$\mathbf{A} = \begin{bmatrix} 0 & -14 & -8 \\ 1 & 3 & 2 \\ -2 & 0 & 6 \end{bmatrix} \qquad \text{(III): } \langle 1 \rangle \longleftrightarrow \langle 2 \rangle$$

$$\Longrightarrow \quad \mathbf{B}_1 = \begin{bmatrix} 1 & 3 & 2 \\ 0 & -14 & -8 \\ -2 & 0 & 6 \end{bmatrix} \qquad \text{(II): } \langle 3 \rangle \longleftarrow 2\langle 1 \rangle + \langle 3 \rangle$$

$$\Longrightarrow \quad \mathbf{B}_2 = \begin{bmatrix} 1 & 3 & 2 \\ 0 & -14 & -8 \\ 0 & 6 & 10 \end{bmatrix} \qquad \text{(I): } \langle 2 \rangle \longleftarrow -\tfrac{1}{14}\langle 2 \rangle$$

$$\Longrightarrow \quad \mathbf{B}_3 = \begin{bmatrix} 1 & 3 & 2 \\ 0 & 1 & \frac{4}{7} \\ 0 & 6 & 10 \end{bmatrix} \qquad \text{(II): } \langle 3 \rangle \longleftarrow -6\langle 2 \rangle + \langle 3 \rangle$$

$$\Longrightarrow \quad \mathbf{B} = \begin{bmatrix} 1 & 3 & 2 \\ 0 & 1 & \frac{4}{7} \\ 0 & 0 & \frac{46}{7} \end{bmatrix}.$$

Because the last matrix $\mathbf{B}$ is in upper triangular form, we stop. (Notice that we do not have to "zero out" the entries above the main diagonal, as we do in calculations of reduced row echelon form.) From part (3) of Theorem 3.4, $|\mathbf{B}| = (1)(1)(\frac{46}{7}) = \frac{46}{7}$. We now work "backward" through the preceding row operations to find $|\mathbf{A}|$. By Theorem 3.5, we have

$$\begin{aligned} |\mathbf{B}| &= |\mathbf{B}_3| && \text{type (II) row operation} \\ &= -\tfrac{1}{14}|\mathbf{B}_2| && \text{type (I) row operation} \\ &= -\tfrac{1}{14}|\mathbf{B}_1| && \text{type (II) row operation} \\ &= \tfrac{1}{14}|\mathbf{A}| && \text{type (III) row operation} \end{aligned}$$

Thus, $|\mathbf{A}| = 14|\mathbf{B}| = 14(\frac{46}{7}) = 92$. ■

A more systematic method of calculating $|\mathbf{A}|$ is to create a variable P with initial value 1 and update P appropriately as each row operation is performed. That is, we replace the current value of P by

$$\begin{cases} P \times (1/c) & \text{for type (I) row operations} \\ P \times (-1) & \text{for type (III) row operations} \end{cases}$$

Of course, row operations of type (II) do not affect the determinant.[†] Then, $|\mathbf{A}| = P|\mathbf{B}|$, where $\mathbf{B}$ is the upper triangular result of the row reduction process. This method is illustrated in the next example.

Example 3

Let us redo the calculation for $|\mathbf{A}|$ in Example 2. We create a variable P and initialize P to 1. Listed below are the row operations used in that example to convert $\mathbf{A}$ into upper triangular form $\mathbf{B}$, with $|\mathbf{B}| = \frac{46}{7}$. After each operation, we update the value of P accordingly:

Operation	Effect	P
(III): $\langle 1 \rangle \longleftrightarrow \langle 2 \rangle$	Multiply P by -1	-1
(II): $\langle 3 \rangle \longleftarrow 2\langle 1 \rangle + \langle 3 \rangle$	No change	1
(I): $\langle 2 \rangle \longleftarrow -\frac{1}{14}\langle 2 \rangle$	Multiply P by -14	14
(II): $\langle 3 \rangle \longleftarrow -6\langle 2 \rangle + \langle 3 \rangle$	No change	14

Then $|\mathbf{A}|$ equals the final value of P times $|\mathbf{B}|$, or $14 \times \frac{46}{7} = 92$. ■

[†]We multiply by $\frac{1}{c}$ (rather than c) when a row operation of type (I) is performed because, although the row reduction goes forward from $\mathbf{A}$ to the upper triangular matrix $\mathbf{B}$, the determinant calculation goes backward from $\mathbf{B}$ to $\mathbf{A}$.

Formal Algorithm for Calculating the Determinant

We now present a formal algorithm for calculating the determinant of an $n \times n$ matrix based on the method shown in Examples 2 and 3. This method is significantly faster than calculating each of the $n!$ terms in the definition when n is large. Still, computing a typical determinant is a tedious task to do by hand if n exceeds 3. Thus, in practice, determinants are usually calculated by computer. If you do not have software that computes determinants, you may want to write your own program based on this algorithm.

ALGORITHM FOR FINDING THE DETERMINANT OF A MATRIX

Suppose that **A** is a given $n \times n$ matrix.

Step 1: Set $P = 1$.

Step 2: Repeat steps 2(a) through 2(e) for each of the n columns of **A** in turn:

(a) If the current column is the ith column, choose the ith row as the current "home row."
(b) If the home row entry of the current column and all entries below it are zero, go to step 4.
(c) If the home row entry of the current column is zero and if there is a later row whose entry in this column is nonzero, use row operation (III) to exchange the home row with any such row. Replace P with $(-1)P$.
(d) Use row operation (I) to multiply the home row by an appropriate scalar c to convert the home row entry of the current column to 1. Replace P with $\frac{1}{c}P$.
(e) Use row operation (II) to add appropriate multiples of the home row to other rows to convert all entries in the current column below the home row to zero.

Step 3: The determinant of **A** is equal to P. Stop.

Step 4: The determinant of **A** is zero.

This algorithm differs in one respect from the method used in Example 3. Step 3 is reached only if all the main diagonal entries have been converted to 1. In this case, the determinant of **A** is the final value of P times these main diagonal entries, which is simply P itself.

Step 4 is reached only if at least one zero main diagonal entry remains after row reduction. Hence, the determinant is zero.

Determinant Criterion for Matrix Singularity

The next theorem gives an alternate way of determining whether the inverse of a given (square) matrix exists.

THEOREM 3.6

An $n \times n$ matrix $\mathbf{A}$ is nonsingular if and only if $|\mathbf{A}| \neq 0$.

Proof of Theorem 3.6

Let $\mathbf{D}$ be the unique matrix in reduced row echelon form for $\mathbf{A}$. Now, using Theorem 3.5, it is easy to see that a single row operation of type (I), (II), or (III) cannot convert a matrix having a nonzero determinant to a matrix having a zero determinant. (Why?) Because $\mathbf{A}$ is converted to $\mathbf{D}$ using a finite number of such row operations, Theorem 3.5 assures us that $|\mathbf{A}|$ and $|\mathbf{D}|$ are either both zero or both nonzero.

Now, if $\mathbf{A}$ is nonsingular (which implies $\mathbf{D} = \mathbf{I}_n$), we know that $|\mathbf{D}| \neq 0$ and therefore $|\mathbf{A}| \neq 0$, and we have done half of the proof.

For the other half, assume that $|\mathbf{A}| \neq 0$. Then $|\mathbf{D}| \neq 0$. Because $\mathbf{D}$ is a square matrix with a staircase pattern of pivots, it is upper triangular. Thus, all main diagonal entries of $\mathbf{D}$ are nonzero, because $|\mathbf{D}| \neq 0$. Hence, they are all pivots, and $\mathbf{D} = \mathbf{I}_n$. Therefore, row reduction transforms $\mathbf{A}$ to $\mathbf{I}_n$, so $\mathbf{A}$ is nonsingular. ■

Notice that Theorem 3.6 agrees with Theorem 2.12 in that an inverse for $\begin{bmatrix} a & b \\ c & d \end{bmatrix}$ exists if and only if $\begin{vmatrix} a & b \\ c & d \end{vmatrix} = ad - bc \neq 0$.

Theorems 2.13 and 3.6 clearly imply the following:

COROLLARY 3.7

Let $\mathbf{A}$ be an $n \times n$ matrix. Then $\text{rank}(\mathbf{A}) = n$ if and only if $|\mathbf{A}| \neq 0$.

Example 4

Consider the matrix $\mathbf{A} = \begin{bmatrix} 1 & 6 \\ -3 & 5 \end{bmatrix}$. Now, $|\mathbf{A}| = 23$. Hence, $\text{rank}(\mathbf{A}) = 2$ by Corollary 3.7. Also, because $\mathbf{A}$ is the coefficient matrix of the system

$$\begin{cases} x + 6y = 20 \\ -3x + 5y = 9 \end{cases}$$

and $|\mathbf{A}| \neq 0$, this system must have a unique solution by Theorem 2.14. In fact, the solution is $(2, 3)$.

On the other hand, the matrix

$$\mathbf{B} = \begin{bmatrix} 1 & 5 & 1 \\ 2 & 1 & -7 \\ -1 & 2 & 6 \end{bmatrix}$$

has determinant zero. Thus, $\text{rank}(\mathbf{B}) < 3$. Also, because $\mathbf{B}$ is the coefficient matrix for the homogeneous system

$$\begin{cases} x_1 + 5x_2 + x_3 = 0 \\ 2x_1 + x_2 - 7x_3 = 0 \\ -x_1 + 2x_2 + 6x_3 = 0 \end{cases},$$

this system has nontrivial solutions by Theorem 2.4. You can easily calculate that its solution set is $\{(4c, -c, c) \,|\, c \in \mathbb{R}\}$. ■

Determinant of a Matrix Product

Our next goal is to prove that the determinant of a product of two matrices $\mathbf{A}$ and $\mathbf{B}$ is equal to the product of their determinants $|\mathbf{A}|$ and $|\mathbf{B}|$. First, let us look at a special case.

LEMMA 3.8

If $\mathbf{E}$ is an elementary $n \times n$ matrix and $\mathbf{K}$ is any $n \times n$ matrix, then $|\mathbf{EK}| = |\mathbf{E}||\mathbf{K}|$.

Proof of Lemma 3.8 (abridged)

Because $\mathbf{E}$ is an elementary $n \times n$ matrix, it is obtained from $\mathbf{I}_n$ by performing a single row operation of type (I), (II), or (III).

Suppose that $\mathbf{E}$ is obtained by the row operation $< i > \leftarrow c < i >$ of type (I). Then, by part (1) of Theorem 3.5, $|\mathbf{E}| = c|\mathbf{I}_n| = c$. Also, by Theorem 2.15, $\mathbf{EK}$ is the matrix obtained by performing this same row operation $< i > \leftarrow c < i >$ on $\mathbf{K}$. Hence, by part (1) of Theorem 3.5, $|\mathbf{EK}| = c|\mathbf{K}|$. Then, $|\mathbf{EK}| = |\mathbf{E}||\mathbf{K}|$.

The cases for type (II) and type (III) operations are handled similarly. ■

A simple induction argument shows that we can replace $\mathbf{E}$ in Lemma 3.8 by the product of any finite number of elementary matrices to obtain $|\mathbf{E}_1\mathbf{E}_2 \cdots \mathbf{E}_k\mathbf{A}| = |\mathbf{E}_1||\mathbf{E}_2| \cdots |\mathbf{E}_k||\mathbf{A}|$. (See Exercise 11(a).) Now we are ready to prove the following theorem.

THEOREM 3.9

If $\mathbf{A}$ and $\mathbf{B}$ are both $n \times n$ matrices, then $|\mathbf{AB}| = |\mathbf{A}||\mathbf{B}|$.

Proof of Theorem 3.9

Suppose first that $\mathbf{A}$ is nonsingular. Then $\mathbf{A}$ is a product of elementary matrices of the form $\mathbf{E}_1\mathbf{E}_2\cdots\mathbf{E}_k$. By the generalized version of Lemma 3.8, $|\mathbf{A}| = |\mathbf{E}_1||\mathbf{E}_2|\cdots|\mathbf{E}_k|$. Hence, $|\mathbf{AB}| = |\mathbf{E}_1\mathbf{E}_2\cdots\mathbf{E}_k\mathbf{B}| = |\mathbf{E}_1||\mathbf{E}_2|\cdots|\mathbf{E}_k||\mathbf{B}| = (|\mathbf{E}_1||\mathbf{E}_2|\cdots|\mathbf{E}_k|)(|\mathbf{B}|) = |\mathbf{A}||\mathbf{B}|$.

Now suppose that $\mathbf{A}$ is singular (that is, $|\mathbf{A}| = 0$). If we can show that $|\mathbf{AB}| = 0$, then $|\mathbf{AB}| = |\mathbf{A}||\mathbf{B}|$, and we will be done. We assume that $|\mathbf{AB}| \neq 0$ and get a contradiction. If $|\mathbf{AB}| \neq 0$, $(\mathbf{AB})^{-1}$ exists, and $\mathbf{I}_n = \mathbf{AB}(\mathbf{AB})^{-1}$. Hence, $\mathbf{B}(\mathbf{AB})^{-1}$ is a right inverse for $\mathbf{A}$. But by Theorem 2.8, $\mathbf{A}^{-1}$ exists, contradicting the fact that $\mathbf{A}$ is singular. ■

Example 5

Let

$$\mathbf{A} = \begin{bmatrix} 3 & 2 & 1 \\ 5 & 0 & -2 \\ -3 & 1 & 4 \end{bmatrix} \quad \text{and} \quad \mathbf{B} = \begin{bmatrix} 1 & -1 & 0 \\ 4 & 2 & -1 \\ -2 & 0 & 3 \end{bmatrix}.$$

It is relatively easy (using "basketweaving" or row reduction) to see that $|\mathbf{A}| = -17$ and that $|\mathbf{B}| = 16$. Therefore, the determinant of

$$\mathbf{AB} = \begin{bmatrix} 9 & 1 & 1 \\ 9 & -5 & -6 \\ -7 & 5 & 11 \end{bmatrix}$$

is $(-17)(16) = -272$. ■

An obvious consequence of Theorem 3.9 is that $|\mathbf{AB}| = 0$ if and only if either $|\mathbf{A}| = 0$ or $|\mathbf{B}| = 0$. (See Exercise 6(a).) Therefore, it follows that $\mathbf{AB}$ is singular if and only if either $\mathbf{A}$ or $\mathbf{B}$ is singular.

Similarly, if $\mathbf{A}$ is nonsingular, Theorem 3.9 and the fact that $\mathbf{AA}^{-1} = \mathbf{I}$ can be used to prove that $|\mathbf{A}^{-1}| = 1/|\mathbf{A}|$. (See Exercise 6(c).)

Determinant of the Transpose

THEOREM 3.10

If $\mathbf{A}$ is an $n \times n$ matrix, then $|\mathbf{A}| = |\mathbf{A}^T|$.

Example 6

Using the "basketweaving" method, we see that

$$\begin{vmatrix} -1 & 4 & 1 \\ 2 & 0 & 3 \\ -1 & -1 & 2 \end{vmatrix} = -33.$$

Hence, by Theorem 3.10,

$$\begin{vmatrix} -1 & 2 & -1 \\ 4 & 0 & -1 \\ 1 & 3 & 2 \end{vmatrix} = -33.$$

■

Proof of Theorem 3.10

First, you can easily show that a type (I) or type (III) elementary matrix equals its own transpose. Also, a type (II) elementary matrix is always upper or lower triangular with 1's on the main diagonal. Thus, so is its transpose. Therefore, if $\mathbf{E}$ is any elementary matrix, $|\mathbf{E}| = |\mathbf{E}^T|$.

Let $\mathbf{A}$ be an $n \times n$ matrix. Also, let $\mathbf{E}_1, \ldots, \mathbf{E}_k$ be elementary matrices associated with a sequence of row operations that reduce $\mathbf{A}$ to an upper triangular matrix $\mathbf{B}$. Hence, $\mathbf{B} = \mathbf{E}_k\mathbf{E}_{k-1}\cdots\mathbf{E}_1\mathbf{A}$. Thus, by Theorem 3.9, $|\mathbf{A}| = |\mathbf{B}|/(|\mathbf{E}_1|\cdots|\mathbf{E}_k|)$. Now, $\mathbf{B}^T = \mathbf{A}^T\mathbf{E}_1^T\cdots\mathbf{E}_k^T$, and therefore $|\mathbf{A}^T| = |\mathbf{B}^T|/(|\mathbf{E}_1^T|\cdots|\mathbf{E}_k^T|)$. But $|\mathbf{E}_i| = |\mathbf{E}_i^T|$, for $1 \le i \le k$. Also, because $\mathbf{B}$ is upper triangular, $\mathbf{B}^T$ is lower triangular with the same diagonal entries as $\mathbf{B}$. Hence, $|\mathbf{B}| = |\mathbf{B}^T|$ by Theorem 3.4, part (3). It clearly follows that $|\mathbf{A}| = |\mathbf{A}^T|$. ■

Theorem 3.10 can be used to prove "column versions" of many earlier results involving determinants. For example, if $\mathbf{A}$ is a square matrix having two equal columns, then we can easily show that $|\mathbf{A}| = 0$. In this case, $\mathbf{A}^T$ has two equal rows, and so $|\mathbf{A}^T| = 0$ by part (2) of Theorem 3.4. Hence, by Theorem 3.10, $|\mathbf{A}| = |\mathbf{A}^T| = 0$.

Also, column operations analogous to the familiar row operations can be defined. For example, a type (I) column operation multiplies all entries of a given column of a matrix by a nonzero scalar. Theorem 3.10 can be combined with Theorem 3.5 to show that a column operation has the same effect on the determinant of a matrix as its corresponding row operation.

Example 7

Let

$$\mathbf{A} = \begin{bmatrix} 2 & 5 & 1 \\ 1 & 2 & 3 \\ -3 & 1 & -1 \end{bmatrix}.$$

Performing the type (II) column operation <col. 2> ⟵ <col. 2> − 3 <col.1 > produces the matrix

$$\mathbf{B} = \begin{bmatrix} 2 & -1 & 1 \\ 1 & -1 & 3 \\ -3 & 10 & -1 \end{bmatrix}.$$

Using "basketweaving," we see that $|\mathbf{A}| = -43 = |\mathbf{B}|$. Thus, this column operation has no effect on the determinant, as we would expect. ■

For reference, we summarize many of the results obtained in Chapters 2 and 3 in Figure 3.4. You should be able to justify each equivalence in the figure by citing a relevant definition or result.

Assume that A is an $n \times n$ matrix. Then the following are all equivalent:	**Assume that A is an $n \times n$ matrix. Then the following are all equivalent:**
A is singular ($\mathbf{A}^{-1}$ does not exist).	**A** is nonsingular ($\mathbf{A}^{-1}$ exists).
Rank(**A**) $\neq n$.	Rank(**A**) $= n$.
$\|\mathbf{A}\| = 0$.	$\|\mathbf{A}\| \neq 0$.
A is not row equivalent to $\mathbf{I}_n$.	**A** is row equivalent to $\mathbf{I}_n$.
$\mathbf{AX} = \mathbf{O}$ has a nontrivial solution for **X**.	$\mathbf{AX} = \mathbf{O}$ has only the trivial solution for **X**.
$\mathbf{AX} = \mathbf{B}$ does not have a unique solution (no solutions or infinitely many solutions).	$\mathbf{AX} = \mathbf{B}$ has a unique solution for **X** ($\mathbf{X} = \mathbf{A}^{-1}\mathbf{B}$).
A is not equal to a (finite) product of elementary matrices.	$\mathbf{A} = \mathbf{E}_1\mathbf{E}_2 \cdots \mathbf{E}_k$, for some elementary matrices $\mathbf{E}_1, \mathbf{E}_2, \ldots, \mathbf{E}_k$.

Figure 3.4
Equivalent conditions for singular and nonsingular matrices

Exercises—Section 3.2

1. Each of the following elementary matrices is obtained from $\mathbf{I}_3$ by performing a single row operation of type (I), (II), or (III). Identify the operation, and use Theorem 3.5 to give the determinant of each matrix:

★(a) $\begin{bmatrix} 1 & -3 & 0 \\ 0 & 1 & 0 \\ 0 & 0 & 1 \end{bmatrix}$ (b) $\begin{bmatrix} 1 & 0 & 0 \\ 0 & 0 & 1 \\ 0 & 1 & 0 \end{bmatrix}$ ★(c) $\begin{bmatrix} 1 & 0 & 0 \\ 0 & 1 & 0 \\ 0 & 0 & -4 \end{bmatrix}$

(d) $\begin{bmatrix} 1 & 0 & 0 \\ 2 & 1 & 0 \\ 0 & 0 & 1 \end{bmatrix}$ (e) $\begin{bmatrix} \frac{1}{2} & 0 & 0 \\ 0 & 1 & 0 \\ 0 & 0 & 1 \end{bmatrix}$ ★(f) $\begin{bmatrix} 0 & 1 & 0 \\ 1 & 0 & 0 \\ 0 & 0 & 1 \end{bmatrix}$

2. Calculate the determinant of each of the following matrices by using row reduction to produce an upper triangular form:

★(a) $\begin{bmatrix} 10 & 4 & 21 \\ 0 & -4 & 3 \\ -5 & -1 & -12 \end{bmatrix}$ (b) $\begin{bmatrix} 18 & -9 & -14 \\ 6 & -3 & -5 \\ -3 & 1 & 2 \end{bmatrix}$

★(c) $\begin{bmatrix} 1 & -1 & 5 & 1 \\ -2 & 1 & -7 & 1 \\ -3 & 2 & -12 & -2 \\ 2 & -1 & 9 & 1 \end{bmatrix}$ (d) $\begin{bmatrix} -8 & 4 & -3 & 2 \\ 2 & 1 & -1 & -1 \\ -3 & -5 & 4 & 0 \\ 2 & -4 & 3 & -1 \end{bmatrix}$

★(e) $\begin{bmatrix} 5 & 3 & -8 & 4 \\ \frac{15}{2} & \frac{1}{2} & -1 & -7 \\ -\frac{5}{2} & \frac{3}{2} & -4 & 1 \\ 10 & -3 & 8 & -8 \end{bmatrix}$ (f) $\begin{bmatrix} 3 & 22 & -1 & -7 & -3 \\ 5 & -1 & -3 & 0 & -5 \\ 2 & 6 & -1 & -2 & -2 \\ 12 & 48 & 5 & -16 & -10 \\ -1 & -3 & 0 & 1 & 1 \end{bmatrix}$

3. By calculating the determinant of each matrix, decide whether it has an inverse. If the inverse exists, what is its determinant?

★(a) $\begin{bmatrix} 5 & 6 \\ -3 & -4 \end{bmatrix}$ (b) $\begin{bmatrix} \cos\theta & -\sin\theta \\ \sin\theta & \cos\theta \end{bmatrix}$

★(c) $\begin{bmatrix} -12 & 7 & -27 \\ 4 & -1 & 2 \\ 3 & 2 & -8 \end{bmatrix}$ (d) $\begin{bmatrix} 31 & -20 & 106 \\ -11 & 7 & -37 \\ -9 & 6 & -32 \end{bmatrix}$

4. By calculating the determinant of the coefficient matrix, decide whether each of the following homogeneous systems has a nontrivial solution. (You do not need to find the actual solutions.) (Hint: Use Corollary 3.7.)

★(a) $\begin{cases} -6x + 3y - 22z = 0 \\ -7x + 4y - 31z = 0 \\ 11x - 6y + 46z = 0 \end{cases}$

(b) $\begin{cases} 4x_1 - x_2 + x_3 = 0 \\ -x_1 + x_2 - 2x_3 = 0 \\ -6x_1 + 9x_2 - 19x_3 = 0 \end{cases}$

(c) $\begin{cases} 2x_1 - 2x_2 + x_3 + 4x_4 = 0 \\ 4x_1 + 2x_2 + x_3 = 0 \\ -x_1 - x_2 - x_4 = 0 \\ -12x_1 - 7x_2 - 5x_3 + 2x_4 = 0 \end{cases}$

5. Let $\mathbf{A}$ and $\mathbf{B}$ be $n \times n$ matrices.

(a) Show that $\mathbf{A}$ is nonsingular if and only if $\mathbf{A}^T$ is nonsingular.

(b) Show that $|\mathbf{AB}| = |\mathbf{BA}|$. (Remember that, in general, $\mathbf{AB} \neq \mathbf{BA}$.)

(c) Show that $|\mathbf{AB}^T| = |\mathbf{A}^T||\mathbf{B}|$.

6. Let **A** and **B** be $n \times n$ matrices.
(a) Show that $|\mathbf{AB}| = 0$ if and only if $|\mathbf{A}| = 0$ or $|\mathbf{B}| = 0$.
(b) Show that if $\mathbf{AB} = -\mathbf{BA}$ and n is odd, then **A** or **B** is singular.
(c) Prove that if **A** is nonsingular, then $|\mathbf{A}^{-1}| = 1/|\mathbf{A}|$.

★ **7.** Give a counterexample to show that, for $n \times n$ matrices **A** and **B**, it is not always true that $|\mathbf{A} + \mathbf{B}| = |\mathbf{A}| + |\mathbf{B}|$.

8. Suppose that $\mathbf{AB} = \mathbf{AC}$ and $|\mathbf{A}| \neq 0$. Show that $\mathbf{B} = \mathbf{C}$.

9. Let **A** be an $n \times n$ matrix. Show that if $\mathbf{A}^k = \mathbf{O}_n$ (the zero $n \times n$ matrix), for some integer $k \geq 1$, then $|\mathbf{A}| = 0$.

10. Show that $|\mathbf{AA}^T| \geq 0$ for any square matrix **A**.

11. Give a proof by induction in each case:
(a) **General form of Lemma 3.8:** Assuming that $|\mathbf{EK}| = |\mathbf{E}||\mathbf{K}|$, whenever **E** is an elementary $n \times n$ matrix and **K** is any $n \times n$ matrix, prove that $|\mathbf{E}_1\mathbf{E}_2 \cdots \mathbf{E}_k\mathbf{K}| = |\mathbf{E}_1||\mathbf{E}_2| \cdots |\mathbf{E}_k||\mathbf{K}|$ for elementary $n \times n$ matrices $\mathbf{E}_1, \mathbf{E}_2, \ldots, \mathbf{E}_k$.
(b) **General form of Theorem 3.9:** Assuming that $|\mathbf{AB}| = |\mathbf{A}||\mathbf{B}|$, prove that $|\mathbf{A}_1\mathbf{A}_2 \cdots \mathbf{A}_k| = |\mathbf{A}_1||\mathbf{A}_2| \cdots |\mathbf{A}_k|$ for any $n \times n$ matrices $\mathbf{A}_1, \mathbf{A}_2, \ldots, \mathbf{A}_k$.
(c) Prove that $|\mathbf{A}^q| = |\mathbf{A}|^q$ for any $n \times n$ matrix **A** and any nonnegative integer q.

12. If $\mathbf{A} = \mathbf{B}_1\mathbf{B}_2 \cdots \mathbf{B}_k$, where each $\mathbf{B}_i$, for $1 \leq i \leq k$, is an $n \times n$ matrix, show that **A** is nonsingular if and only if each $\mathbf{B}_i$ is nonsingular.

13. Let **A** be an $n \times n$ matrix.
(a) Show that if the entries of some row of **A** are proportional to those in another row, then $|\mathbf{A}| = 0$.
(b) Show that if the entries in every row of **A** add up to zero, then $|\mathbf{A}| = 0$. (Hint: Consider the system $\mathbf{AX} = \mathbf{O}$, and note that the $n \times 1$ vector **X** having every entry equal to 1 is a nontrivial solution.)

14. Let **A** be an $n \times n$ skew-symmetric matrix.
(a) If n is odd, show that $|\mathbf{A}| = 0$.
★(b) If n is even, give an example where $|\mathbf{A}| \neq 0$.

15. An **orthogonal** matrix is a (square) matrix **A** with $\mathbf{A}^T = \mathbf{A}^{-1}$.
(a) Why is $\mathbf{I}_n$ obviously orthogonal?
★(b) Find a 3×3 orthogonal matrix other than $\mathbf{I}_3$.
(c) Show that $|\mathbf{A}| = \pm 1$ if **A** is orthogonal.

16. Suppose that $|\mathbf{A}|$ is an integer.
(a) Prove that $|\mathbf{A}^n|$ is not a prime, for $n \geq 2$. (Recall that a **prime** number is an integer greater than 1 with no positive integer divisors except itself and 1.)
(b) Prove that if $\mathbf{A}^n = \mathbf{I}_n$, for all $n \geq 1$, n odd, then $|\mathbf{A}| = 1$.

17. Show that there is no matrix **A** such that

$$\mathbf{A}^2 = \begin{bmatrix} 9 & 0 & -3 \\ 3 & 2 & -1 \\ -6 & 0 & 1 \end{bmatrix}.$$

18. We say that the matrix **B** is **similar** to a matrix **A** if there exists some (nonsingular) matrix **P** such that $\mathbf{PAP}^{-1} = \mathbf{B}$.

(a) Show that if **A** and **B** are similar, then they are both square matrices of the same size.

★(b) Find two different matrices **B** similar to $\mathbf{A} = \begin{bmatrix} 1 & 2 \\ 3 & 4 \end{bmatrix}$.

(c) Show that if **B** is similar to **A**, then **A** is similar to **B**.

(d) Show that if **A** and **B** are similar, then $|\mathbf{A}| = |\mathbf{B}|$.

19. Show that the determinant of

$$\begin{bmatrix} x & -1 & 0 & 0 \\ 0 & x & -1 & 0 \\ 0 & 0 & x & -1 \\ a_0 & a_1 & a_2 & a_3 + x \end{bmatrix}$$

is $x^4 + a_3x^3 + a_2x^2 + a_1x + a_0$. (Hint: Beware! You can't divide by x if $x = 0$.)

20. Use row reduction to show that the determinant of the $n \times n$ matrix symbolically

represented by $\begin{bmatrix} \mathbf{A} & \mathbf{C} \\ \mathbf{O} & \mathbf{B} \end{bmatrix}$ is $|\mathbf{A}||\mathbf{B}|$, where

A is an $m \times m$ submatrix
B is an $(n - m) \times (n - m)$ submatrix
C is an $m \times (n - m)$ submatrix
O is an $(n - m) \times m$ zero submatrix

(As a special case, we have

$$\begin{vmatrix} a & b & * & * \\ c & d & * & * \\ 0 & 0 & e & f \\ 0 & 0 & g & h \end{vmatrix} = \begin{vmatrix} a & b \\ c & d \end{vmatrix}\begin{vmatrix} e & f \\ g & h \end{vmatrix},$$

where the starred entries can be any real numbers.)

21. Prove part (1) of Theorem 3.5 by factoring c out of each term in the determinant formula.

22. This exercise shows that we cannot associate any "other" real number with a matrix and expect it to possess the most important properties of the determinant.

Suppose that $f\colon \mathcal{M}_{nn} \longrightarrow \mathbb{R}$ with $f(\mathbf{I}_n) = 1$, with the property that whenever a single row operation is performed on $\mathbf{A} \in \mathcal{M}_{nn}$ to create $\mathbf{B}$, then

$$f(\mathbf{B}) = \begin{cases} cf(\mathbf{A}) & \text{for a type (I) row operation with } c \neq 0 \\ f(\mathbf{A}) & \text{for a type (II) row operation} \\ -f(\mathbf{A}) & \text{for a type (III) row operation.} \end{cases}$$

Prove that $f(\mathbf{A}) = |\mathbf{A}|$, for all $\mathbf{A} \in \mathcal{M}_{nn}$.

(Hint: If $\mathbf{A}$ is row equivalent to $\mathbf{I}_n$, then the properties of f guarantee that $f(\mathbf{A}) = |\mathbf{A}|$. (Why?) Otherwise, $\mathbf{A}$ is row equivalent to a matrix with a row of zeroes. Then apply a type (I) operation with $c = -1$ to get $f(\mathbf{A}) = 0 = |\mathbf{A}|$.)

3.3 Cofactor Expansion and the Adjoint

In this section, we examine another method for calculating determinants: cofactor expansion. This method is not as fast computationally as the row reduction method outlined in Section 3.2. However, the recursive nature of cofactor expansion makes it interesting. In addition, cofactor expansion is used to define the adjoint matrix, which gives an algebraic formula for the inverse of a matrix. Finally, we discuss Cramer's Rule, a method for solving certain linear systems using determinants.

Cofactors

DEFINITION

Let $\mathbf{A}$ be an $n \times n$ matrix, with $n \geq 2$. The ***(i,j)th* submatrix**, $\mathbf{A}_{ij}$, of $\mathbf{A}$, is the $(n-1) \times (n-1)$ matrix obtained by deleting all entries of the ith row and all entries of the jth column of $\mathbf{A}$. The ***(i,j)th* minor**, $|\mathbf{A}_{ij}|$, of $\mathbf{A}$, is the determinant of the submatrix $\mathbf{A}_{ij}$ of $\mathbf{A}$.

Example 1

Consider the following matrices:

$$\mathbf{A} = \begin{bmatrix} 5 & -2 & 1 \\ 0 & 4 & -3 \\ 2 & -7 & 6 \end{bmatrix} \quad \text{and} \quad \mathbf{B} = \begin{bmatrix} 9 & -1 & 4 & 7 \\ -3 & 2 & 6 & -2 \\ -8 & 0 & 1 & 3 \\ 4 & 7 & -5 & -1 \end{bmatrix}.$$

The submatrix of $\mathbf{A}$ obtained by deleting all entries in the first row and all entries in the third column is $\mathbf{A}_{13} = \begin{bmatrix} 0 & 4 \\ 2 & -7 \end{bmatrix}$, and the submatrix of $\mathbf{B}$ obtained by deleting all entries in the third row and all entries in the fourth column is

$$\mathbf{B}_{34} = \begin{bmatrix} 9 & -1 & 4 \\ -3 & 2 & 6 \\ 4 & 7 & -5 \end{bmatrix}.$$

The corresponding minors for these submatrices are

$$|\mathbf{A}_{13}| = \begin{vmatrix} 0 & 4 \\ 2 & -7 \end{vmatrix} = -8 \quad \text{and} \quad |\mathbf{B}_{34}| = \begin{vmatrix} 9 & -1 & 4 \\ -3 & 2 & 6 \\ 4 & 7 & -5 \end{vmatrix} = -593$$

■

A 3×3 matrix has nine minors. For the matrix $\mathbf{A}$ in Example 1, the minors are:

$$|\mathbf{A}_{11}| = \begin{vmatrix} 4 & -3 \\ -7 & 6 \end{vmatrix} = 3 \qquad |\mathbf{A}_{12}| = \begin{vmatrix} 0 & -3 \\ 2 & 6 \end{vmatrix} = 6 \qquad |\mathbf{A}_{13}| = \begin{vmatrix} 0 & 4 \\ 2 & -7 \end{vmatrix} = -8$$

$$|\mathbf{A}_{21}| = \begin{vmatrix} -2 & 1 \\ -7 & 6 \end{vmatrix} = -5 \qquad |\mathbf{A}_{22}| = \begin{vmatrix} 5 & 1 \\ 2 & 6 \end{vmatrix} = 28 \qquad |\mathbf{A}_{23}| = \begin{vmatrix} 5 & -2 \\ 2 & -7 \end{vmatrix} = -31$$

$$|\mathbf{A}_{31}| = \begin{vmatrix} -2 & 1 \\ 4 & -3 \end{vmatrix} = 2 \qquad |\mathbf{A}_{32}| = \begin{vmatrix} 5 & 1 \\ 0 & -3 \end{vmatrix} = -15 \qquad |\mathbf{A}_{33}| = \begin{vmatrix} 5 & -2 \\ 0 & 4 \end{vmatrix} = 20$$

Similarly, an $n \times n$ matrix has a total of n^2 minors—one for each entry of the matrix.

We now define a "cofactor" for each entry based on its minor.

DEFINITION

> Let $\mathbf{A}$ be an $n \times n$ matrix, with $n \geq 2$. The $(i,j)^{th}$ **cofactor** of $\mathbf{A}$, $\mathcal{A}_{ij}$, is $(-1)^{i+j}|\mathbf{A}_{ij}|$—that is, $(-1)^{i+j}$ times the $(i,j)^{th}$ minor of $\mathbf{A}$.

Example 2

For the matrices $\mathbf{A}$ and $\mathbf{B}$ in Example 1, the cofactor $\mathcal{A}_{13}$ of $\mathbf{A}$ is $(-1)^{1+3}|\mathbf{A}_{13}| = (-1)^4(-8) = -8$, and the cofactor $\mathcal{B}_{34}$ of $\mathbf{B}$ is $(-1)^{3+4}|\mathbf{B}_{34}| = (-1)^7(-593) = 593$.

■

A 3×3 matrix has nine cofactors. For the matrix $\mathbf{A}$ from Example 1, these cofactors are:

$$\begin{aligned}
\mathcal{A}_{11} &= (-1)^{1+1}|\mathbf{A}_{11}| = (-1)^2(3) = 3 \\
\mathcal{A}_{12} &= (-1)^{1+2}|\mathbf{A}_{12}| = (-1)^3(6) = -6 \\
\mathcal{A}_{13} &= (-1)^{1+3}|\mathbf{A}_{13}| = (-1)^4(-8) = -8 \\
\\
\mathcal{A}_{21} &= (-1)^{2+1}|\mathbf{A}_{21}| = (-1)^3(-5) = 5 \\
\mathcal{A}_{22} &= (-1)^{2+2}|\mathbf{A}_{22}| = (-1)^4(28) = 28 \\
\mathcal{A}_{23} &= (-1)^{2+3}|\mathbf{A}_{23}| = (-1)^5(-31) = 31
\end{aligned}$$

$$
\begin{aligned}
\mathcal{A}_{31} &= (-1)^{3+1}|\mathbf{A}_{31}| &= (-1)^4(2) &= 2\\
\mathcal{A}_{32} &= (-1)^{3+2}|\mathbf{A}_{32}| &= (-1)^5(-15) &= 15\\
\mathcal{A}_{33} &= (-1)^{3+3}|\mathbf{A}_{33}| &= (-1)^6(20) &= 20
\end{aligned}
$$

Similarly, an $n \times n$ matrix has n^2 cofactors, one for each matrix entry.

Finding the Determinant by Cofactor Expansion

The next theorem shows an amazing connection between cofactors and the determinant.

THEOREM 3.11

If $\mathbf{A}$ is an $n \times n$ matrix, with $n \geq 2$, then

(1) $a_{i1}\mathcal{A}_{i1} + a_{i2}\mathcal{A}_{i2} + \cdots + a_{in}\mathcal{A}_{in} = |\mathbf{A}|$, for each i, $1 \leq i \leq n$

(2) $a_{i1}\mathcal{A}_{j1} + a_{i2}\mathcal{A}_{j2} + \cdots + a_{in}\mathcal{A}_{jn} = 0$, for $i \neq j$, $1 \leq i, j \leq n$

Part (1) asserts that, if we go along any row of $\mathbf{A}$ multiplying each entry by its corresponding cofactor, the sum of these values equals $|\mathbf{A}|$. Part (2) asserts that, if we go along any row of $\mathbf{A}$, multiplying the entries by the cofactors for a *different* row, the sum of these values equals zero. The proof of this theorem is long and tedious and is omitted.

Example 3

Let us return to the matrix from Example 1:

$$\mathbf{A} = \begin{bmatrix} 5 & -2 & 1 \\ 0 & 4 & -3 \\ 2 & -7 & 6 \end{bmatrix}.$$

It is easy to check that $|\mathbf{A}| = 19$. Now, multiplying every entry of the second row by its cofactor, and summing, we have

$$a_{21}\mathcal{A}_{21} + a_{22}\mathcal{A}_{22} + a_{23}\mathcal{A}_{23} = (0)(5) + (4)(28) + (-3)(31) = 112 - 93 = 19,$$

which equals $|\mathbf{A}|$. Similarly, $a_{11}\mathcal{A}_{11} + a_{12}\mathcal{A}_{12} + a_{13}\mathcal{A}_{13} = 19$. (Try it.) However,

$$a_{21}\mathcal{A}_{31} + a_{22}\mathcal{A}_{32} + a_{23}\mathcal{A}_{33} = (0)(2) + (4)(15) + (-3)(20) = 0,$$

because we are multiplying the second-row elements by the cofactors of elements from a different row (here, the third row). ■

The formula in part (1) of Theorem 3.11 is called the **cofactor expansion** (or the **Laplace expansion**) **along the ith row**. Theorem 3.11 asserts that we can find the determinant by "expanding" along any row of the matrix. The next corollary shows that we can also find the determinant by a similar type of cofactor expansion along any column.

COROLLARY 3.12

If $\mathbf{A}$ is an $n \times n$ matrix, with $n \geq 2$, then

(1) $a_{1j}\mathcal{A}_{1j} + a_{2j}\mathcal{A}_{2j} + \cdots + a_{nj}\mathcal{A}_{nj} = |\mathbf{A}|$, for each j, $1 \leq j \leq n$

(2) $a_{1j}\mathcal{A}_{1k} + a_{2j}\mathcal{A}_{2k} + \cdots + a_{nj}\mathcal{A}_{nk} = 0$, for $j \neq k$, $1 \leq j, k \leq n$

Corollary 3.12 follows easily from Theorem 3.11 using the transpose operation and Theorem 3.10.

Notice that in part (1), each entry of the jth column is multiplied by its corresponding cofactor; in part (2), each entry of the jth column is, instead, multiplied by the cofactor of the related entry from the kth column.

Theorem 3.11 and Corollary 3.12 together show that the determinant can be found by a cofactor expansion along any row or column. The next example shows that the choice of one row or column over another sometimes has advantages in simplifying the computation.

Example 4

Consider the following matrix:

$$\mathbf{A} = \begin{bmatrix} -2 & 5 & 4 & -2 \\ 3 & 1 & 0 & -2 \\ 4 & -4 & 0 & -1 \\ 6 & 1 & -1 & -3 \end{bmatrix}.$$

To find $|\mathbf{A}|$, we could do a cofactor expansion along any row or column, but we choose the third column because it has the largest number of zeroes. Thus, we have only two 3×3 determinants to evaluate instead of four:

$$\begin{aligned} |\mathbf{A}| &= a_{13}\mathcal{A}_{13} + a_{23}\mathcal{A}_{23} + a_{33}\mathcal{A}_{33} + a_{43}\mathcal{A}_{43} \\ &= (4)(-1)^{1+3}|\mathbf{A}_{13}| + (0)(-1)^{2+3}|\mathbf{A}_{23}| + (0)(-1)^{3+3}|\mathbf{A}_{33}| + (-1)(-1)^{4+3}|\mathbf{A}_{43}| \\ &= 4\begin{vmatrix} 3 & -1 & 2 \\ 4 & -4 & -1 \\ 6 & 1 & -3 \end{vmatrix} + \begin{vmatrix} -2 & 5 & -2 \\ 3 & 1 & -2 \\ 4 & -4 & -1 \end{vmatrix}. \end{aligned}$$

At this point, we could use "basketweaving" or row reduction to finish the calculation. Instead, we evaluate each of the remaining determinants using cofactor expansion again to illustrate the recursiveness of the method.

Expanding the first determinant along the second row, we have

$$\begin{vmatrix} 3 & 1 & -2 \\ 4 & -4 & -1 \\ 6 & 1 & -3 \end{vmatrix}$$

$$= (4)(-1)^{2+1}\begin{vmatrix} 1 & -2 \\ 1 & -3 \end{vmatrix} + (-4)(-1)^{2+2}\begin{vmatrix} 3 & -2 \\ 6 & -3 \end{vmatrix} + (-1)(-1)^{2+3}\begin{vmatrix} 3 & 1 \\ 6 & 1 \end{vmatrix}$$

$$= (-4)(-1) + (-4)(3) + (1)(-3) = 4 - 12 - 3 = -11.$$

Expanding the other determinant along the third column, we have

$$\begin{vmatrix} -2 & 5 & -2 \\ 3 & 1 & -2 \\ 4 & -4 & -1 \end{vmatrix}$$

$$= (-2)(-1)^{1+3}\begin{vmatrix} 3 & 1 \\ 4 & -4 \end{vmatrix} + (-2)(-1)^{2+3}\begin{vmatrix} -2 & 5 \\ 4 & -4 \end{vmatrix} + (-1)(-1)^{3+3}\begin{vmatrix} -2 & 5 \\ 3 & 1 \end{vmatrix}$$

$$= (-2)(-16) + (2)(-12) + (-1)(-17) = 32 - 24 + 17 = 25.$$

Thus, $|\mathbf{A}| = (4)(-11) + 25 = -19.$ ■

The Adjoint Matrix

DEFINITION

Let $\mathbf{A}$ be an $n \times n$ matrix, with $n \geq 2$. The **(classical) adjoint** $\mathcal{A}$ of $\mathbf{A}$ is the $n \times n$ matrix whose $(i, j)^{th}$ entry is $\mathcal{A}_{ji}$, the $(j, i)^{th}$ cofactor of $\mathbf{A}$.

Notice that the $(i, j)^{th}$ entry of the adjoint is not the cofactor $\mathcal{A}_{ij}$ of $\mathbf{A}$ but is $\mathcal{A}_{ji}$ instead. Hence, the general form of the adjoint of an $n \times n$ matrix $\mathbf{A}$ is

$$\mathcal{A} = \begin{bmatrix} \mathcal{A}_{11} & \mathcal{A}_{21} & \cdots & \mathcal{A}_{n1} \\ \mathcal{A}_{12} & \mathcal{A}_{22} & \cdots & \mathcal{A}_{n2} \\ \vdots & \vdots & \ddots & \vdots \\ \mathcal{A}_{1n} & \mathcal{A}_{2n} & \cdots & \mathcal{A}_{nn} \end{bmatrix}.$$

Example 5

As you saw after Example 2, the matrix

$$\mathbf{A} = \begin{bmatrix} 5 & -2 & 1 \\ 0 & 4 & -3 \\ 2 & -7 & 6 \end{bmatrix}$$

has cofactors:

$$\begin{array}{lll} \mathcal{A}_{11} = 3 & \mathcal{A}_{12} = -6 & \mathcal{A}_{13} = -8 \\ \mathcal{A}_{21} = 5 & \mathcal{A}_{22} = 28 & \mathcal{A}_{23} = 31 \\ \mathcal{A}_{31} = 2 & \mathcal{A}_{32} = 15 & \mathcal{A}_{33} = 20 \end{array}$$

Therefore, the adjoint matrix of $\mathbf{A}$ is this 3×3 matrix:

$$\mathcal{A} = \begin{bmatrix} 3 & 5 & 2 \\ -6 & 28 & 15 \\ -8 & 31 & 20 \end{bmatrix}.$$

Notice that the cofactors are "transposed"; that is, the cofactors for entries in the same *row* of $\mathbf{A}$ are placed in the same *column* of $\mathcal{A}$. ■

The next theorem shows that the adjoint of $\mathbf{A}$ is "almost" an inverse for $\mathbf{A}$.

THEOREM 3.13

If $\mathbf{A}$ is an $n \times n$ matrix with adjoint matrix $\mathcal{A}$, then

$$\mathbf{A}\mathcal{A} = \mathcal{A}\mathbf{A} = (|\mathbf{A}|)\mathbf{I}_n.$$

Proof of Theorem 3.13

Let $\mathbf{D} = \mathbf{A}\mathcal{A}$. The $(i, j)^{th}$ entry of $\mathbf{D} = d_{ij} = (i^{th}$ row of $\mathbf{A}) \bullet (j^{th}$ column of $\mathcal{A}) = a_{i1}\mathcal{A}_{j1} + a_{i2}\mathcal{A}_{j2} + \cdots + a_{in}\mathcal{A}_{jn}$, because the jth column entries of $\mathcal{A}$ are $\mathcal{A}_{j1}, \mathcal{A}_{j2}, \ldots, \mathcal{A}_{jn}$, respectively. Then, by Theorem 3.11, $d_{ij} = |\mathbf{A}|$ if $i = j$ and $d_{ij} = 0$ if $i \neq j$. Thus $\mathbf{D} = |\mathbf{A}|\mathbf{I}_n$. A similar argument for $\mathcal{A}\mathbf{A}$ concludes the proof (using Corollary 3.12). ■

Example 6

Consider the matrix from Example 5,

$$\mathbf{A} = \begin{bmatrix} 5 & -2 & 1 \\ 0 & 4 & -3 \\ 2 & -7 & 6 \end{bmatrix},$$

whose adjoint is

$$\mathcal{A} = \begin{bmatrix} 3 & 5 & 2 \\ -6 & 28 & 15 \\ -8 & 31 & 20 \end{bmatrix}.$$

$$\text{Then} \qquad \mathbf{A}\mathcal{A} = \mathcal{A}\mathbf{A} = \begin{bmatrix} 19 & 0 & 0 \\ 0 & 19 & 0 \\ 0 & 0 & 19 \end{bmatrix} = |\mathbf{A}|\mathbf{I}_3,$$

as predicted by Theorem 3.13, since $|\mathbf{A}| = 19$ (see Example 3). ∎

Calculating Inverses with the Adjoint Matrix

We have seen that for any $n \times n$ matrix $\mathbf{A}$ and its associated adjoint $\mathcal{A}$, we have $\mathbf{A}\mathcal{A} = (|\mathbf{A}|)\mathbf{I}_n$. Hence, if $|\mathbf{A}| \neq 0$, we can divide by the scalar $|\mathbf{A}|$ to obtain $(1/|\mathbf{A}|)(\mathbf{A}\mathcal{A}) = \mathbf{I}_n$. But then $\mathbf{A}((1/|\mathbf{A}|)\mathcal{A}) = \mathbf{I}_n$. Therefore, the scalar multiple $1/|\mathbf{A}|$ of the adjoint $\mathcal{A}$ must be the inverse matrix of $\mathbf{A}$. Thus we have proved the following:

COROLLARY 3.14

If $\mathbf{A}$ is a nonsingular $n \times n$ matrix with adjoint $\mathcal{A}$, then $\mathbf{A}^{-1} = (1/|\mathbf{A}|)\mathcal{A}$.

This corollary gives an algebraic formula for the inverse of a matrix (when it exists).

Example 7

Let

$$\mathbf{B} = \begin{bmatrix} -2 & 0 & -3 \\ 0 & 1 & 0 \\ 0 & 0 & 4 \end{bmatrix}.$$

The adjoint matrix for $\mathbf{B}$ is

$$\mathcal{B} = \begin{bmatrix} \mathcal{B}_{11} & \mathcal{B}_{21} & \mathcal{B}_{31} \\ \mathcal{B}_{12} & \mathcal{B}_{22} & \mathcal{B}_{32} \\ \mathcal{B}_{13} & \mathcal{B}_{23} & \mathcal{B}_{33} \end{bmatrix},$$

where each $\mathcal{B}_{ij}$ (for $1 \leq i, j \leq 3$) is the $(i, j)^{th}$ cofactor of $\mathbf{B}$. But these cofactors are easily calculated (try it), and we have

$$\mathcal{B} = \begin{bmatrix} 4 & 0 & 3 \\ 0 & -8 & 0 \\ 0 & 0 & -2 \end{bmatrix}.$$

Now, since $|\mathbf{B}| = -8$ (because $\mathbf{B}$ is upper triangular),

$$\mathbf{B}^{-1} = \frac{1}{|\mathbf{B}|}\mathcal{B} = -\frac{1}{8}\begin{bmatrix} 4 & 0 & 3 \\ 0 & -8 & 0 \\ 0 & 0 & -2 \end{bmatrix} = \begin{bmatrix} -\frac{1}{2} & 0 & -\frac{3}{8} \\ 0 & 1 & 0 \\ 0 & 0 & \frac{1}{4} \end{bmatrix}.$$ ∎

Finding the inverse by row reduction is usually quicker than using the adjoint. However, Corollary 3.14 is often a useful tool for proving other results about inverses.

Cramer's Rule

There is an explicit formula, known as **Cramer's Rule**, for the solution to a system of n equations and n variables when it is unique:

THEOREM 3.15: Cramer's Rule

Let $\mathbf{AX} = \mathbf{B}$ be a system of n equations in n variables. Let $\mathbf{A}_i$, for $1 \le i \le n$, be defined as the $n \times n$ matrix obtained by replacing the ith column of $\mathbf{A}$ with $\mathbf{B}$. Then, if $|\mathbf{A}| \neq 0$, the entries of the unique solution $\mathbf{X}$ of $\mathbf{AX} = \mathbf{B}$ are given by

$$x_1 = \frac{|\mathbf{A}_1|}{|\mathbf{A}|}, \quad x_2 = \frac{|\mathbf{A}_2|}{|\mathbf{A}|}, \quad \ldots, \quad x_n = \frac{|\mathbf{A}_n|}{|\mathbf{A}|}.$$

Cramer's Rule cannot be used for a system $\mathbf{AX} = \mathbf{B}$ in which $|\mathbf{A}| = 0$. (Why?) It is frequently used on systems with three equations and three variables having a unique solution, because the determinants involved are relatively easy to calculate by hand.

Example 8

We will solve

$$\begin{cases} 5x_1 - 3x_2 - 10x_3 = -9 \\ 2x_1 + 2x_2 - 3x_3 = 4 \\ -3x_1 - x_2 + 5x_3 = -1 \end{cases}$$

using Cramer's Rule. The coefficient matrix is

$$\mathbf{A} = \begin{bmatrix} 5 & -3 & -10 \\ 2 & 2 & -3 \\ -3 & -1 & 5 \end{bmatrix},$$

and it is easy to compute that $|\mathbf{A}| = -2$. Let

$$\mathbf{A}_1 = \begin{bmatrix} \boxed{-9} & -3 & -10 \\ \boxed{4} & 2 & -3 \\ \boxed{-1} & -1 & 5 \end{bmatrix}, \quad \mathbf{A}_2 = \begin{bmatrix} 5 & \boxed{-9} & -10 \\ 2 & \boxed{4} & -3 \\ -3 & \boxed{-1} & 5 \end{bmatrix}, \quad \text{and} \quad \mathbf{A}_3 = \begin{bmatrix} 5 & -3 & \boxed{-9} \\ 2 & 2 & \boxed{4} \\ -3 & -1 & \boxed{-1} \end{bmatrix}.$$

The matrix $\mathbf{A}_1$ is identical to $\mathbf{A}$, except in the first column, where the entries are taken from the right-hand side of the given system. $\mathbf{A}_2$ and $\mathbf{A}_3$ are created in an analogous manner. It is easy to compute that $|\mathbf{A}_1| = 8$, $|\mathbf{A}_2| = -6$, and $|\mathbf{A}_3| = 4$.

Therefore, $x_1 = |\mathbf{A}_1|/|\mathbf{A}| = 8/(-2) = -4$, $x_2 = |\mathbf{A}_2|/|\mathbf{A}| = -6/(-2) = 3$, and $x_3 = |\mathbf{A}_3|/|\mathbf{A}| = 4/(-2) = -2$. Hence, the unique solution to the given system is $(x_1, x_2, x_3) = (-4, 3, -2)$. ■

Notice that solving the system in Example 8 essentially amounts to calculating the four determinants $|\mathbf{A}|$, $|\mathbf{A}_1|$, $|\mathbf{A}_2|$, and $|\mathbf{A}_3|$.

Proof of Theorem 3.15

The proof for $n = 1$ is trivial, so we give a proof only for $n \geq 2$. Because $|\mathbf{A}| \neq 0$, we have $\mathbf{X} = \mathbf{A}^{-1}\mathbf{B} = \big((1/|\mathbf{A}|)\mathcal{A}\big)\mathbf{B} = (1/|\mathbf{A}|)(\mathcal{A}\mathbf{B})$, by Corollary 3.14. Thus, the kth entry of $\mathbf{X}$ is $(1/|\mathbf{A}|)(b_1\mathcal{A}_{1k} + b_2\mathcal{A}_{2k} + \cdots + b_n\mathcal{A}_{nk})$, for $1 \leq k \leq n$.

Because the kth column cofactors of $\mathbf{A}$ and $\mathbf{A}_k$ agree (why?), the expression $b_1\mathcal{A}_{1k} + b_2\mathcal{A}_{2k} + \cdots + b_n\mathcal{A}_{nk}$ equals the cofactor expansion of the matrix $\mathbf{A}_k$ (as defined in Cramer's Rule) along its kth column. Therefore, by Corollary 3.12, we have $b_1\mathcal{A}_{1k} + b_2\mathcal{A}_{2k} + \cdots + b_n\mathcal{A}_{nk} = |\mathbf{A}_k|$, for $1 \leq k \leq n$. Hence, the kth entry of $\mathbf{X}$ is $|\mathbf{A}_k|/|\mathbf{A}|$. ■

Exercises—Section 3.3

1. Calculate the indicated minors for each given matrix:

★(a) $|\mathbf{A}_{21}|$, for $\mathbf{A} = \begin{bmatrix} -2 & 4 & 3 \\ 3 & -1 & 6 \\ 5 & -2 & 4 \end{bmatrix}$

(b) $|\mathbf{B}_{34}|$, for $\mathbf{B} = \begin{bmatrix} 0 & 2 & -3 & 1 \\ 1 & 4 & 2 & -1 \\ 3 & -2 & 4 & 0 \\ 4 & -1 & 1 & 0 \end{bmatrix}$

★(c) $|\mathbf{C}_{42}|$, for $\mathbf{C} = \begin{bmatrix} -3 & 3 & 0 & 5 \\ 2 & 1 & -1 & 4 \\ 6 & -3 & 4 & 0 \\ -1 & 5 & 1 & -2 \end{bmatrix}$

2. Calculate the indicated cofactors for each given matrix:

★(a) $\mathcal{A}_{22}$, for $\mathbf{A} = \begin{bmatrix} 4 & 1 & -3 \\ 0 & 2 & -2 \\ 9 & 14 & -7 \end{bmatrix}$

(b) $\mathcal{B}_{23}$, for $\mathbf{B} = \begin{bmatrix} -9 & 6 & 7 \\ 2 & -1 & 0 \\ 4 & 3 & -8 \end{bmatrix}$

★(c) $\mathscr{C}_{43}$, for $\mathbf{C} = \begin{bmatrix} -5 & 2 & 2 & 13 \\ -8 & 2 & -5 & 22 \\ -6 & -3 & 0 & -16 \\ 4 & -1 & 7 & -8 \end{bmatrix}$

★(d) $\mathscr{D}_{12}$, for $\mathbf{D} = \begin{bmatrix} x+1 & x & x-7 \\ x-4 & x+5 & x-3 \\ x-1 & x & x+2 \end{bmatrix}$

3. Calculate the adjoint matrix for each of the following by finding the associated cofactor for each entry. Then use the adjoint to find the inverse of the original matrix (if it exists).

★(a) $\begin{bmatrix} 14 & -1 & -21 \\ 2 & 0 & -3 \\ 20 & -2 & -33 \end{bmatrix}$ (b) $\begin{bmatrix} -15 & -6 & -2 \\ 5 & 3 & 2 \\ 5 & 6 & 5 \end{bmatrix}$ ★(c) $\begin{bmatrix} -2 & 1 & 0 & -1 \\ 7 & -4 & 1 & 4 \\ -14 & 11 & -2 & -8 \\ -12 & 10 & -2 & -7 \end{bmatrix}$

(d) $\begin{bmatrix} -4 & 0 & 0 \\ -3 & 2 & 0 \\ 0 & 0 & 3 \end{bmatrix}$ ★(e) $\begin{bmatrix} 3 & -1 & 0 \\ 0 & -3 & 2 \\ 0 & 0 & -1 \end{bmatrix}$ (f) $\begin{bmatrix} 2 & 1 & 0 & 0 \\ 0 & -1 & 1 & 0 \\ 0 & 0 & 1 & -1 \\ 0 & 0 & 0 & -2 \end{bmatrix}$

4. For a general 4×4 matrix $\mathbf{A}$, write out the formula for $|\mathbf{A}|$ using a cofactor expansion along the indicated row or column:

★(a) Third row (b) First row ★(c) Fourth column (d) First column

5. Find the determinant of each of the following matrices by performing a cofactor expansion along the indicated row or column:

★(a) Second row of $\begin{bmatrix} 2 & -1 & 4 \\ 0 & 3 & -2 \\ 5 & -2 & -3 \end{bmatrix}$ (b) Third column of $\begin{bmatrix} 10 & -2 & 7 \\ 3 & 2 & -8 \\ 6 & 5 & -2 \end{bmatrix}$

★(c) First column of $\begin{bmatrix} 4 & -2 & 3 \\ 5 & -1 & -2 \\ 3 & 3 & 2 \end{bmatrix}$ (d) Fourth row of $\begin{bmatrix} 4 & -2 & 0 & -1 \\ -1 & 3 & -3 & 2 \\ 2 & 4 & -4 & -3 \\ 3 & 6 & 0 & -2 \end{bmatrix}$

6. Use Cramer's Rule to solve each of the following systems:

★(a) $\begin{cases} 3x_1 - x_2 - x_3 = -8 \\ 2x_1 - x_2 - x_3 = -4 \\ -9x_1 + x_2 = 39 \end{cases}$

(b) $\begin{cases} -2x_1 + 5x_2 - 4x_3 = -3 \\ 3x_1 - 3x_2 + 4x_3 = 6 \\ 2x_1 - x_2 + 2x_3 = 5 \end{cases}$

(c) $\begin{cases} -5x_1 + 6x_2 + 2x_3 = -16 \\ 3x_1 - 5x_2 - 2x_3 = 8 \\ -3x_1 + 3x_2 + x_3 = -11 \end{cases}$

★(d) $\begin{cases} -5x_1 + 2x_2 - 2x_3 + x_4 = -10 \\ 2x_1 - x_2 + 2x_3 - x_4 = -3 \\ 5x_1 - 2x_2 + 3x_3 - x_4 = 7 \\ -6x_1 + 2x_2 - 2x_3 + x_4 = -14 \end{cases}$

★7. Let **A** and **B** be nonsingular matrices of the same size, with adjoints $\mathcal{A}$ and $\mathcal{B}$. Express $(\mathbf{AB})^{-1}$ in terms of $\mathcal{A}$, $\mathcal{B}$, $|\mathbf{A}|$, and $|\mathbf{B}|$.

8. If all entries of a (square) matrix **A** are integers and $|\mathbf{A}| = \pm 1$, show that all entries of $\mathbf{A}^{-1}$ are integers.

9. If **A** is an $n \times n$ matrix with adjoint $\mathcal{A}$, show that $\mathbf{A}\mathcal{A} = \mathbf{O}_n$ if and only if **A** is singular.

10. Let **A** be an $n \times n$ matrix with adjoint $\mathcal{A}$.
(a) Show that the adjoint of $\mathbf{A}^T$ is $\mathcal{A}^T$.
(b) Show that the adjoint of $k\mathbf{A}$ is $k^{n-1}\mathcal{A}$, for each real number k.

11. (a) Prove that if **A** is symmetric with adjoint matrix $\mathcal{A}$, then $\mathcal{A}$ is symmetric. (Hint: Show that the cofactors $\mathcal{A}_{ij}$ and $\mathcal{A}_{ji}$ of **A** are equal.)
★(b) Give an example to show that part (a) is not necessarily true when *symmetric* is replaced by *skew-symmetric*.

12. If **A** is nonsingular upper triangular, show that $\mathbf{A}^{-1}$ is also upper triangular. (Hint: Use Corollary 3.14.)

13. Let **A** be a matrix with adjoint $\mathcal{A}$.
(a) Prove that if **A** is singular, then $\mathcal{A}$ is singular. (Hint: See Exercise 9.)
(b) Prove that $|\mathcal{A}| = |\mathbf{A}|^{n-1}$. (Hint: Consider the cases $|\mathbf{A}| = 0$ and $|\mathbf{A}| \neq 0$.)

14. Recall that, in Exercise 15 of Section 3.1, we introduced the 3×3 Vandermonde matrix. For $n \geq 3$, the **general $n \times n$ Vandermonde matrix** has this form:

$$\mathbf{V}_n = \begin{bmatrix} 1 & 1 & 1 & \cdots & 1 \\ x_1 & x_2 & x_3 & \cdots & x_n \\ x_1^2 & x_2^2 & x_3^2 & \cdots & x_n^2 \\ \vdots & \vdots & \vdots & \ddots & \vdots \\ x_1^{n-1} & x_2^{n-1} & x_3^{n-1} & \cdots & x_n^{n-1} \end{bmatrix}.$$

If $x_1, x_2, \ldots, x_n$ are distinct real numbers, show that

$$|\mathbf{V}| = (-1)^{n+1}(x_1 - x_n)(x_2 - x_n)\cdots(x_{n-1} - x_n)|\mathbf{V}_{n-1}|.$$

(Hint: Subtract the last column from every other column, and use cofactor expansion along the first row to show that $|\mathbf{V}_n|$ is equal or opposite to the determinant of a matrix $\mathbf{W}$ of size $(n-1) \times (n-1)$. Next, divide each column of $\mathbf{W}$ by the first element of that column, using the "column" version of part (1) of Theorem 3.5 to pull out the factors $x_1 - x_n, x_2 - x_n, \ldots, x_{n-1} - x_n$. [Note that $(x_1^k - x_n^k)/(x_1 - x_n) = x_1^{k-1} + x_1^{k-2}x_n + x_1^{k-3}x_n^2 + \cdots + x_1 x_n^{k-2} + x_n^{k-1}$.] Finally, create $|\mathbf{V}_{n-1}|$ from the resulting matrix by going through each row from 2 to n in *reverse order* and adding $-x_n$ times the previous row to it.)

Summary of Techniques

We conclude this chapter by summarizing many of the techniques developed in Chapters 2 and 3.

Note: Wherever possible, these computations should be done using available software packages if the calculations cannot easily be done by hand.

Techniques for Solving a System $\mathbf{AX} = \mathbf{B}$ of m Linear Equations in n Unknowns

- **Gauss-Jordan row reduction**: Use row operations of types (I), (II), and (III) to find a matrix in reduced row echelon form for $[\mathbf{A}|\mathbf{B}]$. Advantages: easily computerized; relatively efficient. (Section 2.1)
- **Multiplication by inverse matrix:** Use for the case $m = n$ and $|\mathbf{A}| \neq 0$. $\mathbf{X} = \mathbf{A}^{-1}\mathbf{B}$. Advantage: easily computerized. Disadvantage: $\mathbf{A}^{-1}$ must be known or calculated first, and therefore the method is useful only when there are several systems to be solved with the same coefficient matrix $\mathbf{A}$. (Section 2.4)
- **Cramer's Rule**: Use for the case $m = n$ and $|\mathbf{A}| \neq 0$. The solution is $x_1 = |\mathbf{A}_1|/|\mathbf{A}|$, $x_2 = |\mathbf{A}_2|/|\mathbf{A}|, \ldots, x_n = |\mathbf{A}_n|/|\mathbf{A}|$, where $\mathbf{A}_i$ (for $1 \leq i \leq n$) and $\mathbf{A}$ are identical except that the ith column of $\mathbf{A}_i$ equals $\mathbf{B}$. Advantage: can be computerized. Disadvantage: efficient only for small systems because it involves calculating $n + 1$ determinants of size n. (Section 3.3)

Other techniques for solving systems are discussed in Chapter 9. Among these are Gaussian elimination, **LDU** decomposition, and iterative methods, such as the Gauss-Seidel and Jacobi techniques.

Also remember: If $m < n$ and $\mathbf{B} = \mathbf{O}$ (homogeneous case), then there are an infinite number of solutions to $\mathbf{AX} = \mathbf{B}$.

Techniques for Finding the Inverse (If It Exists) of an $n \times n$ Matrix A

- **2×2 case**: The inverse of $\begin{bmatrix} a & b \\ c & d \end{bmatrix}$ exists iff $ad - bc \neq 0$. In that case, the inverse is given by $\left(1/(ad - bc)\right)\begin{bmatrix} d & -b \\ -c & a \end{bmatrix}$. (Section 2.4)
- **Row reduction**: Row reduce $[\mathbf{A}|\mathbf{I}_n]$ to $[\mathbf{I}_n|\mathbf{A}^{-1}]$ (where $\mathbf{A}^{-1}$ does not exist if the process stops prematurely). Advantages: easily computerized; relatively efficient. (Section 2.4)
- **Adjoint matrix**: $\mathbf{A}^{-1} = \left(1/|\mathbf{A}|\right)\mathcal{A}$, where $\mathcal{A}$ is the adjoint matrix of $\mathbf{A}$. Advantage: gives an algebraic formula for $\mathbf{A}^{-1}$. Disadvantage: not very efficient, because $|\mathbf{A}|$ and all n^2 cofactors of $\mathbf{A}$ must be calculated first. (Section 3.3)

Techniques for Finding the Determinant of an $n \times n$ Matrix A

- **2×2 case**: $|\mathbf{A}| = ad - bc$ if $\mathbf{A} = \begin{bmatrix} a & b \\ c & d \end{bmatrix}$. (Sections 2.4 and 3.1)
- **3×3 case**: Use the "basketweaving" technique. (Section 3.1)
- **Row reduction**: Row reduce to an upper triangular form matrix, keeping track of the effect of each row operation on the determinant. Advantages: easily computerized; relatively efficient. (Section 3.2)
- **Cofactor expansion**: Multiply each element along any row or column of $\mathbf{A}$ by its cofactor and sum the results. Advantage: useful for matrices with many zero entries. Disadvantage: not as fast as row reduction. (Section 3.3)

Also remember: $|\mathbf{A}| = 0$ if $\mathbf{A}$ is row equivalent to a matrix with a row or column of zeroes or with two equal rows or with two equal columns.

4

Finite Dimensional Vector Spaces

In Chapter 1, you saw that the operations of addition and scalar multiplication on the set $\mathcal{M}_{mn}$ possessed many of the same algebraic properties as addition and scalar multiplication on the set $\mathbb{R}^n$. In fact, there are many other sets with operations that share these same properties. Instead of studying these sets individually, it is more profitable to study them as a group.

In this chapter, we define vector spaces to be algebraic structures with operations having properties similar to those of addition and scalar multiplication on $\mathbb{R}^n$ and $\mathcal{M}_{mn}$. We then establish many important results relating to vector spaces. Because we are studying vector spaces as a class, this chapter is more abstract than previous chapters have been. However, by working in this more general setting, we generate theorems that apply to all vector spaces, not just $\mathbb{R}^n$ and $\mathcal{M}_{mn}$.

4.1 Introduction to Vector Spaces

Definition of a Vector Space

In Theorems 1.3 and 1.10, we proved eight properties of addition and scalar multiplication in $\mathbb{R}^n$ and $\mathcal{M}_{mn}$. These properties are important, because all other results involving these operations can be derived from them. We now introduce other sets, called **vector spaces**,† that have operations of addition and scalar multiplication with these same eight properties.

†We actually define what are called *real vector spaces*, rather than just vector spaces. The word *real* refers to the fact that the scalars involved in the scalar multiplication are real numbers. In Chapter 7, we will consider *complex vector spaces*, where the scalars are complex numbers. Other types of vector spaces involve more general sets of scalars; they are not considered in this book.

DEFINITION

A **vector space** is a set $\mathcal{V}$ together with an operation called **vector addition** (a rule for adding two elements of $\mathcal{V}$ to obtain a third element of $\mathcal{V}$) and another operation called **scalar multiplication** (a rule for multiplying a real number times an element of $\mathcal{V}$ to obtain a second element of $\mathcal{V}$) on which the following eight properties hold:

For every $\mathbf{u}$, $\mathbf{v}$, and $\mathbf{w}$ in $\mathcal{V}$, and for every a and b in $\mathbb{R}$:

(1)	$\mathbf{u} + \mathbf{v} = \mathbf{v} + \mathbf{u}$	Commutative Law of Addition
(2)	$\mathbf{u} + (\mathbf{v} + \mathbf{w}) = (\mathbf{u} + \mathbf{v}) + \mathbf{w}$	Associative Law of Addition
(3)	There is an element $\mathbf{0}$ of $\mathcal{V}$ so that for every $\mathbf{z}$ in $\mathcal{V}$ we have $\mathbf{0} + \mathbf{z} = \mathbf{z} = \mathbf{z} + \mathbf{0}$.	Existence of Identity Element for Addition
(4)	There is an element $-\mathbf{u}$ in $\mathcal{V}$ such that $\mathbf{u} + (-\mathbf{u}) = \mathbf{0} = (-\mathbf{u}) + \mathbf{u}$.	Existence of Additive Inverse
(5)	$a(\mathbf{u} + \mathbf{v}) = (a\mathbf{u}) + (a\mathbf{v})$	Distributive Laws
(6)	$(a + b)\mathbf{u} = (a\mathbf{u}) + (b\mathbf{u})$	for Scalar Multiplication over Addition
(7)	$(ab)\mathbf{u} = a(b\mathbf{u})$	Associativity of Scalar Multiplication
(8)	$1\mathbf{u} = \mathbf{u}$	Identity Element for Scalar Multiplication

The elements of a vector space $\mathcal{V}$ are called **vectors**.

We have used the standard plus sign, $+$, to indicate both vector addition and the sum of real numbers, two different operations. All sums in properties (1), (2), (3), (4), and (5) are vector sums. In property (6), the $+$ on the left side of the equation represents addition of real numbers; the $+$ on the right side stands for the sum of two vectors. In property (7), the left side of the equation contains one product of real numbers, ab, and one instance of scalar multiplication, ab times $\mathbf{u}$. The right side of property (7) involves two scalar multiplications—first b times $\mathbf{u}$, then a times the vector $b\mathbf{u}$. Usually you can tell from the context which type of operation to use.

In any vector space, the additive identity element is unique (see Exercise 13). Similarly, the additive inverse of each vector in a vector space is unique (see Exercise 14).

Examples of Vector Spaces

Example 1

Let $\mathcal{V} = \mathbb{R}^n$, with addition and scalar multiplication of n-vectors as defined in Section 1.1. By Theorem 1.3, $\mathcal{V} = \mathbb{R}^n$ is a vector space.

Similarly, consider $\mathcal{M}_{mn}$, the set of $m \times n$ matrices. Theorem 1.10 shows that $\mathcal{M}_{mn}$ is a vector space with the usual operations of matrix addition and scalar multiplication. ■

The vector spaces $\mathbb{R}^n$ and $\mathcal{M}_{mn}$ (with the usual operations of addition and scalar multiplication) are representative of most of the vector spaces we consider here. Keep $\mathbb{R}^n$ and $\mathcal{M}_{mn}$ in mind as examples later when we consider theorems involving general vector spaces.

Some vector spaces can have additional operations defined on them. For example, $\mathbb{R}^n$ has the dot product, and $\mathcal{M}_{mn}$ has matrix multiplication and the transpose. But these additional structures are not shared by all vector spaces because they are not included in the definition. We cannot assume the existence of any additional operations in a general discussion of vector spaces. In particular,

> There is no such operation as multiplication or division of one vector by another for general vector spaces.

The only general vector space operation that combines two *vectors* is vector addition.

Example 2

The set $\mathcal{V} = \{\mathbf{0}\}$ is a vector space with the rules for addition and multiplication given by $\mathbf{0} + \mathbf{0} = \mathbf{0}$ and $a\mathbf{0} = \mathbf{0}$ for every scalar (real number) a. It is easy to verify the eight properties for this vector space. This vector space is called the **trivial vector space**, and no smaller vector space is possible. (Why?) ■

The definition of a vector space requires that both the operations of vector addition and scalar multiplication always produce an element of the vector space as a result. These are called the **closure properties** of a vector space. The closure properties must always be verified along with the other eight properties whenever we prove that a set with its operations is a vector space. We did not mention the closure properties in the first two examples because they obviously hold in each case. However, they are not so obvious in the next example, and so, from now on, we will check whether the closure properties hold.

Example 3

Consider $\mathbb{R}^3$ as the set of 3-vectors in three-dimensional space, all with initial points at the origin. Let $\mathcal{P}$ be any plane containing the origin. Consider the set $\mathcal{V}$ of all

3-vectors whose terminal point lies in $\mathcal{P}$ (meaning that the vectors in $\mathcal{V}$ lie entirely in the plane $\mathcal{P}$ when drawn on a graph, since both the initial point and terminal point of each vector lie in $\mathcal{P}$). For example, in Figure 4.1, $\mathcal{P}$ is the plane containing the vectors $\mathbf{u}$ and $\mathbf{v}$ (elements of $\mathcal{V}$); $\mathbf{w}$ is not in $\mathcal{V}$ because its terminal point does not lie in $\mathcal{P}$. Let us prove that $\mathcal{V}$ is a vector space.

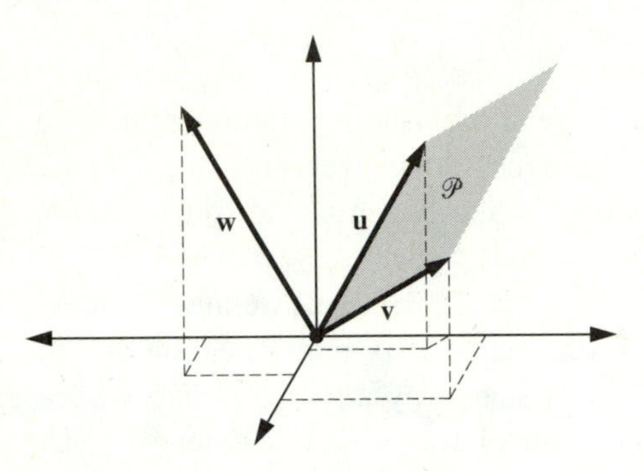

Figure 4.1
A plane $\mathcal{P}$ in $\mathbb{R}^3$ containing the origin

Before we verify the eight vector space properties in $\mathcal{V}$, we must first check the closure properties; that is, we must show that the sum of any two vectors in $\mathcal{V}$ is a vector in $\mathcal{V}$ and that any scalar multiple of a vector in $\mathcal{V}$ also lies in $\mathcal{V}$.

If $\mathbf{x}$ and $\mathbf{y}$ are elements of $\mathcal{V}$, then the parallelogram they form lies entirely in $\mathcal{P}$, because $\mathbf{x}$ and $\mathbf{y}$ do. Hence, the diagonal $\mathbf{x} + \mathbf{y}$ of this parallelogram also lies in $\mathcal{P}$, so $\mathbf{x} + \mathbf{y}$ is in $\mathcal{V}$. This observation verifies that $\mathcal{V}$ is closed under vector addition (that is, the closure property holds for vector addition). Notice that it is not enough to know that the sum of two 3-vectors in $\mathcal{V}$ produces another 3-vector. We have to show that the sum they produce is actually in the set $\mathcal{V}$.

Next consider scalar multiplication. If $\mathbf{x}$ is a vector in $\mathcal{V}$, then any scalar multiple of $\mathbf{x}$, $a\mathbf{x}$, is either parallel to $\mathbf{x}$ or equal to $\mathbf{0}$. Therefore, it lies in any plane through the origin that $\mathbf{x}$ does (in particular, $\mathcal{P}$). Hence, $a\mathbf{x}$ is in $\mathcal{V}$, and $\mathcal{V}$ is closed under scalar multiplication.

We now check that the eight vector space properties hold. Properties (1), (2), (5), (6), (7), and (8) are true for all vectors in $\mathbb{R}^3$ by Theorem 1.3 and thus are true for all vectors in $\mathcal{V}$, since $\mathcal{V} \subseteq \mathbb{R}^3$. However, properties (3) and (4) must be checked separately for $\mathcal{V}$ because they are *existence* properties. We know that the zero vector and additive inverses exist in $\mathbb{R}^3$, but are they in $\mathcal{V}$? First, we make sure that the zero vector is not outside $\mathcal{V}$. But $\mathbf{0} = [0, 0, 0]$ is in $\mathcal{V}$, because the plane $\mathcal{P}$ passes through the origin, thus proving property (3). Finally, we must be sure that the additive inverse of every vector in $\mathcal{V}$ is not outside $\mathcal{V}$. However, the opposite (additive inverse) of any vector lying in the plane $\mathcal{P}$ also lies in $\mathcal{P}$ and is therefore in $\mathcal{V}$, thus proving property (4). Hence, all eight properties and the closure properties are true, so $\mathcal{V}$ is a vector space. ■

Example 4

Let $\mathcal{P}_n$ be the set of polynomials of degree $\leq n$, with real coefficients. The vectors in $\mathcal{P}_n$ have the form $\mathbf{p} = a_n x^n + \cdots + a_1 x + a_0$ for some real numbers $a_0, a_1, \ldots, a_n$. We define addition of polynomials in the usual manner—that is, by adding corresponding coefficients. Then the sum of any two polynomials of degree $\leq n$ also has degree $\leq n$ and so is in $\mathcal{P}_n$. Thus, the closure property of addition holds. Similarly, if b is a real number and $\mathbf{p} = a_n x^n + \cdots + a_1 x + a_0$ is in $\mathcal{P}_n$, we define $b\mathbf{p}$ to be the polynomial $(ba_n)x^n + \cdots + (ba_1)x + ba_0$, which is also obviously in $\mathcal{P}_n$. Hence, the closure property of scalar multiplication holds. Then, if the eight vector space properties hold, $\mathcal{P}_n$ is a vector space under these operations. We verify properties (1), (3), and (4) of the definition and leave the others for you to check.

(1) Commutative Law of Addition: We must show that the order in which two vectors (polynomials) are added makes no difference. Now, by the commutative law of addition for real numbers,

$$\begin{aligned}(a_n x^n + \cdots + a_1 x + a_0) &+ (b_n x^n + \cdots + b_1 x + b_0)\\ &= (a_n + b_n)x^n + \cdots + (a_1 + b_1)x + (a_0 + b_0)\\ &= (b_n + a_n)x^n + \cdots + (b_1 + a_1)x + (b_0 + a_0)\\ &= (b_n x^n + \cdots + b_1 x + b_0) + (a_n x^n + \cdots + a_1 x + a_0).\end{aligned}$$

(3) Existence of Identity Element for Addition: The zero-degree polynomial $\mathbf{z} = 0x^n + \cdots + 0x + 0$ acts as the additive identity element $\mathbf{0}$. That is, adding $\mathbf{z}$ to any vector does not change the vector:

$$\begin{aligned}\mathbf{z} + (a_n x^n + \cdots + a_1 x + a_0) &= (0 + a_n)x^n + \cdots + (0 + a_1)x + (0 + a_0)\\ &= a_n x^n + \cdots + a_1 x + a_0.\end{aligned}$$

(4) Existence of Additive Inverse: We must show that each vector in $\mathcal{P}_n$ has an additive inverse in $\mathcal{P}_n$. Consider the vector $\mathbf{p} = a_n x^n + \cdots + a_1 x + a_0$ in $\mathcal{P}_n$. The vector $-\mathbf{p} = -(a_n x^n + \cdots + a_1 x + a_0) = (-a_n)x^n + \cdots + (-a_1)x + (-a_0)$ has the property that $\mathbf{p} + [-\mathbf{p}] = \mathbf{z}$, the zero vector, and so $-\mathbf{p}$ acts as the additive inverse of $\mathbf{p}$. Because $-\mathbf{p}$ is also in $\mathcal{P}_n$, we are done. ■

The vector space in Example 4 is similar to our prototype $\mathbb{R}^n$. For any polynomial in $\mathcal{P}_n$, consider the sequence of its $n + 1$ coefficients. This sequence completely describes that polynomial and can be thought of as an $(n + 1)$-vector. For example, a polynomial $a_2 x^2 + a_1 x + a_0$ in $\mathcal{P}_2$ can be described by the 3-vector $[a_2, a_1, a_0]$. In this way, the vector space $\mathcal{P}_2$ "resembles" the vector space $\mathbb{R}^3$, and in general, $\mathcal{P}_n$ "resembles" $\mathbb{R}^{n+1}$.

Example 5

The set $\mathcal{P}$ of all polynomials (of all degrees) is easily seen to be a vector space under the usual (term-by-term) operations of addition and scalar multiplication (see Exercise 17). ■

Example 6

Let $\mathcal{V}$ be the set of all real-valued functions defined on $\mathbb{R}$. For example, $\mathbf{f}(x) = \arctan(x)$ is in $\mathcal{V}$. We define addition of functions as usual: $\mathbf{h} = \mathbf{f} + \mathbf{g}$ is the function such that $\mathbf{h}(x) = \mathbf{f}(x) + \mathbf{g}(x)$, for every $x \in \mathbb{R}$. Similarly, if $a \in \mathbb{R}$ and $\mathbf{f}$ is in $\mathcal{V}$, we define the scalar multiple $\mathbf{h} = a\mathbf{f}$ to be the function such that $\mathbf{h}(x) = a\mathbf{f}(x)$, for every $x \in \mathbb{R}$. Clearly the closure properties hold for $\mathcal{V}$, because sums and scalar multiples of real-valued functions produce real-valued functions. To finish verifying that $\mathcal{V}$ is a vector space, we must check that the eight vector space properties hold.

Suppose that $\mathbf{f}$, $\mathbf{g}$, and $\mathbf{h}$ are in $\mathcal{V}$ and that a and b are real numbers.

Property (1): For every x in $\mathbb{R}$, we have $\mathbf{f}(x) + \mathbf{g}(x) = \mathbf{g}(x) + \mathbf{f}(x)$, by the commutative law of addition for real numbers, since $\mathbf{f}(x)$ and $\mathbf{g}(x)$ are both real numbers for each $x \in \mathbb{R}$. Because the formulas $\mathbf{f}(x) + \mathbf{g}(x)$ and $\mathbf{g}(x) + \mathbf{f}(x)$ agree for all $x \in \mathbb{R}$, each must represent the same function of x. Hence, $\mathbf{f} + \mathbf{g} = \mathbf{g} + \mathbf{f}$.

Property (2): For every $x \in \mathbb{R}$, $\mathbf{f}(x) + (\mathbf{g}(x) + \mathbf{h}(x)) = (\mathbf{f}(x) + \mathbf{g}(x)) + \mathbf{h}(x)$, by the associative law of addition for real numbers. Thus, $\mathbf{f} + (\mathbf{g} + \mathbf{h}) = (\mathbf{f} + \mathbf{g}) + \mathbf{h}$.

Property (3): Let $\mathbf{z}$ be the function given by $\mathbf{z}(x) = 0$ for every $x \in \mathbb{R}$. Then, for each x, $\mathbf{z}(x) + \mathbf{f}(x) = 0 + \mathbf{f}(x) = \mathbf{f}(x)$. Hence, $\mathbf{z} + \mathbf{f} = \mathbf{f}$. Similarly, $\mathbf{f} + \mathbf{z} = \mathbf{f}$, so $\mathbf{z}$ acts as the zero vector $\mathbf{0}$ in $\mathcal{V}$.

Property (4): Given $\mathbf{f}$ in $\mathcal{V}$, define $-\mathbf{f}$ by $[-\mathbf{f}](x) = -(\mathbf{f}(x))$ for every $x \in \mathbb{R}$. Then, for all x, $[-\mathbf{f}](x) + \mathbf{f}(x) = -(\mathbf{f}(x)) + \mathbf{f}(x) = 0$. Therefore, $[-\mathbf{f}] + \mathbf{f} = \mathbf{z}$, the zero vector, and so each function in $\mathcal{V}$ has an additive inverse that is also in $\mathcal{V}$.

Properties (5) and (6): For every $x \in \mathbb{R}$, $a(\mathbf{f}(x) + \mathbf{g}(x)) = a\mathbf{f}(x) + a\mathbf{g}(x)$ and $(a + b)\mathbf{f}(x) = a\mathbf{f}(x) + b\mathbf{f}(x)$, by the distributive laws for multiplication of real numbers over addition of real numbers. Hence, $a(\mathbf{f} + \mathbf{g}) = a\mathbf{f} + a\mathbf{g}$ and $(a + b)\mathbf{f} = a\mathbf{f} + b\mathbf{f}$.

Property (7): For every $x \in \mathbb{R}$, $(ab)\mathbf{f}(x) = a(b\mathbf{f}(x))$ follows from the associative law of multiplication for real numbers. Hence, $(ab)\mathbf{f} = a(b\mathbf{f})$.

Property (8): Since $1 \cdot \mathbf{f}(x) = \mathbf{f}(x)$, for every real number x, we have $1 \cdot \mathbf{f} = \mathbf{f}$ in $\mathcal{V}$. ■

Certain classes of functions provide more specialized examples of vector spaces. For example (see Exercise 18), under the same operations of addition and scalar multiplication of functions, both of the following are vector spaces:

The set of all *continuous* real-valued functions on $\mathbb{R}$

The set of all *differentiable* real-valued functions on $\mathbb{R}$

Such function spaces are often useful in solving problems in differential equations and other areas of analysis.

Two Unusual Vector Spaces

The next two examples place unusual operations on familiar sets to create new vector spaces.

Example 7

Let $\mathcal{V}$ be the set $\mathbb{R}^+$ of positive real numbers. This set is not a vector space under the usual operations of addition and scalar multiplication. (Why?) However, we can define new rules for addition and scalar multiplication to make $\mathcal{V}$ a vector space. In defining these new operations, we sometimes think of elements of $\mathbb{R}^+$ as abstract vectors (in which case we use boldface type, such as $\mathbf{v}$) or as the values on the positive real number line they represent (in which case we use italics, such as v).

To define "addition" on $\mathcal{V}$, we use *multiplication* of real numbers. That is,

$$\mathbf{v}_1 \oplus \mathbf{v}_2 = v_1 \cdot v_2$$

for every $\mathbf{v}_1$ and $\mathbf{v}_2$ in $\mathcal{V}$, where we use the symbol $\oplus$ for the "addition" operation on $\mathcal{V}$ to emphasize that this is not addition of real numbers. (The definition of a vector space states only that vector addition must be a rule for combining two vectors to yield a third vector so that properties (1) through (8) hold. There is no stipulation that vector addition be at all similar to ordinary addition of real numbers. You might think that this operation should be called something other than *addition*. However, most of our vector space terminology comes from the motivating example of $\mathbb{R}^n$, where the word *addition* is a natural choice for the name of the operation.)

We next define "scalar multiplication," $\odot$, on $\mathcal{V}$ by

$$a \odot \mathbf{v} = v^a$$

for every $a \in \mathbb{R}$ and $\mathbf{v} \in \mathcal{V}$.

It is easy to see that, if $\mathbf{v}_1$ and $\mathbf{v}_2$ are in $\mathcal{V}$ and a is in $\mathbb{R}$, then both $\mathbf{v}_1 \oplus \mathbf{v}_2$ and $a \odot \mathbf{v}_1$ are in $\mathcal{V}$, thus verifying the two closure properties.

To prove the other eight properties, we assume that $\mathbf{v}_1, \mathbf{v}_2, \mathbf{v}_3 \in \mathcal{V}$ and that $a, b \in \mathbb{R}$. We then have the following:

Property (1): $\mathbf{v}_1 \oplus \mathbf{v}_2 = v_1 \cdot v_2 = v_2 \cdot v_1$ (by the commutative law of multiplication for real numbers) $= \mathbf{v}_2 \oplus \mathbf{v}_1$.

Property (2): $\mathbf{v}_1 \oplus (\mathbf{v}_2 \oplus \mathbf{v}_3) = \mathbf{v}_1 \oplus (v_2 \cdot v_3) = v_1 \cdot (v_2 \cdot v_3) = (v_1 \cdot v_2) \cdot v_3$ (by the associative law of multiplication for real numbers) $= (\mathbf{v}_1 \oplus \mathbf{v}_2) \cdot v_3 = (\mathbf{v}_1 \oplus \mathbf{v}_2) \oplus \mathbf{v}_3$.

Property (3): The number 1 in $\mathbb{R}^+$ acts as the zero vector $\mathbf{0}$ in $\mathcal{V}$. (Why?)

Property (4): The additive inverse of $\mathbf{v}$ in $\mathcal{V}$ is the positive real number $(1/v)$, because $\mathbf{v} \oplus (1/v) = v \cdot (1/v) = 1$, the zero vector in $\mathcal{V}$.

Property (5): $a \odot (\mathbf{v}_1 \oplus \mathbf{v}_2) = a \odot (v_1 \cdot v_2) = (v_1 \cdot v_2)^a = v_1^a \cdot v_2^a = (a \odot \mathbf{v}_1) \cdot (a \odot \mathbf{v}_2) = (a \odot \mathbf{v}_1) \oplus (a \odot \mathbf{v}_2)$.

Property (6): $(a + b) \odot \mathbf{v} = v^{a+b} = v^a \cdot v^b = (a \odot \mathbf{v}) \cdot (b \odot \mathbf{v}) = (a \odot \mathbf{v}) \oplus (b \odot \mathbf{v})$.

Property (7): $(ab) \odot \mathbf{v} = v^{ab} = (v^b)^a = (b \odot \mathbf{v})^a = a \odot (b \odot \mathbf{v})$.

Property (8): $1 \odot \mathbf{v} = v^1 = \mathbf{v}$. ■

Example 8

Let $\mathcal{V} = \mathbb{R}^2$, with addition defined by

$$[x, y] \oplus [w, z] = [x + w + 1, y + z - 2]$$

and scalar multiplication defined by

$$a \odot [x, y] = [ax + a - 1, ay - 2a + 2].$$

The closure properties hold for these operations. (Why?) In fact, $\mathcal{V}$ forms a vector space because the eight vector space properties also hold. We verify properties (2), (3), (4), and (5); you should check the others for yourself.

Property (2):

$$\begin{aligned} [x, y] \oplus ([u, v] \oplus [w, z]) &= [x, y] \oplus [u + w + 1, v + z - 2] \\ &= [x + u + w + 2, y + v + z - 4] \\ &= [x + u + 1, y + v - 2] \oplus [w, z] \\ &= ([x, y] \oplus [u, v]) \oplus [w, z]. \end{aligned}$$

Property (3): The vector $[-1, 2]$ acts as the zero vector, since $[x, y] \oplus [-1, 2] = [x + (-1) + 1, y + 2 - 2] = [x, y]$.

Property (4): The additive inverse of $[x, y]$ is $[-x - 2, -y + 4]$, because $[x, y] \oplus [-x - 2, -y + 4] = [x - x - 2 + 1, y - y + 4 - 2] = [-1, 2]$, the zero vector in $\mathcal{V}$.

Property (5):

$$\begin{aligned} (a + b) \odot [x, y] &= [(a + b)x + (a + b) - 1, (a + b)y - 2(a + b) + 2] \\ &= [(ax + a - 1) + (bx + b - 1) + 1, \\ &\quad (ay - 2a + 2) + (by - 2b + 2) - 2] \\ &= [ax + a - 1, ay - 2a + 2] \oplus [bx + b - 1, by - 2b + 2] \\ &= (a \odot [x, y]) \oplus (b \odot [x, y]). \end{aligned}$$ ■

Some Elementary Properties of Vector Spaces

The next theorem contains several simple results regarding vector spaces. Although these are obviously true in the most familiar examples, we must prove them in general before we know they hold in every possible vector space.

THEOREM 4.1

Let $\mathcal{V}$ be a vector space. Then, for every vector $\mathbf{v}$ in $\mathcal{V}$ and every real number a, we have

(1) $a\mathbf{0} = \mathbf{0}$	Any scalar multiple of the zero vector yields the zero vector.
(2) $0\mathbf{v} = \mathbf{0}$	The scalar zero multiplied by any vector yields the zero vector.
(3) $(-1)\mathbf{v} = -\mathbf{v}$	The scalar -1 multiplied by any vector yields the additive inverse of that vector.
(4) If $a\mathbf{v} = \mathbf{0}$, then, $a = 0$ or $\mathbf{v} = \mathbf{0}$	If a scalar multiplication yields the zero vector, then either the scalar is zero, or the vector is the zero vector, or both.

This theorem must be proved directly from the properties in the definition of a vector space because at this point we have no other known facts about general vector spaces. We prove parts (1), (3), and (4). The proof of part (2) is similar to the proof of part (1) (see Exercise 20).

Proof of Theorem 4.1 (abridged):

Part (1):

$$
\begin{aligned}
a\mathbf{0} &= a\mathbf{0} + \mathbf{0} && \text{by property (3)}\\
&= a\mathbf{0} + (a\mathbf{0} + (-[a\mathbf{0}])) && \text{by property (4)}\\
&= (a\mathbf{0} + a\mathbf{0}) + (-[a\mathbf{0}]) && \text{by property (2)}\\
&= a(\mathbf{0} + \mathbf{0}) + (-[a\mathbf{0}]) && \text{by property (5)}\\
&= a\mathbf{0} + (-[a\mathbf{0}]) && \text{by property (3)}\\
&= \mathbf{0}. && \text{by property (4)}
\end{aligned}
$$

Part (3): This part justifies the notation for the additive inverse in property (4) of the definition of a vector space. After we have proved part (3), we will no longer distinguish between $-\mathbf{v}$ and $(-1)\mathbf{v}$.

$$
\begin{aligned}
-\mathbf{v} &= (-\mathbf{v}) + \mathbf{0} && \text{by property (3)}\\
&= (-\mathbf{v}) + 0\mathbf{v} && \text{by part (2) of Theorem 4.1}\\
&= (-\mathbf{v}) + (1 + (-1))\mathbf{v} && \\
&= (-\mathbf{v}) + (1\mathbf{v} + (-1)\mathbf{v}) && \text{by property (6)}\\
&= ((-\mathbf{v}) + \mathbf{v}) + (-1)\mathbf{v} && \text{by properties (2) and (8)}\\
&= \mathbf{0} + (-1)\mathbf{v} && \text{by property (4)}\\
&= (-1)\mathbf{v}. && \text{by property (3)}
\end{aligned}
$$

Part (4): Assume that $a\mathbf{v} = \mathbf{0}$ and $a \neq 0$. We must show that $\mathbf{v} = \mathbf{0}$. Now,

$$\begin{aligned}
\mathbf{v} &= 1\mathbf{v} && \text{by property (8)} \\
&= (\tfrac{1}{a} \cdot a)\mathbf{v} && \text{because } a \neq 0 \\
&= (\tfrac{1}{a})(a\mathbf{v}) && \text{by property (7)} \\
&= (\tfrac{1}{a})\mathbf{0} && \text{because } a\mathbf{v} = \mathbf{0} \\
&= \mathbf{0}. && \text{by part (1) of Theorem 4.1}
\end{aligned}$$

■

Theorem 4.1 holds even for unusual vector spaces, such as those in Examples 7 and 8. For instance, part (4) of the theorem claims that, in general, $a \odot \mathbf{v} = \mathbf{0}$ implies $a = 0$ or $\mathbf{v} = \mathbf{0}$. This statement can easily be verified for the vector space $\mathcal{V} = \mathbb{R}^+$ with operations $\oplus$ and $\odot$ from Example 7. In this case, $a \odot \mathbf{v} = v^a$, and the zero vector $\mathbf{0}$ is the real number 1. Then, the statement of part (4) is equivalent to the true statement that $v^a = 1$ implies $a = 0$ or $v = 1$.

Applying parts (2) and (3) of Theorem 4.1 to an unusual vector space $\mathcal{V}$ gives us a quick way of finding the zero vector $\mathbf{0}$ of $\mathcal{V}$ and the additive inverse $-\mathbf{v}$ for any vector $\mathbf{v}$ in $\mathcal{V}$. For instance, in Example 8, we have $\mathcal{V} = \mathbb{R}^2$ with scalar multiplication $a \odot [x, y] = [ax + a - 1, ay - 2a + 2]$. To find the zero vector $\mathbf{0}$ in $\mathcal{V}$, we multiply the scalar 0 by any general vector $[x, y]$ in $\mathcal{V}$:

$$\mathbf{0} = 0 \odot [x, y] = [0x + 0 - 1, 0y - 2(0) + 2] = [-1, 2].$$

Similarly, if $[x, y] \in \mathcal{V}$, then $-1 \odot [x, y]$ is the additive inverse of $[x, y]$:

$$\begin{aligned}
-[x, y] &= -1 \odot [x, y] = [-1x + (-1) - 1, -1y - 2(-1) + 2] \\
&= [-x - 2, -y + 4].
\end{aligned}$$

Failure of the Vector Space Conditions

Now that we have examined examples of sets that are vector spaces under various operations, we consider some sets that are not vector spaces and see what can go wrong.

Example 9

The set Φ of continuous real-valued functions, f, defined on the interval $[0, 1]$ such that $f(\frac{1}{2}) = 1$, is not a vector space under the usual operations of function addition and scalar multiplication, because the closure properties do not hold. If f and g are in Φ, then

$$(f + g)(\tfrac{1}{2}) = f(\tfrac{1}{2}) + g(\tfrac{1}{2}) = 1 + 1 = 2 \neq 1,$$

so $f + g$ is not in Φ. Therefore, Φ is not closed under addition and cannot be a vector space. (Is Φ closed under scalar multiplication?) ■

Example 10

Let $\mathbf{A} = \begin{bmatrix} -3 & 1 \\ 5 & -2 \end{bmatrix}$, and let Ψ be the set $\mathbb{R}^2$ with operations

$$\mathbf{v}_1 \oplus \mathbf{v}_2 = \mathbf{v}_1 + \mathbf{v}_2 \quad \text{and} \quad c \odot \mathbf{v} = c(\mathbf{A}\mathbf{v}).$$

With these operations, Ψ is not a vector space. You can easily verify that Ψ is closed under $\oplus$ and $\odot$, but properties (7) and (8) of the definition are not satisfied. For example,

$$1 \odot \begin{bmatrix} 2 \\ 7 \end{bmatrix} = 1\left(\begin{bmatrix} -3 & 1 \\ 5 & -2 \end{bmatrix}\begin{bmatrix} 2 \\ 7 \end{bmatrix}\right) = 1\begin{bmatrix} 1 \\ -4 \end{bmatrix} = \begin{bmatrix} 1 \\ -4 \end{bmatrix} \neq \begin{bmatrix} 2 \\ 7 \end{bmatrix},$$

which shows that property (8) fails. ■

Exercises—Section 4.1

Remember: To verify that a given set with its operations is a vector space, you must prove the closure properties as well as the other eight properties in the definition. To show that a set with operations is *not* a vector space, you need only find an example showing that one of the closure properties or one of the eight properties is not satisfied.

1. Prove that the set of all scalar multiples of the vector $[1, 3, 2]$ in $\mathbb{R}^3$ forms a vector space with the usual operations on 3-vectors.

2. Verify that the set of polynomials $\mathbf{f}$ in $\mathcal{P}_3$, such that $\mathbf{f}(2) = 0$, forms a vector space with the standard operations.

3. Prove that $\mathbb{R}$ is a vector space using the operations $\oplus$ and $\odot$ given by $\mathbf{x} \oplus \mathbf{y} = (x^3 + y^3)^{\frac{1}{3}}$ and $a \odot \mathbf{x} = \sqrt[3]{a}x$.

4. Show that the set of singular 2×2 matrices under the usual operations is not a vector space.

5. Prove that the set of nonsingular $n \times n$ matrices under the usual operations is not a vector space.

6. Consider $\mathbb{R}$ with ordinary addition but with scalar multiplication replaced by $a \odot \mathbf{x} = \mathbf{0}$ for every real number a. Show that $\mathbb{R}$ with these operations is not a vector space.

7. Show that the set $\mathbb{R}$ with the usual scalar multiplication but with addition given by $x \oplus y = 2(x + y)$ is not a vector space.

8. Suppose that Θ is the set $\mathbb{R}^2$ with the usual scalar multiplication but with vector addition replaced by $[x, y] \oplus [w, z] = [x + w, 0]$. Prove that Θ with these operations does not form a vector space.

9. Let $\mathcal{A} = \mathbb{R}$, with the operations $\oplus$ and $\odot$ given by $\mathbf{x} \oplus \mathbf{y} = (x^5 + y^5)^{\frac{1}{5}}$ and $a \odot \mathbf{x} = a\mathbf{x}$. Determine whether $\mathcal{A}$ is a vector space. Prove your answer.

10. Let $\mathbf{A}$ be a fixed $m \times n$ matrix, and let $\mathbf{B}$ be a fixed m-vector (in $\mathbb{R}^m$). Let $\mathcal{V}$ be the set of solutions $\mathbf{X}$ (in $\mathbb{R}^n$) to the matrix equation $\mathbf{AX} = \mathbf{B}$. Endow $\mathcal{V}$ with the usual n-vector operations.
   (a) Show that the closure properties are satisfied in $\mathcal{V}$ if and only if $\mathbf{B} = \mathbf{0}$.
   (b) Explain why properties (1), (2), (5), (6), (7), and (8) in the definition of a vector space have already been proved for $\mathcal{V}$ in Theorem 1.3.
   (c) Prove that property (3) in the definition of a vector space is satisfied if and only if $\mathbf{B} = \mathbf{0}$.
   (d) Explain why property (4) in the definition makes no sense unless property (3) is satisfied. Prove property (4) when $\mathbf{B} = \mathbf{0}$.
   (e) Use parts (a) through (d) of this exercise to determine necessary and sufficient conditions for $\mathcal{V}$ to be a vector space.

11. Property (3) in the definition of a vector space involves two equations: $\mathbf{0} + \mathbf{u} = \mathbf{u}$ and $\mathbf{u} + \mathbf{0} = \mathbf{u}$. Why is it necessary to verify only one of them when proving that a set is a vector space?

12. Property (4) of a vector space involves two equations: $[-\mathbf{u}] + \mathbf{u} = 0$ and $\mathbf{u} + [-\mathbf{u}] = 0$. Why is it necessary to verify only one of them when proving that a set is a vector space?

13. Prove that in a vector space, the identity element for vector addition is unique. (Hint: Use proof by contradiction.)

14. Let $\mathcal{V}$ be a vector space, and let $\mathbf{v} \in \mathcal{V}$. Prove that $\mathbf{v}$ has exactly one additive inverse in $\mathcal{V}$. (Hint: Use proof by contradiction.)

15. The set $\mathbb{R}^2$ with operations $[x, y] \oplus [w, z] = [x + w - 2, y + z + 3]$ and $a \odot [x, y] = [ax - 2a + 2, ay + 3a - 3]$ is a vector space. Use parts (2) and (3) of Theorem 4.1 to find the zero vector $\mathbf{0}$ and the additive inverse of each vector $\mathbf{v} = [x, y]$ for this vector space. Then check your answers.

16. Let $\mathcal{V}$ be a vector space. Prove the following **cancellation laws**:
   (a) If $\mathbf{u}$, $\mathbf{v}$, and $\mathbf{w}$ are vectors in $\mathcal{V}$ for which $\mathbf{u} + \mathbf{v} = \mathbf{w} + \mathbf{v}$, then $\mathbf{u} = \mathbf{w}$.
   (b) If a and b are scalars and $\mathbf{v} \neq \mathbf{0}$ is a vector in $\mathcal{V}$ with $a\mathbf{v} = b\mathbf{v}$, then $a = b$.
   (c) If $a \neq 0$ is a scalar and $\mathbf{v}, \mathbf{w} \in \mathcal{V}$ with $a\mathbf{v} = a\mathbf{w}$, then $\mathbf{v} = \mathbf{w}$.

17. Prove that the set $\mathcal{P}$ of all polynomials with real coefficients forms a vector space under the usual operations of polynomial (term-by-term) addition and scalar multiplication.

18. Show that each of the following sets forms a vector space under the usual operations of function addition and scalar multiplication:
   (a) The set of continuous real-valued functions with domain $\mathbb{R}$
   (b) The set of differentiable real-valued functions with domain $\mathbb{R}$
   (c) The set of all continuous real-valued functions $\mathbf{f}$ defined on the interval [0, 1] such that $\mathbf{f}(\frac{1}{2}) = 0$ (Compare this vector space with the set in Example 9.)

(d) The set of all continuous real-valued functions $\mathbf{f}$ defined on the interval $[0, 1]$ such that $\int_0^1 \mathbf{f}(x)\, dx = 0$

19. Let $\mathbf{v}_1, \ldots, \mathbf{v}_n$ be vectors in a vector space $\mathcal{V}$, and let $a_1, \ldots, a_n$ be any real numbers. Use induction to prove that $\sum_{i=1}^{n} a_i\mathbf{v}_i$ is in $\mathcal{V}$.

20. Prove part (2) of Theorem 4.1.

4.2 Subspaces

Section 4.1 presented many examples in which two vector spaces share the same addition and scalar multiplication operations, with one as a subset of the other. In fact, most of these examples are subsets of either $\mathbb{R}^n$, $\mathcal{M}_{mn}$, or the vector space of real-valued functions defined on some set. As you will see, when a vector space is a subset of a well-known vector space and has the same operations, it becomes easier to handle. These subsets, called **subspaces**, also provide additional information about the larger vector space.

Definition of a Subspace and Examples

DEFINITION

Let $\mathcal{V}$ be a vector space. Then $\mathcal{W}$ is a **subspace** of $\mathcal{V}$ if and only if $\mathcal{W}$ is a subset of $\mathcal{V}$ and $\mathcal{W}$ is itself a vector space with the same operations as $\mathcal{V}$. That is, $\mathcal{W}$ is a vector space inside $\mathcal{V}$ such that, for every a in $\mathbb{R}$ and every $\mathbf{v}$ and $\mathbf{w}$ in $\mathcal{W}$, $\mathbf{v} + \mathbf{w}$ and $a\mathbf{v}$ yield the same vectors when the operations are performed in $\mathcal{W}$ as when they are performed in $\mathcal{V}$.

Example 1

Example 3 of Section 4.1 showed that the set $\mathcal{V}$ of points lying on a plane $\mathcal{P}$ through the origin in $\mathbb{R}^3$ forms a vector space under the usual addition and scalar multiplication in $\mathbb{R}^3$. Hence, $\mathcal{V}$ is a subspace of $\mathbb{R}^3$. ■

Example 2

The set $\mathcal{S}$ of scalar multiples of the vector $[1, 3, 2]$ in $\mathbb{R}^3$ forms a vector space under the usual addition and scalar multiplication in $\mathbb{R}^3$ (see Exercise 1 in Section 4.1). Hence, $\mathcal{S}$ is a subspace of $\mathbb{R}^3$. Notice that $\mathcal{S}$ corresponds geometrically to the set of points lying on the line through the origin in $\mathbb{R}^3$ in the direction of the vector $[1, 3, 2]$ (see Figure 4.2). In the same manner, every line through the origin determines a subspace of $\mathbb{R}^3$, namely the set of scalar multiples of a nonzero vector in the direction of that line.

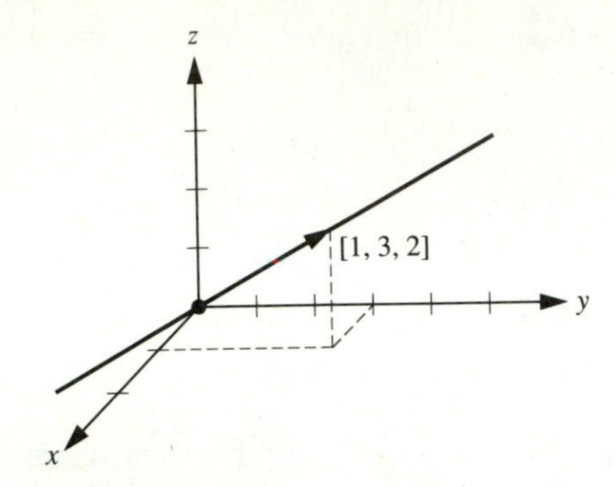

Figure 4.2
Line containing all scalar multiples of [1, 3, 2]

■

Example 3

Let $\mathcal{V}$ be any vector space. Then $\mathcal{V}$ is a subspace of itself. (Why?) Also, if $\mathcal{W}$ is the subset $\{\mathbf{0}\}$ of $\mathcal{V}$, then $\mathcal{W}$ is a vector space under the same operations as $\mathcal{V}$ (see Example 2 of Section 4.1). Therefore,

$$\mathcal{W} = \{\mathbf{0}\}$$

is a subspace of $\mathcal{V}$. ■

Although the subspaces $\mathcal{V}$ and $\{\mathbf{0}\}$ of $\mathcal{V}$ are important, they occasionally complicate matters because they must be considered as special cases in proofs. All subspaces of $\mathcal{V}$ other than $\mathcal{V}$ itself are called **proper subspaces** of $\mathcal{V}$. The subspace $\mathcal{W} = \{\mathbf{0}\}$ is called the **trivial subspace** of $\mathcal{V}$.

A vector space containing at least one nonzero vector has at least two subspaces: the trivial subspace and the vector space itself. In fact, under the usual operations, $\mathbb{R}$ has only these two subspaces (see Exercise 16).

If we consider Examples 1 to 3 in the context of $\mathbb{R}^3$, we find at least four different types of subspaces of $\mathbb{R}^3$: the trivial subspace $\{[0,0,0]\} = \{\mathbf{0}\}$; subspaces like Example 2 that can be geometrically represented as a line (thus "resembling" $\mathbb{R}$); subspaces like Example 1 that can be represented as a plane (thus "resembling" $\mathbb{R}^2$); and the subspace $\mathbb{R}^3$ itself.[†] All but the last are proper subspaces. Later we will show that each subspace of $\mathbb{R}^3$ is, in fact, one of these four types. Similarly, we will show later that all subspaces of $\mathbb{R}^n$ "resemble" one of $\{\mathbf{0}\}$, $\mathbb{R}, \mathbb{R}^2, \mathbb{R}^3, \ldots, \mathbb{R}^{n-1}$, or $\mathbb{R}^n$. We will make this statement mathematically precise later on.

[†] Although some subspaces of $\mathbb{R}^3$ "resemble" $\mathbb{R}$ and $\mathbb{R}^2$ geometrically, note that $\mathbb{R}$ and $\mathbb{R}^2$ are not subspaces of $\mathbb{R}^3$, because they are not subsets of $\mathbb{R}^3$.

Example 4

In Section 4.1, we encountered all the vector spaces (with ordinary function addition and scalar multiplication) in the following chain:

$$\begin{aligned}\mathcal{P}_0 \subset \mathcal{P}_1 \subset \mathcal{P}_2 \subset \cdots \subset\ & \mathcal{P} \\ \subset\ & \{\text{differentiable real-valued functions on } \mathbb{R}\} \\ \subset\ & \{\text{continuous real-valued functions on } \mathbb{R}\} \\ \subset\ & \{\text{all real-valued functions on } \mathbb{R}\}.\end{aligned}$$

Each of these vector spaces is a proper subspace of every vector space after it in the chain. (Why?) ■

When Is a Subset a Subspace?

It is important to note that not every subset of a vector space is a subspace. A subset $\mathcal{S}$ of a vector space $\mathcal{V}$ could fail to be a subspace of $\mathcal{V}$ if $\mathcal{S}$ does not satisfy the properties of a vector space in its own right or if $\mathcal{S}$ does not use the same operations as $\mathcal{V}$.

Example 5

Consider the first quadrant in $\mathbb{R}^2$—that is, the set Ω of all 2-vectors of the form $[x, y]$ where $x \geq 0$ and $y \geq 0$. This subset Ω of $\mathbb{R}^2$ is not a vector space under the normal operations of $\mathbb{R}^2$ because it is not closed under scalar multiplication. (For example, $-2 \cdot [3, 4] = [-6, -8]$ is not in Ω.) Therefore, Ω cannot be a subspace of $\mathbb{R}^2$. ■

Example 6

Consider the vector space $\mathbb{R}$ under the usual operations. Let $\mathcal{W}$ be the subset $\mathbb{R}^+$. By Example 7 of Section 4.1, we know that $\mathcal{W}$ is a vector space under the unusual operations $\oplus$ and $\odot$, where $\oplus$ represents multiplication and $\odot$ represents exponentiation. Although $\mathcal{W}$ is a nonempty subset of $\mathbb{R}$ and is itself a vector space, $\mathcal{W}$ is not a subspace of $\mathbb{R}$, because $\mathcal{W}$ and $\mathbb{R}$ do not share the same operations. ■

The next theorem asserts that, if $\mathcal{W}$ is a (nonempty) subset of a known vector space $\mathcal{V}$ and the closure properties hold, then the eight vector space properties are automatically true in $\mathcal{W}$. Therefore, in the future, to show that $\mathcal{W}$ is a subspace of $\mathcal{V}$, we will need to verify only the closure properties.

THEOREM 4.2

> Let $\mathcal{V}$ be a vector space, and let $\mathcal{W}$ be a nonempty subset of $\mathcal{V}$ using the same operations. Then $\mathcal{W}$ is a subspace of $\mathcal{V}$ if and only if $\mathcal{W}$ is closed under vector addition and scalar multiplication in $\mathcal{V}$.

Notice that this theorem applies only to *nonempty subsets* of a vector space. The empty set, although it is a subset of every vector space, is not a subspace of any vector space, because property (3) in the definition of a vector space fails (the empty set does not contain an additive identity).

Proof of Theorem 4.2

Since this theorem is an "If and only if" statement, the proof has two parts. First we show that if $\mathcal{W}$ is a subspace of $\mathcal{V}$, then it is closed under the two operations. This part is easy since, as a subspace, $\mathcal{W}$ is itself a vector space. Hence, the closure properties hold for $\mathcal{W}$ as they do for any vector space.

For the other part of the proof, we must show that if the closure properties hold for $\mathcal{W}$, then $\mathcal{W}$ is a subspace of $\mathcal{V}$. Thus, we must prove that $\mathcal{W}$ is itself a vector space under the operations in $\mathcal{V}$. Since we assumed that the closure properties hold for $\mathcal{W}$, we must prove only the eight vector space properties for $\mathcal{W}$.

Properties (1), (2), (5), (6), (7), and (8) of the definition are all true in $\mathcal{W}$ because they are true in $\mathcal{V}$, which is known to be a vector space. That is, since these properties hold for all vectors in $\mathcal{V}$, they must be true for all vectors in its subset, $\mathcal{W}$. For example, to prove property (1) for $\mathcal{W}$, let $\mathbf{u}, \mathbf{v} \in \mathcal{W}$. Then,

$$\underbrace{\mathbf{u}+\mathbf{v}}_{\text{addition in } \mathcal{W}} = \underbrace{\mathbf{u}+\mathbf{v}}_{\text{addition in } \mathcal{V}} \quad \text{because } \mathcal{W} \text{ and } \mathcal{V} \text{ share the same operations}$$

$$= \underbrace{\mathbf{v}+\mathbf{u}}_{\text{addition in } \mathcal{V}} \quad \text{because } \mathcal{V} \text{ is a vector space and property (1) holds}$$

$$= \underbrace{\mathbf{v}+\mathbf{u}}_{\text{addition in } \mathcal{W}} \quad \text{because } \mathcal{W} \text{ and } \mathcal{V} \text{ share the same operations}$$

Next we prove property (3), the existence of an additive identity in $\mathcal{W}$. Because $\mathcal{W}$ is assumed to be nonempty, we can choose an element $\mathbf{w}_1$ in $\mathcal{W}$. Now $\mathcal{W}$ is closed under scalar multiplication, so $0\mathbf{w}_1$ must be in $\mathcal{W}$. However, since this is the same operation as in $\mathcal{V}$, which is known to be a vector space, part (2) of Theorem 4.1 implies that $0\mathbf{w}_1 = \mathbf{0}$. Hence, $\mathbf{0}$ is in $\mathcal{W}$. Because $\mathbf{0} + \mathbf{v} = \mathbf{v}$ is true for all $\mathbf{v}$ in $\mathcal{V}$, it is also true that $\mathbf{0} + \mathbf{w} = \mathbf{w}$ holds for all $\mathbf{w}$ in $\mathcal{W}$. Therefore, $\mathcal{W}$ contains an additive identity, namely $\mathbf{0}$, the same additive identity that $\mathcal{V}$ has.

Finally we must prove that property (4), the existence of additive inverses, holds for $\mathcal{W}$. Let $\mathbf{w} \in \mathcal{W}$. Then $\mathbf{w} \in \mathcal{V}$. By property (4) for $\mathcal{V}$, $\mathbf{w}$ has an additive inverse in $\mathcal{V}$. If we can show that this additive inverse is also in $\mathcal{W}$, we will be done. Part (3) of Theorem 4.1 shows that the additive inverse of $\mathbf{w}$ in $\mathcal{V}$ equals $(-1)\mathbf{w}$. But since $\mathcal{W}$ is closed under scalar multiplication, $\mathbf{w} \in \mathcal{W}$ implies that $(-1)\mathbf{w} \in \mathcal{W}$. Therefore, the additive inverse of an element of $\mathcal{W}$ is also in $\mathcal{W}$. ∎

Checking for Subspaces in $\mathcal{M}_{nn}$ and $\mathbb{R}^n$

Let us consider several subsets of the vector spaces $\mathcal{M}_{nn}$ and $\mathbb{R}^n$ and apply Theorem 4.2 to determine whether these subsets are subspaces. Assume that $\mathcal{M}_{nn}$ and $\mathbb{R}^n$ have the usual operations.

Example 7

Consider $\mathcal{U}_n$, the set of upper triangular $n \times n$ matrices. Since $\mathcal{U}_n$ is nonempty, we may apply Theorem 4.2 and check the closure properties for $\mathcal{U}_n$ to see whether it is a subspace of $\mathcal{M}_{nn}$.

Closure of $\mathcal{U}_n$ under vector addition clearly holds, because the sum of any two $n \times n$ upper triangular matrices is again upper triangular. The closure property in $\mathcal{U}_n$ for scalar multiplication also holds, since any scalar multiple of an upper triangular matrix is again upper triangular. Hence, $\mathcal{U}_n$ is a subspace of $\mathcal{M}_{nn}$.

Similar arguments show that $\mathcal{L}_n$ (lower triangular $n \times n$ matrices) and $\mathcal{D}_n$ (diagonal $n \times n$ matrices) are also subspaces of $\mathcal{M}_{nn}$. ∎

The subspace $\mathcal{D}_n$ of $\mathcal{M}_{nn}$ in Example 7 is the intersection of the subspaces $\mathcal{U}_n$ and $\mathcal{L}_n$. In fact, the intersection of subspaces of a vector space always produces a subspace under the same operations (see Exercise 18).

If either closure property fails to hold for a subset, it cannot be a subspace. For this reason, each of the following subsets of $\mathcal{M}_{nn}$, $n \geq 2$, is not a subspace:

(1) The set of nonsingular $n \times n$ matrices
(2) The set of singular $n \times n$ matrices
(3) The set of elementary $n \times n$ matrices
(4) The set of $n \times n$ matrices in reduced row echelon form

You should check that the closure property for vector addition fails in each case and that the closure property for scalar multiplication fails in (1), (3), and (4).

Example 8

Let $\mathcal{V}$ be the set of vectors in $\mathbb{R}^4$ of the form $[a, 0, b, 0]$; that is, 4-vectors whose second and fourth coordinates are zero. We prove that $\mathcal{V}$ is a subspace of $\mathbb{R}^4$ by checking the closure properties.

To prove closure under vector addition, we must add two arbitrary elements of $\mathcal{V}$ and check that the result has the correct form for a vector in $\mathcal{V}$. Now, $[a, 0, b, 0] + [c, 0, d, 0] = [(a + c), 0, (b + d), 0]$. The second and fourth coordinates of the sum are zero, so $\mathcal{V}$ is closed under addition. Similarly, we must check closure under scalar multiplication: $c[a, 0, b, 0] = [ca, 0, cb, 0]$. Since the second and fourth coordinates of the product are zero, $\mathcal{V}$ is closed under scalar multiplication. Hence, by Theorem 4.2, $\mathcal{V}$ is a subspace of $\mathbb{R}^4$. ■

Example 9

Let $\mathcal{W}$ be the set of vectors in $\mathbb{R}^3$ of the form $[a, b, \frac{1}{2}a - 2b]$—that is, 3-vectors whose third coordinate is half the first coordinate minus twice the second coordinate. We show that $\mathcal{W}$ is a subspace of $\mathbb{R}^3$ by checking the closure properties.

Checking closure under vector addition, we have

$$\begin{aligned}[a, b, \tfrac{1}{2}a - 2b] + [c, d, \tfrac{1}{2}c - 2d] &= [a + c, b + d, \tfrac{1}{2}a - 2b + \tfrac{1}{2}c - 2d] \\ &= [a + c, b + d, \tfrac{1}{2}(a + c) - 2(b + d)],\end{aligned}$$

which has the required form, since it equals $[A, B, \frac{1}{2}A - 2B]$, where $A = a + c$ and $B = b + d$.

Checking closure under scalar multiplication, we get

$$c[a, b, \tfrac{1}{2}a - 2b] = [ca, cb, c(\tfrac{1}{2}a - 2b)] = [ca, cb, \tfrac{1}{2}(ca) - 2(cb)],$$

which has the required form. (Why?)

Geometrically, $\mathcal{W}$ is the plane in $\mathbb{R}^3$ containing the points $[0, 0, 0]$, $[1, 0, \frac{1}{2}]$, and $[0, 1, -2]$, shown in Figure 4.3. (Why?) This is the type of subspace of $\mathbb{R}^3$ discussed in Example 1. ■

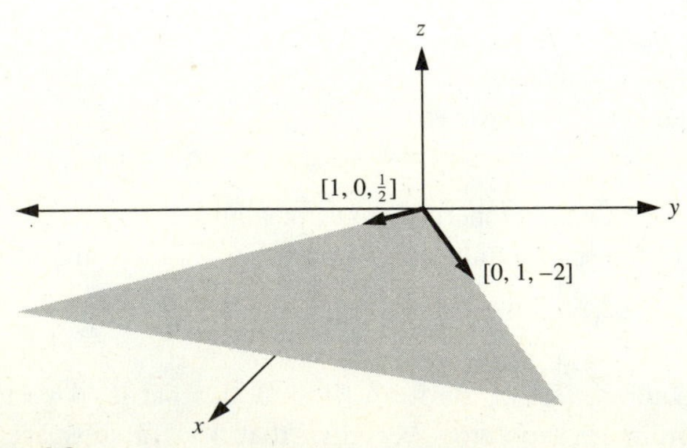

Figure 4.3
The plane containing $[0, 0, 0]$, $[1, 0, \frac{1}{2}]$, and $[0, 1, -2]$

The following subsets of $\mathbb{R}^n$ are not subspaces. In each case, at least one of the two closure properties fails. (Can you determine which ones?)

(1) The set of n-vectors whose first coordinate is nonnegative (in $\mathbb{R}^2$, this set is a half-plane)
(2) The set of unit n-vectors (in $\mathbb{R}^3$, this set is a sphere)
(3) For $n \geq 2$, the set of n-vectors with a zero in at least one coordinate (in $\mathbb{R}^3$, this set is the union of three planes)
(4) The set of n-vectors having all integer coordinates
(5) For $n \geq 2$, the set of all n-vectors whose first two coordinates add up to 3 (in $\mathbb{R}^2$, this is the line $x + y = 3$)

The subsets (2) and (5), which do not contain the additive identity $\mathbf{0}$ of $\mathbb{R}^n$, can quickly be disqualified as subspaces. In general:

> If a subset $\mathcal{S}$ of a vector space $\mathcal{V}$ does not contain the zero vector $\mathbf{0}$ of $\mathcal{V}$, then $\mathcal{S}$ is not a subspace of $\mathcal{V}$.

Checking for the presence of the additive identity is usually easy and thus is a fast way to show that a subset is not a subspace.

The Row Space of a Matrix, Revisited

Section 2.3 introduced the row space of a general $m \times n$ matrix. We now claim it is a subspace of $\mathbb{R}^n$.

THEOREM 4.3

> Let $\mathbf{A}$ be an $m \times n$ matrix. Then the row space of $\mathbf{A}$ forms a subspace of $\mathbb{R}^n$ (under the usual n-vector operations).

Theorem 4.3 is a special case of part (2) of Theorem 4.4 in Section 4.3. However, it can easily be proved at this point (see Exercise 20).

Example 10

Suppose that $\mathbf{A} = \begin{bmatrix} -4 & 3 & -8 \\ 2 & 0 & 1 \end{bmatrix}$. Then $\mathcal{W}$, the row space of $\mathbf{A}$, is the set of all 3-vectors of the form

$$a[-4, 3, -8] + b[2, 0, 1] = [(-4a + 2b), 3a, (-8a + b)].$$

By Theorem 4.3, this is a subspace of $\mathbb{R}^3$. Let us simplify the form of the general vector for $\mathcal{W}$.

Recall that two row equivalent matrices have the same row space (by Theorem 2.7). The reduced row echelon form of **A** is $\mathbf{B} = \begin{bmatrix} 1 & 0 & \frac{1}{2} \\ 0 & 1 & -2 \end{bmatrix}$, so the row space of **B** contains precisely the same vectors as the row space of **A**, which is $\mathcal{W}$. Hence, $\mathcal{W}$ can also be expressed as the subspace containing all 3-vectors of this form:

$$a[1, 0, \tfrac{1}{2}] + b[0, 1, -2] = [a, b, \tfrac{1}{2}a - 2b].$$

This is the same subspace of $\mathbb{R}^3$ discussed in Example 9.

Thus, we now have two interpretations of the row space $\mathcal{W}$: an algebraic one, as the row space of $\begin{bmatrix} 1 & 0 & \frac{1}{2} \\ 0 & 1 & -2 \end{bmatrix}$, and a geometric one, as the plane in 3-space passing through the points $[0, 0, 0]$, $[1, 0, \frac{1}{2}]$, and $[0, 1, -2]$. ∎

The method in Example 10 can be summarized as follows:

> To obtain a simplified form for the row space of an $m \times n$ matrix **A**, find the reduced row echelon form matrix **B** of **A**. Then the row space of **A** is the subspace of $\mathbb{R}^n$ consisting of all linear combinations of the rows of **B**.

Exercises—Section 4.2

Note: From this point onward in the book, use available software packages to avoid tedious calculations.

1. Determine whether each given subset of $\mathbb{R}^2$ is a subspace of $\mathbb{R}^2$ under the usual vector operations. Explain your reasoning. (In these problems, a and b represent arbitrary real numbers.)
   - ⋆(a) The set of unit 2-vectors
   - (b) The set of 2-vectors of the form $[1, a]$
   - ⋆(c) The set of 2-vectors of the form $[a, 2a]$
   - (d) The set of 2-vectors having a zero in at least one coordinate
   - ⋆(e) The set $\{[1, 2]\}$
   - (f) The set of 2-vectors whose second coordinate is zero
   - ⋆(g) The set of 2-vectors of the form $[a, b]$, where $|a| = |b|$
   - (h) The set of points in the plane lying on the line $y = -3x$
   - (i) The set of points in the plane lying on the line $y = 7x - 5$
   - ⋆(j) The set of points lying on the parabola $y = x^2$
   - (k) The set of points in the plane lying above the line $y = 2x - 5$

★(l) The set of points in the plane lying inside the circle of radius 1 centered at the origin

2. Determine whether each given subset of $\mathcal{M}_{22}$ is a subspace of $\mathcal{M}_{22}$ under the usual matrix operations. Explain your reasoning. (In these problems, a and b represent arbitrary real numbers.)

★(a) The set of matrices of the form $\begin{bmatrix} a & -a \\ b & 0 \end{bmatrix}$

(b) The set of 2×2 matrices that have a row of zeroes

★(c) The set of symmetric 2×2 matrices

(d) The set of nonsingular 2×2 matrices

★(e) The set of 2×2 matrices where the sum of the entries is zero

(f) The set of 2×2 matrices whose trace is zero (Recall that the *trace* of a square matrix is the sum of the main diagonal entries.)

★(g) The set of 2×2 matrices $\mathbf{A}$ such that $\mathbf{A}\begin{bmatrix} 1 & 3 \\ -2 & -6 \end{bmatrix} = \begin{bmatrix} 0 & 0 \\ 0 & 0 \end{bmatrix}$

(h) The set of elementary 2×2 matrices

★(i) The set of 2×2 matrices where the product of the entries is zero

3. Determine whether each given subset of $\mathcal{P}_5$ is a subspace of $\mathcal{P}_5$ under the usual operations. Explain your reasoning.

★(a) $\{\mathbf{p} \in \mathcal{P}_5 \mid$ the coefficient of the first-degree term of $\mathbf{p}$ equals the coefficient of the fifth-degree term of $\mathbf{p}\}$

(b) $\{\mathbf{p} \in \mathcal{P}_5 \mid \mathbf{p}(3) = 0\}$

★(c) $\{\mathbf{p} \in \mathcal{P}_5 \mid$ the sum of the coefficients of $\mathbf{p}$ is zero$\}$

(d) $\{\mathbf{p} \in \mathcal{P}_5 \mid \mathbf{p}(3) = \mathbf{p}(5)\}$

★(e) $\{\mathbf{p} \in \mathcal{P}_5 \mid \mathbf{p}$ is an odd-degree polynomial (highest-order nonzero term has odd degree)$\}$

(f) $\{\mathbf{p} \in \mathcal{P}_5 \mid \mathbf{p}$ has a relative maximum at $x = 0\}$

★(g) $\{\mathbf{p} \in \mathcal{P}_5 \mid \mathbf{p}'(4) = 0$, where $\mathbf{p}'$ is the derivative of $\mathbf{p}\}$

(h) $\{\mathbf{p} \in \mathcal{P}_5 \mid \mathbf{p}'(4) = 1$, where $\mathbf{p}'$ is the derivative of $\mathbf{p}\}$

4. Show that the set of vectors of the form $[a, b, 0, c, a - 2b + c]$ in $\mathbb{R}^5$ forms a subspace of $\mathbb{R}^5$ under the usual operations.

5. (a) Show that the set of all 4-vectors of the form $[a + b, a + c, b + c, c]$ is a vector space by showing that it is a subspace of $\mathbb{R}^4$ under the usual operations.

★(b) Express the subspace of part (a) as the row space of some matrix $\mathbf{A}$. (Note that Theorem 4.3 then gives another proof that this is a subspace.)

★(c) Find the reduced row echelon form matrix $\mathbf{B}$ of $\mathbf{A}$.

★(d) Use the matrix $\mathbf{B}$ from part (c) to find a simplified form for the vectors in the subspace of part (a).

6. Repeat Exercise 5 for the set of all vectors of the form $[2a - 3b, a - 5c, a, 4c - b, c]$ in $\mathbb{R}^5$ under the usual operations.

★**7.** (a) Find a general form for a vector in the row space $\mathcal{W}$ of the matrix

$$\mathbf{A} = \begin{bmatrix} 2 & 3 & 1 \\ 1 & 2 & 3 \\ 3 & 1 & -16 \end{bmatrix}.$$

(b) Find the reduced row echelon form matrix **B** of **A**.

(c) Use the matrix **B** from part (b) to find a simplified form for the vectors in $\mathcal{W}$.

(d) Indicate whether $\mathcal{W}$ is all of $\mathbb{R}^3$, a line through the origin in $\mathbb{R}^3$, or a plane through the origin in $\mathbb{R}^3$.

8. (a) Prove that the set of all 3-vectors orthogonal to $[1, -1, 4]$ forms a subspace of $\mathbb{R}^3$.

(b) Is the subspace from part (a) all of $\mathbb{R}^3$, a line passing through the origin in $\mathbb{R}^3$, or a plane passing through the origin in $\mathbb{R}^3$?

9. Suppose that t is a fixed scalar and **A** is an $m \times n$ matrix. Show that the set of all $n \times 1$ vectors **x** satisfying the matrix equation $\mathbf{Ax} = t\mathbf{x}$ is a subspace of $\mathbb{R}^n$. (Do not forget to show that this set is nonempty.)

10. Let $\mathcal{W}$ be the set of differentiable real-valued functions $y = f(x)$ defined on $\mathbb{R}$ that satisfy the differential equation $3(dy/dx) - 2y = 0$. Show that, under the usual function operations, $\mathcal{W}$ is a subspace of the vector space of all differentiable real-valued functions. (Do not forget to show that this set is nonempty.)

11. Show that the set of solutions to the differential equation $y'' + 2y' - 9y = 0$ is a subspace of the vector space of all twice-differentiable real-valued functions defined on $\mathbb{R}$. (Do not forget to show that this set is nonempty.)

12. Prove that the set of discontinuous real-valued functions defined on $\mathbb{R}$ $\left(\text{for example, } f(x) = \begin{cases} 0 & \text{if } x \le 0 \\ 1 & \text{if } x > 0 \end{cases}\right)$ with the usual function operations does not form a vector space.

13. Let **A** be a fixed $n \times n$ matrix, and let $\mathcal{W}$ be the subset of $\mathcal{M}_{nn}$ of all $n \times n$ matrices that commute with **A** under matrix multiplication (that is, $\mathbf{B} \in \mathcal{W}$ if and only if $\mathbf{AB} = \mathbf{BA}$). Show that $\mathcal{W}$ is a subspace of $\mathcal{M}_{nn}$ under the usual vector space operations. (Do not forget to show that $\mathcal{W}$ is nonempty.)

14. (a) A careful reading of the proof of Theorem 4.2 reveals that only closure under scalar multiplication (not closure under addition) is required to prove the eight vector space properties for the nonempty subset $\mathcal{W}$ of the vector space $\mathcal{V}$. Explain, nevertheless, why closure under addition is a necessary condition for $\mathcal{W}$ to be a subspace of $\mathcal{V}$.

(b) Show that the set of singular $n \times n$ matrices is closed under scalar multiplication in $\mathcal{M}_{nn}$.

(c) Use parts (a) and (b) to determine which of the eight vector space properties are true for the set of singular $n \times n$ matrices.

(d) Show that the set of singular $n \times n$ matrices is not closed under vector addition and hence is not a subspace of $\mathcal{M}_{nn}$ $(n \geq 2)$.

★(e) Is the set of nonsingular $n \times n$ matrices closed under scalar multiplication? Why or why not?

15. (a) Prove that the set of all points lying on a line passing through the origin in $\mathbb{R}^2$ is a subspace of $\mathbb{R}^2$ (under the usual operations).

(b) Prove that the set of all points in $\mathbb{R}^2$ lying on a line not passing through the origin does not form a subspace of $\mathbb{R}^2$ (under the usual operations).

16. Prove that $\mathbb{R}$ (under the usual operations) has no subspaces except $\mathbb{R}$ and $\{\mathbf{0}\}$. (Hint: Let $\mathcal{V}$ be a nontrivial subspace of $\mathbb{R}$, and show that $\mathcal{V} = \mathbb{R}$.)

17. Let $\mathcal{W}$ be a subspace of a vector space $\mathcal{V}$. Show that the set $\mathcal{W}' = \{\mathbf{v} \in \mathcal{V} \mid \mathbf{v} \notin \mathcal{W}\}$ is not a subspace of $\mathcal{V}$.

18. Let $\mathcal{V}$ be a vector space, and let $\mathcal{W}_1$ and $\mathcal{W}_2$ be subspaces of $\mathcal{V}$. Prove that $\mathcal{W}_1 \cap \mathcal{W}_2$ is a subspace of $\mathcal{V}$. (If you use Theorem 4.2, do not forget to show first that $\mathcal{W}_1 \cap \mathcal{W}_2$ is nonempty.)

19. Let $\mathcal{V}$ be any vector space, and let $\mathcal{W}$ be a nonempty subset of $\mathcal{V}$.

(a) Prove that $\mathcal{W}$ is a subspace of $\mathcal{V}$ if and only if $a\mathbf{w}_1 + b\mathbf{w}_2$ is an element of $\mathcal{W}$ for every $a, b \in \mathbb{R}$ and every $\mathbf{w}_1$ and $\mathbf{w}_2$ in $\mathcal{W}$. (Hint: One direction of this implication is easy to prove. For the other direction, first consider the case where $a = b = 1$ and then the case where $b = 0$ and a is arbitrary.)

(b) Prove that $\mathcal{W}$ is a subspace of $\mathcal{V}$ if and only if $a\mathbf{w}_1 + \mathbf{w}_2$ is an element of $\mathcal{W}$ for every real number a and every $\mathbf{w}_1$ and $\mathbf{w}_2$ in $\mathcal{W}$.

20. Prove Theorem 4.3. (Hint: Use Lemma 2.6.)

4.3 Span

In this section, we extend the notion of linear combination to general vector spaces. We show that the set of all linear combinations of the vectors in a subset S of $\mathcal{V}$ forms an important subspace of $\mathcal{V}$, called the **span** of S in $\mathcal{V}$.

Linear Combinations

In Chapter 1 we defined a linear combination of a subset of vectors $\{\mathbf{v}_1, \mathbf{v}_2, \ldots, \mathbf{v}_n\}$ in $\mathbb{R}^n$ to be a vector of the form $a_1\mathbf{v}_1 + a_2\mathbf{v}_2 + \cdots + a_n\mathbf{v}_n$. We now define the term *linear combination* for general vector spaces.

DEFINITION

Let S be a nonempty subset of a vector space $\mathcal{V}$. Then a vector $\mathbf{v}$ in $\mathcal{V}$ is said to be a **linear combination** of the vectors in S if and only if there exists a *finite* subset $\{\mathbf{v}_1, \ldots, \mathbf{v}_n\}$ of S such that $\mathbf{v} = a_1\mathbf{v}_1 + \cdots + a_n\mathbf{v}_n$ for some real numbers $a_1, \ldots, a_n$.

Example 1

Consider the subset $S = \{[1, -1, 0], [1, 0, 2], [0, -2, 5]\}$ of $\mathbb{R}^3$. The vector $[1, -2, -2]$ is a linear combination of the vectors in S according to the definition, because $[1, -2, -2] = 2[1, -1, 0] + (-1)[1, 0, 2]$. In this case, the (finite) subset of S used (from the definition) is $\{[1, -1, 0], [1, 0, 2]\}$. However, we could have used all of S to form the linear combination by placing a zero coefficient in front of the remaining vector $[0, -2, 5]$—that is, $[1, -2, -2] = 2[1, -1, 0] + (-1)[1, 0, 2] + 0[0, -2, 5]$. ■

We see from Example 1 that if S is a *finite* subset of a vector space $\mathcal{V}$, any linear combination $\mathbf{v}$ formed using *some* of the vectors in S can always be formed using *all* the vectors in S by placing zero coefficients on the unused vectors.

A linear combination formed from a set $\{\mathbf{v}\}$ containing a single vector is just a scalar multiple $a\mathbf{v}$ of $\mathbf{v}$.

Example 2

Let $S = \{[1, -2, 7]\}$, a subset of $\mathbb{R}^3$ containing a single element. Then the only linear combinations that can be formed from S are scalar multiples of $[1, -2, 7]$, such as $[3, -6, 21]$ and $[-4, 8, -28]$. ■

Next we form linear combinations from an *infinite* subset of a vector space.

Example 3

Consider $\mathcal{P}$, the vector space of polynomials with real coefficients, and let $S = \{1, x^2, x^4, \ldots\}$, the subset of $\mathcal{P}$ consisting of all nonnegative even powers of x (since $x^0 = 1$). We can form linear combinations of vectors in S by first restricting ourselves to a finite subset of S. For example, $\mathbf{p}(x) = 7x^8 - (1/4)x^4 + 10$ is a linear combination formed from S because it is a sum of scalar multiples of elements of a finite subset $\{x^8, x^4, 1\}$ of S.

Clearly, every linear combination of vectors in S is the finite sum of nonnegative even powers of x. ■

Notice that we cannot use all of the elements in an infinite set S when forming a linear combination. Such a linear combination is frequently called a **finite linear combination** to stress that only a finite number of vectors are used at any time.

Example 4

Let $S = \mathcal{U}_2 \cup \mathcal{L}_2$, a subset of $\mathcal{M}_{22}$. (Recall that $\mathcal{U}_2$ and $\mathcal{L}_2$ are, respectively, the sets of upper and lower triangular 2×2 matrices.) The matrix $\mathbf{A} = \begin{bmatrix} 2 & 3 \\ -1 & \frac{1}{2} \end{bmatrix}$ is a linear combination of the elements in S, because

$$\mathbf{A} = \underbrace{\frac{1}{2}\begin{bmatrix} 4 & 6 \\ 0 & 1 \end{bmatrix}}_{\text{in } \mathcal{U}_2} + \underbrace{(-1)\begin{bmatrix} 0 & 0 \\ 1 & 0 \end{bmatrix}}_{\text{in } \mathcal{L}_2}.$$

In fact, there are many other ways to express $\mathbf{A}$ as a finite linear combination of the elements in S. We can add more elements from S with zero coefficients, as in Example 3, but in this case there are further possibilities. For example,

$$\mathbf{A} = 2\underbrace{\begin{bmatrix} 1 & 0 \\ 0 & 0 \end{bmatrix}}_{\substack{\text{in } \mathcal{U}_2 \\ \text{and } \mathcal{L}_2}} + 3\underbrace{\begin{bmatrix} 0 & 1 \\ 0 & 0 \end{bmatrix}}_{\text{in } \mathcal{U}_2} + (-1)\underbrace{\begin{bmatrix} 0 & 0 \\ 1 & 0 \end{bmatrix}}_{\text{in } \mathcal{L}_2} + \frac{1}{2}\underbrace{\begin{bmatrix} 0 & 0 \\ 0 & 1 \end{bmatrix}}_{\substack{\text{in } \mathcal{U}_2 \\ \text{and } \mathcal{L}_2}}.$$

■

Definition of the Span of a Set

Having extended the definition of linear combination to general vector spaces, we can also generalize the notion of a row space in $\mathbb{R}^n$.

DEFINITION

Let S be a nonempty subset of a vector space $\mathcal{V}$. Then the **span** of S in $\mathcal{V}$ is the set of all possible (finite) linear combinations of the vectors in S. We use the notation span(S) to denote the span of S in $\mathcal{V}$.

The span of a set S is analogous to the row space of a matrix; each is just the set of all linear combinations of a set of vectors. In fact,

The row space of a matrix is just the span of the set of rows of the matrix.

Let us consider some examples of span.

Example 5

From Example 3, we see that if $S_1 = \{1, x^2, x^4, \ldots\}$ in $\mathcal{P}$, then span(S_1) is the set of all polynomials containing only even-degree terms. Similarly, let $S_2 = \mathcal{U}_2 \cup \mathcal{L}_2$ in $\mathcal{M}_{22}$, as in Example 4. Then span(S_2) = $\mathcal{M}_{22}$, because every 2×2 matrix can be expressed as a finite linear combination of upper and lower triangular matrices, as follows:

$$\begin{bmatrix} a & b \\ c & d \end{bmatrix} = a\begin{bmatrix} 1 & 0 \\ 0 & 0 \end{bmatrix} + b\begin{bmatrix} 0 & 1 \\ 0 & 0 \end{bmatrix} + c\begin{bmatrix} 0 & 0 \\ 1 & 0 \end{bmatrix} + d\begin{bmatrix} 0 & 0 \\ 0 & 1 \end{bmatrix}.$$

Notice that the span of a given set often (but not always) contains many more vectors than the set itself. ■

Example 5 shows that, with $S = \mathcal{U}_2 \cup \mathcal{L}_2$ and $\mathcal{V} = \mathcal{M}_{22}$, every vector in $\mathcal{V}$ is a linear combination of vectors in S. That is, span(S) = $\mathcal{V}$. When this happens, we say that $\mathcal{V}$ is **spanned by** S or that S **spans** $\mathcal{V}$. Here we are using *span* as a verb to indicate that the span (noun) of a set S equals $\mathcal{V}$. Thus, $\mathcal{M}_{22}$ is spanned (verb) by $\mathcal{U}_2 \cup \mathcal{L}_2$. Similarly, $S_1 = \{1, x^2, x^4, \ldots\}$ spans the vector space of polynomials containing only even-degree terms.

Example 6

Clearly, $\mathbb{R}^3$ is spanned by $S_1 = \{\mathbf{i}, \mathbf{j}, \mathbf{k}\}$, since every 3-vector can be expressed as a linear combination of **i**, **j**, and **k**. (Why?) However, $\mathbb{R}^3$ is not spanned by the smaller set $S_2 = \{\mathbf{i}, \mathbf{j}\}$, since span($S_2$) is the xy-plane in $\mathbb{R}^3$. (Why?) More generally, $\mathbb{R}^n$ is spanned by the set of standard unit vectors $\{\mathbf{e}_1, \ldots, \mathbf{e}_n\}$. Note that no proper subset of $\{\mathbf{e}_1, \ldots, \mathbf{e}_n\}$ will span $\mathbb{R}^n$. ■

Span(*S*) Is the Minimal Subspace Containing *S*

The next theorem completely characterizes the span of a subset of a vector space.

THEOREM 4.4

Let S be a nonempty subset of a vector space $\mathcal{V}$. Then:
(1) $S \subseteq$ span(S).
(2) Span(S) is a subspace of $\mathcal{V}$ (under the same operations as $\mathcal{V}$).
(3) If $\mathcal{W}$ is a subspace of $\mathcal{V}$ with $S \subseteq \mathcal{W}$, then span(S) $\subseteq \mathcal{W}$.
(4) Span(S) is the smallest subspace of $\mathcal{V}$ containing S.

Proof of Theorem 4.4

Part (1): We must show that each vector $\mathbf{w} \in S$ is also in span(S). But $\mathbf{w} \in S$ implies that $\{\mathbf{w}\}$ is a subset of S. Therefore, $\mathbf{w} = 1\mathbf{w}$ is a linear combination of the vectors of S. Hence, $\mathbf{w} \in$ span(S).

Part (2): Since S is nonempty, part (1) shows that span(S) is nonempty. Therefore, by Theorem 4.2, span(S) is a subspace of $\mathcal{V}$ if we can prove the closure properties for span(S).

First, let us verify closure under scalar multiplication. Let $\mathbf{v}$ be in span(S), and let c be a scalar. We must show that $c\mathbf{v} \in$ span(S). Now, since $\mathbf{v} \in$ span(S), a finite subset $\{\mathbf{v}_1, \ldots, \mathbf{v}_n\}$ of S and real numbers $a_1, \ldots, a_n$ exist such that $\mathbf{v} = \sum_{i=1}^{n} a_i \mathbf{v}_i$. Then,

$$c\mathbf{v} = c(a_1\mathbf{v}_1 + \cdots + a_n\mathbf{v}_n) = (ca_1)\mathbf{v}_1 + \cdots + (ca_n)\mathbf{v}_n.$$

Hence, $c\mathbf{v}$ is a linear combination of the finite subset $\{\mathbf{v}_1, \ldots, \mathbf{v}_n\}$ of S, and so $c\mathbf{v} \in \text{span}(S)$.

Finally, we show that span(S) is closed under vector addition. Let $\mathbf{u}$ and $\mathbf{v}$ be two vectors in span(S). Then there are finite subsets $\{\mathbf{u}_1, \ldots, \mathbf{u}_k\}$ and $\{\mathbf{v}_1, \ldots, \mathbf{v}_l\}$ of S such that $\mathbf{u} = a_1\mathbf{u}_1 + \cdots + a_k\mathbf{u}_k$ and $\mathbf{v} = b_1\mathbf{v}_1 + \cdots + b_l\mathbf{v}_l$ for some real numbers $a_1, \ldots, a_k, b_1, \ldots, b_l$. The natural thing to do at this point is to combine the expressions for $\mathbf{u}$ and $\mathbf{v}$ by adding corresponding coefficients. However, either of the subsets $\{\mathbf{u}_1, \ldots, \mathbf{u}_k\}$ or $\{\mathbf{v}_1, \ldots, \mathbf{v}_l\}$ may contain elements not found in the other, so it is difficult to match up their coefficients.

Consider the finite subset $X = \{\mathbf{u}_1, \ldots, \mathbf{u}_k\} \cup \{\mathbf{v}_1, \ldots, \mathbf{v}_l\}$ of S. Renaming the elements of X, we can suppose X is the finite subset $\{\mathbf{w}_1, \ldots, \mathbf{w}_m\}$ of S. Hence, there are real numbers $c_1, \ldots, c_m$ and $d_1, \ldots, d_m$ such that $\mathbf{u} = c_1\mathbf{w}_1 + \cdots + c_m\mathbf{w}_m$ and $\mathbf{v} = d_1\mathbf{w}_1 + \cdots + d_m\mathbf{w}_m$. (Note that $c_i = a_j$ if $\mathbf{w}_i = \mathbf{u}_j$ and that $c_i = 0$ if $\mathbf{w}_i \notin \{\mathbf{u}_1, \ldots, \mathbf{u}_k\}$. A similar formula gives the value of each d_i.) Using this notation, we now get

$$\mathbf{u} + \mathbf{v} = \sum_{i=1}^{m} c_i\mathbf{w}_i + \sum_{i=1}^{m} d_i\mathbf{w}_i = \sum_{i=1}^{m} (c_i + d_i)\mathbf{w}_i.$$

Hence, $\mathbf{u} + \mathbf{v}$ is a linear combination of the elements $\mathbf{w}_i$ in the subset X of S. Thus, $\mathbf{u} + \mathbf{v} \in \text{span}(S)$.

Part (3): Let S be a subset of $\mathcal{V}$, and let $\mathcal{W}$ be a subspace of $\mathcal{V}$ containing S. We must prove that our situation is as shown in Figure 4.4(a), not as shown in Figure 4.4(b). That is, we must show that any finite linear combination of vectors in S can never lie outside of $\mathcal{W}$ (hence, span(S) $\subseteq \mathcal{W}$). We use induction on the number of vectors in the linear combination.

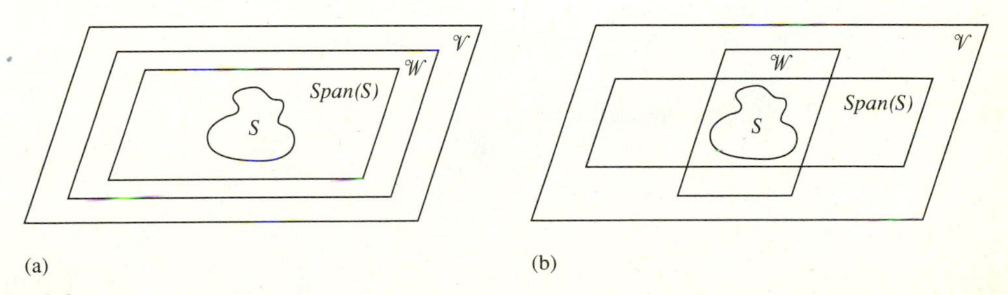

Figure 4.4
(a) Situation that *must* occur if $\mathcal{W}$ is a subspace containing S; (b) Situation that *cannot* occur if $\mathcal{W}$ is a subspace containing S

For the base step, consider a one-element subset $\{\mathbf{v}_1\}$ of S. Now, $S \subseteq \mathcal{W} \Longrightarrow \mathbf{v}_1 \in \mathcal{W}$. Also, any linear combination of $\{\mathbf{v}_1\}$ is a scalar multiple $a_1\mathbf{v}_1$ of $\mathbf{v}_1$. However, $a_1\mathbf{v}_1$ is in the subspace $\mathcal{W}$, because $\mathcal{W}$ is closed under scalar multiplication.

Now consider the inductive step. For the induction hypothesis, assume that if $S \subseteq \mathcal{W}$, then any linear combination of n-vectors in S lies in $\mathcal{W}$. We now use this assumption to prove that every linear combination of $n + 1$ vectors in S also lies in $\mathcal{W}$. Let $\{\mathbf{v}_1, \ldots, \mathbf{v}_{n+1}\}$ be a subset of S, and let $a_1, \ldots, a_{n+1}$ be any real numbers. Then the induction hypothesis assures us that the linear combination $a_1\mathbf{v}_1 + \cdots + a_n\mathbf{v}_n$ is in $\mathcal{W}$, and the proof of the base step shows that $a_{n+1}\mathbf{v}_{n+1}$ is also in $\mathcal{W}$. Hence, the linear combination

$$a_1\mathbf{v}_1 + \cdots + a_{n+1}\mathbf{v}_{n+1} = (a_1\mathbf{v}_1 + \cdots + a_n\mathbf{v}_n) + a_{n+1}\mathbf{v}_{n+1}$$

is in $\mathcal{W}$ by the closure property for vector addition in $\mathcal{W}$.

Part (4): This part is merely a summary of the other three parts. Parts (1) and (2) assert that span(S) is a subspace of $\mathcal{V}$ containing S. Part (3) shows that span(S) is the smallest such subspace, because span(S) must be a subset of, and hence smaller than, any other subspace of $\mathcal{V}$ that contains S. ■

Theorem 4.4 shows that span(S) is created by adding to S precisely those vectors needed to make the closure properties hold. In fact, the whole idea behind span is to "close up" a subset of a vector space to create a subspace.

Example 7

Let $\mathbf{v}_1$ and $\mathbf{v}_2$ be any two vectors in $\mathbb{R}^4$. Then, by Theorem 4.4, span($\{\mathbf{v}_1, \mathbf{v}_2\}$) is the smallest subspace of $\mathbb{R}^4$ containing $\mathbf{v}_1$ and $\mathbf{v}_2$. In particular, if $\mathbf{v}_1 = [1, 3, -2, 5]$ and $\mathbf{v}_2 = [0, -4, 3, -1]$, then span($\{\mathbf{v}_1, \mathbf{v}_2\}$) is the subspace of $\mathbb{R}^4$ consisting of all vectors of the form

$$a[1, 3, -2, 5] + b[0, -4, 3, -1] = [a, 3a - 4b, -2a + 3b, 5a - b].$$

No smaller subspace of $\mathbb{R}^4$ contains $\mathbf{v}_1$ and $\mathbf{v}_2$. ■

The following useful result is easy to prove (see Exercise 18).

THEOREM 4.5

Let $\mathcal{V}$ be a vector space, and let S_1 and S_2 be subsets of $\mathcal{V}$ with $S_1 \subseteq S_2$. Then span(S_1) $\subseteq$ span(S_2).

Calculating Span(S) by the Row Space Technique

For a finite set S of vectors in $\mathbb{R}^n$, we can find span(S) by forming the matrix $\mathbf{A}$ whose rows are the vectors in S. Then, as we observed earlier, span(S) is simply the row space of $\mathbf{A}$.

Example 8

Let S be the subset $\{[1, 4, -1, -5], [2, 8, 5, 4], [-1, -4, 2, 7], [6, 24, -1, -20]\}$ of $\mathbb{R}^4$. By definition, span(S) is the set of all vectors of the form

$$a[1, 4, -1, -5] + b[2, 8, 5, 4] + c[-1, -4, 2, 7] + d[6, 24, -1, -20],$$

for $a, b, c, d \in \mathbb{R}$. To find a simplified form for the vectors in span(S), we create

$$\mathbf{A} = \begin{bmatrix} 1 & 4 & -1 & -5 \\ 2 & 8 & 5 & 4 \\ -1 & -4 & 2 & 7 \\ 6 & 24 & -1 & -20 \end{bmatrix},$$

whose rows are the vectors in S. Then span(S) is the set of all linear combinations of the rows of $\mathbf{A}$, which is precisely the row space of $\mathbf{A}$.

To find the row space of $\mathbf{A}$, we obtain its reduced row echelon form matrix

$$\mathbf{B} = \begin{bmatrix} 1 & 4 & 0 & -3 \\ 0 & 0 & 1 & 2 \\ 0 & 0 & 0 & 0 \\ 0 & 0 & 0 & 0 \end{bmatrix}.$$

The row space of $\mathbf{A}$ is the same as the row space of $\mathbf{B}$, which is the set of all 4-vectors of the form

$$a[1, 4, 0, -3] + b[0, 0, 1, 2] = [a, 4a, b, -3a + 2b].$$

Therefore, span(S) is the subspace of $\mathbb{R}^4$ of 4-vectors of this form. Note, for example, that the vector $[3, 12, -2, -13]$ is in span(S) ($a = 3, b = -2$). However, the vector $[-2, -8, 4, 6]$ is not in span(S), because the following system has no solutions:

$$\begin{cases} a & & = -2 \\ 4a & & = -8 \\ & b & = 4 \\ -3a & + 2b & = 6 \end{cases}.$$

■

The method used in Example 8 works for other vector spaces besides $\mathbb{R}^n$, a fact that will follow from the discussion of isomorphism in Section 5.4. (However, we will not use this fact in proofs of theorems until after Section 5.4.)

Example 9

Let S be the subset $\{5x^3+2x^2+4x-3, -x^2+3x-7, 2x^3+4x^2-8x+5, x^3+2x+5\}$ of $\mathcal{P}_3$. We want to find a simplified form for vectors in span(S).

Consider the coefficients of each polynomial as the coordinates of a vector in $\mathbb{R}^4$, yielding the corresponding set of vectors $T = \{[5, 2, 4, -3], [0, -1, 3, -7], [2, 4, -8, 5], [1, 0, 2, 5]\}$. Using the method of Example 8, we create the following matrix, whose rows are the vectors in T:

$$\mathbf{A} = \begin{bmatrix} 5 & 2 & 4 & -3 \\ 0 & -1 & 3 & -7 \\ 2 & 4 & -8 & 5 \\ 1 & 0 & 2 & 5 \end{bmatrix}.$$

Then span(T) is the row space of the reduced row echelon form of $\mathbf{A}$:

$$\mathbf{B} = \begin{bmatrix} 1 & 0 & 2 & 0 \\ 0 & 1 & -3 & 0 \\ 0 & 0 & 0 & 1 \\ 0 & 0 & 0 & 0 \end{bmatrix}.$$

Taking each row of $\mathbf{B}$ as the coefficients of a polynomial in $\mathcal{P}_3$, we see that span(S) = $\{a(x^3+2x)+b(x^2-3x)+c \mid a, b, c \in \mathbb{R}\} = \{ax^3+bx^2+(2a-3b)x+c \mid a, b, c \in \mathbb{R}\}$.

■

Example 10

Consider the subset S of $\mathcal{M}_{22}$ given by

$$\left\{ \begin{bmatrix} 5 & -3 \\ 1 & -1 \end{bmatrix}, \begin{bmatrix} 4 & -2 \\ 2 & 2 \end{bmatrix}, \begin{bmatrix} 1 & 2 \\ 8 & 18 \end{bmatrix} \right\}.$$

Also consider the entries of each matrix as the coordinates of a vector in $\mathbb{R}^4$. The corresponding subset is $T = \{[5, -3, 1, -1], [4, -2, 2, 2], [1, 2, 8, 18]\}$. Now create the following matrix, whose rows are the vectors in T:

$$\mathbf{A} = \begin{bmatrix} 5 & -3 & 1 & -1 \\ 4 & -2 & 2 & 2 \\ 1 & 2 & 8 & 18 \end{bmatrix}.$$

Then, span(T) is the row space of the reduced row echelon form of $\mathbf{A}$:

$$\mathbf{B} = \begin{bmatrix} 1 & 0 & 2 & 4 \\ 0 & 1 & 3 & 7 \\ 0 & 0 & 0 & 0 \end{bmatrix}.$$

Thus, we have span(S) =

$$\left\{ a\begin{bmatrix} 1 & 0 \\ 2 & 4 \end{bmatrix} + b\begin{bmatrix} 0 & 1 \\ 3 & 7 \end{bmatrix} \middle| a, b \in \mathbb{R} \right\} = \left\{ \begin{bmatrix} a & b \\ 2a + 3b & 4a + 7b \end{bmatrix} \middle| a, b \in \mathbb{R} \right\}.$$ ■

The Span of the Empty Set

Until now, all of our results involving span have specified that the subset S of the vector space $\mathcal{V}$ be nonempty. However, our understanding of span(S) as the smallest subspace of $\mathcal{V}$ containing S allows us to give a meaningful definition for the span of the empty set.

DEFINITION

$$\text{Span}(\{\ \}) = \{\mathbf{0}\}.$$

This definition makes sense because the trivial subspace is the smallest subspace of $\mathcal{V}$, hence the smallest one containing the empty set. Thus, Theorem 4.4 is also true when the set S is empty. Similarly, to maintain consistency, we *define* any linear combination of the empty set of vectors to be $\mathbf{0}$ so that the span of the empty set equals the set of all linear combinations of vectors taken from this set.

Exercises—Section 4.3

1. In each case, use the row space method to find a simplified general form for all the vectors in span(S), where S is the given subset of $\mathbb{R}^n$:

★(a) $S = \{[1, 0, -1], [0, 1, 1]\}$

(b) $S = \{[3, 1, -2], [-3, -1, 2], [6, 2, -4]\}$

★(c) $S = \{[1, -1, 1], [2, -3, 3], [0, 1, -1]\}$

(d) $S = \{[1, 1, 1], [2, 1, 1], [1, 1, 2]\}$

★(e) $S = \{[1, 3, 0, 1], [0, 0, 1, 1], [0, 1, 0, 1], [1, 5, 1, 4]\}$

(f) $S = \{[2, -1, 3, 1], [1, -2, 0, -1], [3, -3, 3, 0], [5, -4, 6, 1], [1, -5, -3, -4]\}$

2. In each case, use the row space method to find a simplified general form for all the vectors in span(S), where S is the given subset of $\mathcal{P}_3$:

★(a) $S = \{x^3 - 1, x^2 - x, x - 1\}$

(b) $S = \{x^3 + 2x^2, 1 - 4x^2, 12 - 5x^3, x^3 - x^2\}$

★(c) $S = \{x^3 - x + 5, 3x^3 - 3x + 10, 5x^3 - 5x - 6, 6x - 6x^3 - 13\}$

3. In each case, use the row space method to find a simplified general form for all the vectors in span(S), where S is the given subset of $\mathcal{M}_{22}$:

★(a) $S = \left\{ \begin{bmatrix} -1 & 1 \\ 0 & 0 \end{bmatrix}, \begin{bmatrix} 0 & 0 \\ 1 & -1 \end{bmatrix}, \begin{bmatrix} -1 & 0 \\ 0 & 1 \end{bmatrix} \right\}$

(b) $S = \left\{ \begin{bmatrix} 4 & -1 \\ 1 & 0 \end{bmatrix}, \begin{bmatrix} 1 & -1 \\ 3 & 0 \end{bmatrix}, \begin{bmatrix} 5 & 1 \\ -7 & 0 \end{bmatrix} \right\}$

★(c) $S = \left\{ \begin{bmatrix} 1 & -1 \\ 3 & 0 \end{bmatrix}, \begin{bmatrix} 2 & -1 \\ 8 & -1 \end{bmatrix}, \begin{bmatrix} -1 & 4 \\ 4 & -1 \end{bmatrix}, \begin{bmatrix} 3 & -4 \\ 5 & 6 \end{bmatrix} \right\}$

4. Prove that the set $S = \{[1, 3, -1], [2, 7, -3], [4, 8, -7]\}$ spans $\mathbb{R}^3$.

5. Prove that the set $S = \{[1, -2, 2], [3, -4, -1], [1, -4, 9], [0, 2, -7]\}$ does not span $\mathbb{R}^3$.

6. Show that the set $\{x^2 + x + 1, x + 1, 1\}$ spans $\mathcal{P}_2$.

7. Prove that the set $\{x^2 + 4x - 3, 2x^2 + x + 5, 7x - 11\}$ does not span $\mathcal{P}_2$.

8. (a) Let $S = \{[1, -2, -2], [3, -5, 1], [-1, 1, -5]\}$. Show that $[-4, 5, -13] \in$ span(S) by expressing it as a linear combination of the vectors in S.
(b) Prove that the set S in part (a) does not span $\mathbb{R}^3$.

★9. Consider the subset $S = \{x^3 - 2x^2 + x - 3, 2x^3 - 3x^2 + 2x + 5, 4x^2 + x - 3, 4x^3 - 7x^2 + 4x - 1\}$ of $\mathcal{P}$. Show that $3x^3 - 8x^2 + 2x + 16$ is in span(S) by expressing it as a linear combination of the elements of S.

10. Prove that the set S of all vectors in $\mathbb{R}^4$ that have zeroes in exactly two coordinates spans $\mathbb{R}^4$. (Hint: Find a subset of S that spans $\mathbb{R}^4$.)

11. Let $\mathbf{a}$ be any nonzero element of $\mathbb{R}$. Prove that span($\{\mathbf{a}\}$) $= \mathbb{R}$.

12.★(a) Suppose that S_1 is the set of symmetric 2×2 matrices and that S_2 is the set of skew-symmetric 2×2 matrices. Prove that span($S_1 \cup S_2$) $= \mathcal{M}_{22}$.
(b) State and prove the corresponding statement for $n \times n$ matrices.

13. Consider the subset $S = \{1 + x^2, x + x^3, 3 - 2x + 3x^2 - 12x^3\}$ of $\mathcal{P}$, and let $\mathcal{W} = \{ax^3 + bx^2 + cx + b \mid a, b, c \in \mathbb{R}\}$. Show that $\mathcal{W} =$ span(S).

14. Let $S_1 = \{\mathbf{v}_1, \ldots, \mathbf{v}_n\}$ be a nonempty subset of a vector space $\mathcal{V}$. Let $S_2 = \{-\mathbf{v}_1, -\mathbf{v}_2, \ldots, -\mathbf{v}_n\}$. Show that span($S_1$) $=$ span(S_2).

15. Let $\mathbf{u}$ and $\mathbf{v}$ be two nonzero vectors in $\mathbb{R}^3$, and let $S = \{\mathbf{u}, \mathbf{v}\}$. Show that span($S$) is a line through the origin if $\mathbf{u} = a\mathbf{v}$ for some real number a but otherwise is a plane through the origin.

16. Let $\mathbf{u} = [u_1, u_2, u_3]$, $\mathbf{v} = [v_1, v_2, v_3]$, and $\mathbf{w} = [w_1, w_2, w_3]$ be three vectors in $\mathbb{R}^3$. Show that $S = \{\mathbf{u}, \mathbf{v}, \mathbf{w}\}$ spans $\mathbb{R}^3$ if and only if

$$\begin{vmatrix} u_1 & u_2 & u_3 \\ v_1 & v_2 & v_3 \\ w_1 & w_2 & w_3 \end{vmatrix} \neq 0.$$

(Hint: Consider the possible reduced row echelon forms of the corresponding matrix.)

17. Let $S = \{\mathbf{p}_1, \ldots, \mathbf{p}_k\}$ be a finite subset of $\mathcal{P}$. Prove that there is some positive integer n such that $\text{span}(S) \subseteq \mathcal{P}_n$.

18. Prove Theorem 4.5.

19. (a) Prove that if S is a nonempty subset of a vector space $\mathcal{V}$, then S is a subspace of $\mathcal{V}$ if and only if $\text{span}(S) = S$.
(b) Use part (a) to show that every subspace $\mathcal{W}$ of a vector space $\mathcal{V}$ has a set of vectors that spans $\mathcal{W}$, namely the set $\mathcal{W}$ itself.
(c) Describe the span of the set of the skew-symmetric matrices in $\mathcal{M}_{33}$.

20. Let S_1 and S_2 be subsets of a vector space $\mathcal{V}$. Prove that $\text{span}(S_1) = \text{span}(S_2)$ if and only if $S_1 \subseteq \text{span}(S_2)$ and $S_2 \subseteq \text{span}(S_1)$.

21. Let S_1 and S_2 be two subsets of a vector space $\mathcal{V}$.
(a) Prove that $\text{span}(S_1 \cap S_2) \subseteq \text{span}(S_1) \cap \text{span}(S_2)$.
★(b) Give an example of distinct subsets S_1 and S_2 of $\mathbb{R}^3$ where the inclusion in part (a) is actually an equality.
★(c) Give an example of subsets S_1 and S_2 of $\mathbb{R}^3$ where the inclusion in part (a) is not an equality.

22. Let S_1 and S_2 be subsets of a vector space $\mathcal{V}$.
(a) Show that $\text{span}(S_1) \cup \text{span}(S_2) \subseteq \text{span}(S_1 \cup S_2)$.
(b) Prove that if $S_1 \subseteq S_2$, then the inclusion in part (a) is an equality.
★(c) Give an example of subsets S_1 and S_2 in $\mathcal{P}_5$ where the inclusion in part (a) is not an equality.

23. Let S be a subset of a vector space $\mathcal{V}$, and let $\mathbf{v} \in \mathcal{V}$. Show that $\text{span}(S) = \text{span}(S \cup \{\mathbf{v}\})$ if and only if $\mathbf{v} \in \text{span}(S)$.

4.4 Linear Independence

In this section, we introduce the concept of a set of linearly independent vectors, a set in which no vector can be expressed as a linear combination of the others. We present several useful methods for determining whether a given set of vectors is linearly independent and show that, in a linearly independent set S, every vector makes an important contribution to $\text{span}(S)$.

Linear Dependence and Independence

DEFINITION

Let S be a subset of a vector space $\mathcal{V}$.

(1) S is a **linearly dependent** subset of $\mathcal{V}$ if and only if some vector $\mathbf{v}$ in S can be expressed as a linear combination of the other vectors in S; that is, there is some vector $\mathbf{v}$ in S such that $\mathbf{v} \in \operatorname{span}(S - \{\mathbf{v}\})$.

(2) S is a **linearly independent** subset of $\mathcal{V}$ if and only if it is not linearly dependent; that is, no vector in S can be expressed as a linear combination of the other vectors in S.

Example 1

The set of vectors $S = \{[1, 2, -1], [0, 1, 2], [2, 7, 4]\}$ in $\mathbb{R}^3$ is linearly dependent because it is possible to express some vector in the set S as a linear combination of the others. For example, $[2, 7, 4] = 2[1, 2, -1] + 3[0, 1, 2]$. From a geometric point of view, the fact that $[2, 7, 4]$ can be expressed as a linear combination of the vectors $[1, 2, -1]$ and $[0, 1, 2]$ means that $[2, 7, 4]$ lies in the plane spanned by $[1, 2, -1]$ and $[0, 1, 2]$, assuming that all three vectors have their initial points at the origin (see Figure 4.5).

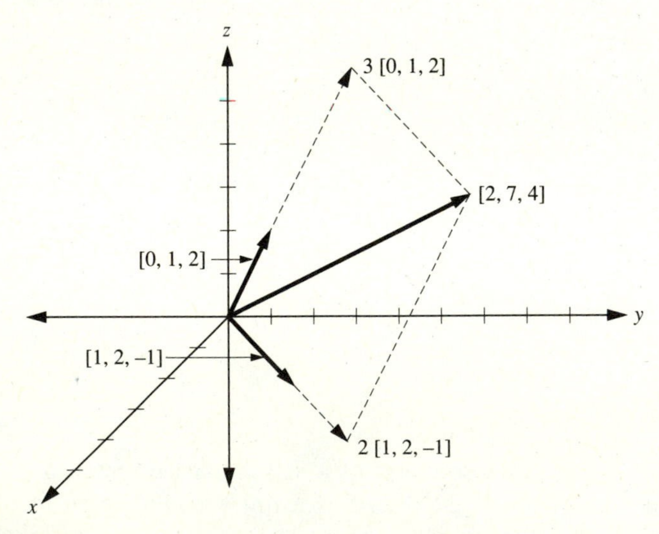

Figure 4.5
The vector [2, 7, 4] in the plane spanned by [1, 2, −1] and [0, 1, 2]

■

Example 2

The set of vectors $S = \{[1, -1, 2], [-3, 3, -6]\}$ in $\mathbb{R}^3$ is linearly dependent since each of the vectors is a linear combination of the other. For example, $[1, -1, 2] = (-1/3)[-3, 3, -6]$. ■

The last example demonstrates that any set of exactly two vectors is linearly dependent precisely when one of the vectors is a scalar multiple of the other—that is, if the vectors are parallel.

Similarly, a set of exactly two vectors is linearly independent precisely when neither is a scalar multiple of the other. For example, the subset $\{[3, -8], [2, 5]\}$ is clearly a linearly independent subset of $\mathbb{R}^2$.

Example 3

The set of vectors $\{\mathbf{i}, \mathbf{j}, \mathbf{k}\}$ in $\mathbb{R}^3$ is linearly independent. None of the three vectors in this set can be expressed as a linear combination of the other two. (Why not?) More generally, the set $\{\mathbf{e}_1, \ldots, \mathbf{e}_n\}$ in $\mathbb{R}^n$ is linearly independent. ■

Example 4

Let S be a subset of a vector space $\mathcal{V}$ containing the zero vector. Then S is linearly dependent. To prove this statement, we will show that $\mathbf{0} \in \text{span}(S - \{\mathbf{0}\})$. Now, by Theorem 4.4, $\text{span}(S - \{\mathbf{0}\})$ is a subspace of $\mathcal{V}$. Because $\mathbf{0}$ is in every subspace, $\mathbf{0}$ is in $\text{span}(S - \{\mathbf{0}\})$. Thus, any subset of a vector space that contains the zero vector is linearly dependent. ■

A Method for Checking Linear Independence

Proving directly from the definition that a set of vectors S is linearly independent takes a lot of effort, because we must show that every vector of S is not a linear combination of all of the other vectors in S. The next theorem gives a faster way to check a finite set S for linear independence.

THEOREM 4.6

Let $\mathcal{V}$ be a vector space, and let $S = \{\mathbf{v}_1, \ldots, \mathbf{v}_n\}$ be a (finite) subset of $\mathcal{V}$. Then S is linearly independent if and only if, for each k with $1 \leq k \leq n$, $\mathbf{v}_k \notin \text{span}(\{\mathbf{v}_1, \ldots, \mathbf{v}_{k-1}\})$.

This theorem says that if S is a finite set of vectors written in some order, then S is linearly independent if and only if *no vector in S can be expressed as a linear* *combination of the vectors preceding it*. It is also important to note the meaning of the statement "$\mathbf{v}_k \notin \text{span}(\{\mathbf{v}_1, \ldots, \mathbf{v}_{k-1}\})$" when $k = 1$. In this case, it means that $\mathbf{v}_1 \notin \text{span}(\{\,\})$; that is, $\mathbf{v}_1$ is nonzero.

We leave this theorem for you to prove in Exercise 21.

Example 5

We use Theorem 4.6 to show that $S = \{1 + x^2,\ x^2 - 2x + 3,\ x - 2\}$ is a linearly independent subset of $\mathcal{P}_3$. We consider each vector in S in turn:

Step 1: $1 + x^2$ is a nonzero vector and hence not in the span of the empty set.

Step 2: $x^2 - 2x + 3 \notin \text{span}(\{1 + x^2\})$ because $x^2 - 2x + 3$ is not a scalar multiple of $1 + x^2$.

Step 3: $x - 2 \notin \text{span}(\{1 + x^2, x^2 - 2x + 3\})$, because elements of span $(\{1 + x^2, x^2 - 2x + 3\})$ are of the form

$$a(1 + x^2) + b(x^2 - 2x + 3) = (a + b)x^2 + (-2b)x + (a + 3b).$$

Setting this last equation equal to $x - 2$ and equating coefficients yields the following system:

$$\begin{cases} a + b = 0 \\ \quad - 2b = 1 \\ a + 3b = -2 \end{cases}.$$

This system has no solutions. (Verify.)

Therefore, none of the vectors $1 + x^2$, $x^2 - 2x + 3$, or $x - 2$ is in the span of the set of preceding vectors. Hence, the set is linearly independent. ■

Example 6

Consider the subset $S = \{[1, 4, 2, 0], [1, -1, 0, 3], [-1, 6, 2, -6], [2, 1, 4, -9]\}$ of $\mathbb{R}^4$. Let us investigate whether S is linearly independent:

Step 1: $[1, 4, 2, 0]$ is a nonzero vector.

Step 2: $[1, -1, 0, 3] \notin \text{span}(\{[1, 4, 2, 0]\})$, because $[1, -1, 0, 3]$ is not a scalar multiple of $[1, 4, 2, 0]$.

Step 3: Is $[-1, 6, 2, -6] \in \text{span}(\{[1, 4, 2, 0], [1, -1, 0, 3]\})$? Suppose that $[-1, 6, 2, -6] = a[1, 4, 2, 0] + b[1, -1, 0, 3]$. Then the corresponding system is

$$\begin{cases} a + b = -1 \\ 4a - b = 6 \\ 2a = 2 \\ 3b = -6 \end{cases}.$$

But this system has a solution, namely $a = 1, b = -2$. Therefore, $[-1, 6, 2, -6]$ is in $\text{span}(\{[1, 4, 2, 0], [1, -1, 0, 3]\})$.

We do not have to proceed to the fourth vector in S, because the test for linear independence from Theorem 4.6 failed at step 3. Hence, S is linearly dependent. ■

A Combination of Linearly Independent Vectors Producing Zero Must Use All Zero Coefficients

The next theorem gives us another characterization of linear independence. In addition to its theoretical importance, it also supplies us with another "fast" method for determining whether a finite subset of a vector space is linearly independent.

THEOREM 4.7

> Let $S = \{\mathbf{v}_1, \ldots, \mathbf{v}_n\}$ be a nonempty finite subset of a vector space $\mathcal{V}$. Then S is linearly independent if and only if the equation $a_1\mathbf{v}_1 + \cdots + a_n\mathbf{v}_n = \mathbf{0}$ implies that $a_1 = a_2 = \cdots = a_n = 0$.

That is, a set S is linearly independent if and only if the zero vector can be expressed as a (nonempty) linear combination of the vectors in S *only* by using all zero coefficients in the linear combination.

To prove Theorem 4.7, we prove instead the following equivalent statement:

Let $S = \{\mathbf{v}_1, \ldots, \mathbf{v}_n\}$ be a nonempty finite subset of a vector space $\mathcal{V}$. Then S is *linearly dependent* if and only if there are real numbers $a_1, \ldots, a_n$ such that $a_1\mathbf{v}_1 + \cdots + a_n\mathbf{v}_n = \mathbf{0}$, where some $a_i \neq 0$.

Proof of Theorem 4.7

To begin, assume that S is linearly dependent. Then there is a vector $\mathbf{v}_i$ in S such that $\mathbf{v}_i \in \text{span}(S - \{\mathbf{v}_i\})$. Therefore, there are real numbers $a_1, \ldots, a_{i-1}, a_{i+1}, \ldots, a_n$ such that

$$\mathbf{v}_i = a_1\mathbf{v}_1 + \cdots + a_{i-1}\mathbf{v}_{i-1} + a_{i+1}\mathbf{v}_{i+1} + \cdots + a_n\mathbf{v}_n.$$

Letting $a_i = -1$, we get $\mathbf{0} = a_1\mathbf{v}_1 + \cdots + a_n\mathbf{v}_n$. Since $a_i \neq 0$, this result completes the first half of the proof.

To prove the reverse implication, assume that we have coefficients $a_1, \ldots, a_n$ such that $\mathbf{0} = a_1\mathbf{v}_1 + \cdots + a_n\mathbf{v}_n$, with $a_i \neq 0$ for some i. Then,

$$\mathbf{v}_i = \left(-\frac{a_1}{a_i}\right)\mathbf{v}_1 + \cdots + \left(-\frac{a_{i-1}}{a_i}\right)\mathbf{v}_{i-1} + \left(-\frac{a_{i+1}}{a_i}\right)\mathbf{v}_{i+1} + \cdots + \left(-\frac{a_n}{a_i}\right)\mathbf{v}_n,$$

which shows that $\mathbf{v}_i \in \text{span}(S - \{\mathbf{v}_i\})$ and hence that S is linearly dependent. ■

Example 7

Consider the subset $S = \{[1, -1, 0, 2], [0, -2, 1, 0], [2, 0, -1, 1]\}$ of $\mathbb{R}^4$. We will investigate whether S is linearly independent using Theorem 4.7.

We proceed by assuming that $a[1, -1, 0, 2] + b[0, -2, 1, 0] + c[2, 0, -1, 1] = [0, 0, 0, 0]$ and solve for a, b, and c to see whether all these coefficients must be zero. That is, we determine whether the following homogeneous system has any nontrivial solutions:

$$\begin{cases} a \phantom{{}-2b} + 2c = 0 \\ -a - 2b \phantom{{}+2c} = 0 \\ \phantom{-a-{}} b - c = 0 \\ 2a \phantom{{}-2b} + c = 0 \end{cases}.$$

Row reduction shows that this system has only the trivial solution. Hence, S is linearly independent. ■

Example 8

Consider the following subset of $\mathcal{M}_{22}$:

$$\left\{ \begin{bmatrix} 2 & 3 \\ -1 & 4 \end{bmatrix}, \begin{bmatrix} -1 & 0 \\ 1 & 1 \end{bmatrix}, \begin{bmatrix} 6 & -1 \\ 3 & 2 \end{bmatrix}, \begin{bmatrix} -11 & 3 \\ -2 & 2 \end{bmatrix} \right\}.$$

Let us determine whether S is linearly independent using Theorem 4.7. We solve the following equation to determine whether the coefficients a, b, c, and d must all be zero:

$$a\begin{bmatrix} 2 & 3 \\ -1 & 4 \end{bmatrix} + b\begin{bmatrix} -1 & 0 \\ 1 & 1 \end{bmatrix} + c\begin{bmatrix} 6 & -1 \\ 3 & 2 \end{bmatrix} + d\begin{bmatrix} -11 & 3 \\ -2 & 2 \end{bmatrix} = \begin{bmatrix} 0 & 0 \\ 0 & 0 \end{bmatrix}.$$

This matrix equation is equivalent to

$$\begin{cases} 2a - b + 6c - 11d = 0 \\ 3a \phantom{{}-b} - c + 3d = 0 \\ -a + b + 3c - 2d = 0 \\ 4a + b + 2c + 2d = 0 \end{cases}.$$

By row reduction, or by noticing that the determinant of the coefficient matrix is zero, we see that this system has nontrivial solutions (for example, $a = -\frac{1}{2}$, $b = -3$, $c = \frac{3}{2}$, $d = 1$). Hence, S is linearly dependent. ■

The systems solved in Examples 7 and 8 are homogeneous. We can use previous results about homogeneous systems to prove the following (see Exercise 13).

COROLLARY 4.8

If S is any set in $\mathbb{R}^n$ containing k vectors, where $k > n$, then S is linearly dependent.

Most cases where we check for linear independence involve a *finite* set S. However, when S is an *infinite* subset of a vector space $\mathcal{V}$, linear independence is proved by checking all of its finite subsets. We state this generalization in the next theorem, which you are asked to prove in Exercise 23.

THEOREM 4.9: Generalization of Theorem 4.7

> Let S be a nonempty subset of a vector space $\mathcal{V}$. Then S is linearly independent if and only if, for every finite subset $\{\mathbf{v}_1, \ldots, \mathbf{v}_n\}$ of S, the equation $a_1\mathbf{v}_1 + \cdots + a_n\mathbf{v}_n = \mathbf{0}$ implies that $a_i = 0$, for every i with $1 \leq i \leq n$.

Uniqueness of Expression of a Vector as a Linear Combination

The next theorem is a very powerful result and the foundation for the rest of this chapter.

THEOREM 4.10

> Let S be a nonempty subset of a vector space $\mathcal{V}$. Then S is linearly independent if and only if every vector $\mathbf{v} \in \text{span}(S)$ can be expressed uniquely as a finite linear combination of the elements of S, if terms with zero coefficients are ignored.

The phrase "if terms with zero coefficients are ignored" means that two linear combinations of the elements of a set S are considered the same when their nonzero coefficient terms agree. For example, with $S = \{\mathbf{v}_1, \mathbf{v}_2, \mathbf{v}_3, \mathbf{v}_4\}$, we consider $a_1\mathbf{v}_1 + a_3\mathbf{v}_3$ to be the same linear combination as $a_1\mathbf{v}_1 + 0\mathbf{v}_2 + a_3\mathbf{v}_3 + 0\mathbf{v}_4$.

To prove Theorem 4.10, we prove the following equivalent statement:

> Let S be a nonempty subset of a vector space $\mathcal{V}$. Then S is linearly dependent if and only if some vector $\mathbf{v} \in \text{span}(S)$ can be represented in more than one way as a finite linear combination of the elements of S (ignoring terms with zero coefficients).

Proof of Theorem 4.10

Suppose that S is linearly dependent. Then there is a vector $\mathbf{v} \in S$ such that $\mathbf{v} \in \text{span}(S - \{\mathbf{v}\})$. Therefore, $\mathbf{v}$ can be expressed as a linear combination of vectors in S other than $\mathbf{v}$. But also, $\mathbf{v} = 1\mathbf{v}$, so there is another way of representing $\mathbf{v}(\in \text{span}(S))$ as a linear combination of elements in S.

Conversely, assume there is some vector $\mathbf{v} \in \text{span}(S)$ such that $\mathbf{v} = a_1\mathbf{v}_1 + \cdots + a_n\mathbf{v}_n$ and $\mathbf{v} = b_1\mathbf{u}_1 + \cdots + b_k\mathbf{u}_k$, with $\mathbf{v}_1, \ldots, \mathbf{v}_n$ and $\mathbf{u}_1, \ldots, \mathbf{u}_k$ vectors in S, are different linear combinations yielding $\mathbf{v}$ (that is, they differ in at least one nonzero term). Then consider the set $X = \{\mathbf{v}_1, \ldots, \mathbf{v}_n, \mathbf{u}_1, \ldots, \mathbf{u}_k\}$. Relabel the distinct vectors in X as $\{\mathbf{x}_1, \ldots, \mathbf{x}_m\}$. Then we can express $\mathbf{v} = a_1\mathbf{v}_1 + \cdots + a_n\mathbf{v}_n$ as $\mathbf{v} = c_1\mathbf{x}_1 + \cdots + c_m\mathbf{x}_m$, and we can express $\mathbf{v} = b_1\mathbf{u}_1 + \cdots + b_k\mathbf{u}_k$ as $\mathbf{v} = d_1\mathbf{x}_1 + \cdots + d_m\mathbf{x}_m$, for some scalars $c_i, d_i, 1 \leq i \leq m$. (Why?) Then $c_1\mathbf{x}_1 + \cdots + c_m\mathbf{x}_m = d_1\mathbf{x}_1 + \cdots + d_m\mathbf{x}_m$, and hence $\mathbf{0} = (c_1 - d_1)\mathbf{x}_1 + \cdots + (c_m - d_m)\mathbf{x}_m$. Because the two linear combinations for $\mathbf{v}$ differ in at least one term, some $c_i - d_i \neq 0$. Thus, we have a nontrivial linear combination of vectors in S producing the zero vector, and by Theorem 4.9, S is linearly dependent. ■

Example 9

Recall the subset $S = \{1 + x^2,\ x^2 - 2x + 3,\ x - 2\}$ of $\mathcal{P}_3$ from Example 5. In that example we proved that S is linearly independent. Notice that the polynomial

$$x^2 + 8x - 11 = 3(1 + x^2) + (-2)(x^2 - 2x + 3) + 4(x - 2)$$

is in span(S). By Theorem 4.10, this is the *only* way to express $x^2 + 8x - 11$ as a linear combination of the elements in S. ■

A Linearly Dependent Set Always Contains a Redundant Vector

The next theorem asserts that every linearly dependent subset S of a vector space has at least one redundant vector, which when removed from S does not affect the span.

THEOREM 4.11

Let S be a subset of a vector space $\mathcal{V}$. Then S is linearly dependent if and only if there is at least one vector $\mathbf{v}$ in S such that $\text{span}(S - \{\mathbf{v}\}) = \text{span}(S)$.

Theorem 4.11 and the definition of linear dependence are similar-looking statements; be sure you understand the difference between them. We leave the details of the proof of Theorem 4.11 for you to provide in Exercise 24.

The next example illustrates why a linearly dependent set must have a redundant vector.

Example 10

Consider the set of vectors $S = \{[1, 2, -1], [0, 1, 2], [2, 7, 4]\}$ in $\mathbb{R}^3$, from Example 1. Recall that S is linearly dependent because $[2, 7, 4] = 2[1, 2, -1] + 3[0, 1, 2]$.

Theorem 4.11 predicts that there is a vector $\mathbf{v}$ in S such that $\text{span}(S-\{\mathbf{v}\}) = \text{span}(S)$. Let us verify this.

Consider $\mathbf{v} = [2,7,4]$, and let $S_1 = S - \{\mathbf{v}\} = \{[1,2,-1],[0,1,2]\}$. We will show that removing $\mathbf{v}$ from S does not affect the span; that is, $\text{span}(S_1) = \text{span}(S)$. Now, since $S_1 \subseteq S$, Theorem 4.5 implies that $\text{span}(S_1) \subseteq \text{span}(S)$. Therefore, we need to show only that $\text{span}(S) \subseteq \text{span}(S_1)$. However, if $\mathbf{w} \in \text{span}(S)$, then

$$\begin{aligned}\mathbf{w} &= a[1,2,-1] + b[0,1,2] + c[2,7,4] \\ &= a[1,2,-1] + b[0,1,2] + c(2[1,2,-1] + 3[0,1,2]) \\ &= (a+2c)[1,2,-1] + (b+3c)[0,1,2],\end{aligned}$$

and so $\mathbf{w} \in \text{span}(S_1)$. Hence, $\text{span}(S) \subseteq \text{span}(S_1)$, and thus $\text{span}(S_1) = \text{span}(S)$. This argument shows that the vector $[2,7,4]$ is redundant in S. ■

Example 11

Let S be the subset $\{[1,1,0],[0,-1,0],[0,1,-1]\}$ of $\mathbb{R}^3$. It is easy to show that $\text{span}(S) = \mathbb{R}^3$ and that S is linearly independent. Thus, Theorem 4.11 predicts that none of the vectors in S is redundant. For example, consider $S_1 = S - \{[0,-1,0]\} = \{[1,1,0],[0,1,-1]\}$. Span$(S_1)$ contains precisely those vectors of the form $d[1,1,0]+e[0,1,-1] = [d, d+e, -e]$. However, $\text{span}(S_1) \neq \text{span}(S) = \mathbb{R}^3$ because it contains only vectors whose first coordinate equals the sum of the other two coordinates. (For example, $[1,1,1]$ is not in $\text{span}(S_1)$.) Thus, $[0,-1,0]$ is not a redundant vector. ■

Summary of Results

This section shows several different ways of determining whether a given set of vectors is linearly independent or dependent. The most important of these methods are summarized in Figure 4.6 (see page 204).

Exercises—Section 4.4

1. Use Theorem 4.6 to determine which of the following subsets of $\mathbb{R}^3$ are linearly independent:

★(a) $\{[1,2,-1],[3,1,-1]\}$

(b) $\{[0,1,1]\}$

★(c) $\{[4,2,1],[-1,3,7],[0,0,0]\}$

(d) $\{[2,-1,3],[4,-1,6],[-2,0,2]\}$

★(e) $\{[1,2,-5],[-2,-4,10]\}$

(f) $\{[1,9,-2],[3,4,5],[-2,5,-7]\}$

★(g) $\{[2,-5,1],[1,1,-1],[0,2,-3],[2,2,6]\}$

2. Use Theorem 4.7 to determine which of the following subsets of $\mathcal{P}_2$ are linearly independent:

Linear Independence of S	Linear Dependence of S	Source
For every $\mathbf{v} \in S$, we have $\mathbf{v} \notin \text{span}(S - \{\mathbf{v}\})$.	There is a $\mathbf{v} \in S$ such that $\mathbf{v} \in \text{span}(S - \{\mathbf{v}\})$.	Definition
Fast Method 1: When $S = \{\mathbf{v}_1, \ldots, \mathbf{v}_n\}$, for each k, $\mathbf{v}_k \notin \text{span}(\{\mathbf{v}_1, \ldots, \mathbf{v}_{k-1}\})$. (Each $\mathbf{v}_k$ is not a linear combination of the previous vectors in S.)	When $S = \{\mathbf{v}_1, \ldots, \mathbf{v}_n\}$, some $\mathbf{v}_k$ can be expressed as $\mathbf{v}_k = a_1\mathbf{v}_1 + \cdots + a_{k-1}\mathbf{v}_{k-1}$. (Some $\mathbf{v}_k$ can be expressed as a linear combination of previous vectors.)	Theorem 4.6
Fast Method 2: $\mathbf{v}_1, \ldots, \mathbf{v}_n \in S$ and $a_1\mathbf{v}_1 + \cdots + a_n\mathbf{v}_n = \mathbf{0}$ imply that $a_1 = a_2 = \cdots = a_n = 0$. (The zero vector requires zero coefficients.)	There are vectors $\mathbf{v}_1, \ldots, \mathbf{v}_n$ in S such that $a_1\mathbf{v}_1 + \cdots + a_n\mathbf{v}_n = \mathbf{0}$, for real coefficients $a_1, a_2, \ldots, a_n$, with some $a_i \neq 0$. (The zero vector does not require all coefficients to be zero.)	Theorems 4.7 and 4.9
For every $\mathbf{v} \in S$, $\text{span}(S - \{\mathbf{v}\})$ does not contain all the vectors of $\text{span}(S)$.	There is a $\mathbf{v} \in S$ such that $\text{span}(S - \{\mathbf{v}\}) = \text{span}(S)$.	Theorem 4.11
Every vector in span(S) can be *uniquely* expressed as a linear combination of the vectors in S. (No vector in S is redundant.)	*Some* vector in span(S) can be expressed in more than one way as a linear combination of the vectors in S. (There is a redundant vector in S.)	Theorem 4.10
Every finite subset of S is linearly independent.	*Some* finite subset of S is linearly dependent.	Exercise 20

Figure 4.6
Equivalent conditions for a subset S of a vector space to be linearly independent or dependent

★(a) $\{x^2 + x + 1, x^2 - 1, x^2 + 1\}$

(b) $\{x^2 - x + 3, 2x^2 - 3x - 1, 5x^2 - 9x - 7\}$

★(c) $\{2x - 6, 7x + 2, 12x - 7\}$

(d) $\{x^2 + ax + b \in \mathcal{P}_2 \mid |a| = |b| = 1\}$

3. Find a redundant vector in each given linearly dependent set:

★(a) $\{[4, -2, 6, 1], [1, 0, -1, 2], [0, 0, 0, 0], [6, -2, 5, 5]\}$

(b) $\{[1, 1, 0, 0], [1, 1, 1, 0], [0, 0, -6, 0]\}$

(c) $\{[x_1, x_2, x_3, x_4] \in \mathbb{R}^4 \mid x_i = \pm 1, \text{for each } i\}$

4. Determine which of the following subsets of $\mathcal{P}$ are linearly independent:

★(a) $\{x^2 - 1, x^2 + 1, x^2 + x\}$

(b) $\{1 + x^2 - x^3, 2x - 1, x + x^3\}$

★(c) $\{4x^2 + 2, x^2 + x - 1, x, x^2 - 5x - 3\}$

★(d) $\{1, x, x^2, x^3, \ldots\}$

(e) $\{3x^3 + 2x + 1, x^3 + x, x - 5, x^3 + x - 10\}$

★(f) $\{1, 1 + x, 1 + x + x^2, 1 + x + x^2 + x^3, \ldots\}$

5. Show that the following is a linearly dependent subset of $\mathcal{M}_{22}$:

$$\left\{\begin{bmatrix} 1 & -2 \\ 0 & 1 \end{bmatrix}, \begin{bmatrix} 3 & 2 \\ -6 & 1 \end{bmatrix}, \begin{bmatrix} 4 & -1 \\ -5 & 2 \end{bmatrix}, \begin{bmatrix} 3 & -3 \\ 0 & 0 \end{bmatrix}\right\}.$$

6. Prove that the following is linearly independent in $\mathcal{M}_{32}$:

$$\left\{\begin{bmatrix} 1 & 2 \\ -1 & 1 \\ 3 & 0 \end{bmatrix}, \begin{bmatrix} 4 & 2 \\ -6 & 1 \\ 0 & 1 \end{bmatrix}, \begin{bmatrix} 0 & 1 \\ 1 & -1 \\ 2 & 2 \end{bmatrix}, \begin{bmatrix} 0 & 7 \\ 5 & 2 \\ -1 & 6 \end{bmatrix}\right\}.$$

7. Let $S = \{[1, 1, 0], [-2, 0, 1]\}$.

(a) Show that S is a linearly independent subset of $\mathbb{R}^3$.

★(b) Find a vector $\mathbf{v}$ in $\mathbb{R}^3$ such that $S \cup \{\mathbf{v}\}$ is also linearly independent.

★(c) Is the vector $\mathbf{v}$ from part (b) unique, or could some other choice for $\mathbf{v}$ have been made? Why or why not?

★(d) Find a nonzero vector $\mathbf{u}$ in $\mathbb{R}^3$ such that $S \cup \{\mathbf{u}\}$ is linearly dependent.

8. Suppose that S is the subset $\{[2, -1, 0, 5], [1, -1, 2, 0], [-1, 0, 1, 1]\}$ of $\mathbb{R}^4$.

(a) Show that S is linearly independent.

(b) Find a linear combination of vectors in S that produces $[-2, 0, 3, -4]$ (an element of span(S)).

(c) Is there a different linear combination of the elements of S that yields $[-2, 0, 3, -4]$? If so, find one. If not, why not?

9. Consider $S = \{2x^3 - x + 3, 3x^3 + 2x - 2, x^3 - 4x + 8, 4x^3 + 5x - 7\} \subseteq \mathcal{P}_3$.

(a) Show that S is linearly dependent.

(b) Show that every three-element subset of S is linearly dependent.

(c) Prove that every subset of S containing exactly two vectors is linearly independent. (Note: There are six possible two-element subsets.)

10. Show that the set of vectors $\{[a_{11}, a_{12}, a_{13}], [a_{21}, a_{22}, a_{23}], [a_{31}, a_{32}, a_{33}]\}$ in $\mathbb{R}^3$ is linearly independent if and only if the determinant

$$\begin{vmatrix} a_{11} & a_{12} & a_{13} \\ a_{21} & a_{22} & a_{23} \\ a_{31} & a_{32} & a_{33} \end{vmatrix} \neq 0.$$

(Hint: Consider an appropriate homogeneous system of linear equations. Then apply Theorem 4.7.) Compare this exercise with Exercise 16 in Section 4.3.

11. For each of the following vector spaces, find a linearly independent subset S containing exactly four elements:

★(a) $\mathbb{R}^4$ (b) $\mathbb{R}^5$ ★ (c) $\mathcal{P}_3$ (d) $\mathcal{M}_{23}$

★(e) $\mathcal{V}$ = set of all symmetric matrices in $\mathcal{M}_{33}$

12. Let $S_1 = \{\mathbf{v}_1, \ldots, \mathbf{v}_n\}$ be a subset of a vector space $\mathcal{V}$, let c be a nonzero real number, and let $S_2 = \{c\mathbf{v}_1, \ldots, c\mathbf{v}_n\}$. Show that S_1 is linearly independent if and only if S_2 is linearly independent.

13. Prove Corollary 4.8. (Hint: Use Theorem 4.7. Construct an appropriate homogeneous system of linear equations; then determine whether the system has a nontrivial solution.)

14. Let $\mathcal{V}$ be a vector space. Prove that the empty set is a linearly independent subset of $\mathcal{V}$.

15. Let $\mathbf{f}$ be a polynomial with more than one nonzero term. Prove that the set $\{\mathbf{f}(x), x\mathbf{f}'(x)\}$ (where $\mathbf{f}'$ is the derivative of $\mathbf{f}$) is linearly independent in $\mathcal{P}$.

16. Let $\mathbf{f}$ be an nth degree polynomial in $\mathcal{P}$, and let $\mathbf{f}^{(i)}$ be the ith derivative of $\mathbf{f}$. Show that $\{\mathbf{f}, \mathbf{f}^{(1)}, \mathbf{f}^{(2)}, \ldots, \mathbf{f}^{(n)}\}$ is a linearly independent subset of $\mathcal{P}$. (Hint: Reverse the order of the elements, and use Theorem 4.6.)

17. Let $\mathcal{V}$ be a vector space, $\mathcal{W}$ a subspace of $\mathcal{V}$, S a linearly independent subset of $\mathcal{W}$, and $\mathbf{v} \in \mathcal{V} - \mathcal{W}$. Prove that $S \cup \{\mathbf{v}\}$ is linearly independent.

18. Let $\mathbf{A}$ be an $n \times m$ matrix, let $S = \{\mathbf{v}_1, \ldots, \mathbf{v}_k\}$ be a finite subset of $\mathbb{R}^m$, and let $T = \{\mathbf{A}\mathbf{v}_1, \ldots, \mathbf{A}\mathbf{v}_k\}$, a subset of $\mathbb{R}^n$.
(a) Prove that if T is a linearly independent subset of $\mathbb{R}^n$, then S is a linearly independent subset of $\mathbb{R}^m$. (Hint: Prove the contrapositive.)
★(b) Find a matrix $\mathbf{A}$ for which the converse to part (a) is false.

19. Prove that every subset of a linearly independent set is linearly independent.

20. Suppose that S is a subset of a vector space $\mathcal{V}$. Show that if every finite subset of S is linearly independent, then S itself is linearly independent.

21. Prove Theorem 4.6. (Hint: Half of the proof is done by contrapositive. For this half, assume that S is linearly dependent, and use computations similar to those in the second half of the proof of Theorem 4.7.)

22. Let S be a subset of a vector space $\mathcal{V}$.
(a) Prove that S is linearly independent if and only if *some* vector $\mathbf{v}$ in span(S) has a unique expression as a linear combination of the vectors in S (ignoring zero coefficients).
(b) The contrapositive of the statement in part (a) gives necessary and sufficient conditions for S to be linearly dependent. What are these conditions?

23. Prove Theorem 4.9 by proving the following equivalent statement:

Let S be a nonempty subset of a vector space $\mathcal{V}$. Then S is linearly dependent if and only if there is some finite subset $\{\mathbf{v}_1, \mathbf{v}_2, \ldots, \mathbf{v}_n\}$ of S such that $a_1\mathbf{v}_1 + a_2\mathbf{v}_2 + \cdots + a_n\mathbf{v}_n = \mathbf{0}$, for coefficients $a_1, a_2, \ldots, a_n$, with some $a_i \neq 0$.

Your proof should have the same general outline as the proof of Theorem 4.7. However, it will have several technical differences, because S could be infinite.

24. Prove Theorem 4.11.

4.5 Basis and Dimension

Suppose that S is a subset of a vector space $\mathcal{V}$ and that $\mathbf{v}$ is some vector in $\mathcal{V}$. We can ask two fundamental questions about S and $\mathbf{v}$:

Existence: Is there a linear combination of vectors in S equal to $\mathbf{v}$?
Uniqueness: If so, is this the only such linear combination?

The interplay between existence and uniqueness questions is a pervasive theme throughout mathematics. We studied similar questions in examining solutions to linear equations in Chapter 2. Answering the existence question is equivalent to determining whether $\mathbf{v} \in \text{span}(S)$. Answering the uniqueness question is equivalent (by Theorem 4.10) to determining whether S is linearly independent.

We are most interested in cases where both existence and uniqueness occur. In this section, we tie together these concepts by examining those subsets of vector spaces that simultaneously span and are linearly independent. Such a subset is called a **basis**.

Definition of Basis

DEFINITION

Let $\mathcal{V}$ be a vector space, and let B be a subset of $\mathcal{V}$. Then B is a **basis** for $\mathcal{V}$ if and only if both of the following are true:
(1) B spans $\mathcal{V}$.
(2) B is linearly independent.

Example 1

We show that $B = \{[1, 2, 1], [2, 3, 1], [-1, 2, -3]\}$ is a basis for $\mathbb{R}^3$ by showing that it both spans $\mathbb{R}^3$ and is linearly independent.

First we use the row space method from Section 4.3 to show that B spans $\mathbb{R}^3$. Row reducing the matrix

$$\begin{bmatrix} 1 & 2 & 1 \\ 2 & 3 & 1 \\ -1 & 2 & -3 \end{bmatrix} \quad \text{yields} \quad \begin{bmatrix} 1 & 0 & 0 \\ 0 & 1 & 0 \\ 0 & 0 & 1 \end{bmatrix},$$

which proves that $\text{span}(B) = \{a[1,0,0] + b[0,1,0] + c[0,0,1] \mid a, b, c \in \mathbb{R}\} = \mathbb{R}^3$.

Next we must show that B is linearly independent. It is easy to show that the vector equation $a[1, 2, 1] + b[2, 3, 1] + c[-1, 2, -3] = [0, 0, 0]$ has only the trivial solution $a = b = c = 0$. Therefore, B is linearly independent by Theorem 4.7. Since B has already been shown to span $\mathbb{R}^3$, B is a basis for $\mathbb{R}^3$. (B is not the only basis for $\mathbb{R}^3$, as we show in the next example.) ■

Example 2

The vector space $\mathbb{R}^n$ has $\{\mathbf{e}_1, \ldots, \mathbf{e}_n\}$ as a basis. Although $\mathbb{R}^n$ has other bases as well, the basis $\{\mathbf{e}_1, \ldots, \mathbf{e}_n\}$ is the most useful for general applications and is therefore referred to as the **standard basis** for $\mathbb{R}^n$. Thus, we refer to $\{\mathbf{i}, \mathbf{j}\}$ and $\{\mathbf{i}, \mathbf{j}, \mathbf{k}\}$ as the standard bases for $\mathbb{R}^2$ and $\mathbb{R}^3$, respectively. ■

Each of our basic examples of vector spaces also has a "standard basis."

Example 3

The standard basis in $\mathcal{M}_{32}$ is defined as the set

$$\left\{\begin{bmatrix}1&0\\0&0\\0&0\end{bmatrix}, \begin{bmatrix}0&1\\0&0\\0&0\end{bmatrix}, \begin{bmatrix}0&0\\1&0\\0&0\end{bmatrix}, \begin{bmatrix}0&0\\0&1\\0&0\end{bmatrix}, \begin{bmatrix}0&0\\0&0\\1&0\end{bmatrix}, \begin{bmatrix}0&0\\0&0\\0&1\end{bmatrix}\right\}.$$

More generally, we define the standard basis in $\mathcal{M}_{mn}$ to be the set of $m \cdot n$ different matrices

$$\{\mathbf{A}_{ij} \mid 1 \leq i \leq m, 1 \leq j \leq n\},$$

where $\mathbf{A}_{ij}$ is the $m \times n$ matrix with 1 in the $(i, j)^{th}$ position and zeroes elsewhere. You should check that these $m \cdot n$ matrices are linearly independent and do span $\mathcal{M}_{mn}$. In addition to the standard basis, $\mathcal{M}_{mn}$ has many other bases as well. ■

Example 4

We define $\{1, x, x^2, x^3\}$ to be the standard basis for $\mathcal{P}_3$. More generally, the standard basis for $\mathcal{P}_n$ is defined to be the set $\{1, x, x^2, \ldots, x^n\}$, containing $n + 1$ elements. Similarly, we define the infinite set $\{1, x, x^2, \ldots\}$ to be the standard basis for $\mathcal{P}$. Again, note that in each case these sets both span and are linearly independent.

Of course, the polynomial spaces also have other bases. For example, the following is a basis for $\mathcal{P}_4$:

$$\{x^4, x^4 - x^3, x^4 - x^3 + x^2, x^4 - x^3 + x^2 - x, x^3 - 1\}.$$

In Exercise 3, you are asked to verify that this is a basis. ■

Since a basis B for a vector space $\mathcal{V}$ spans $\mathcal{V}$, every vector in $\mathcal{V}$ is a linear combination of the vectors in B. Also, by Theorem 4.10, the fact that B is linearly independent implies that each different linear combination of vectors in B produces a different vector in $\mathcal{V}$. That is, every vector in $\mathcal{V}$ can be expressed in a unique way using the basis B for $\mathcal{V}$.

Example 5

Example 1 showed that $B = \{[1, 2, 1], [2, 3, 1], [-1, 2, -3]\}$ is a basis for $\mathbb{R}^3$. Therefore, every vector in $\mathbb{R}^3$ can be expressed in one and only one way as a linear combination of the vectors in B. For example, the vector $[-5, 6, -7]$ can be expressed in terms of B as

$$[-5, 6, -7] = 6[1, 2, 1] - 4[2, 3, 1] + 3[-1, 2, -3],$$

and no other linear combination of the vectors in B equals $[-5, 6, -7]$. ■

Finite Dimension

Most of the results in the remainder of this book will apply only to vector spaces that have some finite basis. Thus, we give this property a name.

DEFINITION

A vector space $\mathcal{V}$ is said to be **finite dimensional** if and only if $\mathcal{V}$ contains some finite subset that is a basis for $\mathcal{V}$.

$\mathbb{R}^n$, $\mathcal{P}_n$, and $\mathcal{M}_{mn}$ are all finite dimensional vector spaces, because the standard bases for these spaces are all finite. At the end of this section, we prove that all subspaces of these vector spaces are also finite dimensional. Thus, most of our basic examples have finite dimension. However, the vector space $\mathcal{P}$ is, instead, **infinite dimensional**, because no finite subset of $\mathcal{P}$ forms a basis for $\mathcal{P}$ (see Exercise 18). Be careful! Many results that are true in the finite dimensional case do not hold for infinite dimensional spaces.

The next lemma essentially says that a large enough linearly independent subset of a finite dimensional vector space must also span that vector space.

LEMMA 4.12

Let $\mathcal{V}$ be a finite dimensional vector space spanned by the n-element set $\{\mathbf{v}_1, \ldots, \mathbf{v}_n\}$, and let $S = \{\mathbf{w}_1, \ldots, \mathbf{w}_n\}$ be a set of n linearly independent vectors in $\mathcal{V}$. Then $\text{span}(S) = \mathcal{V}$, and hence S is a basis for $\mathcal{V}$.

Proof of Lemma 4.12

We must show that $S = \{\mathbf{w}_1, \ldots, \mathbf{w}_n\}$ spans $\mathcal{V}$. To do this, we will construct a sequence of n-element sets $T_0, T_1, \ldots, T_n$, with each T_i of the form $T_i = X_i \cup Y_i$, where X_i is an i-element subset of S and where Y_i is an $(n - i)$-element subset of $\{\mathbf{v}_1, \ldots, \mathbf{v}_n\}$. In particular, $T_0 = \{\mathbf{v}_1, \ldots, \mathbf{v}_n\}$ and $T_n = S = \{\mathbf{w}_1, \ldots, \mathbf{w}_n\}$. While constructing these sets, we prove inductively that each T_i spans $\mathcal{V}$. Thus, it will follow that $T_n = S$ spans $\mathcal{V}$, and the lemma will be proved.

(1) Base step: Let $X_0 = \{\ \}$ and $Y_0 = \{\mathbf{v}_1, \ldots, \mathbf{v}_n\}$. Then $T_0 = X_0 \cup Y_0 = \{\mathbf{v}_1, \ldots, \mathbf{v}_n\}$. Clearly, T_0 spans $\mathcal{V}$.

(2) Inductive step: We assume as the inductive hypothesis that, for some $k < n$, we have X_k, a k-element subset of $\{\mathbf{w}_1, \ldots, \mathbf{w}_n\}$, and Y_k, an $(n - k)$-element subset of $\{\mathbf{v}_1, \ldots, \mathbf{v}_n\}$, such that $T_k = X_k \cup Y_k$ spans $\mathcal{V}$. Our goal is to construct X_{k+1} and Y_{k+1} (having the proper form) so that $T_{k+1} = X_{k+1} \cup Y_{k+1}$ spans $\mathcal{V}$.

If Y_k contains a vector $\mathbf{w}_\alpha$ in $\{\mathbf{w}_1, \ldots, \mathbf{w}_n\}$, then let $X_{k+1} = X_k \cup \{\mathbf{w}_\alpha\}$ and $Y_{k+1} = Y_k - \{\mathbf{w}_\alpha\}$. Thus, $T_{k+1} = X_{k+1} \cup Y_{k+1} = (X_k \cup \{\mathbf{w}_\alpha\}) \cup (Y_k - \{\mathbf{w}_\alpha\}) = X_k \cup Y_k = T_k$. Hence, T_{k+1} spans $\mathcal{V}$ because T_k does, and the inductive step will be done. Therefore, we can assume that Y_k contains no $\mathbf{w}_i$.

Since X_k contains only k vectors, $k < n$, and Y_k contains none of the vectors $\mathbf{w}_1, \ldots, \mathbf{w}_n$, some $\mathbf{w}_i$ is not in T_k. Call it $\mathbf{w}_\beta$. Let $Z = T_k \cup \{\mathbf{w}_\beta\}$. Because T_k spans $\mathcal{V}$, $\mathbf{w}_\beta \in \text{span}(T_k) = \text{span}(Z - \{\mathbf{w}_\beta\})$. Thus, Z is linearly dependent.

Next, reorder the elements of Z by first listing the vectors in X_k, then $\mathbf{w}_\beta$, and finally listing the vectors in Y_k. By Theorem 4.6, some vector $\mathbf{v}$ in Z can be expressed as a linear combination of those preceding it. However, $\mathbf{v}$ cannot be among the vectors in $X_k \cup \{\mathbf{w}_\beta\}$, because Theorem 4.6 would contradict the linear independence of the set $\{\mathbf{w}_1, \ldots, \mathbf{w}_k\}$. Hence, $\mathbf{v} \in Y_k$.

Let $X_{k+1} = X_k \cup \{\mathbf{w}_\beta\}$ and $Y_{k+1} = Y_k - \{\mathbf{v}\}$. Then X_{k+1} contains $k + 1$ vectors from $\{\mathbf{w}_1, \ldots, \mathbf{w}_n\}$, and Y_{k+1} contains $n - k - 1 = n - (k + 1)$ vectors from $\{\mathbf{v}_1, \ldots, \mathbf{v}_n\}$. To finish, we show that $T_{k+1} = X_{k+1} \cup Y_{k+1}$ spans $\mathcal{V}$.

Now, $Z \subseteq \text{span}(T_{k+1})$, since every element of Z except $\mathbf{v}$ is in T_{k+1}, and $\mathbf{v}$ is a linear combination of other vectors in Z. Therefore, $\text{span}(Z) \subseteq \text{span}(T_{k+1})$, by Theorem 4.4. We also know that $\mathcal{V} = \text{span}(T_k) \subseteq \text{span}(Z)$, so we have $\mathcal{V} \subseteq \text{span}(T_{k+1})$, proving that T_{k+1} spans $\mathcal{V}$ and completing the proof.

■

Dimension

We can now prove the main result of this section, which states that all bases for the same finite dimensional vector space have the same number of elements. In what follows, if S is a finite set, we let the notation $|S|$ represent the number of elements in S.

THEOREM 4.13

Let $\mathcal{V}$ be a finite dimensional vector space, and let B be a basis for $\mathcal{V}$. Then B has finitely many elements. Furthermore, every basis for $\mathcal{V}$ has size $|B|$.

Proof of Theorem 4.13

Because $\mathcal{V}$ is finite dimensional, some finite subset A of $\mathcal{V}$ is a basis for $\mathcal{V}$. Let $|A| = n$. If B is any other basis for $\mathcal{V}$, we must show that $|B| = |A|$.

Suppose that $|B| > |A|$. Then there is some subset $\{\mathbf{w}_1, \ldots, \mathbf{w}_{n+1}\}$ of B containing $n+1$ vectors. Since B is linearly independent and every subset of a linearly independent subset is linearly independent (see Exercise 19 in Section 4.4), $\{\mathbf{w}_1, \ldots, \mathbf{w}_{n+1}\}$ is linearly independent. Similarly, $\{\mathbf{w}_1, \ldots, \mathbf{w}_n\}$ is linearly independent. But A spans $\mathcal{V}$ and $|A| = n$, so Lemma 4.12 implies that $\{\mathbf{w}_1, \ldots, \mathbf{w}_n\}$ is a basis for $\mathcal{V}$. Hence, $\mathbf{w}_{n+1} \in \mathcal{V} = \text{span}(\{\mathbf{w}_1, \ldots, \mathbf{w}_n\})$. This statement contradicts the linear independence of $\{\mathbf{w}_1, \ldots, \mathbf{w}_{n+1}\}$. Therefore, $|B| \leq |A|$.

A similar argument with the roles of A and B reversed shows that $|A| \leq |B|$. Hence, $|B| = |A|$. ∎

DEFINITION

Let $\mathcal{V}$ be a finite dimensional vector space. Then the **dimension** of $\mathcal{V}$, $\dim(\mathcal{V})$, is the number of elements in any basis for $\mathcal{V}$.

Theorem 4.13 had to be proved before we could formally make this definition. Without that result, the dimension of a vector space would not be a well-defined quantity. That is, Theorem 4.13 assures us that two different bases for $\mathcal{V}$ could not produce different values for $\dim(\mathcal{V})$.

Example 6

The dimension of $\mathbb{R}^3$ is clearly 3, because $\mathbb{R}^3$ has the (standard) basis $\{\mathbf{i}, \mathbf{j}, \mathbf{k}\}$. Theorem 4.13 then implies that every other basis for $\mathbb{R}^3$ also has exactly three elements. More generally, we can see that $\dim(\mathbb{R}^n) = n$, since $\mathbb{R}^n$ has the basis $\{\mathbf{e}_1, \ldots, \mathbf{e}_n\}$. ∎

Example 7

Because the standard basis $\{1, x, x^2, x^3\}$ for $\mathcal{P}_3$ has four elements, $\dim(\mathcal{P}_3) = 4$. Every other basis for $\mathcal{P}_3$, such as $\{x^3 - x, x^2 + x + 1, x^3 + x - 5, 2x^3 + x^2 + x - 3\}$, also has four elements. (Verify that this set is a basis for $\mathcal{P}_3$.)

Also, $\dim(\mathcal{P}_n) = n+1$, since $\mathcal{P}_n$ has the basis $\{1, x, x^2, \ldots, x^n\}$, containing $n+1$ elements. Be careful! Many students erroneously try to determine the dimension of $\mathcal{P}_n$ just by looking at the subscript n. ■

Example 8

The standard basis for $\mathcal{M}_{22}$ contains four elements. Hence, $\dim(\mathcal{M}_{22}) = 4$. In general, from the size of the standard basis for $\mathcal{M}_{mn}$, we see that $\dim(\mathcal{M}_{mn}) = m \cdot n$. ■

Example 9

Let $\mathcal{V} = \{\mathbf{0}\}$ be the trivial vector space. Then $\dim(\mathcal{V}) = 0$, because the empty set, which contains no elements, is a basis for $\mathcal{V}$. (Why?) ■

Example 10

Consider the subset $S = \{x^2 - 3x + 5, 3x^3 + 4x - 8, 6x^3 - x^2 + 11x - 21, 2x^5 - 7x^3 + 5x\}$ of $\mathcal{P}_5$. It is easy to show that the subset $B = \{x^2 - 3x + 5, 3x^3 + 4x - 8, 2x^5 - 7x^3 + 5x\}$ of S is linearly independent. (Verify.) You can also check that the polynomial $6x^3 - x^2 + 11x - 21$ in S is in span(B), implying that $S \subseteq \text{span}(B)$. Thus, the three-element set B is a basis for $\mathcal{W} = \text{span}(S)$ and so $\mathcal{W}$ has dimension 3.

Although B is a basis for $\mathcal{W}$, B is not a basis for $\mathcal{P}_5$ (even though B is a linearly independent subset of $\mathcal{P}_5$). Every basis for $\mathcal{P}_5$ must have $\dim(\mathcal{P}_5) = 6$ elements. However, $|B| = 3$. Thus, $\mathcal{W}$ is a proper subspace of $\mathcal{P}_5$. ■

Using Row Reduction to Construct a Basis

We now present a common method for finding a basis for a subspace of $\mathbb{R}^n$, which we will refer to as the "row reduction method." This method can be adapted to other vector spaces, as in Example 12.

Method for Finding a Basis Using Row Reduction

Suppose that S is a finite nonempty subset of $\mathbb{R}^n$ containing k vectors, and let $\mathcal{V} = \text{span}(S)$.

Step 1: Form a $k \times n$ matrix $\mathbf{A}$ by using the vectors in S as the rows of $\mathbf{A}$.
Step 2: Let $\mathbf{C}$ be the reduced row echelon form matrix for $\mathbf{A}$.
Step 3: The set B of vectors comprising the nonzero rows of $\mathbf{C}$ is a basis for $\mathcal{V}$.

An analysis of the reduced row echelon form will convince you that the rows of $\mathbf{C}$ are linearly independent. Theorem 2.7 implies that they span $\mathcal{V}$. Thus, B must be a basis for $\mathcal{V}$.

Example 11

Let S be the subset $\{[2, -2, 3, 5, 5], [-1, 1, 4, 14, -8], [4, -4, -2, -14, 18], [3, -3, -1, -9, 13]\}$ of $\mathbb{R}^5$. We can use the row reduction method to find a basis B for $\mathcal{V} = \text{span}(S)$.

Step 1: We construct the following matrix:

$$\mathbf{A} = \begin{bmatrix} 2 & -2 & 3 & 5 & 5 \\ -1 & 1 & 4 & 14 & -8 \\ 4 & -4 & -2 & -14 & 18 \\ 3 & -3 & -1 & -9 & 13 \end{bmatrix}.$$

Step 2: Using row reduction on $\mathbf{A}$ yields

$$\mathbf{C} = \begin{bmatrix} 1 & -1 & 0 & -2 & 4 \\ 0 & 0 & 1 & 3 & -1 \\ 0 & 0 & 0 & 0 & 0 \\ 0 & 0 & 0 & 0 & 0 \end{bmatrix}. \quad \text{(Verify.)}$$

Step 3: The desired basis for $\mathcal{V}$ is the set $B = \{[1, -1, 0, -2, 4], [0, 0, 1, 3, -1]\}$, the collection of the nonzero rows of $\mathbf{C}$. Thus, $\dim(\mathcal{V}) = 2$. ■

In general, the row reduction method for creating a basis produces vectors whose form is simpler than the vectors in the original spanning set. This effect occurs because a reduced row echelon form matrix has the simplest form of all matrices that are row equivalent to it.

Example 12

Recall from Example 10 the subset $S = \{x^2 - 3x + 5, 3x^3 + 4x - 8, 6x^3 - x^2 + 11x - 21, 2x^5 - 7x^3 + 5x\}$ of $\mathcal{P}_5$. In that example, we found that $\mathcal{W} = \text{span}(S)$ has dimension 3 by finding the three-element basis $B = \{x^2 - 3x + 5, 3x^3 + 4x - 8, 2x^5 - 7x^3 + 5x\}$. Now we use the row reduction method to find a different basis for $\mathcal{W}$.

Since S is a subset of $\mathcal{P}_5$ instead of $\mathbb{R}^n$, we must alter our method slightly. We cannot use the polynomials in S themselves as rows of a matrix, so we "peel off" the coefficients of the polynomials to create four 6-vectors, which we use as the rows of the matrix. That is, we take

$$\begin{array}{cccccc} x^5 & x^4 & x^3 & x^2 & x & 1 \end{array}$$

$$\mathbf{A} = \begin{bmatrix} 0 & 0 & 0 & 1 & -3 & 5 \\ 0 & 0 & 3 & 0 & 4 & -8 \\ 0 & 0 & 6 & -1 & 11 & -21 \\ 2 & 0 & -7 & 0 & 5 & 0 \end{bmatrix}.$$

Row reducing this matrix produces

$$\mathbf{C} = \begin{array}{c} \begin{matrix} x^5 & x^4 & x^3 & x^2 & x & 1 \end{matrix} \\ \left[\begin{matrix} 1 & 0 & 0 & 0 & \frac{43}{6} & -\frac{28}{3} \\ 0 & 0 & 1 & 0 & \frac{4}{3} & -\frac{8}{3} \\ 0 & 0 & 0 & 1 & -3 & 5 \\ 0 & 0 & 0 & 0 & 0 & 0 \end{matrix}\right] \end{array}.$$

This matrix yields an alternate three-element basis for $\mathcal{W}$: $D = \{x^5 + \frac{43}{6}x - \frac{28}{3}, x^3 + \frac{4}{3}x - \frac{8}{3}, x^2 - 3x + 5\}$. ■

Subspaces of a Finite Dimensional Vector Space

We conclude this section by showing that every subspace of a finite dimensional vector space is also finite dimensional. This result is important: it tells us that the theorems we have developed apply to all subspaces of our basic examples $\mathbb{R}^n$, $\mathcal{M}_{mn}$, and $\mathcal{P}_n$.

THEOREM 4.14

> Let $\mathcal{V}$ be a finite dimensional vector space, and let $\mathcal{W}$ be a subspace of $\mathcal{V}$. Then $\mathcal{W}$ is also finite dimensional with $\dim(\mathcal{W}) \le \dim(\mathcal{V})$. Moreover, $\dim(\mathcal{W}) = \dim(\mathcal{V})$ if and only if $\mathcal{W} = \mathcal{V}$.

Before proving this theorem, let us consider some examples.

Example 13

Consider the nested sequence of subspaces of $\mathbb{R}^3$ given by $\{\mathbf{0}\} \subset$ {scalar multiples of $[4, -7, 0]\} \subset xy$-plane $\subset \mathbb{R}^3$. Their respective dimensions are 0, 1, 2, and 3. (Why?) Hence, the dimensions of each pair of these subspaces satisfy the inequality given in Theorem 4.14. ■

Example 14

It is easy to show that $B = \{x^3 + 2x^2 - 4x + 18, 3x^2 + 4x - 4, x^3 + 5x^2 - 3, 3x + 2\}$ is a linearly independent subset of $\mathcal{P}_3$. Therefore, B is a basis for $\mathcal{W} = \text{span}(B)$. (Why?) Clearly, $\dim(\mathcal{W}) = 4$. But since $\mathcal{W}$ is a subspace of $\mathcal{P}_3$ and $\dim(\mathcal{P}_3) = 4$, Theorem 4.14 implies that $\mathcal{W} = \mathcal{P}_3$. Hence, B is a basis for $\mathcal{P}_3$. ■

The following lemma is needed for the proof of Theorem 4.14.

LEMMA 4.15

Let $\mathcal{V}$ be a vector spacc with spanning set S, and let B be a maximal linearly independent subset of S. Then B is a basis for $\mathcal{V}$.

The phrase "B is a maximal linearly independent subset of S" means that all of the following are true:

B is a subset of S.
B is linearly independent.
If $C \subseteq S$ with $B \subset C$ and $B \neq C$, then C is linearly dependent.

Proof of Lemma 4.15

Let $\mathcal{V}$ be a vector space with spanning set S and maximal linearly independent subset B. Now, since B is linearly independent, we need only show that it spans $\mathcal{V}$ in order to show that it is a basis for $\mathcal{V}$. If we can prove that $S \subseteq \text{span}(B)$, then $\mathcal{V} = \text{span}(S) \subseteq \text{span}(B)$ by part (3) of Theorem 4.4, and B spans $\mathcal{V}$. Thus, it is sufficient to show that $S \subseteq \text{span}(B)$.

Let $\mathbf{w} \in S$. We must show that $\mathbf{w} \in \text{span}(B)$. This is clearly true if $\mathbf{w} \in B$. If, however, $\mathbf{w} \notin B$, let $C = B \cup \{\mathbf{w}\}$. Because B is a *maximal* linearly independent subset of S, C must be linearly dependent. Thus by Theorem 4.9, for some finite set $\{\mathbf{v}_1, \mathbf{v}_2, \ldots, \mathbf{v}_k\}$ of vectors in C, $b_1\mathbf{v}_1 + b_2\mathbf{v}_2 + \cdots + b_k\mathbf{v}_k = \mathbf{0}$ with at least one nonzero coefficient. Now, some $\mathbf{v}_i = \mathbf{w}$; if not, all $\mathbf{v}_i$ are in B, which would contradict the linear independence of B. If we reorder the vectors so that $\mathbf{w} = \mathbf{v}_1$, then $b_1\mathbf{w} + b_2\mathbf{v}_2 + \cdots + b_k\mathbf{v}_k = \mathbf{0}$. We must have $b_1 \neq 0$, or the linear independence of B is contradicted again. We can now solve for $\mathbf{w}$, obtaining

$$\mathbf{w} = -\frac{b_2}{b_1}\mathbf{v}_2 - \cdots - \frac{b_k}{b_1}\mathbf{v}_k,$$

thus showing that $\mathbf{w} \in \text{span}(B)$, as required. ■

Example 15

Consider the subset $S = \{[1, -2, 1], [3, 1, -2], [5, -3, 0], [5, 4, -5], [0, 0, 0]\}$ of $\mathbb{R}^3$ and the subset $B = \{[1, -2, 1], [5, -3, 0]\}$ of S. We show that B is a maximal linearly independent subset of S and hence, by Lemma 4.15, that it is a basis for span(S).

Clearly, $B \subseteq S$. Also, B is obviously linearly independent. The following equations show that, if any of the remaining vectors of S are added to B, the set is no longer linearly independent:

$$\begin{array}{rcrcr}[3,1,-2] & = & -2\,[1,-2,1] & + & [5,-3,0]\,, \\ [5,4,-5] & = & -5\,[1,-2,1] & + & 2\,[5,-3,0]\,, \\ [0,0,0] & = & 0\,[1,-2,1] & + & 0\,[5,-3,0]\,.\end{array}$$

Thus, B is a maximal linearly independent subset of S and so is a basis for span(S). ■

Proof of Theorem 4.14

We first show that $\mathcal{W}$ is finite dimensional by proving that $\mathcal{W}$ has a finite basis. (This is trickier than you might suspect, because it is not clear that $\mathcal{W}$ has any basis at all.)

Consider the set A of nonnegative integers k for which there is a linearly independent subset T of $\mathcal{W}$ of k elements. That is, $A = \{k \mid \text{a set } T \text{ exists with } |T| = k,\ T \subseteq \mathcal{W}, \text{ and } T \text{ linearly independent}\}$. Now, $0 \in A$, since the empty set is a linearly independent subset of $\mathcal{W}$.

Next we notice that no number larger than $\dim(\mathcal{V})$ can be in A. This is because a linearly independent subset of $\mathcal{W}$ is also a linearly independent subset of $\mathcal{V}$, and no such subset has size larger than $\dim(\mathcal{V})$ by an argument similar to that in the proof of Theorem 4.13. Thus, A is a finite, nonempty set of integers.

Let n represent the largest number in A. Then there is a linearly independent set $T \subseteq \mathcal{W}$ with $|T| = n$. Now, T is a maximal linearly independent subset of $\mathcal{W}$. (Why?) Thus, T is a (finite) basis for $\mathcal{W}$ by Lemma 4.15 (where $\mathcal{W}$ itself plays the role of the spanning set in that lemma). Hence, $\mathcal{W}$ is finite dimensional.

Because $n \in A$ and no element of A is larger than $\dim(\mathcal{V})$, we have $\dim(\mathcal{W}) = |T| = n \leq \dim(\mathcal{V})$.

Finally, if $\dim(\mathcal{W}) = \dim(\mathcal{V})$, then T is a linearly independent subset of $\mathcal{V}$ with $|T| = \dim(\mathcal{V})$, and so T is a basis for $\mathcal{V}$ by Lemma 4.12. Thus, $\mathcal{W} = \text{span}(T) = \mathcal{V}$. Conversely, if $\mathcal{W} = \mathcal{V}$, it is clearly true that $\dim(\mathcal{W}) = \dim(\mathcal{V})$. ■

Exercises—Section 4.5

1. Prove that each of the following subsets of $\mathbb{R}^4$ is a basis for $\mathbb{R}^4$ by showing that it both spans $\mathbb{R}^4$ and is linearly independent:

   (a) $\{[2,1,0,0],\ [0,1,1,-1], [0,-1,2,-2], [3,1,0,-2]\}$

   (b) $\{[6,1,1,-1],\ [1,0,0,9], [-2,3,2,4],\ [2,2,5,-5]\}$

   (c) $\{[1,1,1,1],\ [1,1,1,-1], [1,1,-1,-1],\ [1,-1,-1,-1]\}$

   (d) $\{[\frac{15}{2}, 5, \frac{12}{5}, 1], [2, \frac{1}{2}, \frac{3}{4}, 1], [\frac{-13}{2}, 1, 0, 4], [\frac{18}{5}, 0, \frac{1}{5}, -\frac{1}{5}]\}$

2. Prove that the following set is a basis for $\mathcal{M}_{22}$ by showing that it spans $\mathcal{M}_{22}$ and is linearly independent:

$$\left\{\begin{bmatrix}1 & 4\\ 2 & 0\end{bmatrix}, \begin{bmatrix}0 & 2\\ 1 & 0\end{bmatrix}, \begin{bmatrix}-3 & 1\\ -1 & 0\end{bmatrix}, \begin{bmatrix}5 & -2\\ 0 & -3\end{bmatrix}\right\}.$$

3. Show that the subset $\{x^4, x^4 - x^3, x^4 - x^3 + x^2, x^4 - x^3 + x^2 - x, x^3 - 1\}$ of $\mathcal{P}_4$ is a basis for $\mathcal{P}_4$.

4. Determine which of the following subsets of $\mathbb{R}^4$ form a basis for $\mathbb{R}^4$:

★(a) $S = \{[7, 1, 2, 0], [8, 0, 1, -1], [1, 0, 0, -2]\}$

(b) $S = \{[1, 3, 2, 0], [-2, 0, 6, 7], [0, 6, 10, 7]\}$

★(c) $S = \{[7, 1, 2, 0], [8, 0, 1, -1], [1, 0, 0, -2], [3, 0, 1, -1]\}$

(d) $S = \{[1, 3, 2, 0], [-2, 0, 6, 7], [0, 6, 10, 7], [2, 10, -3, 1]\}$

★(e) $S = \{[2, 17, 3, -1], [1, -1, 0, 3], [5, 0, -7, 19], [12, 0, -8, 3], [-9, 7, 2, 0]\}$

5. For each of the given subsets S of $\mathbb{R}^5$, find a basis for $\mathcal{V} = \text{span}(S)$ by row reducing an appropriate matrix:

★(a) $S = \{[1, 2, 3, -1, 0], [3, 6, 8, -2, 0], [-1, -1, -3, 1, 1], [-2, -3, -5, 1, 1]\}$

(b) $S = \{[3, 2, -1, 0, 1], [1, -1, 0, 3, 1], [4, 1, -1, 3, 2], [3, 7, -2, -9, -1], [-1, -4, 1, 6, 1]\}$

(c) $S = \{[0, 1, 1, 0, 6], [2, -1, 0, -2, 1], [-1, 2, 1, 1, 2], [3, -2, 0, -2, -3], [1, 1, 1, -1, 4], [2, -1, -1, 1, 3]\}$

★(d) $S = \{[1, 1, 1, 1, 1], [1, 2, 3, 4, 5], [0, 1, 2, 3, 4], [0, 0, 4, 0, -1]\}$

★6. Adapt the row reduction method to find a basis for the subspace of $\mathcal{P}_3$ spanned by $S = \{x^3 - 3x^2 + 2, 2x^3 - 7x^2 + x - 3, 4x^3 - 13x^2 + x + 5\}$.

★7. Adapt the row reduction method to find a basis for the subspace of $\mathcal{M}_{32}$ spanned by

$$S = \left\{ \begin{bmatrix} 1 & 4 \\ 0 & -1 \\ 2 & 2 \end{bmatrix}, \begin{bmatrix} 2 & 5 \\ 1 & -1 \\ 4 & 9 \end{bmatrix}, \begin{bmatrix} 1 & 7 \\ -1 & -2 \\ 2 & -3 \end{bmatrix}, \begin{bmatrix} 3 & 6 \\ 2 & -1 \\ 6 & 12 \end{bmatrix} \right\}.$$

8. (a) Show that $B = \{[2, 3, 0, -1], [-1, 1, 1, -1]\}$ is a maximal linearly independent subset of $S = \{[1, 4, 1, -2], [-1, 1, 1, -1], [3, 2, -1, 0], [2, 3, 0, -1]\}$.

★(b) Calculate $\dim(\text{span}(S))$.

★(c) Does $\text{span}(S) = \mathbb{R}^4$? Why or why not?

9. (a) Show that $B = \{x^3 - x^2 + 2x + 1, 2x^3 + 4x - 7, 3x^3 - x^2 - 6x + 6\}$ is a maximal linearly independent subset of $S = \{x^3 - x^2 + 2x + 1, x - 1, 2x^3 + 4x - 7, x^3 - 3x^2 - 22x + 34, 3x^3 - x^2 - 6x + 6\}$.

(b) Calculate $\dim(\text{span}(S))$.

(c) Does $\text{span}(S) = \mathcal{P}_3$? Why or why not?

10. Let $\mathcal{W}$ be the solution set to the matrix equation $\mathbf{AX} = \mathbf{O}$, where

$$\mathbf{A} = \begin{bmatrix} 1 & 2 & 1 & 0 & -1 \\ 2 & -1 & 0 & 1 & 3 \\ 1 & -3 & -1 & 1 & 4 \\ 2 & 9 & 4 & -1 & -7 \end{bmatrix}.$$

(a) Show that $\mathcal{W}$ is a subspace of $\mathbb{R}^5$.
(b) Find a basis for $\mathcal{W}$.
(c) Show that $\dim(\mathcal{W}) + \text{rank}(\mathbf{A}) = 5$.

11. Prove that every proper nontrivial subspace of $\mathbb{R}^3$ can be thought of, from a geometric point of view, as either a line through the origin or a plane through the origin.

12. Let $\mathbf{f}$ be a polynomial of degree n. Show that the set $\{\mathbf{f}, \mathbf{f}^{(1)}, \mathbf{f}^{(2)}, \ldots, \mathbf{f}^{(n)}\}$ is a basis for $\mathcal{P}_n$ (where $\mathbf{f}^{(i)}$ denotes the ith derivative of $\mathbf{f}$). (Hint: See Exercise 16 in Section 4.4.)

13. (a) Let $\mathbf{A}$ be a 2×2 matrix. Prove that there are real numbers $a_0, a_1, \ldots, a_4$, not all zero, such that $a_4\mathbf{A}^4 + a_3\mathbf{A}^3 + a_2\mathbf{A}^2 + a_1\mathbf{A} + a_0\mathbf{I}_2 = \mathbf{O}_2$. (Hint: You can assume that $\mathbf{A}$, $\mathbf{A}^2$, $\mathbf{A}^3$, $\mathbf{A}^4$, $\mathbf{I}_2$ are all distinct, since the statement is clearly true if they are not.)
(b) Suppose $\mathbf{B}$ is an $n \times n$ matrix. Show that there must be a nonzero polynomial $\mathbf{p} \in \mathcal{P}_{n^2}$ such that $\mathbf{p}(\mathbf{B}) = \mathbf{O}_n$.

14. (a) Show that $B = \{(x-2), x(x-2), x^2(x-2), x^3(x-2), x^4(x-2)\}$ is a basis for $\mathcal{V} = \{\mathbf{p} \in \mathcal{P}_5 \mid \mathbf{p}(2) = 0\}$.
★(b) What is $\dim(\mathcal{V})$?
★(c) Find a basis for $\mathcal{W} = \{\mathbf{p} \in \mathcal{P}_5 \mid \mathbf{p}(2) = \mathbf{p}(3) = 0\}$.
★(d) Calculate $\dim(\mathcal{W})$.

15. (a) Let S be a finite subset of $\mathbb{R}^n$. Prove that the row reduction method applied to S produces the standard basis for $\mathbb{R}^n$ if and only if $\text{span}(S) = \mathbb{R}^n$.
(b) Let $B \subseteq \mathbb{R}^n$ with $|B| = n$, and let $\mathbf{A}$ be the $n \times n$ matrix whose rows are the vectors in B. Prove that B is a basis for $\mathbb{R}^n$ if and only if $|\mathbf{A}| \neq 0$.

16. Show that the rank of an $n \times m$ matrix $\mathbf{A}$ is the same as the dimension of its row space.

17. Let $\mathbf{A}$ be a nonsingular $n \times n$ matrix, and let B be a basis for $\mathbb{R}^n$.
(a) Show that $B_1 = \{\mathbf{A}\mathbf{v} \mid \mathbf{v} \in B\}$ is also a basis for $\mathbb{R}^n$. (Think of the vectors in B as column vectors.)
(b) Show that $B_2 = \{\mathbf{v}\mathbf{A} \mid \mathbf{v} \in B\}$ is also a basis for $\mathbb{R}^n$. (Think of the vectors in B as row vectors.)
(c) Letting B be the standard basis for $\mathbb{R}^n$, use the result of part (a) to show that the columns of $\mathbf{A}$ form a basis for $\mathbb{R}^n$.
(d) Prove that the rows of $\mathbf{A}$ form a basis for $\mathbb{R}^n$.

18. Prove that $\mathcal{P}$ is infinite dimensional by showing that no finite subset S of $\mathcal{P}$ can span $\mathcal{P}$, as follows:
(a) Let S be a finite subset of $\mathcal{P}$. Show that $S \subseteq \mathcal{P}_n$, for some n.
(b) Use part (a) to prove that $\text{span}(S) \subseteq \mathcal{P}_n$.
(c) Conclude that S cannot span $\mathcal{P}$.

19. (a) Prove that if a vector space $\mathcal{V}$ has an infinite linearly independent subset, then $\mathcal{V}$ is not finite dimensional.

(b) Use part (a) to prove that any vector space having $\mathcal{P}$ as a subspace is not finite dimensional.

20. Let S be a subset of a finite dimensional vector space $\mathcal{V}$. Prove that S is a basis for $\mathcal{V}$ if and only if S is a minimal spanning set for $\mathcal{V}$.

21. Let $\mathcal{V}$ be a subspace of $\mathbb{R}^n$ with $\dim(\mathcal{V}) = n - 1$. (Such a subspace is called a **hyperplane** in $\mathbb{R}^n$.) Prove that there is a nonzero $\mathbf{x} \in \mathbb{R}^n$ such that $\mathcal{V} = \{\mathbf{v} \in \mathbb{R}^n \mid \mathbf{x} \cdot \mathbf{v} = 0\}$. (Hint: Use a basis for $\mathcal{V}$ to set up a homogeneous system of equations whose coefficient matrix has the vectors in this basis as its rows. Then notice that this system has at least one nontrivial solution, say $\mathbf{x}$.)

22. Let $\alpha_1, \ldots, \alpha_n$ and $\beta_1, \ldots, \beta_n$ be any real numbers, with $n > 2$. Consider the $n \times n$ matrix $\mathbf{A}$ whose $(i, j)^{th}$ term is $a_{ij} = \sin(\alpha_i + \beta_j)$. Prove that $|\mathbf{A}| = 0$. (Hint: Consider $\mathbf{x}_1 = [\sin\beta_1, \sin\beta_2, \ldots, \sin\beta_n]$, $\mathbf{x}_2 = [\cos\beta_1, \cos\beta_2, \ldots, \cos\beta_n]$. Show that $\text{span}(\{\mathbf{x}_1, \mathbf{x}_2\})$ contains the row space of $\mathbf{A}$.)

4.6 Constructing Special Bases

In this section, we present two additional methods for finding a basis for a given finite dimensional vector space. In particular, we show that, for a finite dimensional vector space, a spanning set can always be "shrunk" to a basis and a linearly independent set can always be "enlarged" to a basis.

Every Spanning Set for a Finite Dimensional Vector Space Contains a Basis

The next theorem asserts that any spanning set of $\mathcal{V}$, including an infinite one, must contain a basis for $\mathcal{V}$.

THEOREM 4.16

If S is any spanning set for a finite dimensional vector space $\mathcal{V}$, then there is a set $B \subseteq S$ that is a basis for $\mathcal{V}$.

The proof of this theorem is very similar to the first part of the proof of Theorem 4.14. We leave it for you to do in Exercise 13. Theorem 4.16 has an important corollary.

COROLLARY 4.17

Let $\mathcal{V}$ be a finite dimensional vector space, and let S be a subset of $\mathcal{V}$ that spans $\mathcal{V}$. Then, if S is finite, $\dim(\mathcal{V}) \leq |S|$. Additionally, S is a finite spanning set with $|S| = \dim(\mathcal{V})$ if and only if S is a basis for $\mathcal{V}$.

Proof of Corollary 4.17

Suppose that S is finite. Then Theorem 4.16 implies that there is a basis B for $\mathcal{V}$ with $B \subseteq S$. Therefore, $\dim(\mathcal{V}) = |B| \leq |S|$.

Next, assume that S is a finite spanning set with $\dim(\mathcal{V}) = |S|$. We want to prove that S is a basis for $\mathcal{V}$. Again, Theorem 4.16 gives us a basis B for $\mathcal{V}$ with $B \subseteq S$. However, $|B| = \dim(\mathcal{V}) = |S|$ implies that B and S must be the same size. Since S is finite, $B = S$. Therefore, S is a basis for $\mathcal{V}$.

Finally, suppose that S is a basis for a finite dimensional vector space $\mathcal{V}$. Then, by definition, S is a finite spanning set for $\mathcal{V}$ with $|S| = \dim(\mathcal{V})$. ■

Example 1

Let $S = \{[1, 3, -2], [2, 1, 4], [0, 5, -8], [1, -7, 14]\}$, and let $\mathcal{V} = \text{span}(S)$. Corollary 4.17 predicts that $\dim(\mathcal{V}) \leq |S| = 4$. In fact, $\dim(\mathcal{V}) \leq \dim(\mathbb{R}^3) = 3$, by Theorem 4.14. Also, it is easy to show that the subset $B = \{[1, 3, -2], [2, 1, 4]\}$ is a maximal linearly independent subset of S. Hence, B is a basis for $\mathcal{V}$ contained in S, thus illustrating Theorem 4.16. ■

Shrinking a Finite Spanning Set to a Basis

We now examine another method for finding a basis for a finite dimensional vector space $\mathcal{V}$. This method assumes that we begin with a finite spanning set S. Our goal is to find a basis B for $\mathcal{V}$ with $B \subseteq S$. This technique resembles a proof by induction in that there is a "base" step followed by an "inductive" step that is repeated until the desired basis is found.†

METHOD FOR SHRINKING A FINITE SPANNING SET TO A BASIS

Let S be a finite set of vectors spanning a vector space $\mathcal{V}$.

(1) Base step: Choose some nonzero vector in S. Call this vector $\mathbf{v}_1$.

Repeat the following step as many times as possible:

†We assume that S has at least one nonzero vector. Otherwise, $\mathcal{V}$ would be the trivial vector space, because S spans $\mathcal{V}$. In this case, the desired basis for $\mathcal{V}$ is the empty set, { }.

(2) Inductive step: Assuming that vectors $\mathbf{v}_1, \ldots, \mathbf{v}_{k-1}$ have been chosen from S, choose another vector $\mathbf{v}_k \in S$ such that $\mathbf{v}_k \notin \operatorname{span}(\{\mathbf{v}_1, \ldots, \mathbf{v}_{k-1}\})$.

The final set constructed is a basis for $\mathcal{V}$.

We will refer to this method as the "shrinking method." The idea behind this technique is to construct the largest linearly independent subset of S that we can. We choose only those vectors of the spanning set S that are not redundant, thus maintaining linear independence throughout the process. The method stops when we have constructed a set of vectors $\{\mathbf{v}_1, \ldots, \mathbf{v}_n\}$ such that $\operatorname{span}(\{\mathbf{v}_1, \ldots, \mathbf{v}_n\}) = \operatorname{span}(S) = \mathcal{V}$. Sometimes this happens when there are no more vectors remaining in S to choose from. In any event, the process stops after a finite number of steps. The final set $\{\mathbf{v}_1, \ldots, \mathbf{v}_n\}$ is the desired maximal linearly independent subset of S and hence a basis for $\mathcal{V}$ by Lemma 4.15.

Example 2

Let $S = \{[1,3,2,-1], [-2,-6,-4,2], [2,4,-1,0], [0,1,0,-1], [1,3,1,-1]\}$, a subset of $\mathbb{R}^4$. Let $\mathcal{V} = \operatorname{span}(S)$, a subspace of $\mathbb{R}^4$. We use the "shrinking method" to find a subset B of S that is a basis for $\mathcal{V}$.

The base step is to choose $\mathbf{v}_1$, a nonzero vector in S. We let $\mathbf{v}_1 = [1,3,2,-1]$.

Moving on to the inductive step, we look for $\mathbf{v}_2$ in S so that $\mathbf{v}_2 \notin \operatorname{span}(\{\mathbf{v}_1\})$. Hence, $\mathbf{v}_2$ may not be a scalar multiple of $\mathbf{v}_1$. Therefore, we may not choose $[-2,-6,-4,2]$. Instead, we choose $\mathbf{v}_2 = [2,4,-1,0]$.

Repeating the inductive step, we look for $\mathbf{v}_3$ in S so that $\mathbf{v}_3 \notin \operatorname{span}(\{\mathbf{v}_1, \mathbf{v}_2\})$. We try $[0,1,0,-1]$ as a candidate for $\mathbf{v}_3$. To show that $[0,1,0,-1] \notin \operatorname{span}(\{\mathbf{v}_1, \mathbf{v}_2\})$, we must determine whether there are any solutions a and b to the equation

$$[0,1,0,-1] = a\mathbf{v}_1 + b\mathbf{v}_2 = a[1,3,2,-1] + b[2,4,-1,0].$$

To do so, we create the system

$$\begin{cases} a + 2b = 0 \\ 3a + 4b = 1 \\ 2a - b = 0 \\ -a = -1 \end{cases}.$$

A simple computation shows that this system has no solutions. Therefore, we choose $\mathbf{v}_3 = [0,1,0,-1]$.

Next we look for $\mathbf{v}_4 \in S$ so that $\mathbf{v}_4 \notin \operatorname{span}(\{\mathbf{v}_1, \mathbf{v}_2, \mathbf{v}_3\})$. We try $[1,3,1,-1]$, the only remaining vector, as a candidate for $\mathbf{v}_4$. Again, we determine whether there are any solutions a, b, and c to the equation

$$[1,3,1,-1] = a\mathbf{v}_1 + b\mathbf{v}_2 + c\mathbf{v}_3 = a[1,3,2,-1] + b[2,4,-1,0] + c[0,1,0,-1].$$

We get the system

$$\begin{cases} a + 2b = 1 \\ 3a + 4b + c = 3 \\ 2a - b = 1 \\ -a - c = -1 \end{cases}.$$

You can easily see that $a = \frac{3}{5}$, $b = \frac{1}{5}$, $c = \frac{2}{5}$ is a solution to this system. Hence, $[1, 3, 1, -1]$ is not a possible choice for $\mathbf{v}_4$, because $[1, 2, 1, -1] \in \text{span}(\{\mathbf{v}_1, \mathbf{v}_2, \mathbf{v}_3\})$. However, there are no more vectors in S for us to try, so the induction process must terminate here. Therefore, $B = \{\mathbf{v}_1, \mathbf{v}_2, \mathbf{v}_3\} = \{[1, 3, 2, -1], [2, 4, -1, 0], [0, 1, 0, -1]\}$ is the desired basis for $\mathcal{V}$. ■

Incidentally, we can now show that in Example 2, $\mathcal{V} = \text{span}(B)$ is not all of $\mathbb{R}^4$, because $\dim(\mathcal{V}) = 3 \neq \dim(\mathbb{R}^4)$. You can verify, for example, that the vector $[1, 0, 0, 0] \in \mathbb{R}^4$ cannot be expressed as a linear combination of the vectors in B and hence is not in $\mathcal{V} = \text{span}(B)$.

Shrinking a Spanning Set Using Row Reduction

Next we describe a more efficient way to carry out the "shrinking method." In Example 2, the augmented matrices corresponding to the systems we solved are

$$\left[\begin{array}{cc|c} 1 & 2 & 0 \\ 3 & 4 & 1 \\ 2 & -1 & 0 \\ -1 & 0 & -1 \end{array}\right] \quad \text{and} \quad \left[\begin{array}{ccc|c} 1 & 2 & 0 & 1 \\ 3 & 4 & 1 & 3 \\ 2 & -1 & 0 & 1 \\ -1 & 0 & -1 & -1 \end{array}\right].$$

The columns in these matrices are just the vectors in S. When we row reduce either of these matrices, we duplicate some of the row operations performed on the other, since the initial columns are the same. In general, it is easier to row reduce the single matrix whose columns are all the vectors in S. For instance, in Example 2 we can row reduce

$$\mathbf{A} = \begin{bmatrix} 1 & -2 & 2 & 0 & 1 \\ 3 & -6 & 4 & 1 & 3 \\ 2 & -4 & -1 & 0 & 1 \\ -1 & 2 & 0 & -1 & -1 \end{bmatrix} \quad \text{to obtain} \quad \mathbf{C} = \begin{bmatrix} 1 & -2 & 0 & 0 & \frac{3}{5} \\ 0 & 0 & 1 & 0 & \frac{1}{5} \\ 0 & 0 & 0 & 1 & \frac{2}{5} \\ 0 & 0 & 0 & 0 & 0 \end{bmatrix}.$$

The columns with nonzero pivots in $\mathbf{C}$ (first, third, and fourth) correspond to the columns in $\mathbf{A}$ that form the basis for span(S). Thus, $\mathbf{B} = \{$ first column of $\mathbf{A}$, third column of $\mathbf{A}$, fourth column of $\mathbf{A}$ $\} = \{[1, 3, 2, -1], [2, 4, -1, 0], [0, 1, 0, -1]\}$, just as in Example 2.

We will refer to this technique as the "efficient shrinking method." A proof of its validity is outlined in Exercise 16. To summarize:

EFFICIENT SHRINKING METHOD

> Let S be a finite set of vectors spanning a vector space $\mathcal{V}$.
>
> Step 1: Create the matrix $\mathbf{A}$ whose columns are the vectors in S.
> Step 2: Find $\mathbf{C}$, the reduced row echelon form of $\mathbf{A}$.
> Step 3: Let B be the set of vectors in S corresponding to the columns of $\mathbf{C}$ that have nonzero pivots.
>
> Then B is a basis for $\mathcal{V} = \text{span}(S)$.

Example 3

Let $S = \{x^3 - 3x^2 + 1, 2x^2 + x, 2x^3 + 3x + 2, 4x - 5\} \subseteq \mathcal{P}_3$. We use the "efficient shrinking method" to find a subset B of S that is a basis for $\mathcal{V} = \text{span}(S)$.

Steps 1 and 2: Let

$$\mathbf{A} = \begin{bmatrix} 1 & 0 & 2 & 0 \\ -3 & 2 & 0 & 0 \\ 0 & 1 & 3 & 4 \\ 1 & 0 & 2 & -5 \end{bmatrix} \quad \text{which reduces to} \quad \mathbf{C} = \begin{bmatrix} 1 & 0 & 2 & 0 \\ 0 & 1 & 3 & 0 \\ 0 & 0 & 0 & 1 \\ 0 & 0 & 0 & 0 \end{bmatrix}.$$

Step 3: $B = \{x^3 - 3x^2 + 1, 2x^2 + x, 4x - 5\}$, the first, second, and fourth vectors in S, because we have nonzero pivots in the first, second, and fourth columns of $\mathbf{C}$. B is the desired basis for $\mathcal{V}$. ■

Shrinking an Infinite Spanning Set to a Basis

The "shrinking method" can sometimes be used successfully when the spanning set S is infinite.

Example 4

Let $\mathcal{V}$ be the subspace of $\mathcal{M}_{22}$ consisting of all 2×2 symmetric matrices, and let S be the set of nonsingular matrices in $\mathcal{V}$. We reduce S to a basis for $\mathcal{W} = \text{span}(S)$ using the "shrinking method," even though S is infinite. (We prove later that $\mathcal{W} = \mathcal{V}$, and so the basis we construct is actually a basis for $\mathcal{V}$.)

The strategy is to guess a finite subset T of S that spans $\mathcal{W}$. We then use the "shrinking method" on T to find the desired basis. We try to pick vectors for T whose forms are as simple as possible to make computation easier and to produce a "nice" basis. In this case, we choose the set of all nonsingular symmetric 2×2 matrices having only zeroes and ones as entries. That is,

$$T = \left\{ \begin{bmatrix} 1 & 0 \\ 0 & 1 \end{bmatrix}, \begin{bmatrix} 1 & 1 \\ 1 & 0 \end{bmatrix}, \begin{bmatrix} 0 & 1 \\ 1 & 0 \end{bmatrix}, \begin{bmatrix} 0 & 1 \\ 1 & 1 \end{bmatrix} \right\}.$$

(Verify.) We now use the "efficient shrinking method" on T:
Steps 1 and 2:

$$\mathbf{A} = \begin{bmatrix} 1 & 1 & 0 & 0 \\ 0 & 1 & 1 & 1 \\ 0 & 1 & 1 & 1 \\ 1 & 0 & 0 & 1 \end{bmatrix} \quad \text{reduces to} \quad \mathbf{C} = \begin{bmatrix} 1 & 0 & 0 & 1 \\ 0 & 1 & 0 & -1 \\ 0 & 0 & 1 & 2 \\ 0 & 0 & 0 & 0 \end{bmatrix}.$$

Step 3: The desired basis is

$$B = \left\{ \begin{bmatrix} 1 & 0 \\ 0 & 1 \end{bmatrix}, \begin{bmatrix} 1 & 1 \\ 1 & 0 \end{bmatrix}, \begin{bmatrix} 0 & 1 \\ 1 & 0 \end{bmatrix} \right\}.$$

To finish, we must ensure that span$(B) = \mathcal{W}$, since we only guessed at a finite spanning set T for $\mathcal{W}$, and we must verify that this guess is valid. To do so, we show that every symmetric matrix is in span(B) by finding real numbers x, y, and z so that

$$\begin{bmatrix} a & b \\ b & c \end{bmatrix} = x \begin{bmatrix} 1 & 0 \\ 0 & 1 \end{bmatrix} + y \begin{bmatrix} 1 & 1 \\ 1 & 0 \end{bmatrix} + z \begin{bmatrix} 0 & 1 \\ 1 & 0 \end{bmatrix}.$$

Thus, we must prove that the system

$$\begin{cases} x + y = a \\ y + z = b \\ y + z = b \\ x = c \end{cases}$$

has solutions for x, y, and z in terms of a, b, and c. But $x = c, y = a - c, z = b - (a - c)$ obviously satisfies the system. Hence, $\mathcal{V} \subseteq \text{span}(B)$. Since span$(B) \subseteq \mathcal{V}$, we have span$(B) = \mathcal{V} = \mathcal{W}$. ■

The "shrinking method" is not guaranteed to work when the spanning set S has infinitely many elements, because our choice for the finite set T might not have the same span as S. When this happens, try choosing a larger set for T.

Every Linearly Independent Set in a Finite Dimensional Vector Space Is Contained in Some Basis

Suppose that $T = \{\mathbf{t}_1, \ldots, \mathbf{t}_k\}$ is a linearly independent set of vectors in a finite dimensional vector space $\mathcal{V}$. Because $\mathcal{V}$ is finite dimensional, it has a finite basis, say $A = \{\mathbf{a}_1, \ldots, \mathbf{a}_n\}$. Consider the set $T \cup A$. Now, $T \cup A$ certainly spans $\mathcal{V}$ (A alone spans $\mathcal{V}$). We can therefore apply the "shrinking method" to $T \cup A$ to produce a basis B for $\mathcal{V}$. If we order the vectors in $T \cup A$ so that all the vectors in T are listed first, then none of these vectors will be eliminated during the process, since no vector in T is a linear combination of earlier listed vectors in T. In this manner we construct a basis B for $\mathcal{V}$ that contains T.

We have just proved the following:

THEOREM 4.18

Let T be a linearly independent subset of a finite dimensional vector space $\mathcal{V}$. Then $\mathcal{V}$ has a basis B with $T \subseteq B$.

Compare this result with Theorem 4.16. Theorem 4.18 also has an important corollary, which you should compare with Corollary 4.17.

COROLLARY 4.19

Let T be a linearly independent subset of a finite dimensional vector space $\mathcal{V}$. Then $|T| \leq \dim(\mathcal{V})$. Moreover, $|T| = \dim(\mathcal{V})$ if and only if T is a basis for $\mathcal{V}$.

Proof of Corollary 4.19

By Theorem 4.18, there is a basis B for $\mathcal{V}$ containing T. Hence, $|T| \leq |B| = \dim(\mathcal{V})$.

Now, suppose that $|T| = \dim(\mathcal{V})$. If B is any basis for $\mathcal{V}$, then $|T| = |B|$. Hence, Lemma 4.12 shows that T is a basis for $\mathcal{V}$.

Conversely, if T is a basis for $\mathcal{V}$, then clearly $|T| = \dim(\mathcal{V})$ by definition. ■

Enlarging a Linearly Independent Subset to a Basis

We modify slightly the method outlined before Theorem 4.18 to find a basis for a finite dimensional vector space $\mathcal{V}$ containing a given linearly independent subset T of $\mathcal{V}$.

METHOD FOR ENLARGING A LINEARLY INDEPENDENT SET TO A BASIS

Suppose that $T = \{\mathbf{v}_1, \ldots, \mathbf{v}_k\}$ is a linearly independent subset of a finite dimensional vector space $\mathcal{V}$.

Step 1: Find a finite spanning set $A = \{\mathbf{a}_1, \ldots, \mathbf{a}_n\}$ for $\mathcal{V}$.
Step 2: Form the ordered spanning set $S = \{\mathbf{v}_1, \ldots, \mathbf{v}_k, \mathbf{a}_1, \ldots, \mathbf{a}_n\}$ for $\mathcal{V}$.
Step 3: Use either "shrinking method" on S to produce a subset B of S.

Then B is a basis for $\mathcal{V}$ containing T.

We will refer to this technique as the "enlarging method." The basis produced by this method is easier to use if the additional vectors in the set A have a simple form. Ideally, we choose A to be some standard basis for $\mathcal{V}$.

Example 5

Consider the linearly independent subset $T = \{[2, 0, 4, -12], [0, -1, -3, 9]\}$ of $\mathcal{V} = \mathbb{R}^4$. We use the "enlarging method" to find a basis for $\mathbb{R}^4$ that contains T.

Step 1: We choose A to be the standard basis $\{\mathbf{e}_1, \mathbf{e}_2, \mathbf{e}_3, \mathbf{e}_4\}$ for $\mathbb{R}^4$.

Step 2: We create

$$S = \{[2, 0, 4, -12], [0, -1, -3, 9], [1, 0, 0, 0], [0, 1, 0, 0], [0, 0, 1, 0], [0, 0, 0, 1]\}.$$

Step 3: The matrix

$$\begin{bmatrix} 2 & 0 & 1 & 0 & 0 & 0 \\ 0 & -1 & 0 & 1 & 0 & 0 \\ 4 & -3 & 0 & 0 & 1 & 0 \\ -12 & 9 & 0 & 0 & 0 & 1 \end{bmatrix} \quad \text{reduces to} \quad \begin{bmatrix} 1 & 0 & 0 & -\frac{3}{4} & 0 & -\frac{1}{12} \\ 0 & 1 & 0 & -1 & 0 & 0 \\ 0 & 0 & 1 & \frac{3}{2} & 0 & \frac{1}{6} \\ 0 & 0 & 0 & 0 & 1 & \frac{1}{3} \end{bmatrix}.$$

Since columns 1, 2, 3, and 5 have nonzero pivots, the "efficient shrinking method" indicates that the set $B = \{[2, 0, 4, -12], [0, -1, -3, 9], [1, 0, 0, 0], [0, 0, 1, 0]\}$ is a basis for $\mathbb{R}^4$ containing T. ■

In general, we can use the "enlarging method" only when we already know of a basis or a finite spanning set to use for A. If one is not known, we can make an intelligent guess, just as we did when using the "shrinking method" on an infinite spanning set. However, we must then verify that the resulting set actually spans the vector space.

Span and Linear Independence, Revisited

We conclude by stating the following useful result, whose proof follows immediately from Corollaries 4.17 and 4.19:

THEOREM 4.20

Let $\mathcal{V}$ be an n-dimensional vector space, and let S be a subset of $\mathcal{V}$ containing exactly n elements. Then S spans $\mathcal{V}$ if and only if S is linearly independent.

Exercises—Section 4.6

1. For each given subset S of $\mathbb{R}^3$, find a subset B of S that is a basis for $\mathcal{V} = \text{span}(S)$:

★(a) $S = \{[1, 3, -2], [2, 1, 4], [3, -6, 18], [0, 1, -1], [-2, 1, -6]\}$

(b) $S = \{[1, 4, -2], [-2, -8, 4], [2, -8, 5], [0, -7, 2]\}$

★(c) $S = \{[3, -2, 2], [1, 2, -1], [3, -2, 7], [-1, -10, 6]\}$

(d) $S = \{[3, 1, 0], [2, -1, 7], [0, 0, 0], [0, 5, -21], [6, 2, 0], [1, 5, 7]\}$

(e) $S =$ the set of all 3-vectors whose second coordinate is zero

★(f) $S =$ the set of all 3-vectors whose second coordinate is -3 times its first coordinate plus its third coordinate

2. For each given subset S of $\mathcal{P}_3$, find a subset B of S that is a basis for $\mathcal{V} = \text{span}(S)$:

★(a) $S = \{x^3 - 8x^2 + 1, 3x^3 - 2x^2 + x, 4x^3 + 2x - 10, x^3 - 20x^2 - x + 12, x^3 + 24x^2 + 2x - 13, x^3 + 14x^2 - 7x + 18\}$

(b) $S = \{-2x^3 + x + 2, 3x^3 - x^2 + 4x + 6, 8x^3 + x^2 + 6x + 10, -4x^3 - 3x^2 + 3x + 4, -3x^3 - 4x^2 + 8x + 12\}$

★(c) $S =$ the set of all polynomials in $\mathcal{P}_3$ with a zero constant term

(d) $S = \mathcal{P}_2$

★(e) $S =$ the set of all polynomials in $\mathcal{P}_3$ with the coefficient of the x^2 term equal to the coefficient of the x^3 term

(f) $S =$ the set of all polynomials in $\mathcal{P}_3$ with the coefficient of the x^3 term equal to 8

3. For each given subset S of $\mathcal{M}_{33}$, find a subset B of S that is a basis for $\mathcal{V} = \text{span}(S)$:

★(a) $S = \{\mathbf{A} \in \mathcal{M}_{33} |\ \text{each } a_{ij} \text{ is either 0 or 1}\}$

(b) $S = \{\mathbf{A} \in \mathcal{M}_{33} |\ \text{each } a_{ij} \text{ is either 1 or } -1\}$

★(c) $S =$ the set of all symmetric 3×3 matrices

(d) $S =$ the set of all nonsingular 3×3 matrices

4. Enlarge each of the given linearly independent subsets T of $\mathbb{R}^5$ to a basis B for $\mathbb{R}^5$ containing T:

★(a) $T = \{[1, -3, 0, 1, 4], [2, 2, 1, -3, 1]\}$

(b) $T = \{[1, 1, 1, 1, 1], [0, 1, 1, 1, 1], [0, 0, 1, 1, 1]\}$

★(c) $T = \{[1, 0, -1, 0, 0], [0, 1, -1, 1, 0], [2, 3, -8, -1, 0]\}$

5. Enlarge each of the given linearly independent subsets T of $\mathcal{P}_4$ to a basis B for $\mathcal{P}_4$ that contains T:

★(a) $T = \{x^3 - x^2, x^4 - 3x^3 + 5x^2 - x\}$

(b) $T = \{3x - 2, x^3 - 6x + 4\}$

★(c) $T = \{x^4 - x^3 + x^2 - x + 1, x^3 - x^2 + x - 1, x^2 - x + 1\}$

6. Enlarge each of the given linearly independent subsets T of $\mathcal{M}_{32}$ to a basis B for $\mathcal{M}_{32}$ that contains T:

★(a) $T = \left\{ \begin{bmatrix} 1 & -1 \\ -1 & 1 \\ 0 & 0 \end{bmatrix}, \begin{bmatrix} 0 & 0 \\ 1 & -1 \\ -1 & 1 \end{bmatrix} \right\}$

(b) $T = \left\{ \begin{bmatrix} 0 & -2 \\ 1 & 0 \\ -1 & 2 \end{bmatrix}, \begin{bmatrix} 0 & -3 \\ 0 & 1 \\ 3 & -6 \end{bmatrix}, \begin{bmatrix} 0 & 1 \\ 1 & 1 \\ -4 & 8 \end{bmatrix} \right\}$

★(c) $T = \left\{ \begin{bmatrix} 3 & 0 \\ -1 & 7 \\ 0 & 1 \end{bmatrix}, \begin{bmatrix} -1 & 0 \\ 1 & 3 \\ 0 & -2 \end{bmatrix}, \begin{bmatrix} 2 & 0 \\ 3 & 1 \\ 0 & -1 \end{bmatrix}, \begin{bmatrix} 6 & 0 \\ 0 & 1 \\ 0 & -1 \end{bmatrix} \right\}$

★**7.** Find a basis for the vector space $\mathcal{U}_4$ consisting of all 4×4 upper triangular matrices.

8. In each case, find the dimension of $\mathcal{V}$ by using an appropriate method to find a basis:

(a) $\mathcal{V} = \text{span}(\{[5, 2, 1, 0, -1], [3, 0, 1, 1, 0], [0, 0, 0, 0, 0], [-2, 4, -2, -4, -2], [0, 12, -4, -10, -6], [-6, 0, -2, -2, 0]\})$, a subspace of $\mathbb{R}^5$

★(b) $\mathcal{V} = \{\mathbf{A} \in \mathcal{M}_{33} \mid \text{trace}(\mathbf{A}) = 0\}$, a subspace of $\mathcal{M}_{33}$ (Recall that the trace of a matrix is the sum of the terms on the main diagonal.)

(c) $\mathcal{V} = \text{span}(\{x^4 - x^3 + 2x^2, 2x^4 + x - 5, 2x^3 - 4x^2 + x - 4, 6, x^2 - 1\})$

★(d) $\mathcal{V} = \{\mathbf{p} \in \mathcal{P}_6 \mid \mathbf{p} = ax^6 - bx^5 + ax^4 - cx^3 + (a + b + c)x^2 - (a - c)x + (3a - 2b + 16c)$, for some real numbers a, b, and $c\}$

9. (a) Show that each of these subspaces of $\mathcal{M}_{nn}$ has dimension $(n^2 + n)/2$:

(i) The set of upper triangular $n \times n$ matrices

(ii) The set of lower triangular $n \times n$ matrices

(iii) The set of symmetric $n \times n$ matrices

★(b) What is the dimension of the set of skew-symmetric $n \times n$ matrices?

★**10.** Let $\mathcal{V}$ be a finite dimensional vector space.

(a) Let S be a subset of $\mathcal{V}$ with $\dim(\mathcal{V}) \leq |S|$. Find an example to show that S need not span $\mathcal{V}$.

(b) Let T be a subset of $\mathcal{V}$ with $|T| \leq \dim(\mathcal{V})$. Find an example to show that T need not be linearly independent.

11. Let S be a subset of a finite dimensional vector space $\mathcal{V}$ such that $|S| = \dim(\mathcal{V})$. If S is not a basis for $\mathcal{V}$, prove that S neither spans $\mathcal{V}$ nor is linearly independent.

12. Let **A** be an $m \times n$ matrix.
   (a) Prove that $S_{\mathbf{A}} = \{\mathbf{X} \in \mathbb{R}^n \mid \mathbf{AX} = \mathbf{0}\}$, the solution set of the homogeneous system $\mathbf{AX} = \mathbf{0}$, is a subspace of $\mathbb{R}^n$.
   (b) Prove that $\dim(S_{\mathbf{A}}) + \text{rank}(\mathbf{A}) = n$. (Hint: First consider the case where **A** is in reduced row echelon form.)

13. Prove Theorem 4.16.

14. Let $\mathcal{W}$ be a subspace of a finite dimensional vector space $\mathcal{V}$.
   (a) Show that $\mathcal{V}$ has some basis B with a subset B' that is a basis for $\mathcal{W}$.
   ★(b) If B is any given basis for $\mathcal{V}$, must some subset B' of B be a basis for $\mathcal{W}$? Prove that your answer is correct.
   ★(c) If B is any given basis for $\mathcal{V}$ and B' is some subset of B, is there necessarily a subspace $\mathcal{Y}$ of $\mathcal{V}$ such that B' is a basis for $\mathcal{Y}$? Why or why not?

15. Let $\mathcal{V}$ be a finite dimensional vector space, and let $\mathcal{W}$ be a subspace of $\mathcal{V}$.
   (a) Prove that $\mathcal{V}$ has a subspace $\mathcal{W}'$ such that every vector in $\mathcal{V}$ can be uniquely expressed as a sum of a vector in $\mathcal{W}$ and a vector in $\mathcal{W}'$. (In other words, show that there is a subspace $\mathcal{W}'$ so that, for every $\mathbf{v}$ in $\mathcal{V}$, there are unique vectors $\mathbf{w} \in \mathcal{W}$ and $\mathbf{w}' \in \mathcal{W}'$ such that $\mathbf{v} = \mathbf{w} + \mathbf{w}'$.)
   ★(b) Give an example of a subspace $\mathcal{W}$ of some finite dimensional vector space $\mathcal{V}$ for which the subspace $\mathcal{W}'$ from part (a) is not unique.

16. Let $S = \{\mathbf{v}_1, \ldots, \mathbf{v}_k\}$ be a finite subset of $\mathbb{R}^n$, and let **A** be the matrix whose columns are the vectors in S in the order $\mathbf{v}_1, \mathbf{v}_2, \ldots, \mathbf{v}_k$. Let **B** be the reduced row echelon form matrix that is row equivalent to **A**.
   (a) Prove that if the ith column of **B** contains a nonzero pivot, then $\mathbf{v}_i \notin \text{span}(\{\mathbf{v}_1, \ldots, \mathbf{v}_{i-1}\})$.
   (b) Prove that if the ith column of **B** has no nonzero pivot, then $\mathbf{v}_i \in \text{span}(\{\mathbf{v}_1, \ldots, \mathbf{v}_{i-1}\})$.
   (c) Use parts (a) and (b) together with Theorem 4.6 and Lemma 4.15 to show that the "efficient shrinking method" works correctly to produce a basis for $\text{span}(S)$.

4.7 Coordinatization

If B is a basis for a vector space $\mathcal{V}$, then we know that every vector in $\mathcal{V}$ has a unique expression as a linear combination of the vectors in B. For example, the set $\{\mathbf{e}_1, \ldots, \mathbf{e}_n\}$ is a basis (the standard one) for $\mathbb{R}^n$, and we are accustomed to writing the vector $[a_1, \ldots, a_n]$ in $\mathbb{R}^n$ as $a_1\mathbf{e}_1 + \cdots + a_n\mathbf{e}_n$. Dealing with the standard basis in $\mathbb{R}^n$ is easy, because the coefficients in the linear combination are the same as the coordinates of the vector. However, this property is not necessarily true for other bases.

In this section, we develop a notation for representing any vector in a finite dimensional vector space in terms of its coefficients with respect to a given basis. Such a representation is known as the coordinatization of the vector. We also determine how the coordinatization changes whenever we switch bases.

Coordinates with Respect to an Ordered Basis

DEFINITION

An **ordered basis** for a vector space $\mathcal{V}$ is an ordered n-tuple of vectors $(\mathbf{v}_1, \ldots, \mathbf{v}_n)$ such that the set $\{\mathbf{v}_1, \ldots, \mathbf{v}_n\}$ is a basis for $\mathcal{V}$.

The new idea in this definition is that the basis elements for $\mathcal{V}$ are written in a specific order. Thus, two ordered bases for $\mathcal{V}$ are the same only if they contain the same vectors in the same order. Thus, $(\mathbf{i}, \mathbf{j}, \mathbf{k})$ and $(\mathbf{j}, \mathbf{i}, \mathbf{k})$ are different ordered bases for $\mathbb{R}^3$.

By Theorem 4.10, if $B = (\mathbf{v}_1, \mathbf{v}_2, \ldots, \mathbf{v}_n)$ is an ordered basis for $\mathcal{V}$, then for every vector $\mathbf{w} \in \mathcal{V}$, there are unique scalars $a_1, a_2, \ldots, a_n$ such that $\mathbf{w} = a_1\mathbf{v}_1 + a_2\mathbf{v}_2 + \cdots + a_n\mathbf{v}_n$. We use these scalars $a_1, a_2, \ldots, a_n$ to **coordinatize** the vector $\mathbf{w}$ as follows:

DEFINITION

Let $B = (\mathbf{v}_1, \mathbf{v}_2, \ldots, \mathbf{v}_n)$ be an ordered basis for a vector space $\mathcal{V}$. Suppose that $\mathbf{w} = a_1\mathbf{v}_1 + a_2\mathbf{v}_2 + \cdots + a_n\mathbf{v}_n \in \mathcal{V}$. Then $[\mathbf{w}]_B$, the **coordinatization of w with respect to** B, is the n-vector $[a_1, a_2, \ldots, a_n]$.

The vector $[\mathbf{w}]_B = [a_1, a_2, \ldots, a_n]$ is frequently referred to as **w expressed in** B**-coordinates**. When necessary, we will express $[\mathbf{w}]_B$ as a column vector.

Example 1

Let $B = (x^3, x^2, x, 1)$, an ordered basis for $\mathcal{P}_3$. Then $[6x^3 - 2x + 18]_B = [6, 0, -2, 18]$, and $[4 - 3x + 9x^2 - 7x^3]_B = [-7, 9, -3, 4]$. ■

Example 2

The following set is an ordered basis for the vector space of skew-symmetric 3×3 matrices:

$$B = \left(\begin{bmatrix} 0 & 1 & 0 \\ -1 & 0 & 0 \\ 0 & 0 & 0 \end{bmatrix}, \begin{bmatrix} 0 & 0 & 1 \\ 0 & 0 & 0 \\ -1 & 0 & 0 \end{bmatrix}, \begin{bmatrix} 0 & 0 & 0 \\ 0 & 0 & 1 \\ 0 & -1 & 0 \end{bmatrix} \right).$$

Since a general 3×3 skew-symmetric matrix can be expressed as

$$\begin{bmatrix} 0 & a & b \\ -a & 0 & c \\ -b & -c & 0 \end{bmatrix} = a\begin{bmatrix} 0 & 1 & 0 \\ -1 & 0 & 0 \\ 0 & 0 & 0 \end{bmatrix} + b\begin{bmatrix} 0 & 0 & 1 \\ 0 & 0 & 0 \\ -1 & 0 & 0 \end{bmatrix} + c\begin{bmatrix} 0 & 0 & 0 \\ 0 & 0 & 1 \\ 0 & -1 & 0 \end{bmatrix},$$

we see that

$$\begin{bmatrix} 0 & a & b \\ -a & 0 & c \\ -b & -c & 0 \end{bmatrix}_B = [a, b, c].$$ ■

Notice in Example 2 that the coordinatized vector $[a, b, c]$ is more "compact" than the full matrix of nine entries but still contains the same essential information.

The next theorem shows that the coordinatization of a vector can play as full a functional role as the original vector itself as far as the vector space operations of addition and scalar multiplication are concerned:

THEOREM 4.21

Let $B = (\mathbf{v}_1, \ldots, \mathbf{v}_n)$ be an ordered basis for a vector space $\mathcal{V}$. Suppose $\mathbf{w}_1, \ldots, \mathbf{w}_k \in \mathcal{V}$ and $a_1, \ldots, a_k$ are scalars. Then,

(1) $[\mathbf{w}_1 + \mathbf{w}_2]_B = [\mathbf{w}_1]_B + [\mathbf{w}_2]_B$
(2) $[a_1\mathbf{w}_1]_B = a_1[\mathbf{w}_1]_B$
(3) $[a_1\mathbf{w}_1 + a_2\mathbf{w}_2 + \cdots + a_k\mathbf{w}_k]_B = a_1[\mathbf{w}_1]_B + a_2[\mathbf{w}_2]_B + \cdots + a_k[\mathbf{w}_k]_B$

This theorem says that finding a linear combination of vectors in $\mathcal{V}$ and then putting that answer into B-coordinates is equivalent to first finding the B-coordinates of each vector individually and then calculating the linear combination in $\mathbb{R}^n$.

The proof of Theorem 4.21 is very easy (see Exercise 10).

Example 3

Consider the subspace $\mathcal{V}$ of $\mathbb{R}^5$ spanned by the ordered basis $C = ([-4, 5, -1, 0, -1], [1, -3, 2, 2, 5], [1, -2, 1, 1, 3])$. Notice that the vectors in $\mathcal{V}$ can be put into C-coordinates by solving an appropriate system. For example, to find $[-23, 30, -7, -1, -7]_C$, we solve the equation

$$[-23, 30, -7, -1, -7] = a[-4, 5, -1, 0, -1] + b[1, -3, 2, 2, 5] + c[1, -2, 1, 1, 3].$$

The equivalent system is

$$\begin{cases} -4a + b + c = -23 \\ 5a - 3b - 2c = 30 \\ -a + 2b + c = -7 \\ \quad\; 2b + c = -1 \\ -a + 5b + 3c = -7 \end{cases}.$$

Since the (unique) solution for this system is $a = 6, b = -2, c = 3$, we see that $[-23, 30, -7, -1, -7]_C = [6, -2, 3]$. Similarly, you can see that

$$\begin{aligned} [-4, 5, -1, 0, -1]_C &= [1, 0, 0], \\ [1, -3, 2, 2, 5]_C &= [0, 1, 0], \\ [1, -2, 1, 1, 3]_C &= [0, 0, 1]. \end{aligned}$$

Clearly, vectors in $\mathbb{R}^5$ that are not in $\mathcal{V}$ cannot be expressed in C-coordinates. For example, the vector $[1, 2, 3, 4, 5]$ is not in $\mathcal{V}$, because the system

$$\begin{cases} -4a + b + c = 1 \\ 5a - 3b - 2c = 2 \\ -a + 2b + c = 3 \\ 2b + c = 4 \\ -a + 5b + 3c = 5 \end{cases}$$

has no solutions.

Finally, notice that we can easily convert back from C-coordinates to a 5-vector in $\mathcal{V}$ merely by calculating the appropriate linear combination. For example,

$$\begin{aligned} [3, 7, -1] &= 3[1, 0, 0] + 7[0, 1, 0] - 1[0, 0, 1] \\ &= 3[-4, 5, -1, 0, -1]_C + 7[1, -3, 2, 2, 5]_C - [1, -2, 1, 1, 3]_C \\ &= [-12, 15, -3, 0, -3]_C + [7, -21, 14, 14, 35]_C + [-1, 2, -1, -1, -3]_C \\ &= [-6, -4, 10, 13, 29]_C. \end{aligned}$$

Thus, $[3, 7, -1]$ is the coordinatization of $[-6, -4, 10, 13, 29]$ in $\mathcal{V}$. ■

Example 4

Let $B = ([4, 2], [1, 3])$ be an ordered basis for $\mathbb{R}^2$. From a geometric viewpoint, converting to B-coordinates in $\mathbb{R}^2$ creates a new coordinate grid in $\mathbb{R}^2$ using parallelograms whose sides are the vectors in B. For example, $[11, 13]$ equals $[2, 3]$ when expressed in B-coordinates, because $[11, 13] = 2[4, 2] + 3[1, 3]$. The vector $[11, 13]$ is graphed on this new grid in Figure 4.7. ■

Changing Coordinates

Our next goal is to determine how the coordinates of a vector change when we convert from one ordered basis for a vector space to another.

DEFINITION

Suppose that $\mathcal{V}$ is an n-dimensional vector space with ordered bases B and C. Let $\mathbf{P}$ be the $n \times n$ matrix whose i^{th} column, for $1 \leq i \leq n$, equals $[\mathbf{b}_i]_C$, where $\mathbf{b}_i$ is the ith basis vector in B. Then $\mathbf{P}$ is called the **transition matrix from B-coordinates to C-coordinates**.

We often refer to the matrix $\mathbf{P}$ in this definition as the "transition matrix from B to C."

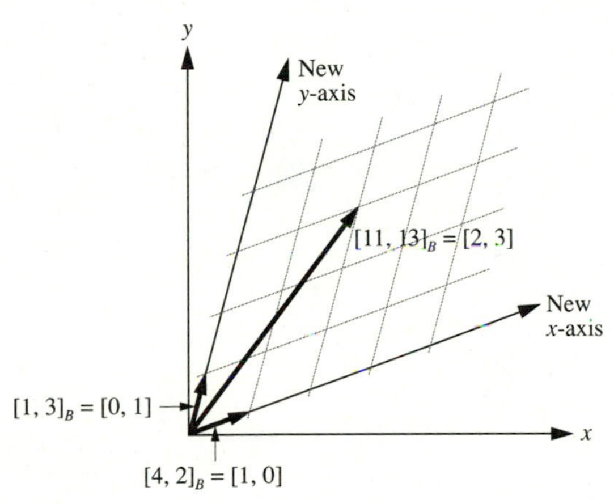

Figure 4.7
A B-coordinate grid in $\mathbb{R}^2$: picturing [2, 3] in B-coordinates

Example 5

Consider these ordered bases for $\mathcal{U}_2$:

$$B = \left(\begin{bmatrix} 7 & 3 \\ 0 & 0 \end{bmatrix}, \begin{bmatrix} 1 & 2 \\ 0 & -1 \end{bmatrix}, \begin{bmatrix} 1 & -1 \\ 0 & 1 \end{bmatrix} \right) \quad \text{and} \quad C = \left(\begin{bmatrix} 1 & 0 \\ 0 & 0 \end{bmatrix}, \begin{bmatrix} 0 & 1 \\ 0 & 0 \end{bmatrix}, \begin{bmatrix} 0 & 0 \\ 0 & 1 \end{bmatrix} \right).$$

Since

$$\begin{bmatrix} 7 & 3 \\ 0 & 0 \end{bmatrix}_C = \begin{bmatrix} 7 \\ 3 \\ 0 \end{bmatrix}, \quad \begin{bmatrix} 1 & 2 \\ 0 & -1 \end{bmatrix}_C = \begin{bmatrix} 1 \\ 2 \\ -1 \end{bmatrix}, \quad \text{and} \quad \begin{bmatrix} 1 & -1 \\ 0 & 1 \end{bmatrix}_C = \begin{bmatrix} 1 \\ -1 \\ 1 \end{bmatrix},$$

by definition, the transition matrix $\mathbf{P}$ from B to C is given by

$$\begin{bmatrix} 7 & 1 & 1 \\ 3 & 2 & -1 \\ 0 & -1 & 1 \end{bmatrix}.$$

■

The next theorem shows that if the B-coordinatization of a vector $\mathbf{v}$ is known, then its C-coordinatization can be found by multiplying the transition matrix from B to C times the B-coordinate vector.

THEOREM 4.22

Suppose that B and C are ordered bases for an n-dimensional vector space $\mathcal{V}$, and let $\mathbf{P}$ be an $n \times n$ matrix. Then $\mathbf{P}$ is the transition matrix from B to C if and only if for every $\mathbf{v} \in \mathcal{V}$, $\mathbf{P}[\mathbf{v}]_B = [\mathbf{v}]_C$.

Proof of Theorem 4.22

Let B and C be ordered bases for a vector space $\mathcal{V}$, with $B = (\mathbf{b}_1, \ldots, \mathbf{b}_n)$. First, suppose that $\mathbf{P}$ is the transition matrix from B to C. We show that, for every $\mathbf{v} \in \mathcal{V}$, $\mathbf{P}[\mathbf{v}]_B = [\mathbf{v}]_C$.

Suppose that $\mathbf{v} \in \mathcal{V}$ and $[\mathbf{v}]_B = [a_1, \ldots, a_n]$. Then $\mathbf{v} = a_1\mathbf{b}_1 + \cdots + a_n\mathbf{b}_n$. Hence,

$$\mathbf{P}[\mathbf{v}]_B = \begin{bmatrix} p_{11} & \cdots & p_{1n} \\ \vdots & \ddots & \vdots \\ p_{n1} & \cdots & p_{nn} \end{bmatrix} \begin{bmatrix} a_1 \\ a_2 \\ \vdots \\ a_n \end{bmatrix}$$

$$= a_1 \begin{bmatrix} p_{11} \\ p_{21} \\ \vdots \\ p_{n1} \end{bmatrix} + a_2 \begin{bmatrix} p_{12} \\ p_{22} \\ \vdots \\ p_{n2} \end{bmatrix} + \cdots + a_n \begin{bmatrix} p_{1n} \\ p_{2n} \\ \vdots \\ p_{nn} \end{bmatrix}.$$

However, $\mathbf{P}$ is the transition matrix from B to C, so the ith column of $\mathbf{P}$ equals $[\mathbf{b}_i]_C$. Therefore,

$$\begin{aligned} \mathbf{P}[\mathbf{v}]_B &= a_1[\mathbf{b}_1]_C + a_2[\mathbf{b}_2]_C + \cdots + a_n[\mathbf{b}_n]_C \\ &= [a_1\mathbf{b}_1 + a_2\mathbf{b}_2 + \cdots + a_n\mathbf{b}_n]_C \qquad \text{by Theorem 4.21} \\ &= [\mathbf{v}]_C. \end{aligned}$$

Conversely, suppose that $\mathbf{P}$ is an $n \times n$ matrix and that $\mathbf{P}[\mathbf{v}]_B = [\mathbf{v}]_C$ for every $\mathbf{v} \in \mathcal{V}$. We show that $\mathbf{P}$ is the transition matrix from B to C. By definition, it is enough to show that the ith column of $\mathbf{P}$ is equal to $[\mathbf{b}_i]_C$. Consider the case when $\mathbf{v} = \mathbf{b}_i$. Then $[\mathbf{v}]_B = \mathbf{e}_i$, the ith standard basis vector for $\mathbb{R}^n$. Since $\mathbf{P}[\mathbf{v}]_B = [\mathbf{v}]_C$, we have $\mathbf{P}\mathbf{e}_i = [\mathbf{b}_i]_C$. But $\mathbf{P}\mathbf{e}_i = i^{th}$ column of $\mathbf{P}$, and so the ith column of $\mathbf{P} = [\mathbf{b}_i]_C$. ■

Example 6

Consider the ordered bases $B = ([12, 9], [15, -1])$ and $C = ([3, 4], [6, 1])$ for $\mathbb{R}^2$. Solving the systems

$$\begin{cases} 3a + 6b = 12 \\ 4a + b = 9 \end{cases} \quad \text{and} \quad \begin{cases} 3a + 6b = 15 \\ 4a + b = -1 \end{cases}$$

gives $[12, 9]_C = [2, 1]$ and $[15, -1]_C = [-1, 3]$. Hence, the transition matrix from B to C is $\mathbf{P} = \begin{bmatrix} 2 & -1 \\ 1 & 3 \end{bmatrix}$.

Now, consider $\mathbf{v} = [3, 39]$. You can easily verify that $[\mathbf{v}]_B = [4, -3]$. Then, by Theorem 4.22,

$$[\mathbf{v}]_C = \mathbf{P}[\mathbf{v}]_B = \begin{bmatrix} 2 & -1 \\ 1 & 3 \end{bmatrix}\begin{bmatrix} 4 \\ -3 \end{bmatrix} = \begin{bmatrix} 11 \\ -5 \end{bmatrix}.$$

You should check that this is the correct answer for $[\mathbf{v}]_C$. ■

Calculating the Transition Matrix by Row Reduction

The next example motivates the general technique for calculating a transition matrix using row reduction.

Example 7

Let B and C be the ordered bases for $\mathcal{P}_3$ given by

$$\begin{aligned} B &= (3x^3 + 10x^2 + 4x - 3, 2x^3 - x^2 - 2x + 3, x^3 + 6x^2 + 4x - 7, \\ &\quad -4x^3 - 7x^2 - x - 2), \\ C &= (7x^3 + 7x^2 + 1, 4x^3 + 2x^2 - x + 1, -3x^3 + 5x^2 + 4x - 2, -x^3 - 2x^2 - 2). \end{aligned}$$

We can find the columns of the transition matrix from B to C by expressing each polynomial in B in C-coordinates. We do so by solving the four systems

$$\mathbf{DX} = \mathbf{F}_1, \quad \mathbf{DX} = \mathbf{F}_2, \quad \mathbf{DX} = \mathbf{F}_3, \quad \mathbf{DX} = \mathbf{F}_4,$$

where $\mathbf{D}$ is the matrix whose columns are the coefficients of the polynomials in C and where $\mathbf{F}_i$, for $1 \le i \le 4$, is the column containing the coefficients of the ith polynomial in B. All four systems can be solved simultaneously by row reducing the augmented matrix

$$[\mathbf{D} \mid \mathbf{F}_1\mathbf{F}_2\mathbf{F}_3\mathbf{F}_4] = \left[\begin{array}{cccc|cccc} 7 & 4 & -3 & -1 & 3 & 2 & 1 & -4 \\ 7 & 2 & 5 & -2 & 10 & -1 & 6 & -7 \\ 0 & -1 & 4 & 0 & 4 & -2 & 4 & -1 \\ 1 & 1 & -2 & -2 & -3 & 3 & -7 & -2 \end{array}\right].$$

This matrix reduces to

$$\left[\begin{array}{cccc|cccc} 1 & 0 & 0 & 0 & 1 & -1 & 1 & -1 \\ 0 & 1 & 0 & 0 & 0 & 2 & 0 & 1 \\ 0 & 0 & 1 & 0 & 1 & 0 & 1 & 0 \\ 0 & 0 & 0 & 1 & 1 & -1 & 3 & 1 \end{array}\right].$$

Thus, the desired transition matrix from B to C is

$$\mathbf{P} = \begin{bmatrix} 1 & -1 & 1 & -1 \\ 0 & 2 & 0 & 1 \\ 1 & 0 & 1 & 0 \\ 1 & -1 & 3 & 1 \end{bmatrix}.$$

■

We summarize this method as follows:

CALCULATING THE TRANSITION MATRIX

To find the transition matrix $\mathbf{P}$ from B to C, where B and C are ordered bases for a k-dimensional subspace of $\mathbb{R}^n$, use row reduction on

$$\left[\begin{array}{cccc|cccc} 1^{st} & 2^{nd} & & k^{th} & 1^{st} & 2^{nd} & & k^{th} \\ \text{vector} & \text{vector} & \cdots & \text{vector} & \text{vector} & \text{vector} & \cdots & \text{vector} \\ \text{in} & \text{in} & & \text{in} & \text{in} & \text{in} & & \text{in} \\ C & C & & C & B & B & & B \end{array}\right]$$

$$\text{to produce} \left[\begin{array}{c|c} \mathbf{I}_k & \mathbf{P} \\ \hline \text{rows of} & \text{zeroes} \end{array}\right].$$

In Exercise 8 you are asked to show that, in the special cases where either B or C is the standard basis in $\mathbb{R}^n$, there are simple formulas for the transition matrix from B to C.

Example 8

Recall from Example 3 the subspace $\mathcal{V}$ of $\mathbb{R}^5$ that is spanned by the ordered basis $C = ([-4, 5, -1, 0, -1], [1, -3, 2, 2, 5], [1, -2, 1, 1, 3])$. Using the row reduction method, we can determine that $B = ([1, 0, -1, 0, 4], [0, 1, -1, 0, 3], [0, 0, 0, 1, 5])$ is also an ordered basis for $\mathcal{V}$. To find the transition matrix from B to C, we must solve for the C-coordinates of each vector in B. In essence, we are solving three systems of linear equations simultaneously by row reducing the corresponding augmented matrix

$$\left[\begin{array}{rrr|rrr} -4 & 1 & 1 & 1 & 0 & 0 \\ 5 & -3 & -2 & 0 & 1 & 0 \\ -1 & 2 & 1 & -1 & -1 & 0 \\ 0 & 2 & 1 & 0 & 0 & 1 \\ -1 & 5 & 3 & 4 & 3 & 5 \end{array}\right]$$

to produce

$$\left[\begin{array}{ccc|ccc} 1 & 0 & 0 & 1 & 1 & 1 \\ 0 & 1 & 0 & -5 & -4 & -3 \\ 0 & 0 & 1 & 10 & 8 & 7 \\ 0 & 0 & 0 & 0 & 0 & 0 \\ 0 & 0 & 0 & 0 & 0 & 0 \end{array}\right].$$

We ignore the last two rows because they contribute no information to the solutions to the linear systems. The columns on the right now show how to express each vector in B as a linear combination of the vectors in C. For example,

$$[1, 0, -1, 0, 4] = 1[-4, 5, -1, 0, -1] - 5[1, -3, 2, 2, 5] + 10[1, -2, 1, 1, 3].$$

Therefore, the entries to the right of the bar lead to the transition matrix from B to C, namely

$$\mathbf{P} = \begin{bmatrix} 1 & 1 & 1 \\ -5 & -4 & -3 \\ 10 & 8 & 7 \end{bmatrix}.$$

■

Composition of Transitions

The next theorem shows that the cumulative effect of two transitions between bases is represented by the product of the transition matrices in reverse order.

THEOREM 4.23

Suppose that B, C, and D are ordered bases for a finite dimensional vector space $\mathcal{V}$. Let $\mathbf{P}$ be the transition matrix from B to C, and let $\mathbf{Q}$ be the transition matrix from C to D. Then $\mathbf{QP}$ is the transition matrix from B to D.

The proof of this theorem is left for you to do in Exercise 11.

Example 9

Recall the ordered bases B and C given for $\mathcal{P}_3$ in Example 7:

$$B = (3x^3 + 10x^2 + 4x - 3, 2x^3 - x^2 - 2x + 3, x^3 + 6x^2 + 4x - 7,$$
$$- 4x^3 - 7x^2 - x - 2),$$
$$C = (7x^3 + 7x^2 + 1, 4x^3 + 2x^2 - x + 1, -3x^3 + 5x^2 + 4x - 2,$$
$$- x^3 - 2x^2 - 2).$$

Also consider the standard basis $S = (x^3, x^2, x, 1)$ for $\mathcal{P}_3$.

In Example 7, we found that the transition matrix from B to C is

$$\mathbf{P} = \begin{bmatrix} 1 & -1 & 1 & -1 \\ 0 & 2 & 0 & 1 \\ 1 & 0 & 1 & 0 \\ 1 & -1 & 3 & 1 \end{bmatrix}.$$

Because it is easy to express each vector in C in S-coordinates, we can quickly calculate that the transition matrix from C to S is

$$\mathbf{Q} = \begin{bmatrix} 7 & 4 & -3 & -1 \\ 7 & 2 & 5 & -2 \\ 0 & -1 & 4 & 0 \\ 1 & 1 & -2 & -2 \end{bmatrix}.$$

Then, by Theorem 4.23, the product

$$\mathbf{QP} = \begin{bmatrix} 3 & 2 & 1 & -4 \\ 10 & -1 & 6 & -7 \\ 4 & -2 & 4 & -1 \\ -3 & 3 & -7 & -2 \end{bmatrix}$$

is the transition matrix from B to S. This matrix is clearly correct, since the columns of $\mathbf{QP}$ are the vectors of B expressed in S-coordinates. ■

Reversing the Order of Transition

If we know the transition matrix for changing from B-coordinates to C-coordinates, we can convert from C-coordinates to B-coordinates using the next result. You are asked to prove this theorem in Exercise 12.

THEOREM 4.24

Let B and C be ordered bases for a finite dimensional vector space $\mathcal{V}$, and let $\mathbf{P}$ be the transition matrix from B to C. Then $\mathbf{P}$ is nonsingular, and $\mathbf{P}^{-1}$ is the transition matrix from C to B.

Example 10

Let us consider again the vector space $\mathcal{V}$ of Examples 3 and 8, with ordered bases

$$B = ([1, 0, -1, 0, 4], [0, 1, -1, 0, 3], [0, 0, 0, 1, 5]),$$
$$C = ([-4, 5, -1, 0, -1], [1, -3, 2, 2, 5], [1, -2, 1, 1, 3]).$$

In Example 8 we calculated the transition matrix $\mathbf{P}$ from B to C. Now, by Theorem 4.24, $\mathbf{P}^{-1}$ is the transition matrix from C to B. Because

$$\mathbf{P} = \begin{bmatrix} 1 & 1 & 1 \\ -5 & -4 & -3 \\ 10 & 8 & 7 \end{bmatrix}, \quad \text{we get} \quad \mathbf{P}^{-1} = \begin{bmatrix} -4 & 1 & 1 \\ 5 & -3 & -2 \\ 0 & 2 & 1 \end{bmatrix}.$$

We can verify this matrix by computing the transition matrix from C to B directly. To do this, we express the vectors in C in B-coordinates. The simple form of the vectors in B makes this calculation easy:

$$\begin{aligned} [-4, 5, -1, 0, -1]_B &= [-4, 5, 0], \\ [1, -3, 2, 2, 5]_B &= [1, -3, 2], \\ [1, -2, 1, 1, 3]_B &= [1, -2, 1]. \end{aligned}$$

These are the columns of $\mathbf{P}^{-1}$, as expected. ■

Finally, let us return to the situation in Example 9 and use the inverses of the transition matrices to find the B-coordinates of a polynomial in $\mathcal{P}_3$.

Example 11

Consider again the bases B, C, and S in Example 9 and the transition matrices $\mathbf{P}$ from B to C and $\mathbf{Q}$ from C to S. The transition matrices from C to B and from S to C, respectively, are

$$\mathbf{P}^{-1} = \begin{bmatrix} -\frac{3}{2} & -1 & 3 & -\frac{1}{2} \\ 1 & 1 & -1 & 0 \\ \frac{3}{2} & 1 & -2 & \frac{1}{2} \\ -2 & -1 & 2 & 0 \end{bmatrix} \quad \text{and} \quad \mathbf{Q}^{-1} = \begin{bmatrix} -22 & 24 & -53 & -13 \\ 48 & -52 & 115 & 28 \\ 12 & -13 & 29 & 7 \\ 1 & -1 & 2 & 0 \end{bmatrix}.$$

Now,

$$[\mathbf{v}]_B = \mathbf{P}^{-1}[\mathbf{v}]_C = \mathbf{P}^{-1}(\mathbf{Q}^{-1}[\mathbf{v}]_S) = (\mathbf{P}^{-1}\mathbf{Q}^{-1})[\mathbf{v}]_S,$$

and so $\mathbf{P}^{-1}\mathbf{Q}^{-1}$ acts as the transition matrix from S to B (see Figure 4.8). For example, if $\mathbf{v} = 2x^3 - 36x^2 - 21x + 16$, then

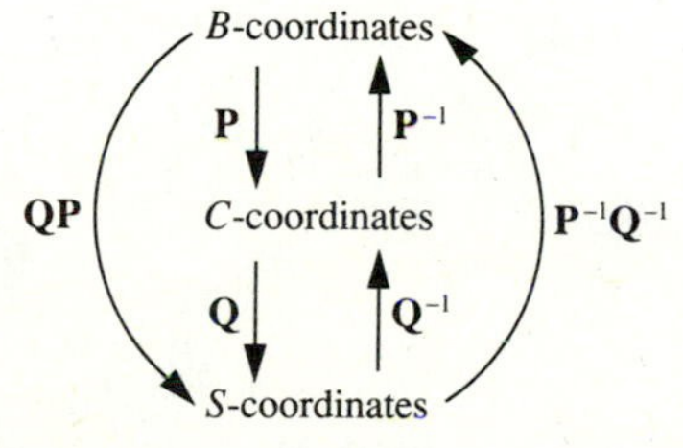

Figure 4.8
Transition matrices used to convert between B-, C-, and S-coordinates in $\mathcal{P}_3$

$$[\mathbf{v}]_B = \mathbf{P}^{-1}\mathbf{Q}^{-1}[\mathbf{v}]_S = \begin{bmatrix} -\frac{3}{2} & -1 & 3 & -\frac{1}{2} \\ 1 & 1 & -1 & 0 \\ \frac{3}{2} & 1 & -2 & \frac{1}{2} \\ -2 & -1 & 2 & 0 \end{bmatrix} \begin{bmatrix} -22 & 24 & -53 & -13 \\ 48 & -52 & 115 & 28 \\ 12 & -13 & 29 & 7 \\ 1 & -1 & 2 & 0 \end{bmatrix} \begin{bmatrix} 2 \\ -36 \\ -21 \\ 16 \end{bmatrix}$$

$$= \begin{bmatrix} -\frac{19}{2} \\ 3 \\ \frac{9}{2} \\ -5 \end{bmatrix}.$$

(Verify.) ■

♦ **Application:** You have now covered the prerequisites for Section 8.6, "Function Spaces"; and Section 8.7, "Rotation of Axes."

Exercises—Section 4.7

1. In each part, let B represent an ordered basis for a subspace $\mathcal{V}$ of $\mathbb{R}^n$, $\mathcal{P}_n$, or $\mathcal{M}_{mn}$. Find $[\mathbf{v}]_B$, for the given $\mathbf{v} \in \mathcal{V}$, by solving the appropriate linear system:

★(a) $B = ([1, -4, 1], [5, -7, 2], [0, -4, 1])$; $\mathbf{v} = [2, -1, 0]$

(b) $B = ([4, 6, 0, 1], [5, 1, -1, 0], [0, 15, 1, 3], [1, 5, 0, 1])$; $\mathbf{v} = [0, -9, 1, -2]$

★(c) $B = ([2, 3, 1, -2, 2], [4, 3, 3, 1, -1], [1, 2, 1, -1, 1])$; $\mathbf{v} = [7, -4, 5, 13, -13]$

(d) $B = ([-3, 1, -2, 5, -1], [6, 1, 2, -1, 0], [9, 2, 1, -4, 2], [3, 1, 0, -2, 1])$; $\mathbf{v} = [3, 16, -12, 41, -7]$

★(e) $B = (3x^2 - x + 2, x^2 + 2x - 3, 2x^2 + 3x - 1)$; $\mathbf{v} = 13x^2 - 5x + 20$

(f) $B = (4x^2 + 3x - 1, 2x^2 - x + 4, x^2 - 2x + 3)$; $\mathbf{v} = -5x^2 - 17x + 20$

★(g) $B = (2x^3 - x^2 + 3x - 1, x^3 + 2x^2 - x + 3, -3x^3 - x^2 + x + 1)$; $\mathbf{v} = 8x^3 + 11x^2 - 9x + 11$

★(h) $B = \left(\begin{bmatrix} 1 & -2 \\ 0 & 1 \end{bmatrix}, \begin{bmatrix} 2 & -1 \\ 1 & 0 \end{bmatrix}, \begin{bmatrix} 1 & -1 \\ 3 & 1 \end{bmatrix}\right)$; $\mathbf{v} = \begin{bmatrix} -3 & -2 \\ 0 & 3 \end{bmatrix}$

(i) $B = \left(\begin{bmatrix} -2 & 3 \\ 0 & 2 \end{bmatrix}, \begin{bmatrix} 1 & 1 \\ -1 & 2 \end{bmatrix}, \begin{bmatrix} 0 & -3 \\ 2 & 1 \end{bmatrix}\right)$; $\mathbf{v} = \begin{bmatrix} -8 & 35 \\ -14 & 8 \end{bmatrix}$

★(j) $B = \left(\begin{bmatrix} 1 & 3 & -1 \\ 2 & 1 & 4 \end{bmatrix}, \begin{bmatrix} -3 & 1 & 7 \\ 1 & 2 & 5 \end{bmatrix}\right)$; $\mathbf{v} = \begin{bmatrix} 11 & 13 & -19 \\ 8 & 1 & 10 \end{bmatrix}$

2. In each part, ordered bases B and C are given for a subspace of $\mathbb{R}^n$, $\mathcal{P}_n$, or $\mathcal{M}_{mn}$. Find the transition matrix from B-coordinates to C-coordinates:

★(a) $B = ([1, 0, 0], [0, 1, 0], [0, 0, 1])$; $C = ([1, 5, 1], [1, 6, -6], [1, 3, 14])$

(b) $B = ([1, 2, 2], [3, 5, 1], [2, 1, 6])$; $C = ([1, 4, 1], [0, 41, 8], [0, 5, 1])$

★(c) $B = (2x^2 + 3x - 1, 8x^2 + x + 1, x^2 + 6)$;
$C = (x^2 + 3x + 1, 3x^2 + 4x + 1, 10x^2 + 17x + 5)$

★(d) $B = \left(\begin{bmatrix} 1 & 3 \\ 5 & 1 \end{bmatrix}, \begin{bmatrix} 2 & 1 \\ 0 & 4 \end{bmatrix}, \begin{bmatrix} 3 & 1 \\ 1 & 0 \end{bmatrix}, \begin{bmatrix} 0 & 2 \\ -4 & 1 \end{bmatrix}\right)$;

$C = \left(\begin{bmatrix} -1 & 1 \\ 3 & -1 \end{bmatrix}, \begin{bmatrix} 1 & 0 \\ 0 & 1 \end{bmatrix}, \begin{bmatrix} 3 & -4 \\ -7 & 4 \end{bmatrix}, \begin{bmatrix} 1 & -1 \\ -2 & 1 \end{bmatrix}\right)$

(e) $B = ([1, 3, -2, 0, 1, 4], [-6, 2, 7, -5, -11, -14])$;
$C = ([3, 1, -4, 2, 5, 8], [4, 0, -5, 3, 7, 10])$

★(f) $B = (6x^4 + 20x^3 + 7x^2 + 19x - 4, x^4 + 5x^3 + 7x^2 - x + 6,$
$5x^3 + 17x^2 - 10x + 19)$;
$C = (x^4 + 3x^3 + 4x - 2, 2x^4 + 7x^3 + 4x^2 + 3x + 1,$
$2x^4 + 5x^3 - 3x^2 + 8x - 7)$

(g) $B = \left(\begin{bmatrix} -1 & 4 \\ 8 & 4 \\ -9 & 0 \end{bmatrix}, \begin{bmatrix} 3 & 7 \\ 2 & 10 \\ -7 & 3 \end{bmatrix}, \begin{bmatrix} -9 & -1 \\ 20 & 0 \\ 3 & 1 \end{bmatrix}\right)$;

$C = \left(\begin{bmatrix} 1 & 2 \\ 0 & 4 \\ 1 & 2 \end{bmatrix}, \begin{bmatrix} 2 & 3 \\ -1 & 5 \\ -1 & 2 \end{bmatrix}, \begin{bmatrix} -1 & 2 \\ 5 & 3 \\ -2 & 1 \end{bmatrix}\right)$

3. Draw the B-coordinate grid in $\mathbb{R}^2$ as in Example 4, where $B = ([3, 2], [-2, 1])$. Plot the point $(2, 6)$. Convert this point to B-coordinates, and show that it is at the proper place on the B-coordinate grid.

4. In each part of this exercise, ordered bases B, C, and D are given for $\mathbb{R}^n$ or $\mathcal{P}_n$. Calculate independently

(i) The transition matrix $\mathbf{P}$ from B to C
(ii) The transition matrix $\mathbf{Q}$ from C to D
(iii) The transition matrix $\mathbf{T}$ from B to D

Then verify Theorem 4.23 by showing that $\mathbf{T} = \mathbf{QP}$.

★(a) $B = ([3, 1], [7, 2])$; $C = ([3, 7], [2, 5])$; $D = ([5, 2], [2, 1])$

(b) $B = ([8, 1, 0], [2, 11, 5], [-1, 2, 1])$; $C = ([2, 11, 5], [-1, 2, 1], [8, 1, 0])$;
$D = ([-1, 2, 1], [2, 11, 5], [8, 1, 0])$

★(c) $B = (x^2 + 2x + 2, 3x^2 + 7x + 8, 3x^2 + 9x + 13)$; $C = (x^2 + 4x + 1,$
$2x^2 + x, x^2)$; $D = (7x^2 - 3x + 2, x^2 + 7x - 3, x^2 - 2x + 1)$

(d) $B = (4x^3 + x^2 + 5x + 2, 2x^3 - 2x^2 + x + 1, 3x^3 - x^2 + 7x + 3, x^3 - x^2 + 2x + 1)$;
$C = (x^3 + x + 3, x^2 + 2x - 1, x^3 + 2x^2 + 6x + 6, 3x^3 - x^2 + 6x + 36)$;
$D = (x^3, x^2, x, 1)$

5. In each part of this exercise, an ordered basis B is given for a subspace $\mathcal{V}$ of $\mathbb{R}^n$. Perform the following steps:

(i) Use the row reduction method of Section 4.5 to find a second ordered basis C.
(ii) Find the transition matrix $\mathbf{P}$ from B to C.
(iii) Use Theorem 4.24 to find the transition matrix $\mathbf{Q}$ from C to B.
(iv) For the given vector $\mathbf{v} \in \mathcal{V}$, independently calculate $[\mathbf{v}]_B$ and $[\mathbf{v}]_C$.
(v) Check your answer to step (iv) by using $\mathbf{Q}$ and $[\mathbf{v}]_C$ to calculate $[\mathbf{v}]_B$.

⋆(a) $B = ([1, -4, 1, 2, 1], [6, -24, 5, 8, 3], [3, -12, 3, 6, 2])$;
$\mathbf{v} = [2, -8, -2, -12, 3]$

(b) $B = ([1, -5, 2, 0, -4], [3, -14, 9, 2, -3], [1, -4, 5, 3, 7])$;
$\mathbf{v} = [2, -9, 7, 5, 7]$

⋆(c) $B = ([3, -1, 4, 6], [6, 7, -3, -2], [-4, -3, 3, 4], [-2, 0, 1, 2])$;
$\mathbf{v} = [10, 14, 3, 12]$

⋆**6.** Consider the ordered basis $B = ([-2, 1, 3], [1, 0, 2], [-13, 5, 10])$ for $\mathbb{R}^3$. Suppose that C is another ordered basis for $\mathbb{R}^3$ and that the transition matrix from B to C is given by

$$\begin{bmatrix} 1 & 9 & -1 \\ 2 & 13 & -11 \\ -1 & -8 & 3 \end{bmatrix}.$$

Find C.

7. ⋆(a) Let $\mathbf{u} = [-5, 9, -1]$, $\mathbf{v} = [3, -9, 2]$, and $\mathbf{w} = [2, -5, 1]$. Find the transition matrix from the ordered basis $B = (\mathbf{u}, \mathbf{v}, \mathbf{w})$ to each of the following ordered bases: $C_1 = (\mathbf{v}, \mathbf{w}, \mathbf{u})$, $C_2 = (\mathbf{w}, \mathbf{u}, \mathbf{v})$, $C_3 = (\mathbf{u}, \mathbf{w}, \mathbf{v})$, $C_4 = (\mathbf{v}, \mathbf{u}, \mathbf{w})$, $C_5 = (\mathbf{w}, \mathbf{v}, \mathbf{u})$.

(b) Let B be an ordered basis for an n-dimensional vector space $\mathcal{V}$. Let C be another ordered basis for $\mathcal{V}$ with the same vectors as B but listed in a different order. Prove that the transition matrix from B to C is obtained by permuting rows of $\mathbf{I}_n$ in the same fashion that vectors in B are permuted to obtain C.

8. Let B and C be ordered bases for $\mathbb{R}^n$.

(a) Show that if B is the standard basis in $\mathbb{R}^n$, then the transition matrix from B to C is given by

$$\begin{bmatrix} 1^{st} & 2^{nd} & & n^{th} \\ \text{vector} & \text{vector} & \cdots & \text{vector} \\ \text{in} & \text{in} & & \text{in} \\ C & C & & C \end{bmatrix}^{-1}.$$

(b) Show that if C is the standard basis in $\mathbb{R}^n$, then the transition matrix from B to C is given by

$$\begin{bmatrix} 1^{st} & 2^{nd} & & n^{th} \\ \text{vector} & \text{vector} & \cdots & \text{vector} \\ \text{in} & \text{in} & & \text{in} \\ B & B & & B \end{bmatrix}.$$

9. Let B and C be ordered bases for $\mathbb{R}^n$. Let $\mathbf{P}$ be the matrix whose columns are the vectors in B, and let $\mathbf{Q}$ be the matrix whose columns are the vectors in C. Prove that the transition matrix from B to C equals $\mathbf{Q}^{-1}\mathbf{P}$.

10. Prove Theorem 4.21. (Hint: Use a proof by induction for part (3).)

11. Prove Theorem 4.23.

12. Prove Theorem 4.24. (Hint: Let $\mathbf{Q}$ be the transition matrix from C to B. Prove that $\mathbf{QP} = \mathbf{I}$ by using Theorems 4.22 and 4.23.)

5

Linear Transformations and Orthogonality

In this chapter, we study functions whose domain and codomain are vector spaces; that is, we study functions that map the vectors in one vector space to those in another. We concentrate on a special class of these functions, known as linear transformations. You are already familiar with several types of linear transformations from geometry: reflections, rotations, and projections. In Section 5.2, we show that the effect of any linear transformation is equivalent to multiplication by a corresponding matrix. In Section 5.3, we examine an important relationship between the dimensions of the domain and the range of a linear transformation, known as the Dimension Theorem. In Section 5.4, we establish that all n-dimensional vector spaces are in some sense equivalent.

Finally, in Sections 5.5 and 5.6 we discuss orthogonality of vectors in $\mathbb{R}^n$ in more detail and show that each subspace of $\mathbb{R}^n$ has an important subspace orthogonal to it, known as its orthogonal complement.

5.1 Introduction to Linear Transformations

In this section, we introduce linear transformations and examine their elementary properties.

Functions

If you are not familiar with the terms *domain*, *codomain*, *range*, *image*, and *pre-image* in the context of functions, read Appendix B before proceeding. The following example illustrates some of these terms.

Example 1

Let $f\colon \mathcal{M}_{23} \longrightarrow \mathcal{M}_{22}$ be given by

$$f\left(\begin{bmatrix} a & b & c \\ d & e & f \end{bmatrix}\right) = \begin{bmatrix} a & b \\ 0 & 0 \end{bmatrix}.$$

Then, f is a function that maps one vector space to another. The domain of f is $\mathcal{M}_{23}$, the codomain of f is $\mathcal{M}_{22}$, and the range of f is the set of all 2×2 matrices with second-row entries equal to zero. The image of $\begin{bmatrix} 1 & 2 & 3 \\ 4 & 5 & 6 \end{bmatrix}$ under f is $\begin{bmatrix} 1 & 2 \\ 0 & 0 \end{bmatrix}$. The matrix $\begin{bmatrix} 1 & 2 & 10 \\ 11 & 12 & 13 \end{bmatrix}$ is one of the pre-images of $\begin{bmatrix} 1 & 2 \\ 0 & 0 \end{bmatrix}$ under f. Also, the image under f of the set S of all matrices of the form $\begin{bmatrix} 7 & * & * \\ * & * & * \end{bmatrix}$ (where $*$ represents any real number) is the set $f(S)$ containing all matrices of the form $\begin{bmatrix} 7 & * \\ 0 & 0 \end{bmatrix}$. Finally, the pre-image under f of the set T of all matrices of the form $\begin{bmatrix} a & a+2 \\ 0 & 0 \end{bmatrix}$ is the set $f^{-1}(T)$ consisting of all matrices of the form $\begin{bmatrix} a & a+2 & * \\ * & * & * \end{bmatrix}$. ■

Linear Transformations

DEFINITION

Let $\mathcal{V}$ and $\mathcal{W}$ be vector spaces, and let $f\colon \mathcal{V} \longrightarrow \mathcal{W}$ be a function from $\mathcal{V}$ to $\mathcal{W}$. (That is, for each vector $\mathbf{v} \in \mathcal{V}$, $f(\mathbf{v})$ represents exactly one vector of $\mathcal{W}$.) Then f is a **linear transformation** if and only if both of the following are true:

(1) $f(\mathbf{v}_1 + \mathbf{v}_2) = f(\mathbf{v}_1) + f(\mathbf{v}_2)$, for all $\mathbf{v}_1, \mathbf{v}_2 \in \mathcal{V}$
(2) $f(c\mathbf{v}_1) = cf(\mathbf{v}_1)$, for all $c \in \mathbb{R}$ and all $\mathbf{v}_1 \in \mathcal{V}$

Thus, a linear transformation is a function between vector spaces that "preserves" the operations that give structure to the spaces.

To determine whether a given function f from a vector space $\mathcal{V}$ to a vector space $\mathcal{W}$ is a linear transformation, we need only verify properties (1) and (2) in the definition.

Example 2

Consider the mapping $f\colon \mathcal{M}_{mn} \longrightarrow \mathcal{M}_{nm}$ given by $f(\mathbf{A}) = \mathbf{A}^T$, for any $m \times n$ matrix $\mathbf{A}$. We will show that f is a linear transformation.

(1) We must show that $f(\mathbf{A}_1 + \mathbf{A}_2) = f(\mathbf{A}_1) + f(\mathbf{A}_2)$, for matrices $\mathbf{A}_1, \mathbf{A}_2 \in \mathcal{M}_{mn}$. However, $f(\mathbf{A}_1 + \mathbf{A}_2) = (\mathbf{A}_1 + \mathbf{A}_2)^T = \mathbf{A}_1^T + \mathbf{A}_2^T$ (by part (2) of Theorem 1.11) $= f(\mathbf{A}_1) + f(\mathbf{A}_2)$.

(2) We must show that $f(c\mathbf{A}) = cf(\mathbf{A})$, for all $c \in \mathbb{R}$ and for all $\mathbf{A} \in \mathcal{M}_{mn}$. However, $f(c\mathbf{A}) = (c\mathbf{A})^T = c(\mathbf{A}^T)$ (by part (3) of Theorem 1.11) $= cf(\mathbf{A})$.

Hence, f is a linear transformation. ■

Example 3

Consider the function $g: \mathcal{P}_n \longrightarrow \mathcal{P}_{n-1}$ given by $g(\mathbf{p}) = \mathbf{p}'$, the derivative of $\mathbf{p}$. We will show that g is a linear transformation.

(1) We must show that $g(\mathbf{p}_1 + \mathbf{p}_2) = g(\mathbf{p}_1) + g(\mathbf{p}_2)$, for all $\mathbf{p}_1, \mathbf{p}_2 \in \mathcal{P}_n$. Now, $g(\mathbf{p}_1 + \mathbf{p}_2) = (\mathbf{p}_1 + \mathbf{p}_2)'$. From calculus we know that the derivative of a sum is the sum of the derivatives, so $(\mathbf{p}_1 + \mathbf{p}_2)' = \mathbf{p}_1' + \mathbf{p}_2' = g(\mathbf{p}_1) + g(\mathbf{p}_2)$.

(2) We must show that $g(c\mathbf{p}) = cg(\mathbf{p})$, for all $\mathbf{p} \in \mathcal{P}_n$. Now, $g(c\mathbf{p}) = (c\mathbf{p})'$. Again, from calculus we know that the derivative of a constant times a function is equal to the constant times the derivative of the function, so $(c\mathbf{p})' = c(\mathbf{p}') = cg(\mathbf{p})$.

Hence, g is a linear transformation. ■

Example 4

Let $\mathcal{V}$ be a finite dimensional vector space, and let $B = (\mathbf{v}_1, \mathbf{v}_2, \ldots, \mathbf{v}_n)$ be an ordered basis for $\mathcal{V}$. Then every element $\mathbf{v} \in \mathcal{V}$ has its coordinatization $[\mathbf{v}]_B$ with respect to B. Consider the mapping $f: \mathcal{V} \longrightarrow \mathbb{R}^n$ given by $f(\mathbf{v}) = [\mathbf{v}]_B$. We will show that f is a linear transformation.

By Theorem 4.21, $[\mathbf{v}_1 + \mathbf{v}_2]_B = [\mathbf{v}_1]_B + [\mathbf{v}_2]_B$. Hence, we have

$$f(\mathbf{v}_1 + \mathbf{v}_2) = [\mathbf{v}_1 + \mathbf{v}_2]_B = [\mathbf{v}_1]_B + [\mathbf{v}_2]_B = f(\mathbf{v}_1) + f(\mathbf{v}_2).$$

Next, let $c \in \mathbb{R}$. Again by Theorem 4.21, $[c\mathbf{v}_1]_B = c[\mathbf{v}_1]_B$. Hence, we have

$$f(c\mathbf{v}_1) = [c\mathbf{v}_1]_B = c[\mathbf{v}_1]_B = cf(\mathbf{v}_1).$$

Thus, f is a linear transformation from $\mathcal{V}$ to $\mathbb{R}^n$. ■

Not every function between vector spaces is a linear transformation. For example, consider the function $h: \mathbb{R}^2 \longrightarrow \mathbb{R}^2$ given by $h([x, y]) = [x + 1, y - 2] = [x, y] + [1, -2]$. In this case, h merely adds $[1, -2]$ to each vector $[x, y]$ (see Figure 5.1). This type of mapping is called a **translation**. However, h is not a linear transformation. To show that it is not, we have to produce a counterexample to

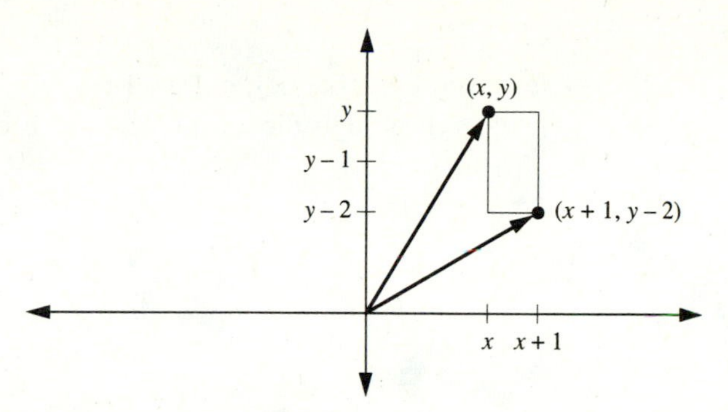

Figure 5.1
A translation in $\mathbb{R}^2$

verify that either property (1) or property (2) of the definition fails. In fact, property (1) fails: $h([1,2] + [3,4]) = h([4,6]) = [5,4]$, while $h([1,2]) + h([3,4]) = [2,0] + [4,2] = [6,2]$.

In general, when given a function f between vector spaces, we do not always know right away whether f is a linear transformation. If you suspect that either property (1) or (2) does not hold for f, then look for a counterexample.

Linear Operators and Some Geometric Examples

An important type of linear transformation is one that maps a vector space to itself:

DEFINITION

Let $\mathcal{V}$ be a vector space. A **linear operator** on $\mathcal{V}$ is a linear transformation whose domain and codomain are both $\mathcal{V}$.

Example 5

If $\mathcal{V}$ is any vector space, then it is easy to show that the mapping $i: \mathcal{V} \longrightarrow \mathcal{V}$ given by $i(\mathbf{v}) = \mathbf{v}$, for all $\mathbf{v} \in \mathcal{V}$, is a linear operator known as the **identity linear operator**. Also, the constant mapping $z: \mathcal{V} \longrightarrow \mathcal{V}$ given by $z(\mathbf{v}) = \mathbf{0}$, where $\mathbf{0}$ is the zero vector of $\mathcal{V}$, is a linear operator known as the **zero linear operator** (see Exercise 2). ■

The next few examples exhibit operators that are especially important in computer graphics, where figures are manipulated geometrically on a terminal screen. In these examples, assume that all vectors begin at the origin.

Example 6: Reflections

Consider the mapping $f: \mathbb{R}^3 \longrightarrow \mathbb{R}^3$ given by $f([a_1, a_2, a_3]) = [a_1, a_2, -a_3]$. This mapping "reflects" the vector $[a_1, a_2, a_3]$ through the xy-plane, which acts like a "mirror" (see Figure 5.2).

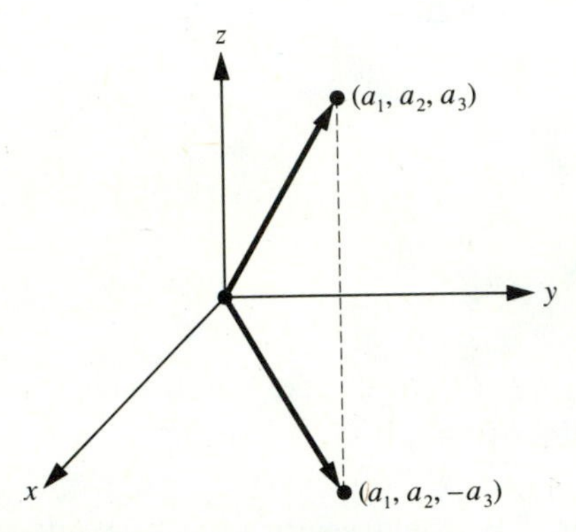

Figure 5.2
Reflection in $\mathbb{R}^3$ through the *xy*-plane

Now, since $f([a_1, a_2, a_3] + [b_1, b_2, b_3]) = f([a_1 + b_1, a_2 + b_2, a_3 + b_3]) = [a_1 + b_1, a_2 + b_2, -(a_3 + b_3)] = [a_1, a_2, -a_3] + [b_1, b_2, -b_3] = f([a_1, a_2, a_3]) + f([b_1, b_2, b_3])$ and $f(c[a_1, a_2, a_3]) = f([ca_1, ca_2, ca_3]) = [ca_1, ca_2, -ca_3] = c[a_1, a_2, -a_3] = cf([a_1, a_2, a_3])$, we see that f is a linear operator. Similarly, reflection through the xz-plane or the yz-plane is also a linear operator on $\mathbb{R}^3$ (see Exercise 4). ■

Example 7: Contractions and Dilations

Consider the mapping $g: \mathbb{R}^n \longrightarrow \mathbb{R}^n$ given by scalar multiplication by k, where $k \in \mathbb{R}$; that is, $g(\mathbf{v}) = k\mathbf{v}$, for $\mathbf{v} \in \mathbb{R}^n$. Then g is easily seen to be a linear operator (see Exercise 3). If $|k| > 1$, g represents a **dilation** (lengthening) of the vectors in $\mathbb{R}^n$; if $|k| < 1$, g represents a **contraction** (shrinking). ■

Example 8: Projections

Consider the mapping $h: \mathbb{R}^3 \longrightarrow \mathbb{R}^3$ given by $h([a_1, a_2, a_3]) = [a_1, a_2, 0]$. This mapping takes each vector in $\mathbb{R}^3$ to a corresponding vector in the xy-plane (see Figure 5.3).

Similarly, consider the mapping $j: \mathbb{R}^4 \longrightarrow \mathbb{R}^4$ given by $j([a_1, a_2, a_3, a_4]) = [0, a_2, 0, a_4]$. This mapping takes each vector in $\mathbb{R}^4$ to a corresponding vector whose first and third coordinates are zero. The functions h and j are both linear operators (see Exercise 5). Such mappings, where one or more of the coordinates is "zeroed out," are examples of **projection mappings**. All such mappings are easily seen to

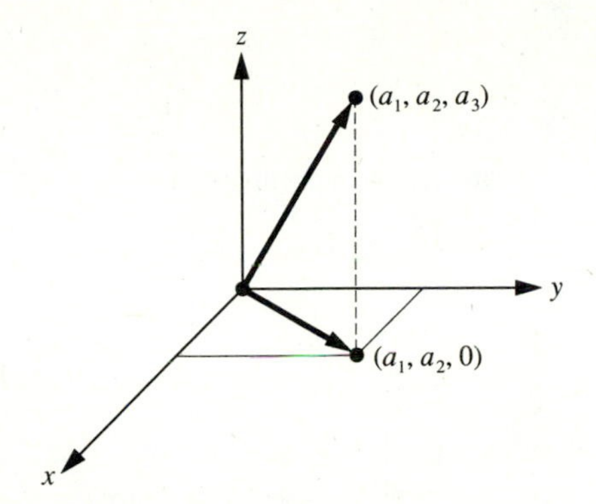

Figure 5.3
Projection of $[a_1, a_2, a_3]$ to the xy-plane

be linear operators. (Other types of projection mappings are illustrated in Exercises 6 and 7.) ■

Example 9: Rotations

Let θ be a fixed angle in $\mathbb{R}^2$, and let $l: \mathbb{R}^2 \longrightarrow \mathbb{R}^2$ be given by

$$l\left(\begin{bmatrix} x \\ y \end{bmatrix}\right) = \begin{bmatrix} \cos\theta & -\sin\theta \\ \sin\theta & \cos\theta \end{bmatrix}\begin{bmatrix} x \\ y \end{bmatrix} = \begin{bmatrix} x\cos\theta - y\sin\theta \\ x\sin\theta + y\cos\theta \end{bmatrix}.$$

In Exercise 8 you are asked to show that l rotates $[x, y]$ counterclockwise through the angle θ (see Figure 5.4).

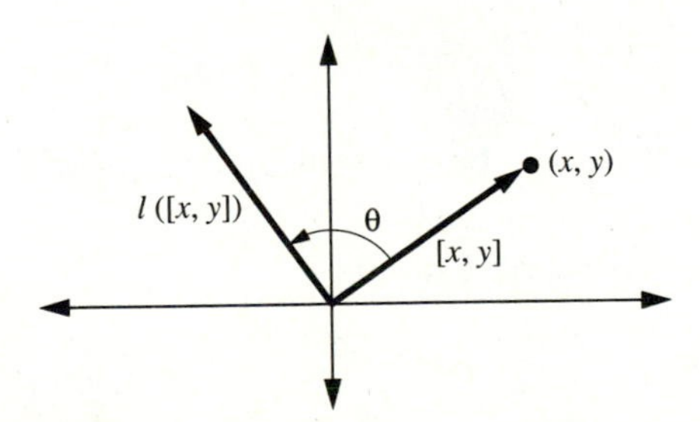

Figure 5.4
Counterclockwise rotation of a vector by l through an angle θ in $\mathbb{R}^2$

Now, let $\mathbf{v}_1 = [x_1, y_1]$ and $\mathbf{v}_2 = [x_2, y_2]$ be two vectors in $\mathbb{R}^2$. Then,

$$\begin{aligned} l(\mathbf{v}_1 + \mathbf{v}_2) &= \begin{bmatrix} \cos\theta & -\sin\theta \\ \sin\theta & \cos\theta \end{bmatrix}(\mathbf{v}_1 + \mathbf{v}_2) \\ &= \begin{bmatrix} \cos\theta & -\sin\theta \\ \sin\theta & \cos\theta \end{bmatrix}\mathbf{v}_1 + \begin{bmatrix} \cos\theta & -\sin\theta \\ \sin\theta & \cos\theta \end{bmatrix}\mathbf{v}_2 \\ &= l(\mathbf{v}_1) + l(\mathbf{v}_2). \end{aligned}$$

Similarly, $l(c\mathbf{v}) = cl(\mathbf{v})$, for any $c \in \mathbb{R}$ and $\mathbf{v} \in \mathbb{R}^2$. Hence, l is a linear operator. ■

Beware! Not all geometric operations are linear operators. Recall that the translation function is not a linear operator!

Multiplication Transformation

The linear operator in Example 9 is actually a special case of the next example, which shows that multiplication by an $m \times n$ matrix is always a linear transformation from $\mathbb{R}^n$ to $\mathbb{R}^m$.

Example 10

Let $\mathbf{A}$ be a given $m \times n$ matrix. It is easy to show that the function $f: \mathbb{R}^n \longrightarrow \mathbb{R}^m$ defined by $f(\mathbf{x}) = \mathbf{Ax}$, for all $\mathbf{x} \in \mathbb{R}^n$, is a linear transformation. Let $\mathbf{x}_1, \mathbf{x}_2 \in \mathbb{R}^n$ and $c \in \mathbb{R}$. Then $f(\mathbf{x}_1 + \mathbf{x}_2) = \mathbf{A}(\mathbf{x}_1 + \mathbf{x}_2) = \mathbf{Ax}_1 + \mathbf{Ax}_2 = f(\mathbf{x}_1) + f(\mathbf{x}_2)$, and $f(c\mathbf{x}_1) = \mathbf{A}(c\mathbf{x}_1) = c(\mathbf{Ax}_1) = cf(\mathbf{x}_1)$. Hence, f is a linear transformation. ■

As a specific example, consider the matrix $\mathbf{A} = \begin{bmatrix} -1 & 4 & 2 \\ 5 & 6 & -3 \end{bmatrix}$. The mapping given by

$$f\left(\begin{bmatrix} x_1 \\ x_2 \\ x_3 \end{bmatrix}\right) = \begin{bmatrix} -1 & 4 & 2 \\ 5 & 6 & -3 \end{bmatrix}\begin{bmatrix} x_1 \\ x_2 \\ x_3 \end{bmatrix} = \begin{bmatrix} -x_1 + 4x_2 + 2x_3 \\ 5x_1 + 6x_2 - 3x_3 \end{bmatrix}$$

is then a linear transformation from $\mathbb{R}^3$ to $\mathbb{R}^2$. In the next section, we will show that the converse of Example 10 also holds: every linear transformation from $\mathbb{R}^n$ to $\mathbb{R}^m$ is equivalent to multiplication by an appropriate matrix.

Elementary Properties of Linear Transformations

We conclude this section by proving some basic properties of linear transformations. From here on, we usually use italicized capital letters, such as L, to represent linear transformations.

THEOREM 5.1

Let $\mathcal{V}$ and $\mathcal{W}$ be vector spaces, and let $L: \mathcal{V} \longrightarrow \mathcal{W}$ be a linear transformation. Let $\mathbf{0}_{\mathcal{V}}$ be the zero vector in $\mathcal{V}$ and $\mathbf{0}_{\mathcal{W}}$ be the zero vector in $\mathcal{W}$. Then,

(1) $L(\mathbf{0}_{\mathcal{V}}) = \mathbf{0}_{\mathcal{W}}$
(2) $L(-\mathbf{v}) = -L(\mathbf{v})$, for all $\mathbf{v} \in \mathcal{V}$
(3) $L(a_1\mathbf{v}_1 + a_2\mathbf{v}_2 + \cdots + a_n\mathbf{v}_n) = a_1L(\mathbf{v}_1) + a_2L(\mathbf{v}_2) + \cdots + a_nL(\mathbf{v}_n)$, for all $a_1, \ldots, a_n \in \mathbb{R}$, and $\mathbf{v}_1, \ldots, \mathbf{v}_n \in \mathcal{V}$, for $n \geq 2$.

Proof of Theorem 5.1

Part (1):

$$\begin{aligned} L(\mathbf{0}_{\mathcal{V}}) &= L(0\mathbf{0}_{\mathcal{V}}) && \text{Theorem 4.1, part (2), in } \mathcal{V} \\ &= 0L(\mathbf{0}_{\mathcal{V}}) && \text{property (2) of linear transformation} \\ &= \mathbf{0}_{\mathcal{W}} && \text{Theorem 4.1, part (2), in } \mathcal{W} \end{aligned}$$

Part (2):

$$\begin{aligned} L(-\mathbf{v}) &= L(-1\mathbf{v}) && \text{Theorem 4.1, part (3), in } \mathcal{V} \\ &= -1(L(\mathbf{v})) && \text{property (2) of linear transformation} \\ &= -L(\mathbf{v}) && \text{Theorem 4.1, part (3), in } \mathcal{W} \end{aligned}$$

Part (3) (abridged): This part is proved by induction. We prove the base step ($n = 2$) here and leave the inductive step for you to do in Exercise 28. Now we must show that $L(a_1\mathbf{v}_1 + a_2\mathbf{v}_2) = a_1L(\mathbf{v}_1) + a_2L(\mathbf{v}_2)$:

$$\begin{aligned} L(a_1\mathbf{v}_1 + a_2\mathbf{v}_2) &= L(a_1\mathbf{v}_1) + L(a_2\mathbf{v}_2) && \text{property (1) of linear transformation} \\ &= a_1L(\mathbf{v}_1) + a_2L(\mathbf{v}_2) && \text{property (2) of linear transformation} \end{aligned}$$

■

The next theorem asserts that the composition $L_2 \circ L_1$ of linear transformations L_1 and L_2 is again a linear transformation (see Appendix B for a review of composition of functions).

THEOREM 5.2

Let $\mathcal{V}_1$, $\mathcal{V}_2$, and $\mathcal{V}_3$ be vector spaces. Let $L_1\colon \mathcal{V}_1 \longrightarrow \mathcal{V}_2$ and $L_2\colon \mathcal{V}_2 \longrightarrow \mathcal{V}_3$ be linear transformations. Then $L_2 \circ L_1\colon \mathcal{V}_1 \longrightarrow \mathcal{V}_3$ given by $(L_2 \circ L_1)(\mathbf{v}) = L_2(L_1(\mathbf{v}))$, for all $\mathbf{v} \in \mathcal{V}_1$, is a linear transformation.

Proof of Theorem 5.2 (abridged)

To show that $L_2 \circ L_1$ is a linear transformation, we must show that it satisfies these two properties for all $c \in \mathbb{R}$ and $\mathbf{v}_1, \mathbf{v}_2 \in \mathcal{V}$:

$$(L_2 \circ L_1)(\mathbf{v}_1 + \mathbf{v}_2) = (L_2 \circ L_1)(\mathbf{v}_1) + (L_2 \circ L_1)(\mathbf{v}_2) \quad \text{and} \quad (L_2 \circ L_1)(c\mathbf{v}_1) = c(L_2 \circ L_1)(\mathbf{v}_1).$$

However,

$$
\begin{aligned}
(L_2 \circ L_1)(\mathbf{v}_1 + \mathbf{v}_2) &= L_2(L_1(\mathbf{v}_1 + \mathbf{v}_2)) \\
&= L_2(L_1(\mathbf{v}_1) + L_1(\mathbf{v}_2)) && \text{because } L_1 \text{ is a linear transformation} \\
&= L_2(L_1(\mathbf{v}_1)) + L_2(L_1(\mathbf{v}_2)) && \text{because } L_2 \text{ is a linear transformation} \\
&= (L_2 \circ L_1)(\mathbf{v}_1) + (L_2 \circ L_1)(\mathbf{v}_2).
\end{aligned}
$$

■

We leave the second property of a linear transformation for you to prove in Exercise 32.

Example 11

Let L_1 represent the rotation of vectors in $\mathbb{R}^2$ through a fixed angle θ, and let L_2 represent the reflection of vectors in $\mathbb{R}^2$ through the x-axis. That is, if $\mathbf{v} = [v_1, v_2]$, then

$$L_1(\mathbf{v}) = \begin{bmatrix} \cos\theta & -\sin\theta \\ \sin\theta & \cos\theta \end{bmatrix}\begin{bmatrix} v_1 \\ v_2 \end{bmatrix} \quad \text{and} \quad L_2(\mathbf{v}) = \begin{bmatrix} v_1 \\ -v_2 \end{bmatrix}.$$

Because L_1 and L_2 are both linear transformations, Theorem 5.2 asserts that the following is also a linear transformation:

$$L_2(L_1(\mathbf{v})) = L_2\left(\begin{bmatrix} v_1 \cos\theta - v_2 \sin\theta \\ v_1 \sin\theta + v_2 \cos\theta \end{bmatrix}\right) = \begin{bmatrix} v_1 \cos\theta - v_2 \sin\theta \\ -v_1 \sin\theta - v_2 \cos\theta \end{bmatrix}.$$

This transformation represents a rotation of $\mathbf{v}$ through θ followed by a reflection through the x-axis. ■

Theorem 5.2 easily generalizes to more than two linear transformations. That is, if $L_1, L_2, \ldots, L_k$ are linear transformations and the composition $L_k \circ \cdots \circ L_2 \circ L_1$ is defined, then $L_k \circ \cdots \circ L_2 \circ L_1$ is also a linear transformation.

The next theorem assures us that, under a linear transformation $L: \mathcal{V} \longrightarrow \mathcal{W}$, subspaces of $\mathcal{V}$ "correspond" to subspaces of $\mathcal{W}$, and vice versa.

THEOREM 5.3

Let $L: \mathcal{V} \longrightarrow \mathcal{W}$ be a linear transformation. Then
(1) If $\mathcal{V}'$ is a subspace of $\mathcal{V}$, then the image of $\mathcal{V}'$ in $\mathcal{W}$ is a subspace of $\mathcal{W}$. In particular, the range of L is a subspace of $\mathcal{W}$.
(2) If $\mathcal{W}'$ is a subspace of $\mathcal{W}$, then the pre-image of $\mathcal{W}'$ in $\mathcal{V}$ is a subspace of $\mathcal{V}$.

We prove part (1) and leave part (2) for you (see Exercise 30).

Proof of Theorem 5.3, Part (1)

Suppose that $L: \mathcal{V} \longrightarrow \mathcal{W}$ is a linear transformation and that $\mathcal{V}'$ is a subspace of $\mathcal{V}$. Let $\mathcal{W}' = L(\mathcal{V}')$ be the image of $\mathcal{V}'$ in $\mathcal{V}$ (see Figure 5.5). Now, $\mathcal{W}'$ is clearly nonempty. (Why?) Hence, to show that $\mathcal{W}'$ is a subspace of $\mathcal{W}$, we must prove that $\mathcal{W}'$ is closed under addition and scalar multiplication.

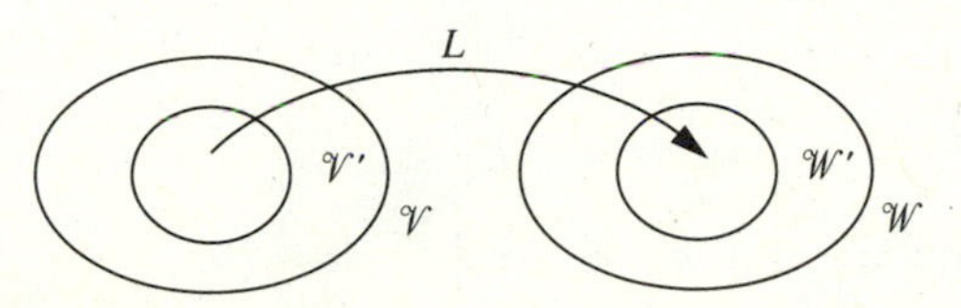

Figure 5.5
Subspaces of $\mathcal{V}$ correspond to subspaces of $\mathcal{W}$ under a linear transformation $L: \mathcal{V} \longrightarrow \mathcal{W}$.

First, suppose that $\mathbf{w}_1, \mathbf{w}_2 \in \mathcal{W}'$. Then, by definition of $\mathcal{W}'$, we have $\mathbf{w}_1 = L(\mathbf{v}_1)$ and $\mathbf{w}_2 = L(\mathbf{v}_2)$, for some $\mathbf{v}_1, \mathbf{v}_2 \in \mathcal{V}'$. Then, $\mathbf{w}_1 + \mathbf{w}_2 = L(\mathbf{v}_1) + L(\mathbf{v}_2) = L(\mathbf{v}_1 + \mathbf{v}_2)$, because L is a linear transformation. However, since $\mathcal{V}'$ is a subspace of $\mathcal{V}$, $(\mathbf{v}_1 + \mathbf{v}_2) \in \mathcal{V}'$. Thus, $(\mathbf{w}_1 + \mathbf{w}_2)$ is the image of $(\mathbf{v}_1 + \mathbf{v}_2)$ in $\mathcal{V}'$, and so $(\mathbf{w}_1 + \mathbf{w}_2) \in \mathcal{W}'$. Hence, $\mathcal{W}'$ is closed under addition.

Next, suppose that $c \in \mathbb{R}$ and $\mathbf{w}_1 \in \mathcal{W}'$. By definition of $\mathcal{W}'$, $L(\mathbf{v}_1) = \mathbf{w}_1$, for some $\mathbf{v}_1 \in \mathcal{V}'$. Then, $c\mathbf{w}_1 = cL(\mathbf{v}_1) = L(c\mathbf{v}_1)$, since L is a linear transformation. Now, $c\mathbf{v}_1 \in \mathcal{V}'$, because $\mathcal{V}'$ is a subspace of $\mathcal{V}$. Thus, $c\mathbf{w}_1$ is the image of $c\mathbf{v}_1$ in $\mathcal{V}'$, and so $c\mathbf{w}_1 \in \mathcal{W}'$. Hence, $\mathcal{W}'$ is closed under scalar multiplication. ■

Example 12

Let $L: \mathcal{M}_{22} \longrightarrow \mathbb{R}^3$, where $L\left(\begin{bmatrix} a & b \\ c & d \end{bmatrix}\right) = [b, 0, c]$. It is easy to show that L is a linear transformation. (Verify.) By Theorem 5.3, the range of any linear transformation is a subspace of the codomain. Hence, the range of $L = \{[b, 0, c] \mid b, c \in \mathbb{R}\}$ is a subspace of $\mathbb{R}^3$.

Also, consider the subspace $\mathcal{U}_2 = \left\{\begin{bmatrix} a & b \\ 0 & d \end{bmatrix} \middle|\, a, b, d \in \mathbb{R}\right\}$ of $\mathcal{M}_{22}$. Then the image of $\mathcal{U}_2$ under L is $\{[b, 0, 0] \mid b \in \mathbb{R}\}$. This image is a subspace of $\mathbb{R}^3$, as Theorem 5.3 asserts. Finally, consider the subspace $\mathcal{W} = \{[b, e, 2b] \mid b, e \in \mathbb{R}\}$ of $\mathbb{R}^3$. The pre-image of $\mathcal{W}$ consists of all matrices in $\mathcal{M}_{22}$ of the form $\begin{bmatrix} a & b \\ 2b & d \end{bmatrix}$. Notice that this pre-image is a subspace of $\mathcal{M}_{22}$, as claimed by Theorem 5.3. ■

Exercises—Section 5.1

1. Determine which of the following functions are linear transformations and which are not. Prove that your answers are correct. Which are linear operators?

★(a) f: $\mathbb{R}^2 \longrightarrow \mathbb{R}^2$ given by $f([x, y]) = [3x - 4y, -x + 2y]$

★(b) h: $\mathbb{R}^4 \longrightarrow \mathbb{R}^4$ given by $h([x_1, x_2, x_3, x_4]) = [x_1 + 2, x_2 - 1, x_3, -3]$

(c) k: $\mathbb{R}^3 \longrightarrow \mathbb{R}^3$ given by $k([x_1, x_2, x_3]) = [x_2, x_3, x_1]$

★(d) l: $\mathcal{M}_{22} \longrightarrow \mathcal{M}_{22}$ given by $l\left(\begin{bmatrix} a & b \\ c & d \end{bmatrix}\right) = \begin{bmatrix} a - 2c + d & 3b - c \\ -4a & b + c - 3d \end{bmatrix}$

(e) n: $\mathcal{M}_{22} \longrightarrow \mathbb{R}$ given by $n\left(\begin{bmatrix} a & b \\ c & d \end{bmatrix}\right) = ad - bc$

★(f) r: $\mathcal{P}_3 \longrightarrow \mathcal{P}_2$ given by $r(ax^3 + bx^2 + cx + d) = \sqrt{a}x^2 - b^2x + c$

(g) s: $\mathbb{R}^3 \longrightarrow \mathbb{R}^3$ given by $s([x_1, x_2, x_3]) = [\cos x_1, \sin x_2, e^{x_3}]$

★(h) t: $\mathcal{P}_3 \longrightarrow \mathbb{R}$ given by $t(a_3x^3 + a_2x^2 + a_1x + a_0) = a_3 + a_2 + a_1 + a_0$

(i) u: $\mathbb{R}^4 \longrightarrow \mathbb{R}$ given by $u([x_1, x_2, x_3, x_4]) = |x_2|$

★(j) v: $\mathcal{P}_2 \longrightarrow \mathbb{R}$ given by $v(ax^2 + bx + c) = abc$

(k) g: $\mathcal{M}_{32} \longrightarrow \mathcal{P}_4$ given by $g\left(\begin{bmatrix} a_{11} & a_{12} \\ a_{21} & a_{22} \\ a_{31} & a_{32} \end{bmatrix}\right) = a_{11}x^4 - a_{21}x^2 + a_{31}$

★(l) e: $\mathbb{R}^2 \longrightarrow \mathbb{R}$ given by $e([x, y]) = \sqrt{x^2 + y^2}$

2. Let $\mathcal{V}$ and $\mathcal{W}$ be vector spaces.

(a) Show that the identity mapping $i: \mathcal{V} \longrightarrow \mathcal{V}$ given by $i(\mathbf{v}) = \mathbf{v}$, for all $\mathbf{v} \in \mathcal{V}$, is a linear operator.

(b) Show that the zero mapping $z: \mathcal{V} \longrightarrow \mathcal{W}$ given by $z(\mathbf{v}) = \mathbf{0}_{\mathcal{W}}$, for all $\mathbf{v} \in \mathcal{V}$, is a linear transformation.

3. Let k be a fixed scalar in $\mathbb{R}$. Show that the mapping f: $\mathbb{R}^n \longrightarrow \mathbb{R}^n$ given by $f([x_1, x_2, \ldots, x_n]) = k[x_1, x_2, \ldots, x_n]$ is a linear operator.

4. (a) Show that f: $\mathbb{R}^3 \longrightarrow \mathbb{R}^3$ given by $f([x, y, z]) = [-x, y, z]$ (reflection of a vector through the yz-plane) is a linear operator.

(b) What mapping from $\mathbb{R}^3$ to $\mathbb{R}^3$ would reflect a vector through the xz-plane? Is it a linear operator? Why or why not?

(c) What mapping from $\mathbb{R}^2$ to $\mathbb{R}^2$ would reflect a vector through the y-axis? through the x-axis? Are these linear operators? Why or why not?

5. Show that the projection mappings h: $\mathbb{R}^3 \longrightarrow \mathbb{R}^3$ given by $h([a_1, a_2, a_3]) = [a_1, a_2, 0]$ and j: $\mathbb{R}^4 \longrightarrow \mathbb{R}^4$ given by $j([a_1, a_2, a_3, a_4]) = [0, a_2, 0, a_4]$ in Example 8 are linear operators.

6. The mapping f: $\mathbb{R}^n \longrightarrow \mathbb{R}$ given by $f([x_1, x_2, \ldots, x_i, \ldots, x_n]) = x_i$ is another type of projection mapping. Show that f is a linear transformation.

7. Let $\mathbf{x}$ be a nonzero vector in $\mathbb{R}^3$. Show that the mapping g: $\mathbb{R}^3 \longrightarrow \mathbb{R}^3$ given by $g(\mathbf{y}) = \mathbf{proj_x y}$ is a linear operator.

8. Let θ be a fixed angle in the xy-plane. Show that the linear operator L: $\mathbb{R}^2 \longrightarrow \mathbb{R}^2$ given by $L\left(\begin{bmatrix} x \\ y \end{bmatrix}\right) = \begin{bmatrix} \cos\theta & -\sin\theta \\ \sin\theta & \cos\theta \end{bmatrix}\begin{bmatrix} x \\ y \end{bmatrix}$ rotates the vector $[x, y]$ counterclockwise through the angle θ in the plane. (Hint: Consider the vector $[x', y']$, obtained by rotating $[x, y]$ counterclockwise through the angle θ. Let $r = \sqrt{x^2 + y^2}$. Then $x = r\cos\alpha$ and $y = r\sin\alpha$, where α is the angle shown in Figure 5.6. Notice that $x' = r(\cos(\theta + \alpha))$ and $y' = r(\sin(\theta + \alpha))$. Then show that $L([x, y]) = [x', y']$.)

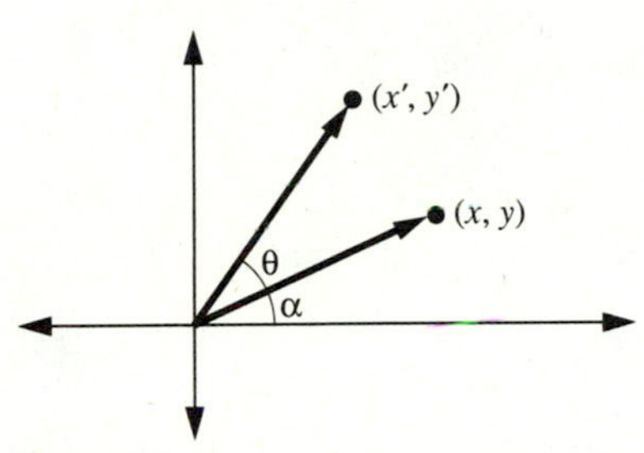

Figure 5.6
The vectors $[x, y]$ and $[x', y']$

9. (a) Explain why the mapping $L : \mathbb{R}^3 \longrightarrow \mathbb{R}^3$ given by

$$L\left(\begin{bmatrix} x \\ y \\ z \end{bmatrix}\right) = \begin{bmatrix} \cos\theta & -\sin\theta & 0 \\ \sin\theta & \cos\theta & 0 \\ 0 & 0 & 1 \end{bmatrix}\begin{bmatrix} x \\ y \\ z \end{bmatrix}$$

is a linear operator.

(b) Show that the mapping L in part (a) rotates the vector $[x, y, z]$ about the z-axis through an angle of θ in the xy-plane.

★(c) What matrix should be multiplied times $[x, y, z]$ to create the linear operator that rotates this vector about the y-axis through an angle ϕ in the xz-plane?

10. Shears: Let f_1, f_2: $\mathbb{R}^2 \longrightarrow \mathbb{R}^2$ be given by

$$f_1\left(\begin{bmatrix} x \\ y \end{bmatrix}\right) = \begin{bmatrix} 1 & k \\ 0 & 1 \end{bmatrix}\begin{bmatrix} x \\ y \end{bmatrix} = \begin{bmatrix} x + ky \\ y \end{bmatrix}$$

and

$$f_2\left(\begin{bmatrix} x \\ y \end{bmatrix}\right) = \begin{bmatrix} 1 & 0 \\ k & 1 \end{bmatrix}\begin{bmatrix} x \\ y \end{bmatrix} = \begin{bmatrix} x \\ kx + y \end{bmatrix}.$$

The mapping f_1 is called a **shear in the *x*-direction with factor *k***; f_2 is called a **shear in the *y*-direction with factor *k***. The effect of these functions (for $k > 1$) on the vector $[1, 1]$ is shown in Figure 5.7. Show that f_1 and f_2 are linear operators directly, without using Example 10.

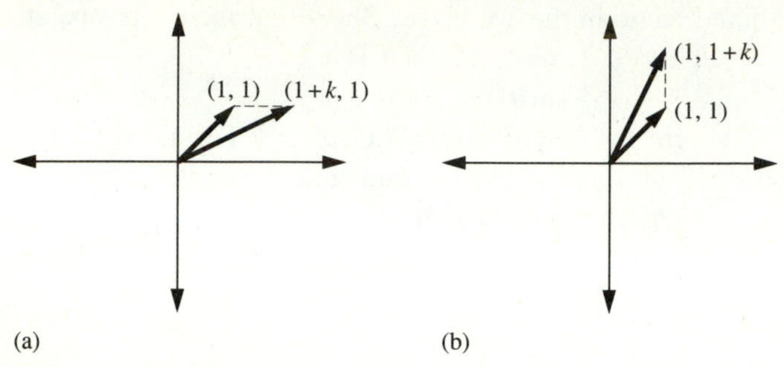

Figure 5.7
(a) Shear in the x-direction; (b) Shear in the y-direction (both for $k > 1$)

11. Let $f: \mathcal{M}_{nn} \longrightarrow \mathbb{R}$ be given by $f(\mathbf{A}) = \text{trace}(\mathbf{A})$. (The trace is defined in Exercise 14 of Section 1.4.) Prove that f is a linear transformation.

12. Show that the mappings g, h: $\mathcal{M}_{nn} \longrightarrow \mathcal{M}_{nn}$ given by $g(\mathbf{A}) = \mathbf{A} + \mathbf{A}^T$ and $h(\mathbf{A}) = \mathbf{A} - \mathbf{A}^T$ are linear operators on $\mathcal{M}_{nn}$.

13. (a) Show that if $\mathbf{p} \in \mathcal{P}_n$, then the (indefinite integration) function $f: \mathcal{P}_n \longrightarrow \mathcal{P}_{n+1}$ given by $f(\mathbf{p}) = \int \mathbf{p}\,dx$, with constant of integration equal to zero, is a linear transformation.

(b) Show that if $\mathbf{p} \in \mathcal{P}_n$, then the (definite integral) function $g: \mathcal{P}_n \longrightarrow \mathbb{R}$ given by $g(\mathbf{p}) = \int_a^b \mathbf{p}\,dx$ is a linear transformation, for any fixed $a, b \in \mathbb{R}$.

14. Let $\mathcal{V}$ be the vector space of all functions f from $\mathbb{R}$ to $\mathbb{R}$ that are infinitely differentiable (that is, for which $f^{(n)}$, the nth derivative of f, exists for every integer $n \geq 1$). Use Theorem 5.2 to show that for any given integer $k \geq 1$, the mapping $L: \mathcal{V} \longrightarrow \mathcal{V}$ given by $L(f) = f^{(k)}$ is a linear operator.

15. Consider the function $f: \mathcal{M}_{nn} \longrightarrow \mathcal{M}_{nn}$ given by $f(\mathbf{A}) = \mathbf{BA}$, where $\mathbf{B}$ is some fixed $n \times n$ matrix. Show that f is a linear operator.

16. Let $\mathbf{B}$ be a fixed nonsingular matrix in $\mathcal{M}_{nn}$. Show that the mapping $f: \mathcal{M}_{nn} \longrightarrow \mathcal{M}_{nn}$ given by $f(\mathbf{A}) = \mathbf{B}^{-1}\mathbf{AB}$ is a linear operator.

17. Let a be a fixed real number.

(a) Let $L: \mathcal{P}_n \longrightarrow \mathbb{R}$ be given by $L(\mathbf{p}) = \mathbf{p}(a)$. (That is, L evaluates polynomials in $\mathcal{P}_n$ at $x = a$.) Show that L is a linear transformation.

(b) Let $L: \mathcal{P}_n \longrightarrow \mathcal{P}_n$ be given by $(L(\mathbf{p}))(x) = \mathbf{p}(x + a)$. (For example, when a is positive, L shifts the graph of $\mathbf{p}(x)$ to the *left* a units.) Prove that L is a linear operator.

18. Let $\mathbf{A}$ be a fixed matrix in $\mathcal{M}_{nn}$. Define $f: \mathcal{P}_n \longrightarrow \mathcal{M}_{nn}$ by $f(a_nx^n + a_{n-1}x^{n-1} + \cdots + a_1x + a_0) = a_n\mathbf{A}^n + a_{n-1}\mathbf{A}^{n-1} + \cdots + a_1\mathbf{A} + a_0\mathbf{I}_n$. Show that f is a linear transformation.

19. Let $\mathcal{V}$ be the unusual vector space from Example 7 in Section 4.1. Show that $L: \mathcal{V} \longrightarrow \mathbb{R}$ given by $L(x) = \ln(x)$ is a linear transformation.

20. Let $\mathcal{V}$ be a vector space, with $\mathbf{x} \neq \mathbf{0}$ in $\mathcal{V}$. Prove that the translation function $f: \mathcal{V} \longrightarrow \mathcal{V}$ given by $f(\mathbf{v}) = \mathbf{v} + \mathbf{x}$ is not a linear transformation.

21. Show that if $\mathbf{A}$ is a fixed matrix in $\mathcal{M}_{mn}$ and $\mathbf{y} \neq \mathbf{0}$ is a vector in $\mathbb{R}^m$, then the mapping $f: \mathbb{R}^n \longrightarrow \mathbb{R}^m$ given by $f(\mathbf{x}) = \mathbf{Ax} + \mathbf{y}$ is not a linear transformation by showing that part (1) of Theorem 5.1 is contradicted for f.

22. Prove that $f: \mathcal{M}_{33} \longrightarrow \mathbb{R}$ given by $f(\mathbf{A}) = |\mathbf{A}|$ is not a linear transformation. (A similar result is true for $\mathcal{M}_{nn}$, for $n > 1$.)

23. Suppose that $L_1: \mathcal{V} \longrightarrow \mathcal{W}$ is a linear transformation and that $L_2: \mathcal{V} \longrightarrow \mathcal{W}$ is defined by $L_2(\mathbf{v}) = L_1(2\mathbf{v})$. Show that L_2 is a linear transformation.

24. Suppose that $L: \mathbb{R}^3 \longrightarrow \mathbb{R}^3$ is a linear operator and that $L([1, 0, 0]) = [-2, 1, 0]$, $L([0, 1, 0]) = [3, -2, 1]$, and $L([0, 0, 1]) = [0, -1, 3]$. Find $L([-3, 2, 4])$. Give a formula for $L([x, y, z])$, for any $[x, y, z] \in \mathbb{R}^3$.

★**25.** Suppose that $L: \mathbb{R}^2 \longrightarrow \mathbb{R}^2$ is a linear operator and that $L(\mathbf{i} + \mathbf{j}) = \mathbf{i} - 3\mathbf{j}$ and $L(-2\mathbf{i} + 3\mathbf{j}) = -4\mathbf{i} + 2\mathbf{j}$. Find $L(\mathbf{i})$ and $L(\mathbf{j})$ as linear combinations of $\mathbf{i}$ and $\mathbf{j}$.

26. Let $L: \mathcal{V} \longrightarrow \mathcal{W}$ be a linear transformation. Show that $L(\mathbf{x} - \mathbf{y}) = L(\mathbf{x}) - L(\mathbf{y})$, for all vectors $\mathbf{x}, \mathbf{y} \in \mathcal{V}$.

27. Part (3) of Theorem 5.1 assures us that if $L: \mathcal{V} \longrightarrow \mathcal{W}$ is a linear transformation, then $L(a\mathbf{v}_1 + b\mathbf{v}_2) = aL(\mathbf{v}_1) + bL(\mathbf{v}_2)$, for all $\mathbf{v}_1, \mathbf{v}_2 \in \mathcal{V}$ and all $a, b \in \mathbb{R}$. Prove that the converse of this statement is true. (Hint: Consider two cases: first $a = b = 1$ and then $b = 0$.)

28. Finish the proof of part (3) of Theorem 5.1 by doing the inductive step.

29. (a) Suppose that $L: \mathcal{V} \longrightarrow \mathcal{W}$ is a linear transformation. Suppose also that $\{L(\mathbf{v}_1), L(\mathbf{v}_2), \ldots, L(\mathbf{v}_n)\}$ is a linearly independent set in $\mathcal{W}$, for some vectors $\mathbf{v}_1, \ldots, \mathbf{v}_n \in \mathcal{V}$. Show that $\{\mathbf{v}_1, \mathbf{v}_2, \ldots, \mathbf{v}_n\}$ is a linearly independent set in $\mathcal{V}$.

★(b) Find a counterexample to the converse of part (a).

30. Finish the proof of Theorem 5.3 by showing that if $L: \mathcal{V} \longrightarrow \mathcal{W}$ is a linear transformation and if $\mathcal{W}'$ is a subspace of $\mathcal{W}$ with pre-image $L^{-1}(\mathcal{W}')$, then $L^{-1}(\mathcal{W}')$ is a subspace of $\mathcal{V}$.

31. Show that every linear operator $L: \mathbb{R} \longrightarrow \mathbb{R}$ has the form $L(\mathbf{x}) = c\mathbf{x}$, for some $c \in \mathbb{R}$.

32. Finish the proof of Theorem 5.2 by proving property (2) of a linear transformation for $L_2 \circ L_1$.

33. Let $L_1, L_2: \mathcal{V} \longrightarrow \mathcal{W}$ be linear transformations. Define $(L_1 \oplus L_2): \mathcal{V} \longrightarrow \mathcal{W}$ by $(L_1 \oplus L_2)(\mathbf{v}) = L_1(\mathbf{v}) + L_2(\mathbf{v})$ (where the latter addition takes place in $\mathcal{W}$). Also define $(c \odot L_1): \mathcal{V} \longrightarrow \mathcal{W}$ by $(c \odot L_1)\mathbf{v} = c(L_1(\mathbf{v}))$ (where the latter scalar multiplication takes place in $\mathcal{W}$).

(a) Show that $(L_1 \oplus L_2)$ and $(c \odot L_1)$ are linear transformations.

(b) Use the results in part (a) above and part (b) of Exercise 2 to show that the set of all linear transformations from $\mathcal{V}$ to $\mathcal{W}$ is a vector space under the operations $\oplus$ and $\odot$.

34. Let $L\colon \mathbb{R}^2 \longrightarrow \mathbb{R}^2$ be a nonzero linear operator. Show that L maps a line to a line.

5.2 The Matrix of a Linear Transformation

In this section we show that the behavior of any linear transformation $L\colon \mathcal{V} \longrightarrow \mathcal{W}$ is determined by its effect on a basis for $\mathcal{V}$. When $\mathcal{V}$ and $\mathcal{W}$ are finite dimensional and ordered bases for $\mathcal{V}$ and $\mathcal{W}$ are chosen, we can obtain a matrix corresponding to the linear transformation. Multiplying this matrix by the coordinatization of any vector in $\mathcal{V}$ is equivalent to performing the linear transformation on that vector. Finally, we investigate how the matrix for the linear transformation changes as we change the bases for $\mathcal{V}$ and $\mathcal{W}$.

A Linear Transformation Is Determined by Its Action on a Basis

It is easy to see that, if the action of a linear transformation $L\colon \mathcal{V} \longrightarrow \mathcal{W}$ on a basis for $\mathcal{V}$ is known, then the action of L can be computed for all elements of $\mathcal{V}$.

Example 1

You can easily verify that

$$B = \left(\begin{bmatrix} 1 & 0 \\ 0 & 0 \end{bmatrix}, \begin{bmatrix} 0 & 0 \\ 0 & 1 \end{bmatrix}, \begin{bmatrix} 0 & 1 \\ 1 & 0 \end{bmatrix}, \begin{bmatrix} 0 & 1 \\ -1 & 0 \end{bmatrix} \right)$$

is an ordered basis for $\mathcal{M}_{22}$. Now suppose that $L\colon \mathcal{M}_{22} \longrightarrow \mathcal{P}_3$ is a linear transformation for which

$$L\left(\begin{bmatrix} 1 & 0 \\ 0 & 0 \end{bmatrix}\right) = x^3, \qquad L\left(\begin{bmatrix} 0 & 0 \\ 0 & 1 \end{bmatrix}\right) = x^2,$$

$$L\left(\begin{bmatrix} 0 & 1 \\ 1 & 0 \end{bmatrix}\right) = x^3 + x, \quad \text{and} \quad L\left(\begin{bmatrix} 0 & 1 \\ -1 & 0 \end{bmatrix}\right) = x^2 - x.$$

We can use the values of L on B to compute L for other vectors in $\mathcal{M}_{22}$. For example,

$$\begin{aligned} L\left(\begin{bmatrix} -9 & -5 \\ 11 & 1 \end{bmatrix}\right) &= L\left(-9\begin{bmatrix} 1 & 0 \\ 0 & 0 \end{bmatrix} + 1\begin{bmatrix} 0 & 0 \\ 0 & 1 \end{bmatrix} + 3\begin{bmatrix} 0 & 1 \\ 1 & 0 \end{bmatrix} - 8\begin{bmatrix} 0 & 1 \\ -1 & 0 \end{bmatrix}\right) \\ &= -9L\left(\begin{bmatrix} 1 & 0 \\ 0 & 0 \end{bmatrix}\right) + 1L\left(\begin{bmatrix} 0 & 0 \\ 0 & 1 \end{bmatrix}\right) + 3L\left(\begin{bmatrix} 0 & 1 \\ 1 & 0 \end{bmatrix}\right) - 8L\left(\begin{bmatrix} 0 & 1 \\ -1 & 0 \end{bmatrix}\right) \\ &= -9x^3 + x^2 + 3(x^3 + x) - 8(x^2 - x) \\ &= -6x^3 - 7x^2 + 11x. \end{aligned}$$

In general, if $\mathbf{K} \in \mathcal{M}_{22}$ and $[\mathbf{K}]_B = [k_1, k_2, k_3, k_4]$, then

$$\begin{aligned} L(\mathbf{K}) &= k_1(x^3) + k_2(x^2) + k_3(x^3 + x) + k_4(x^2 - x) \\ &= (k_1 + k_3)x^3 + (k_2 + k_4)x^2 + (k_3 - k_4)x. \end{aligned}$$

Thus, we have derived a general formula for L from its effect on the basis B. ■

Example 1 illustrates the uniqueness assertion in the next theorem.

THEOREM 5.4

Let $B = (\mathbf{v}_1, \mathbf{v}_2, \ldots, \mathbf{v}_n)$ be an ordered basis for a vector space $\mathcal{V}$. Let $\mathcal{W}$ be a vector space, and let $\mathbf{w}_1, \mathbf{w}_2, \ldots, \mathbf{w}_n$ be any n vectors in $\mathcal{W}$. Then there is a unique linear transformation $L: \mathcal{V} \longrightarrow \mathcal{W}$ such that $L(\mathbf{v}_1) = \mathbf{w}_1, L(\mathbf{v}_2) = \mathbf{w}_2, \ldots, L(\mathbf{v}_n) = \mathbf{w}_n$.

Proof of Theorem 5.4 (abridged)

Let $\mathbf{v} \in \mathcal{V}$. Then $\mathbf{v} = a_1\mathbf{v}_1 + \cdots + a_n\mathbf{v}_n$, for some unique a_i's in $\mathbb{R}$. Let $\mathbf{w}_1, \ldots, \mathbf{w}_n$ be any vectors in $\mathcal{W}$. Define $L: \mathcal{V} \longrightarrow \mathcal{W}$ by $L(\mathbf{v}) = a_1\mathbf{w}_1 + a_2\mathbf{w}_2 + \cdots + a_n\mathbf{w}_n$. Notice that $L(\mathbf{v})$ is well defined since the a_i's are unique.

Next, to show that L is a linear transformation, we must prove that $L(\mathbf{x}_1 + \mathbf{x}_2) = L(\mathbf{x}_1) + L(\mathbf{x}_2)$ and $L(c\mathbf{x}_1) = cL(\mathbf{x}_1)$, for all $\mathbf{x}_1, \mathbf{x}_2 \in \mathcal{V}$ and all $c \in \mathbb{R}$. Suppose that $\mathbf{x}_1 = d_1\mathbf{v}_1 + \cdots + d_n\mathbf{v}_n$ and $\mathbf{x}_2 = e_1\mathbf{v}_1 + \cdots + e_n\mathbf{v}_n$. Then, by definition, $L(\mathbf{x}_1) = d_1\mathbf{w}_1 + \cdots + d_n\mathbf{w}_n$ and $L(\mathbf{x}_2) = e_1\mathbf{w}_1 + \cdots + e_n\mathbf{w}_n$. However,

$$\mathbf{x}_1 + \mathbf{x}_2 = (d_1 + e_1)\mathbf{v}_1 + \cdots + (d_n + e_n)\mathbf{v}_n,$$

$$\text{so} \qquad L(\mathbf{x}_1 + \mathbf{x}_2) = (d_1 + e_1)\mathbf{w}_1 + \cdots + (d_n + e_n)\mathbf{w}_n.$$

Hence, $L(\mathbf{x}_1) + L(\mathbf{x}_2) = L(\mathbf{x}_1 + \mathbf{x}_2)$.

Similarly, $c\mathbf{x}_1 = cd_1\mathbf{v}_1 + \cdots + cd_n\mathbf{v}_n$, and so $L(c\mathbf{x}_1) = cd_1\mathbf{w}_1 + \cdots + cd_n\mathbf{w}_n = cL(\mathbf{x}_1)$. Hence, L is a linear transformation.

Finally, the uniqueness assertion is easily verified using the fact that the effect of L on a basis is known. ■

The Matrix of a Linear Transformation

We can express the linear transformation in Example 1 as a matrix multiplication. When creating such a matrix, we must choose ordered bases for the domain and codomain. Let us again use the basis B in Example 1 for the domain $\mathcal{M}_{22}$ and suppose that $\mathbf{K} \in \mathcal{M}_{22}$ has the coordinatization $[k_1, k_2, k_3, k_4]$ with respect to B. Also, suppose we select the standard basis $C = (x^3, x^2, x, 1)$ for $\mathcal{P}_3$. Then the polynomial $L(\mathbf{K}) = (k_1 + k_3)x^3 + (k_2 + k_4)x^2 + (k_3 - k_4)x$ that we obtained in

Example 1 has the representation $[k_1 + k_3, k_2 + k_4, k_3 - k_4, 0]$ in C-coordinates. Notice that

$$\text{if}\quad \mathbf{A} = \begin{bmatrix} 1 & 0 & 1 & 0 \\ 0 & 1 & 0 & 1 \\ 0 & 0 & 1 & -1 \\ 0 & 0 & 0 & 0 \end{bmatrix}, \quad \text{then}\quad \mathbf{A}\begin{bmatrix} k_1 \\ k_2 \\ k_3 \\ k_4 \end{bmatrix} = \begin{bmatrix} k_1 + k_3 \\ k_2 + k_4 \\ k_3 - k_4 \\ 0 \end{bmatrix}.$$

The matrix $\mathbf{A}$ contains all of the information needed for carrying out the linear transformation L in Example 1.

A similar process can be used for any linear transformation between finite dimensional vector spaces.

THEOREM 5.5

Let $\mathcal{V}$ and $\mathcal{W}$ be vector spaces, with $\dim(\mathcal{V}) = n$ and $\dim(\mathcal{W}) = m$. Let $B = (\mathbf{v}_1, \mathbf{v}_2, \ldots, \mathbf{v}_n)$ and $C = (\mathbf{w}_1, \mathbf{w}_2, \ldots, \mathbf{w}_m)$ be ordered bases for $\mathcal{V}$ and $\mathcal{W}$, respectively. Let $L\colon \mathcal{V} \longrightarrow \mathcal{W}$ be a linear transformation. Then there is a unique $m \times n$ matrix $\mathbf{A}_{BC}$ such that $\mathbf{A}_{BC}[\mathbf{v}]_B = [L(\mathbf{v})]_C$, for all $\mathbf{v} \in \mathcal{V}$. (That is, $\mathbf{A}_{BC}$ times the coordinatization of $\mathbf{v}$ with respect to B gives the coordinatization of $L(\mathbf{v})$ with respect to C.)

Furthermore, for $1 \le i \le n$, the ith column of $\mathbf{A}_{BC} = [L(\mathbf{v}_i)]_C$.

This theorem asserts that, once ordered bases for $\mathcal{V}$ and $\mathcal{W}$ have been selected, *each linear transformation $L\colon \mathcal{V} \longrightarrow \mathcal{W}$ is equivalent to multiplication by a unique corresponding matrix*. The matrix $\mathbf{A}_{BC}$ in Theorem 5.5 is known as the **matrix of the linear transformation L with respect to the ordered bases B** (for $\mathcal{V}$) **and C** (for $\mathcal{W}$). The subscripts B and C on $\mathbf{A}$ are sometimes omitted when the bases being used are clear from context. Beware! If different ordered bases are chosen for $\mathcal{V}$ or $\mathcal{W}$, the matrix for the linear transformation will probably change.

Proof of Theorem 5.5

Consider the $m \times n$ matrix $\mathbf{A}_{BC}$ whose i^{th} column equals $[L(\mathbf{v}_i)]_C$, for $1 \le i \le n$. Let $\mathbf{v} \in \mathcal{V}$. We first prove that $\mathbf{A}_{BC}[\mathbf{v}]_B = [L(\mathbf{v})]_C$.

Suppose that $[\mathbf{v}]_B = [k_1, k_2, \ldots, k_n]$. Then $\mathbf{v} = k_1\mathbf{v}_1 + k_2\mathbf{v}_2 + \cdots + k_n\mathbf{v}_n$, and $L(\mathbf{v}) = k_1L(\mathbf{v}_1) + k_2L(\mathbf{v}_2) + \cdots + k_nL(\mathbf{v}_n)$, by Theorem 5.1. Hence,

$$\begin{aligned} [L(\mathbf{v})]_C &= [k_1L(\mathbf{v}_1) + k_2L(\mathbf{v}_2) + \cdots + k_nL(\mathbf{v}_n)]_C \\ &= k_1[L(\mathbf{v}_1)]_C + k_2[L(\mathbf{v}_2)]_C + \cdots + k_n[L(\mathbf{v}_n)]_C \quad \text{by Theorem 4.21} \\ &= k_1(1^{\text{st}} \text{ column of } \mathbf{A}_{BC}) + k_2(2^{\text{nd}} \text{ column of } \mathbf{A}_{BC}) \\ &\qquad + \cdots + k_n(n^{th} \text{ column of } \mathbf{A}_{BC}) \end{aligned}$$

$$= \mathbf{A}_{BC}\begin{bmatrix} k_1 \\ k_2 \\ \vdots \\ k_n \end{bmatrix} = \mathbf{A}_{BC}[\mathbf{v}]_B.$$

To complete the proof, we need to establish the uniqueness of $\mathbf{A}_{BC}$. Suppose that $\mathbf{H}$ is an $m \times n$ matrix such that $\mathbf{H}[\mathbf{v}]_B = [L(\mathbf{v})]_C$, for all $\mathbf{v} \in \mathcal{V}$. We will show that $\mathbf{H} = \mathbf{A}_{BC}$. It is enough to show that the ith column of $\mathbf{H}$ equals the ith column of $\mathbf{A}_{BC}$, for $1 \leq i \leq n$. Consider the ith vector, $\mathbf{v}_i$, of the ordered basis B for $\mathcal{V}$. Since $[\mathbf{v}_i]_B = \mathbf{e}_i$, we have i^{th} column of $\mathbf{H} = \mathbf{H}\mathbf{e}_i = \mathbf{H}[\mathbf{v}_i]_B = [L(\mathbf{v}_i)]_C = \mathbf{A}_{BC}[\mathbf{v}_i]_B = \mathbf{A}_{BC}\mathbf{e}_i = i^{th}$ column of $\mathbf{A}_{BC}$. ■

Because a linear transformation can be represented by multiplication by an appropriate matrix, we can solve problems involving linear transformations by performing matrix multiplications, which can easily be done by computer.

Example 2

The table in Figure 5.8 lists the matrices corresponding to some geometric linear operators on $\mathbb{R}^3$, with respect to the standard basis. By Theorem 5.5, the columns of each matrix are the images of the basis elements $\mathbf{e}_1$, $\mathbf{e}_2$, and $\mathbf{e}_3$ in each case.

For example, to calculate the effect of the reflection L_1 in Figure 5.8 (page 262) on the vector $[3, -4, 2]$, we simply multiply by its corresponding matrix to get

$$\begin{bmatrix} 1 & 0 & 0 \\ 0 & 1 & 0 \\ 0 & 0 & -1 \end{bmatrix}\begin{bmatrix} 3 \\ -4 \\ 2 \end{bmatrix} = \begin{bmatrix} 3 \\ -4 \\ -2 \end{bmatrix}.$$

■

Example 3

We will find the matrix for the linear transformation L: $\mathcal{P}_3 \longrightarrow \mathbb{R}^3$ given by $L(a_3x^3 + a_2x^2 + a_1x + a_0) = [a_0 + a_1, 2a_2, a_3 - a_0]$ with respect to the standard ordered bases $B = (x^3, x^2, x, 1)$ for $\mathcal{P}_3$ and $C = (\mathbf{e}_1, \mathbf{e}_2, \mathbf{e}_3)$ for $\mathbb{R}^3$. We need to find $L(\mathbf{v})$, for each $\mathbf{v} \in B$. By definition of L, we have

$$L(x^3) = [0, 0, 1], \quad L(x^2) = [0, 2, 0], \quad L(x) = [1, 0, 0], \quad \text{and} \quad L(1) = [1, 0, -1].$$

Each of these images in $\mathbb{R}^3$ is its own coordinatization (because we are using the standard basis C for $\mathbb{R}^3$). By Theorem 5.5, the matrix $\mathbf{A}_{BC}$ for L is the matrix whose columns are these images; that is,

$$\mathbf{A}_{BC} = \overset{\begin{matrix} L(x^3) & L(x^2) & L(x) & L(1) \end{matrix}}{\begin{bmatrix} 0 & 0 & 1 & 1 \\ 0 & 2 & 0 & 0 \\ 1 & 0 & 0 & -1 \end{bmatrix}}.$$

Transformation	Formula	Matrix
Reflection (through xy-plane)	$L_1\left(\begin{bmatrix} a_1 \\ a_2 \\ a_3 \end{bmatrix}\right) = \begin{bmatrix} a_1 \\ a_2 \\ -a_3 \end{bmatrix}$	$\begin{array}{ccc} L_1(\mathbf{e}_1) & L_1(\mathbf{e}_2) & L_1(\mathbf{e}_3) \end{array}$ $\begin{bmatrix} 1 & 0 & 0 \\ 0 & 1 & 0 \\ 0 & 0 & -1 \end{bmatrix}$
Contraction or dilation	$L_2\left(\begin{bmatrix} a_1 \\ a_2 \\ a_3 \end{bmatrix}\right) = \begin{bmatrix} ca_1 \\ ca_2 \\ ca_3 \end{bmatrix}$, for $c \in \mathbb{R}$	$\begin{array}{ccc} L_2(\mathbf{e}_1) & L_2(\mathbf{e}_2) & L_2(\mathbf{e}_3) \end{array}$ $\begin{bmatrix} c & 0 & 0 \\ 0 & c & 0 \\ 0 & 0 & c \end{bmatrix}$
Projection (onto xy-plane)	$L_3\left(\begin{bmatrix} a_1 \\ a_2 \\ a_3 \end{bmatrix}\right) = \begin{bmatrix} a_1 \\ a_2 \\ 0 \end{bmatrix}$	$\begin{array}{ccc} L_3(\mathbf{e}_1) & L_3(\mathbf{e}_2) & L_3(\mathbf{e}_3) \end{array}$ $\begin{bmatrix} 1 & 0 & 0 \\ 0 & 1 & 0 \\ 0 & 0 & 0 \end{bmatrix}$
Rotation (about z-axis through angle θ in xy-plane)	$L_4\left(\begin{bmatrix} a_1 \\ a_2 \\ a_3 \end{bmatrix}\right) = \begin{bmatrix} a_1\cos\theta - a_2\sin\theta \\ a_1\sin\theta + a_2\cos\theta \\ a_3 \end{bmatrix}$	$\begin{array}{ccc} L_4(\mathbf{e}_1) & L_4(\mathbf{e}_2) & L_4(\mathbf{e}_3) \end{array}$ $\begin{bmatrix} \cos\theta & -\sin\theta & 0 \\ \sin\theta & \cos\theta & 0 \\ 0 & 0 & 1 \end{bmatrix}$
Shear (in the z-direction with factor k) (see Exercise 10 in Section 5.1)	$L_5\left(\begin{bmatrix} a_1 \\ a_2 \\ a_3 \end{bmatrix}\right) = \begin{bmatrix} a_1 + ka_3 \\ a_2 + ka_3 \\ a_3 \end{bmatrix}$	$\begin{array}{ccc} L_5(\mathbf{e}_1) & L_5(\mathbf{e}_2) & L_5(\mathbf{e}_3) \end{array}$ $\begin{bmatrix} 1 & 0 & k \\ 0 & 1 & k \\ 0 & 0 & 1 \end{bmatrix}$

Figure 5.8
Matrices for several geometric linear operators on $\mathbb{R}^3$

We will compute $L(5x^3 - x^2 + 3x + 2)$ using this matrix. Now, $[5x^3 - x^2 + 3x + 2]_B = [5, -1, 3, 2]$. Then, multiplication by $\mathbf{A}_{BC}$ gives

$$[L(5x^3 - x^2 + 3x + 2)]_C = \begin{bmatrix} 0 & 0 & 1 & 1 \\ 0 & 2 & 0 & 0 \\ 1 & 0 & 0 & -1 \end{bmatrix} \begin{bmatrix} 5 \\ -1 \\ 3 \\ 2 \end{bmatrix} = \begin{bmatrix} 5 \\ -2 \\ 3 \end{bmatrix},$$

which is easily seen to be the correct answer. ■

Example 4

We will find the matrix for the linear transformation $L: \mathcal{P}_3 \longrightarrow \mathbb{R}^3$ of Example 3 with respect to the different ordered bases

$$D = (x^3 + x^2, x^2 + x, x + 1, 1) \quad \text{and} \quad E = ([-2, 1, -3], [1, -3, 0], [3, -6, 2]).$$

You should verify that D and E are bases for $\mathcal{P}_3$ and $\mathbb{R}^3$, respectively.

We need to find $L(\mathbf{v})$, for each $\mathbf{v} \in D$. By definition of L, we have $L(x^3 + x^2) = [0, 2, 1]$, $L(x^2 + x) = [1, 2, 0]$, $L(x + 1) = [2, 0, -1]$, and $L(1) = [1, 0, -1]$. Now we must find the coordinatization of each of these images in terms of the basis E for $\mathbb{R}^3$. Since we must solve for the coordinates of many vectors, it is quicker to use the transition matrix $\mathbf{Q}$ from the standard basis C to the basis E for $\mathbb{R}^3$. From Theorem 4.24, $\mathbf{Q}$ is the inverse of the matrix whose columns are the vectors in E; that is,

$$\mathbf{Q} = \begin{bmatrix} -2 & 1 & 3 \\ 1 & -3 & -6 \\ -3 & 0 & 2 \end{bmatrix}^{-1} = \begin{bmatrix} -6 & -2 & 3 \\ 16 & 5 & -9 \\ -9 & -3 & 5 \end{bmatrix}.$$

Now, multiplying $\mathbf{Q}$ times each of these images, we get

$$[L(x^3 + x^2)]_E = \mathbf{Q}\begin{bmatrix} 0 \\ 2 \\ 1 \end{bmatrix} = \begin{bmatrix} -1 \\ 1 \\ -1 \end{bmatrix}, \quad [L(x^2 + x)]_E = \mathbf{Q}\begin{bmatrix} 1 \\ 2 \\ 0 \end{bmatrix} = \begin{bmatrix} -10 \\ 26 \\ -15 \end{bmatrix},$$

$$[L(x + 1)]_E = \mathbf{Q}\begin{bmatrix} 2 \\ 0 \\ -1 \end{bmatrix} = \begin{bmatrix} -15 \\ 41 \\ -23 \end{bmatrix}, \quad \text{and} \quad [L(1)]_E = \mathbf{Q}\begin{bmatrix} 1 \\ 0 \\ -1 \end{bmatrix} = \begin{bmatrix} -9 \\ 25 \\ -14 \end{bmatrix}.$$

By Theorem 5.5, the matrix $\mathbf{A}_{DE}$ for L is the matrix whose columns are the above products:

$$\mathbf{A}_{DE} = \begin{bmatrix} -1 & -10 & -15 & -9 \\ 1 & 26 & 41 & 25 \\ -1 & -15 & -23 & -14 \end{bmatrix}.$$

We will compute $L(5x^3 - x^2 + 3x + 2)$ using this matrix. We must first find the representation for $5x^3 - x^2 + 3x + 2$ in terms of the basis D. Solving $5x^3 - x^2 + 3x + 2 = a(x^3 + x^2) + b(x^2 + x) + c(x + 1) + d(1)$ for a, b, c, and d, we get the unique solution $a = 5$, $b = -6$, $c = 9$, and $d = -7$. (Verify.) Hence, $[5x^3 - x^2 + 3x + 2]_D = [5, -6, 9, -7]$. Then

$$[L(5x^3 - x^2 + 3x + 2)]_E = \begin{bmatrix} -1 & -10 & -15 & -9 \\ 1 & 26 & 41 & 25 \\ -1 & -15 & -23 & -14 \end{bmatrix}\begin{bmatrix} 5 \\ -6 \\ 9 \\ -7 \end{bmatrix} = \begin{bmatrix} -17 \\ 43 \\ -24 \end{bmatrix}.$$

This matrix represents a coordinate vector in terms of the basis E, and so

$$L(5x^3 - x^2 + 3x + 2) = -17\begin{bmatrix} -2 \\ 1 \\ -3 \end{bmatrix} + 43\begin{bmatrix} 1 \\ -3 \\ 0 \end{bmatrix} - 24\begin{bmatrix} 3 \\ -6 \\ 2 \end{bmatrix} = \begin{bmatrix} 5 \\ -2 \\ 3 \end{bmatrix},$$

the correct answer. ■

Finding the New Matrix for a Linear Transformation after a Change of Basis

The next theorem indicates precisely how the matrix for a linear transformation changes when you alter the bases for the vector spaces.

THEOREM 5.6

Let $\mathcal{V}$ and $\mathcal{W}$ be two finite dimensional vector spaces with ordered bases B and C, respectively. Let $L: \mathcal{V} \longrightarrow \mathcal{W}$ be a linear transformation with matrix $\mathbf{A}_{BC}$ in terms of bases B and C. Suppose that D and E are other ordered bases for $\mathcal{V}$ and $\mathcal{W}$, respectively. Let $\mathbf{P}$ be the transition matrix from B to D, and let $\mathbf{Q}$ be the transition matrix from C to E. Then the matrix $\mathbf{A}_{DE}$ for L in terms of bases D and E is given by $\mathbf{A}_{DE} = \mathbf{Q}\mathbf{A}_{BC}\mathbf{P}^{-1}$.

The situation in Theorem 5.6 is summarized in Figure 5.9.

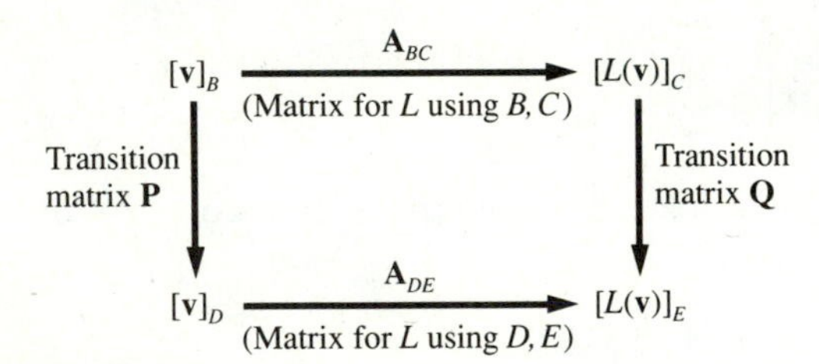

Figure 5.9
Relationship between matrices $\mathbf{A}_{BC}$ and $\mathbf{A}_{DE}$ for a linear transformation under a change of basis

Proof of Theorem 5.6

For all $\mathbf{v} \in \mathcal{V}$,

$$\mathbf{A}_{BC}[\mathbf{v}]_B = [L(\mathbf{v})]_C$$ because $\mathbf{A}_{BC}$ is the matrix for L using bases B and C

$$\Longrightarrow \quad \mathbf{Q}\mathbf{A}_{BC}[\mathbf{v}]_B = \mathbf{Q}[L(\mathbf{v})]_C$$

$$\Longrightarrow \quad \mathbf{Q}\mathbf{A}_{BC}[\mathbf{v}]_B = [L(\mathbf{v})]_E$$ because $\mathbf{Q}$ is the transition matrix from C to E

$$\Longrightarrow \quad \mathbf{Q}\mathbf{A}_{BC}\mathbf{P}^{-1}[\mathbf{v}]_D = [L(\mathbf{v})]_E$$ because $\mathbf{P}^{-1}$ is the transition matrix from D to B

However, $\mathbf{A}_{DE}$ is the unique matrix such that $\mathbf{A}_{DE}[\mathbf{v}]_D = [L(\mathbf{v})]_E$, for all $\mathbf{v} \in \mathcal{V}$. Hence, $\mathbf{A}_{DE} = \mathbf{Q}\mathbf{A}_{BC}\mathbf{P}^{-1}$. ■

Theorem 5.6 gives us an alternate method for finding the matrix of a linear transformation with respect to one pair of bases when the matrix for another pair of bases is known.

Example 5

Recall the linear transformation $L: \mathcal{P}_3 \longrightarrow \mathbb{R}^3$ from Examples 3 and 4, given by

$$L(a_3x^3 + a_2x^2 + a_1x + a_0) = [a_0 + a_1, 2a_2, a_3 - a_0].$$

Example 3 shows that the matrix for L using the standard bases B (for $\mathcal{P}_3$) and C (for $\mathbb{R}^3$) is

$$\mathbf{A}_{BC} = \begin{bmatrix} 0 & 0 & 1 & 1 \\ 0 & 2 & 0 & 0 \\ 1 & 0 & 0 & -1 \end{bmatrix}.$$

Also, in Example 4, we computed directly to find the matrix $\mathbf{A}_{DE}$ for the ordered bases $D = (x^3+x^2, x^2+x, x+1, 1)$ for $\mathcal{P}_3$ and $E = ([-2, 1, -3], [1, -3, 0], [3, -6, 2])$ for $\mathbb{R}^3$. Instead, we now use Theorem 5.6 to calculate $\mathbf{A}_{DE}$. Recall from Example 4 that the transition matrix $\mathbf{Q}$ from bases C to E is

$$\mathbf{Q} = \begin{bmatrix} -6 & -2 & 3 \\ 16 & 5 & -9 \\ -9 & -3 & 5 \end{bmatrix}.$$

Also, the transition matrix $\mathbf{P}^{-1}$ from bases D to B is easily seen to be

$$\mathbf{P}^{-1} = \begin{bmatrix} 1 & 0 & 0 & 0 \\ 1 & 1 & 0 & 0 \\ 0 & 1 & 1 & 0 \\ 0 & 0 & 1 & 1 \end{bmatrix}. \quad \text{(Verify.)}$$

Hence,

$$\mathbf{A}_{DE} = \mathbf{Q}\mathbf{A}_{BC}\mathbf{P}^{-1} = \begin{bmatrix} -6 & -2 & 3 \\ 16 & 5 & -9 \\ -9 & -3 & 5 \end{bmatrix} \begin{bmatrix} 0 & 0 & 1 & 1 \\ 0 & 2 & 0 & 0 \\ 1 & 0 & 0 & -1 \end{bmatrix} \begin{bmatrix} 1 & 0 & 0 & 0 \\ 1 & 1 & 0 & 0 \\ 0 & 1 & 1 & 0 \\ 0 & 0 & 1 & 1 \end{bmatrix}$$

$$= \begin{bmatrix} -1 & -10 & -15 & -9 \\ 1 & 26 & 41 & 25 \\ -1 & -15 & -23 & -14 \end{bmatrix},$$

which agrees with the calculation of $\mathbf{A}_{DE}$ in Example 4. ■

Matrix for the Composition of Linear Transformations

The next theorem shows how to find the corresponding matrix for the composition of linear transformations. The proof is easy (see Exercise 14).

THEOREM 5.7

Let $\mathcal{V}_1$, $\mathcal{V}_2$, and $\mathcal{V}_3$ be vector spaces with ordered bases B, C, and D, respectively. Let $L_1: \mathcal{V}_1 \longrightarrow \mathcal{V}_2$ be a linear transformation with matrix $\mathbf{A}_{BC}$ with respect to bases B and C, and let $L_2: \mathcal{V}_2 \longrightarrow \mathcal{V}_3$ be a linear transformation with matrix $\mathbf{A}_{CD}$ with respect to bases C and D. Then the matrix for the composite linear transformation $L_2 \circ L_1: \mathcal{V}_1 \longrightarrow \mathcal{V}_3$ with respect to bases B and D is the product $\mathbf{A}_{CD}\mathbf{A}_{BC}$.

Theorem 5.7 can easily be generalized to compositions of several linear transformations, as in the next example.

Example 6

Let $L_1, L_2, \ldots, L_5$ be the geometric linear operators on $\mathbb{R}^3$ given in Example 2. Let $\mathbf{A}_1, \ldots, \mathbf{A}_5$ be the matrices for these operators using the standard basis for $\mathbb{R}^3$. (These matrices are listed in Figure 5.8.) Then, the matrix for the composition $L_4 \circ L_5$ is

$$\mathbf{A}_4\mathbf{A}_5 = \begin{bmatrix} \cos\theta & -\sin\theta & 0 \\ \sin\theta & \cos\theta & 0 \\ 0 & 0 & 1 \end{bmatrix}\begin{bmatrix} 1 & 0 & k \\ 0 & 1 & k \\ 0 & 0 & 1 \end{bmatrix} = \begin{bmatrix} \cos\theta & -\sin\theta & k\cos\theta - k\sin\theta \\ \sin\theta & \cos\theta & k\sin\theta + k\cos\theta \\ 0 & 0 & 1 \end{bmatrix}.$$

Similarly, the matrix for the composition $L_2 \circ L_3 \circ L_1 \circ L_5$ is

$$\mathbf{A}_2\mathbf{A}_3\mathbf{A}_1\mathbf{A}_5 = \begin{bmatrix} c & 0 & 0 \\ 0 & c & 0 \\ 0 & 0 & c \end{bmatrix}\begin{bmatrix} 1 & 0 & 0 \\ 0 & 1 & 0 \\ 0 & 0 & 0 \end{bmatrix}\begin{bmatrix} 1 & 0 & 0 \\ 0 & 1 & 0 \\ 0 & 0 & -1 \end{bmatrix}\begin{bmatrix} 1 & 0 & k \\ 0 & 1 & k \\ 0 & 0 & 1 \end{bmatrix}$$

$$= \begin{bmatrix} c & 0 & kc \\ 0 & c & kc \\ 0 & 0 & 0 \end{bmatrix}.$$

■

Similar Matrices

Suppose L is a linear operator on a finite dimensional vector space $\mathcal{V}$. If B is a basis for $\mathcal{V}$, then there is some matrix $\mathbf{A}_{BB}$ for L with respect to B. Also, if C is another basis for $\mathcal{V}$, then there is some matrix $\mathbf{A}_{CC}$ for L with respect to C. Let $\mathbf{P}$ be the transition matrix from B to C (see Figure 5.10). Notice that by Theorem 5.6 we have $\mathbf{A}_{CC} = \mathbf{P}\mathbf{A}_{BB}\mathbf{P}^{-1}$, which leads us to the following definition.

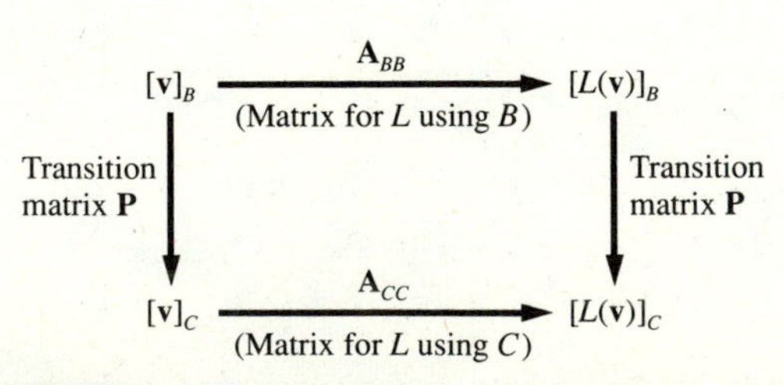

Figure 5.10
Relationship between matrices $\mathbf{A}_{BB}$ and $\mathbf{A}_{CC}$ for a linear operator under a change of basis

DEFINITION

Let $\mathbf{X}$ and $\mathbf{Y}$ be $n \times n$ matrices. Then $\mathbf{Y}$ is **similar** to $\mathbf{X}$ if and only if $\mathbf{Y} = \mathbf{PXP}^{-1}$, for some nonsingular $n \times n$ matrix $\mathbf{P}$.

Notice that if $\mathbf{X}$ is similar to $\mathbf{Y}$, then $\mathbf{Y} = \mathbf{PXP}^{-1}$, and so $\mathbf{X} = \mathbf{P}^{-1}\mathbf{YP} = \mathbf{P}^{-1}\mathbf{Y}(\mathbf{P}^{-1})^{-1}$, which means that $\mathbf{Y}$ is similar to $\mathbf{X}$. Thus we usually speak of such matrices $\mathbf{X}$ and $\mathbf{Y}$ as being similar to each other.

The discussion before the definition shows that any two matrices for the same linear operator are similar. In fact, the converse of this statement is also true (see Exercise 20). Hence, we have the following result:

THEOREM 5.8

Let $\mathcal{V}$ be an n-dimensional vector space. Then, $\mathbf{X}$ and $\mathbf{Y}$ are similar $n \times n$ matrices if and only if $\mathbf{X}$ and $\mathbf{Y}$ are both matrices for the same linear operator $L: \mathcal{V} \longrightarrow \mathcal{V}$ with respect to (possibly) different bases (that is, there are bases B and C for $\mathcal{V}$ such that $\mathbf{X}$ is the matrix $\mathbf{A}_{BB}$ for some linear operator L and $\mathbf{Y}$ is the matrix $\mathbf{A}_{CC}$ for L).

Example 7

Let B be the standard basis for $\mathbb{R}^2$, and consider $L: \mathbb{R}^2 \longrightarrow \mathbb{R}^2$ with matrix $\mathbf{A}_{BB} = \begin{bmatrix} 1 & -2 \\ -1 & 3 \end{bmatrix}$. Also, let $\mathbf{P} = \begin{bmatrix} 4 & 3 \\ 9 & 7 \end{bmatrix}$. Then $\mathbf{P}^{-1} = \begin{bmatrix} 7 & -3 \\ -9 & 4 \end{bmatrix}$. (Why?) Finally, let

$$\mathbf{Y} = \mathbf{PA}_{BB}\mathbf{P}^{-1} = \begin{bmatrix} 4 & 3 \\ 9 & 7 \end{bmatrix}\begin{bmatrix} 1 & -2 \\ -1 & 3 \end{bmatrix}\begin{bmatrix} 7 & -3 \\ -9 & 4 \end{bmatrix} = \begin{bmatrix} -2 & 1 \\ -13 & 6 \end{bmatrix}.$$

Then $\mathbf{Y}$ is similar to $\mathbf{A}_{BB}$. By Theorem 5.8, there is some basis C for $\mathbb{R}^2$ so that $\mathbf{Y}$ is the matrix for L with respect to C; that is, $\mathbf{Y} = \mathbf{A}_{CC}$.

We find the basis C and calculate $\mathbf{A}_{CC}$ to verify this result. We must choose C so that $\mathbf{P}$ is the transition matrix from B to C. Because B is the standard basis, the vectors in C are the columns of $\mathbf{P}^{-1}$ by Exercise 8 in Section 4.7. That is, $C = ([7, -9], [-3, 4])$. Note that in standard coordinates,

$$L\left(\begin{bmatrix} 7 \\ -9 \end{bmatrix}\right) = \begin{bmatrix} 1 & -2 \\ -1 & 3 \end{bmatrix}\begin{bmatrix} 7 \\ -9 \end{bmatrix} = \begin{bmatrix} 25 \\ -34 \end{bmatrix},$$

which equals $[-2, -13]$ in C-coordinates. (Why?) Also,

$$L\left(\begin{bmatrix} -3 \\ 4 \end{bmatrix}\right) = \begin{bmatrix} 1 & -2 \\ -1 & 3 \end{bmatrix}\begin{bmatrix} -3 \\ 4 \end{bmatrix} = \begin{bmatrix} -11 \\ 15 \end{bmatrix},$$

which equals [1, 6] in C-coordinates. Hence, $\mathbf{A}_{CC}$ is the matrix whose columns are $\begin{bmatrix} -2 \\ -13 \end{bmatrix}$ and $\begin{bmatrix} 1 \\ 6 \end{bmatrix}$, which is $\mathbf{Y}$, as expected. ■

Some useful properties of similar matrices are listed in Exercises 15 through 18.

Exercises—Section 5.2

1. Verify that the correct matrix is given for each of the geometric linear operators in Figure 5.8.

2. For each of the following linear transformations $L: \mathcal{V} \longrightarrow \mathcal{W}$, find the matrix for L with respect to the standard bases for $\mathcal{V}$ and $\mathcal{W}$.
   ★(a) $L: \mathbb{R}^3 \longrightarrow \mathbb{R}^3$ given by $L([x, y, z]) = [-6x + 4y - z, -2x + 3y - 5z, 3x - y + 7z]$
   (b) $L: \mathbb{R}^4 \longrightarrow \mathbb{R}^2$ given by $L([x, y, z, w]) = [3x - 5y + z - 2w, 5x + y - 2z + 8w]$
   ★(c) $L: \mathcal{P}_3 \longrightarrow \mathbb{R}^3$ given by $L(ax^3 + bx^2 + cx + d) = [4a - b + 3c + 3d, a + 3b - c + 5d, -2a - 7b + 5c - d]$
   (d) $L: \mathcal{P}_3 \longrightarrow \mathcal{M}_{22}$ given by
   $$L(ax^3 + bx^2 + cx + d) = \begin{bmatrix} -3a - 2c & -b + 4d \\ 4b - c + 3d & -6a - b + 2d \end{bmatrix}$$

3. For each of the following linear transformations $L: \mathcal{V} \longrightarrow \mathcal{W}$, find the matrix $\mathbf{A}_{BC}$ for L with respect to the given bases B for $\mathcal{V}$ and C for $\mathcal{W}$ using the method of Theorem 5.5.
   ★(a) $L: \mathbb{R}^3 \longrightarrow \mathbb{R}^2$ given by $L([x, y, z]) = [-2x + 3z, x + 2y - z]$ with $B = ([1, -3, 2], [-4, 13, -3], [2, -3, 20])$ and $C = ([-2, -1], [5, 3])$
   (b) $L: \mathbb{R}^2 \longrightarrow \mathbb{R}^3$ given by $L([x, y]) = [2x - y, x + 3y, -x]$ with $B = ([2, 3], [-3, -4])$ and $C = ([-1, 3, 2], [-4, -1, 0], [1, -1, 1])$
   ★(c) $L: \mathbb{R}^2 \longrightarrow \mathcal{P}_2$ given by $L([a, b]) = (-a + 5b)x^2 + (3a - b)x + 2b$ with $B = ([5, 3], [3, 2])$ and $C = (3x^2 - 2x, -2x^2 + 2x - 1, x^2 - x + 1)$
   (d) $L: \mathcal{M}_{22} \longrightarrow \mathbb{R}^3$ given by $L\left(\begin{bmatrix} a & b \\ c & d \end{bmatrix}\right) = [a - c + 2d, 2a + b - d, -2c + d]$ with $B = \left(\begin{bmatrix} 2 & 5 \\ 2 & -1 \end{bmatrix}, \begin{bmatrix} -2 & -2 \\ 0 & 1 \end{bmatrix}, \begin{bmatrix} -3 & -4 \\ 1 & 2 \end{bmatrix}, \begin{bmatrix} -1 & -3 \\ 0 & 1 \end{bmatrix}\right)$ and $C = ([7, 0, -3], [2, -1, -2], [-2, 0, 1])$
   ★(e) $L: \mathcal{P}_2 \longrightarrow \mathcal{M}_{23}$ given by
   $$L(ax^2 + bx + c) = \begin{bmatrix} -a & 2b + c & 3a - c \\ a + b & c & -2a + b - c \end{bmatrix}$$
   with $B = (-5x^2 - x - 1, -6x^2 + 3x + 1, 2x + 1)$ and $C = \left(\begin{bmatrix} 1 & 0 & 0 \\ 0 & 0 & 0 \end{bmatrix}, \begin{bmatrix} 0 & -1 & 0 \\ 0 & 0 & 0 \end{bmatrix}, \begin{bmatrix} 0 & 1 & 1 \\ 0 & 0 & 0 \end{bmatrix}, \begin{bmatrix} 0 & 0 & 0 \\ -1 & 0 & 0 \end{bmatrix}, \begin{bmatrix} 0 & 0 & 0 \\ 0 & 1 & 1 \end{bmatrix}, \begin{bmatrix} 0 & 0 & 0 \\ 0 & 0 & 1 \end{bmatrix}\right)$

4. In each case, find the matrix $\mathbf{A}_{DE}$ for the given linear transformation $L: \mathcal{V} \longrightarrow \mathcal{W}$ with respect to the given bases D and E by first finding the matrix for L with respect to the standard bases B and C for $\mathcal{V}$ and $\mathcal{W}$, respectively, and then using the method of Theorem 5.6.

★(a) $L: \mathbb{R}^3 \longrightarrow \mathbb{R}^3$ given by $L([a, b, c]) = [-2a + b, -b - c, a + 3c]$ with $D = ([15, -6, 4], [2, 0, 1], [3, -1, 1])$ and $E = ([1, -3, 1], [0, 3, -1], [2, -2, 1])$

★(b) $L: \mathcal{M}_{22} \longrightarrow \mathbb{R}^2$ given by

$$L\left(\begin{bmatrix} a & b \\ c & d \end{bmatrix}\right) = [6a - b + 3c - 2d, -2a + 3b - c + 4d]$$

with $D = \left(\begin{bmatrix} 2 & 1 \\ 0 & 1 \end{bmatrix}, \begin{bmatrix} 0 & 2 \\ 1 & 1 \end{bmatrix}, \begin{bmatrix} 1 & 1 \\ 2 & 1 \end{bmatrix}, \begin{bmatrix} 1 & 1 \\ 1 & 1 \end{bmatrix}\right)$

and $E = ([-2, 5], [-1, 2])$

(c) $L: \mathcal{M}_{22} \longrightarrow \mathcal{P}_2$ given by

$$L\left(\begin{bmatrix} a & b \\ c & d \end{bmatrix}\right) = (b - c)x^2 + (3a - d)x + (4a - 2c + d)$$

with $D = \left(\begin{bmatrix} 3 & -4 \\ 1 & -1 \end{bmatrix}, \begin{bmatrix} -2 & 1 \\ 1 & 1 \end{bmatrix}, \begin{bmatrix} 2 & -2 \\ 1 & -1 \end{bmatrix}, \begin{bmatrix} -2 & 1 \\ 0 & 1 \end{bmatrix}\right)$

and $E = (2x - 1, -5x^2 + 3x - 1, x^2 - 2x + 1)$

5. Verify that the same matrix is obtained for L in Exercise 3(d) by first finding the matrix for L with respect to the standard bases and then using the method of Theorem 5.6.

6. In each case, find the matrix $\mathbf{A}_{BB}$ for each of the given linear operators $L: \mathcal{V} \longrightarrow \mathcal{V}$ with respect to the given basis B by using the method of Theorem 5.5. Then check your answer by calculating the matrix for L using the standard basis and applying the method of Theorem 5.6.

★(a) $L: \mathbb{R}^2 \longrightarrow \mathbb{R}^2$ given by $L([x, y]) = [2x - y, x - 3y]$ with $B = ([4, -1], [-7, 2])$

★(b) $L: \mathcal{P}_2 \longrightarrow \mathcal{P}_2$ given by $L(ax^2 + bx + c) = (b - 2c)x^2 + (2a + c)x + (a - b - c)$ with $B = (2x^2 + 2x - 1, x, -3x^2 - 2x + 1)$

(c) $L: \mathcal{M}_{22} \longrightarrow \mathcal{M}_{22}$ given by

$$L\left(\begin{bmatrix} a & b \\ c & d \end{bmatrix}\right) = \begin{bmatrix} 2a - c + d & a - b \\ -3b - 2d & -a - 2c + 3d \end{bmatrix}$$

with

$$B = \left(\begin{bmatrix} -2 & -1 \\ 0 & 1 \end{bmatrix}, \begin{bmatrix} 3 & 1 \\ 0 & -1 \end{bmatrix}, \begin{bmatrix} -2 & 0 \\ 0 & 1 \end{bmatrix}, \begin{bmatrix} 1 & -1 \\ 1 & -1 \end{bmatrix}\right)$$

7.★(a) Let $L: \mathcal{P}_3 \longrightarrow \mathcal{P}_2$ be given by $L(\mathbf{p}) = \mathbf{p}'$, for $\mathbf{p} \in \mathcal{P}_3$. Find the matrix for L with respect to the standard bases for $\mathcal{P}_3$ and $\mathcal{P}_2$. Use this matrix to calculate $L(4x^3 - 5x^2 + 6x - 7)$ by matrix multiplication.

(b) Let $L: \mathcal{P}_2 \longrightarrow \mathcal{P}_3$ be the integration linear transformation; that is, $L(\mathbf{p}) = \int \mathbf{p}(x)\,dx$, with constant of integration equal to zero. Find the matrix for L with respect to the standard bases for $\mathcal{P}_2$ and $\mathcal{P}_3$. Use this matrix to calculate $L(2x^2 - x + 5)$ by matrix multiplication.

8. Let $L: \mathbb{R}^2 \longrightarrow \mathbb{R}^2$ be the linear operator that performs a counterclockwise rotation through an angle of $\frac{\pi}{6}$ radians (30°).

★(a) Find the matrix for L with respect to the standard basis for $\mathbb{R}^2$.

(b) Find the matrix for L with respect to the basis $B = ([4, -3], [3, -2])$.

9. Let $L: \mathcal{M}_{23} \longrightarrow \mathcal{M}_{32}$ be given by $L(\mathbf{A}) = \mathbf{A}^T$.

(a) Find the matrix for L with respect to the standard bases.

★(b) Find the matrix for L with respect to the bases

$$B = \left(\begin{bmatrix} 1 & 0 & 0 \\ 0 & 0 & 0 \end{bmatrix}, \begin{bmatrix} 0 & 1 & -1 \\ 0 & 0 & 0 \end{bmatrix}, \begin{bmatrix} 0 & 1 & 0 \\ 0 & 0 & 0 \end{bmatrix}, \begin{bmatrix} 0 & 0 & 0 \\ -1 & 0 & 0 \end{bmatrix}, \begin{bmatrix} 0 & 0 & 0 \\ 0 & -1 & -1 \end{bmatrix}, \begin{bmatrix} 0 & 0 & 0 \\ 0 & 0 & 1 \end{bmatrix}\right)$$

for $\mathcal{M}_{23}$ and

$$C = \left(\begin{bmatrix} 1 & 1 \\ 0 & 0 \\ 0 & 0 \end{bmatrix}, \begin{bmatrix} 1 & -1 \\ 0 & 0 \\ 0 & 0 \end{bmatrix}, \begin{bmatrix} 0 & 0 \\ 1 & 1 \\ 0 & 0 \end{bmatrix}, \begin{bmatrix} 0 & 0 \\ 1 & -1 \\ 0 & 0 \end{bmatrix}, \begin{bmatrix} 0 & 0 \\ 0 & 0 \\ 1 & 1 \end{bmatrix}, \begin{bmatrix} 0 & 0 \\ 0 & 0 \\ 1 & -1 \end{bmatrix}\right)$$

for $\mathcal{M}_{32}$.

★**10.** Let B be a basis for $\mathcal{V}_1$, C be a basis for $\mathcal{V}_2$, and D be a basis for $\mathcal{V}_3$. Suppose that $L_1: \mathcal{V}_1 \longrightarrow \mathcal{V}_2$ is represented by the matrix

$$\mathbf{A}_{BC} = \begin{bmatrix} -2 & 3 & -1 \\ 4 & 0 & -2 \end{bmatrix}$$

and that $L_2: \mathcal{V}_2 \longrightarrow \mathcal{V}_3$ is represented by the matrix

$$\mathbf{A}_{CD} = \begin{bmatrix} 4 & -1 \\ 2 & 0 \\ -1 & -3 \end{bmatrix}.$$

Find the matrix $\mathbf{A}_{BD}$ representing the composition $L_2 \circ L_1: \mathcal{V}_1 \longrightarrow \mathcal{V}_3$.

11. Let $L_1: \mathbb{R}^3 \longrightarrow \mathbb{R}^4$ be given by $L_1([x, y, z]) = [x - y - z, 2y + 3z, x + 3y, -2x + z]$, and let $L_2: \mathbb{R}^4 \longrightarrow \mathbb{R}^2$ be given by $L_2([x, y, z, w]) = [2y - 2z + 3w, x - z + w]$.

(a) Find the matrix for L_1 with respect to the standard bases for $\mathbb{R}^3$ and $\mathbb{R}^4$ and the matrix for L_2 with respect to the standard bases for $\mathbb{R}^4$ and $\mathbb{R}^2$.

(b) Find the matrix for $L_2 \circ L_1$ with respect to the standard bases for $\mathbb{R}^3$ and $\mathbb{R}^2$ using Theorem 5.7.

(c) Check your answer to part (b) by computing $(L_2 \circ L_1)([x, y, z])$ and finding the matrix for $L_2 \circ L_1$ directly from this result.

12. Let $\mathbf{A} = \begin{bmatrix} \cos\theta & -\sin\theta \\ \sin\theta & \cos\theta \end{bmatrix}$, the matrix representing the counterclockwise rotation of $\mathbb{R}^2$ about the origin through an angle θ.

(a) Use Theorem 5.7 to show that $\mathbf{A}^2 = \begin{bmatrix} \cos 2\theta & -\sin 2\theta \\ \sin 2\theta & \cos 2\theta \end{bmatrix}$.

(b) Generalize the result of part (a) to show that, for any integer $n \geq 1$,

$$\mathbf{A}^n = \begin{bmatrix} \cos n\theta & -\sin n\theta \\ \sin n\theta & \cos n\theta \end{bmatrix}.$$

13. Let $B = (\mathbf{v}_1, \mathbf{v}_2, \ldots, \mathbf{v}_n)$ be an ordered basis for a vector space $\mathcal{V}$. Find the matrix with respect to B for each of the following linear operators.

★(a) $L: \mathcal{V} \longrightarrow \mathcal{V}$ given by $L(\mathbf{v}_1) = \mathbf{v}_1, L(\mathbf{v}_2) = \mathbf{v}_2, \ldots, L(\mathbf{v}_n) = \mathbf{v}_n$ (identity linear operator)

(b) $L: \mathcal{V} \longrightarrow \mathcal{V}$ given by $L(\mathbf{v}_1) = L(\mathbf{v}_2) = \cdots = L(\mathbf{v}_n) = \mathbf{0}$ (zero linear operator)

★(c) $L: \mathcal{V} \longrightarrow \mathcal{V}$ given by $L(\mathbf{v}_1) = c\mathbf{v}_1, L(\mathbf{v}_2) = c\mathbf{v}_2, \ldots, L(\mathbf{v}_n) = c\mathbf{v}_n$, for some fixed $c \in \mathbb{R}$ (scalar linear operator)

(d) $L: \mathcal{V} \longrightarrow \mathcal{V}$ given by $L(\mathbf{v}_1) = \mathbf{v}_2, L(\mathbf{v}_2) = \mathbf{v}_3, \ldots, L(\mathbf{v}_{n-1}) = \mathbf{v}_n, L(\mathbf{v}_n) = \mathbf{v}_1$ (forward rotation of basis vectors)

★(e) $L: \mathcal{V} \longrightarrow \mathcal{V}$ given by $L(\mathbf{v}_1) = \mathbf{v}_n, L(\mathbf{v}_2) = \mathbf{v}_1, \ldots, L(\mathbf{v}_n) = \mathbf{v}_{n-1}$ (reverse rotation of basis vectors)

14. Prove Theorem 5.7.

15. Let $\mathbf{A}$, $\mathbf{B}$, and $\mathbf{C}$ be $n \times n$ matrices. Show that if $\mathbf{A}$ is similar to $\mathbf{B}$ and $\mathbf{B}$ is similar to $\mathbf{C}$, then $\mathbf{A}$ is similar to $\mathbf{C}$.

16. Let $\mathbf{A}$ and $\mathbf{B}$ be similar $n \times n$ matrices. Prove each of the following:

(a) $\mathbf{A}^k$ is similar to $\mathbf{B}^k$, for each positive integer k.

(b) $\mathbf{A}^T$ is similar to $\mathbf{B}^T$.

(c) $|\mathbf{A}| = |\mathbf{B}|$.

(d) $\mathbf{A}$ is nonsingular if and only if $\mathbf{B}$ is nonsingular.

(e) If $\mathbf{A}$ and $\mathbf{B}$ are both nonsingular, then $\mathbf{A}^{-1}$ is similar to $\mathbf{B}^{-1}$.

(f) $\mathbf{A} + \mathbf{I}_n$ is similar to $\mathbf{B} + \mathbf{I}_n$.

(g) Trace($\mathbf{A}$) = trace($\mathbf{B}$). (Hint: Use Exercise 21(c) in Section 1.5.)

17. Show that if $\mathbf{A}$ is a nonsingular $n \times n$ matrix and $\mathbf{B}$ is any $n \times n$ matrix, then $\mathbf{AB}$ is similar to $\mathbf{BA}$.

18. Show that a 2×2 matrix similar only to itself must be a diagonal matrix with all main diagonal entries equal.

19. Let L be a linear operator on a vector space $\mathcal{V}$ with ordered basis $B = (\mathbf{v}_1, \ldots, \mathbf{v}_n)$. Suppose that k is a nonzero real number, and let C be the ordered basis $(k\mathbf{v}_1, \ldots, k\mathbf{v}_n)$ for $\mathcal{V}$. Show that $\mathbf{A}_{BB} = \mathbf{A}_{CC}$.

20. Let $\mathcal{V}$ be an n-dimensional vector space, and let $\mathbf{X}$ and $\mathbf{Y}$ be similar matrices. Prove that there is a linear operator $L: \mathcal{V} \longrightarrow \mathcal{V}$ and bases B and C such that $\mathbf{X}$ is the matrix for L with respect to B and $\mathbf{Y}$ is the matrix for L with respect to C. (Hint: Suppose that $\mathbf{Y} = \mathbf{PXP}^{-1}$. Choose any basis B for $\mathcal{V}$. Then create the linear operator $L: \mathcal{V} \longrightarrow \mathcal{V}$ whose matrix with respect to B is $\mathbf{X}$. Let $\mathbf{v}_i$ be the vector so that $[\mathbf{v}_i]_B = i^{th}$ column of $\mathbf{P}^{-1}$. Define C to be $(\mathbf{v}_1, \ldots, \mathbf{v}_n)$. Prove that C is a basis for $\mathcal{V}$. Then show that $\mathbf{P}$ is the transition matrix from B to C and that $\mathbf{Y}$ is the matrix for L with respect to C.)

21. Let $B = ([a, b], [c, d])$ be a basis for $\mathbb{R}^2$. Then $ad - bc \neq 0$. (Why?) Let $L: \mathbb{R}^2 \longrightarrow \mathbb{R}^2$ be a linear operator such that $L([a, b]) = [c, d]$ and $L([c, d]) = [a, b]$. Show that the matrix for L with respect to the standard basis for $\mathbb{R}^2$ is

$$\frac{1}{ad - bc}\begin{bmatrix} cd - ab & a^2 - c^2 \\ d^2 - b^2 & ab - cd \end{bmatrix}.$$

22. Let $L: \mathbb{R}^2 \longrightarrow \mathbb{R}^2$ be the linear transformation where $L(\mathbf{v})$ is the reflection of $\mathbf{v}$ through the line $y = mx$. (Assume that the initial point of $\mathbf{v}$ is the origin.) Show that the matrix for L with respect to the standard basis for $\mathbb{R}^2$ is

$$\frac{1}{1 + m^2}\begin{bmatrix} 1 - m^2 & 2m \\ 2m & m^2 - 1 \end{bmatrix}.$$

23. Find the set of all matrices with respect to the standard basis for $\mathbb{R}^2$ for all linear operators that:
(a) Take all vectors of the form $[0, y]$ to vectors of the form $[0, y']$
(b) Take all vectors of the form $[x, 0]$ to vectors of the form $[x', 0]$
(c) Satisfy both parts (a) and (b) simultaneously

24. In this exercise, we show that using a nonstandard (that is, more complicated) basis can sometimes lead to a diagonal matrix for a given linear transformation. (We will study this idea in greater depth in Section 6.2.) Let $L: \mathbb{R}^3 \longrightarrow \mathbb{R}^3$ be given by $L([x, y, z]) = [-4y - 13z, -6x + 5y + 6z, 2x - 2y - 3z]$.
(a) What is the matrix for L with respect to the standard basis for $\mathbb{R}^3$?
(b) What is the matrix for L with respect to the basis $B = ([-1, -6, 2], [3, 4, -1], [-1, -3, 1])$?

25. Let $\mathcal{V}$ and $\mathcal{W}$ be finite dimensional vector spaces, and let $\mathcal{Y}$ be a subspace of $\mathcal{V}$. Suppose that $L: \mathcal{Y} \longrightarrow \mathcal{W}$ is a linear transformation. Prove that there is a linear transformation $L': \mathcal{V} \longrightarrow \mathcal{W}$ such that $L'(\mathbf{y}) = L(\mathbf{y})$ for every $\mathbf{y} \in \mathcal{Y}$. ($L'$ is called an **extension** of L to $\mathcal{V}$.)

5.3 The Dimension Theorem

In this section, we explore the properties of two special subspaces associated with a linear transformation $L: \mathcal{V} \longrightarrow \mathcal{W}$: the kernel of L (a subspace of $\mathcal{V}$) and the range of L (a subspace of $\mathcal{W}$). We will illustrate techniques for calculating bases for both the kernel and the range and show that their dimensions are related to the rank of the associated matrix for the linear transformation. Finally, we will show that any matrix and its transpose have the same rank.

Kernel and Range

DEFINITION

Let $L: \mathcal{V} \longrightarrow \mathcal{W}$ be a linear transformation. The **kernel of** L, denoted $\ker(L)$, is the subset of all vectors in $\mathcal{V}$ that map to $\mathbf{0}_{\mathcal{W}}$, the zero vector in $\mathcal{W}$.[†] That is, $\ker(L) = \{\mathbf{v} \in \mathcal{V} \mid L(\mathbf{v}) = \mathbf{0}_{\mathcal{W}}\}$. The **range of** L, or $\text{range}(L)$, is the subset of all vectors in $\mathcal{W}$ that are the image of some vector in $\mathcal{V}$. That is, $\text{range}(L) = \{L(\mathbf{v}) \mid \mathbf{v} \in \mathcal{V}\}$.

Remember that the kernel is a subset of the *domain* and the range is a subset of the *codomain*. Since the kernel of $L: \mathcal{V} \longrightarrow \mathcal{W}$ is the pre-image of the subspace $\{\mathbf{0}_{\mathcal{W}}\}$ of $\mathcal{W}$, it must be a subspace of $\mathcal{V}$ by Theorem 5.3. This theorem also assures us that the range of L is a subspace of $\mathcal{W}$. Hence, we have the following:

THEOREM 5.9

If $L: \mathcal{V} \longrightarrow \mathcal{W}$ is a linear transformation, then the kernel of L is a subspace of $\mathcal{V}$ and the range of L is a subspace of $\mathcal{W}$.

Example 1: Projection

For $n \geq 3$, consider the linear operator $L: \mathbb{R}^n \longrightarrow \mathbb{R}^n$ given by $L([a_1, a_2, \ldots, a_n]) = [a_1, a_2, 0, \ldots, 0]$. Now, $\ker(L)$ consists of those elements of the domain that map to $[0, 0, \ldots, 0]$, the zero vector of the codomain. As a result, for vectors in the kernel, $a_1 = a_2 = 0$, but $a_3, \ldots, a_n$ can have any values. Thus, $\ker(L) = \{[0, 0, a_3, \ldots, a_n] \mid a_3, \ldots, a_n \in \mathbb{R}\}$. Notice that $\ker(L)$ is a subspace of the domain and that $\dim(\ker(L)) = n - 2$, because the standard basis vectors $\mathbf{e}_3, \ldots, \mathbf{e}_n$ of $\mathbb{R}^n$ span $\ker(L)$.

Also, $\text{range}(L)$ consists of those elements of the codomain that are images of domain elements. Hence, $\text{range}(L) = \{[a_1, a_2, 0, \ldots, 0] \mid a_1, a_2 \in \mathbb{R}\}$. Notice that $\text{range}(L)$ is a subspace of the codomain and that $\dim(\text{range}(L)) = 2$, since the standard basis vectors $\mathbf{e}_1$ and $\mathbf{e}_2$ span $\text{range}(L)$. ∎

[†] Some textbooks refer to the kernel of L as the **nullspace** of L.

Example 2: Differentiation

Consider the linear transformation $L: \mathcal{P}_3 \longrightarrow \mathcal{P}_2$ given by $L(ax^3 + bx^2 + cx + d) = 3ax^2 + 2bx + c$. Now, ker($L$) consists of the polynomials in $\mathcal{P}_3$ that map to the zero polynomial in $\mathcal{P}_2$. However, for $3ax^2 + 2bx + c$ to equal zero, we must have $a = b = c = 0$. Hence, ker(L) $= \{0x^3 + 0x^2 + 0x + d \mid d \in \mathbb{R}\}$; that is, ker($L$) is just the subset of all constant polynomials. Notice that ker(L) is a subspace of $\mathcal{P}_3$ and that dim(ker(L)) $= 1$, because the single polynomial 1 spans ker(L).

Also, range(L) consists of all polynomials in the codomain $\mathcal{P}_2$ of the form $3ax^2 + 2bx + c$. Since every polynomial $Ax^2 + Bx + C$ of degree 2 or less can be expressed in this form (take $a = A/3$, $b = B/2$, $c = C$), range(L) is all of $\mathcal{P}_2$. Obviously, range(L) is a subspace of $\mathcal{P}_2$, and dim(range(L)) $= 3$. ■

Example 3: Rotation

Recall that the linear transformation $L: \mathbb{R}^2 \longrightarrow \mathbb{R}^2$ given by

$$L\left(\begin{bmatrix} x \\ y \end{bmatrix}\right) = \begin{bmatrix} \cos\theta & -\sin\theta \\ \sin\theta & \cos\theta \end{bmatrix} \begin{bmatrix} x \\ y \end{bmatrix},$$

for some (fixed) angle θ, represents the rotation of any vector $[x, y]$ with initial point at the origin counterclockwise through the angle θ.

Now, ker(L) consists of all vectors in the domain $\mathbb{R}^2$ that map to $[0, 0]$ in the codomain $\mathbb{R}^2$. However, only $[0, 0]$ itself is sent by L to the zero vector. Hence, ker(L) $= \{[0, 0]\}$. In this case, ker(L) is obviously a subspace of $\mathbb{R}^2$, and dim(ker(L)) $= 0$.

Also, range(L) is all of the codomain $\mathbb{R}^2$, because every nonzero vector $\mathbf{v}$ in $\mathbb{R}^2$ is the image of the vector of the same length at the angle θ *clockwise* from $\mathbf{v}$. Thus, range(L) $= \mathbb{R}^2$ and, obviously, range(L) is a subspace of $\mathbb{R}^2$, with dim(range(L)) $= 2$. ■

The Dimension Theorem

Notice that in Examples 1, 2, and 3, the sum of the dimensions of the kernel and the range is equal to the dimension of the domain. This is no coincidence.

THEOREM 5.10: Dimension Theorem

If $L: \mathcal{V} \longrightarrow \mathcal{W}$ is a linear transformation and $\mathcal{V}$ is finite dimensional, then

$$\dim(\ker(L)) + \dim(\text{range}(L)) = \dim(\mathcal{V}).$$

Proof of Theorem 5.10

Let $L: \mathcal{V} \longrightarrow \mathcal{W}$ be a linear transformation with $\dim(\mathcal{V}) = n$. Suppose that $\{\mathbf{v}_1, \ldots, \mathbf{v}_n\}$ is a basis for $\mathcal{V}$. Now, $\ker(L)$ is a subspace of $\mathcal{V}$, so $\dim(\ker(L)) \leq \dim(\mathcal{V}) = n$. Suppose that $\dim(\ker(L)) = s$. Let $\{\mathbf{k}_1, \ldots, \mathbf{k}_s\}$ be a basis for $\ker(L)$.

Now, range(L) is a subspace of $\mathcal{W}$. If $\mathbf{v} \in \mathcal{V}$, then $\mathbf{v} = a_1\mathbf{v}_1 + \cdots + a_n\mathbf{v}_n$, for some $a_1, \ldots, a_n \in \mathbb{R}$. Hence, the range element $L(\mathbf{v}) = L(a_1\mathbf{v}_1 + \cdots + a_n\mathbf{v}_n) = a_1L(\mathbf{v}_1) + \cdots + a_nL(\mathbf{v}_n)$. Thus, range$(L)$ is spanned by $\{L(\mathbf{v}_1), \ldots, L(\mathbf{v}_n)\}$, and range$(L)$ is also finite dimensional. Suppose that $\dim(\text{range}(L)) = t$. Let $\{\mathbf{r}_1, \ldots, \mathbf{r}_t\}$ be a basis for range(L).

Our goal is to show that $s + t = n$, for then the proof will be complete. Now, each of $\mathbf{r}_1, \ldots, \mathbf{r}_t$ is the image under L of a vector in $\mathcal{V}$. Let $\mathbf{q}_1, \ldots, \mathbf{q}_t$ be pre-images under L for $\mathbf{r}_1, \ldots, \mathbf{r}_t$, respectively. That is, $L(\mathbf{q}_i) = \mathbf{r}_i$, for $1 \leq i \leq t$. If we can show that $B = \{\mathbf{k}_1, \ldots, \mathbf{k}_s, \mathbf{q}_1, \ldots, \mathbf{q}_t\}$ is a basis for $\mathcal{V}$, then $n = \dim(\mathcal{V}) = s + t$, and we will be done. It is enough to show that B spans $\mathcal{V}$ and is linearly independent.

First we show that B spans $\mathcal{V}$. Let $\mathbf{v} \in \mathcal{V}$. Since $L(\mathbf{v}) \in \text{range}(L)$, we have $L(\mathbf{v}) = c_1\mathbf{r}_1 + \cdots + c_t\mathbf{r}_t$, for some $c_1, \ldots, c_t \in \mathbb{R}$. Then

$$L(\mathbf{v}) = c_1L(\mathbf{q}_1) + \cdots + c_tL(\mathbf{q}_t) = L(c_1\mathbf{q}_1 + \cdots + c_t\mathbf{q}_t).$$

Hence, $L(\mathbf{v}) - L(c_1\mathbf{q}_1 + \cdots + c_t\mathbf{q}_t) = \mathbf{0}_{\mathcal{W}}$, implying that $L(\mathbf{v} - c_1\mathbf{q}_1 - \cdots - c_t\mathbf{q}_t) = \mathbf{0}_{\mathcal{W}}$. Therefore, $v - c_1\mathbf{q}_1 - \cdots - c_t\mathbf{q}_t \in \ker(L)$. Hence,

$$\mathbf{v} - c_1\mathbf{q}_1 - \cdots - c_t\mathbf{q}_t = d_1\mathbf{k}_1 + \cdots + d_s\mathbf{k}_s,$$

for some $d_1, \ldots, d_s \in \mathbb{R}$. As a result, $\mathbf{v} = d_1\mathbf{k}_1 + \cdots + d_s\mathbf{k}_s + c_1\mathbf{q}_1 + \cdots + c_t\mathbf{q}_t$, and so $B = \{\mathbf{k}_1, \ldots, \mathbf{k}_s, \mathbf{q}_1, \ldots, \mathbf{q}_t\}$ spans $\mathcal{V}$.

Finally, we show that B is linearly independent. Assume that

$$a_1\mathbf{k}_1 + \cdots + a_s\mathbf{k}_s + b_1\mathbf{q}_1 + \cdots + b_t\mathbf{q}_t = \mathbf{0}_{\mathcal{V}}.$$

We will prove that $a_1 = \cdots = a_s = b_1 = \cdots = b_t = 0$. Now,

$$L(a_1\mathbf{k}_1 + \cdots + a_s\mathbf{k}_s + b_1\mathbf{q}_1 + \cdots + b_t\mathbf{q}_t) = L(\mathbf{0}_{\mathcal{V}}) = \mathbf{0}_{\mathcal{W}}$$

$$\Longrightarrow a_1L(\mathbf{k}_1) + \cdots + a_sL(\mathbf{k}_s) + b_1L(\mathbf{q}_1) + \cdots + b_tL(\mathbf{q}_t) = \mathbf{0}_{\mathcal{W}}$$

$$\Longrightarrow a_1(\mathbf{0}_{\mathcal{W}}) + \cdots + a_s(\mathbf{0}_{\mathcal{W}}) + b_1\mathbf{r}_1 + \cdots + b_t\mathbf{r}_t = \mathbf{0}_{\mathcal{W}} \quad \text{because } \mathbf{k}_1, \ldots, \mathbf{k}_s \in \ker(L)$$

$$\Longrightarrow b_1\mathbf{r}_1 + \cdots + b_t\mathbf{r}_t = \mathbf{0}_{\mathcal{W}}.$$

Since $\{\mathbf{r}_1, \ldots, \mathbf{r}_t\}$ is a basis for range(L), it is linearly independent, so we must have $b_1 = \cdots = b_t = 0$. Then the equation $a_1\mathbf{k}_1 + \cdots + a_s\mathbf{k}_s + b_1\mathbf{q}_1 + \cdots + b_t\mathbf{q}_t = \mathbf{0}_{\mathcal{V}}$ reduces to $a_1\mathbf{k}_1 + \cdots + a_s\mathbf{k}_s = \mathbf{0}_{\mathcal{V}}$. But because $\{\mathbf{k}_1, \ldots, \mathbf{k}_s\}$ is a basis for $\ker(L)$, it is linearly independent, and so $a_1 = \cdots = a_s = 0$. Hence, all of $a_1, \ldots, a_s, b_1, \ldots, b_t$ are zero. ■

An obvious corollary of the Dimension Theorem is the following:

COROLLARY 5.11

If $L: \mathcal{V} \longrightarrow \mathcal{W}$ is a linear transformation and $\mathcal{V}$ is finite dimensional, then $\dim(\ker(L)) \leq \dim(\mathcal{V})$ and $\dim(\text{range}(L)) \leq \dim(\mathcal{V})$.

Beware! If $L: \mathcal{V} \longrightarrow \mathcal{W}$ is a linear transformation, it is not necessarily true that $\dim(\mathcal{W}) \leq \dim(\mathcal{V})$. In fact, let $\mathcal{V} = \mathbb{R}^3$ and $\mathcal{W} = \mathbb{R}^4$ and consider the linear transformation $L: \mathbb{R}^3 \longrightarrow \mathbb{R}^4$ given by $L([x, y, z]) = [x, y, z, 0]$. Then, $\dim(\mathcal{W}) = \dim(\mathbb{R}^4) = 4 > 3 = \dim(\mathbb{R}^3) = \dim(\mathcal{V})$.

Example 4

Consider $L: \mathcal{M}_{nn} \longrightarrow \mathcal{M}_{nn}$ given by $L(\mathbf{A}) = \mathbf{A} + \mathbf{A}^T$. Now, $\ker(L) = \{\mathbf{A} \mid \mathbf{A} \in \mathcal{M}_{nn}$ and $\mathbf{A} + \mathbf{A}^T = \mathbf{O}_n\}$. However, $\mathbf{A} + \mathbf{A}^T = \mathbf{O}_n$ implies that $\mathbf{A} = -\mathbf{A}^T$. Hence, $\ker(L)$ is precisely the set of all skew-symmetric $n \times n$ matrices.

The range of L is the set of all matrices $\mathbf{B}$ of the form $\mathbf{A} + \mathbf{A}^T$, for some $n \times n$ matrix $\mathbf{A}$. However, if $\mathbf{B} = \mathbf{A} + \mathbf{A}^T$, then $\mathbf{B}^T = (\mathbf{A} + \mathbf{A}^T)^T = \mathbf{A}^T + \mathbf{A} = \mathbf{B}$, so $\mathbf{B}$ is symmetric. Thus, $\text{range}(L) \subseteq \{\text{symmetric } n \times n \text{ matrices}\}$.

In Exercise 9 of Section 4.6, you were asked to show that dim(skew-symmetric $n \times n$ matrices) $= (n^2 - n)/2$ and that dim(symmetric $n \times n$ matrices) $= (n^2 + n)/2$. Thus, the Dimension Theorem implies that $\dim(\ker(L)) + \dim(\text{range}(L)) = \dim(\mathcal{M}_{nn})$, or $((n^2 - n)/2) + \dim(\text{range}(L)) = n^2$. Solving for $\dim(\text{range}(L))$ yields $\dim(\text{range}(L)) = (n^2 + n)/2$. Hence, Theorem 4.14 implies that $\text{range}(L) =$ symmetric $n \times n$ matrices. ■

Finding the Kernel and Range from the Matrix of a Linear Transformation

Consider the linear transformation $L: \mathbb{R}^n \longrightarrow \mathbb{R}^m$ given by $L(\mathbf{X}) = \mathbf{AX}$, where $\mathbf{A}$ is a (fixed) $m \times n$ matrix and $\mathbf{X} \in \mathbb{R}^n$. Now, $\ker(L)$ is the subspace of all vectors $\mathbf{X}$ in the domain $\mathbb{R}^n$ that are solutions of the homogeneous system $\mathbf{AX} = \mathbf{O}$. Therefore, since $\mathbf{AX} = \mathbf{O}$ has the same solution set as $\mathbf{BX} = \mathbf{O}$, where $\mathbf{B}$ is the reduced row echelon form for $\mathbf{A}$, we have the following:

If $L: \mathbb{R}^n \longrightarrow \mathbb{R}^m$ is a linear transformation having matrix $\mathbf{A}$ with respect to the standard bases, then $\ker(L)$ is the solution set of $\mathbf{BX} = \mathbf{O}$, where $\mathbf{B}$ is the reduced row echelon form of $\mathbf{A}$.

This observation enables us to prove the next theorem, which relates the dimensions of the kernel and the range to the rank of the matrix for a linear transformation from $\mathbb{R}^n$ to $\mathbb{R}^m$.

THEOREM 5.12

If $L: \mathbb{R}^n \longrightarrow \mathbb{R}^m$ is a linear transformation with matrix **A** with respect to any bases for $\mathbb{R}^n$ and $\mathbb{R}^m$, then
(1) $\dim(\ker(L)) = n - \text{rank}(\mathbf{A})$
(2) $\dim(\text{range}(L)) = \text{rank}(\mathbf{A})$

Proof of Theorem 5.12

From the observation preceding the theorem, $\ker(L)$ is the solution set of the homogeneous system $\mathbf{BX} = \mathbf{O}$, where **B** is the reduced row echelon form for **A**. (These solutions are expressed in coordinates with respect to the basis used in $\mathbb{R}^n$ when calculating **A**.) Now, consider how we use the matrix **B** to describe the complete solution set of the homogeneous system. Each basis vector for this set is found by setting one of the independent variables in the system equal to 1 while setting the rest of the independent variables equal to zero. We finish calculating this basis vector by solving for the values of the dependent variables in terms of the independent variables. Since the number of independent variables equals the number of nonpivot columns in **B**, $\dim(\ker(L)) =$ number of basis vectors for the solution set $= n - \text{rank}(\mathbf{B}) = n - \text{rank}(\mathbf{A})$.

Finally, by the Dimension Theorem, $\dim(\text{range}(L)) = \dim(\mathbb{R}^n) - \dim(\ker(L)) = \text{rank}(\mathbf{A})$. ■

Notice that the range of L is the subspace of the codomain $\mathbb{R}^m$ consisting of all images of the form **AX**. However,

$$\mathbf{AX} = \begin{bmatrix} a_{11} & a_{12} & \cdots & a_{1n} \\ a_{21} & a_{22} & \cdots & a_{2n} \\ \vdots & \vdots & \ddots & \vdots \\ a_{m1} & a_{m2} & \cdots & a_{mn} \end{bmatrix} \begin{bmatrix} x_1 \\ x_2 \\ \vdots \\ x_n \end{bmatrix} = \begin{bmatrix} a_{11}x_1 + a_{12}x_2 + \cdots + a_{1n}x_n \\ a_{21}x_1 + a_{22}x_2 + \cdots + a_{2n}x_n \\ \vdots \\ a_{m1}x_1 + a_{m2}x_2 + \cdots + a_{mn}x_n \end{bmatrix}$$

$$= x_1 \begin{bmatrix} a_{11} \\ a_{21} \\ \vdots \\ a_{m1} \end{bmatrix} + x_2 \begin{bmatrix} a_{12} \\ a_{22} \\ \vdots \\ a_{m2} \end{bmatrix} + \cdots + x_n \begin{bmatrix} a_{1n} \\ a_{2n} \\ \vdots \\ a_{mn} \end{bmatrix}.$$

Hence, the elements of the range are precisely the linear combinations of the columns of **A**. That is, range(L) = span({columns of **A**}). This subspace of the codomain is frequently called the **column space** of **A**.

Example 5

Consider the linear transformation $L: \mathbb{R}^5 \longrightarrow \mathbb{R}^4$ given by

$$L\left(\begin{bmatrix} x_1 \\ x_2 \\ x_3 \\ x_4 \\ x_5 \end{bmatrix}\right) = \begin{bmatrix} 1 & -3 & 2 & -4 & 8 \\ 3 & -9 & 6 & -12 & 24 \\ -2 & 6 & -5 & 11 & -18 \\ 1 & -3 & 6 & -16 & 16 \end{bmatrix}\begin{bmatrix} x_1 \\ x_2 \\ x_3 \\ x_4 \\ x_5 \end{bmatrix}.$$

Let **A** be the matrix of this transformation using the standard bases for $\mathbb{R}^5$ and $\mathbb{R}^4$. The reduced row echelon form matrix for **A** is

$$\mathbf{B} = \begin{bmatrix} \textcircled{1} & -3 & 0 & 2 & 4 \\ 0 & 0 & \textcircled{1} & -3 & 2 \\ 0 & 0 & 0 & 0 & 0 \\ 0 & 0 & 0 & 0 & 0 \end{bmatrix},$$

where the nonzero pivot elements are circled. Now, ker(L) is the solution set of the homogeneous system **BX** = **O**. Thus,

$$\begin{aligned} \ker(L) &= \{[3a - 2b - 4c, a, 3b - 2c, b, c] \mid a, b, c \in \mathbb{R}\} \\ &= \{a[3, 1, 0, 0, 0] + b[-2, 0, 3, 1, 0] + c[-4, 0, -2, 0, 1] \mid a, b, c \in \mathbb{R}\}, \end{aligned}$$

and a basis for ker(L) is $\{[3, 1, 0, 0, 0], [-2, 0, 3, 1, 0], [-4, 0, -2, 0, 1]\}$. Thus, dim(ker($L$)) = 3.

The range of L is the span of the columns of **A**; that is,

$$\begin{aligned} \text{range}(L) = \text{span}(\{&[1, 3, -2, 1], [-3, -9, 6, -3], [2, 6, -5, 6], \\ &[-4, -12, 11, -16], [8, 24, -18, 16]\}). \end{aligned}$$

However, by Theorem 5.12, dim(range(L)) = rank(**A**) = 2. Since we have already row reduced **A**, we can use the "efficient shrinking method" of Section 4.6 to obtain the basis $\{[1, 3, -2, 1], [2, 6, -5, 6]\}$ for range(L). ■

In Example 5, we illustrated how to use the "efficient shrinking method" to find a basis for range(L). In general, we have the following:

> If $L: \mathbb{R}^n \longrightarrow \mathbb{R}^m$ is a linear transformation having matrix **A** with respect to the standard bases, then a basis for range(L) is obtained by taking those columns of **A** that become pivot columns in the row reduction process.

Rank of the Transpose

COROLLARY 5.13

If $\mathbf{A}$ is any matrix, then $\text{rank}(\mathbf{A}) = \text{rank}(\mathbf{A}^T)$.

Proof of Corollary 5.13

Let $\mathbf{A}$ be an $m \times n$ matrix. Consider the linear transformation $L: \mathbb{R}^n \longrightarrow \mathbb{R}^m$ with associated matrix $\mathbf{A}$ (using the standard bases). By the observations made before Example 5, range(L) is the span of the column vectors of $\mathbf{A}$. Hence, range(L) is the span of the row vectors of $\mathbf{A}^T$—that is, the row space of $\mathbf{A}^T$. Thus, $\dim(\text{range}(L)) = \text{rank}(\mathbf{A}^T)$. But by Theorem 5.12, $\dim(\text{range}(L)) = \text{rank}(\mathbf{A})$. Hence, $\text{rank}(\mathbf{A}) = \text{rank}(\mathbf{A}^T)$. ■

In some textbooks, rank($\mathbf{A}$) is called the **row rank** of $\mathbf{A}$ and rank($\mathbf{A}^T$) is called the **column rank** of $\mathbf{A}$. With this terminology, Corollary 5.13 asserts that the row rank of $\mathbf{A}$ equals the column rank of $\mathbf{A}$. Equivalently, dim(row space of $\mathbf{A}$) = dim(column space of $\mathbf{A}$). Be careful! This statement does not imply that these *spaces* are equal, only that their *dimensions* are equal. In fact, unless $\mathbf{A}$ is square, they contain vectors of different sizes.

Corollary 5.13 is illustrated in Example 5, which showed that dim(row space of $\mathbf{A}$) = (row) rank($\mathbf{A}$) = 2. In Example 5 we also calculated a basis for the column space of $\mathbf{A}$ containing two vectors, and so dim(column space of $\mathbf{A}$) = column rank($\mathbf{A}$) = 2 as well.

Exercises—Section 5.3

1. Let $L: \mathbb{R}^3 \longrightarrow \mathbb{R}^3$ be given by

$$L\left(\begin{bmatrix} x_1 \\ x_2 \\ x_3 \end{bmatrix}\right) = \begin{bmatrix} 5 & 1 & -1 \\ -3 & 0 & 1 \\ 1 & -1 & -1 \end{bmatrix}\begin{bmatrix} x_1 \\ x_2 \\ x_3 \end{bmatrix}.$$

★(a) Is $[1, -2, 3]$ in ker(L)? Why or why not?
(b) Is $[2, -1, 4]$ in ker(L)? Why or why not?
★(c) Is $[2, -1, 4]$ in range(L)? Why or why not?
(d) Is $[-16, 12, -8]$ in range(L)? Why or why not?

2. Let $L: \mathcal{P}_3 \longrightarrow \mathcal{P}_3$ be given by $L(ax^3+bx^2+cx+d) = 2cx^3-(a+b)x+(d-c)$.
★(a) Is $x^3 - 5x^2 + 3x - 6$ in ker(L)? Why or why not?
(b) Is $4x^3 - 4x^2$ in ker(L)? Why or why not?
★(c) Is $8x^3 - x - 1$ in range(L)? Why or why not?
(d) Is $4x^3 - 3x^2 + 7$ in range(L)? Why or why not?

3. For each of the following linear transformations $L: \mathcal{V} \longrightarrow \mathcal{W}$, find a basis for $\ker(L)$ and a basis for $\text{range}(L)$, and verify that $\dim(\ker(L)) + \dim(\text{range}(L)) = \dim(\mathcal{V})$.

★(a) $L: \mathbb{R}^3 \longrightarrow \mathbb{R}^2$ given by $L([x_1, x_2, x_3]) = [0, x_2]$

(b) $L: \mathbb{R}^2 \longrightarrow \mathbb{R}^3$ given by $L([x_1, x_2]) = [x_1, x_1 + x_2, x_2]$

(c) $L: \mathcal{M}_{22} \longrightarrow \mathcal{M}_{32}$ given by $L\left(\begin{bmatrix} a_{11} & a_{12} \\ a_{21} & a_{22} \end{bmatrix}\right) = \begin{bmatrix} 0 & -a_{12} \\ -a_{21} & 0 \\ 0 & 0 \end{bmatrix}$

★(d) $L: \mathcal{P}_4 \longrightarrow \mathcal{P}_2$ given by $L(ax^4 + bx^3 + cx^2 + dx + e) = cx^2 + dx + e$

(e) $L: \mathcal{P}_2 \longrightarrow \mathcal{P}_3$ given by $L(ax^2 + bx + c) = cx^3 + bx^2 + ax$

★(f) $L: \mathbb{R}^3 \longrightarrow \mathbb{R}^3$ given by $L([x_1, x_2, x_3]) = [x_1, 0, x_1 - x_2 + x_3]$

★(g) $L: \mathcal{M}_{22} \longrightarrow \mathcal{M}_{22}$ given by $L(\mathbf{A}) = \mathbf{A}^T$

(h) $L: \mathcal{M}_{33} \longrightarrow \mathcal{M}_{33}$ given by $L(\mathbf{A}) = \mathbf{A} - \mathbf{A}^T$

★(i) $L: \mathcal{P}_2 \longrightarrow \mathbb{R}^2$ given by $L(\mathbf{p}) = [\mathbf{p}(1), \mathbf{p}'(1)]$

(j) $L: \mathcal{P}_4 \longrightarrow \mathbb{R}^3$ given by $L(\mathbf{p}) = [\mathbf{p}(-1), \mathbf{p}(0), \mathbf{p}(1)]$

4. For each of the following linear transformations $L: \mathcal{V} \longrightarrow \mathcal{W}$, find a basis for $\ker(L)$ and a basis for $\text{range}(L)$ using the method of Example 5. Verify that $\dim(\ker(L)) + \dim(\text{range}(L)) = \dim(\mathcal{V})$.

★(a) $L: \mathbb{R}^3 \longrightarrow \mathbb{R}^3$ given by $L\left(\begin{bmatrix} x_1 \\ x_2 \\ x_3 \end{bmatrix}\right) = \begin{bmatrix} 1 & -1 & 5 \\ -2 & 3 & -13 \\ 3 & -3 & 15 \end{bmatrix}\begin{bmatrix} x_1 \\ x_2 \\ x_3 \end{bmatrix}$

(b) $L: \mathbb{R}^3 \longrightarrow \mathbb{R}^4$ given by $L\left(\begin{bmatrix} x_1 \\ x_2 \\ x_3 \end{bmatrix}\right) = \begin{bmatrix} 4 & -2 & 8 \\ 7 & 1 & 5 \\ -2 & -1 & 0 \\ 3 & -2 & 7 \end{bmatrix}\begin{bmatrix} x_1 \\ x_2 \\ x_3 \end{bmatrix}$

(c) $L: \mathbb{R}^3 \longrightarrow \mathbb{R}^2$ given by $L\left(\begin{bmatrix} x_1 \\ x_2 \\ x_3 \end{bmatrix}\right) = \begin{bmatrix} 3 & 2 & 11 \\ 2 & 1 & 8 \end{bmatrix}\begin{bmatrix} x_1 \\ x_2 \\ x_3 \end{bmatrix}$

★(d) $L: \mathbb{R}^4 \longrightarrow \mathbb{R}^5$ given by $L\left(\begin{bmatrix} x_1 \\ x_2 \\ x_3 \\ x_4 \end{bmatrix}\right) = \begin{bmatrix} -14 & -8 & -10 & 2 \\ -4 & -1 & 1 & -2 \\ -6 & 2 & 12 & -10 \\ 3 & -7 & -24 & 17 \\ 4 & 2 & 2 & 0 \end{bmatrix}\begin{bmatrix} x_1 \\ x_2 \\ x_3 \\ x_4 \end{bmatrix}$

5. (a) Suppose that $L: \mathcal{V} \longrightarrow \mathcal{W}$ is the linear transformation given by $L(\mathbf{v}) = \mathbf{0}_{\mathcal{W}}$, for all $\mathbf{v} \in \mathcal{V}$. What is $\ker(L)$? What is $\text{range}(L)$?

(b) Suppose that $L: \mathcal{V} \longrightarrow \mathcal{V}$ is the linear transformation given by $L(\mathbf{v}) = \mathbf{v}$, for all $\mathbf{v} \in \mathcal{V}$. What is $\ker(L)$? What is $\text{range}(L)$?

★**6.** Consider the mapping $L: \mathcal{M}_{33} \longrightarrow \mathbb{R}$ given by $L(\mathbf{A}) = \text{trace}(\mathbf{A})$ (see Exercise 14 in Section 1.4). Show that L is a linear transformation. What is $\ker(L)$? What is $\text{range}(L)$? Calculate $\dim(\ker(L))$ and $\dim(\text{range}(L))$.

7. Let $\mathcal{V}$ be a vector space with fixed basis $B = \{\mathbf{v}_1, \ldots, \mathbf{v}_n\}$. Define $L: \mathcal{V} \longrightarrow \mathcal{V}$ by $L(\mathbf{v}_1) = \mathbf{v}_2, L(\mathbf{v}_2) = \mathbf{v}_3, \ldots, L(\mathbf{v}_{n-1}) = \mathbf{v}_n, L(\mathbf{v}_n) = \mathbf{v}_1$. Find range($L$). What is ker($L$)?

★8. Consider $L: \mathcal{P}_2 \longrightarrow \mathcal{P}_4$ given by $L(\mathbf{p}) = x^2\mathbf{p}$. What is ker($L$)? What is range($L$)? Verify that $\dim(\ker(L)) + \dim(\text{range}(L)) = \dim(\mathcal{P}_2)$.

9. Consider $L: \mathcal{P}_4 \longrightarrow \mathcal{P}_2$ given by $L(\mathbf{p}) = \mathbf{p}''$. What is ker($L$)? What is range($L$)? Verify that $\dim(\ker(L)) + \dim(\text{range}(L)) = \dim(\mathcal{P}_4)$.

★10. Consider $L: \mathcal{P}_n \longrightarrow \mathcal{P}_n$ given by $L(\mathbf{p}) = \mathbf{p}^{(k)}$ (the kth derivative of $\mathbf{p}$), where $k \leq n$. What is dim(ker(L))? What is dim(range(L))? What happens when $k > n$?

11. Let a be a fixed real number. Consider $L: \mathcal{P}_n \longrightarrow \mathbb{R}$ given by $L(\mathbf{p}(x)) = \mathbf{p}(a)$ (that is, the evaluation of $\mathbf{p}$ at $x = a$). (Recall from Exercise 17 in Section 5.1 that L is a linear transformation.) Show that $\{x - a, x^2 - a^2, \ldots, x^n - a^n\}$ is a basis for ker(L). (Hint: What is range(L)?)

★12. Suppose that $L: \mathbb{R}^n \longrightarrow \mathbb{R}^n$ is a linear operator given by $L(\mathbf{X}) = \mathbf{AX}$, where $|\mathbf{A}| \neq 0$. What is ker(L)? What is range(L)?

13. Let $\mathcal{V}$ be a finite dimensional vector space, and let $L: \mathcal{V} \longrightarrow \mathcal{V}$ be a linear operator. Show that $\ker(L) = \{\mathbf{0}_{\mathcal{V}}\}$ if and only if $\text{range}(L) = \mathcal{V}$.

14. Let $L: \mathcal{V} \longrightarrow \mathcal{W}$ be a linear transformation, where $\{\mathbf{v}_1, \mathbf{v}_2, \ldots, \mathbf{v}_n\}$ spans $\mathcal{V}$. Prove that $\{L(\mathbf{v}_1), L(\mathbf{v}_2), \ldots, L(\mathbf{v}_n)\}$ spans range(L).

15. Let $L: \mathcal{V} \longrightarrow \mathcal{W}$ be a linear transformation. Prove directly that ker(L) is a subspace of $\mathcal{V}$ and that range(L) is a subspace of $\mathcal{W}$ using Theorem 4.2—that is, without invoking Theorem 5.9.

16. Let $L_1: \mathcal{V} \longrightarrow \mathcal{W}$ and $L_2: \mathcal{W} \longrightarrow \mathcal{X}$ be linear transformations.
(a) Show that $\ker(L_1) \subseteq \ker(L_2 \circ L_1)$.
(b) Show that $\text{range}(L_2 \circ L_1) \subseteq \text{range}(L_2)$.
(c) Show that $\dim(\text{range}(L_2 \circ L_1)) \leq \dim(\text{range}(L_1))$.

★17. Give an example of a linear operator $L: \mathbb{R}^2 \longrightarrow \mathbb{R}^2$ such that $\ker(L) = \text{range}(L)$.

18. Let $L: \mathbb{R}^3 \longrightarrow \mathbb{R}^3$ be a linear operator. Show that L takes a plane through the origin to one of the following: a plane through the origin, a line through the origin, or the origin itself. (Hint: Use Corollary 5.11.)

5.4 Isomorphism

In this section, we examine methods for determining whether two vector spaces are equivalent, or isomorphic. Isomorphism is important because, if a certain algebraic result is true in one of the vector spaces, a corresponding result holds in the other as well.

We first study two special types of linear transformations: one-to-one and onto. You will see that, if there is a linear transformation from a vector space $\mathcal{V}$ to a vector space $\mathcal{W}$ that is both one-to-one and onto, then $\mathcal{V}$ and $\mathcal{W}$ are isomorphic vector spaces.

One-to-One and Onto Linear Transformations

One-to-one functions and onto functions are defined and discussed in Appendix B. In particular, Appendix B contains the usual methods for proving that a given function is or is not one-to-one or onto. However, we are interested primarily in linear transformations here, so we restate the definitions of *one-to-one* and *onto* specifically as they apply to this type of function.

DEFINITION

Let $\mathcal{V}$ and $\mathcal{W}$ be vector spaces:

(1) A linear transformation $L: \mathcal{V} \longrightarrow \mathcal{W}$ is **one-to-one** if and only if distinct vectors in $\mathcal{V}$ have different images in $\mathcal{W}$. That is, L is **one-to-one** if and only if, for all $\mathbf{v}_1, \mathbf{v}_2 \in \mathcal{V}$, $L(\mathbf{v}_1) = L(\mathbf{v}_2)$ implies $\mathbf{v}_1 = \mathbf{v}_2$.

(2) A linear transformation $L: \mathcal{V} \longrightarrow \mathcal{W}$ is **onto** if and only if every vector in the codomain $\mathcal{W}$ is the image of some vector in the domain $\mathcal{V}$. That is, L is **onto** if and only if, for every $\mathbf{w} \in \mathcal{W}$, there is some $\mathbf{v} \in \mathcal{V}$ such that $L(\mathbf{v}) = \mathbf{w}$.

Example 1: Rotation

Recall the rotation linear operator $L: \mathbb{R}^2 \longrightarrow \mathbb{R}^2$ from Example 9 in Section 5.1 given by $L(\mathbf{v}) = \mathbf{A}\mathbf{v}$, where $\mathbf{A} = \begin{bmatrix} \cos\theta & -\sin\theta \\ \sin\theta & \cos\theta \end{bmatrix}$. We will show that L is *both one-to-one and onto*.

To show that L is one-to-one, we take any two arbitrary vectors $\mathbf{v}_1$ and $\mathbf{v}_2$ in the domain $\mathbb{R}^2$, assume that $L(\mathbf{v}_1) = L(\mathbf{v}_2)$, and prove that $\mathbf{v}_1 = \mathbf{v}_2$. Now, if $L(\mathbf{v}_1) = L(\mathbf{v}_2)$, then $\mathbf{A}\mathbf{v}_1 = \mathbf{A}\mathbf{v}_2$. Because $\mathbf{A}$ is nonsingular, we can multiply both sides on the left by its inverse to obtain $\mathbf{v}_1 = \mathbf{v}_2$.† Hence, L is one-to-one.

To show that L is onto, we must take any arbitrary vector $\mathbf{w}$ in the codomain $\mathbb{R}^2$ and show that there is some vector $\mathbf{v}$ in the domain $\mathbb{R}^2$ that maps to $\mathbf{w}$. Consider

---

† $\mathbf{A}^{-1} = \begin{bmatrix} \cos\theta & \sin\theta \\ -\sin\theta & \cos\theta \end{bmatrix} = \begin{bmatrix} \cos(-\theta) & -\sin(-\theta) \\ \sin(-\theta) & \cos(-\theta) \end{bmatrix}$ represents a *clockwise* rotation through the angle θ.

the vector $\mathbf{v} = \mathbf{A}^{-1}\mathbf{w} \in \mathbb{R}^2$. Then, $L(\mathbf{v}) = \mathbf{A}(\mathbf{A}^{-1}\mathbf{w}) = \mathbf{w}$. Since the vector $\mathbf{v}$ in the domain maps to $\mathbf{w}$ under L, L is onto. ■

Example 2

Consider the linear transformation L: $\mathbb{R}^2 \longrightarrow \mathbb{R}^3$ given by $L([x, y]) = [x, x + y, x - y]$. We will show that L is *one-to-one but not onto*.

To show that L is one-to-one, we must take two arbitrary vectors in the domain, say $[x_1, y_1]$ and $[x_2, y_2]$; assume that $L([x_1, y_1]) = L([x_2, y_2])$; and try to prove that $[x_1, y_1] = [x_2, y_2]$. However, if $L([x_1, y_1]) = L([x_2, y_2])$, then $[x_1, x_1 + y_1, x_1 - y_1] = [x_2, x_2 + y_2, x_2 - y_2]$. Equating the first entries, we have $x_1 = x_2$. Equating the middle entries, we have $x_1 + y_1 = x_2 + y_2$. Because $x_1 = x_2$, we must have $y_1 = y_2$. Hence, $[x_1, y_1] = [x_2, y_2]$, and L is one-to-one.

To show that L is not onto, we must find some vector $\mathbf{w}$ in the codomain $\mathbb{R}^3$ that is not the image under L of a vector in the domain $\mathbb{R}^2$. Consider $[0, 1, 0]$. You can easily show that there are no values of x and y such that $L([x, y]) = [0, 1, 0]$. (Verify.) Hence, L is not onto. ■

Example 3: Differentiation

Consider the linear transformation L: $\mathcal{P}_3 \longrightarrow \mathcal{P}_2$ given by $L(\mathbf{p}) = \mathbf{p}'$. We will show that L is *onto but not one-to-one*.

To show that L is not one-to-one, we must find two different vectors $\mathbf{p}_1$ and $\mathbf{p}_2$ in the domain $\mathcal{P}_3$ that have the same image. Consider $\mathbf{p}_1 = x + 1$ and $\mathbf{p}_2 = x + 2$. Since $L(\mathbf{p}_1) = L(\mathbf{p}_2) = 1$, L is not one-to-one.

To show that L is onto, we must take an arbitrary vector $\mathbf{q}$ in $\mathcal{P}_2$ and find some vector $\mathbf{p}$ in $\mathcal{P}_3$ such that $L(\mathbf{p}) = \mathbf{q}$. Consider the vector $\mathbf{p} = \int \mathbf{q}(x)\,dx$ with zero constant term. Because $L(\mathbf{p}) = \mathbf{q}$, we see that L is onto. ■

Example 4

Consider the linear operator L: $\mathcal{M}_{22} \longrightarrow \mathcal{M}_{22}$ given by $L\left(\begin{bmatrix} a & b \\ c & d \end{bmatrix}\right) = \begin{bmatrix} a & -b \\ 0 & d \end{bmatrix}$. We will show that L is *neither one-to-one nor onto*.

Consider $\mathbf{v}_1 = \begin{bmatrix} 1 & 1 \\ 1 & 1 \end{bmatrix}$ and $\mathbf{v}_2 = \begin{bmatrix} 1 & 1 \\ 2 & 1 \end{bmatrix}$. Then $L(\mathbf{v}_1) = L(\mathbf{v}_2) = \begin{bmatrix} 1 & -1 \\ 0 & 1 \end{bmatrix}$, and so L is not one-to-one. Next, consider $\mathbf{w} = \begin{bmatrix} 2 & 2 \\ 2 & 2 \end{bmatrix}$. Because the (2, 1) entry of $\mathbf{w}$ is $2 \neq 0$, $\mathbf{w}$ is never the image of any vector $\mathbf{v}$ under L. Hence L is not onto. ■

Examples 1 through 4 show that the concepts of one-to-one and onto are independent: there are linear transformations that have either property with or without the other.

Theorem B.1 in Appendix B shows that the composition of two one-to-one linear transformations is also one-to-one. Similarly, the composition of onto linear transformations is onto.

Example 5

Recall from Example 1 the rotation linear operator $L: \mathbb{R}^2 \longrightarrow \mathbb{R}^2$ given by $L\left(\begin{bmatrix} x \\ y \end{bmatrix}\right) = \begin{bmatrix} \cos\theta & -\sin\theta \\ \sin\theta & \cos\theta \end{bmatrix}\begin{bmatrix} x \\ y \end{bmatrix}$. Consider also the operator $R: \mathbb{R}^2 \longrightarrow \mathbb{R}^2$ given by $R\left(\begin{bmatrix} x \\ y \end{bmatrix}\right) = \begin{bmatrix} 0 & 1 \\ 1 & 0 \end{bmatrix}\begin{bmatrix} x \\ y \end{bmatrix} = \begin{bmatrix} y \\ x \end{bmatrix}$, which reflects a given vector (beginning at the origin) through the line $y = x$. We have already seen that L is both one-to-one and onto. It is easy to check that R is also both one-to-one and onto. (Verify.) Then, by Theorem B.1, the composition $R \circ L: \mathbb{R}^2 \longrightarrow \mathbb{R}^2$ given by $R \circ L\left(\begin{bmatrix} x \\ y \end{bmatrix}\right) = \begin{bmatrix} \sin\theta & \cos\theta \\ \cos\theta & -\sin\theta \end{bmatrix}\begin{bmatrix} x \\ y \end{bmatrix}$ is both one-to-one and onto. ■

The Kernel of a One-to-One Linear Transformation

The next theorem gives an alternate way of proving that a linear transformation is one-to-one.

THEOREM 5.14

Let $\mathcal{V}$ and $\mathcal{W}$ be vector spaces, and let $L: \mathcal{V} \longrightarrow \mathcal{W}$ be a linear transformation. Then L is one-to-one if and only if $\ker(L) = \{\mathbf{0}_{\mathcal{V}}\}$ (or, equivalently, if and only if $\dim(\ker(L)) = 0$).

Thus, a linear transformation whose kernel contains a nonzero vector cannot be one-to-one.

Proof of Theorem 5.14

First suppose that L is one-to-one, and let $\mathbf{v} \in \ker(L)$. We must show that $\mathbf{v} = \mathbf{0}_{\mathcal{V}}$. Now, $L(\mathbf{v}) = \mathbf{0}_{\mathcal{W}}$. However, by Theorem 5.1, $L(\mathbf{0}_{\mathcal{V}}) = \mathbf{0}_{\mathcal{W}}$. Because $L(\mathbf{v}) = L(\mathbf{0}_{\mathcal{V}})$ and L is one-to-one, we must have $\mathbf{v} = \mathbf{0}_{\mathcal{V}}$.

Conversely, suppose that $\ker(L) = \{\mathbf{0}_{\mathcal{V}}\}$. We must show that L is one-to-one. Let $\mathbf{v}_1, \mathbf{v}_2 \in \mathcal{V}$, with $L(\mathbf{v}_1) = L(\mathbf{v}_2)$. We must show that $\mathbf{v}_1 = \mathbf{v}_2$. Now, $L(\mathbf{v}_1) - L(\mathbf{v}_2) = \mathbf{0}_{\mathcal{W}}$, implying that $L(\mathbf{v}_1 - \mathbf{v}_2) = \mathbf{0}_{\mathcal{W}}$. However, then $\mathbf{v}_1 - \mathbf{v}_2 \in \ker(L)$, by definition of the kernel. Since $\ker(L) = \{\mathbf{0}_{\mathcal{V}}\}$, $\mathbf{v}_1 - \mathbf{v}_2 = \mathbf{0}_{\mathcal{V}}$, and so $\mathbf{v}_1 = \mathbf{v}_2$. ■

Example 6

Consider the linear transformation $L\colon \mathcal{M}_{22} \longrightarrow \mathcal{M}_{32}$ given by $L\left(\begin{bmatrix} a & b \\ c & d \end{bmatrix}\right) = \begin{bmatrix} a-b & 0 & c-d \\ c+d & a+b & 0 \end{bmatrix}$. If $\begin{bmatrix} a & b \\ c & d \end{bmatrix}$ is in $\ker(L)$, then $a - b = c - d = c + d = a + b = 0$. Solving these equations yields $a = b = c = d = 0$, and so $\ker(L)$ contains only the zero matrix $\begin{bmatrix} 0 & 0 \\ 0 & 0 \end{bmatrix}$. By Theorem 5.14, L is one-to-one.

On the other hand, consider $M\colon \mathcal{M}_{32} \longrightarrow \mathcal{M}_{22}$ given by $M\left(\begin{bmatrix} a & b & c \\ d & e & f \end{bmatrix}\right) = \begin{bmatrix} a+b & a+c \\ d+e & d+f \end{bmatrix}$. Notice that $M\left(\begin{bmatrix} 1 & -1 & -1 \\ 1 & -1 & -1 \end{bmatrix}\right) = \begin{bmatrix} 0 & 0 \\ 0 & 0 \end{bmatrix}$, and so $\ker(M)$ contains a nonzero vector. Therefore, M is not one-to-one. ■

Linear Independence and Span, Revisited

Another important property of one-to-one linear transformations is that they are the only transformations that preserve linear independence:

THEOREM 5.15

> Let $L\colon \mathcal{V} \longrightarrow \mathcal{W}$ be a linear transformation between vector spaces. Then L is one-to-one if and only if for every linearly independent set T in $\mathcal{V}$, the set $L(T)$ is linearly independent in $\mathcal{W}$.

Proof of Theorem 5.15

Let $L\colon \mathcal{V} \longrightarrow \mathcal{W}$ be a linear transformation. First, suppose that L is one-to-one and T is a linearly independent set of vectors in $\mathcal{V}$. We want to show that the image $L(T)$ in $\mathcal{W}$ is also linearly independent.

Suppose that $\mathbf{w}_1, \ldots, \mathbf{w}_n \in L(T)$ and $a_1\mathbf{w}_1 + \cdots + a_n\mathbf{w}_n = \mathbf{0}_{\mathcal{W}}$, for some scalars $a_1, \ldots, a_n$. Our goal is to show that $a_1 = \cdots = a_n = 0$. Now, since $\mathbf{w}_1, \ldots, \mathbf{w}_n \in L(T)$, each $\mathbf{w}_i$ equals $L(\mathbf{v}_i)$, for some $\mathbf{v}_i \in T$, $1 \le i \le n$. Then $a_1L(\mathbf{v}_1) + \cdots + a_nL(\mathbf{v}_n) = \mathbf{0}_{\mathcal{W}}$. However, because L is a linear transformation, $L(a_1\mathbf{v}_1 + \cdots + a_n\mathbf{v}_n) = \mathbf{0}_{\mathcal{W}}$. Hence, $a_1\mathbf{v}_1 + \cdots + a_n\mathbf{v}_n \in \ker(L)$. Now, since L is one-to-one, $\ker(L) = \{\mathbf{0}_{\mathcal{V}}\}$, and so $a_1\mathbf{v}_1 + \cdots + a_n\mathbf{v}_n = \mathbf{0}_{\mathcal{V}}$. However, $\mathbf{v}_1, \ldots, \mathbf{v}_n \in T$, a linearly independent set. Hence, by Theorem 4.9, $a_1 = \cdots = a_n = 0$, and $L(T)$ is linearly independent.

Conversely, suppose that the image in $\mathcal{W}$ of every linearly independent set in $\mathcal{V}$ is also linearly independent. We want to show that L is one-to-one. By Theorem 5.14, it is enough to show that $\ker(L) = \{\mathbf{0}_{\mathcal{V}}\}$. We prove this by contradiction.

Suppose that $\ker(L) \neq \{\mathbf{0}_{\mathcal{V}}\}$. Let $\mathbf{v}_1 \neq \mathbf{0}_{\mathcal{V}}$ be in $\ker(L)$. Then, $L(\mathbf{v}_1) = \mathbf{0}_{\mathcal{W}}$. Because $\{\mathbf{v}_1\}$ is a linearly independent set (why?), its image $\{\mathbf{0}_{\mathcal{W}}\}$ must also be a linearly independent set. However, $\{\mathbf{0}_{\mathcal{W}}\}$ contains the zero vector and so is linearly dependent. This contradiction shows that $\ker(L) = \{\mathbf{0}_{\mathcal{V}}\}$, and so L is one-to-one. ■

Example 7

Consider the linear operator $L: \mathcal{P}_2 \longrightarrow \mathcal{P}_2$ given by $L(ax^2 + bx + c) = (2a - b)x^2 + (a - c)x - b$. Now, $\ker(L) = \{ax^2 + bx + c \mid 2a - b = 0, a - c = 0, -b = 0\}$. It is easy to show that these conditions force $\ker(L) = \{0x^2 + 0x + 0\}$. Hence, L is one-to-one, by Theorem 5.14. Then, by Theorem 5.15, the image of any linearly independent set in $\mathcal{P}_2$ is again linearly independent. For example, because $\{x^2, x, 1\}$ is linearly independent, it follows that $\{L(x^2), L(x), L(1)\} = \{2x^2 + x, -x^2 - 1, -x\}$ is linearly independent. (Verify.) ■

It is easy to see that a linear transformation $L: \mathcal{V} \longrightarrow \mathcal{W}$ between finite dimensional vector spaces is onto if and only if $\dim(\text{range}(L)) = \dim(\mathcal{W})$. (Why?) Also, just as one-to-one linear transformations preserve linear independence, onto linear transformations preserve spanning. That is:

THEOREM 5.16

Let $L: \mathcal{V} \longrightarrow \mathcal{W}$ be a linear transformation between vector spaces, and let S be a spanning set for $\mathcal{V}$. Then L is onto if and only if $L(S)$ is a spanning set for $\mathcal{W}$.

The proof of Theorem 5.16 is easy and is left for you to do in Exercise 13. Notice that if L is not onto, the image of a spanning set for $\mathcal{V}$ will not span all of $\mathcal{W}$ but will span only range(L).

Example 8

Consider the linear operator $L: \mathbb{R}^3 \longrightarrow \mathbb{R}^3$ given by $L([x, y, z]) = [x+y+z, y+z, z]$. Now, L is onto because every vector $[a, b, c]$ in the codomain $\mathbb{R}^3$ can be expressed in the form $[x + y + z, y + z, z]$ by letting $x = a - b$, $y = b - c$, and $z = c$. Thus, $\text{range}(L) = \mathbb{R}^3$, and $\dim(\text{range}(L)) = 3$.

Because L is onto, Theorem 5.16 asserts that the image of any spanning set for the domain also spans the codomain $\mathcal{W}$. For example, the image of the standard basis for $\mathbb{R}^3$ is $\{[1, 0, 0], [1, 1, 0], [1, 1, 1]\}$. This is easily seen to be a spanning set (and, in fact, a basis) for the codomain $\mathbb{R}^3$. ■

Isomorphisms: Invertible Linear Transformations

We restate here the definition from Appendix B for the inverse of a function as it applies to linear transformations:

DEFINITION

> Let $\mathcal{V}$ and $\mathcal{W}$ be vector spaces, and let $L: \mathcal{V} \longrightarrow \mathcal{W}$ be a linear transformation. Then L is an **invertible linear transformation** if and only if there is a function $M: \mathcal{W} \longrightarrow \mathcal{V}$ such that $(M \circ L)(\mathbf{v}) = \mathbf{v}$, for all $\mathbf{v} \in \mathcal{V}$, and $(L \circ M)(\mathbf{w}) = \mathbf{w}$, for all $\mathbf{w} \in \mathcal{W}$. Such a function M is called an **inverse** of L.

If the inverse M of $L: \mathcal{V} \longrightarrow \mathcal{W}$ exists, then it is unique by Theorem B.3 and is usually denoted by $L^{-1}: \mathcal{W} \longrightarrow \mathcal{V}$.

DEFINITION

> Let $\mathcal{V}$ and $\mathcal{W}$ be vector spaces. Then a linear transformation $L: \mathcal{V} \longrightarrow \mathcal{W}$ that is both one-to-one and onto is called an **isomorphism** from $\mathcal{V}$ to $\mathcal{W}$.

The next result shows that the last two definitions actually refer to the same class of linear transformations:

THEOREM 5.17

> Let $\mathcal{V}$ and $\mathcal{W}$ be vector spaces, and let $L: \mathcal{V} \longrightarrow \mathcal{W}$ be a linear transformation. Then L is an isomorphism if and only if L is an invertible linear transformation. Moreover, if L is invertible, then L^{-1} is also a linear transformation.

Theorem 5.17 shows us that whenever L is an isomorphism, L^{-1} is also an isomorphism, because the inverse of an invertible function is also invertible.

Proof of Theorem 5.17

The "if and only if" part of Theorem 5.17 follows directly from Theorem B.2. Now, suppose that $L: \mathcal{V} \longrightarrow \mathcal{W}$ is invertible (and thus, an isomorphism) with inverse L^{-1}. We must show that both of the following properties hold:

(1) $L^{-1}(\mathbf{w}_1 + \mathbf{w}_2) = L^{-1}(\mathbf{w}_1) + L^{-1}(\mathbf{w}_2)$, for all $\mathbf{w}_1, \mathbf{w}_2 \in \mathcal{W}$

(2) $L^{-1}(c\mathbf{w}) = cL^{-1}(\mathbf{w})$, for all $c \in \mathbb{R}$, $\mathbf{w} \in \mathcal{W}$

Property (1): Because L is an isomorphism, L is onto. Hence, if $\mathbf{w}_1, \mathbf{w}_2 \in \mathcal{W}$, there are vectors $\mathbf{v}_1, \mathbf{v}_2 \in \mathcal{V}$ such that $L(\mathbf{v}_1) = \mathbf{w}_1$ and $L(\mathbf{v}_2) = \mathbf{w}_2$. However, then $L^{-1}(\mathbf{w}_1) = \mathbf{v}_1$ and $L^{-1}(\mathbf{w}_2) = \mathbf{v}_2$. (Why?) Therefore,

$$\begin{aligned} L^{-1}(\mathbf{w}_1 + \mathbf{w}_2) &= L^{-1}(L(\mathbf{v}_1) + L(\mathbf{v}_2)) \\ &= L^{-1}(L(\mathbf{v}_1 + \mathbf{v}_2)) && \text{because } L \text{ is a linear transformation} \\ &= (L^{-1} \circ L)(\mathbf{v}_1 + \mathbf{v}_2) \\ &= \mathbf{v}_1 + \mathbf{v}_2 && \text{because } L \text{ and } L^{-1} \text{ are inverses} \\ &= L^{-1}(\mathbf{w}_1) + L^{-1}(\mathbf{w}_2). \end{aligned}$$

Property (2): Again, because L is an isomorphism, L is onto. Hence, if $\mathbf{w} \in \mathcal{W}$, there is a vector $\mathbf{v} \in \mathcal{V}$ such that $L(\mathbf{v}) = \mathbf{w}$. But then $L^{-1}(\mathbf{w}) = \mathbf{v}$. Thus,

$$L^{-1}(c\mathbf{w}) = L^{-1}(cL(\mathbf{v})) = L^{-1}(L(c\mathbf{v})) = (L^{-1} \circ L)(c\mathbf{v}) = c\mathbf{v} = cL^{-1}(\mathbf{w}).$$

Because both properties (1) and (2) hold, L^{-1} is a linear transformation. ■

Example 9

Recall again the rotation linear operator $L: \mathbb{R}^2 \longrightarrow \mathbb{R}^2$ with

$$L\left(\begin{bmatrix} x \\ y \end{bmatrix}\right) = \begin{bmatrix} \cos\theta & -\sin\theta \\ \sin\theta & \cos\theta \end{bmatrix}\begin{bmatrix} x \\ y \end{bmatrix}$$

given in Example 1. We have shown that L is both one-to-one and onto. Hence, L is an isomorphism and has an inverse, L^{-1}. Because L represents a counterclockwise rotation of vectors through the angle θ, L^{-1} must represent a counterclockwise rotation through the angle $-\theta$. Thus,

$$L^{-1}\left(\begin{bmatrix} x \\ y \end{bmatrix}\right) = \begin{bmatrix} \cos(-\theta) & -\sin(-\theta) \\ \sin(-\theta) & \cos(-\theta) \end{bmatrix}\begin{bmatrix} x \\ y \end{bmatrix} = \begin{bmatrix} \cos\theta & \sin\theta \\ -\sin\theta & \cos\theta \end{bmatrix}\begin{bmatrix} x \\ y \end{bmatrix}.$$

Notice that L^{-1} is also an isomorphism. ■

Example 10

The reflection linear operator given in Example 5, $R: \mathbb{R}^2 \longrightarrow \mathbb{R}^2$ with $R\left(\begin{bmatrix} x \\ y \end{bmatrix}\right) = \begin{bmatrix} 0 & 1 \\ 1 & 0 \end{bmatrix}\begin{bmatrix} x \\ y \end{bmatrix} = \begin{bmatrix} y \\ x \end{bmatrix}$, is also both one-to-one and onto. Hence, R is an isomorphism, and so R has an inverse. In fact, R is its own inverse; that is, $R^{-1} = R$. (Why?) ■

Theorem B.4 shows that the composition of isomorphisms results in an isomorphism. The inverse of the composition $L_2 \circ L_1$ is $L_1^{-1} \circ L_2^{-1}$. That is, the transformations must be undone in *reverse* order to arrive at the correct inverse. (Compare this statement with part (3) of Theorem 2.10 for matrix multiplication.)

Isomorphism, Dimension, and Rank

Not every pair of vector spaces is isomorphic. We now show that vector spaces $\mathcal{V}$ and $\mathcal{W}$ must have the same dimension for an isomorphism to exist between them.

THEOREM 5.18

Let $\mathcal{V}$ be a finite dimensional vector space, and let $L: \mathcal{V} \longrightarrow \mathcal{W}$ be an isomorphism. Then $\mathcal{W}$ is finite dimensional and $\dim(\mathcal{V}) = \dim(\mathcal{W})$.

Proof of Theorem 5.18

Since L is onto, Theorem 5.16 clearly implies that $\mathcal{W}$ is finite dimensional.

Because L is one-to-one, $\dim(\ker(L)) = 0$ by Theorem 5.14. Since L is onto, $\dim(\text{range}(L)) = \dim(\mathcal{W})$. Then, by the Dimension Theorem, $\dim(\mathcal{V}) = \dim(\ker(L)) + \dim(\text{range}(L))$, so $\dim(\mathcal{V}) = 0 + \dim(\mathcal{W}) = \dim(\mathcal{W})$. ■

Theorem 5.18 implies that there is no possible isomorphism from, say, $\mathbb{R}^3$ to $\mathcal{P}_4$ or from $\mathcal{M}_{22}$ to $\mathbb{R}^3$, because the dimensions of the spaces do not agree. In addition, the corresponding matrix for any isomorphism $L: \mathcal{V} \longrightarrow \mathcal{W}$ (with respect to any bases for $\mathcal{V}$ and $\mathcal{W}$) must be a square matrix, since $\dim(\mathcal{V}) = \dim(\mathcal{W})$.

The next theorem gives a simple method for determining whether a linear transformation between finite dimensional vector spaces is an isomorphism.

THEOREM 5.19

Let $\mathcal{V}$ and $\mathcal{W}$ both be n-dimensional vector spaces with ordered bases B and C, respectively, and let $L: \mathcal{V} \longrightarrow \mathcal{W}$ be a linear transformation. Then L is an isomorphism if and only if the matrix representation $\mathbf{A}_{BC}$ for L with respect to B and C is nonsingular.

Theorem 5.19 is proved by noting that if L is an isomorphism, then the matrix representation for L^{-1} with respect to C and B is the inverse of $\mathbf{A}_{BC}$. An analogous observation proves the converse. We leave the details for you to provide in Exercise 14.

Also note that Theorem 5.19 implies L is an isomorphism if and only if $\text{rank}(\mathbf{A}_{BC}) = n$.

Example 11

Consider $L: \mathbb{R}^3 \longrightarrow \mathbb{R}^3$ given by $L(\mathbf{v}) = \mathbf{A}\mathbf{v}$, where

$$\mathbf{A} = \begin{bmatrix} 1 & 0 & 3 \\ 0 & 1 & 3 \\ 0 & 0 & 1 \end{bmatrix}.$$

Now $\mathbf{A}$ is nonsingular ($|\mathbf{A}| = 1 \neq 0$). Hence, by Theorem 5.19, L is an isomorphism. Geometrically, L represents a shear in the z-direction (see Figure 5.8 in Section 5.2). ■

The next theorem gives another important property of isomorphisms.

THEOREM 5.20

Let $L: \mathcal{V} \longrightarrow \mathcal{W}$ be a linear transformation between vector spaces. Let B be a basis for $\mathcal{V}$, such that the images under L of vectors in B are distinct vectors in $\mathcal{W}$. Then L is an isomorphism if and only if $L(B)$ is a basis for $\mathcal{W}$.

Proof of Theorem 5.20

First, suppose that L is an isomorphism. By Theorem 5.15, $L(B)$ is linearly independent in $\mathcal{W}$. Also, by Theorem 5.16, $L(B)$ is a spanning set for $\mathcal{W}$. Hence, $L(B)$ is a basis for $\mathcal{W}$.

Conversely, suppose that $L(B)$ is a basis for $\mathcal{W}$. Then Theorem 5.16 clearly shows that L is onto. We must show that L is one-to-one to finish the proof that L is an isomorphism.

Suppose that $L(\mathbf{v}) = \mathbf{0}_{\mathcal{W}}$, for some $\mathbf{v} \in \mathcal{V}$. We will show that $\mathbf{v} = \mathbf{0}_{\mathcal{V}}$, thus completing the proof by Theorem 5.14. Now, $\mathbf{v} = a_1\mathbf{u}_1 + \cdots + a_k\mathbf{u}_k$, for some $\mathbf{u}_1, \ldots, \mathbf{u}_k \in B$, and $a_1, \ldots, a_k \in \mathbb{R}$. Thus, $\mathbf{0}_{\mathcal{W}} = L(\mathbf{v}) = L(a_1\mathbf{u}_1 + \cdots + a_k\mathbf{u}_k) = a_1L(\mathbf{u}_1) + \cdots + a_kL(\mathbf{u}_k)$. But $\{L(\mathbf{u}_1), \ldots, L(\mathbf{u}_k)\}$ is a linearly independent set of k distinct vectors, because it is a subset of $L(B)$. Hence, $a_1 = \cdots = a_k = 0$. Therefore, $\mathbf{v} = \mathbf{0}_{\mathcal{V}}$. ■

Example 12

Recall from Example 8 the onto linear operator $L: \mathbb{R}^3 \longrightarrow \mathbb{R}^3$ given by $L([x, y, z]) = [x + y + z, y + z, z]$. That example shows that the image under L of the standard basis for the domain $\mathbb{R}^3$ is a basis for the codomain $\mathbb{R}^3$. Hence, by Theorem 5.20, L is an isomorphism. ■

Suppose that $L: \mathcal{V} \longrightarrow \mathcal{W}$ is a linear transformation. If $\dim(\mathcal{V}) = \dim(\mathcal{W})$, as in Example 12, the next result asserts that we need only check that L is either one-to-one or onto to prove that L is an isomorphism.

THEOREM 5.21

Let $\mathcal{V}$ and $\mathcal{W}$ be finite dimensional vector spaces with $\dim(\mathcal{V}) = \dim(\mathcal{W})$. Let $L: \mathcal{V} \longrightarrow \mathcal{W}$ be a linear transformation. Then L is one-to-one if and only if L is onto.

Proof of Theorem 5.21

Let $\mathcal{V}$ and $\mathcal{W}$ be vector spaces with $\dim(\mathcal{V}) = \dim(\mathcal{W})$. Let $L: \mathcal{V} \longrightarrow \mathcal{W}$ be a linear transformation. Then

	L is one-to-one	
$\Longleftrightarrow$	$\dim(\ker(L)) = 0$	by Theorem 5.14
$\Longleftrightarrow$	$\dim(\mathcal{V}) = \dim(\text{range}(L))$	by the Dimension Theorem
$\Longleftrightarrow$	$\dim(\mathcal{W}) = \dim(\text{range}(L))$	because $\dim(\mathcal{V}) = \dim(\mathcal{W})$
$\Longleftrightarrow$	L is onto	by Theorem 4.14 ■

Example 13

Consider $L: \mathbb{R}^3 \longrightarrow \mathcal{P}_2$ given by $L([a, b, c]) = (a + 2b)x^2 - bx - 3c$. Now, $\dim(\mathbb{R}^3) = \dim(\mathcal{P}_2) = 3$. Hence, by Theorem 5.21, L is an isomorphism if L is either one-to-one or onto.

We will show that L is one-to-one. If $L([a, b, c]) = 0$, then $a + 2b = b = -3c = 0$. Clearly, these equations imply that $a = b = c = 0$, showing that L is one-to-one by Theorem 5.14. Thus, by Theorem 5.21, L is an isomorphism. ■

All n-Dimensional Vector Spaces Are Isomorphic

DEFINITION

Let $\mathcal{V}$ and $\mathcal{W}$ be vector spaces. Then $\mathcal{V}$ is **isomorphic** to $\mathcal{W}$, denoted $\mathcal{V} \cong \mathcal{W}$, if and only if there exists an isomorphism $L: \mathcal{V} \longrightarrow \mathcal{W}$.

If $\mathcal{V} \cong \mathcal{W}$, there is some isomorphism $L: \mathcal{V} \longrightarrow \mathcal{W}$. Then by Theorem 5.17, $L^{-1}: \mathcal{W} \longrightarrow \mathcal{V}$ is also an isomorphism, so $\mathcal{W} \cong \mathcal{V}$. Hence, we usually speak of such $\mathcal{V}$ and $\mathcal{W}$ as being *isomorphic to each other*.

Also notice that if $\mathcal{V} \cong \mathcal{W}$ and $\mathcal{W} \cong \mathcal{X}$, then there are isomorphisms $L_1: \mathcal{V} \longrightarrow \mathcal{W}$ and $L_2: \mathcal{W} \longrightarrow \mathcal{X}$. But then $L_2 \circ L_1: \mathcal{V} \longrightarrow \mathcal{X}$ is an isomorphism, and so $\mathcal{V} \cong \mathcal{X}$. In other words, two vector spaces such as $\mathcal{V}$ and $\mathcal{X}$ that are both isomorphic to the same vector space $\mathcal{W}$ are isomorphic to each other.

Example 14

Consider $L_1: \mathbb{R}^4 \longrightarrow \mathcal{P}_3$ given by $L_1([a, b, c, d]) = ax^3 + bx^2 + cx + d$ and $L_2: \mathcal{M}_{22} \longrightarrow \mathcal{P}_3$ given by $L_2\left(\begin{bmatrix} a & b \\ c & d \end{bmatrix}\right) = ax^3 + bx^2 + cx + d$. L_1 and L_2 are obviously both isomorphisms. Hence, $\mathbb{R}^4 \cong \mathcal{P}_3$ and $\mathcal{M}_{22} \cong \mathcal{P}_3$. Thus, the composition $L_2^{-1} \circ L_1: \mathbb{R}^4 \longrightarrow \mathcal{M}_{22}$ is also an isomorphism, and so $\mathbb{R}^4 \cong \mathcal{M}_{22}$. Notice that all of these vector spaces have dimension 4. ■

Example 14 hints that any two finite dimensional vector spaces of the same dimension are isomorphic. This result, which is one of the most important in all linear algebra, is a corollary of the next theorem:

THEOREM 5.22

If $\mathcal{V}$ is any n-dimensional vector space, then $\mathcal{V} \cong \mathbb{R}^n$.

Proof of Theorem 5.22

Suppose that $\mathcal{V}$ is a vector space with $\dim(\mathcal{V}) = n$. If we can find an isomorphism $L: \mathcal{V} \longrightarrow \mathbb{R}^n$, then $\mathcal{V} \cong \mathbb{R}^n$, and we will be done. Now let $B = (\mathbf{v}_1, \ldots, \mathbf{v}_n)$ be an ordered basis for $\mathcal{V}$. Consider the mapping $L(\mathbf{v}) = [\mathbf{v}]_B$, for all $\mathbf{v} \in \mathcal{V}$. We know that L is a linear transformation (see Example 4 in Section 5.1). Now, $L(\mathbf{v}_i) = \mathbf{e}_i$, for each i, $1 \le i \le n$. Thus, $L(B)$ is the standard basis for $\mathbb{R}^n$. Hence, L is an isomorphism by Theorem 5.20. ■

In particular, Theorem 5.22 tells us that $\mathcal{P}_n \cong \mathbb{R}^{n+1}$ and that $\mathcal{M}_{mn} \cong \mathbb{R}^{mn}$. Also, the proof of Theorem 5.22 illustrates that coordinatization of vectors in an n-dimensional vector space $\mathcal{V}$ gives an isomorphism of $\mathcal{V}$ with $\mathbb{R}^n$.

By the remarks before Example 14, Theorem 5.22 implies the following:

COROLLARY 5.23

Any two n-dimensional vector spaces $\mathcal{V}$ and $\mathcal{W}$ are isomorphic. That is, if $\dim(\mathcal{V}) = \dim(\mathcal{W})$, then $\mathcal{V} \cong \mathcal{W}$.

For example, suppose that $\mathcal{V}$ and $\mathcal{W}$ are both vector spaces with $\dim(\mathcal{V}) = \dim(\mathcal{W}) = 47$. Then by Corollary 5.23, $\mathcal{V} \cong \mathcal{W}$, and both $\mathcal{V}$ and $\mathcal{W}$ are isomorphic to $\mathbb{R}^{47}$.

Although we do not prove so here, whenever two vector spaces $\mathcal{V}$ and $\mathcal{W}$ are isomorphic, the elements of $\mathcal{V}$ behave "identically" to those of $\mathcal{W}$, except that their "names" are different. This observation enables us to extend many of the results we have proved for $\mathbb{R}^n$ to other n-dimensional spaces. For example, in Theorem 5.12 we proved that for a linear transformation $L: \mathbb{R}^n \longrightarrow \mathbb{R}^m$ with matrix $\mathbf{A}$, we have $\dim(\text{range}(L)) = \text{rank}(\mathbf{A})$ and $\dim(\ker(L)) = n - \text{rank}(\mathbf{A})$. A corresponding result holds with other finite dimensional vector spaces, as in the next example.

Example 15

Suppose that $L: \mathcal{P}_4 \longrightarrow \mathcal{M}_{22}$ is a linear transformation with matrix

$$\mathbf{A} = \begin{bmatrix} 3 & -2 & 1 & 13 & -9 \\ -3 & 3 & -1 & -16 & 11 \\ 2 & -2 & 1 & 11 & -8 \\ -5 & 3 & -3 & -22 & 17 \end{bmatrix},$$

using the standard bases for $\mathcal{P}_4$ and $\mathcal{M}_{22}$. Now, $\mathcal{P}_4 \cong \mathbb{R}^5$ and $\mathcal{M}_{22} \cong \mathbb{R}^4$. A natural pair of isomorphisms to use here is the coordinatization of $\mathcal{P}_4$ and $\mathcal{M}_{22}$ with respect to their standard bases. Therefore, we can replace every polynomial $a_4x^4 + a_3x^3 + a_2x^2 + a_1x + a_0$ in $\mathcal{P}_4$ with the vector $[a_4, a_3, a_2, a_1, a_0]$ in $\mathbb{R}^5$ and every matrix $\begin{bmatrix} a & b \\ c & d \end{bmatrix}$ in $\mathcal{M}_{22}$ with the vector $[a, b, c, d]$ in $\mathbb{R}^4$. Hence, for all practical purposes, we can replace the given mapping $L: \mathcal{P}_4 \longrightarrow \mathcal{M}_{22}$ with an equivalent function $L': \mathbb{R}^5 \longrightarrow \mathbb{R}^4$. Then $\mathbf{A}$ is also the matrix for the equivalent mapping L'. (Why?)

Now the reduced row echelon form for $\mathbf{A}$ is

$$\mathbf{B} = \begin{bmatrix} 1 & 0 & 0 & 2 & -1 \\ 0 & 1 & 0 & -3 & 2 \\ 0 & 0 & 1 & 1 & -2 \\ 0 & 0 & 0 & 0 & 0 \end{bmatrix}.$$

By Theorem 5.12, $\dim(\text{range}(L')) = \text{rank}(\mathbf{A}) = \text{rank}(\mathbf{B}) = 3$, and $\dim(\ker(L')) = \dim(\mathbb{R}^5) - \text{rank}(\mathbf{A}) = 5 - 3 = 2$. Hence, $\dim(\text{range}(L)) = 3$ and $\dim(\ker(L)) = 2$.

We can also find bases for range(L) and ker(L) by first finding bases for range(L') and ker(L'). By taking the columns of $\mathbf{A}$ that become pivot columns of $\mathbf{B}$, we have

$$\text{Basis for range}(L') = \{[3, -3, 2, -5], [-2, 3, -2, 3], [1, -1, 1, -3]\}.$$

By replacing these vectors in $\mathbb{R}^4$ with the corresponding 2×2 matrices. we have

$$\text{Basis for range}(L) = \left\{ \begin{bmatrix} 3 & -3 \\ 2 & -5 \end{bmatrix}, \begin{bmatrix} -2 & 3 \\ -2 & 3 \end{bmatrix}, \begin{bmatrix} 1 & -1 \\ 1 & -3 \end{bmatrix} \right\}.$$

Also, ker(L') is the solution set of $\mathbf{BX} = \mathbf{O}$, which is $\{a[-2, 3, -1, 1, 0] + b[1, -2, 2, 0, 1] \mid a, b \in \mathbb{R}\}$. Hence, a basis for ker(L') is $\{[-2, 3, -1, 1, 0], [1, -2, 2, 0, 1]\}$. By replacing these vectors in $\mathbb{R}^5$ with the corresponding polynomials in $\mathcal{P}_4$, we find that a basis for ker(L) is $\{-2x^4 + 3x^3 - x^2 + x, x^4 - 2x^3 + 2x^2 + 1\}$. ■

So far, we have proved many important results concerning the concepts of one-to-one, onto, and isomorphism. For convenience, these are summarized in Figure 5.11.

Let $L: \mathcal{V} \longrightarrow \mathcal{W}$ be a linear transformation between vector spaces, and let B be a basis for $\mathcal{V}$.

Then L is *one-to-one:*	
$\Longleftrightarrow$ $\ker(L) = \{\mathbf{0}_{\mathcal{W}}\}$	Theorem 5.14
$\Longleftrightarrow$ $\dim(\ker(L)) = 0$	Theorem 5.14
$\Longleftrightarrow$ the image of every linearly independent set in $\mathcal{V}$ is linearly independent in $\mathcal{W}$	Theorem 5.15
Then L is *onto:*	
$\Longleftrightarrow$ range $(L) = \mathcal{W}$	Definition
$\Longleftrightarrow$ $\dim(\text{range}(L)) = \dim(\mathcal{W})$	Theorems 4.14 and 5.9†
$\Longleftrightarrow$ the image of every spanning set of $\mathcal{V}$ is a spanning set for $\mathcal{W}$	Theorem 5.16
$\Longleftrightarrow$ the image of some spanning set of $\mathcal{V}$ is a spanning set for $\mathcal{W}$	Theorem 5.16
Then L is an *isomorphism:*	
$\Longleftrightarrow$ L is both one-to-one and onto	Definition
$\Longleftrightarrow$ L is invertible (that is, $L^{-1}: \mathcal{W} \longrightarrow \mathcal{V}$ exists)	Theorem 5.17
$\Longleftrightarrow$ the matrix for L (with respect to any pair of ordered bases for $\mathcal{V}$ and $\mathcal{W}$) is nonsingular	Theorem 5.19†
$\Longleftrightarrow$ the matrix for L (with respect to some pair of ordered bases for $\mathcal{V}$ and $\mathcal{W}$) is nonsingular	Theorem 5.19†
$\Longleftrightarrow$ the images of vectors in B are distinct and $L(B)$ is a basis for $\mathcal{W}$	Theorem 5.20
$\Longleftrightarrow$ L is one-to-one and $\dim(\mathcal{V}) = \dim(\mathcal{W})$	Theorem 5.21†
$\Longleftrightarrow$ L is onto and $\dim(\mathcal{V}) = \dim(\mathcal{W})$	Theorem 5.21†
Furthermore, if $L: \mathcal{V} \longrightarrow \mathcal{W}$ is an isomorphism, then	
(1) $\dim(\mathcal{V}) = \dim(\mathcal{W})$	Theorem 5.18†
(2) L^{-1} is an isomorphism from $\mathcal{W}$ to $\mathcal{V}$	Theorem 5.17
(3) $L(B)$ is a basis for $\mathcal{W}$	Theorem 5.20

† True only in the finite dimensional case

Figure 5.11
Conditions on linear transformations that are one-to-one, onto, or isomorphisms

Exercises—Section 5.4

1. Which of the following linear transformations are one-to-one? onto? isomorphisms? Justify your answers without using row reduction.

★(a) $L: \mathbb{R}^3 \longrightarrow \mathbb{R}^4$ given by $L([x, y, z]) = [y, z, -y, 0]$

(b) $L: \mathbb{R}^3 \longrightarrow \mathbb{R}^2$ given by $L([x, y, z]) = [x + y, y + z]$

★(c) $L: \mathbb{R}^3 \longrightarrow \mathbb{R}^3$ given by $L([x, y, z]) = [2x, x + y + z, -y]$

(d) $L: \mathcal{P}_3 \longrightarrow \mathcal{P}_2$ given by $L(ax^3 + bx^2 + cx + d) = ax^2 + bx + c$

★(e) $L: \mathcal{P}_2 \longrightarrow \mathcal{P}_2$ given by $L(ax^2 + bx + c) = (a + b)x^2 + (b + c)x + (a + c)$

(f) $L: \mathcal{M}_{22} \longrightarrow \mathcal{M}_{22}$ given by $L\left(\begin{bmatrix} a & b \\ c & d \end{bmatrix}\right) = \begin{bmatrix} d & b+c \\ b-c & a \end{bmatrix}$

★(g) $L: \mathcal{M}_{32} \longrightarrow \mathcal{M}_{22}$ given by $L\left(\begin{bmatrix} a & b & c \\ d & e & f \end{bmatrix}\right) = \begin{bmatrix} a & -c \\ 2e & d+f \end{bmatrix}$

★(h) $L: \mathcal{P}_2 \longrightarrow \mathcal{M}_{22}$ given by $L(ax^2 + bx + c) = \begin{bmatrix} a+c & 0 \\ b-c & -3a \end{bmatrix}$

2. Which of the following linear transformations are one-to-one? onto? isomorphisms? Justify your answers by using row reduction to determine the dimensions of the kernel and range.

★(a) $L: \mathbb{R}^2 \longrightarrow \mathbb{R}^2$ given by $L\left(\begin{bmatrix} x_1 \\ x_2 \end{bmatrix}\right) = \begin{bmatrix} -4 & -3 \\ 2 & 2 \end{bmatrix}\begin{bmatrix} x_1 \\ x_2 \end{bmatrix}$

★(b) $L: \mathbb{R}^2 \longrightarrow \mathbb{R}^3$ given by $L\left(\begin{bmatrix} x_1 \\ x_2 \end{bmatrix}\right) = \begin{bmatrix} -3 & 4 \\ -6 & 9 \\ 7 & -8 \end{bmatrix}\begin{bmatrix} x_1 \\ x_2 \end{bmatrix}$

★(c) $L: \mathbb{R}^3 \longrightarrow \mathbb{R}^3$ given by $L\left(\begin{bmatrix} x_1 \\ x_2 \\ x_3 \end{bmatrix}\right) = \begin{bmatrix} -7 & 4 & -2 \\ 16 & -7 & 2 \\ 4 & -3 & 2 \end{bmatrix}\begin{bmatrix} x_1 \\ x_2 \\ x_3 \end{bmatrix}$

(d) $L: \mathbb{R}^4 \longrightarrow \mathbb{R}^3$ given by $L\left(\begin{bmatrix} x_1 \\ x_2 \\ x_3 \\ x_4 \end{bmatrix}\right) = \begin{bmatrix} -5 & 3 & 1 & 18 \\ -2 & 1 & 1 & 6 \\ -7 & 3 & 4 & 19 \end{bmatrix}\begin{bmatrix} x_1 \\ x_2 \\ x_3 \\ x_4 \end{bmatrix}$

3. In each of the following cases, the matrix for a linear transformation with respect to some ordered bases for the domain and codomain is given. Which of these linear transformations are one-to-one? onto? isomorphisms? Justify your answers by using row reduction to determine the dimensions of the kernel and range.

★(a) $L: \mathcal{P}_2 \longrightarrow \mathcal{P}_2$ having matrix $\begin{bmatrix} 1 & -3 & 0 \\ -4 & 13 & -1 \\ 8 & -25 & 2 \end{bmatrix}$

(b) $L: \mathcal{M}_{22} \longrightarrow \mathcal{M}_{22}$ having matrix $\begin{bmatrix} 6 & -9 & 2 & 8 \\ 10 & -6 & 12 & 4 \\ -3 & 3 & -4 & -4 \\ 8 & -9 & 9 & 11 \end{bmatrix}$

★(c) $L: \mathcal{M}_{22} \longrightarrow \mathcal{P}_3$ having matrix $\begin{bmatrix} 2 & 3 & -1 & 1 \\ 5 & 2 & -4 & 7 \\ 1 & 7 & 1 & -4 \\ -2 & 19 & 7 & -19 \end{bmatrix}$

★**4.** (a) Suppose that $L: \mathbb{R}^6 \longrightarrow \mathcal{P}_5$ is a linear transformation and that L is not onto. Is L one-to-one? Why or why not?

(b) Suppose that L: $\mathcal{M}_{22} \longrightarrow \mathcal{P}_3$ is a linear transformation and that L is not one-to-one. Is L onto? Why or why not?

5. Suppose that $m > n$.

(a) Show there is no onto linear transformation from $\mathbb{R}^n$ to $\mathbb{R}^m$.

(b) Show there is no one-to-one linear transformation from $\mathbb{R}^m$ to $\mathbb{R}^n$.

6. Let $\mathbf{A}$ be a fixed $n \times n$ matrix, and consider L: $\mathcal{M}_{nn} \longrightarrow \mathcal{M}_{nn}$ given by $L(\mathbf{B}) = \mathbf{AB} - \mathbf{BA}$.

(a) Show that L is not one-to-one. (Hint: Consider $L(\mathbf{A})$.)

(b) Use part (a) to show that L is not onto.

7. Each part of this exercise gives matrices for linear operators L_1 and L_2 on $\mathbb{R}^3$ with respect to the standard basis. For each part, do the following:

(i) Show that L_1 and L_2 are isomorphisms.

(ii) Find L_1^{-1} and L_2^{-1}.

(iii) Calculate $L_2 \circ L_1$ directly.

(iv) Calculate $(L_2 \circ L_1)^{-1}$ by inverting the appropriate matrix.

(v) Calculate $L_1^{-1} \circ L_2^{-1}$ directly from your answer to (ii) and verify that the answer agrees with the result you obtained in (iv).

★(a) L_1: $\begin{bmatrix} 0 & -2 & 1 \\ 0 & -1 & 0 \\ 1 & 0 & 0 \end{bmatrix}$, $\quad L_2$: $\begin{bmatrix} 1 & 0 & 0 \\ -2 & 0 & 1 \\ 0 & -3 & 0 \end{bmatrix}$

(b) L_1: $\begin{bmatrix} -4 & 0 & 1 \\ 0 & 1 & 0 \\ 1 & 2 & 0 \end{bmatrix}$, $\quad L_2$: $\begin{bmatrix} 0 & 3 & -1 \\ 1 & 0 & 0 \\ 0 & -2 & 1 \end{bmatrix}$

★(c) L_1: $\begin{bmatrix} -9 & 2 & 1 \\ -6 & 1 & 1 \\ 5 & 0 & -2 \end{bmatrix}$, $\quad L_2$: $\begin{bmatrix} -4 & 2 & 1 \\ -3 & 1 & 0 \\ -5 & 2 & 1 \end{bmatrix}$

8. Show that L: $\mathcal{M}_{mn} \longrightarrow \mathcal{M}_{nm}$ given by $L(\mathbf{A}) = \mathbf{A}^T$ is an isomorphism.

9. Let $\mathbf{A}$ be a fixed nonsingular $n \times n$ matrix.

(a) Show that the linear operator L_1: $\mathcal{M}_{nn} \longrightarrow \mathcal{M}_{nn}$ given by $L_1(\mathbf{B}) = \mathbf{AB}$ is an isomorphism. (Hint: Be sure to show first that L_1 is a linear operator.)

(b) Show that the linear operator L_2: $\mathcal{M}_{nn} \longrightarrow \mathcal{M}_{nn}$ given by $L_2(\mathbf{B}) = \mathbf{ABA}^{-1}$ is an isomorphism.

10. Show that L: $\mathcal{P}_n \longrightarrow \mathcal{P}_n$ given by $L(\mathbf{p}) = \mathbf{p} + \mathbf{p}'$ is an isomorphism. (Hint: First show that L is a linear operator.)

11. Prove that the change of basis process is essentially an isomorphism; that is, if B and C are two different finite bases for a vector space $\mathcal{V}$, with $\dim(\mathcal{V}) = n$, then the mapping L: $\mathbb{R}^n \longrightarrow \mathbb{R}^n$ given by $L([\mathbf{v}]_B) = [\mathbf{v}]_C$ is an isomorphism. (Hint: First show that L is a linear operator.)

12. Let $\mathcal{V}$, $\mathcal{W}$, and $\mathcal{X}$ be vector spaces. Let $L_1: \mathcal{V} \longrightarrow \mathcal{W}$ and $L_2: \mathcal{V} \longrightarrow \mathcal{W}$ be linear transformations. Let $M: \mathcal{W} \longrightarrow \mathcal{X}$ be an isomorphism. If $M \circ L_1 = M \circ L_2$, show that $L_1 = L_2$.

13. Prove Theorem 5.16.

14. Prove Theorem 5.19.

15. Use an argument similar to the proof of Theorem 5.20 to prove the following: let $L: \mathcal{V} \longrightarrow \mathcal{W}$ be a linear transformation, and let B be a basis for $\mathcal{V}$ whose images under L are all distinct. Then L is one-to-one if and only if $L(B)$ is a basis for range(L).

16. (a) Explain why $\mathcal{M}_{mn} \cong \mathcal{M}_{nm}$.
(b) Explain why $\mathcal{P}_{4n+3} \cong \mathcal{M}_{4,n+1}$.
(c) Explain why the subspace of upper triangular matrices in $\mathcal{M}_{nn}$ is isomorphic to $\mathbb{R}^{(n(n+1))/2}$. Is the subspace still isomorphic to $\mathbb{R}^{(n(n+1))/2}$ if *upper* is replaced by *lower*?

17. Let $\mathcal{V}$ be a vector space. Show that a linear operator $L: \mathcal{V} \longrightarrow \mathcal{V}$ is an isomorphism if and only if $L \circ L$ is an isomorphism.

18. Let $\mathcal{V}$ be a nontrivial vector space. Suppose that $L: \mathcal{V} \longrightarrow \mathcal{V}$ is a linear operator.
(a) If $L \circ L$ is the zero transformation, show that L is not an isomorphism.
(b) If $L \circ L = L$ and L is not the identity transformation, show that L is not an isomorphism.

19. Let $L: \mathbb{R}^n \longrightarrow \mathbb{R}^n$ be a linear operator with matrix $\mathbf{A}$ (using the standard basis for $\mathbb{R}^n$). Prove that L is an isomorphism if and only if the columns of $\mathbf{A}$ are linearly independent.

20. We show in this exercise that any isomorphism from $\mathbb{R}^2$ to $\mathbb{R}^2$ is the composition of certain types of reflections, contractions/dilations, and shears. (See Exercise 10 in Section 5.1 for the definition of a shear.) Recall from Corollary 2.18 that any nonsingular 2×2 matrix is the product of a finite number of elementary 2×2 matrices.
(a) Show that any elementary 2×2 matrix is one of the following, for $k \in \mathbb{R}$:
$$\begin{bmatrix} k & 0 \\ 0 & 1 \end{bmatrix}, \begin{bmatrix} 1 & 0 \\ 0 & k \end{bmatrix}, \begin{bmatrix} 1 & 0 \\ k & 1 \end{bmatrix}, \begin{bmatrix} 1 & k \\ 0 & 1 \end{bmatrix}, \begin{bmatrix} 0 & 1 \\ 1 & 0 \end{bmatrix}.$$
(b) Show that when $k \geq 0$, multiplication of either of the first two matrices in part (a) times the vector $[x, y]$ represents a contraction/dilation along the x-coordinate or the y-coordinate.
(c) Show that when $k < 0$, multiplication of either of the first two matrices in part (a) times the vector $[x, y]$ represents a contraction/dilation along the x-coordinate or the y-coordinate, followed by a reflection through one of the axes. $\left(\text{Hint: } \begin{bmatrix} k & 0 \\ 0 & 1 \end{bmatrix} = \begin{bmatrix} -1 & 0 \\ 0 & 1 \end{bmatrix}\begin{bmatrix} -k & 0 \\ 0 & 1 \end{bmatrix}.\right)$

(d) Explain why multiplication by either of the third or fourth matrices in part (a) times $[x, y]$ represents a shear.

(e) Explain why multiplication of the last matrix in part (a) times $[x, y]$ represents a reflection through the line $y = x$.

(f) Using parts (a) through (e), show that any isomorphism from $\mathbb{R}^2$ to $\mathbb{R}^2$ is the composition of a finite number of linear operators of the following types: reflection through an axis, reflection through the line $y = x$, contraction/dilation of x-coordinate or y-coordinate, shear in the x- or y-direction.

21. Express the linear transformation $L: \mathbb{R}^2 \longrightarrow \mathbb{R}^2$ that rotates the plane 45° in a counterclockwise direction as a composition of the transformations described in part (f) of Exercise 20.

5.5 Orthogonality and the Gram-Schmidt Process

In this section, we deal only with vectors in $\mathbb{R}^n$. We begin by investigating orthogonality of vectors in more detail. Our main goal is the Gram-Schmidt Process, a method for constructing a basis of mutually orthogonal vectors for any nontrivial subspace of $\mathbb{R}^n$.

Orthogonal and Orthonormal Vectors

DEFINITION

Let $\{\mathbf{v}_1, \mathbf{v}_2, \ldots, \mathbf{v}_k\}$ be a subset of k distinct vectors of $\mathbb{R}^n$. Then $\{\mathbf{v}_1, \mathbf{v}_2, \ldots, \mathbf{v}_k\}$ is an **orthogonal set of vectors** if and only if the dot product of any two distinct vectors in this set is zero—that is, if and only if $\mathbf{v}_i \cdot \mathbf{v}_j = 0$, for $1 \leq i, j \leq k, i \neq j$. Also, $\{\mathbf{v}_1, \mathbf{v}_2, \ldots, \mathbf{v}_k\}$ is an **orthonormal set of vectors** if and only if it is an orthogonal set and all its vectors are unit vectors (that is, $\|\mathbf{v}_i\| = 1$, for $1 \leq i \leq k$).

In particular, a set containing a single vector is orthogonal, and a set containing a single unit vector is orthonormal.

Example 1

In $\mathbb{R}^3$, $\{\mathbf{i}, \mathbf{j}, \mathbf{k}\}$ is an orthogonal set because $\mathbf{i} \cdot \mathbf{j} = \mathbf{j} \cdot \mathbf{k} = \mathbf{k} \cdot \mathbf{i} = 0$. In fact, this is an orthonormal set since we also have $\|\mathbf{i}\| = \|\mathbf{j}\| = \|\mathbf{k}\| = 1$.

In $\mathbb{R}^4$, $\{[1, 0, -1, 0], [3, 0, 3, 0]\}$ is an orthogonal set because $[1, 0, -1, 0] \cdot [3, 0, 3, 0] = 0$. If we divide each of these vectors by its length, we create the orthonormal set of vectors

$$\left\{\left[\frac{1}{\sqrt{2}}, 0, -\frac{1}{\sqrt{2}}, 0\right], \left[\frac{1}{\sqrt{2}}, 0, \frac{1}{\sqrt{2}}, 0\right]\right\}.$$

■

The next theorem is proved in the same manner as Result 7 in Section 1.3.

THEOREM 5.24

Let $T = \{\mathbf{v}_1, \ldots, \mathbf{v}_k\}$ be an orthogonal set of nonzero vectors in $\mathbb{R}^n$. Then T is a linearly independent set.

Notice that the orthogonal sets in Example 1 are indeed linearly independent sets.

Orthogonal and Orthonormal Bases

Theorem 5.24 assures us that any orthogonal set of nonzero vectors in $\mathbb{R}^n$ is linearly independent, so any such set forms a basis for some subspace of $\mathbb{R}^n$.

DEFINITION

A set B of vectors in a subspace $\mathcal{W}$ of $\mathbb{R}^n$ is an **orthogonal basis** for $\mathcal{W}$ if and only if B is an orthogonal set and B is a basis for $\mathcal{W}$. Similarly, B is an **orthonormal basis** for $\mathcal{W}$ if and only if B is an orthonormal set and B is a basis for $\mathcal{W}$.

The following corollary follows immediately from Theorem 5.24:

COROLLARY 5.25

If B is an orthogonal set of n nonzero vectors in $\mathbb{R}^n$, then B is an orthogonal basis for $\mathbb{R}^n$. Similarly, if B is an orthonormal set of n vectors in $\mathbb{R}^n$, then B is an orthonormal basis for $\mathbb{R}^n$.

Example 2

Consider the following subset of $\mathbb{R}^3$: $\{[1, 0, -1], [-1, 4, -1], [2, 1, 2]\}$. Because every pair of distinct vectors in this set is orthogonal (verify), this is an orthogonal set. By Corollary 5.25, this is also an orthogonal basis for $\mathbb{R}^3$. Normalizing each vector, we obtain an orthonormal basis for $\mathbb{R}^3$:

$$\left\{\left[\frac{1}{\sqrt{2}}, 0, -\frac{1}{\sqrt{2}}\right], \left[-\frac{1}{3\sqrt{2}}, \frac{4}{3\sqrt{2}}, -\frac{1}{3\sqrt{2}}\right], \left[\frac{2}{3}, \frac{1}{3}, \frac{2}{3}\right]\right\}.$$ ■

One of the advantages of using an orthogonal or orthonormal basis is that it is easy to coordinatize vectors with respect to that basis.

THEOREM 5.26

If $B = (\mathbf{v}_1, \mathbf{v}_2, \ldots, \mathbf{v}_k)$ is an orthogonal ordered basis for a subspace $\mathcal{W}$ of $\mathbb{R}^n$, and if $\mathbf{v}$ is any vector in $\mathcal{W}$, then $[\mathbf{v}]_B = \left[(\mathbf{v} \cdot \mathbf{v}_1)/\|\mathbf{v}_1\|^2, (\mathbf{v} \cdot \mathbf{v}_2)/\|\mathbf{v}_2\|^2, \ldots, (\mathbf{v} \cdot \mathbf{v}_k)/\|\mathbf{v}_k\|^2\right]$. In particular, if B is an orthonormal ordered basis for $\mathcal{W}$, then $[\mathbf{v}]_B = [\mathbf{v} \cdot \mathbf{v}_1, \mathbf{v} \cdot \mathbf{v}_2, \ldots, \mathbf{v} \cdot \mathbf{v}_k]$.

Proof of Theorem 5.26

Suppose that $[\mathbf{v}]_B = [a_1, a_2, \ldots, a_k]$, where $a_1, a_2, \ldots, a_k \in \mathbb{R}$. We must show that $a_i = (\mathbf{v} \cdot \mathbf{v}_i)/\|\mathbf{v}_i\|^2$, for $1 \le i \le k$. Now, $\mathbf{v} = a_1\mathbf{v}_1 + a_2\mathbf{v}_2 + \cdots + a_k\mathbf{v}_k$. Hence,

$$\begin{aligned} \mathbf{v} \cdot \mathbf{v}_i &= (a_1\mathbf{v}_1 + a_2\mathbf{v}_2 + \cdots + a_i\mathbf{v}_i + \cdots + a_k\mathbf{v}_k) \cdot \mathbf{v}_i \\ &= a_1(\mathbf{v}_1 \cdot \mathbf{v}_i) + a_2(\mathbf{v}_2 \cdot \mathbf{v}_i) + \cdots + a_i(\mathbf{v}_i \cdot \mathbf{v}_i) + \cdots + a_k(\mathbf{v}_k \cdot \mathbf{v}_i) \\ &= a_1(0) + a_2(0) + \cdots + a_i\|\mathbf{v}_i\|^2 + \cdots + a_n(0) \quad \text{because } B \text{ is orthogonal} \\ &= a_i\|\mathbf{v}_i\|^2. \end{aligned}$$

Thus, $a_i = (\mathbf{v} \cdot \mathbf{v}_i)/\|\mathbf{v}_i\|^2$. In the special case when B is orthonormal, $\|\mathbf{v}_i\| = 1$, for $1 \le i \le k$, and so $a_i = \mathbf{v} \cdot \mathbf{v}_i$. ■

Example 3

Consider the ordered orthonormal basis $B = (\mathbf{v}_1, \mathbf{v}_2, \mathbf{v}_3)$ for $\mathbb{R}^3$, where $\mathbf{v}_1 = [1/\sqrt{2}, 0, -1/\sqrt{2}]$, $\mathbf{v}_2 = [-1/(3\sqrt{2}), 4/(3\sqrt{2}), -1/(3\sqrt{2})]$, and $\mathbf{v}_3 = [\frac{2}{3}, \frac{1}{3}, \frac{2}{3}]$ are the vectors from Example 2. Let $\mathbf{v} = [-1, 5, 3]$. We will use Theorem 5.26 to find the coordinatization of $\mathbf{v}$ with respect to B. Now, $\mathbf{v} \cdot \mathbf{v}_1 = (-1/\sqrt{2}) - (3/\sqrt{2}) = -2\sqrt{2}$, $\mathbf{v} \cdot \mathbf{v}_2 = (1/(3\sqrt{2})) + (20/(3\sqrt{2})) - (3/(3\sqrt{2})) = 3\sqrt{2}$, and $\mathbf{v} \cdot \mathbf{v}_3 = -\frac{2}{3} + \frac{5}{3} + 2 = 3$. Hence, $[\mathbf{v}]_B = [-2\sqrt{2}, 3\sqrt{2}, 3]$. These coordinates can easily be verified by checking that

$$[-1, 5, 3] = -2\sqrt{2}\left[\frac{1}{\sqrt{2}}, 0, -\frac{1}{\sqrt{2}}\right] + 3\sqrt{2}\left[-\frac{1}{3\sqrt{2}}, \frac{4}{3\sqrt{2}}, -\frac{1}{3\sqrt{2}}\right] + 3\left[\frac{2}{3}, \frac{1}{3}, \frac{2}{3}\right].$$ ■

The Gram-Schmidt Process for Finding an Orthogonal Basis for a Subspace of $\mathbb{R}^n$

Suppose that $\mathcal{W}$ is a subspace of $\mathbb{R}^n$ with basis $B = \{\mathbf{w}_1, \ldots, \mathbf{w}_k\}$. There is a straightforward way to replace B with an orthogonal basis for $\mathcal{W}$, known as the Gram-Schmidt Process:

GRAM-SCHMIDT PROCESS

Let $\{\mathbf{w}_1, \ldots, \mathbf{w}_k\}$ be a linearly independent subset of $\mathbb{R}^n$. We create a new set $\{\mathbf{v}_1, \ldots, \mathbf{v}_k\}$ of vectors as follows:

Let $\mathbf{v}_1 = \mathbf{w}_1.$

Let $$\mathbf{v}_2 = \mathbf{w}_2 - \left(\frac{\mathbf{w}_2 \cdot \mathbf{v}_1}{\mathbf{v}_1 \cdot \mathbf{v}_1}\right)\mathbf{v}_1.$$

Let $$\mathbf{v}_3 = \mathbf{w}_3 - \left(\frac{\mathbf{w}_3 \cdot \mathbf{v}_1}{\mathbf{v}_1 \cdot \mathbf{v}_1}\right)\mathbf{v}_1 - \left(\frac{\mathbf{w}_3 \cdot \mathbf{v}_2}{\mathbf{v}_2 \cdot \mathbf{v}_2}\right)\mathbf{v}_2.$$

$$\vdots$$

Let $$\mathbf{v}_k = \mathbf{w}_k - \left(\frac{\mathbf{w}_k \cdot \mathbf{v}_1}{\mathbf{v}_1 \cdot \mathbf{v}_1}\right)\mathbf{v}_1 - \left(\frac{\mathbf{w}_k \cdot \mathbf{v}_2}{\mathbf{v}_2 \cdot \mathbf{v}_2}\right)\mathbf{v}_2 - \cdots - \left(\frac{\mathbf{w}_k \cdot \mathbf{v}_{k-1}}{\mathbf{v}_{k-1} \cdot \mathbf{v}_{k-1}}\right)\mathbf{v}_{k-1}.$$

The next theorem asserts that the Gram-Schmidt Process works.

THEOREM 5.27

Let $B = \{\mathbf{w}_1, \ldots, \mathbf{w}_k\}$ be a basis for a subspace $\mathcal{W}$ of $\mathbb{R}^n$. Then the set $\{\mathbf{v}_1, \ldots, \mathbf{v}_k\}$ obtained by applying the Gram-Schmidt Process to B is an orthogonal basis for $\mathcal{W}$.

Hence, any nontrivial subspace $\mathcal{W}$ of $\mathbb{R}^n$ has an orthogonal basis.

Proof of Theorem 5.27

Let $\mathcal{W}$ be a subspace of $\mathbb{R}^n$ having basis $B = \{\mathbf{w}_1, \ldots, \mathbf{w}_k\}$. Let $T = \{\mathbf{v}_1, \ldots, \mathbf{v}_k\}$ be the set obtained from applying the Gram-Schmidt Process to the basis B. We first show $T \subseteq \mathcal{W}$ by proving by induction that each $\mathbf{v}_i \in T$ is in span($\{\mathbf{w}_1, \mathbf{w}_2, \ldots, \mathbf{w}_i\}$), which is obviously in $\mathcal{W}$.

For the base step, notice that the first vector $\mathbf{v}_1 \in T$ is equal to $\mathbf{w}_1 \in B$, so $\mathbf{v}_1 \in$ span($\{\mathbf{w}_1\}$). For the inductive step, we assume that, for all $\mathbf{v}_j$ with $1 \leq j < i$, $\mathbf{v}_j \in$ span($\{\mathbf{w}_1, \ldots, \mathbf{w}_j\}$), and try to prove that $\mathbf{v}_i \in$ span($\{\mathbf{w}_1, \ldots, \mathbf{w}_i\}$). But $\mathbf{v}_i$ is a linear combination of $\mathbf{w}_i$ and $\mathbf{v}_1, \ldots, \mathbf{v}_{i-1}$, all of which are in span($\{\mathbf{w}_1, \ldots, \mathbf{w}_i\}$) by our inductive hypothesis. Hence, $\mathbf{v}_i \in$ span($\{\mathbf{w}_1, \ldots, \mathbf{w}_i\}$), which is in $\mathcal{W}$. Thus, $T \subseteq \mathcal{W}$.

Now, because $|T| = k$ and $\dim(\mathcal{W}) = k$, Theorem 4.19 shows that if T is a linearly independent set, then T is a basis for $\mathcal{W}$. So by Theorem 5.24, if we show that T is an orthogonal set of nonzero vectors, then T is an orthogonal basis for $\mathcal{W}$.

First we show that all vectors in T are nonzero. Now, if some vector $\mathbf{v}_i \in T$ is $\mathbf{0}$, then by definition of $\mathbf{v}_i$, it follows that

$$\mathbf{w}_i = \left(\frac{\mathbf{w}_i \cdot \mathbf{v}_1}{\mathbf{v}_1 \cdot \mathbf{v}_1}\right)\mathbf{v}_1 + \left(\frac{\mathbf{w}_i \cdot \mathbf{v}_2}{\mathbf{v}_2 \cdot \mathbf{v}_2}\right)\mathbf{v}_2 + \cdots + \left(\frac{\mathbf{w}_i \cdot \mathbf{v}_{i-1}}{\mathbf{v}_{i-1} \cdot \mathbf{v}_{i-1}}\right)\mathbf{v}_{i-1}.$$

However, then $\mathbf{w}_i \in \text{span}(\{\mathbf{v}_1, \ldots, \mathbf{v}_{i-1}\}) \subseteq \text{span}(\{\mathbf{w}_1, \ldots, \mathbf{w}_{i-1}\})$, from our earlier induction argument. This result contradicts the fact that B is a basis and hence a linearly independent set. Thus, all vectors in T are nonzero.

Finally, we need to show that T is orthogonal. We give another proof by induction. For the base step, notice that $\{\mathbf{v}_1\}$ is orthogonal, since $\mathbf{v}_1 = \mathbf{w}_1 \neq \mathbf{0}$. For the inductive step, we assume that $\{\mathbf{v}_1, \ldots, \mathbf{v}_{i-1}\}$ is orthogonal and show that $\mathbf{v}_i$ is orthogonal to each of $\mathbf{v}_1, \ldots, \mathbf{v}_{i-1}$. Now,

$$\mathbf{v}_i = \mathbf{w}_i - \left(\frac{\mathbf{w}_i \cdot \mathbf{v}_1}{\mathbf{v}_1 \cdot \mathbf{v}_1}\right)\mathbf{v}_1 - \left(\frac{\mathbf{w}_i \cdot \mathbf{v}_2}{\mathbf{v}_2 \cdot \mathbf{v}_2}\right)\mathbf{v}_2 - \cdots - \left(\frac{\mathbf{w}_i \cdot \mathbf{v}_{i-1}}{\mathbf{v}_{i-1} \cdot \mathbf{v}_{i-1}}\right)\mathbf{v}_{i-1}.$$

Notice that

$$\begin{aligned}
\mathbf{v}_i \cdot \mathbf{v}_1 &= \mathbf{w}_i \cdot \mathbf{v}_1 - \left(\frac{\mathbf{w}_i \cdot \mathbf{v}_1}{\mathbf{v}_1 \cdot \mathbf{v}_1}\right)(\mathbf{v}_1 \cdot \mathbf{v}_1) - \left(\frac{\mathbf{w}_i \cdot \mathbf{v}_2}{\mathbf{v}_2 \cdot \mathbf{v}_2}\right)(\mathbf{v}_2 \cdot \mathbf{v}_1) \\
&\quad - \cdots - \left(\frac{\mathbf{w}_i \cdot \mathbf{v}_{i-1}}{\mathbf{v}_{i-1} \cdot \mathbf{v}_{i-1}}\right)(\mathbf{v}_{i-1} \cdot \mathbf{v}_1) \\
&= \mathbf{w}_i \cdot \mathbf{v}_1 - \left(\frac{\mathbf{w}_i \cdot \mathbf{v}_1}{\mathbf{v}_1 \cdot \mathbf{v}_1}\right)(\mathbf{v}_1 \cdot \mathbf{v}_1) - \left(\frac{\mathbf{w}_i \cdot \mathbf{v}_2}{\mathbf{v}_2 \cdot \mathbf{v}_2}\right)(0) \\
&\quad - \cdots - \left(\frac{\mathbf{w}_i \cdot \mathbf{v}_{i-1}}{\mathbf{v}_{i-1} \cdot \mathbf{v}_{i-1}}\right)(0) \quad \text{inductive hypothesis} \\
&= \mathbf{w}_i \cdot \mathbf{v}_1 - \mathbf{w}_i \cdot \mathbf{v}_1 = 0.
\end{aligned}$$

Similar arguments show that $\mathbf{v}_i \cdot \mathbf{v}_2 = \mathbf{v}_i \cdot \mathbf{v}_3 = \cdots = \mathbf{v}_i \cdot \mathbf{v}_{i-1} = 0$. Hence, for each i, $\{\mathbf{v}_1, \ldots, \mathbf{v}_i\}$ is an orthogonal set. In particular, when $i = k$, the entire set $T = \{\mathbf{v}_1, \ldots, \mathbf{v}_k\}$ is orthogonal. ■

Once we have an orthogonal basis for a subspace $\mathcal{W}$ of $\mathbb{R}^n$, we can easily convert it to an orthonormal basis for $\mathcal{W}$ by normalizing each vector. Also, a little thought will convince you that in the Gram-Schmidt Process, any of the newly created vectors $\mathbf{v}_i$ could be replaced with a nonzero scalar multiple of itself, and the proof of Theorem 5.27 still holds. Hence, in applying the Gram-Schmidt Process, we can often replace these $\mathbf{v}_i$'s with appropriate multiples to avoid fractions. The next example illustrates these techniques.

Example 4

You can easily check that $B = \{[2, 1, 0, -1], [1, 0, 2, -1], [0, -2, 1, 0]\}$ is a linearly independent set in $\mathbb{R}^4$. Let $\mathcal{W} = \text{span}(B)$. Because B is not an orthogonal basis for $\mathcal{W}$, we will replace it with an orthogonal basis. Let $\mathbf{w}_1 = [2, 1, 0, -1]$, $\mathbf{w}_2 = [1, 0, 2, -1]$, and $\mathbf{w}_3 = [0, -2, 1, 0]$. Beginning the Gram-Schmidt Process, we obtain $\mathbf{v}_1 = \mathbf{w}_1 = [2, 1, 0, -1]$ and

$$\begin{aligned}\mathbf{v}_2 &= \mathbf{w}_2 - \left(\frac{\mathbf{w}_2 \cdot \mathbf{v}_1}{\mathbf{v}_1 \cdot \mathbf{v}_1}\right)\mathbf{v}_1 \\ &= [1, 0, 2, -1] - \left(\frac{[1, 0, 2, -1] \cdot [2, 1, 0, -1]}{[2, 1, 0, -1] \cdot [2, 1, 0, -1]}\right)[2, 1, 0, -1] \\ &= [1, 0, 2, -1] - \left(\tfrac{3}{6}\right)[2, 1, 0, -1] \\ &= [1, 0, 2, -1] - \tfrac{1}{2}[2, 1, 0, -1] = [0, -\tfrac{1}{2}, 2, -\tfrac{1}{2}].\end{aligned}$$

To avoid fractions in what follows, we replace this vector with an appropriate scalar multiple. Multiplying by 2, we get $\mathbf{v}_2 = [0, -1, 4, -1]$. Notice that $\mathbf{v}_2$ is orthogonal to $\mathbf{v}_1$. Finally,

$$\begin{aligned}\mathbf{v}_3 &= \mathbf{w}_3 - \left(\frac{\mathbf{w}_3 \cdot \mathbf{v}_1}{\mathbf{v}_1 \cdot \mathbf{v}_1}\right)\mathbf{v}_1 - \left(\frac{\mathbf{w}_3 \cdot \mathbf{v}_2}{\mathbf{v}_2 \cdot \mathbf{v}_2}\right)\mathbf{v}_2 \\ &= [0, -2, 1, 0] - \left(\frac{[0, -2, 1, 0] \cdot [2, 1, 0, -1]}{[2, 1, 0, -1] \cdot [2, 1, 0, -1]}\right)[2, 1, 0, -1] \\ &\quad - \left(\frac{[0, -2, 1, 0] \cdot [0, -1, 4, -1]}{[0, -1, 4, -1] \cdot [0, -1, 4, -1]}\right)[0, -1, 4, -1] \\ &= [0, -2, 1, 0] - \left(\tfrac{-2}{6}\right)[2, 1, 0, -1] - \left(\tfrac{6}{18}\right)[0, -1, 4, -1] \\ &= [0, -2, 1, 0] + \left(\tfrac{1}{3}\right)[2, 1, 0, -1] - \left(\tfrac{1}{3}\right)[0, -1, 4, -1] = \left[\tfrac{2}{3}, -\tfrac{4}{3}, -\tfrac{1}{3}, 0\right].\end{aligned}$$

To avoid fractions, we multiply this vector by 3, yielding $\mathbf{v}_3 = [2, -4, -1, 0]$. Notice that $\mathbf{v}_3$ is orthogonal to both $\mathbf{v}_1$ and $\mathbf{v}_2$.

Hence, $\{\mathbf{v}_1, \mathbf{v}_2, \mathbf{v}_3\} = \{[2, 1, 0, -1], [0, -1, 4, -1], [2, -4, -1, 0]\}$ is an orthogonal basis for $\mathcal{W}$. To find an orthonormal basis for $\mathcal{W}$, we normalize $\mathbf{v}_1$, $\mathbf{v}_2$, and $\mathbf{v}_3$ to obtain

$$\left\{\left[\frac{2}{\sqrt{6}}, \frac{1}{\sqrt{6}}, 0, -\frac{1}{\sqrt{6}}\right], \left[0, -\frac{1}{3\sqrt{2}}, \frac{4}{3\sqrt{2}}, -\frac{1}{3\sqrt{2}}\right], \left[\frac{2}{\sqrt{21}}, -\frac{4}{\sqrt{21}}, -\frac{1}{\sqrt{21}}, 0\right]\right\}.$$ ■

Suppose that $T = \{\mathbf{w}_1, \ldots, \mathbf{w}_k\}$ is an orthogonal set of nonzero vectors in a subspace $\mathcal{W}$ of $\mathbb{R}^n$. By Theorem 5.24, T is linearly independent. Hence, by Theorem 4.18, we can enlarge T to an ordered basis $(\mathbf{w}_1, \ldots, \mathbf{w}_k, \mathbf{w}_{k+1}, \ldots, \mathbf{w}_l)$ for $\mathcal{W}$. Applying the Gram-Schmidt Process to this enlarged basis gives an ordered orthogonal basis $B = (\mathbf{v}_1, \ldots, \mathbf{v}_k, \mathbf{v}_{k+1}, \ldots, \mathbf{v}_l)$ for $\mathcal{W}$. However, because $(\mathbf{w}_1, \ldots, \mathbf{w}_k)$ is already orthogonal, the first k vectors, $\mathbf{v}_1, \ldots, \mathbf{v}_k$, created by the Gram-Schmidt

Process will be equal to $\mathbf{w}_1, \ldots, \mathbf{w}_k$, respectively. (Why?) Hence, B is an ordered orthogonal basis for $\mathcal{W}$ that contains T. Similarly, if the original set $T = (\mathbf{w}_1, \ldots, \mathbf{w}_k)$ is *orthonormal*, T can be enlarged to an *orthonormal* basis for $\mathcal{W}$. (Why?) These remarks prove the following:

THEOREM 5.28

Let $\mathcal{W}$ be a subspace of $\mathbb{R}^n$. Then any orthogonal [orthonormal] set of nonzero vectors in $\mathcal{W}$ is contained in (can be enlarged to) an orthogonal [orthonormal] basis for $\mathcal{W}$.

Example 5

We will find an orthogonal basis B for $\mathbb{R}^4$ that contains the orthogonal set $T = \{[2,1,0,-1], [0,-1,4,-1], [2,-4,-1,0]\}$ from Example 4. You can easily check that $[1,0,0,0]$ is not a linear combination of the given vectors $[2,1,0,-1], [0,-1,4,-1], [2,-4,-1,0]$. Hence, $\{[2,1,0,-1], [0,-1,4,-1], [2,-4,-1,0], [1,0,0,0]\}$ is a basis for $\mathbb{R}^4$. Now we use the Gram-Schmidt Process to convert this basis to an orthogonal basis for $\mathbb{R}^4$.

Let $\mathbf{w}_1 = [2,1,0,-1]$, $\mathbf{w}_2 = [0,-1,4,-1]$, $\mathbf{w}_3 = [2,-4,-1,0]$, and $\mathbf{w}_4 = [1,0,0,0]$. The first few steps of the Gram-Schmidt Process give $\mathbf{v}_1 = \mathbf{w}_1$, $\mathbf{v}_2 = \mathbf{w}_2$, and $\mathbf{v}_3 = \mathbf{w}_3$. (Why?) Finally,

$$\begin{aligned}
\mathbf{v}_4 &= \mathbf{w}_4 - \left(\frac{\mathbf{w}_4 \cdot \mathbf{v}_1}{\mathbf{v}_1 \cdot \mathbf{v}_1}\right)\mathbf{v}_1 - \left(\frac{\mathbf{w}_4 \cdot \mathbf{v}_2}{\mathbf{v}_2 \cdot \mathbf{v}_2}\right)\mathbf{v}_2 - \left(\frac{\mathbf{w}_4 \cdot \mathbf{v}_3}{\mathbf{v}_3 \cdot \mathbf{v}_3}\right)\mathbf{v}_3 \\
&= [1,0,0,0] - \left(\frac{[1,0,0,0] \cdot [2,1,0,-1]}{[2,1,0,-1] \cdot [2,1,0,-1]}\right)[2,1,0,-1] \\
&\quad - \left(\frac{[1,0,0,0] \cdot [0,-1,4,-1]}{[0,-1,4,-1] \cdot [0,-1,4,-1]}\right)[0,-1,4,-1] \\
&\quad - \left(\frac{[1,0,0,0] \cdot [2,-4,-1,0]}{[2,-4,-1,0] \cdot [2,-4,-1,0]}\right)[2,-4,-1,0] \\
&= [1,0,0,0] - \tfrac{1}{3}[2,1,0,-1] - \tfrac{2}{21}[2,-4,-1,0] = \left[\tfrac{1}{7}, \tfrac{1}{21}, \tfrac{2}{21}, \tfrac{1}{3}\right].
\end{aligned}$$

To avoid fractions, we multiply this vector by 21 to obtain $\mathbf{v}_4 = [3,1,2,7]$. Notice that $\mathbf{v}_4$ is orthogonal to $\mathbf{v}_1$, $\mathbf{v}_2$, and $\mathbf{v}_3$. Hence, $\{\mathbf{v}_1, \mathbf{v}_2, \mathbf{v}_3, \mathbf{v}_4\}$ is an orthogonal basis for $\mathbb{R}^4$ containing T. ■

Orthogonal Matrices

DEFINITION

A square (nonsingular) matrix $\mathbf{A}$ is **orthogonal** if and only if $\mathbf{A}^T = \mathbf{A}^{-1}$.

Notice that if $\mathbf{A}$ is orthogonal, then $|\mathbf{A}| = \pm 1$, because $|\mathbf{A}^T| = |\mathbf{A}^{-1}| \Longrightarrow |\mathbf{A}| = 1/|\mathbf{A}| \Longrightarrow |\mathbf{A}|^2 = 1 \Longrightarrow |\mathbf{A}| = \pm 1$. (Beware! The converse is not true: if $|\mathbf{A}| = \pm 1$, then $\mathbf{A}$ is not necessarily orthogonal.)

The proof of the next theorem is very easy (see Exercise 9).

THEOREM 5.29

If $\mathbf{A}$ and $\mathbf{B}$ are orthogonal matrices of the same size, then $\mathbf{AB}$ is orthogonal.

The next theorem characterizes all orthogonal matrices.

THEOREM 5.30

Let $\mathbf{A}$ be an $n \times n$ matrix. Then $\mathbf{A}$ is orthogonal
(1) if and only if the rows of $\mathbf{A}$ form an orthonormal basis for $\mathbb{R}^n$
(2) if and only if the columns of $\mathbf{A}$ form an orthonormal basis for $\mathbb{R}^n$

Theorem 5.30 suggests that it is probably more appropriate to refer to orthogonal matrices as "orthonormal matrices." Unfortunately, the term *orthogonal* has become traditional usage in linear algebra.

Proof of Theorem 5.30 (abridged)

We prove half of part (1) and leave the rest of the theorem for you to prove in Exercise 15.

Suppose that $\mathbf{A}$ is an orthogonal $n \times n$ matrix. Then we have $\mathbf{AA}^T = \mathbf{I}_n$. (Why?) Hence, for $1 \le i, j \le n$, with $i \ne j$, we have $[i^{th}$ row of $\mathbf{A}] \cdot [j^{th}$ column of $\mathbf{A}^T] = 0$. Therefore $[i^{th}$ row of $\mathbf{A}] \cdot [j^{th}$ row of $\mathbf{A}] = 0$, which shows that distinct rows of $\mathbf{A}$ are orthogonal. Again, because $\mathbf{AA}^T = \mathbf{I}_n$, for each $i, 1 \le i \le n$, we have $[i^{th}$ row of $\mathbf{A}] \cdot [i^{th}$ column of $\mathbf{A}^T] = 1$. But then $[i^{th}$ row of $\mathbf{A}] \cdot [i^{th}$ row of $\mathbf{A}] = 1$, which shows that each row of $\mathbf{A}$ is a unit vector. Thus, the n rows of $\mathbf{A}$ form an orthonormal set and, hence, an orthonormal basis for $\mathbb{R}^n$. ■

$\mathbf{I}_n$ is obviously an orthogonal matrix, for any $n \ge 1$. In the next example, we show how Theorem 5.30 can be used to find other orthogonal matrices.

Example 6

Consider the orthonormal basis $\{\mathbf{v}_1, \mathbf{v}_2, \mathbf{v}_3\}$ for $\mathbb{R}^3$ from Example 2, where

$$\mathbf{v}_1 = \left[\frac{1}{\sqrt{2}}, 0, -\frac{1}{\sqrt{2}}\right], \quad \mathbf{v}_2 = \left[-\frac{1}{3\sqrt{2}}, \frac{4}{3\sqrt{2}}, -\frac{1}{3\sqrt{2}}\right], \quad \text{and} \quad \mathbf{v}_3 = \left[\frac{2}{3}, \frac{1}{3}, \frac{2}{3}\right].$$

By parts (1) and (2) of Theorem 5.30, respectively,

$$\mathbf{A} = \begin{bmatrix} \frac{1}{\sqrt{2}} & 0 & -\frac{1}{\sqrt{2}} \\ -\frac{1}{3\sqrt{2}} & \frac{4}{3\sqrt{2}} & -\frac{1}{3\sqrt{2}} \\ \frac{2}{3} & \frac{1}{3} & \frac{2}{3} \end{bmatrix} \quad \text{and} \quad \mathbf{B} = \begin{bmatrix} \frac{1}{\sqrt{2}} & -\frac{1}{3\sqrt{2}} & \frac{2}{3} \\ 0 & \frac{4}{3\sqrt{2}} & \frac{1}{3} \\ -\frac{1}{\sqrt{2}} & -\frac{1}{3\sqrt{2}} & \frac{2}{3} \end{bmatrix}$$

are both orthogonal matrices. You can easily verify that **A** and **B** are orthogonal by checking that $\mathbf{A}^T\mathbf{A} = \mathbf{B}\mathbf{B}^T = \mathbf{I}_3$. ■

One use for orthogonal matrices is given in the next theorem.

THEOREM 5.31

Let B and C be ordered orthonormal bases for $\mathbb{R}^n$. Then the transition matrix from B to C is an orthogonal matrix.

Proof of Theorem 5.31

Let S be the standard basis for $\mathbb{R}^n$. The matrix **P** whose columns are the vectors in B is the transition matrix from B to S. Similarly, the matrix **Q**, whose columns are the vectors in C, is the transition matrix from C to S. Both **P** and **Q** are orthogonal matrices by Theorem 5.30. But then $\mathbf{Q}^{-1}$ is also orthogonal, because $\mathbf{Q}^{-1} = \mathbf{Q}^T$, whose *rows* are the set of orthogonal vectors in C. Finally, by Theorems 4.23 and 4.24, $\mathbf{Q}^{-1}\mathbf{P}$ is the transition matrix from B to C, which is orthogonal by Theorem 5.29 (see Figure 5.12). ■

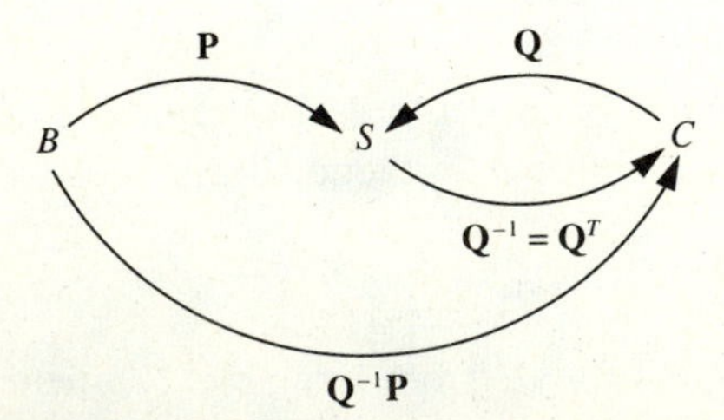

Figure 5.12
Visualizing $\mathbf{Q}^{-1}\mathbf{P}$ as the transition matrix from B to C

Example 7

Consider these ordered orthonormal bases for $\mathbb{R}^2$:

$$B = \left(\left[\frac{\sqrt{2}}{2}, \frac{\sqrt{2}}{2}\right], \left[\frac{\sqrt{2}}{2}, -\frac{\sqrt{2}}{2}\right]\right) \quad \text{and} \quad C = \left(\left[\frac{\sqrt{3}}{2}, \frac{1}{2}\right], \left[-\frac{1}{2}, \frac{\sqrt{3}}{2}\right]\right).$$

By Theorem 5.31, the transition matrix from B to C is orthogonal. To verify that it is, we can use Theorem 5.26 to obtain

$$\left[\frac{\sqrt{2}}{2}, \frac{\sqrt{2}}{2}\right] = \left(\frac{\sqrt{6}+\sqrt{2}}{4}\right)\left[\frac{\sqrt{3}}{2}, \frac{1}{2}\right] + \left(\frac{\sqrt{6}-\sqrt{2}}{4}\right)\left[-\frac{1}{2}, \frac{\sqrt{3}}{2}\right] \quad \text{and}$$

$$\left[\frac{\sqrt{2}}{2}, -\frac{\sqrt{2}}{2}\right] = \left(\frac{\sqrt{6}-\sqrt{2}}{4}\right)\left[\frac{\sqrt{3}}{2}, \frac{1}{2}\right] + \left(\frac{-\sqrt{6}-\sqrt{2}}{4}\right)\left[-\frac{1}{2}, \frac{\sqrt{3}}{2}\right].$$

Hence, the transition matrix from B to C is

$$\mathbf{A} = \frac{1}{4}\begin{bmatrix} \sqrt{6}+\sqrt{2} & \sqrt{6}-\sqrt{2} \\ \sqrt{6}-\sqrt{2} & -\sqrt{6}-\sqrt{2} \end{bmatrix}.$$

Because $\mathbf{AA}^T = \mathbf{I}_2$ (verify), $\mathbf{A}$ is an orthogonal matrix. ■

The final theorem of this section can be used to prove that multiplying two n-vectors by an orthogonal matrix $\mathbf{A}$ does not change the angle between them (see Exercise 16).

THEOREM 5.32

Let $\mathbf{A}$ be an $n \times n$ orthogonal matrix, and let $\mathbf{v}$ and $\mathbf{w}$ be vectors in $\mathbb{R}^n$. Then $\mathbf{v} \cdot \mathbf{w} = \mathbf{Av} \cdot \mathbf{Aw}$.

Proof of Theorem 5.32

Recall that the dot product $\mathbf{x} \cdot \mathbf{y}$ of two column vectors $\mathbf{x}$ and $\mathbf{y}$ can be written in matrix multiplication form as $\mathbf{x}^T\mathbf{y}$. Let $\mathbf{v}, \mathbf{w} \in \mathbb{R}^n$, and let $\mathbf{A}$ be an $n \times n$ orthogonal matrix. Then

$$\mathbf{v} \cdot \mathbf{w} = \mathbf{v}^T\mathbf{w} = \mathbf{v}^T\mathbf{I}_n\mathbf{w} = \mathbf{v}^T\mathbf{A}^T\mathbf{A}\mathbf{w} = (\mathbf{Av})^T\mathbf{Aw} = \mathbf{Av} \cdot \mathbf{Aw}.$$

■

Exercises—Section 5.5

1. Which of the following sets of vectors are orthogonal? orthonormal?

★(a) $\{[3, -2], [4, 6]\}$

(b) $\left\{\left[-\frac{1}{\sqrt{5}}, \frac{2}{\sqrt{5}}\right], \left[\frac{2}{\sqrt{5}}, \frac{1}{\sqrt{5}}\right]\right\}$

★(c) $\left\{\left[\frac{3}{\sqrt{13}}, -\frac{2}{\sqrt{13}}\right], \left[\frac{1}{\sqrt{10}}, -\frac{3}{\sqrt{10}}\right]\right\}$

(d) $\left\{\left[\frac{1}{3}, \frac{2}{3}, \frac{2}{3}\right], \left[\frac{2}{3}, \frac{1}{3}, -\frac{2}{3}\right], \left[\frac{2}{3}, -\frac{2}{3}, \frac{1}{3}\right]\right\}$

(e) $\{[\frac{3}{5}, 0, -\frac{4}{5}]\}$

★(f) $\{[2, -3, 1, 2], [-1, 2, 8, 0], [6, -1, 1, -8]\}$

(g) $\{[\frac{1}{4}, \frac{1}{4}, \frac{1}{4}, -\frac{1}{2}, \frac{3}{4}], [\frac{1}{6}, \frac{1}{6}, -\frac{1}{2}, \frac{2}{3}, \frac{1}{2}]\}$

2. Which of the following matrices are orthogonal?

★(a) $\begin{bmatrix} \frac{\sqrt{3}}{2} & \frac{1}{2} \\ -\frac{1}{2} & \frac{\sqrt{3}}{2} \end{bmatrix}$ (b) $\begin{bmatrix} 3 & -2 \\ 2 & 3 \end{bmatrix}$ ★(c) $\begin{bmatrix} 3 & 0 & 10 \\ -1 & 3 & 3 \\ 3 & 1 & -9 \end{bmatrix}$

(d) $\begin{bmatrix} \frac{2}{15} & \frac{5}{15} & \frac{14}{15} \\ \frac{10}{15} & \frac{10}{15} & -\frac{5}{15} \\ \frac{11}{15} & -\frac{10}{15} & \frac{2}{15} \end{bmatrix}$ ★(e) $\begin{bmatrix} \frac{2}{3} & \frac{2}{3} & 0 & \frac{1}{3} \\ \frac{2}{3} & -\frac{2}{3} & -\frac{1}{3} & 0 \\ \frac{1}{3} & 0 & \frac{2}{3} & -\frac{2}{3} \\ 0 & \frac{1}{3} & -\frac{2}{3} & -\frac{2}{3} \end{bmatrix}$

3. In each of the following cases, verify that the given ordered basis B is orthonormal. Then, for the given vector $\mathbf{v}$, find $[\mathbf{v}]_B$, using the method of Theorem 5.26.

★(a) $\mathbf{v} = [-2, 3], B = \left(\left[-\frac{\sqrt{3}}{2}, \frac{1}{2}\right], \left[\frac{1}{2}, \frac{\sqrt{3}}{2}\right]\right)$

(b) $\mathbf{v} = [4, -1, 2], B = \left([\frac{3}{7}, -\frac{6}{7}, -\frac{2}{7}], [\frac{2}{7}, \frac{3}{7}, -\frac{6}{7}], [\frac{6}{7}, \frac{2}{7}, \frac{3}{7}]\right)$

★(c) $\mathbf{v} = [8, 4, -3, 5], B = \left([\frac{1}{2}, -\frac{1}{2}, \frac{1}{2}, \frac{1}{2}], \left[\frac{3}{2\sqrt{3}}, \frac{1}{2\sqrt{3}}, -\frac{1}{2\sqrt{3}}, -\frac{1}{2\sqrt{3}}\right],\right.$
$\left.\left[0, \frac{2}{\sqrt{6}}, \frac{1}{\sqrt{6}}, \frac{1}{\sqrt{6}}\right], \left[0, 0, -\frac{1}{\sqrt{2}}, \frac{1}{\sqrt{2}}\right]\right)$

4. Each of the following represents a basis for a subspace of $\mathbb{R}^n$, for some n. Use the Gram-Schmidt Process to find an orthogonal basis for the subspace.

★(a) $\{[5, -1, 2], [2, -1, -4]\}$ in $\mathbb{R}^3$

(b) $\{[2, -1, 3, 1], [-3, 0, -1, 4]\}$ in $\mathbb{R}^4$

★(c) $\{[2, 1, 0, -1], [1, 1, 1, -1], [1, -2, 1, 1]\}$ in $\mathbb{R}^4$

5. Enlarge each of the following orthogonal sets to an orthogonal basis for $\mathbb{R}^n$. (Avoid fractions by using appropriate scalar multiples.)

★(a) $\{[2, 2, -3]\}$ (b) $\{[1, -4, 3]\}$

★(c) $\{[1, -3, 1], [2, 5, 13]\}$ (d) $\{[3, 1, -2], [5, -3, 6]\}$

★(e) $\{[2, 1, -2, 1]\}$ (f) $\{[2, 1, 0, -3], [0, 3, 2, 1]\}$

6. (a) Show that if $\{\mathbf{v}_1, \ldots, \mathbf{v}_k\}$ is an orthogonal set in $\mathbb{R}^n$ and $c_1, \ldots, c_k$ are nonzero scalars, then $\{c_1\mathbf{v}_1, \ldots, c_k\mathbf{v}_k\}$ is also an orthogonal set.

★(b) Is part (a) still true if *orthogonal* is replaced by *orthonormal* everywhere?

7. Suppose that $\{\mathbf{u}_1, \ldots, \mathbf{u}_n\}$ is an orthonormal basis for $\mathbb{R}^n$.

(a) If $\mathbf{v}, \mathbf{w} \in \mathbb{R}^n$, show that

$$\mathbf{v} \cdot \mathbf{w} = (\mathbf{v} \cdot \mathbf{u}_1)(\mathbf{w} \cdot \mathbf{u}_1) + (\mathbf{v} \cdot \mathbf{u}_2)(\mathbf{w} \cdot \mathbf{u}_2) + \cdots + (\mathbf{v} \cdot \mathbf{u}_n)(\mathbf{w} \cdot \mathbf{u}_n).$$

(b) If $\mathbf{v} \in \mathbb{R}^n$, use part (a) to prove **Parseval's Equality**:

$$\|\mathbf{v}\|^2 = (\mathbf{v} \cdot \mathbf{u}_1)^2 + (\mathbf{v} \cdot \mathbf{u}_2)^2 + \cdots + (\mathbf{v} \cdot \mathbf{u}_n)^2.$$

(Parseval's Equality is a special case of Bessel's Inequality, which appears in Exercise 8.)

8. Let $\{\mathbf{u}_1, \ldots, \mathbf{u}_k\}$ be an orthonormal set of vectors in $\mathbb{R}^n$. For any vector $\mathbf{v} \in \mathbb{R}^n$, prove **Bessel's Inequality**:

$$|\mathbf{v} \cdot \mathbf{u}_1|^2 + \cdots + |\mathbf{v} \cdot \mathbf{u}_k|^2 \leq \|\mathbf{v}\|^2.$$

(Hint: Let $\mathcal{W}$ be the subspace spanned by $\{\mathbf{u}_1, \ldots, \mathbf{u}_k\}$. Enlarge $\{\mathbf{u}_1, \ldots, \mathbf{u}_k\}$ to an orthonormal basis for $\mathbb{R}^n$. Then use Theorem 5.26.)

9. Prove Theorem 5.29.

10. Let $\mathbf{A}$ be an $n \times n$ matrix with $\mathbf{A}^2 = \mathbf{I}_n$. Prove that $\mathbf{A}$ is symmetric if and only if $\mathbf{A}$ is orthogonal.

11. Show that no orthogonal $n \times n$ matrix $\mathbf{A}$ is skew-symmetric when n is odd. (Hint: Suppose $\mathbf{A}$ is both orthogonal and skew-symmetric. Show that $\mathbf{A}^2 = -\mathbf{I}_n$, and then use determinants.)

12. If $\mathbf{A}$ is an $n \times n$ orthogonal matrix with $|\mathbf{A}| = -1$, show that $\mathbf{A} + \mathbf{I}_n$ has no inverse. (Hint: Show that $\mathbf{A} + \mathbf{I}_n = \mathbf{A}(\mathbf{A} + \mathbf{I}_n)^T$, and then use determinants.)

13. Suppose that $\mathbf{A}$ is a 3×3 upper triangular orthogonal matrix. Show that $\mathbf{A}$ is diagonal and that all main diagonal entries of $\mathbf{A}$ equal ± 1. (Note: This result is true for any $n \times n$ upper triangular orthogonal matrix.)

14. (a) If $\mathbf{u}$ is any unit vector in $\mathbb{R}^n$, explain why there exists an $n \times n$ orthogonal matrix with $\mathbf{u}$ as its first row. (Hint: Consider Theorem 5.28.)

★(b) Find an orthogonal matrix whose first row is $\frac{1}{\sqrt{6}}[1, 2, 1]$.

15. Finish the proof of Theorem 5.30.

16. Suppose that $\mathbf{A}$ is an $n \times n$ orthogonal matrix.

(a) Prove that for every $\mathbf{v} \in \mathbb{R}^n$, $\|\mathbf{v}\| = \|\mathbf{A}\mathbf{v}\|$.

(b) Prove that for all $\mathbf{v}, \mathbf{w} \in \mathbb{R}^n$, the angle between $\mathbf{v}$ and $\mathbf{w}$ equals the angle between $\mathbf{A}\mathbf{v}$ and $\mathbf{A}\mathbf{w}$.

17. Let B be an ordered orthonormal basis for a k-dimensional subspace $\mathcal{V}$ of $\mathbb{R}^n$. Prove that for all $\mathbf{v}_1, \mathbf{v}_2 \in \mathcal{V}$, $\mathbf{v}_1 \cdot \mathbf{v}_2 = [\mathbf{v}_1]_B \cdot [\mathbf{v}_2]_B$, where the first dot product takes place in $\mathbb{R}^n$ and the second takes place in $\mathbb{R}^k$.

5.6 Orthogonal Complements

As in Section 5.5, we deal here only with vectors in $\mathbb{R}^n$. For each subspace $\mathcal{W}$ of $\mathbb{R}^n$, there is a corresponding subspace consisting of the vectors that are orthogonal to all vectors in $\mathcal{W}$, called the orthogonal complement of $\mathcal{W}$. In this section we study many elementary properties of orthogonal complements and then investigate the orthogonal projection of a vector onto a subspace of $\mathbb{R}^n$.

Orthogonal Complements

DEFINITION

Let $\mathcal{W}$ be a subspace of $\mathbb{R}^n$. Then the **orthogonal complement**, $\mathcal{W}^\perp$, of $\mathcal{W}$ in $\mathbb{R}^n$, is the set of all vectors $\mathbf{x} \in \mathbb{R}^n$ with the property that $\mathbf{x} \cdot \mathbf{w} = 0$, for all $\mathbf{w} \in \mathcal{W}$. That is, $\mathcal{W}^\perp$ contains those vectors of $\mathbb{R}^n$ orthogonal to every vector in $\mathcal{W}$.

The proof of the next theorem is easy and is left for you to do in Exercise 13.

THEOREM 5.33

If $\mathcal{W}$ is a subspace of $\mathbb{R}^n$, then a vector $\mathbf{v}$ is in $\mathcal{W}^\perp$ if and only if $\mathbf{v}$ is orthogonal to every vector in a spanning set for $\mathcal{W}$.

Example 1

Consider the subspace $\mathcal{W} = \{[a, b, 0] \mid a, b \in \mathbb{R}\}$ of $\mathbb{R}^3$. Now, $\mathcal{W}$ is spanned by $\{[1, 0, 0], [0, 1, 0]\}$. By Theorem 5.33, a vector $[x, y, z]$ is in $\mathcal{W}^\perp$, the orthogonal complement of $\mathcal{W}$, if and only if it is orthogonal to both $[1, 0, 0]$ and $[0, 1, 0]$ (why?)—that is, if and only if $x = y = 0$. Hence, $\mathcal{W}^\perp = \{[0, 0, z] \mid z \in \mathbb{R}\}$. Notice that $\mathcal{W}^\perp$ is a subspace of $\mathbb{R}^3$ of dimension 1 and that $\dim(\mathcal{W}) + \dim(\mathcal{W}^\perp) = \dim(\mathbb{R}^3)$. ■

Example 2

Consider the subspace $\mathcal{W} = \{a[-3, 2, 4] \mid a \in \mathbb{R}\}$ of $\mathbb{R}^3$. Since $[-3, 2, 4]$ spans $\mathcal{W}$, the orthogonal complement $\mathcal{W}^\perp$ of $\mathcal{W}$ is the set of all vectors $[x, y, z]$ in $\mathbb{R}^3$ such that $[x, y, z] \cdot [-3, 2, 4] = 0$. That is, $\mathcal{W}^\perp$ is precisely the set of all vectors $[x, y, z]$ lying in the plane $-3x + 2y + 4z = 0$. Notice that $\mathcal{W}^\perp$ is a subspace of $\mathbb{R}^3$ of dimension 2 and that $\dim(\mathcal{W}) + \dim(\mathcal{W}^\perp) = \dim(\mathbb{R}^3)$. ■

Example 3

The orthogonal complement of $\mathbb{R}^n$ itself is just the trivial subspace $\{\mathbf{0}\}$, since $\mathbf{0}$ is the only vector orthogonal to each of $\mathbf{e}_1, \mathbf{e}_2, \ldots, \mathbf{e}_n \in \mathbb{R}^n$. (Why?)

Conversely, the orthogonal complement of the trivial subspace in $\mathbb{R}^n$ is all of $\mathbb{R}^n$, because every vector in $\mathbb{R}^n$ is orthogonal to the zero vector.

Hence, in $\mathbb{R}^n$, $\{\mathbf{0}\}$ and $\mathbb{R}^n$ itself are orthogonal complements of each other. Notice that the dimensions of these two subspaces add up to $\dim(\mathbb{R}^n)$. ■

Properties of Orthogonal Complements

Examples 1, 2, and 3 suggest that the orthogonal complement $\mathcal{W}^\perp$ of a subspace $\mathcal{W}$ is always a subspace and that $\dim(\mathcal{W}^\perp)$ is related to $\dim(\mathcal{W})$. These results are proved in Theorem 5.34 and Corollary 5.36, which follow. Also, Example 3 suggests that $\mathcal{W}$ and $\mathcal{W}^\perp$ are orthogonal complements of each other; that is, $(\mathcal{W}^\perp)^\perp = \mathcal{W}$. This statement is proved in Corollary 5.37.

THEOREM 5.34

Let $\mathcal{W}$ be a subspace of $\mathbb{R}^n$. Then $\mathcal{W}^\perp$ is a subspace of $\mathbb{R}^n$, and $\mathcal{W} \cap \mathcal{W}^\perp = \{\mathbf{0}\}$.

Proof of Theorem 5.34

$\mathcal{W}^\perp$ is nonempty, because $\mathbf{0} \in \mathcal{W}^\perp$. (Why?) Thus, to show that $\mathcal{W}^\perp$ is a subspace, we need only verify the closure properties for $\mathcal{W}^\perp$.

Suppose, then, that $\mathbf{x}_1, \mathbf{x}_2 \in \mathcal{W}^\perp$. We want to show that $\mathbf{x}_1 + \mathbf{x}_2 \in \mathcal{W}^\perp$. However, for all $\mathbf{w} \in \mathcal{W}$, $(\mathbf{x}_1 + \mathbf{x}_2) \cdot \mathbf{w} = (\mathbf{x}_1 \cdot \mathbf{w}) + (\mathbf{x}_2 \cdot \mathbf{w}) = 0 + 0 = 0$, since $\mathbf{x}_1, \mathbf{x}_2 \in \mathcal{W}^\perp$. Hence, $\mathbf{x}_1 + \mathbf{x}_2 \in \mathcal{W}^\perp$. Next, suppose that $\mathbf{x} \in \mathcal{W}^\perp$ and $c \in \mathbb{R}$. We want to show that $c\mathbf{x} \in \mathcal{W}^\perp$. However, for all $\mathbf{w} \in \mathcal{W}$, $(c\mathbf{x}) \cdot \mathbf{w} = c(\mathbf{x} \cdot \mathbf{w}) = c(0) = 0$, since $\mathbf{x} \in \mathcal{W}^\perp$. Hence, $c\mathbf{x} \in \mathcal{W}^\perp$. Thus, $\mathcal{W}^\perp$ is a subspace of $\mathbb{R}^n$.

Finally, suppose that $\mathbf{w} \in \mathcal{W} \cap \mathcal{W}^\perp$. Then $\mathbf{w} \in \mathcal{W}$ and $\mathbf{w} \in \mathcal{W}^\perp$, so $\mathbf{w}$ is orthogonal to itself. Hence, $\mathbf{w} \cdot \mathbf{w} = 0$, and so $\mathbf{w} = \mathbf{0}$. ■

The next theorem shows how we can obtain an orthogonal basis for $\mathcal{W}^\perp$.

THEOREM 5.35

Let $\mathcal{W}$ be a subspace of $\mathbb{R}^n$. Let $\{\mathbf{v}_1, \ldots, \mathbf{v}_k\}$ be an orthogonal basis for $\mathcal{W}$ contained in an orthogonal basis $\{\mathbf{v}_1, \ldots, \mathbf{v}_k, \mathbf{v}_{k+1}, \ldots, \mathbf{v}_n\}$ for $\mathbb{R}^n$. Then $\{\mathbf{v}_{k+1}, \ldots, \mathbf{v}_n\}$ is an orthogonal basis for $\mathcal{W}^\perp$.

Proof of Theorem 5.35

Let $\{\mathbf{v}_1, \ldots, \mathbf{v}_n\}$ be an orthogonal basis for $\mathbb{R}^n$, with $\mathcal{W} = \text{span}(\{\mathbf{v}_1, \ldots, \mathbf{v}_k\})$. Let $\mathcal{X} = \text{span}(\{\mathbf{v}_{k+1}, \ldots, \mathbf{v}_n\})$. We want to prove that $\mathcal{X} = \mathcal{W}^\perp$. We will show that $\mathcal{X} \subseteq \mathcal{W}^\perp$ and $\mathcal{W}^\perp \subseteq \mathcal{X}$.

To show $\mathcal{X} \subseteq \mathcal{W}^\perp$, we must prove that any vector $\mathbf{x}$ of the form $d_{k+1}\mathbf{v}_{k+1} + \cdots + d_n\mathbf{v}_n$ (for some scalars $d_{k+1}, \ldots, d_n$) is orthogonal to every vector

$\mathbf{w} \in \mathcal{W}$. Now if $\mathbf{w} \in \mathcal{W}$, then $\mathbf{w} = c_1\mathbf{v}_1 + \cdots + c_k\mathbf{v}_k$, for some scalars $c_1, \ldots, c_k$. Hence,

$$\mathbf{x} \cdot \mathbf{w} = (d_{k+1}\mathbf{v}_{k+1} + \cdots + d_n\mathbf{v}_n) \cdot (c_1\mathbf{v}_1 + \cdots + c_k\mathbf{v}_k),$$

which equals zero when expanded because all vectors in $\{\mathbf{v}_{k+1}, \ldots, \mathbf{v}_n\}$ are orthogonal to all vectors in $\{\mathbf{v}_1, \ldots, \mathbf{v}_k\}$. Hence, $\mathbf{x} \in \mathcal{W}^\perp$, and so $\mathcal{X} \subseteq \mathcal{W}^\perp$.

To show $\mathcal{W}^\perp \subseteq \mathcal{X}$, we must show that any vector $\mathbf{x}$ in $\mathcal{W}^\perp$ is also in $\text{span}(\{\mathbf{v}_{k+1}, \ldots, \mathbf{v}_n\})$. Let $\mathbf{x} \in \mathcal{W}^\perp$. Since $\{\mathbf{v}_1, \ldots, \mathbf{v}_n\}$ is an orthogonal basis for $\mathbb{R}^n$, Theorem 5.26 tells us that

$$\mathbf{x} = \frac{(\mathbf{x} \cdot \mathbf{v}_1)}{\|\mathbf{v}_1\|^2}\mathbf{v}_1 + \cdots + \frac{(\mathbf{x} \cdot \mathbf{v}_k)}{\|\mathbf{v}_k\|^2}\mathbf{v}_k + \frac{(\mathbf{x} \cdot \mathbf{v}_{k+1})}{\|\mathbf{v}_{k+1}\|^2}\mathbf{v}_{k+1} + \cdots + \frac{(\mathbf{x} \cdot \mathbf{v}_n)}{\|\mathbf{v}_n\|^2}\mathbf{v}_n.$$

However, since each of $\mathbf{v}_1, \ldots, \mathbf{v}_k$ is in $\mathcal{W}$, we know that $\mathbf{x} \cdot \mathbf{v}_1 = \cdots = \mathbf{x} \cdot \mathbf{v}_k = 0$. Hence,

$$\mathbf{x} = \frac{(\mathbf{x} \cdot \mathbf{v}_{k+1})}{\|\mathbf{v}_{k+1}\|^2}\mathbf{v}_{k+1} + \cdots + \frac{(\mathbf{x} \cdot \mathbf{v}_n)}{\|\mathbf{v}_n\|^2}\mathbf{v}_n,$$

and so $\mathbf{x} \in \text{span}(\{\mathbf{v}_{k+1}, \ldots, \mathbf{v}_n\})$. Thus, $\mathcal{W}^\perp \subseteq \mathcal{X}$. ■

Example 4

Consider the subspace $\mathcal{W} = \text{span}(\{[2, -1, 0, 1], [-1, 3, 1, -1]\})$ of $\mathbb{R}^4$. We want to find an orthogonal basis for $\mathcal{W}^\perp$. We start by finding an orthogonal basis for $\mathcal{W}$.

Let $\mathbf{w}_1 = [2, -1, 0, 1]$ and $\mathbf{w}_2 = [-1, 3, 1, -1]$. Performing the Gram-Schmidt Process yields $\mathbf{v}_1 = \mathbf{w}_1 = [2, -1, 0, 1]$ and $\mathbf{v}_2 = \mathbf{w}_2 - ((\mathbf{w}_2 \cdot \mathbf{v}_1)/(\mathbf{v}_1 \cdot \mathbf{v}_1))\mathbf{v}_1 = [1, 2, 1, 0]$. Hence, $\{\mathbf{v}_1, \mathbf{v}_2\} = \{[2, -1, 0, 1], [1, 2, 1, 0]\}$ is an orthogonal basis for $\mathcal{W}$.

We now expand this basis for $\mathcal{W}$ to a basis for all of $\mathbb{R}^4$. Continuing the Gram-Schmidt Process with $\mathbf{w}_3 = [1, 0, 0, 0]$ and $\mathbf{w}_4 = [0, 1, 0, 0]$ produces $\mathbf{v}_3 = [1, 0, -1, -2]$ and $\mathbf{v}_4 = [0, 1, -2, 1]$. (Verify.) Then $\{\mathbf{v}_1, \mathbf{v}_2, \mathbf{v}_3, \mathbf{v}_4\}$ is an orthogonal basis for $\mathbb{R}^4$. Since $\{\mathbf{v}_1, \mathbf{v}_2\}$ is an orthogonal basis for $\mathcal{W}$, Theorem 5.35 tells us that $\{\mathbf{v}_3, \mathbf{v}_4\} = \{[1, 0, -1, -2], [0, 1, -2, 1]\}$ is an orthogonal basis for $\mathcal{W}^\perp$. ■

An important corollary of Theorem 5.35 is the following:

COROLLARY 5.36

Let $\mathcal{W}$ be a subspace of $\mathbb{R}^n$. Then $\dim(\mathcal{W}) + \dim(\mathcal{W}^\perp) = n = \dim(\mathbb{R}^n)$.

Proof of Corollary 5.36

Let $\mathcal{W}$ be a subspace of $\mathbb{R}^n$ of dimension k. By Theorem 5.27, $\mathcal{W}$ has an orthogonal basis $\{\mathbf{v}_1, \ldots, \mathbf{v}_k\}$. By Theorem 5.28, we can expand this basis for $\mathcal{W}$ to an orthogonal basis $\{\mathbf{v}_1, \ldots, \mathbf{v}_k, \mathbf{v}_{k+1}, \ldots, \mathbf{v}_n\}$ for all of $\mathbb{R}^n$. Then, by Theorem 5.35, $\{\mathbf{v}_{k+1}, \ldots, \mathbf{v}_n\}$ is a basis for $\mathcal{W}^{\perp}$, and so $\dim(\mathcal{W}^{\perp}) = n - k$. Hence, $\dim(\mathcal{W}) + \dim(\mathcal{W}^{\perp}) = n$. ■

Example 5

If $\mathcal{W}$ is a one-dimensional subspace of $\mathbb{R}^n$, then $\mathcal{W}$ is the set of all multiples of a single nonzero vector. Corollary 5.36 asserts that $\dim(\mathcal{W}^{\perp}) = n - 1$. For example, in $\mathbb{R}^2$, the orthogonal complement of the one-dimensional subspace $\mathcal{W} = \text{span}(\{[a, b]\})$, where $[a, b] \neq \mathbf{0}$, is also one-dimensional. It is easy to see that $\mathcal{W}^{\perp} = \text{span}(\{[b, -a]\})$ (see Figure 5.13(a)). That is, $\mathcal{W}^{\perp}$ is the set of all vectors on the line through the origin perpendicular to $[a, b]$.

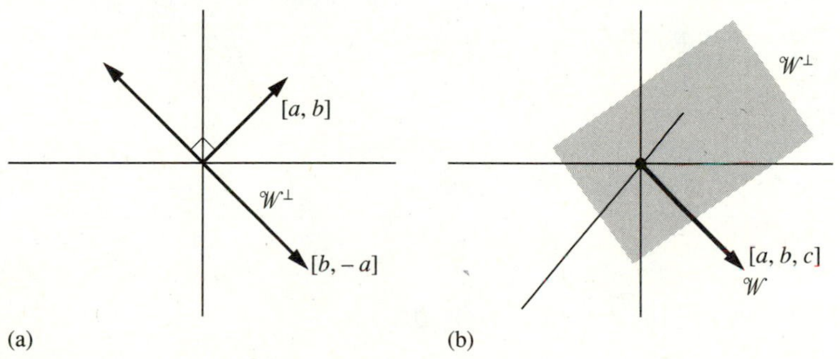

Figure 5.13
(a) The orthogonal complement of $\mathcal{W} = \text{span}(\{[a, b]\})$ in $\mathbb{R}^2$, a one-dimensional line through the origin perpendicular to $[a, b]$, when $[a, b] \neq \mathbf{0}$; (b) The orthogonal complement of $\mathcal{W} = \text{span}(\{[a, b, c]\})$ in $\mathbb{R}^3$, a two-dimensional plane through the origin perpendicular to $[a, b, c]$, when $[a, b, c] \neq \mathbf{0}$.

In $\mathbb{R}^3$, the orthogonal complement of the one-dimensional subspace $\mathcal{W} = \text{span}(\{[a, b, c]\})$, where $[a, b, c] \neq [0, 0, 0]$, is two-dimensional. A little thought will convince you that $\mathcal{W}^{\perp}$ is the plane through the origin perpendicular to $[a, b, c]$—that is, the plane $ax + by + cz = 0$ (see Figure 5.13(b)). ■

If $\mathcal{W}$ is a subspace of $\mathbb{R}^n$, Corollary 5.36 indicates that the dimensions of $\mathcal{W}$ and $\mathcal{W}^{\perp}$ add up to n. For this reason, many students get the mistaken impression that every vector in $\mathbb{R}^n$ either lies in $\mathcal{W}$ or in $\mathcal{W}^{\perp}$. But $\mathcal{W}$ and $\mathcal{W}^{\perp}$ are not "setwise" complements of each other; a more accurate depiction is given in Figure 5.14. For example, recall the subspace $\mathcal{W} = \{[a, b, 0] \mid a, b \in \mathbb{R}\}$ of Example 1. We showed that $\mathcal{W}^{\perp} = \{[0, 0, z] \mid z \in \mathbb{R}\}$. Yet $[1, 1, 1]$ is in neither $\mathcal{W}$ nor $\mathcal{W}^{\perp}$, even though $\dim(\mathcal{W}) + \dim(\mathcal{W}^{\perp}) = \dim(\mathbb{R}^3)$.

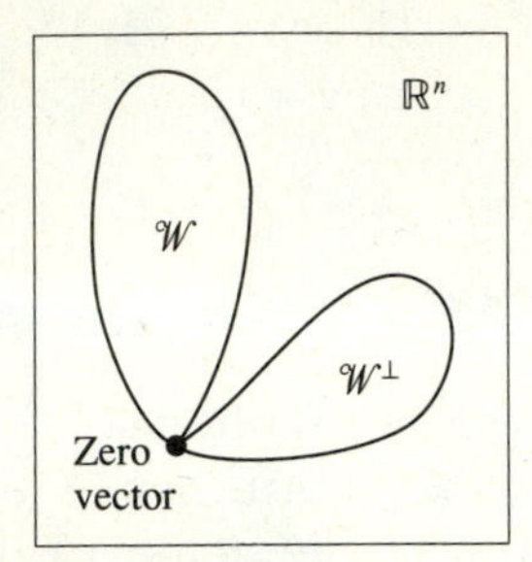

Figure 5.14
Symbolic depiction of $\mathcal{W}$ and $\mathcal{W}^\perp$

The next corollary asserts that each subspace $\mathcal{W}$ of $\mathbb{R}^n$ is, in fact, the orthogonal complement of $\mathcal{W}^\perp$. Hence, $\mathcal{W}$ and $\mathcal{W}^\perp$ are orthogonal complements of each other. You are asked to prove this result in Exercise 14.

COROLLARY 5.37

Let $\mathcal{W}$ be a subspace of $\mathbb{R}^n$. Then $(\mathcal{W}^\perp)^\perp = \mathcal{W}$.

Orthogonal Projection onto a Subspace

Next we present the Projection Theorem, a generalization of Theorem 1.9. Recall that Theorem 1.9 states that every nonzero vector in $\mathbb{R}^n$ can be decomposed into the sum of two vectors, one parallel to a given vector $\mathbf{a}$ and another orthogonal to $\mathbf{a}$.

THEOREM 5.38: Projection Theorem

Let $\mathcal{W}$ be a subspace of $\mathbb{R}^n$. Then every vector $\mathbf{v} \in \mathbb{R}^n$ can be expressed in a unique way as $\mathbf{w}_1 + \mathbf{w}_2$, where $\mathbf{w}_1 \in \mathcal{W}$ and $\mathbf{w}_2 \in \mathcal{W}^\perp$.

Proof of Theorem 5.38

Let $\mathcal{W}$ be a subspace of $\mathbb{R}^n$, and let $\mathbf{v} \in \mathbb{R}^n$. We first show that $\mathbf{v}$ can be expressed as $\mathbf{w}_1 + \mathbf{w}_2$, where $\mathbf{w}_1 \in \mathcal{W}$, $\mathbf{w}_2 \in \mathcal{W}^\perp$. Then we will show that there is a unique pair $\mathbf{w}_1, \mathbf{w}_2$ for each $\mathbf{v}$.

Let $\{\mathbf{u}_1, \ldots, \mathbf{u}_k\}$ be an orthonormal basis for $\mathcal{W}$. Expand $\{\mathbf{u}_1, \ldots, \mathbf{u}_k\}$ to an orthonormal basis $\{\mathbf{u}_1, \ldots, \mathbf{u}_k, \mathbf{u}_{k+1}, \ldots, \mathbf{u}_n\}$ for $\mathbb{R}^n$. Then, by Theorem 5.26, $\mathbf{v} = (\mathbf{v} \cdot \mathbf{u}_1)\mathbf{u}_1 + \cdots + (\mathbf{v} \cdot \mathbf{u}_n)\mathbf{u}_n$. Let $\mathbf{w}_1 = (\mathbf{v} \cdot \mathbf{u}_1)\mathbf{u}_1 + \cdots + (\mathbf{v} \cdot \mathbf{u}_k)\mathbf{u}_k$ and

$\mathbf{w}_2 = (\mathbf{v} \cdot \mathbf{u}_{k+1})\mathbf{u}_{k+1} + \cdots + (\mathbf{v} \cdot \mathbf{u}_n)\mathbf{u}_n$. Clearly, $\mathbf{v} = \mathbf{w}_1 + \mathbf{w}_2$. Also, Theorem 5.35 implies that $\mathbf{w}_1 \in \mathcal{W}$ and $\mathbf{w}_2$ is in $\mathcal{W}^\perp$.

Finally, we want to show uniqueness of decomposition. Suppose that $\mathbf{v} = \mathbf{w}_1 + \mathbf{w}_2$ and $\mathbf{v} = \mathbf{w}_1' + \mathbf{w}_2'$, where $\mathbf{w}_1, \mathbf{w}_1' \in \mathcal{W}$ and $\mathbf{w}_2, \mathbf{w}_2' \in \mathcal{W}^\perp$. We want to show that $\mathbf{w}_1 = \mathbf{w}_1'$ and $\mathbf{w}_2 = \mathbf{w}_2'$. Now, $\mathbf{w}_1 - \mathbf{w}_1' = \mathbf{w}_2' - \mathbf{w}_2$. (Why?) Also, $\mathbf{w}_1 - \mathbf{w}_1' \in \mathcal{W}$, but $\mathbf{w}_2' - \mathbf{w}_2 \in \mathcal{W}^\perp$. Thus, $\mathbf{w}_1 - \mathbf{w}_1' = \mathbf{w}_2' - \mathbf{w}_2 \in \mathcal{W} \cap \mathcal{W}^\perp$. By Theorem 5.34, $\mathbf{w}_1 - \mathbf{w}_1' = \mathbf{w}_2' - \mathbf{w}_2 = \mathbf{0}$. Hence, $\mathbf{w}_1 = \mathbf{w}_1'$ and $\mathbf{w}_2 = \mathbf{w}_2'$. ■

We give a special name to the vector $\mathbf{w}_1$ in the proof of Theorem 5.38:

DEFINITION

Let $\mathcal{W}$ be a subspace of $\mathbb{R}^n$ with orthonormal basis $\{\mathbf{u}_1, \ldots, \mathbf{u}_k\}$, and let $\mathbf{v} \in \mathbb{R}^n$. Then, $\mathbf{proj}_{\mathcal{W}}\mathbf{v}$, the **orthogonal projection** of $\mathbf{v}$ onto $\mathcal{W}$, is the vector $(\mathbf{v} \cdot \mathbf{u}_1)\mathbf{u}_1 + \cdots + (\mathbf{v} \cdot \mathbf{u}_k)\mathbf{u}_k$.

Notice that the choice of orthonormal basis for $\mathcal{W}$ in this definition does not matter, because Theorem 5.38 guarantees the uniqueness of $\mathbf{w}_1 = \mathbf{proj}_{\mathcal{W}}\mathbf{v}$. That is, if $\{\mathbf{z}_1, \ldots, \mathbf{z}_k\}$ is any other orthonormal basis for $\mathcal{W}$, then the same vector $\mathbf{proj}_{\mathcal{W}}\mathbf{v}$ is obtained from $(\mathbf{v} \cdot \mathbf{z}_1)\mathbf{z}_1 + \cdots + (\mathbf{v} \cdot \mathbf{z}_k)\mathbf{z}_k$. This fact is illustrated in the next example.

Example 6

You can easily check that

$$\mathbf{A} = \begin{bmatrix} 8 & -1 & -4 \\ 4 & 4 & 7 \end{bmatrix} \quad \text{and} \quad \mathbf{B} = \begin{bmatrix} 4 & 1 & 1 \\ 4 & -5 & -11 \end{bmatrix}$$

both have reduced row echelon form $\begin{bmatrix} 1 & 0 & -\frac{1}{4} \\ 0 & 1 & 2 \end{bmatrix}$. Hence, the row spaces of $\mathbf{A}$ and $\mathbf{B}$ are equal. Let $\mathcal{W}$ be this common row space. Also notice that the rows of $\mathbf{A}$ are orthogonal and that the rows of $\mathbf{B}$ are orthogonal. Hence, after normalizing, we obtain these orthonormal bases for $\mathcal{W}$:

$$\{\mathbf{u}_1, \mathbf{u}_2\} = \left\{\left[\tfrac{8}{9}, -\tfrac{1}{9}, -\tfrac{4}{9}\right], \left[\tfrac{4}{9}, \tfrac{4}{9}, \tfrac{7}{9}\right]\right\}$$

$$\text{and} \quad \{\mathbf{z}_1, \mathbf{z}_2\} = \left\{\left[\frac{4}{3\sqrt{2}}, \frac{1}{3\sqrt{2}}, \frac{1}{3\sqrt{2}}\right], \left[\frac{4}{9\sqrt{2}}, -\frac{5}{9\sqrt{2}}, -\frac{11}{9\sqrt{2}}\right]\right\}.$$

Let $\mathbf{v} = [1, 2, 3]$. We will verify that the same vector for $\mathbf{proj}_{\mathcal{W}}\mathbf{v}$ is obtained if either $\{\mathbf{u}_1, \mathbf{u}_2\}$ or $\{\mathbf{z}_1, \mathbf{z}_2\}$ is used as the orthonormal basis for $\mathcal{W}$. Now,

$$(\mathbf{v} \cdot \mathbf{u}_1)\mathbf{u}_1 + (\mathbf{v} \cdot \mathbf{u}_2)\mathbf{u}_2 = -\tfrac{2}{3}\left[\tfrac{8}{9}, -\tfrac{1}{9}, -\tfrac{4}{9}\right] + \tfrac{11}{3}\left[\tfrac{4}{9}, \tfrac{4}{9}, \tfrac{7}{9}\right] = \left[\tfrac{28}{27}, \tfrac{46}{27}, \tfrac{85}{27}\right].$$

Similarly,

$$
\begin{aligned}
(\mathbf{v} \cdot \mathbf{z}_1)\mathbf{z}_1 + (\mathbf{v} \cdot \mathbf{z}_2)\mathbf{z}_2 &= \frac{3}{\sqrt{2}}\left[\frac{4}{3\sqrt{2}}, \frac{1}{3\sqrt{2}}, \frac{1}{3\sqrt{2}}\right] + \left(-\frac{13}{3\sqrt{2}}\right)\left[\frac{4}{9\sqrt{2}}, -\frac{5}{9\sqrt{2}}, -\frac{11}{9\sqrt{2}}\right] \\
&= \left[\frac{28}{27}, \frac{46}{27}, \frac{85}{27}\right].
\end{aligned}
$$

Hence, with either orthonormal basis we obtain $\mathbf{proj}_{\mathcal{W}}\mathbf{v} = \left[\frac{28}{27}, \frac{46}{27}, \frac{85}{27}\right]$. ■

The proof of Theorem 5.38 illustrates the following:

> If $\mathcal{W}$ is a subspace of $\mathbb{R}^n$ and $\mathbf{v} \in \mathbb{R}^n$, then $\mathbf{v}$ can be uniquely expressed as $\mathbf{w}_1 + \mathbf{w}_2$, where $\mathbf{w}_1 = \mathbf{proj}_{\mathcal{W}}\mathbf{v} \in \mathcal{W}$ and $\mathbf{w}_2 = \mathbf{v} - \mathbf{proj}_{\mathcal{W}}\mathbf{v} \in \mathcal{W}^{\perp}$. Moreover, $\mathbf{w}_2$ can also be expressed as $\mathbf{proj}_{\mathcal{W}^{\perp}}\mathbf{v}$.

The vector $\mathbf{w}_1$ is the generalization of the projection vector $\mathbf{proj}_{\mathbf{a}}\mathbf{b}$ from Section 1.2 (see Exercise 12).

Example 7

Let $\mathcal{W}$ be the subspace of $\mathbb{R}^3$ whose vectors (beginning at the origin) lie in the plane $\mathcal{L}$ with equation $2x + y + z = 0$. Let $\mathbf{v} = [-6, 10, 5]$. (Notice that $\mathbf{v} \notin \mathcal{W}$.) We will find $\mathbf{proj}_{\mathcal{W}}\mathbf{v}$.

First, notice that $[-1, 0, 2]$ and $[-2, 3, 1]$ are two linearly independent vectors in $\mathcal{W}$. By using the Gram-Schmidt Process on these vectors, we obtain the orthogonal basis $\{[-1, 0, 2], [-2, 5, -1]\}$ for $\mathcal{W}$. (Verify.) After normalizing, we have the orthonormal basis $\{\mathbf{u}_1, \mathbf{u}_2\}$ for $\mathcal{W}$, where

$$
\mathbf{u}_1 = \left[-\frac{1}{\sqrt{5}}, 0, \frac{2}{\sqrt{5}}\right] \quad \text{and} \quad \mathbf{u}_2 = \left[-\frac{2}{\sqrt{30}}, \frac{5}{\sqrt{30}}, -\frac{1}{\sqrt{30}}\right].
$$

Now,

$$
\begin{aligned}
\mathbf{proj}_{\mathcal{W}}\mathbf{v} &= (\mathbf{v} \cdot \mathbf{u}_1)\mathbf{u}_1 + (\mathbf{v} \cdot \mathbf{u}_2)\mathbf{u}_2 \\
&= \frac{16}{\sqrt{5}}\left[-\frac{1}{\sqrt{5}}, 0, \frac{2}{\sqrt{5}}\right] + \frac{57}{\sqrt{30}}\left[-\frac{2}{\sqrt{30}}, \frac{5}{\sqrt{30}}, -\frac{1}{\sqrt{30}}\right] \\
&= \left[-\frac{16}{5}, 0, \frac{32}{5}\right] + \left[-\frac{114}{30}, \frac{285}{30}, -\frac{57}{30}\right] \\
&= \left[-7, \frac{19}{2}, \frac{9}{2}\right].
\end{aligned}
$$

Notice that this vector is in $\mathcal{W}$. Finally, $\mathbf{v} - \mathbf{proj}_{\mathcal{W}}\mathbf{v} = [1, \frac{1}{2}, \frac{1}{2}]$, which is indeed in $\mathcal{W}^{\perp}$ because it is orthogonal to both $\mathbf{u}_1$ and $\mathbf{u}_2$. (Verify.) Hence, we have decomposed

$\mathbf{v} = [-6, 10, 5]$ as the sum of two vectors $[-7, \frac{19}{2}, \frac{9}{2}]$ and $[1, \frac{1}{2}, \frac{1}{2}]$, where the first is in $\mathcal{W}$ and the second is in $\mathcal{W}^{\perp}$. ■

You can think of the orthogonal projection vector $\mathbf{proj}_{\mathcal{W}}\mathbf{v}$ in Example 7 as the "shadow" that $\mathbf{v}$ casts on the plane $\mathcal{L}$ as light falls directly onto $\mathcal{L}$ from a light source above and parallel to $\mathcal{L}$. This concept is illustrated in Figure 5.15.

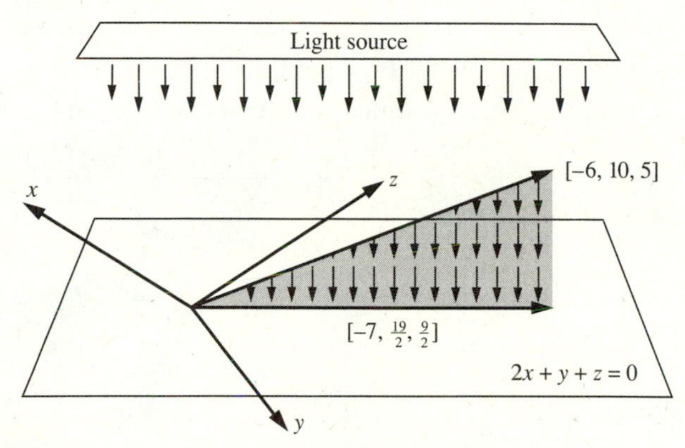

Figure 5.15
The orthogonal projection vector $[-7, \frac{19}{2}, \frac{9}{2}]$ of $[-6, 10, 5]$ onto the plane $2x + y + z = 0$, pictured as a shadow cast by $\mathbf{v}$ from a light source above and parallel to the plane

There are two special cases of the Projection Theorem. First, if $\mathbf{v} \in \mathcal{W}$, then $\mathbf{proj}_{\mathcal{W}}\mathbf{v}$ simply equals $\mathbf{v}$ itself. Also, if $\mathbf{v} \in \mathcal{W}^{\perp}$, then $\mathbf{proj}_{\mathcal{W}}\mathbf{v}$ equals $\mathbf{0}$. You are asked to verify these results in Exercise 8.

The next theorem assures us that orthogonal projection onto a subspace of $\mathbb{R}^n$ is a linear operator on $\mathbb{R}^n$. The proof is easy (see Exercise 15).

THEOREM 5.39

Let $\mathcal{W}$ be a subspace of $\mathbb{R}^n$. Then the mapping $L: \mathbb{R}^n \longrightarrow \mathbb{R}^n$ given by $L(\mathbf{v}) = \mathbf{proj}_{\mathcal{W}}\mathbf{v}$ is a linear operator with $\ker(L) = \mathcal{W}^{\perp}$.

Application: Distance from a Point to a Subspace

DEFINITION

Let $\mathcal{W}$ be a subspace of $\mathbb{R}^n$, and let P be any point in n-dimensional space. Then the **minimum distance** from P to $\mathcal{W}$ is the smallest distance between P and the terminal point of any vector in $\mathcal{W}$.

Suppose that $\mathcal{W}$ is a subspace of $\mathbb{R}^n$ and P is a point in n-dimensional space. Consider the vector $\mathbf{v}$ from the origin to P and the vector $\mathbf{proj}_{\mathcal{W}}\mathbf{v}$. Let S be the terminal point of $\mathbf{proj}_{\mathcal{W}}\mathbf{v}$. Then the distance from P to S is $\|\mathbf{v} - \mathbf{proj}_{\mathcal{W}}\mathbf{v}\|$. The next theorem asserts that this is the smallest distance from P to $\mathcal{W}$ (see Figure 5.16). You are asked to prove this theorem in Exercise 18.

THEOREM 5.40

> Let $\mathcal{W}$ be a subspace of $\mathbb{R}^n$, and let P be a point in n-dimensional space. If $\mathbf{v}$ is the vector from the origin to P, then the minimum distance from P to $\mathcal{W}$ is $\|\mathbf{v} - \mathbf{proj}_{\mathcal{W}}\mathbf{v}\|$.

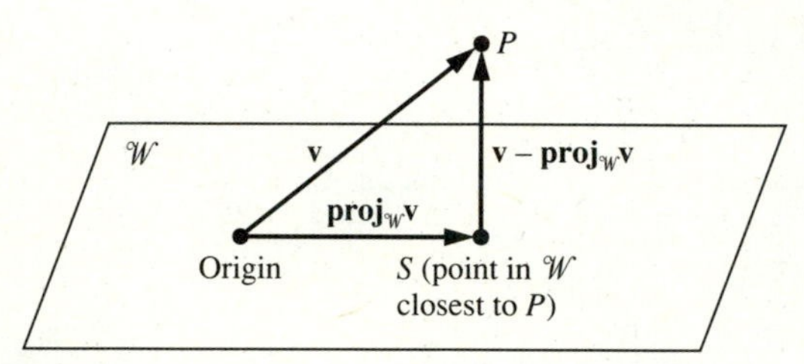

Figure 5.16
The minimum distance from P to $\mathcal{W}$, $\|\mathbf{v} - \mathbf{proj}_{\mathcal{W}}\mathbf{v}\|$

Example 8

Consider the subspace $\mathcal{W}$ of $\mathbb{R}^3$ from Example 7, whose vectors lie in the plane $2x + y + z = 0$. In that example, for $\mathbf{v} = [-6, 10, 5]$, we calculated that $\mathbf{v} - \mathbf{proj}_{\mathcal{W}}\mathbf{v} = [1, \frac{1}{2}, \frac{1}{2}]$. Hence, the minimum distance from $P = (-6, 10, 5)$ to $\mathcal{W}$ is $\|\mathbf{v} - \mathbf{proj}_{\mathcal{W}}\mathbf{v}\| = \sqrt{1^2 + (\frac{1}{2})^2 + (\frac{1}{2})^2} = \sqrt{\frac{3}{2}} \approx 1.2247$. ■

Exercises—Section 5.6

1. For each of the following subspaces $\mathcal{W}$ of $\mathbb{R}^n$, find a basis for $\mathcal{W}^\perp$, and verify Corollary 5.36.
   ★(a) In $\mathbb{R}^2$, $\mathcal{W} = \text{span}(\{[3, -2]\})$
   (b) In $\mathbb{R}^3$, $\mathcal{W} = \text{span}(\{[1, -2, 1]\})$
   ★(c) In $\mathbb{R}^3$, $\mathcal{W} = \text{span}(\{[1, 4, -2], [2, 1, -1]\})$
   (d) In $\mathbb{R}^3$, $\mathcal{W} =$ the plane $3x - y + 4z = 0$
   ★(e) In $\mathbb{R}^3$, $\mathcal{W} =$ the plane $-2x + 5y - z = 0$
   ★(f) In $\mathbb{R}^4$, $\mathcal{W} = \text{span}(\{[1, -1, 0, 2], [0, 1, 2, -1]\})$
   (g) In $\mathbb{R}^4$, $\mathcal{W} = \{(x, y, z, w) \mid 3x - 2y + 4z + w = 0\}$

2. For each of the following subspaces $\mathcal{W}$ of $\mathbb{R}^n$ and for the given $\mathbf{v} \in \mathbb{R}^n$, find $\mathbf{proj}_{\mathcal{W}}\mathbf{v}$, and decompose $\mathbf{v}$ into $\mathbf{w}_1 + \mathbf{w}_2$, where $\mathbf{w}_1 \in \mathcal{W}$ and $\mathbf{w}_2 \in \mathcal{W}^\perp$. (Hint: You may need to find an orthonormal basis for $\mathcal{W}$ first.)

★(a) In $\mathbb{R}^3$, $\mathcal{W} = \text{span}(\{[1, -2, -1], [3, -1, 0]\})$, $\mathbf{v} = [-1, 3, 2]$

★(b) In $\mathbb{R}^3$, $\mathcal{W} =$ the plane $2x - 2y + z = 0$, $\mathbf{v} = [1, -4, 3]$

(c) In $\mathbb{R}^3$, $\mathcal{W} = \text{span}(\{[-1, 3, 2]\})$, $\mathbf{v} = [2, 2, -3]$

(d) In $\mathbb{R}^4$, $\mathcal{W} = \text{span}(\{[2, -1, 1, 0], [1, -1, 2, 2]\})$, $\mathbf{v} = [-1, 3, 3, 2]$

3. Let $\mathbf{v} = [a, b, c]$. If $\mathcal{W}$ is the xy-plane, verify that $\mathbf{proj}_{\mathcal{W}}\mathbf{v} = [a, b, 0]$.

4. In each of the following cases, find the minimum distance between the given point P and the given subspace $\mathcal{W}$ of $\mathbb{R}^n$:

★(a) $P = (-2, 3, 1)$, $\mathcal{W} = \text{span}(\{[-1, 4, 4], [2, -1, 0]\})$ in $\mathbb{R}^3$

(b) $P = (4, -1, 2)$, $\mathcal{W} = \text{span}(\{[-2, 3, -3]\})$ in $\mathbb{R}^3$

(c) $P = (2, 3, -3, 1)$, $\mathcal{W} = \text{span}(\{[-1, 2, -1, 1], [2, -1, 1, -1]\})$ in $\mathbb{R}^4$

★(d) $P = (-1, 4, -2, 2)$, $\mathcal{W} = \{(x, y, z, w) \mid 2x - 3z + 2w = 0\}$ in $\mathbb{R}^4$

5. By Theorem 5.39, the mapping $L\colon \mathbb{R}^n \longrightarrow \mathbb{R}^n$ given by $L(\mathbf{v}) = \mathbf{proj}_{\mathcal{W}}\mathbf{v}$ is a linear operator for any subspace $\mathcal{W}$ of $\mathbb{R}^n$. In each of the following cases, find the matrix representation of this operator with respect to the standard basis for $\mathbb{R}^n$:

★(a) In $\mathbb{R}^3$, $\mathcal{W} = \text{span}(\{[2, -1, 1], [1, 0, -3]\})$

(b) In $\mathbb{R}^3$, $\mathcal{W} =$ the plane $3x - 2y + 2z = 0$

★(c) In $\mathbb{R}^4$, $\mathcal{W} = \text{span}(\{[1, 2, 1, 0], [-1, 0, -2, 1]\})$

(d) In $\mathbb{R}^4$, $\mathcal{W} = \text{span}(\{[-3, -1, 1, 2]\})$

6. Prove that if $\mathcal{W}_1$ and $\mathcal{W}_2$ are subspaces of $\mathbb{R}^n$ with $\mathcal{W}_1^\perp = \mathcal{W}_2^\perp$, then $\mathcal{W}_1 = \mathcal{W}_2$.

7. Prove that if $\mathcal{W}_1$ and $\mathcal{W}_2$ are subspaces of $\mathbb{R}^n$ with $\mathcal{W}_1 \subseteq \mathcal{W}_2$, then $\mathcal{W}_2^\perp \subseteq \mathcal{W}_1^\perp$.

8. Let $\mathcal{W}$ be a subspace of $\mathbb{R}^n$.

(a) Show that if $\mathbf{v} \in \mathcal{W}$, then $\mathbf{proj}_{\mathcal{W}}\mathbf{v} = \mathbf{v}$.

(b) Show that if $\mathbf{v} \in \mathcal{W}^\perp$, then $\mathbf{proj}_{\mathcal{W}}\mathbf{v} = \mathbf{0}$.

9. Let $\mathcal{W}$ be a subspace of $\mathbb{R}^n$. Suppose that $\mathbf{v}$ is a nonzero vector with initial point at the origin and terminal point P. Prove that $\mathbf{v} \in \mathcal{W}^\perp$ if and only if the distance between P and $\mathcal{W}$ is $\|\mathbf{v}\|$.

10. Let $\mathcal{W}$ be a subspace of $\mathbb{R}^n$, and let $\mathbf{v}_1$ and $\mathbf{v}_2$ be vectors in $\mathbb{R}^n$. Suppose that $\mathbf{p}_1 = \mathbf{proj}_{\mathcal{W}}\mathbf{v}_1$ and $\mathbf{p}_2 = \mathbf{proj}_{\mathcal{W}}\mathbf{v}_2$.

(a) What is $\mathbf{proj}_{\mathcal{W}}(\mathbf{v}_1 + \mathbf{v}_2)$? Prove your answer.

(b) If $c \in \mathbb{R}$, what is $\mathbf{proj}_{\mathcal{W}}(c\mathbf{v}_1)$? Prove your answer.

11. You can think of matrices in $\mathcal{M}_{nn}$ as n^2-vectors by using their coordinatization with respect to the standard basis. Use this technique to prove that the orthogonal complement of the subspace $\mathcal{V}$ of symmetric matrices in $\mathcal{M}_{nn}$ is the subspace $\mathcal{W}$ of $n \times n$ skew-symmetric matrices. (Hint: First show that $\mathcal{W} \subseteq \mathcal{V}^\perp$. Then prove equality by showing that $\dim(\mathcal{W}) = n^2 - \dim(\mathcal{V})$.)

12. Show that if $\mathcal{W}$ is a one-dimensional subspace of $\mathbb{R}^n$ spanned by $\mathbf{a}$ and if $\mathbf{b} \in \mathbb{R}^n$, then the value of $\mathbf{proj}_{\mathcal{W}}\mathbf{b}$ agrees with the definition for $\mathbf{proj}_{\mathbf{a}}\mathbf{b}$ in Section 1.2.

13. Prove Theorem 5.33.

14. Prove Corollary 5.37. (Hint: Note that $\mathcal{W} \subseteq (\mathcal{W}^\perp)^\perp$. Use Corollary 5.36 to show that $\dim(\mathcal{W}) = \dim((\mathcal{W}^\perp)^\perp)$, and apply Theorem 4.16.)

15. Prove Theorem 5.39. (Hint: To prove $\ker(L) = \mathcal{W}^\perp$, first show that $\text{range}(L) = \mathcal{W}$. Hence, $\dim(\ker(L)) = n - \dim(\mathcal{W}) = \dim(\mathcal{W}^\perp)$. (Why?) Finally, show $\mathcal{W}^\perp \subseteq \ker(L)$, and apply Theorem 4.16.)

16. Let $L: \mathbb{R}^n \longrightarrow \mathbb{R}^m$ be a linear transformation with matrix $\mathbf{A}$ (with respect to the standard basis). Show that $\ker(L)$ is the orthogonal complement of the row space of $\mathbf{A}$.

17. Let $L: \mathbb{R}^n \longrightarrow \mathbb{R}^m$ be a linear transformation. Consider the mapping $L'\colon (\ker(L))^\perp \longrightarrow \mathbb{R}^m$ given by $L'(\mathbf{v}) = L(\mathbf{v})$, for all $\mathbf{v} \in (\ker(L))^\perp$. ($L'$ is the **restriction** of L to $(\ker(L))^\perp$.) Prove that L' is one-to-one.

18. Prove Theorem 5.40. (Hint: Suppose that T is any point in $\mathcal{W}$ and $\mathbf{w}$ is the vector from the origin to T. We need to show that $\|\mathbf{v} - \mathbf{w}\| \geq \|\mathbf{v} - \mathbf{proj}_{\mathcal{W}}\mathbf{v}\|$—that is, that the distance from P to T is never smaller than the distance from P to the terminal point of $\mathbf{proj}_{\mathcal{W}}\mathbf{v}$. Let $\mathbf{a} = \mathbf{v} - \mathbf{proj}_{\mathcal{W}}\mathbf{v}$ and $\mathbf{b} = \mathbf{proj}_{\mathcal{W}}\mathbf{v} - \mathbf{w}$. Show that $\mathbf{a} \in \mathcal{W}^\perp$, $\mathbf{b} \in \mathcal{W}$, and $\|\mathbf{v} - \mathbf{w}\|^2 = \|\mathbf{a}\|^2 + \|\mathbf{b}\|^2$.)

19. Let $\mathcal{X}$ be a subspace of $\mathbb{R}^n$, and let $\mathcal{W}$ be a subspace of $\mathcal{X}$. We define the orthogonal complement of $\mathcal{W}$ in $\mathcal{X}$ to be the set of all vectors in $\mathcal{X}$ that are orthogonal to every vector in $\mathcal{W}$.

(a) Prove that the orthogonal complement of $\mathcal{W}$ in $\mathcal{X}$ is a subspace of $\mathcal{X}$.

(b) Prove that the dimensions of $\mathcal{W}$ and its orthogonal complement in $\mathcal{X}$ add up to the dimension of $\mathcal{X}$. (Hint: Let B be an orthonormal basis for $\mathcal{W}$. First enlarge B to an orthonormal basis for $\mathcal{X}$, and then enlarge this basis to an orthonormal basis for $\mathbb{R}^n$.)

6

Eigenvalues and Eigenvectors

Trying to understand the effects of a specific linear operator on a vector space $\mathcal{V}$ can often be quite puzzling. Simply looking at the entries in a matrix representation for the operator may provide no insight at all into the action of the operator. Frequently, however, the correct change of basis for $\mathcal{V}$ reveals a simple matrix representation for the operator that can easily be understood. Because diagonal matrices are one of the simplest types of matrices, we search for a change of basis, if it exists, that will give a diagonal matrix for the linear operator. Eigenvalues and eigenvectors are the tools necessary for finding such a diagonal representation.

6.1 Introduction to Eigenvalues

In this section, we define eigenvalues and eigenvectors and establish their basic properties.

Defining Eigenvalues

DEFINITION

Let $\mathcal{V}$ be a finite dimensional vector space, and let L: $\mathcal{V} \longrightarrow \mathcal{V}$ be a linear operator. A real number λ is said to be an **eigenvalue** of L if and only if there is a nonzero vector $\mathbf{v} \in \mathcal{V}$ such that $L(\mathbf{v}) = \lambda\mathbf{v}$. Also, any nonzero vector $\mathbf{v}$ such that $L(\mathbf{v}) = \lambda\mathbf{v}$ is said to be an **eigenvector** corresponding to the eigenvalue λ.

In some textbooks, eigenvalues are called **characteristic values** and eigenvectors are called **characteristic vectors**.

If $\mathbf{v}$ is an eigenvector associated with an eigenvalue λ for a linear operator L, then by the preceding definitions, L acts on $\mathbf{v}$ by scalar multiplication; that is, $L(\mathbf{v})$ produces the same result as the product $\lambda\mathbf{v}$. If we are working in $\mathbb{R}^n$, the image $L(\mathbf{v})$ is parallel to the vector $\mathbf{v}$, dilating (or lengthening) $\mathbf{v}$ if $|\lambda| > 1$ and contracting (or shortening) $\mathbf{v}$ if $|\lambda| < 1$. In the case where $\lambda = 0$, $L(\mathbf{v}) = 0\mathbf{v} = \mathbf{0}$, and thus $\mathbf{v} \in \ker(L)$.

Notice that an eigenvalue can be zero. However, by definition, an eigenvector is never the zero vector.

Example 1

Suppose that $L\colon \mathbb{R}^3 \longrightarrow \mathbb{R}^3$ is the linear operator given by

$$L\left(\begin{bmatrix} x \\ y \\ z \end{bmatrix}\right) = \begin{bmatrix} 3 & -11 & 16 \\ 2 & -8 & 8 \\ 1 & -3 & 2 \end{bmatrix}\begin{bmatrix} x \\ y \\ z \end{bmatrix} = \begin{bmatrix} 3x - 11y + 16z \\ 2x - 8y + 8z \\ x - 3y + 2z \end{bmatrix}.$$

Then -3 is an eigenvalue for L, since

$$L\left(\begin{bmatrix} 1 \\ 2 \\ 1 \end{bmatrix}\right) = \begin{bmatrix} 3 & -11 & 16 \\ 2 & -8 & 8 \\ 1 & -3 & 2 \end{bmatrix}\begin{bmatrix} 1 \\ 2 \\ 1 \end{bmatrix} = \begin{bmatrix} -3 \\ -6 \\ -3 \end{bmatrix} = -3\begin{bmatrix} 1 \\ 2 \\ 1 \end{bmatrix}.$$

Notice that $[1, 2, 1]$ is an eigenvector corresponding to the eigenvalue -3. In fact, any nonzero scalar multiple of $[1, 2, 1]$ is also an eigenvector corresponding to -3, since $L(c[1, 2, 1]) = c(L([1, 2, 1])) = c(-3[1, 2, 1]) = -3(c[1, 2, 1])$. Therefore, there are infinitely many eigenvectors corresponding to this eigenvalue. In fact, adding $\mathbf{0}$ to the set of all these eigenvectors forms a subspace of $\mathbb{R}^3$, as we prove later in this section. The action of L on any vector in this subspace is to reverse its direction and multiply its length by 3. ■

The Characteristic Polynomial of a Matrix

Our next goal is to find a method for determining the eigenvalues and eigenvectors of a linear operator. Suppose that $\mathbf{A}$ is an $n \times n$ matrix and $L\colon \mathbb{R}^n \longrightarrow \mathbb{R}^n$ is the linear operator given by $L(\mathbf{v}) = \mathbf{A}\mathbf{v}$. For instance, the operator in Example 1 is of this type. If $\mathbf{v}$ is an eigenvector for L corresponding to the eigenvalue λ, then we have

$$\mathbf{A}\mathbf{v} = \lambda\mathbf{v} = \lambda\mathbf{I}_n\mathbf{v}, \quad \text{or} \quad (\lambda\mathbf{I}_n - \mathbf{A})\mathbf{v} = \mathbf{0}.$$

Therefore, $\mathbf{v}$ is a nontrivial solution to the homogeneous system whose coefficient matrix is $\lambda\mathbf{I}_n - \mathbf{A}$. Corollary 3.7 then shows that $|\lambda\mathbf{I}_n - \mathbf{A}| = 0$. Since all of the steps in this argument are reversible, we have proved the following:

THEOREM 6.1

> Let $\mathbf{A}$ be an $n \times n$ matrix, and let $L: \mathbb{R}^n \longrightarrow \mathbb{R}^n$ be the linear operator given by $L(\mathbf{v}) = \mathbf{Av}$. Then λ is an eigenvalue for L if and only if $|\lambda\mathbf{I}_n - \mathbf{A}| = 0$. The eigenvectors corresponding to λ are the nontrivial solutions of the homogeneous system $(\lambda\mathbf{I}_n - \mathbf{A})\mathbf{v} = \mathbf{0}$.

Eigenvalues and eigenvectors for an operator of the type given in Theorem 6.1 are also called **eigenvalues** and **eigenvectors for the matrix A**.

Example 2

Recall the linear operator $L: \mathbb{R}^3 \longrightarrow \mathbb{R}^3$ from Example 1 given by $L(\mathbf{v}) = \mathbf{Av}$, where

$$\mathbf{A} = \begin{bmatrix} 3 & -11 & 16 \\ 2 & -8 & 8 \\ 1 & -3 & 2 \end{bmatrix}.$$

We discovered that $\lambda_1 = -3$ is an eigenvalue for $\mathbf{A}$. Notice that

$$|-3\mathbf{I}_3 - \mathbf{A}| = \left| \begin{bmatrix} -3 & 0 & 0 \\ 0 & -3 & 0 \\ 0 & 0 & -3 \end{bmatrix} - \begin{bmatrix} 3 & -11 & 16 \\ 2 & -8 & 8 \\ 1 & -3 & 2 \end{bmatrix} \right| = \begin{vmatrix} -6 & 11 & -16 \\ -2 & 5 & -8 \\ -1 & 3 & -5 \end{vmatrix}.$$

This determinant equals zero. (Verify.) We can find the eigenvectors corresponding to $\lambda_1 = -3$ by solving the homogeneous system

$$\begin{bmatrix} -6 & 11 & -16 \\ -2 & 5 & -8 \\ -1 & 3 & -5 \end{bmatrix} \begin{bmatrix} x \\ y \\ z \end{bmatrix} = \begin{bmatrix} 0 \\ 0 \\ 0 \end{bmatrix},$$

producing the solution set $\{[c, 2c, c] \mid c \in \mathbb{R}\} = \{c[1, 2, 1] \mid c \in \mathbb{R}\}$. The nonzero vectors in this set are the eigenvectors for $\lambda_1 = -3$. Setting $c = 1$, we get the eigenvector $[1, 2, 1]$ given in Example 1.

To find other eigenvalues for $\mathbf{A}$, Theorem 6.1 suggests that we set $|\lambda\mathbf{I}_3 - \mathbf{A}| = 0$ and solve for λ:

$$0 = \left| \begin{bmatrix} \lambda & 0 & 0 \\ 0 & \lambda & 0 \\ 0 & 0 & \lambda \end{bmatrix} - \begin{bmatrix} 3 & -11 & 16 \\ 2 & -8 & 8 \\ 1 & -3 & 2 \end{bmatrix} \right| = \begin{vmatrix} \lambda - 3 & 11 & -16 \\ -2 & \lambda + 8 & -8 \\ -1 & 3 & \lambda - 2 \end{vmatrix}.$$

After some simplification, this equation becomes $0 = \lambda^3 + 3\lambda^2 - 4\lambda - 12 = (\lambda + 3)(\lambda - 2)(\lambda + 2)$, which yields these three solutions: $\lambda_1 = -3$, $\lambda_2 = 2$, and $\lambda_3 = -2$. We were already aware of $\lambda_1 = -3$ as a solution, but we have now discovered two other eigenvalues for $\mathbf{A}$, namely $\lambda_2 = 2$ and $\lambda_3 = -2$. We can find eigenvectors corresponding to λ_2 and λ_3 by solving appropriate homogeneous systems, just as we found eigenvectors for λ_1. In particular, for $\lambda_2 = 2$, we solve $(2\mathbf{I}_3 - \mathbf{A})\mathbf{v} = \mathbf{0}$, or

$$\begin{bmatrix} -1 & 11 & -16 \\ -2 & 10 & -8 \\ -1 & 3 & 0 \end{bmatrix} \begin{bmatrix} x \\ y \\ z \end{bmatrix} = \begin{bmatrix} 0 \\ 0 \\ 0 \end{bmatrix},$$

which has the solution set $\{[6c, 2c, c] \mid c \in \mathbb{R}\} = \{c[6, 2, 1] \mid c \in \mathbb{R}\}$. Therefore, $\mathbf{v}_2 = [6, 2, 1]$ is an eigenvector for $\mathbf{A}$ corresponding to λ_2. We can check this result by noting that

$$L(\mathbf{v}_2) = \mathbf{A}\mathbf{v}_2 = \begin{bmatrix} 3 & -11 & 16 \\ 2 & -8 & 8 \\ 1 & -3 & 2 \end{bmatrix} \begin{bmatrix} 6 \\ 2 \\ 1 \end{bmatrix} = \begin{bmatrix} 12 \\ 4 \\ 2 \end{bmatrix} = 2 \begin{bmatrix} 6 \\ 2 \\ 1 \end{bmatrix} = \lambda_2 \mathbf{v}_2.$$

Similarly, solving the homogeneous system $(-2\mathbf{I}_3 - \mathbf{A})\mathbf{v} = \mathbf{0}$ gives eigenvectors corresponding to $\lambda_3 = -2$. This solution set is $\{[-c, c, c] \mid c \in \mathbb{R}\} = \{c[-1, 1, 1] \mid c \in \mathbb{R}\}$, and thus $\mathbf{v}_3 = [-1, 1, 1]$ is an eigenvector for $\mathbf{A}$ corresponding to the eigenvalue $\lambda_3 = -2$. You should verify that $\mathbf{v}_3$ is such an eigenvector by checking that $L(\mathbf{v}_3) = \mathbf{A}\mathbf{v}_3 = -2\mathbf{v}_3 = \lambda_3\mathbf{v}_3$. ■

Theorem 6.1 points out that we can find the eigenvalues for an $n \times n$ matrix $\mathbf{A}$ by solving the equation $|\lambda\mathbf{I}_n - \mathbf{A}| = 0$, as we did in Example 2.

DEFINITION

If $\mathbf{A}$ is an $n \times n$ matrix, then the **characteristic polynomial** of $\mathbf{A}$ is the degree n polynomial $p_{\mathbf{A}}(x) = |x\mathbf{I}_n - \mathbf{A}|$.

Using this terminology, we can rephrase Theorem 6.1 as

The eigenvalues of an $n \times n$ matrix $\mathbf{A}$ are precisely the real roots of the characteristic polynomial $p_{\mathbf{A}}(x)$.

Example 3

The characteristic polynomial of $\mathbf{A} = \begin{bmatrix} 12 & -51 \\ 2 & -11 \end{bmatrix}$ is

$$p_{\mathbf{A}}(x) = \left| \begin{bmatrix} x & 0 \\ 0 & x \end{bmatrix} - \begin{bmatrix} 12 & -51 \\ 2 & -11 \end{bmatrix} \right| = \begin{vmatrix} x - 12 & 51 \\ -2 & x + 11 \end{vmatrix}$$

$$= (x - 12)(x + 11) + 102 = x^2 - x - 30 = (x - 6)(x + 5).$$

Therefore, the eigenvalues of $\mathbf{A}$ are the solutions to $p_{\mathbf{A}}(x) = 0$, or $\lambda_1 = 6$ and $\lambda_2 = -5$.

Similarly, the characteristic polynomial of

$$\mathbf{B} = \begin{bmatrix} 7 & 1 & -1 \\ -11 & -3 & 2 \\ 18 & 2 & -4 \end{bmatrix} \quad \text{is} \quad p_{\mathbf{B}}(x) = \begin{vmatrix} x-7 & -1 & 1 \\ 11 & x+3 & -2 \\ -18 & -2 & x+4 \end{vmatrix},$$

which simplifies to $p_{\mathbf{B}}(x) = x^3 - 12x - 16$, or $p_{\mathbf{B}}(x) = (x+2)^2(x-4)$. Hence, $\alpha_1 = -2$ and $\alpha_2 = 4$ are the eigenvalues for $\mathbf{B}$. ■

Calculating the characteristic polynomial of a 4×4 or larger matrix can be tedious. Computing the roots of the characteristic polynomial may be just as difficult. Thus, in practice, you should use a computer with appropriate software to compute the eigenvalues of a matrix. Numerical techniques for finding eigenvalues without using the characteristic polynomial are discussed in Section 9.3.

The Characteristic Polynomial of a Linear Operator

Now that we have a method for finding the eigenvalues for a given $n \times n$ matrix, we can use it to find the eigenvalues for a linear operator L on a finite dimensional vector space $\mathcal{V}$. We find an ordered basis B for $\mathcal{V}$, then solve for the matrix representation $\mathbf{A}$ of L with respect to B. Then L is defined by $[L(\mathbf{v})]_B = \mathbf{A}[\mathbf{v}]_B$. Thus, finding the eigenvalues of $\mathbf{A}$ gives the eigenvalues of L.

Example 4

Let $L: \mathbb{R}^2 \longrightarrow \mathbb{R}^2$ be the linear operator given by $L([a, b]) = [b, a]$—that is, a reflection about the line $y = x$. We will calculate the eigenvalues for L two ways: first using the standard basis for $\mathbb{R}^2$ and then using a nonstandard basis.

Since $L(\mathbf{i}) = \mathbf{j}$ and $L(\mathbf{j}) = \mathbf{i}$, the matrix for L with respect to the standard basis is $\mathbf{A} = \begin{bmatrix} 0 & 1 \\ 1 & 0 \end{bmatrix}$. Then $p_{\mathbf{A}}(x) = \begin{vmatrix} x & -1 \\ -1 & x \end{vmatrix} = x^2 - 1 = (x-1)(x+1)$. Hence, the eigenvalues for $\mathbf{A}$ (and L) are $\lambda_1 = 1$ and $\lambda_2 = -1$. Solving the homogeneous system $(1\mathbf{I}_2 - \mathbf{A})\mathbf{v} = \mathbf{0}$ yields $\mathbf{v}_1 = [1, 1]$ as an eigenvector corresponding to $\lambda_1 = 1$. Similarly, $\mathbf{v}_2 = [1, -1]$, for $\lambda_2 = -1$.

Notice that this result makes sense geometrically: $\mathbf{v}_1$ runs parallel to the axis of reflection and thus is unchanged by L, since $L(\mathbf{v}_1) = \lambda_1\mathbf{v}_1 = 1\mathbf{v}_1 = \mathbf{v}_1$. On the other hand, $\mathbf{v}_2$ is perpendicular to the axis of reflection, and so L reverses its direction: $L(\mathbf{v}_2) = \lambda_2\mathbf{v}_2 = -\mathbf{v}_2$.

Now, instead of using the standard basis in $\mathbb{R}^2$, let us find the matrix representation of L with respect to $B = (\mathbf{v}_1, \mathbf{v}_2)$. Since $[L(\mathbf{v}_1)]_B = [1, 0]$ and $[L(\mathbf{v}_2)]_B = [0, -1]$ (why?), the matrix for L with respect to B is

$$\mathbf{C} = \begin{array}{cc} [L(\mathbf{v}_1)]_B & [L(\mathbf{v}_2)]_B \end{array} \\ \begin{bmatrix} 1 & 0 \\ 0 & -1 \end{bmatrix},$$

a diagonal matrix with the eigenvalues for L on the main diagonal. Notice that

$$p_{\mathbf{C}}(x) = \begin{vmatrix} x-1 & 0 \\ 0 & x+1 \end{vmatrix} = (x-1)(x+1) = p_{\mathbf{A}}(x),$$

giving us (of course) the same eigenvalues $\lambda_1 = 1$ and $\lambda_2 = -1$ for L. ■

Example 4 illustrates how two different matrix representations for the same linear operator (using different ordered bases) produce the same characteristic polynomial, as the next theorem asserts:

THEOREM 6.2

Let L be a linear operator on a finite dimensional vector space $\mathcal{V}$, and suppose that $\mathbf{A}$ and $\mathbf{B}$ are matrix representations for L with respect to two different ordered bases for $\mathcal{V}$. Then $p_{\mathbf{A}}(x) = p_{\mathbf{B}}(x)$.

By Theorem 5.8, if two $n \times n$ matrices $\mathbf{A}$ and $\mathbf{B}$ represent the same linear operator with respect to different ordered bases, then there is a nonsingular matrix $\mathbf{P}$ such that $\mathbf{B} = \mathbf{PAP}^{-1}$; that is, $\mathbf{B}$ and $\mathbf{A}$ are similar. The matrix $\mathbf{P}$ is the transition matrix between the two ordered bases. Therefore, Theorem 6.2 is proved if we can establish the following lemma:

LEMMA 6.3

Let $\mathbf{A}$ and $\mathbf{B}$ be similar $n \times n$ matrices. Then $p_{\mathbf{A}}(x) = p_{\mathbf{B}}(x)$.

Proof of Lemma 6.3

Let $\mathbf{A}$ and $\mathbf{B}$ be similar $n \times n$ matrices, and let $\mathbf{P}$ be a nonsingular matrix such that $\mathbf{B} = \mathbf{PAP}^{-1}$. Then,

$$\begin{aligned} p_{\mathbf{B}}(x) &= |x\mathbf{I}_n - \mathbf{B}| = |x\mathbf{I}_n - \mathbf{PAP}^{-1}| \\ &= |x(\mathbf{PI}_n\mathbf{P}^{-1}) - \mathbf{PAP}^{-1}| = |(x\mathbf{PI}_n - \mathbf{PA})\mathbf{P}^{-1}| \\ &= |\mathbf{P}(x\mathbf{I}_n - \mathbf{A})\mathbf{P}^{-1}| = |\mathbf{P}||x\mathbf{I}_n - \mathbf{A}||\mathbf{P}^{-1}| \\ &= |\mathbf{P}||x\mathbf{I}_n - \mathbf{A}|\frac{1}{|\mathbf{P}|} = |x\mathbf{I}_n - \mathbf{A}| = p_{\mathbf{A}}(x). \end{aligned}$$

■

Theorem 6.2 allows us to define the characteristic polynomial of a linear operator, since this polynomial is independent of the ordered basis used.

DEFINITION

Let L be a linear operator on a finite dimensional vector space $\mathcal{V}$. Suppose $\mathbf{A}$ is the matrix representation of L with respect to some ordered basis for $\mathcal{V}$. Then the **characteristic polynomial of** L, $p_L(x)$, is defined to be $p_{\mathbf{A}}(x)$.

Example 5

Consider $L\colon \mathcal{P}_2 \longrightarrow \mathcal{P}_2$ determined by $L(f(x)) = x^2 f''(x) + (3x-2)f'(x) + 5f(x)$. You can easily check that $L(x^2) = 13x^2 - 4x$, $L(x) = 8x - 2$, and $L(1) = 5$. Thus, the matrix representation of L with respect to the standard basis $S = (x^2, x, 1)$ is

$$\mathbf{A} = \begin{bmatrix} 13 & 0 & 0 \\ -4 & 8 & 0 \\ 0 & -2 & 5 \end{bmatrix}.$$

Hence,

$$p_L(x) = p_{\mathbf{A}}(x) = \begin{vmatrix} x-13 & 0 & 0 \\ 4 & x-8 & 0 \\ 0 & 2 & x-5 \end{vmatrix} = (x-13)(x-8)(x-5),$$

since this is the determinant of a lower triangular matrix. The eigenvalues of L are the roots of $p_L(x)$, namely $\lambda_1 = 13$, $\lambda_2 = 8$, and $\lambda_3 = 5$. Solving each of the homogeneous systems $(13\mathbf{I}_3 - \mathbf{A})\mathbf{v} = \mathbf{0}$, $(8\mathbf{I}_3 - \mathbf{A})\mathbf{v} = \mathbf{0}$, and $(5\mathbf{I}_3 - \mathbf{A})\mathbf{v} = \mathbf{0}$ yields respective eigenvectors for λ_1, λ_2, and λ_3: $\mathbf{v}_1 = 5x^2 - 4x + 1$, $\mathbf{v}_2 = 3x - 2$, and $\mathbf{v}_3 = 1$. (Verify.)

It is easy to show that the set $\{\mathbf{v}_1, \mathbf{v}_2, \mathbf{v}_3\}$ is linearly independent, and thus, since $\dim(\mathcal{P}_2) = 3$, $B = (\mathbf{v}_1, \mathbf{v}_2, \mathbf{v}_3)$ is an ordered basis for $\mathcal{P}_2$. Because $L(\mathbf{v}_1) = 13\mathbf{v}_1$, $L(\mathbf{v}_2) = 8\mathbf{v}_2$, and $L(\mathbf{v}_3) = 5\mathbf{v}_3$, the matrix representation of L with respect to B is

$$\begin{array}{ccc} [L(\mathbf{v}_1)]_B & [L(\mathbf{v}_2)]_B & [L(\mathbf{v}_3)]_B \end{array}$$
$$\begin{bmatrix} 13 & 0 & 0 \\ 0 & 8 & 0 \\ 0 & 0 & 5 \end{bmatrix},$$

a diagonal matrix, with the eigenvalues for L on the main diagonal. ■

Analyzing Geometric Operators Using Eigenvalues

We now present two examples that illustrate the use of eigenvalues to study operators that can be described geometrically.

Example 6

Consider the linear operator L on $\mathbb{R}^2$ that rotates the plane counterclockwise through an angle θ. We have seen that the matrix representation of L with respect to the standard basis is $\begin{bmatrix} \cos\theta & -\sin\theta \\ \sin\theta & \cos\theta \end{bmatrix}$. Thus,

$$p_L(x) = \begin{vmatrix} (x - \cos\theta) & \sin\theta \\ -\sin\theta & (x - \cos\theta) \end{vmatrix} = x^2 - (2\cos\theta)x + 1.$$

Using the quadratic formula to solve for eigenvalues yields

$$\lambda = \frac{2\cos\theta \pm \sqrt{4\cos^2\theta - 4}}{2} = \cos\theta \pm \sqrt{(-\sin^2\theta)}.$$

Thus, there are no eigenvalues, unless $\theta = 0$ or $\theta = \pi$.

In retrospect, this result makes perfect sense. If we rotate the plane through an angle θ other than 0 or π, all vectors will change their direction, and no nonzero vector will end up parallel to where it started. Thus, there cannot be any eigenvectors (or eigenvalues) in this case. When $\theta = 0$, no rotation is actually performed, and L is the identity transformation. The only eigenvalue in this case is 1, and every nonzero vector is an eigenvector. However, if $\theta = \pi$, the direction of every vector is reversed. Hence, -1 is the only eigenvalue in this case, and every nonzero vector is an eigenvector. ■

Example 7

Let $L: \mathbb{R}^3 \longrightarrow \mathbb{R}^3$ be the orthogonal projection onto the plane $2x + y - 3z = 0$. Our goal is to use eigenvalues and eigenvectors to find the matrix representation of L with respect to the standard basis.

The equation of the plane can be expressed as $[2, 1, -3] \cdot [x, y, z] = 0$. Thus, the plane consists of all vectors perpendicular to $\mathbf{v}_1 = [2, 1, -3]$; that is, $\mathbf{v}_1$ is orthogonal to the plane. But then $L(\mathbf{v}_1) = \mathbf{0}$, since all vectors orthogonal to the plane that we are projecting onto are sent to $\mathbf{0}$. Thus, $L(\mathbf{v}_1) = 0\mathbf{v}_1$, and so $\mathbf{v}_1$ is an eigenvector for L corresponding to the eigenvalue $\lambda_1 = 0$.

Next, any vector in the plane $2x + y - 3z = 0$ is mapped to itself by L. We can find such vectors by choosing arbitrary values for x and y and solving for z—for example, $\mathbf{v}_2 = [1, 1, 1]$ and $\mathbf{v}_3 = [2, -1, 1]$. Notice that $\{\mathbf{v}_2, \mathbf{v}_3\}$ is linearly independent. Since the plane is two-dimensional, $\{\mathbf{v}_2, \mathbf{v}_3\}$ spans the plane. Both of these are eigenvectors for L corresponding to the eigenvalue $\lambda_2 = 1$, since $L(\mathbf{v}_2) = 1\mathbf{v}_2$ and $L(\mathbf{v}_3) = 1\mathbf{v}_3$.

Thus, $B = (\mathbf{v}_1, \mathbf{v}_2, \mathbf{v}_3)$ is an ordered basis for $\mathbb{R}^3$. The matrix representation for L with respect to B is

$$\begin{array}{ccc} [L(\mathbf{v}_1)]_B & [L(\mathbf{v}_2)]_B & [L(\mathbf{v}_3)]_B \end{array}$$
$$\begin{bmatrix} 0 & 0 & 0 \\ 0 & 1 & 0 \\ 0 & 0 & 1 \end{bmatrix},$$

a diagonal matrix with the eigenvalues for L on the main diagonal. Hence, the matrix representation of L with respect to the standard basis is $\mathbf{PAP}^{-1}$, where $\mathbf{P}$ is the transition matrix from B to standard coordinates—that is, the matrix whose columns are the vectors $\mathbf{v}_1$, $\mathbf{v}_2$, and $\mathbf{v}_3$. Solving for $\mathbf{P}^{-1}$ and then computing, we get the desired matrix representation for L:

$$\mathbf{PAP}^{-1} = \begin{bmatrix} 2 & 1 & 2 \\ 1 & 1 & -1 \\ -3 & 1 & 1 \end{bmatrix} \begin{bmatrix} 0 & 0 & 0 \\ 0 & 1 & 0 \\ 0 & 0 & 1 \end{bmatrix} \begin{bmatrix} \frac{1}{7} & \frac{1}{14} & -\frac{3}{14} \\ \frac{1}{7} & \frac{4}{7} & \frac{2}{7} \\ \frac{2}{7} & -\frac{5}{14} & \frac{1}{14} \end{bmatrix} = \frac{1}{14}\begin{bmatrix} 10 & -2 & 6 \\ -2 & 13 & 3 \\ 6 & 3 & 5 \end{bmatrix}. \blacksquare$$

We will return to such orthogonal projections in Section 6.2.

Properties of Eigenvectors

In Examples 4 and 5 we formed a basis for the given vector space $\mathcal{V}$ by choosing one eigenvector for each eigenvalue of the linear operator. The next theorem guarantees that, in general, such a set of eigenvectors is linearly independent, although, as we will see later, it might not span $\mathcal{V}$.

THEOREM 6.4

Let L be a linear operator on a finite dimensional vector space $\mathcal{V}$, and let $\lambda_1, \ldots, \lambda_n$ be distinct eigenvalues for L. If $\mathbf{v}_1, \ldots, \mathbf{v}_n$ are eigenvectors for L corresponding to $\lambda_1, \ldots, \lambda_n$, respectively, then the set $\{\mathbf{v}_1, \ldots, \mathbf{v}_n\}$ is linearly independent. That is, eigenvectors corresponding to distinct eigenvalues are linearly independent.

Proof of Theorem 6.4

We proceed by induction on n:

Base step: Suppose that $n = 1$. Any eigenvector $\mathbf{v}_1$ for λ_1 is nonzero, so $\{\mathbf{v}_1\}$ is linearly independent.

Inductive step: Let $\lambda_1, \ldots, \lambda_{k+1}$ be distinct eigenvalues for L, and let $\mathbf{v}_1, \ldots, \mathbf{v}_{k+1}$ be corresponding eigenvectors. Our inductive hypothesis is that the set $\{\mathbf{v}_1, \ldots, \mathbf{v}_k\}$ is linearly independent. We must prove that $\{\mathbf{v}_1, \ldots, \mathbf{v}_k, \mathbf{v}_{k+1}\}$ is linearly independent. Suppose that $a_1\mathbf{v}_1 + \cdots + a_k\mathbf{v}_k + a_{k+1}\mathbf{v}_{k+1} = \mathbf{0}_{\mathcal{V}}$. Showing that $a_1 = a_2 = \cdots = a_k = a_{k+1} = 0$ will finish the proof by Theorem 4.7. Now,

$$\begin{aligned} L(a_1\mathbf{v}_1 + \cdots + a_k\mathbf{v}_k + a_{k+1}\mathbf{v}_{k+1}) &= L(\mathbf{0}_{\mathcal{V}}) \\ \Longrightarrow \quad a_1L(\mathbf{v}_1) + \cdots + a_kL(\mathbf{v}_k) + a_{k+1}L(\mathbf{v}_{k+1}) &= L(\mathbf{0}_{\mathcal{V}}) \\ \Longrightarrow \quad a_1\lambda_1\mathbf{v}_1 + \cdots + a_k\lambda_k\mathbf{v}_k + a_{k+1}\lambda_{k+1}\mathbf{v}_{k+1} &= \mathbf{0}_{\mathcal{V}}. \end{aligned}$$

Multiplying both sides of the equation $a_1\mathbf{v}_1 + \cdots + a_k\mathbf{v}_k + a_{k+1}\mathbf{v}_{k+1} = \mathbf{0}_{\mathcal{V}}$ by λ_{k+1} yields

$$a_1\lambda_{k+1}\mathbf{v}_1 + \cdots + a_k\lambda_{k+1}\mathbf{v}_k + a_{k+1}\lambda_{k+1}\mathbf{v}_{k+1} = \mathbf{0}_{\mathcal{V}}.$$

Subtracting this equation from the previous equation gives

$$a_1(\lambda_1 - \lambda_{k+1})\mathbf{v}_1 + \cdots + a_k(\lambda_k - \lambda_{k+1})\mathbf{v}_k = \mathbf{0}_{\mathcal{V}}.$$

Hence, our induction hypothesis, together with Theorem 4.7, implies that

$$a_1(\lambda_1 - \lambda_{k+1}) = \cdots = a_k(\lambda_k - \lambda_{k+1}) = 0.$$

Since the eigenvalues $\lambda_1, \ldots, \lambda_{k+1}$ are distinct, none of the factors $\lambda_i - \lambda_{k+1}$ in these equations can equal zero, for $1 \le i \le k$. Thus, $a_1 = a_2 = \cdots = a_k = 0$. Finally, plugging these values into the earlier equation $a_1\mathbf{v}_1 + \cdots + a_k\mathbf{v}_k + a_{k+1}\mathbf{v}_{k+1} = \mathbf{0}_{\mathcal{V}}$ gives $a_{k+1}\mathbf{v}_{k+1} = \mathbf{0}_{\mathcal{V}}$. Since $\mathbf{v}_{k+1} \neq \mathbf{0}_{\mathcal{V}}$, we must have $a_{k+1} = 0$ as well. ■

Example 8

Consider $L: \mathbb{R}^4 \longrightarrow \mathbb{R}^4$ given by

$$L\left(\begin{bmatrix} x_1 \\ x_2 \\ x_3 \\ x_4 \end{bmatrix}\right) = \begin{bmatrix} 5 & -1 & 0 & 9 \\ -5 & 5 & 1 & -21 \\ 5 & -2 & 2 & 17 \\ -2 & 1 & 0 & -6 \end{bmatrix}\begin{bmatrix} x_1 \\ x_2 \\ x_3 \\ x_4 \end{bmatrix}.$$

In Exercise 3, you are asked to verify that the characteristic polynomial of L is $p_L(x) = (x-2)^2(x-3)(x+1)$. Therefore, $\lambda_1 = 2$, $\lambda_2 = 3$, and $\lambda_3 = -1$ are distinct eigenvalues for L. Exercise 3 also asks you to show that $\mathbf{v}_1 = [1, -6, 2, -1]$, $\mathbf{v}_2 = [1, 2, 1, 0]$, and $\mathbf{v}_3 = [1, -3, 2, -1]$ are eigenvectors corresponding to λ_1, λ_2, and λ_3, respectively. Therefore $\{\mathbf{v}_1, \mathbf{v}_2, \mathbf{v}_3\}$ is linearly independent by Theorem 6.4. You can check this conclusion by row reducing the following matrix to observe that its rank equals 3:

$$\begin{bmatrix} 1 & -6 & 2 & -1 \\ 1 & 2 & 1 & 0 \\ 1 & -3 & 2 & -1 \end{bmatrix}.$$

■

DEFINITION

Let $\mathcal{V}$ be a vector space, and let $L: \mathcal{V} \longrightarrow \mathcal{V}$ be a linear operator on $\mathcal{V}$. Let λ be an eigenvalue for L. Then E_λ, the **eigenspace of** λ, is defined to be the set of all eigenvectors for L corresponding to λ, together with the zero vector $\mathbf{0}_{\mathcal{V}}$ of $\mathcal{V}$. That is, $E_\lambda = \{\mathbf{v} \in \mathcal{V} \mid L(\mathbf{v}) = \lambda\mathbf{v}\}$.

Although eigenvectors corresponding to *different* eigenvalues for an operator on a given vector space are linearly independent, the set of all eigenvectors for the *same* eigenvalue (together with the zero vector) always forms a subspace of the vector space.

THEOREM 6.5

Let L be a linear operator on a finite dimensional vector space $\mathcal{V}$, and let λ be an eigenvalue for L. Then the eigenspace E_λ of λ is a subspace of $\mathcal{V}$.

Proof of Theorem 6.5

Let $\mathcal{V}$ be an n-dimensional vector space, and let L, λ, and E_λ be as given in the statement of the theorem. Consider the linear operator $T: \mathcal{V} \longrightarrow \mathcal{V}$ given by $T(\mathbf{v}) = \lambda\mathbf{v} - L(\mathbf{v})$. Then, clearly, $E_\lambda = \ker(T)$, which is a subspace of $\mathcal{V}$ by Theorem 5.9. ∎

Example 9

Recall from Example 7 the orthogonal projection $L: \mathbb{R}^3 \longrightarrow \mathbb{R}^3$ onto the plane $2x + y - 3z = 0$. The eigenvalues for L are $\lambda_1 = 0$ and $\lambda_2 = 1$. Then $E_{\lambda_1} = E_0 = \{c[2, 1, -3] \mid c \in \mathbb{R}\}$, the one-dimensional subspace of $\mathbb{R}^3$ spanned by a vector perpendicular to the plane. Also, $E_{\lambda_2} = E_1 = \{[x, y, z] \mid 2x + y - 3z = 0\}$, the set of vectors lying in the plane itself. As Example 7 shows, E_1 is a two-dimensional subspace of $\mathbb{R}^3$ spanned by $\{[1, 1, 1], [2, -1, 1]\}$. ∎

Geometric and Algebraic Multiplicity

DEFINITION

Let L be a linear operator on a finite dimensional vector space, and let λ be an eigenvalue for L. Then the dimension of the eigenspace E_λ is called the **geometric multiplicity of** λ.

In Example 9, the projection operator L had two eigenvalues, $\lambda_1 = 0$ and $\lambda_2 = 1$. The dimensions of the eigenspaces corresponding to these eigenvalues are $\dim(E_{\lambda_1}) = 1$ and $\dim(E_{\lambda_2}) = 2$. Thus, the geometric multiplicity of λ_1 is 1, and the geometric multiplicity of λ_2 is 2.

Let us consider this same operator from a different point of view. In Example 7 we found its matrix representation with respect to the ordered basis $B = ([2, 1, -3], [1, 1, 1], [2, -1, 1])$ to be

$$\mathbf{A} = \begin{bmatrix} 0 & 0 & 0 \\ 0 & 1 & 0 \\ 0 & 0 & 1 \end{bmatrix}.$$

Using this matrix to find the characteristic polynomial for L quickly yields $p_L(x) = x(x-1)^2$. Notice that the linear factor x, corresponding to the eigenvalue $\lambda_1 = 0$, appears *once* in the factored form of $p_L(x)$. The linear factor $x - 1$, corresponding to the eigenvalue $\lambda_2 = 1$, appears *twice* (that is, to the second power). The number of times these linear factors appear in the characteristic polynomial is also important, and so we have the following:

DEFINITION

Let L be a linear operator on a finite dimensional vector space, and let λ be an eigenvalue for L. Suppose that $p_L(x) = (x - \lambda)^k g(x)$, where $g(\lambda) \neq 0$. (That is, $(x - \lambda)^k$ is the highest power of $(x - \lambda)$ that divides $p_L(x)$.) Then k is called the **algebraic multiplicity of λ.**

For the projection operator of Examples 7 and 9, the algebraic multiplicity of λ_1 is 1, and the algebraic multiplicity of λ_2 is 2. For this operator, the algebraic multiplicities of the eigenvalues equal their geometric multiplicities. However, they are not always equal, as we show in the next example.

Example 10

Consider the linear operator $L: \mathbb{R}^4 \longrightarrow \mathbb{R}^4$ given by

$$L\left(\begin{bmatrix} x_1 \\ x_2 \\ x_3 \\ x_4 \end{bmatrix}\right) = \begin{bmatrix} 5 & 2 & 0 & 1 \\ -2 & 1 & 0 & -1 \\ 4 & 4 & 3 & 2 \\ 16 & 0 & -8 & -5 \end{bmatrix} \begin{bmatrix} x_1 \\ x_2 \\ x_3 \\ x_4 \end{bmatrix}.$$

In Exercise 4 you are asked to verify that $p_L(x) = (x-3)^3(x+5)$. Thus, the eigenvalues for L are $\lambda_1 = 3$ and $\lambda_2 = -5$. Also, the algebraic multiplicity of λ_1 is 3, and the algebraic multiplicity of λ_2 is 1.

Next we find the eigenspaces of λ_1 and λ_2 by solving appropriate homogeneous systems. For $\lambda_1 = 3$, we solve $(3\mathbf{I}_4 - L)\mathbf{v} = \mathbf{0}$ by row reducing

$$\left[\begin{array}{cccc|c} -2 & -2 & 0 & -1 & 0 \\ 2 & 2 & 0 & 1 & 0 \\ -4 & -4 & 0 & -2 & 0 \\ -16 & 0 & 8 & 8 & 0 \end{array}\right] \quad \text{to obtain} \quad \left[\begin{array}{cccc|c} 1 & 0 & -\frac{1}{2} & -\frac{1}{2} & 0 \\ 0 & 1 & \frac{1}{2} & 1 & 0 \\ 0 & 0 & 0 & 0 & 0 \\ 0 & 0 & 0 & 0 & 0 \end{array}\right].$$

The eigenspace E_3 is the solution space to this system, and thus $E_3 = \{[\frac{1}{2}a + \frac{1}{2}b, -\frac{1}{2}a - b, a, b] \,|\, a, b \in \mathbb{R}\}$, which is the two-dimensional subspace of $\mathbb{R}^4$ with basis $\{[\frac{1}{2}, -\frac{1}{2}, 1, 0], [\frac{1}{2}, -1, 0, 1]\}$. Hence, the geometric multiplicity of λ_1 is 2, which is less than its algebraic multiplicity.

In Exercise 4, you are asked to solve an appropriate system to show that the eigenspace for $\lambda_2 = -5$ has dimension 1, with $\{[1, -1, 2, -8]\}$ being a basis for E_{-5}. Thus, the geometric multiplicity of λ_2 is 1. For this eigenvalue, the geometric and algebraic multiplicities are equal. ■

Example 10 illustrates the following theorem:

THEOREM 6.6

Let L be a linear operator on a finite dimensional vector space $\mathcal{V}$, and let λ be an eigenvalue for L. Then

$$1 \leq (\text{geometric multiplicity of } \lambda) \leq (\text{algebraic multiplicity of } \lambda).$$

The proof of Theorem 6.6 uses the following lemma:

LEMMA 6.7

Let $\mathbf{A}$ be an $n \times n$ matrix symbolically represented by $\mathbf{A} = \begin{bmatrix} \mathbf{B} & \mathbf{C} \\ \mathbf{O} & \mathbf{D} \end{bmatrix}$, where $\mathbf{B}$ is an $m \times m$ submatrix, $\mathbf{C}$ is an $m \times (n - m)$ submatrix, $\mathbf{O}$ is an $(n - m) \times m$ zero submatrix, and $\mathbf{D}$ is an $(n - m) \times (n - m)$ submatrix. Then, $|\mathbf{A}| = |\mathbf{B}| \cdot |\mathbf{D}|$.

Lemma 6.7 is true from Exercise 20 in Section 3.2. (We suggest you work out that exercise if you have not done so previously.)

Proof of Theorem 6.6

Let $\mathcal{V}$, L, and λ be as given in the statement of the theorem, and let k represent the geometric multiplicity of λ. Clearly, by definition, the eigenspace E_λ must contain at least one nonzero vector, and thus $k = \dim(E_\lambda) \geq 1$. Thus, the first inequality in the theorem is proved.

Next, choose a basis $\{\mathbf{v}_1, \ldots, \mathbf{v}_k\}$ for E_λ and expand it to an ordered basis $B = (\mathbf{v}_1, \ldots, \mathbf{v}_k, \mathbf{v}_{k+1}, \ldots, \mathbf{v}_n)$ for $\mathcal{V}$. Let $\mathbf{A}$ be the matrix representation for L with respect to B. First we show that $\mathbf{A}$ has the form

$$\mathbf{A} = \begin{bmatrix} \lambda\mathbf{I}_k & \mathbf{C} \\ \mathbf{O} & \mathbf{D} \end{bmatrix},$$

where $\mathbf{C}$ is a $k \times (n-k)$ submatrix, $\mathbf{O}$ is an $(n-k) \times k$ zero submatrix, and $\mathbf{D}$ is an $(n-k) \times (n-k)$ submatrix. Now, $\mathbf{A}$ has such a form if its ith column equals $\lambda\mathbf{e}_i$, for $1 \leq i \leq k$. This fact follows from

$$i^{th} \text{ column of } \mathbf{A} = [L(\mathbf{v}_i)]_B = [\lambda\mathbf{v}_i]_B = \lambda[\mathbf{v}_i]_B = \lambda\mathbf{e}_i.$$

The form of $\mathbf{A}$ now makes it easy to calculate the characteristic polynomial of L. In particular,

$$\begin{aligned} p_L(x) = p_\mathbf{A}(x) = |x\mathbf{I}_n - \mathbf{A}| &= \left| x\mathbf{I}_n - \begin{bmatrix} \lambda\mathbf{I}_k & \mathbf{C} \\ \mathbf{O} & \mathbf{D} \end{bmatrix} \right| \\ &= \begin{vmatrix} (x-\lambda)\mathbf{I}_k & \mathbf{C} \\ \mathbf{O} & x\mathbf{I}_{n-k} - \mathbf{D} \end{vmatrix} \\ &= (x-\lambda)^k \cdot p_\mathbf{D}(x) \qquad \text{by Lemma 6.7} \end{aligned}$$

Let l be the number of factors of $x - \lambda$ in $p_\mathbf{D}(x)$; that is, $p_\mathbf{D}(x) = (x-\lambda)^l \cdot g(x)$, where $g(\lambda) \neq 0$. Note that $l \geq 0$, with $l = 0$ if $p_\mathbf{D}(\lambda) \neq 0$. Then, $p_L(x) = (x-\lambda)^{k+l} \cdot g(x)$, where $g(\lambda) \neq 0$. Hence,

$$\text{geometric multiplicity of } \lambda = k \leq k + l = \text{algebraic multiplicity of } \lambda.$$

■

The Cayley-Hamilton Theorem

We conclude this section with an interesting relationship between a matrix and its characteristic polynomial:

THEOREM 6.8: Cayley-Hamilton Theorem

> Let $\mathbf{A}$ be an $n \times n$ matrix, and let $p_\mathbf{A}(x)$ be its characteristic polynomial. Then $p_\mathbf{A}(\mathbf{A}) = \mathbf{O}_n$.

Example 11

Let $\mathbf{A} = \begin{bmatrix} 3 & 2 \\ 4 & -1 \end{bmatrix}$. Then $p_\mathbf{A}(x) = x^2 - 2x - 11$. The Cayley-Hamilton Theorem states that $p_\mathbf{A}(\mathbf{A}) = \mathbf{O}_2$. To verify this statement, we calculate

$$p_{\mathbf{A}}(\mathbf{A}) = \mathbf{A}^2 - 2\mathbf{A} - 11\mathbf{I}_2 = \begin{bmatrix} 17 & 4 \\ 8 & 9 \end{bmatrix} - \begin{bmatrix} 6 & 4 \\ 8 & -2 \end{bmatrix} - \begin{bmatrix} 11 & 0 \\ 0 & 11 \end{bmatrix} = \begin{bmatrix} 0 & 0 \\ 0 & 0 \end{bmatrix}.$$

■

Although the Cayley-Hamilton Theorem is an important result in advanced linear algebra, it is not central to this chapter. Therefore, we have placed its proof in Appendix A for the interested reader.

♦ **Application:** You have now covered the prerequisites for Section 9.3, "The Power Method for Finding Eigenvalues."

Exercises—Section 6.1

1. Find the characteristic polynomial of each given matrix. (Hint: For (e), do a cofactor expansion along the third row.)

★(a) $\begin{bmatrix} 3 & 1 \\ -2 & 4 \end{bmatrix}$ (b) $\begin{bmatrix} 2 & 5 & 8 \\ 0 & -1 & 9 \\ 0 & 0 & 5 \end{bmatrix}$ ★(c) $\begin{bmatrix} 2 & 1 & -1 \\ -6 & 6 & 0 \\ 3 & 0 & 0 \end{bmatrix}$

(d) $\begin{bmatrix} 5 & 1 & 4 \\ 1 & 2 & 3 \\ 3 & -1 & 1 \end{bmatrix}$ ★(e) $\begin{bmatrix} 0 & -1 & 0 & 1 \\ -5 & 2 & -1 & 2 \\ 0 & 1 & 1 & 0 \\ 4 & -1 & 3 & 0 \end{bmatrix}$

2. Find a basis for the eigenspace E_λ corresponding to the given eigenvalue λ for each of the following matrices.

★(a) $\begin{bmatrix} 1 & 1 \\ -2 & 4 \end{bmatrix}$, $\lambda = 2$ (b) $\begin{bmatrix} 1 & -1 & -1 \\ 1 & 3 & 2 \\ -3 & -3 & -2 \end{bmatrix}$, $\lambda = 2$

★(c) $\begin{bmatrix} -5 & 2 & 0 \\ -8 & 3 & 0 \\ 4 & -2 & -1 \end{bmatrix}$, $\lambda = -1$

3. Consider the linear operator $L: \mathbb{R}^4 \longrightarrow \mathbb{R}^4$ from Example 8.
(a) Verify that $p_L(x) = (x - 2)^2(x - 3)(x + 1) = x^4 - 6x^3 + 9x^2 + 4x - 12$. (Hint: Use a cofactor expansion along the third column of the matrix representation for L.)
(b) Verify that $\mathbf{v}_1 = [1, -6, 2, -1]$, $\mathbf{v}_2 = [1, 2, 1, 0]$, and $\mathbf{v}_3 = [1, -3, 2, -1]$ are eigenvectors corresponding to the eigenvalues $\lambda_1 = 2$, $\lambda_2 = 3$, and $\lambda_3 = -1$, respectively. (Hint: Just calculate $L(\mathbf{v}_1)$, $L(\mathbf{v}_2)$, and $L(\mathbf{v}_3)$.)
★(c) Calculate the algebraic and geometric multiplicities of $\lambda_1 = 2$.

4. Consider the linear operator $L: \mathbb{R}^4 \longrightarrow \mathbb{R}^4$ from Example 10.
(a) Verify that $p_L(x) = (x - 3)^3(x + 5) = x^4 - 4x^3 - 18x^2 + 108x - 135$. (Hint: Use a cofactor expansion along the third column of the matrix representation for L.)

(b) Show that $\{[1, -1, 2, -8]\}$ is a basis for the eigenspace E_{-5} for L by solving an appropriate homogeneous system.

5. Find all eigenvalues corresponding to each given matrix, and find a basis for the eigenspace corresponding to each eigenvalue. Compare the geometric and algebraic multiplicities of each eigenvalue.

★(a) $\begin{bmatrix} 1 & 3 \\ 0 & 1 \end{bmatrix}$ (b) $\begin{bmatrix} 2 & -1 \\ 0 & 3 \end{bmatrix}$ ★(c) $\begin{bmatrix} 1 & 0 & 1 \\ 0 & 2 & -3 \\ 0 & 0 & -5 \end{bmatrix}$

(d) $\begin{bmatrix} 8 & -21 \\ 3 & -8 \end{bmatrix}$ ★(e) $\begin{bmatrix} 4 & 0 & -2 \\ 6 & 2 & -6 \\ 4 & 0 & -2 \end{bmatrix}$ (f) $\begin{bmatrix} 3 & 4 & 12 \\ 4 & -12 & 3 \\ 12 & 3 & -4 \end{bmatrix}$

(g) $\begin{bmatrix} 2 & 1 & -2 & -4 \\ -2 & -4 & 4 & 10 \\ 3 & 4 & -5 & -12 \\ -2 & -3 & 4 & 9 \end{bmatrix}$ ★(h) $\begin{bmatrix} 3 & -1 & 4 & -1 \\ 0 & 3 & -3 & 3 \\ -6 & 2 & -8 & 2 \\ -6 & -4 & -2 & -4 \end{bmatrix}$

6. The linear operator represented by a 3×3 orthogonal matrix with determinant 1 (with respect to the standard basis) is always a rotation about some axis in $\mathbb{R}^3$. (We do not prove this statement here.) The axis of rotation is parallel to an eigenvector corresponding to the eigenvalue $\lambda = 1$.

Thus, each of the following orthogonal matrices represents a rotation about an axis in $\mathbb{R}^3$. Solve for a vector in the direction of the axis of rotation.

★(a) $\frac{1}{11}\begin{bmatrix} 2 & 6 & -9 \\ -9 & 6 & 2 \\ 6 & 7 & 6 \end{bmatrix}$ (b) $\frac{1}{17}\begin{bmatrix} 12 & 1 & 12 \\ 8 & 12 & -9 \\ -9 & 12 & 8 \end{bmatrix}$

★(c) $\frac{1}{7}\begin{bmatrix} 6 & 2 & 3 \\ 3 & -6 & -2 \\ 2 & 3 & -6 \end{bmatrix}$ (d) $\frac{1}{15}\begin{bmatrix} 2 & 14 & 5 \\ 10 & -5 & 10 \\ 11 & 2 & -10 \end{bmatrix}$

★**7.** Let $L: \mathbb{R}^3 \longrightarrow \mathbb{R}^3$ be the orthogonal projection onto the plane $2x - y + 2z = 0$. Use eigenvalues and eigenvectors to find the matrix representation of L with respect to the standard basis. (Hint: Follow the method of Example 7.)

8. Let $L: \mathbb{R}^3 \longrightarrow \mathbb{R}^3$ be the reflection through the plane $3x - y + 2z = 0$. Use eigenvalues and eigenvectors to find the matrix representation of L with respect to the standard basis. (Hint: Use a technique similar to that used in Example 7.)

9. Find the characteristic polynomial for each of the given linear operators. Also, solve for all eigenvalues for each operator. Finally, find a basis for each eigenspace. (Hint: For (e), (f), and (g), find the characteristic polynomial after finding the eigenvalues.)

★(a) L: $\mathcal{P}_2 \longrightarrow \mathcal{P}_2$ given by $L(f(x)) = (x-3)^2 f''(x) + xf'(x) - 5f(x)$

(b) L: $\mathcal{P}_2 \longrightarrow \mathcal{P}_2$ given by $L(f(x)) = x^2 f''(x) + f'(x) - 3f(x)$

★(c) L: $\mathbb{R}^2 \longrightarrow \mathbb{R}^2$ such that L is the counterclockwise rotation about the origin through an angle of $\frac{\pi}{3}$ radians

(d) L: $\mathbb{R}^4 \longrightarrow \mathbb{R}^4$ given by $L([x_1, x_2, x_3, x_4]) = [x_2, x_1, x_4, x_3]$

★(e) L: $\mathbb{R}^3 \longrightarrow \mathbb{R}^3$ such that L is the orthogonal projection onto the plane $4x - 3y + 2z = 0$

(f) L: $\mathbb{R}^3 \longrightarrow \mathbb{R}^3$ such that L is the orthogonal projection onto the line through the origin spanned by $[4, -1, 3]$

★(g) L: $\mathbb{R}^3 \longrightarrow \mathbb{R}^3$ such that L is the reflection through the plane $3x + 5y - z = 0$

10. Let L: $\mathcal{P}_2 \longrightarrow \mathcal{P}_2$ be the translation operator given by $L(f(x)) = f(x + a)$, for some (fixed) real number a.

★(a) Find all eigenvalues for L when $a = 1$, and find a basis for each eigenspace.

(b) Find all eigenvalues for L when a is an arbitrary nonzero number, and find a basis for each eigenspace.

11. Prove that $\begin{bmatrix} a & b \\ c & d \end{bmatrix}$ has two distinct eigenvalues if $(a - d)^2 + 4bc > 0$, one distinct eigenvalue if $(a - d)^2 + 4bc = 0$, and no eigenvalues if $(a - d)^2 + 4bc < 0$.

12. Let $\mathbf{A}$ be an $n \times n$ matrix, and let k be a positive integer.

(a) Prove that if λ is an eigenvalue of $\mathbf{A}$, then λ^k is an eigenvalue of $\mathbf{A}^k$.

★(b) Give a 2×2 matrix $\mathbf{A}$ and an integer k that provide a counterexample to the converse of part (a). (Hint: Consider a rotation in $\mathbb{R}^2$.)

13. (a) Suppose that L is a linear operator on a finite dimensional vector space. Prove that L is an isomorphism if and only if 0 is not an eigenvalue for L.

(b) Let L be an isomorphism from a finite dimensional vector space to itself. Suppose that λ is an eigenvalue for L having a corresponding eigenvector $\mathbf{v}$. Prove that $\mathbf{v}$ is an eigenvector for L^{-1} corresponding to the eigenvalue $1/\lambda$.

14. Suppose that $\mathbf{A}$ is a nonsingular $n \times n$ matrix. Prove that

$$p_{\mathbf{A}^{-1}}(x) = (-x)^n |\mathbf{A}^{-1}| p_{\mathbf{A}}\left(\tfrac{1}{x}\right).$$

(Hint: Expand the right-hand side, then collect the result into one determinant.)

15. Let $\mathbf{A}$ be an upper triangular $n \times n$ matrix. (Note: Parts (a) and (b) are also true if $\mathbf{A}$ is a lower triangular matrix.)

(a) Prove that λ is an eigenvalue for $\mathbf{A}$ if and only if λ appears on the main diagonal of $\mathbf{A}$.

(b) Show that the algebraic multiplicity of an eigenvalue λ of $\mathbf{A}$ equals the number of times λ appears on the main diagonal.

★(c) Give an example of a 3×3 upper triangular matrix having an eigenvalue λ with algebraic multiplicity 3 and geometric multiplicity 2.

★(d) Give an example of a 3×3 upper triangular matrix, one of whose eigenvalues has algebraic multiplicity 2 and geometric multiplicity 2.

16. Let $\mathbf{A}$ be an $n \times n$ matrix. Prove that $\mathbf{A}$ and $\mathbf{A}^T$ have the same characteristic polynomial and hence the same eigenvalues.

17. (Note: You must have covered the material in Section 8.4 in order to do this exercise.) Suppose that $\mathbf{A}$ is a stochastic $n \times n$ matrix. Prove that $\lambda = 1$ is an eigenvalue for $\mathbf{A}$. (Hint: Let $\mathbf{v} = [1, 1, \ldots, 1]$, and consider $\mathbf{A}^T\mathbf{v}$. Then use Exercise 16.) (This exercise implies that every stochastic matrix has a steady state solution. However, not all initial conditions reach this steady state, as demonstrated in Example 3 in Section 8.4.)

18. (a) Show that the characteristic polynomial of a 2×2 matrix $\mathbf{A}$ is $x^2 - (\text{trace}(\mathbf{A}))x + |\mathbf{A}|$.

(b) Prove that the characteristic polynomial of an $n \times n$ matrix always has degree n, with the coefficient of x^n equal to 1.

(c) If $\mathbf{A}$ is an $n \times n$ matrix, show that the constant term of $p_{\mathbf{A}}(x)$ is $(-1)^n|\mathbf{A}|$.

(d) If $\mathbf{A}$ is an $n \times n$ matrix, show that the coefficient of x^{n-1} in $p_{\mathbf{A}}(x)$ is $-\text{trace}(\mathbf{A})$.

6.2 Diagonalization

In Examples 4 and 5 of Section 6.1, you saw that certain linear operators can be represented by a diagonal matrix, as long as we use the correct ordered basis. Using a diagonal matrix makes working with a linear operator much easier. Our goal in this section is to determine which operators have a diagonal form and to present a method for finding the correct basis when such a form exists.

Throughout the remainder of the book, we will frequently use the characteristic polynomial to find the eigenvalues of an operator. However, calculating the characteristic polynomial, and finding its roots, is often difficult. In these cases, you should use appropriate computer software to find the eigenvalues instead of computing them by hand.

Diagonalizable Operators

DEFINITION

An $n \times n$ matrix $\mathbf{A}$ is **diagonalizable** if and only if there is a nonsingular matrix $\mathbf{P}$ such that $\mathbf{P}^{-1}\mathbf{A}\mathbf{P}$ is a diagonal matrix.

A linear operator L on a finite dimensional vector space $\mathcal{V}$ is **diagonalizable** if and only if the matrix representation of L with respect to some ordered basis for $\mathcal{V}$ is a diagonal matrix.

Note that a diagonalizable matrix is one that is similar to a diagonal matrix.

Example 1

The matrix

$$\begin{bmatrix} 11 & -10 & 4 \\ 12 & -15 & 8 \\ 12 & -18 & 11 \end{bmatrix}$$

is diagonalizable, since

$$\begin{bmatrix} 1 & -1 & -1 \\ 1 & -2 & -2 \\ 1 & -2 & -3 \end{bmatrix}^{-1} \begin{bmatrix} 11 & -10 & 4 \\ 12 & -15 & 8 \\ 12 & -18 & 11 \end{bmatrix} \begin{bmatrix} 1 & -1 & -1 \\ 1 & -2 & -2 \\ 1 & -2 & -3 \end{bmatrix}$$
$$= \begin{bmatrix} 2 & -1 & 0 \\ 1 & -2 & 1 \\ 0 & 1 & -1 \end{bmatrix} \begin{bmatrix} 11 & -10 & 4 \\ 12 & -15 & 8 \\ 12 & -18 & 11 \end{bmatrix} \begin{bmatrix} 1 & -1 & -1 \\ 1 & -2 & -2 \\ 1 & -2 & -3 \end{bmatrix} = \begin{bmatrix} 5 & 0 & 0 \\ 0 & -1 & 0 \\ 0 & 0 & 3 \end{bmatrix},$$

a diagonal matrix. ■

Example 2

The linear operator on $\mathbb{R}^3$ that performs an orthogonal projection onto the xy-plane is diagonalizable, since its matrix representation with respect to the standard basis is the following diagonal matrix:

$$\begin{bmatrix} 1 & 0 & 0 \\ 0 & 1 & 0 \\ 0 & 0 & 0 \end{bmatrix}.$$

■

The following theorem explains the connection between diagonalizable matrices and diagonalizable operators:

THEOREM 6.9

Let $\mathcal{V}$ be a finite dimensional vector space and let L be a linear operator on $\mathcal{V}$. Let B be any ordered basis for $\mathcal{V}$. Then L is a diagonalizable linear operator if and only if the matrix representation $\mathbf{A}$ of L with respect to B is a diagonalizable matrix.

Essentially, this theorem says that diagonalizable matrices are precisely those that represent diagonalizable operators. Thus, when discussing diagonalization, we can refer interchangeably to either the linear operator itself or its matrix representation.

The proof of Theorem 6.9, which follows quickly from Theorem 5.6, is outlined in Exercise 9.

Criterion for Diagonalization

The next result indicates precisely which linear operators (and hence, which matrices) are diagonalizable:

THEOREM 6.10

Let L be a linear operator on an n-dimensional vector space $\mathcal{V}$. Then L is diagonalizable if and only if there is a set of n linearly independent eigenvectors for L.

Proof of Theorem 6.10

Suppose that L is diagonalizable. Then there is an ordered basis $B = (\mathbf{v}_1, \ldots, \mathbf{v}_n)$ for $\mathcal{V}$ such that the matrix representation for L with respect to B is a diagonal matrix $\mathbf{D}$. Now, B is a linearly independent set. If we can show that each vector $\mathbf{v}_i$ in B, for $1 \leq i \leq n$, is an eigenvector corresponding to some eigenvalue for L, then B will be a set of n linearly independent eigenvectors for L.

Now, for each $\mathbf{v}_i$, with $1 \leq i \leq n$, we have $[L(\mathbf{v}_i)]_B = \mathbf{D}[\mathbf{v}_i]_B = \mathbf{D}\mathbf{e}_i = d_{ii}\mathbf{e}_i = d_{ii}[\mathbf{v}_i]_B$, where d_{ii} is the (i, i) entry of $\mathbf{D}$. Thus, $L(\mathbf{v}_i) = d_{ii}\mathbf{v}_i$, and so $\mathbf{v}_i$ is an eigenvector for L corresponding to the eigenvalue d_{ii}.

Conversely, suppose that $B = \{\mathbf{w}_1, \ldots, \mathbf{w}_n\}$ is a set of n linearly independent eigenvectors for L, corresponding to the (not necessarily distinct) eigenvalues $\lambda_1, \ldots, \lambda_n$, respectively. Since B contains $n = \dim(\mathcal{V})$ linearly independent vectors, B is a basis for $\mathcal{V}$. We show that the matrix $\mathbf{A}$ for L with respect to B is, in fact, diagonal (and hence diagonalizable). Now,

$$i^{th} \text{ column of } \mathbf{A} = [L(\mathbf{w}_i)]_B = [\lambda_i \mathbf{w}_i]_B = \lambda_i[\mathbf{w}_i]_B = \lambda_i \mathbf{e}_i.$$

Thus, $\mathbf{A}$ is a diagonal matrix. ■

Example 3

Suppose that $L: \mathbb{R}^3 \longrightarrow \mathbb{R}^3$ is given by

$$L\left(\begin{bmatrix} x_1 \\ x_2 \\ x_3 \end{bmatrix}\right) = \begin{bmatrix} 18 & -6 & -30 \\ -25 & 10 & 45 \\ 17 & -6 & -29 \end{bmatrix}\begin{bmatrix} x_1 \\ x_2 \\ x_3 \end{bmatrix}.$$

Then $p_L(x) = x^3 + x^2 - 2x = x(x+2)(x-1)$, and so $\lambda_1 = 0$, $\lambda_2 = -2$, and $\lambda_3 = 1$ are eigenvalues for L. Solving for eigenvectors corresponding to these eigenvalues yields $\mathbf{v}_1 = [1, -2, 1]$, $\mathbf{v}_2 = [3, -5, 3]$, and $\mathbf{v}_3 = [0, -5, 1]$, respectively.

Theorem 6.4 assures us that the set $\{\mathbf{v}_1, \mathbf{v}_2, \mathbf{v}_3\}$ is linearly independent, since λ_1, λ_2, and λ_3 are distinct eigenvalues. Thus, we have a set of three linearly independent eigenvectors for L, and so L is diagonalizable by Theorem 6.10.

The proof of Theorem 6.10 shows that the matrix representation of L with respect to $B = (\mathbf{v}_1, \mathbf{v}_2, \mathbf{v}_3)$ is the diagonal matrix

$$\begin{bmatrix} \lambda_1 & 0 & 0 \\ 0 & \lambda_2 & 0 \\ 0 & 0 & \lambda_3 \end{bmatrix} = \begin{bmatrix} 0 & 0 & 0 \\ 0 & -2 & 0 \\ 0 & 0 & 1 \end{bmatrix}.$$

We can check this result either by computing this matrix representation directly (which is easy to do) or instead by doing the following: First, form the matrix $\mathbf{P}$ whose columns are the vectors $\mathbf{v}_1, \mathbf{v}_2$, and $\mathbf{v}_3$. Recall that $\mathbf{P}$ is the transition matrix from B-coordinates to standard coordinates. Then, by Theorem 5.6, the matrix representation of L with respect to B-coordinates is

$$\begin{aligned} \mathbf{P}^{-1}\begin{bmatrix} 18 & -6 & -30 \\ -25 & 10 & 45 \\ 17 & -6 & -29 \end{bmatrix}\mathbf{P} &= \begin{bmatrix} 10 & -3 & -15 \\ -3 & 1 & 5 \\ -1 & 0 & 1 \end{bmatrix}\begin{bmatrix} 18 & -6 & -30 \\ -25 & 10 & 45 \\ 17 & -6 & -29 \end{bmatrix}\begin{bmatrix} 1 & 3 & 0 \\ -2 & -5 & -5 \\ 1 & 3 & 1 \end{bmatrix} \\ &= \begin{bmatrix} 0 & 0 & 0 \\ 0 & -2 & 0 \\ 0 & 0 & 1 \end{bmatrix}. \end{aligned}$$

■

As illustrated in Example 3, Theorems 6.4 and 6.10 combine to prove the following:

COROLLARY 6.11

If L is a linear operator on an n-dimensional vector space and L has n distinct eigenvalues, then L is diagonalizable.

The converse to this corollary is false, since it is possible to get n linearly independent eigenvectors from fewer than n eigenvalues provided that the geometric multiplicities of the eigenvalues are high enough (see Exercise 10). We will also see a diagonalizable operator on $\mathbb{R}^4$ with fewer than four eigenvalues in Example 4.

Method for Diagonalizing a Linear Operator

Next we formally outline the method for diagonalizing a given linear operator L, when possible. That is, we show how to find a basis B so that the matrix for L with respect to B is diagonal.

METHOD FOR DIAGONALIZING A LINEAR OPERATOR (WHEN POSSIBLE)

Let $L: \mathcal{V} \longrightarrow \mathcal{V}$ be a linear operator on an n dimensional vector space $\mathcal{V}$.

Step 1: Find a basis C for $\mathcal{V}$. (If $\mathcal{V} = \mathbb{R}^n$, you can use the standard basis.)
Step 2: Calculate the matrix representation $\mathbf{A}$ of L with respect to C.
Step 3: Solve for all of the eigenvalues $\lambda_1, \ldots, \lambda_k$ of $\mathbf{A}$ (which can often be done by using the characteristic polynomial of $\mathbf{A}$ or appropriate software).
Step 4: Find a basis for each eigenspace E_{λ_i} of $\mathbf{A}$ (which can often be done by solving an appropriate homogeneous system).
Step 5: If the union of the bases from step 4 contains fewer than n elements, then L is not diagonalizable. Otherwise, let $B = (\mathbf{v}_1, \ldots, \mathbf{v}_n)$ be the ordered basis for $\mathcal{V}$ formed by taking this union.
Step 6: The matrix representation for L with respect to B is the diagonal matrix $\mathbf{D}$ whose (i, i) entry d_{ii} equals the eigenvalue for L corresponding to $\mathbf{v}_i$.
Step 7: In most practical situations, the transition matrix $\mathbf{P}$ from B to C coordinates is useful; $\mathbf{P}$ is the $n \times n$ matrix whose columns are $[\mathbf{v}_1]_C, \ldots, [\mathbf{v}_n]_C$. Note that $\mathbf{D} = \mathbf{P}^{-1}\mathbf{AP}$.

Step 5 of this algorithm assumes that the union of bases for distinct eigenspaces yields a linearly independent set. This fact is a generalization of Theorem 6.4, the proof of which is outlined in Exercise 17.

Example 4

We use the method outlined above to diagonalize the operator $L: \mathbb{R}^4 \longrightarrow \mathbb{R}^4$ given by $L(\mathbf{v}) = \mathbf{Av}$, where

$$\mathbf{A} = \begin{bmatrix} 5 & 0 & -8 & 8 \\ 8 & 1 & -16 & 16 \\ -4 & 0 & 9 & -8 \\ -8 & 0 & 16 & -15 \end{bmatrix}.$$

Steps 1 and 2: Since $\mathcal{V} = \mathbb{R}^4$, we let C be the standard basis for $\mathbb{R}^4$. Then the matrix representation for L with respect to C is the matrix $\mathbf{A}$ given above.

Step 3: A lengthy computation produces the characteristic polynomial

$$p_{\mathbf{A}}(x) = x^4 - 6x^2 + 8x - 3 = (x-1)^3(x+3).$$

Thus, the eigenvalues for $\mathbf{A}$ are $\lambda_1 = 1$ and $\lambda_2 = -3$.

Step 4: Solving the homogeneous system $(\lambda_1\mathbf{I}_4 - \mathbf{A})\mathbf{v} = \mathbf{0}$ yields a basis for the eigenspace E_{λ_1}. Since $\lambda_1 = 1$, solving this system amounts to row reducing

$$\left[\begin{array}{rrrr|r} -4 & 0 & 8 & -8 & 0 \\ -8 & 0 & 16 & -16 & 0 \\ 4 & 0 & -8 & 8 & 0 \\ 8 & 0 & -16 & 16 & 0 \end{array}\right] \text{ to get } \left[\begin{array}{rrrr|r} 1 & 0 & -2 & 2 & 0 \\ 0 & 0 & 0 & 0 & 0 \\ 0 & 0 & 0 & 0 & 0 \\ 0 & 0 & 0 & 0 & 0 \end{array}\right].$$

The solution space of this system is three-dimensional and is spanned by $\{[0, 1, 0, 0], [2, 0, 1, 0], [-2, 0, 0, 1]\}$. (Why?) Therefore, this set is a basis for E_{λ_1}.

Working similarly to find a basis for E_{λ_2} yields

$$\text{Basis for } E_{\lambda_2} = \{[1, 2, -1, -2]\}.$$

Step 5: Set $\mathbf{v}_1 = [0, 1, 0, 0]$, $\mathbf{v}_2 = [2, 0, 1, 0]$, $\mathbf{v}_3 = [-2, 0, 0, 1]$, and $\mathbf{v}_4 = [1, 2, -1, -2]$. Then $B = (\mathbf{v}_1, \mathbf{v}_2, \mathbf{v}_3, \mathbf{v}_4)$ is an ordered basis for $\mathbb{R}^4$.

Step 6: The matrix representation of L with respect to B is the 4×4 diagonal matrix $\mathbf{D}$ with d_{ii} equal to the eigenvalue corresponding to $\mathbf{v}_i$, for $1 \leq i \leq 4$. In particular,

$$\mathbf{D} = \begin{bmatrix} 1 & 0 & 0 & 0 \\ 0 & 1 & 0 & 0 \\ 0 & 0 & 1 & 0 \\ 0 & 0 & 0 & -3 \end{bmatrix}.$$

Step 7: The transition matrix $\mathbf{P}$ from B-coordinates to standard coordinates is formed by using $\mathbf{v}_1$, $\mathbf{v}_2$, $\mathbf{v}_3$, and $\mathbf{v}_4$ as columns. Hence,

$$\mathbf{P} = \begin{bmatrix} 0 & 2 & -2 & 1 \\ 1 & 0 & 0 & 2 \\ 0 & 1 & 0 & -1 \\ 0 & 0 & 1 & -2 \end{bmatrix}.$$

After calculating $\mathbf{P}^{-1}$, you can easily verify that

$$\mathbf{P}^{-1}\mathbf{A}\mathbf{P} = \begin{bmatrix} 2 & 1 & -4 & 4 \\ -1 & 0 & 3 & -2 \\ -2 & 0 & 4 & -3 \\ -1 & 0 & 2 & -2 \end{bmatrix} \begin{bmatrix} 5 & 0 & -8 & 8 \\ 8 & 1 & -16 & 16 \\ -4 & 0 & 9 & -8 \\ -8 & 0 & 16 & -15 \end{bmatrix} \begin{bmatrix} 0 & 2 & -2 & 1 \\ 1 & 0 & 0 & 2 \\ 0 & 1 & 0 & -1 \\ 0 & 0 & 1 & -2 \end{bmatrix} = \mathbf{D}.$$ ■

In the next example, the linear operator is not originally defined by matrix multiplication, and so step 2 of the process requires some work.

Example 5

Let $L: \mathcal{P}_3 \longrightarrow \mathcal{P}_3$ be given by $L(f(x)) = xf'(x) + f(x+1)$. We want to find an ordered basis B for $\mathcal{P}_3$ such that the matrix representation of L with respect to B is diagonal.

Step 1: Let $C = (x^3, x^2, x, 1)$, the standard basis for $\mathcal{P}_3$.

Step 2: Next, we find the matrix for L with respect to C. Calculating directly, we get

$$\begin{aligned} L(x^3) &= x(3x^2) + (x+1)^3 &&= 4x^3 + 3x^2 + 3x + 1, \\ L(x^2) &= x(2x) + (x+1)^2 &&= 3x^2 + 2x + 1, \\ L(x) &= x(1) + (x+1) &&= 2x + 1, \\ \text{and} \quad L(1) &= x(0) + 1 &&= 1. \end{aligned}$$

Thus, the matrix for L with respect to C is

$$\mathbf{A} = \begin{bmatrix} 4 & 0 & 0 & 0 \\ 3 & 3 & 0 & 0 \\ 3 & 2 & 2 & 0 \\ 1 & 1 & 1 & 1 \end{bmatrix}.$$

Step 3: The characteristic polynomial of $\mathbf{A}$ is clearly $p_{\mathbf{A}}(x) = (x-4)(x-3)(x-2)(x-1)$, since $\mathbf{A}$ is lower triangular. Thus, the eigenvalues for $\mathbf{A}$ are $\lambda_1 = 4$, $\lambda_2 = 3$, $\lambda_3 = 2$, and $\lambda_4 = 1$.

Step 4: Solving for a basis for each eigenspace of $\mathbf{A}$ gives: basis for $E_{\lambda_1} = \{[6, 18, 27, 17]\}$, basis for $E_{\lambda_2} = \{[0, 2, 4, 3]\}$, basis for $E_{\lambda_3} = \{[0, 0, 1, 1]\}$, and basis for $E_{\lambda_4} = \{[0, 0, 0, 1]\}$.

Step 5: An ordered basis $B = (\mathbf{v}_1, \mathbf{v}_2, \mathbf{v}_3, \mathbf{v}_4)$ for $\mathcal{P}_3$ is found by taking the polynomials corresponding to the basis vectors in step 2 — namely, $\mathbf{v}_1 = 6x^3 + 18x^2 + 27x + 17$, $\mathbf{v}_2 = 2x^2 + 4x + 3$, $\mathbf{v}_3 = x + 1$, and $\mathbf{v}_4 = 1$.

Step 6: The diagonal matrix

$$\mathbf{D} = \begin{bmatrix} 4 & 0 & 0 & 0 \\ 0 & 3 & 0 & 0 \\ 0 & 0 & 2 & 0 \\ 0 & 0 & 0 & 1 \end{bmatrix}$$

is the matrix representation of L in B-coordinates and has the eigenvalues of L appearing on the main diagonal.

Step 7: The transition matrix $\mathbf{P}$ from B-coordinates to C-coordinates is

$$\mathbf{P} = \begin{bmatrix} 6 & 0 & 0 & 0 \\ 18 & 2 & 0 & 0 \\ 27 & 4 & 1 & 0 \\ 17 & 3 & 1 & 1 \end{bmatrix}.$$

It can easily be verified that $\mathbf{D} = \mathbf{P}^{-1}\mathbf{A}\mathbf{P}$. ■

Application: Large Powers of a Matrix

If $\mathbf{D}$ is a diagonal matrix, any positive (integer) power of $\mathbf{D}$ can be obtained by merely raising the diagonal entries of $\mathbf{D}$ to that power. For example,

$$\begin{bmatrix} 3 & 0 \\ 0 & -2 \end{bmatrix}^{12} = \begin{bmatrix} 3^{12} & 0 \\ 0 & (-2)^{12} \end{bmatrix} = \begin{bmatrix} 531441 & 0 \\ 0 & 4096 \end{bmatrix}.$$

Now, suppose that $\mathbf{A}$ is an $n \times n$ diagonalizable matrix. Then there exists a nonsingular matrix $\mathbf{P}$ such that $\mathbf{P}^{-1}\mathbf{AP} = \mathbf{D}$, a diagonal matrix. Thus, $\mathbf{A} = \mathbf{PDP}^{-1}$. But then,

$$\mathbf{A}^2 = \mathbf{AA} = (\mathbf{PDP}^{-1})(\mathbf{PDP}^{-1}) = \mathbf{PD}(\mathbf{P}^{-1}\mathbf{P})\mathbf{DP}^{-1} = \mathbf{PDI}_n\mathbf{DP}^{-1} = \mathbf{PD}^2\mathbf{P}^{-1}.$$

More generally, a straightforward proof by induction can be used to show that, for all positive integers k, $\mathbf{A}^k = \mathbf{PD}^k\mathbf{P}^{-1}$ (see Exercise 12). Hence, if $\mathbf{P}$ and $\mathbf{D}$ are known, calculating positive powers of $\mathbf{A}$ is much easier.

Example 6

Consider the following matrix from Example 3:

$$\mathbf{A} = \begin{bmatrix} 18 & -6 & -30 \\ -25 & 10 & 45 \\ 17 & -6 & -29 \end{bmatrix}.$$

We will calculate $\mathbf{A}^{11}$. Example 3 showed that $\mathbf{P}^{-1}\mathbf{AP} = \mathbf{D}$, where

$$\mathbf{P} = \begin{bmatrix} 1 & 3 & 0 \\ -2 & -5 & -5 \\ 1 & 3 & 1 \end{bmatrix}, \ \mathbf{P}^{-1} = \begin{bmatrix} 10 & -3 & -15 \\ -3 & 1 & 5 \\ -1 & 0 & 1 \end{bmatrix}, \ \text{and } \mathbf{D} = \begin{bmatrix} 0 & 0 & 0 \\ 0 & -2 & 0 \\ 0 & 0 & 1 \end{bmatrix}.$$

Therefore, $\mathbf{A} = \mathbf{PDP}^{-1}$. We then have

$$\begin{aligned} \mathbf{A}^{11} &= \mathbf{PD}^{11}\mathbf{P}^{-1} \\ &= \begin{bmatrix} 1 & 3 & 0 \\ -2 & -5 & -5 \\ 1 & 3 & 1 \end{bmatrix} \begin{bmatrix} 0 & 0 & 0 \\ 0 & -2048 & 0 \\ 0 & 0 & 1 \end{bmatrix} \begin{bmatrix} 10 & -3 & -15 \\ -3 & 1 & 5 \\ -1 & 0 & 1 \end{bmatrix} \\ &= \begin{bmatrix} 18432 & -6144 & -30720 \\ -30715 & 10240 & 51195 \\ 18431 & -6144 & -30719 \end{bmatrix}. \end{aligned}$$

■

Application: Orthogonal Projections and Reflections in $\mathbb{R}^3$

Suppose that L: $\mathbb{R}^3 \longrightarrow \mathbb{R}^3$ represents the orthogonal projection onto a certain plane through the origin. Let $\mathbf{A}$ be the matrix for L relative to the standard basis for $\mathbb{R}^3$. Now, if $\mathbf{v}_1$ is a vector normal (perpendicular) to this plane, then $\mathbf{v}_1$ spans ker(L). Also, if $\mathbf{v}_2$ and $\mathbf{v}_3$ are linearly independent vectors in this plane, then $\{\mathbf{v}_2, \mathbf{v}_3\}$ spans the plane. Thus, $L(\mathbf{v}_1) = \mathbf{0}$, $L(\mathbf{v}_2) = \mathbf{v}_2$, and $L(\mathbf{v}_3) = \mathbf{v}_3$. Hence, $\mathbf{v}_1$ is an eigenvector corresponding to $\lambda_1 = 0$, and $\{\mathbf{v}_2, \mathbf{v}_3\}$ is a basis for the eigenspace corresponding to $\lambda_2 = 1$. Therefore, the matrix representation with respect to the ordered basis $B = (\mathbf{v}_1, \mathbf{v}_2, \mathbf{v}_3)$ is

$$\mathbf{D} = \begin{bmatrix} 0 & 0 & 0 \\ 0 & 1 & 0 \\ 0 & 0 & 1 \end{bmatrix}.$$

Now, $\mathbf{D} = \mathbf{P}^{-1}\mathbf{A}\mathbf{P}$, where $\mathbf{P}$ is the transition matrix from B to the standard basis. (Why?) Notice that, since the columns of $\mathbf{P}$ are equal to $\mathbf{v}_1$, $\mathbf{v}_2$, and $\mathbf{v}_3$, respectively, and since $\mathbf{v}_1 \perp \mathbf{v}_2$ and $\mathbf{v}_1 \perp \mathbf{v}_3$, the transition matrix $\mathbf{P}$ must have its first column orthogonal to the other columns.

Every orthogonal projection operator onto a plane in $\mathbb{R}^3$ is diagonalizable in this fashion. That is,

> A linear operator on $\mathbb{R}^3$ represents an orthogonal projection onto a plane through the origin if and only if its matrix with respect to the standard basis can be diagonalized to the matrix $\mathbf{D}$ given above and the transition matrix used in the diagonalization has its first column orthogonal to the other columns.

A diagonalized matrix for this orthogonal projection might have its diagonal entries in a different order if the basis used does not have the normal vector to the plane listed first. Not listing the normal vector first can also change which pairs of columns of the transition matrix are orthogonal. Reordering the basis will make the diagonal matrix and the transition matrix fit the given description.

Example 7

The linear operator from Example 3 diagonalizes as

$$\begin{bmatrix} 0 & 0 & 0 \\ 0 & -2 & 0 \\ 0 & 0 & 1 \end{bmatrix}.$$

Since one of the diagonal entries is -2, this operator is not an orthogonal projection onto a plane through the origin.

Similarly, you can verify that

$$\mathbf{A}_1 = \begin{bmatrix} -3 & 1 & -1 \\ 16 & -3 & 4 \\ 28 & -7 & 8 \end{bmatrix} \quad \text{diagonalizes to} \quad \mathbf{D}_1 = \begin{bmatrix} 0 & 0 & 0 \\ 0 & 1 & 0 \\ 0 & 0 & 1 \end{bmatrix}.$$

Now, $\mathbf{D}_1$ clearly has the proper form. However, the transition matrix $\mathbf{P}_1$ used in the diagonalization is found to be

$$\mathbf{P}_1 = \begin{bmatrix} -1 & 0 & 1 \\ 4 & -1 & -6 \\ 7 & -1 & -10 \end{bmatrix}.$$

Since the first column of $\mathbf{P}_1$ is not orthogonal to the other two columns of $\mathbf{P}_1$, $\mathbf{A}_1$ does not represent an orthogonal projection onto a plane through the origin.

In contrast, the matrix

$$\mathbf{A}_2 = \frac{1}{14}\begin{bmatrix} 5 & -3 & -6 \\ -3 & 13 & -2 \\ -6 & -2 & 10 \end{bmatrix} \quad \text{diagonalizes to} \quad \mathbf{D}_2 = \begin{bmatrix} 0 & 0 & 0 \\ 0 & 1 & 0 \\ 0 & 0 & 1 \end{bmatrix}$$

with transition matrix

$$\mathbf{P}_2 = \begin{bmatrix} 3 & -4 & 1 \\ 1 & 2 & -1 \\ 2 & 5 & -1 \end{bmatrix}.$$

Clearly, $\mathbf{D}_2$ has the correct form, as does $\mathbf{P}_2$, since the first column of $\mathbf{P}_2$ is orthogonal to both its second and third columns. Hence, $\mathbf{A}_2$ does represent an orthogonal projection onto a plane through the origin in $\mathbb{R}^3$. In fact, it is the orthogonal projection onto the plane $3x + y + 2z = 0$—that is, all vectors $[x, y, z]$ orthogonal to the first column of $\mathbf{P}_2$. ■

We can discover which linear operators are **orthogonal reflections** through a plane through the origin in $\mathbb{R}^3$ in a similar manner. The only difference is that the vector $\mathbf{v}_1$ normal to the plane now corresponds to the eigenvalue $\lambda_1 = -1$ (instead of $\lambda_1 = 0$), since $\mathbf{v}_1$ reflects through the plane into $-\mathbf{v}_1$.

Example 8

The matrix

$$\mathbf{A} = \frac{1}{35}\begin{bmatrix} -15 & 10 & -30 \\ 10 & 33 & 6 \\ -30 & 6 & 17 \end{bmatrix}$$

represents an orthogonal reflection through the plane $5x - y + 3z = 0$, because $\mathbf{A}$ diagonalizes to

$$\mathbf{D} = \begin{bmatrix} -1 & 0 & 0 \\ 0 & 1 & 0 \\ 0 & 0 & 1 \end{bmatrix} \quad \text{with transition matrix} \quad \mathbf{P} = \begin{bmatrix} 5 & 2 & -1 \\ -1 & -2 & 4 \\ 3 & -4 & 3 \end{bmatrix}.$$

Notice that $\mathbf{D}$ has the right form and that the first column of $\mathbf{P}$ (corresponding to the -1 diagonal entry of $\mathbf{D}$) is orthogonal to the other columns of $\mathbf{P}$. ■

Nondiagonalizable Operators

Remember that not every matrix is diagonalizable—only those that have n linearly independent eigenvectors. For example, as we have seen, a clockwise rotation of $\mathbb{R}^2$ through an angle θ (where $\theta \neq 0, \pi$) has no eigenvalues and hence no eigenvectors. Thus, a rotation of the plane is not generally diagonalizable. Similarly, a rotation about an axis in $\mathbb{R}^3$ (again with $\theta \neq 0, \pi$) has only one eigenvalue ($\lambda = 1$) with a one-dimensional eigenspace. (Why?) Thus it is impossible to find three linearly independent eigenvectors, and so a rotation about an axis in $\mathbb{R}^3$ is also generally not diagonalizable.

Many other types of operators have corresponding matrices that are not diagonalizable. For instance, in Example 10 of Section 6.1, we were able to find only three linearly independent eigenvectors for the 4×4 matrix

$$\mathbf{A} = \begin{bmatrix} 5 & 2 & 0 & 1 \\ -2 & 1 & 0 & -1 \\ 4 & 4 & 3 & 2 \\ 16 & 0 & -8 & -5 \end{bmatrix},$$

and so $\mathbf{A}$ is not diagonalizable.

♦ **Application:** You have now covered the prerequisites for Section 8.8, "Differential Equations."

Exercises—Section 6.2

Note: Use appropriate software (when needed) to help solve for eigenvalues and corresponding eigenvectors of operators in these exercises.

1. Each of the following matrices $\mathbf{A}$ is the representation of a linear operator $L: \mathbb{R}^n \longrightarrow \mathbb{R}^n$ with respect to the standard basis. Follow steps 3 through 5 of the method given in the text to determine whether L is diagonalizable. If L is diagonalizable, finish the method by performing steps 6 and 7. In particular, find:

(i) An ordered basis B for $\mathbb{R}^n$ consisting of eigenvectors for L
(ii) The transition matrix $\mathbf{P}$ from B to standard coordinates
(iii) The diagonal matrix $\mathbf{D}$ that is the matrix representation of L with respect to B

Finally, check your work by verifying that $\mathbf{D} = \mathbf{P}^{-1}\mathbf{A}\mathbf{P}$.

★(a) $\mathbf{A} = \begin{bmatrix} 19 & -48 \\ 8 & -21 \end{bmatrix}$ (b) $\mathbf{A} = \begin{bmatrix} -18 & 40 \\ -8 & 18 \end{bmatrix}$ ★(c) $\mathbf{A} = \begin{bmatrix} 13 & -34 \\ 5 & -13 \end{bmatrix}$

★(d) $\mathbf{A} = \begin{bmatrix} -13 & -3 & 18 \\ -20 & -4 & 26 \\ -14 & -3 & 19 \end{bmatrix}$ (e) $\mathbf{A} = \begin{bmatrix} -3 & 3 & -1 \\ 2 & 2 & 4 \\ 6 & -3 & 4 \end{bmatrix}$

★(f) $\mathbf{A} = \begin{bmatrix} 5 & -8 & -12 \\ -2 & 3 & 4 \\ 4 & -6 & -9 \end{bmatrix}$ ★(g) $\mathbf{A} = \begin{bmatrix} 2 & 0 & 0 \\ -3 & 4 & 1 \\ 3 & -2 & 1 \end{bmatrix}$

(h) $\mathbf{A} = \begin{bmatrix} -5 & 18 & 6 \\ -2 & 7 & 2 \\ 1 & -3 & 0 \end{bmatrix}$ ★(i) $\mathbf{A} = \begin{bmatrix} 3 & 1 & -6 & -2 \\ 4 & 0 & -6 & -4 \\ 2 & 0 & -3 & -2 \\ 0 & 1 & -2 & 1 \end{bmatrix}$

2. Each of the following represents a linear operator L on a vector space $\mathcal{V}$. Follow steps 1 through 5 of the method given in the text to determine whether L is diagonalizable. Let C be the standard basis in each case, and let $\mathbf{A}$ be the matrix representation of L with respect to C. If L is diagonalizable, finish the method by performing steps 6 and 7. In particular, find:

(i) An ordered basis B of $\mathcal{V}$ consisting of eigenvectors for L
(ii) The transition matrix $\mathbf{P}$ from B to C
(iii) The diagonal matrix $\mathbf{D}$ that is the matrix representation of L with respect to B

Finally, check your work by verifying that $\mathbf{D} = \mathbf{P}^{-1}\mathbf{AP}$.

★(a) $L: \mathcal{P}_2 \longrightarrow \mathcal{P}_2$ given by $L(p(x)) = (x-1)p'(x)$
(b) $L: \mathcal{P}_3 \longrightarrow \mathcal{P}_3$ given by $L(p(x)) = x^2 p''(x-1) - p(x) + 2p'(1)$
(c) $L: \mathcal{M}_{22} \longrightarrow \mathcal{M}_{22}$ given by $L(\mathbf{K}) = \mathbf{K}^T$
★(d) $L: \mathcal{M}_{22} \longrightarrow \mathcal{M}_{22}$ given by $L(\mathbf{K}) = \begin{bmatrix} -4 & 3 \\ -10 & 7 \end{bmatrix}\mathbf{K}$

3. Use diagonalization to calculate the indicated powers of $\mathbf{A}$ in each case.

★(a) $\mathbf{A}^{15}$, where $\mathbf{A} = \begin{bmatrix} 4 & -6 \\ 3 & -5 \end{bmatrix}$

(b) $\mathbf{A}^{30}$, where $\mathbf{A} = \begin{bmatrix} 11 & -6 & -12 \\ 13 & -6 & -16 \\ 5 & -3 & -5 \end{bmatrix}$

★(c) $\mathbf{A}^{49}$, where $\mathbf{A}$ is the matrix of part (b)

(d) $\mathbf{A}^{11}$, where $\mathbf{A} = \begin{bmatrix} 4 & -4 & 6 \\ -1 & 2 & -1 \\ -1 & 4 & -3 \end{bmatrix}$

★(e) $\mathbf{A}^{10}$, where $\mathbf{A} = \begin{bmatrix} 7 & 9 & -12 \\ 10 & 16 & -22 \\ 8 & 12 & -16 \end{bmatrix}$

4. In each part, let L be the linear operator on $\mathbb{R}^3$ with the given matrix representation with respect to the standard basis. Determine whether L is

(i) An orthogonal projection onto a plane through the origin
(ii) An orthogonal reflection through a plane through the origin
(iii) Neither

If L is of type (i) or (ii), give the equation of the plane involved.

★(a) $\frac{1}{11}\begin{bmatrix} 2 & -3 & -3 \\ -3 & 10 & -1 \\ -3 & -1 & 10 \end{bmatrix}$

(b) $\frac{1}{9}\begin{bmatrix} 7 & 4 & 4 \\ 4 & 1 & -8 \\ 4 & -8 & 1 \end{bmatrix}$

(c) $\frac{1}{3}\begin{bmatrix} 11 & 49 & -77 \\ -18 & -66 & 99 \\ -10 & -35 & 52 \end{bmatrix}$

★(d) $\frac{1}{15}\begin{bmatrix} 7 & -2 & -14 \\ -4 & 14 & -7 \\ -12 & -3 & -6 \end{bmatrix}$

5. Let $\mathbf{A}$ be an $n \times n$ matrix. Suppose that $(\mathbf{v}_1, \ldots, \mathbf{v}_n)$ is a basis for $\mathbb{R}^n$ of eigenvectors for $\mathbf{A}$ with corresponding eigenvalues $\lambda_1, \lambda_2, \ldots, \lambda_n$. Show that $|\mathbf{A}| = \lambda_1\lambda_2 \cdots \lambda_n$.

6. Let $\mathbf{A}$ be a diagonalizable $n \times n$ matrix.

(a) Show that $\mathbf{A}$ has a cube root—that is, that there is a matrix $\mathbf{B}$ such that $\mathbf{B}^3 = \mathbf{A}$.

★(b) Give a sufficient condition for $\mathbf{A}$ to have a square root. Prove that your condition works.

7. Let $\mathbf{A}$ be an $n \times n$ upper triangular matrix with all main diagonal entries distinct. Prove that $\mathbf{A}$ is diagonalizable.

8. Let $\mathbf{A}$ be an $n \times n$ upper triangular matrix with all main diagonal entries equal. Show that $\mathbf{A}$ is diagonalizable if and only if $\mathbf{A}$ is a diagonal matrix.

9. This exercise outlines the proof of Theorem 6.9. Let L be a linear operator on a finite dimensional vector space $\mathcal{V}$, and let B be an ordered basis for $\mathcal{V}$. Also, let $\mathbf{A}$ be the matrix for L with respect to B.

(a) Assume that L is diagonalizable, and let C be an ordered basis for $\mathcal{V}$ such that the matrix representation of L with respect to C is diagonal. If $\mathbf{P}$ is the transition matrix from C to B, prove that $\mathbf{P}^{-1}\mathbf{AP}$ is a diagonal matrix and hence that $\mathbf{A}$ is diagonalizable.

(b) Assume that $\mathbf{A}$ is diagonalizable. Prove that there is an ordered basis C for $\mathcal{V}$ such that the matrix representation of L with respect to C is diagonal and hence that L is a diagonalizable operator.

10. Let L be a linear operator on an n-dimensional vector space, with $\{\lambda_1, \ldots, \lambda_k\}$ equal to the set of all distinct eigenvalues for L.
  (a) Show that $n \geq \sum_{i=1}^{k}$ (geometric multiplicity of λ_i).
  (b) Show that $n = \sum_{i=1}^{k}$ (geometric multiplicity of λ_i) if and only if L is diagonalizable.

11. Let L be a linear operator on a finite dimensional vector space $\mathcal{V}$.
  (a) Prove that if L is diagonalizable, then the algebraic multiplicity of λ equals the geometric multiplicity of λ for every eigenvalue λ of L.
  ★(b) Disprove the converse to part (a) by finding a linear operator L on $\mathbb{R}^3$ that provides a counterexample. (Hint: Try a rotation about an axis.)
  (c) Show that if L is diagonalizable, then every root of $p_L(x)$ is real.

12. Let **A**, **P**, and **D** be $n \times n$ matrices with **P** nonsingular and $\mathbf{P}^{-1}\mathbf{AP} = \mathbf{D}$. Use a proof by induction to show that $\mathbf{A}^k = \mathbf{PD}^k\mathbf{P}^{-1}$, for every integer $k > 0$.

13. Let **A** and **B** be commuting $n \times n$ matrices.
  (a) Show that if λ is an eigenvalue for **A** and $\mathbf{v} \in E_\lambda$ (the eigenspace for **A** associated with λ), then $\mathbf{Bv} \in E_\lambda$.
  (b) Prove that if **A** has n distinct eigenvalues, then **B** is diagonalizable.

14. Let **A** be a diagonalizable matrix. Prove that $\mathbf{A}^T$ is diagonalizable.

15. Let **A** be a nonsingular diagonalizable matrix. Prove that $\mathbf{A}^{-1}$ is diagonalizable.

16. (a) Let **A** be a fixed 2×2 matrix with distinct eigenvalues λ_1 and λ_2. Consider the linear operator $L: \mathcal{M}_{22} \longrightarrow \mathcal{M}_{22}$ given by $L(\mathbf{K}) = \mathbf{AK}$. Show that L is diagonalizable with eigenvalues λ_1 and λ_2, each having multiplicity 2. (Hint: Use eigenvectors for **A** to help create eigenvectors for L.)
  (b) Generalize part (a) as follows: let **A** be a fixed diagonalizable $n \times n$ matrix with distinct eigenvalues $\lambda_1, \ldots, \lambda_k$. Consider the linear operator $L: \mathcal{M}_{nn} \longrightarrow \mathcal{M}_{nn}$ given by $L(\mathbf{K}) = \mathbf{AK}$. Show that L is diagonalizable with eigenvalues $\lambda_1, \ldots, \lambda_k$. In addition, show that, for each i, the geometric multiplicity of λ_i for L is n times the geometric multiplicity of λ_i for **A**.

17. Generalization of Theorem 6.4: Let $L: \mathcal{V} \longrightarrow \mathcal{V}$ be a linear operator on a finite dimensional vector space $\mathcal{V}$. Suppose that $\lambda_1, \ldots, \lambda_n$ are distinct eigenvalues for L and that $B_i = \{\mathbf{v}_{i1}, \ldots, \mathbf{v}_{ik_i}\}$ is a basis for the eigenspace E_{λ_i}, for $1 \leq i \leq n$. The goal of this exercise is to show that $B = \bigcup_{i=1}^{n} B_i$ is linearly independent. Suppose that $\sum_{i=1}^{n} \sum_{j=1}^{k_i} a_{ij}\mathbf{v}_{ij} = \mathbf{0}$.
  (a) Let $\mathbf{u}_i = \sum_{j=1}^{k_i} a_{ij}\mathbf{v}_{ij}$. Show that $\mathbf{u}_i \in E_{\lambda_i}$.
  (b) Explain why $\sum_{i=1}^{n} \mathbf{u}_i = \mathbf{0}$.
  (c) Use Theorem 6.4 and a proof by contradiction to show that $\mathbf{u}_i = \mathbf{0}$, for $1 \leq i \leq n$.
  (d) Conclude that $a_{ij} = 0$, for $1 \leq i \leq n$ and $1 \leq j \leq k_i$.
  (e) Explain why parts (a) through (d) prove that B is linearly independent.

6.3 Orthogonal Diagonalization

In this section we study only linear operators on subspaces of $\mathbb{R}^n$. Our goal is to determine which operators on $\mathbb{R}^n$ have an orthonormal basis B of eigenvectors. Such operators are said to be orthogonally diagonalizable. For this type of operator, the transition matrix $\mathbf{P}$ from B-coordinates to standard coordinates is an orthogonal matrix. Such a transition preserves much of the geometric structure of $\mathbb{R}^n$, including lengths of vectors and the angles between them. Essentially, then, an orthogonally diagonalizable operator is one for which we can find a diagonal form without losing other geometric properties of the operator.

We begin by defining symmetric operators and studying their properties. Then we show that precisely these operators are orthogonally diagonalizable. In addition, we present a method for orthogonally diagonalizing an operator analogous to the step-by-step method in Section 6.2 for ordinary diagonalization.

Symmetric Operators

DEFINITION

Let $\mathcal{V}$ be a subspace of $\mathbb{R}^n$. A linear operator $L: \mathcal{V} \longrightarrow \mathcal{V}$ is a **symmetric operator** on $\mathcal{V}$ if and only if $L(\mathbf{v}_1) \cdot \mathbf{v}_2 = \mathbf{v}_1 \cdot L(\mathbf{v}_2)$, for every $\mathbf{v}_1, \mathbf{v}_2 \in \mathcal{V}$.

Example 1

The operator on $\mathbb{R}^3$ given by $L([a, b, c]) = [b, a, -c]$ is symmetric since

$$L([a, b, c]) \cdot [d, e, f] = [b, a, -c] \cdot [d, e, f] = bd + ae - cf$$

and $$[a, b, c] \cdot L([d, e, f]) = [a, b, c] \cdot [e, d, -f] = ae + bd - cf.$$

■

You can easily verify that the matrix representation for the operator in Example 1 with respect to the standard basis is

$$\begin{bmatrix} 0 & 1 & 0 \\ 1 & 0 & 0 \\ 0 & 0 & -1 \end{bmatrix},$$

a symmetric matrix. The next theorem asserts that an operator on a subspace $\mathcal{V}$ of $\mathbb{R}^n$ is symmetric if and only if its matrix representation with respect to any orthonormal basis for $\mathcal{V}$ is symmetric.

THEOREM 6.12

Let $\mathcal{V}$ be a nontrivial subspace of $\mathbb{R}^n$, and let L be a linear operator on $\mathcal{V}$. Let B be an ordered orthonormal basis for $\mathcal{V}$, and let $\mathbf{A}$ be the matrix for L with respect to B. Then L is a symmetric operator if and only if $\mathbf{A}$ is a symmetric matrix.

Theorem 6.12 gives a quick way of recognizing symmetric operators just by looking at their matrix representations. Such operators occur frequently in applications. (For example, see the discussion of quadratic forms in Section 8.9.)

The proof of Theorem 6.12 is long, and so we have placed it in Appendix A for the interested reader.

A Symmetric Operator Always Has an Eigenvalue

The following lemma is needed for the proof of Theorem 6.15, the main theorem of this section.

LEMMA 6.13

Let L be a symmetric operator on a nontrivial subspace $\mathcal{V}$ of $\mathbb{R}^n$. Then L has at least one eigenvalue.

Simpler proofs of Lemma 6.13 exist than the one given below, but they involve complex vector spaces, which will be discussed in Section 7.1. Nevertheless, the following proof is interesting, since it brings together a variety of topics already developed and some familiar theorems from algebra.

Proof of Lemma 6.13

Suppose that L is a symmetric operator on a nontrivial subspace $\mathcal{V}$ of $\mathbb{R}^n$ with $\dim(\mathcal{V}) = k$, and let B be an orthonormal basis for $\mathcal{V}$. By Theorem 6.12, the matrix representation $\mathbf{A}$ for L with respect to B is a symmetric matrix.

Let $p_{\mathbf{A}}(x) = x^k + \alpha_{k-1}x^{k-1} + \cdots + \alpha_1 x + \alpha_0$ be the characteristic polynomial for $\mathbf{A}$. From algebra, we know that $p_{\mathbf{A}}(x)$ can be factored into a product of linear terms and irreducible (nonfactorable) quadratic terms. Since $k \times k$ matrices follow the same laws of algebra as real numbers, with the exception of the commutative law for multiplication, and since $\mathbf{A}$ commutes with itself and $\mathbf{I}_k$, it follows that the polynomial $p_{\mathbf{A}}(\mathbf{A}) = \mathbf{A}^k + \alpha_{k-1}\mathbf{A}^{k-1} + \cdots + \alpha_1\mathbf{A} + \alpha_0\mathbf{I}_k$ can also be factored into linear and irreducible quadratic factors. Hence, $p_{\mathbf{A}}(\mathbf{A}) = \mathbf{F}_1\mathbf{F}_2\cdots\mathbf{F}_j$, where each factor $\mathbf{F}_i$ is either of the form $a_i\mathbf{A} + b_i\mathbf{I}_k$, with $a_i \neq 0$, or of the form $a_i\mathbf{A}^2 + b_i\mathbf{A} + c_i\mathbf{I}_k$, with $a_i \neq 0$ and $b_i^2 - 4a_ic_i < 0$ (where the latter condition makes this quadratic irreducible).

Now, by the Cayley-Hamilton Theorem, $p_{\mathbf{A}}(\mathbf{A}) = \mathbf{F}_1\mathbf{F}_2\cdots\mathbf{F}_j = \mathbf{O}_k$. Hence, the determinant $|\mathbf{F}_1\mathbf{F}_2\cdots\mathbf{F}_j| = 0$. Since $|\mathbf{F}_1\mathbf{F}_2\cdots\mathbf{F}_j| = |\mathbf{F}_1||\mathbf{F}_2|\cdots|\mathbf{F}_j|$, some $\mathbf{F}_i$ must have a zero determinant. There are two possible cases:

Case 1: Suppose that $\mathbf{F}_i = a_i\mathbf{A} + b_i\mathbf{I}_k$. Since $|a_i\mathbf{A} + b_i\mathbf{I}_k| = 0$, there is a nonzero vector $\mathbf{u}$ with $(a_i\mathbf{A} + b_i\mathbf{I}_k)\mathbf{u} = \mathbf{0}$. Thus, $\mathbf{Au} = -(b_i/a_i)\mathbf{u}$, and $\mathbf{u}$ is an eigenvector for $\mathbf{A}$ with eigenvalue $-b_i/a_i$. Hence, $-b_i/a_i$ is an eigenvalue for L.

Case 2: Suppose that $\mathbf{F}_i = a_i\mathbf{A}^2 + b_i\mathbf{A} + c_i\mathbf{I}_k$. We show that this case cannot occur by exhibiting a contradiction. As in case 1, there is a nonzero vector $\mathbf{u}$ with $(a_i\mathbf{A}^2 + b_i\mathbf{A} + c_i\mathbf{I}_k)\mathbf{u} = \mathbf{0}$. Completing the square yields

$$a_i\left(\left(\mathbf{A} + \frac{b_i}{2a_i}\mathbf{I}_k\right)^2 - \left(\frac{b_i^2 - 4a_ic_i}{4a_i^2}\right)\mathbf{I}_k\right)\mathbf{u} = \mathbf{0}.$$

Let $\mathbf{C} = \mathbf{A}+(b_i/2a_i)\mathbf{I}_k$ and $d = -(b_i^2-4a_ic_i)/4a_i^2$. Then $\mathbf{C}$ is a symmetric matrix since it is the sum of symmetric matrices, and $d > 0$ since $b_i^2 - 4a_ic_i < 0$. These substitutions simplify the preceding equation to $a_i(\mathbf{C}^2 + d\mathbf{I}_k)\mathbf{u} = \mathbf{0}$, or $\mathbf{C}^2\mathbf{u} = -d\mathbf{u}$. Thus,

$$\begin{aligned} 0 &\leq (\mathbf{Cu}) \cdot (\mathbf{Cu}) \\ &= \mathbf{u} \cdot (\mathbf{C}^2\mathbf{u}) && \text{since } \mathbf{C} \text{ is symmetric} \\ &= \mathbf{u} \cdot (-d\mathbf{u}) \\ &= -d(\mathbf{u} \cdot \mathbf{u}) \\ &< 0 && \text{since } d > 0 \end{aligned}$$

However, $0 < 0$ is a contradiction. Hence, case 2 cannot occur. ■

Example 2

The operator $L([a, b, c]) = [b, a, -c]$ on $\mathbb{R}^3$ is symmetric, as shown in Example 1. Lemma 6.13 then states that L has at least one eigenvalue. In fact, L has two eigenvalues: $\lambda_1 = 1$ and $\lambda_2 = -1$. The eigenspaces E_{λ_1} and E_{λ_2} have bases $\{[1, 1, 0]\}$ and $\{[1, -1, 0], [0, 0, 1]\}$, respectively. ■

Orthogonally Diagonalizable Operators

DEFINITION

Let $\mathcal{V}$ be a subspace of $\mathbb{R}^n$, and let $L: \mathcal{V} \longrightarrow \mathcal{V}$ be a linear operator. Then L is an **orthogonally diagonalizable operator** if and only if there is an ordered orthonormal basis B for $\mathcal{V}$ such that the matrix for L with respect to B is a diagonal matrix.

A $k \times k$ matrix $\mathbf{A}$ is an **orthogonally diagonalizable matrix** if and only if there is an orthogonal matrix $\mathbf{P}$ such that $\mathbf{D} = \mathbf{P}^{-1}\mathbf{AP}$ is a diagonal matrix.

An argument similar to the proof of Theorem 6.9 (see Exercise 9 in Section 6.2) proves the following theorem, which connects the concept of an orthogonally diagonalizable operator with that of an orthogonally diagonalizable matrix.

THEOREM 6.14

Let $\mathcal{V}$ be a nontrivial subspace of $\mathbb{R}^n$, let L be a linear operator on $\mathcal{V}$, and let B be any ordered orthonormal basis for $\mathcal{V}$. Then L is an orthogonally diagonalizable operator if and only if the matrix for L with respect to B is orthogonally diagonalizable.

Equivalence of Symmetric and Orthogonally Diagonalizable Operators

We are now ready to show that symmetric operators and orthogonally diagonalizable operators are, in reality, the same:

THEOREM 6.15

Let $\mathcal{V}$ be a nontrivial subspace of $\mathbb{R}^n$, and let L be a linear operator on $\mathcal{V}$. Then L is orthogonally diagonalizable if and only if L is symmetric.

Proof of Theorem 6.15

Suppose that L is a linear operator on a nontrivial subspace $\mathcal{V}$ of $\mathbb{R}^n$. First we show that if L is orthogonally diagonalizable, then L is symmetric. The definition of an orthogonally diagonalizable operator states that there is an ordered orthonormal basis B for $\mathcal{V}$ such that the matrix representation $\mathbf{A}$ for L with respect to B is diagonal. Since every diagonal matrix is also symmetric, L is a symmetric operator by Theorem 6.12.

To finish the proof, we must show that if L is a symmetric operator, then L is orthogonally diagonalizable. Suppose L has an ordered orthonormal basis B consisting entirely of eigenvectors of L. Then, clearly, the matrix for L with respect to B is a diagonal matrix (having the eigenvalues corresponding to the eigenvectors in B along its main diagonal), and so L is orthogonally diagonalizable. Therefore, our goal is to find an orthonormal basis of eigenvectors for L.

We use a proof by induction on $\dim(\mathcal{V})$. For the base step, we assume that $\dim(\mathcal{V}) = 1$. Now, any subset $\{\mathbf{u}\}$ of $\mathcal{V}$ consisting of a single unit vector is an orthonormal basis for $\mathcal{V}$. Since $L(\mathbf{u}) \in \mathcal{V}$ and $\{\mathbf{u}\}$ is a basis for $\mathcal{V}$, we must have $L(\mathbf{u}) = \lambda\mathbf{u}$, for some real number λ. Hence, $\{\mathbf{u}\}$ is an orthonormal basis of eigenvectors for $\mathcal{V}$, thus completing the base step.

We now perform the inductive step. The inductive hypothesis is as follows:

If $\mathcal{W}$ is a subspace of $\mathbb{R}^n$ with dimension k and T is any symmetric operator on $\mathcal{W}$, then $\mathcal{W}$ has an orthonormal basis of eigenvectors for T.

We must prove the following:

> If $\mathcal{V}$ is a subspace of $\mathbb{R}^n$ with dimension $k+1$ and L is a symmetric operator on $\mathcal{V}$, then $\mathcal{V}$ has an orthonormal basis of eigenvectors for L.

Now, L has at least one eigenvalue λ, by Lemma 6.13. Take any eigenvector for L corresponding to λ and normalize it to create a unit eigenvector $\mathbf{v}$. Let $\mathcal{Y} = \text{span}(\{\mathbf{v}\})$. Now, we want to enlarge $\{\mathbf{v}\}$ to an orthonormal basis of eigenvectors for L in $\mathcal{V}$. We look for these basis vectors in the orthogonal complement $\mathcal{Y}^\perp$ of $\mathcal{Y}$ in $\mathbb{R}^n$.

Consider the subspace $\mathcal{W}$ of $\mathcal{V}$ given by $\mathcal{W} = \mathcal{Y}^\perp \cap \mathcal{V}$. Now, $\mathbf{w} \in \mathcal{W}$ if and only if $\mathbf{w} \in \mathcal{V}$ and $\mathbf{v} \cdot \mathbf{w} = 0$. Thus, $\mathcal{W}$ is essentially the orthogonal complement of $\mathcal{Y}$ in $\mathcal{V}$ (see Exercise 19 in Section 5.6). In particular, if $\{\mathbf{v}, \mathbf{v}_1, \ldots, \mathbf{v}_k\}$ is an orthonormal basis for $\mathcal{V}$ containing $\mathbf{v}$, then $\{\mathbf{v}_1, \ldots, \mathbf{v}_k\}$ is an orthonormal set in $\mathcal{W}$. Hence, $\dim(\mathcal{W}) \geq k$. But $\mathbf{v} \notin \mathcal{W}$ implies $\dim(\mathcal{W}) < \dim(\mathcal{V}) = k+1$, and so $\dim(\mathcal{W}) = k$.

Next, we claim that for every $\mathbf{w} \in \mathcal{W}$, we have $L(\mathbf{w}) \in \mathcal{W}$. This statement is true since

$$\begin{aligned}
\mathbf{v} \cdot L(\mathbf{w}) &= L(\mathbf{v}) \cdot \mathbf{w} && \text{since } L \text{ is symmetric} \\
&= (\lambda \mathbf{v}) \cdot \mathbf{w} \\
&= \lambda(\mathbf{v} \cdot \mathbf{w}) \\
&= \lambda(0) = 0,
\end{aligned}$$

which shows that $L(\mathbf{w})$ is orthogonal to $\mathbf{v}$ and hence is in $\mathcal{W}$. Therefore, we can define a linear operator $T: \mathcal{W} \longrightarrow \mathcal{W}$ by $T(\mathbf{w}) = L(\mathbf{w})$. ($T$ is the **restriction** of L to $\mathcal{W}$.) Now, T is a symmetric operator on $\mathcal{W}$ since, for every $\mathbf{w}_1, \mathbf{w}_2 \in \mathcal{W}$,

$$\begin{aligned}
T(\mathbf{w}_1) \cdot \mathbf{w}_2 &= L(\mathbf{w}_1) \cdot \mathbf{w}_2 && \text{definition of } T \\
&= \mathbf{w}_1 \cdot L(\mathbf{w}_2) && \text{since } L \text{ is symmetric} \\
&= \mathbf{w}_1 \cdot T(\mathbf{w}_2) && \text{definition of } T
\end{aligned}$$

Since $\mathcal{W}$ is k-dimensional, the inductive hypothesis implies that $\mathcal{W}$ has an orthonormal basis $\{\mathbf{u}_1, \ldots, \mathbf{u}_k\}$ of eigenvectors for T. Then, by definition of T, $\{\mathbf{u}_1, \ldots, \mathbf{u}_k\}$ is also a set of eigenvectors for L, all of which are orthogonal to $\mathbf{v}$ (since they are in $\mathcal{W}$). Hence, $B = \{\mathbf{v}, \mathbf{u}_1, \ldots, \mathbf{u}_k\}$ is an orthonormal basis for $\mathcal{V}$ of eigenvectors for L, and we have finished the proof of the inductive step. ■

Method for Orthogonally Diagonalizing a Linear Operator

We now state a method for orthogonally diagonalizing a symmetric operator, based on Theorem 6.15. You should compare this method to the method for diagonalizing a linear operator given in Section 6.2. Notice that the following method assumes that eigenvectors of a symmetric operator corresponding to different eigenvalues are orthogonal. (You are asked to prove this statement in Exercise 11.)

METHOD FOR ORTHOGONALLY DIAGONALIZING A SYMMETRIC OPERATOR

Let $L: \mathcal{V} \longrightarrow \mathcal{V}$ be a symmetric operator on a subspace $\mathcal{V}$ of $\mathbb{R}^n$.

Step 1: Find an ordered orthonormal basis C for $\mathcal{V}$. (When $\mathcal{V} = \mathbb{R}^n$, you can use the standard basis for $\mathbb{R}^n$.)

Step 2: Calculate the matrix representation $\mathbf{A}$ for L with respect to C (which should be a symmetric matrix).

Step 3: Solve for all the eigenvalues $\lambda_1, \ldots, \lambda_k$ of $\mathbf{A}$ (which can often be done by using the characteristic polynomial of $\mathbf{A}$ or appropriate software).

Step 4: Find a basis for each eigenspace E_{λ_i} of $\mathbf{A}$ (which can often be done by solving an appropriate homogeneous system).

Step 5: Perform the Gram-Schmidt Process on the basis for each E_{λ_i} from step 4. Normalize to get an orthonormal basis B_i for each E_{λ_i}.

Step 6: Let $B = (\mathbf{v}_1, \ldots, \mathbf{v}_l)$ be the ordered orthonormal basis for $\mathcal{V}$ formed by taking the union of the bases from step 5.

Step 7: The matrix representation for L with respect to B is the diagonal matrix $\mathbf{D}$, whose (i, i) entry d_{ii} equals the eigenvalue for L corresponding to $\mathbf{v}_i$.

Step 8: The transition matrix $\mathbf{P}$ from B to C coordinates is the $l \times l$ matrix whose columns are equal to $[\mathbf{v}_1]_C, \ldots, [\mathbf{v}_l]_C$. Also, $\mathbf{P}$ is an orthogonal matrix, and $\mathbf{D} = \mathbf{P}^{-1}\mathbf{A}\mathbf{P}$.

The following example illustrates this method.

Example 3

Consider the operator $L: \mathbb{R}^4 \longrightarrow \mathbb{R}^4$ given by $L(\mathbf{v}) = \mathbf{A}\mathbf{v}$, where

$$\mathbf{A} = \frac{1}{7}\begin{bmatrix} 15 & -21 & -3 & -5 \\ -21 & 35 & -7 & 0 \\ -3 & -7 & 23 & 15 \\ -5 & 0 & 15 & 39 \end{bmatrix}.$$

L is clearly symmetric, since its matrix with respect to the standard basis C for $\mathbb{R}^4$ is $\mathbf{A}$, which is symmetric. We find an orthonormal basis B such that the matrix for L with respect to B is diagonal.

Steps 1 and 2: With C as the standard basis for $\mathbb{R}^4$, we have already seen that $\mathbf{A}$ is the matrix for L with respect to C.

Step 3: A lengthy calculation yields

$$p_{\mathbf{A}}(x) = x^4 - 16x^3 + 77x^2 - 98x = x(x-2)(x-7)^2,$$

giving us the eigenvalues $\lambda_1 = 0$, $\lambda_2 = 2$, and $\lambda_3 = 7$.

Step 4: Solving the appropriate homogeneous systems to find bases for the eigenspaces corresponding to the eigenvalues in step 3 produces:

$$\begin{aligned}\text{Basis for } E_{\lambda_1} &= \{[3, 2, 1, 0]\}\\ \text{Basis for } E_{\lambda_2} &= \{[1, 0, -3, 2]\}\\ \text{Basis for } E_{\lambda_3} &= \{[-2, 3, 0, 1], [3, -5, 1, 0]\}\end{aligned}$$

Step 5: There is no need to perform the Gram-Schmidt Process on the bases for E_{λ_1} and E_{λ_2}, since each of these eigenspaces is one-dimensional. Normalizing the basis vectors yields:

$$\begin{aligned}\text{Orthonormal basis for } E_{\lambda_1} &= \left\{\frac{1}{\sqrt{14}}[3, 2, 1, 0]\right\}\\ \text{Orthonormal basis for } E_{\lambda_2} &= \left\{\frac{1}{\sqrt{14}}[1, 0, -3, 2]\right\}\end{aligned}$$

Now we must perform the Gram-Schmidt Process on the basis for E_{λ_3}. Let $\mathbf{w}_1 = \mathbf{v}_1 = [-2, 3, 0, 1]$ and $\mathbf{w}_2 = [3, -5, 1, 0]$. Then

$$\begin{aligned}\mathbf{v}_2 &= [3, -5, 1, 0] - \left(\frac{[3, -5, 1, 0] \cdot [-2, 3, 0, 1]}{[-2, 3, 0, 1] \cdot [-2, 3, 0, 1]}\right)[-2, 3, 0, 1]\\ &= \left[0, -\frac{1}{2}, 1, \frac{3}{2}\right].\end{aligned}$$

Finally, normalizing $\mathbf{v}_1$ and $\mathbf{v}_2$, we obtain:

$$\text{Orthonormal basis for } E_{\lambda_3} = \left\{\frac{1}{\sqrt{14}}[-2, 3, 0, 1], \frac{1}{\sqrt{14}}[0, -1, 2, 3]\right\}.$$

Step 6: Letting $\mathbf{v}_1 = \left(1/\sqrt{14}\right)[3, 2, 1, 0]$, $\mathbf{v}_2 = \left(1/\sqrt{14}\right)[1, 0, -3, 2]$, $\mathbf{v}_3 = \left(1/\sqrt{14}\right)[-2, 3, 0, 1]$, and $\mathbf{v}_4 = \left(1/\sqrt{14}\right)[0, -1, 2, 3]$ gives us an ordered orthonormal basis $B = (\mathbf{v}_1, \mathbf{v}_2, \mathbf{v}_3, \mathbf{v}_4)$.

Steps 7 and 8: The matrix representation $\mathbf{D}$ of L with respect to B is

$$\mathbf{D} = \begin{bmatrix} \lambda_1 & 0 & 0 & 0\\ 0 & \lambda_2 & 0 & 0\\ 0 & 0 & \lambda_3 & 0\\ 0 & 0 & 0 & \lambda_3 \end{bmatrix} = \begin{bmatrix} 0 & 0 & 0 & 0\\ 0 & 2 & 0 & 0\\ 0 & 0 & 7 & 0\\ 0 & 0 & 0 & 7 \end{bmatrix}.$$

The transition matrix $\mathbf{P}$ from B to C is the orthogonal matrix

$$\mathbf{P} = \frac{1}{\sqrt{14}}\begin{bmatrix} 3 & 1 & -2 & 0 \\ 2 & 0 & 3 & -1 \\ 1 & -3 & 0 & 2 \\ 0 & 2 & 1 & 3 \end{bmatrix}.$$

Since $\mathbf{P}^{-1} = \mathbf{P}^T$, we can show $\mathbf{P}^{-1}\mathbf{AP} = \mathbf{D}$ by checking that $\mathbf{P}^T\mathbf{AP} = \mathbf{D}$:

$$\left(\frac{1}{\sqrt{14}}\begin{bmatrix} 3 & 2 & 1 & 0 \\ 1 & 0 & -3 & 2 \\ -2 & 3 & 0 & 1 \\ 0 & -1 & 2 & 3 \end{bmatrix}\right)\left(\frac{1}{7}\begin{bmatrix} 15 & -21 & -3 & -5 \\ -21 & 35 & -7 & 0 \\ -3 & -7 & 23 & 15 \\ -5 & 0 & 15 & 39 \end{bmatrix}\right)\left(\frac{1}{\sqrt{14}}\begin{bmatrix} 3 & 1 & -2 & 0 \\ 2 & 0 & 3 & -1 \\ 1 & -3 & 0 & 2 \\ 0 & 2 & 1 & 3 \end{bmatrix}\right)$$
$$= \begin{bmatrix} 0 & 0 & 0 & 0 \\ 0 & 2 & 0 & 0 \\ 0 & 0 & 7 & 0 \\ 0 & 0 & 0 & 7 \end{bmatrix}.$$

■

We conclude by examining a symmetric operator whose domain is a proper subspace of $\mathbb{R}^n$.

Example 4

Consider the operators L_1, L_2, and L_3 on $\mathbb{R}^3$ given by

L_1: orthogonal projection onto the plane $x + y + z = 0$
L_2: orthogonal projection onto the plane $x + y - z = 0$
L_3: orthogonal projection onto the xy-plane (that is, $z = 0$)

Let $L: \mathbb{R}^3 \longrightarrow \mathbb{R}^3$ be given by $L = L_3 \circ L_2 \circ L_1$, and let $\mathcal{V}$ be the xy-plane in $\mathbb{R}^3$. Then, since range$(L_3) = \mathcal{V}$, we see that range$(L) \subseteq \mathcal{V}$. Thus, restricting the domain of L to $\mathcal{V}$, we can think of L as a linear operator on $\mathcal{V}$.

First we calculate the matrix representation $\mathbf{A}$ of L with respect to the ordered orthonormal basis $C = ([1, 0, 0], [0, 1, 0])$ for $\mathcal{V}$. Using the orthonormal basis $\left\{\left(1/\sqrt{2}\right)[1, -1, 0], \left(1/\sqrt{6}\right)[1, 1, -2]\right\}$ for the plane $x + y + z = 0$, the basis $\left\{\left(1/\sqrt{2}\right)[1, -1, 0], \left(1/\sqrt{6}\right)[1, 1, 2]\right\}$ for the plane $x + y - z = 0$, and the basis C for the xy-plane, we can use the method of Example 7 in Section 5.6 to compute the required orthogonal projections. These computations yield

$$L([1, 0, 0]) = L_3(L_2(L_1([1, 0, 0]))) = L_3(L_2([\tfrac{2}{3}, -\tfrac{1}{3}, -\tfrac{1}{3}]))$$
$$= L_3([\tfrac{4}{9}, -\tfrac{5}{9}, -\tfrac{1}{9}]) = \tfrac{1}{9}[4, -5, 0]$$

and $$L([0, 1, 0]) = L_3(L_2(L_1([0, 1, 0]))) = L_3(L_2([-\tfrac{1}{3}, \tfrac{2}{3}, -\tfrac{1}{3}]))$$
$$= L_3([-\tfrac{5}{9}, \tfrac{4}{9}, -\tfrac{1}{9}]) = \tfrac{1}{9}[-5, 4, 0].$$

Expressing these vectors in C-coordinates, we see that the matrix representation of L with respect to C is $\mathbf{A} = \frac{1}{9}\begin{bmatrix} 4 & -5 \\ -5 & 4 \end{bmatrix}$, a symmetric matrix. Thus, by Theorem 6.12, L is a symmetric operator on $\mathcal{V}$.[†] We will orthogonally diagonalize L:

Steps 1 and 2: We have already chosen the orthonormal basis C for $\mathcal{V}$ and found the matrix representation $\mathbf{A}$ for L with respect to C.

Step 3: The characteristic polynomial for $\mathbf{A}$ is $p_{\mathbf{A}}(x) = x^2 - \frac{8}{9}x - \frac{1}{9} = (x-1)(x+\frac{1}{9})$, giving eigenvalues $\lambda_1 = 1$ and $\lambda_2 = -\frac{1}{9}$.

Step 4: Solving the appropriate homogeneous systems to find bases for the eigenspaces of $\mathbf{A}$ yields

$$\text{Basis for } E_{\lambda_1} = \{[1,-1]\}, \quad \text{basis for } E_{\lambda_2} = \{[1,1]\}.$$

Notice that we expressed the bases in C-coordinates.

Step 5: Since the eigenspaces are one-dimensional, there is no need to perform the Gram-Schmidt Process on the bases for E_{λ_1} and E_{λ_2}. Normalizing the basis vectors produces:

$$\text{Orthonormal basis for } E_{\lambda_1} = \left\{\frac{1}{\sqrt{2}}[1,-1]\right\}$$

$$\text{Orthonormal basis for } E_{\lambda_2} = \left\{\frac{1}{\sqrt{2}}[1,1]\right\}.$$

Step 6: Expressed in C-coordinates, the orthonormal basis B for $\mathcal{V}$ is $\{\frac{1}{\sqrt{2}}[1,-1], \frac{1}{\sqrt{2}}[1,1]\}$. Writing these vectors in standard coordinates in $\mathbb{R}^3$, we get $B = \{\frac{1}{\sqrt{2}}[1,-1,0], \frac{1}{\sqrt{2}}[1,1,0]\}$.

Steps 7 and 8: The matrix $\mathbf{D} = \begin{bmatrix} 1 & 0 \\ 0 & -\frac{1}{9} \end{bmatrix}$ is the matrix representation of L with respect to B. The transition matrix $\mathbf{P} = \left(1/\sqrt{2}\right)\begin{bmatrix} 1 & 1 \\ -1 & 1 \end{bmatrix}$ from B- to C-coordinates is the orthogonal matrix whose columns are the vectors in B expressed in C-coordinates. Notice that

$$\begin{aligned} \mathbf{P}^{-1}\mathbf{A}\mathbf{P} &= \left(\frac{1}{\sqrt{2}}\begin{bmatrix} 1 & -1 \\ 1 & 1 \end{bmatrix}\right)\left(\frac{1}{9}\begin{bmatrix} 4 & -5 \\ -5 & 4 \end{bmatrix}\right)\left(\frac{1}{\sqrt{2}}\begin{bmatrix} 1 & 1 \\ -1 & 1 \end{bmatrix}\right) \\ &= \begin{bmatrix} 1 & 0 \\ 0 & -\frac{1}{9} \end{bmatrix} = \mathbf{D}. \end{aligned}$$

■

♦ **Application:** You have now covered the prerequisites for Section 8.9, "Quadratic Forms."

[†] You can easily verify that L is not a symmetric operator on all of $\mathbb{R}^3$, even though it is symmetric on the subspace $\mathcal{V}$.

Exercises—Section 6.3

Note: Use a computer with appropriate software to help (when needed) in solving for eigenvalues and eigenvectors and to help in performing the Gram-Schmidt Process.

1. Determine which of the following linear operators are symmetric. Explain why each is or is not symmetric.

★(a) $L: \mathbb{R}^2 \longrightarrow \mathbb{R}^2$ given by $L\left(\begin{bmatrix} x \\ y \end{bmatrix}\right) = \begin{bmatrix} 3x + 2y \\ 2x + 5y \end{bmatrix}$

(b) $L: \mathbb{R}^2 \longrightarrow \mathbb{R}^2$ given by $L\left(\begin{bmatrix} x \\ y \end{bmatrix}\right) = \begin{bmatrix} 5x - 7y \\ 7x + 6y \end{bmatrix}$

(c) $L: \mathbb{R}^3 \longrightarrow \mathbb{R}^3$ given by the orthogonal projection onto the plane $x + y + z = 0$

★(d) $L: \mathbb{R}^3 \longrightarrow \mathbb{R}^3$ given by the orthogonal projection onto the plane $ax + by + cz = 0$

★(e) $L: \mathbb{R}^3 \longrightarrow \mathbb{R}^3$ given by a counterclockwise rotation through an angle of $\frac{\pi}{3}$ radians about the line through the origin in the direction $[1, 1, -1]$

(f) $L: \mathbb{R}^3 \longrightarrow \mathbb{R}^3$ given by the orthogonal reflection through the plane $4x - 3y + 5z = 0$

★(g) $L: \mathbb{R}^4 \longrightarrow \mathbb{R}^4$ given by $L = L_1^{-1} \circ L_2 \circ L_1$, where $L_1: \mathbb{R}^4 \longrightarrow \mathcal{M}_{22}$ is given by $L_1([a, b, c, d]) = \begin{bmatrix} a & b \\ c & d \end{bmatrix}$, $L_2: \mathcal{M}_{22} \longrightarrow \mathcal{M}_{22}$ is given by $L_2(\mathbf{K}) = \begin{bmatrix} 4 & 3 \\ 3 & 9 \end{bmatrix}\mathbf{K}$

2. In each part, find a symmetric matrix having the given eigenvalues and the given bases for their associated eigenspaces. (Be careful: the given bases may not be orthonormal.)

★(a) $\lambda_1 = 1$, $\lambda_2 = -1$, $E_{\lambda_1} = \text{span}(\{\frac{1}{5}[3, 4]\})$, $E_{\lambda_2} = \text{span}(\{\frac{1}{5}[4, -3]\})$.

(b) $\lambda_1 = 0$, $\lambda_2 = 1$, $\lambda_3 = 2$, $E_{\lambda_1} = \text{span}(\{\frac{1}{11}[6, 2, -9]\})$, $E_{\lambda_2} = \text{span}(\{\frac{1}{11}[7, 6, 6]\})$, $E_{\lambda_3} = \text{span}(\{\frac{1}{11}[6, -9, 2]\})$

(c) $\lambda_1 = 1$, $\lambda_2 = 2$, $E_{\lambda_1} = \text{span}(\{[6, 3, 2], [8, -3, 5]\})$, $E_{\lambda_2} = \text{span}(\{[3, -2, -6]\})$

★(d) $\lambda_1 = -1$, $\lambda_2 = 1$, $E_{\lambda_1} = \text{span}(\{[12, 3, 4, 0], [12, -1, 7, 12]\})$, $E_{\lambda_2} = \text{span}(\{[-3, 12, 0, 4], [-2, 24, -12, 11]\})$

3. In each part of this exercise, the matrix $\mathbf{A}$ with respect to the standard basis for a symmetric linear operator on $\mathbb{R}^n$ is given. Orthogonally diagonalize each operator by following steps 3 through 8 of the method given in the text. Your answers should include the ordered orthonormal basis B, the orthogonal matrix $\mathbf{P}$, and the diagonal matrix $\mathbf{D}$. Check your work by verifying that $\mathbf{D} = \mathbf{P}^{-1}\mathbf{A}\mathbf{P}$. (Hint: For (c), use $[-1, 2, 2]$ as the first eigenvector for the two-dimensional eigenspace in step 4. For (e), $p_{\mathbf{A}}(x) = (x - 2)^2(x + 3)(x - 5)$.)

★(a) $\mathbf{A} = \begin{bmatrix} 144 & -60 \\ -60 & 25 \end{bmatrix}$

(b) $\mathbf{A} = \frac{1}{25}\begin{bmatrix} 39 & 48 \\ 48 & 11 \end{bmatrix}$

★(c) $\mathbf{A} = \frac{1}{9}\begin{bmatrix} 17 & 8 & -4 \\ 8 & 17 & -4 \\ -4 & -4 & 11 \end{bmatrix}$ (d) $\mathbf{A} = \frac{1}{27}\begin{bmatrix} -13 & -40 & -16 \\ -40 & 176 & -124 \\ -16 & -124 & -1 \end{bmatrix}$

★(e) $\mathbf{A} = \frac{1}{14}\begin{bmatrix} 23 & 0 & 15 & -10 \\ 0 & 31 & -6 & -9 \\ 15 & -6 & -5 & 48 \\ -10 & -9 & 48 & 35 \end{bmatrix}$

(f) $\mathbf{A} = \begin{bmatrix} 3 & 4 & 12 \\ 4 & -12 & 3 \\ 12 & 3 & -4 \end{bmatrix}$ ★(g) $\mathbf{A} = \begin{bmatrix} 11 & 2 & -10 \\ 2 & 14 & 5 \\ -10 & 5 & -10 \end{bmatrix}$

4. In each part of this exercise, a symmetric linear operator L is defined on a subspace $\mathcal{V}$ of $\mathbb{R}^n$. Orthogonally diagonalize each operator by following all eight steps of the method given in the text. Your answers should include the ordered orthonormal basis C for $\mathcal{V}$, the matrix $\mathbf{A}$ for L with respect to C, the ordered orthonormal basis B for $\mathcal{V}$, the orthogonal matrix $\mathbf{P}$, and the diagonal matrix $\mathbf{D}$. Check your work by verifying that $\mathbf{D} = \mathbf{P}^{-1}\mathbf{AP}$.

★(a) $L: \mathcal{V} \longrightarrow \mathcal{V}$, where $\mathcal{V}$ is the plane $6x + 10y - 15z = 0$ in $\mathbb{R}^3$, $L([-10, 15, 6]) = [50, -18, 8]$, and $L([15, 6, 10]) = [-5, 36, 22]$

(b) $L: \mathcal{V} \longrightarrow \mathcal{V}$, where $\mathcal{V}$ is the subspace of $\mathbb{R}^4$ spanned by $\{[1, -1, 1, 1], [-1, 1, 1, 1], [1, 1, 1, -1]\}$ and L is given by

$$L\left(\begin{bmatrix} w \\ x \\ y \\ z \end{bmatrix}\right) = \begin{bmatrix} 1 & -2 & 1 & 1 \\ -1 & 2 & 0 & 2 \\ 2 & 2 & 1 & 2 \\ 1 & 1 & 1 & -2 \end{bmatrix}\begin{bmatrix} w \\ x \\ y \\ z \end{bmatrix}$$

5. In each case, use orthogonal diagonalization to find a symmetric matrix $\mathbf{A}$ such that:

★(a) $\mathbf{A}^3 = \frac{1}{25}\begin{bmatrix} 119 & -108 \\ -108 & 56 \end{bmatrix}$ (b) $\mathbf{A}^2 = \begin{bmatrix} 481 & -360 \\ -360 & 964 \end{bmatrix}$

★(c) $\mathbf{A}^2 = \begin{bmatrix} 17 & 16 & -16 \\ 16 & 41 & -32 \\ -16 & -32 & 41 \end{bmatrix}$

★**6.** Give an example of a 3×3 matrix that is diagonalizable but not orthogonally diagonalizable.

★**7.** Find the diagonal matrix $\mathbf{D}$ to which $\begin{bmatrix} a & b \\ b & c \end{bmatrix}$ is similar by an orthogonal change of coordinates. (Hint: Think! The full method for orthogonal diagonalization is not needed.)

8. Let L be a symmetric linear operator on a subspace $\mathcal{V}$ of $\mathbb{R}^n$.
(a) If 1 is the only eigenvalue for L, prove that L is the identity operator.
★(b) What must be true about L if zero is its only eigenvalue? Prove it.

9. Let L_1 and L_2 be symmetric operators on $\mathbb{R}^n$. Prove that $L_2 \circ L_1$ is symmetric if and only if $L_2 \circ L_1 = L_1 \circ L_2$.

10. Two $n \times n$ matrices $\mathbf{A}$ and $\mathbf{B}$ are said to be **orthogonally similar** if and only if there is an orthogonal matrix $\mathbf{P}$ such that $\mathbf{A} = \mathbf{PBP}^{-1}$. Prove that the following statements are equivalent for all $n \times n$ symmetric matrices $\mathbf{A}$ and $\mathbf{B}$:

(i) $\mathbf{A}$ and $\mathbf{B}$ are similar.
(ii) $\mathbf{A}$ and $\mathbf{B}$ have the same characteristic polynomial.
(iii) $\mathbf{A}$ and $\mathbf{B}$ are orthogonally similar.

(Hint: Show that (i) $\Longrightarrow$ (ii) $\Longrightarrow$ (iii) $\Longrightarrow$ (i).)

11. Let L be a symmetric operator on a subspace $\mathcal{V}$ of $\mathbb{R}^n$. Suppose that λ_1 and λ_2 are distinct eigenvalues for L with corresponding eigenvectors $\mathbf{v}_1$ and $\mathbf{v}_2$. Prove that $\mathbf{v}_1 \perp \mathbf{v}_2$. (Hint: Use the definition of a symmetric operator to show that $(\lambda_2 - \lambda_1)(\mathbf{v}_1 \cdot \mathbf{v}_2) = 0$.)

12. Let $\mathbf{A}$ be an $n \times n$ symmetric matrix. Prove that $\mathbf{A}$ is orthogonal if and only if all eigenvalues for $\mathbf{A}$ are either 1 or -1. (Hint: For one half of the proof, use Theorem 5.32. For the other half, orthogonally diagonalize to help calculate $\mathbf{A}^2 = \mathbf{AA}^T$.)

7

Complex Vector Spaces and General Inner Products

In this chapter, we extend many previous results to more complicated algebraic structures. In some cases we must work with the larger set of complex numbers rather than with just real numbers. In Section 7.1, we examine complex vector spaces and consider their similarities to and differences from real vector spaces. In Section 7.2, we discuss inner product spaces, which possess an additional operation analogous to the dot product on $\mathbb{R}^n$.

We use the complex number system throughout this chapter, and we assume you are familiar with its basic operations. For quick reference, Appendix C lists the definition of a complex number and the rules for complex addition, multiplication, conjugation, absolute value, and reciprocal.

7.1 Complex Vector Spaces

Until now, our set of scalars has always been the real numbers. In this section, however, we use the complex numbers to define and study complex vector spaces. Most of the results obtained with real vector spaces have complex counterparts. However, in this section, we emphasize the differences between real and complex vector spaces.

Complex Vectors

DEFINITION

A **complex n-vector** is an ordered sequence (or ordered n-tuple) of n complex numbers. The set of all complex n-vectors is denoted by $\mathbb{C}^n$.

For example, $[3-2i, 4+3i, -i]$ is a vector in $\mathbb{C}^3$. We often write $\mathbf{z} = [z_1, z_2, \ldots, z_n]$ (where $z_1, z_2, \ldots, z_n \in \mathbb{C}$) to represent an arbitrary vector in $\mathbb{C}^n$.

When dealing with complex vectors, we often need to extend our definition of *scalar* to include all complex numbers instead of only real numbers. In what follows, it will always be clear from context whether we are using complex scalars or real scalars.

Scalar multiplication and addition of complex vectors are defined coordinate-wise, just as for real vectors. For example, $(-2+i)[4+i, -1-2i] + [-3-2i, -2+i] = [-9+2i, 4+3i] + [-3-2i, -2+i] = [-12, 2+4i]$. It is easy to show that all the properties in Theorem 1.3 carry over to complex vectors (with real or complex scalars). If we restrict ourselves to real scalars, these properties show that $\mathbb{C}^n$ is another example of a *real vector space*.

The **complex conjugate** of a vector $\mathbf{z} = [z_1, z_2, \ldots, z_n] \in \mathbb{C}^n$ is defined, using the complex conjugate operation, to be $\overline{\mathbf{z}} = [\overline{z_1}, \overline{z_2}, \ldots, \overline{z_n}]$. For example, if $\mathbf{z} = [3-2i, -5-4i, -2i]$, then $\overline{\mathbf{z}} = [3+2i, -5+4i, 2i]$.

The **complex dot product** of two vectors $\mathbf{z} = [z_1, z_2, \ldots, z_n]$ and $\mathbf{w} = [w_1, w_2, \ldots, w_n]$ in $\mathbb{C}^n$ is defined as follows:

$$\mathbf{z} \cdot \mathbf{w} = z_1\overline{w_1} + z_2\overline{w_2} + \cdots + z_n\overline{w_n}.$$

Notice that if $\mathbf{z}$ and $\mathbf{w}$ are both real vectors, then $\mathbf{z} \cdot \mathbf{w}$ is the same as the familiar dot product in $\mathbb{R}^n$. The next example illustrates the complex dot product.

Example 1

Let $\mathbf{z} = [3-2i, -2+i, -4-3i]$ and $\mathbf{w} = [-2+4i, 5-i, -2i]$. Then

$$\begin{aligned}\mathbf{z} \cdot \mathbf{w} &= (3-2i)\overline{(-2+4i)} + (-2+i)\overline{(5-i)} + (-4-3i)\overline{(-2i)}\\ &= (3-2i)(-2-4i) + (-2+i)(5+i) + (-4-3i)(+2i) = -19-13i.\end{aligned}$$

However,

$$\begin{aligned}\mathbf{w} \cdot \mathbf{z} &= (-2+4i)\overline{(3-2i)} + (5-i)\overline{(-2+i)} + (-2i)\overline{(-4-3i)}\\ &= (-2+4i)(3+2i) + (5-i)(-2-i) + (-2i)(-4+3i) = -19+13i.\end{aligned}$$

Notice that $\mathbf{z} \cdot \mathbf{w} = \overline{\mathbf{w} \cdot \mathbf{z}}$. This statement is true in general, as you will see shortly. ■

Notice that if $\mathbf{z} = [z_1, \ldots, z_n]$, then $\mathbf{z} \cdot \mathbf{z} = z_1\overline{z_1} + \cdots + z_n\overline{z_n} = |z_1|^2 + \cdots + |z_n|^2$, a real number. We define the **length** of a complex vector $\mathbf{z} = [z_1, \ldots, z_n]$ as $\|\mathbf{z}\| = \sqrt{\mathbf{z} \cdot \mathbf{z}}$. For example, if $\mathbf{z} = [3-i, -2i, 4+3i]$, then

$$\|\mathbf{z}\| = \sqrt{(3-i)(3+i) + (-2i)(2i) + (4+3i)(4-3i)} = \sqrt{10+4+25} = \sqrt{39}.$$

The following theorem lists the most important properties of the complex dot product. You are asked to prove parts of this theorem in Exercise 2. Notice the use of the complex conjugate in parts (1) and (5).

THEOREM 7.1

Let $\mathbf{z}_1$, $\mathbf{z}_2$, and $\mathbf{z}_3$ be vectors in $\mathbb{C}^n$, and let $k \in \mathbb{C}$ be any scalar. Then

(1) $\mathbf{z}_1 \cdot \mathbf{z}_2 = \overline{\mathbf{z}_2 \cdot \mathbf{z}_1}$ — Conjugate-Commutativity of Complex Dot Product

(2) $\mathbf{z}_1 \cdot \mathbf{z}_1 = \|\mathbf{z}_1\|^2 \geq 0$
(3) $\mathbf{z}_1 \cdot \mathbf{z}_1 = 0$ if and only if $\mathbf{z}_1 = \mathbf{0}$ — Relationships between Complex Dot Product and Length

(4) $k(\mathbf{z}_1 \cdot \mathbf{z}_2) = (k\mathbf{z}_1) \cdot \mathbf{z}_2$
(5) $\overline{k}(\mathbf{z}_1 \cdot \mathbf{z}_2) = \mathbf{z}_1 \cdot (k\mathbf{z}_2)$ — Relationships between Scalar Multiplication and Complex Dot Product

(6) $\mathbf{z}_1 \cdot (\mathbf{z}_2 + \mathbf{z}_3) = (\mathbf{z}_1 \cdot \mathbf{z}_2) + (\mathbf{z}_1 \cdot \mathbf{z}_3)$
(7) $(\mathbf{z}_1 + \mathbf{z}_2) \cdot \mathbf{z}_3 = (\mathbf{z}_1 \cdot \mathbf{z}_3) + (\mathbf{z}_2 \cdot \mathbf{z}_3)$ — Distributive Laws of Complex Dot Product over Addition

Complex Matrices

DEFINITION

An $m \times n$ **complex matrix** is a rectangular array of complex numbers arranged in m rows and n columns. The set of all $m \times n$ complex matrices is denoted as $\mathcal{M}_{mn}^{\mathbb{C}}$, or **complex** $\mathcal{M}_{mn}$.

Addition and scalar multiplication of matrices are defined entrywise in the usual manner, and the properties in Theorem 1.10 also hold for complex matrices. When we restrict ourselves to real scalars, these properties tell us that $\mathcal{M}_{mn}^{\mathbb{C}}$ is another example of a real vector space.

We next define multiplication of complex matrices. Beware! Complex matrices are multiplied the same way as real matrices. We do not take complex conjugates of entries in the second matrix as we do with the complex dot product. If $\mathbf{Z}$ is an $m \times n$ matrix and $\mathbf{W}$ is an $n \times r$ matrix, then $\mathbf{ZW}$ is the $m \times r$ matrix whose (i, j) entry equals

$$(\mathbf{ZW})_{ij} = z_{i1}w_{1j} + z_{i2}w_{2j} + \cdots + z_{in}w_{nj}.$$

Example 2

Let $\mathbf{Z} = \begin{bmatrix} 1-i & 2i & -2+i \\ -3i & 3-2i & -1-i \end{bmatrix}$ and $\mathbf{W} = \begin{bmatrix} -2i & 1-4i \\ -1+3i & 2-3i \\ -2+i & -4+i \end{bmatrix}$.

Then the (1, 1) entry of **ZW** is

$$\begin{aligned}(1-i)(-2i)+(2i)(-1+3i)+(-2+i)(-2+i) &= -2i-2-2i-6+3-4i \\ &= -5-8i.\end{aligned}$$

You can easily check that the entire product is $\mathbf{ZW} = \begin{bmatrix} -5-8i & 10-7i \\ 12i & -7-13i \end{bmatrix}$. ■

The familiar properties of matrix multiplication and matrix inverses all carry over to the complex case.

The **complex conjugate** $\overline{\mathbf{Z}}$ of a complex matrix $\mathbf{Z} = [z_{ij}]$ is the matrix whose (i, j) entry is $\overline{z_{ij}}$. The **transpose** $\mathbf{Z}^T$ of an $m \times n$ complex matrix $\mathbf{Z} = [z_{ij}]$ is the $n \times m$ matrix whose (j, i) entry is z_{ij}. These two operations are combined as follows: we define the **conjugate transpose** $\mathbf{Z}^*$ of a complex matrix to be

$$\mathbf{Z}^* = (\overline{\mathbf{Z}})^T = \overline{(\mathbf{Z}^T)}.$$

It is easy to see that, in fact, $(\overline{\mathbf{Z}})^T = \overline{(\mathbf{Z}^T)}$ for any complex matrix **Z**.

Example 3

If $\mathbf{Z} = \begin{bmatrix} 2-3i & -i & 5 \\ 4i & 1+2i & -2-4i \end{bmatrix}$, then $\overline{\mathbf{Z}} = \begin{bmatrix} 2+3i & i & 5 \\ -4i & 1-2i & -2+4i \end{bmatrix}$, and

$$\mathbf{Z}^* = (\overline{\mathbf{Z}})^T = \begin{bmatrix} 2+3i & -4i \\ i & 1-2i \\ 5 & -2+4i \end{bmatrix}.$$

■

The following theorem lists the most important properties of the complex conjugate and conjugate transpose operations.

THEOREM 7.2

Let **Z** and **Y** be $m \times n$ complex matrices, and let **W** be an $n \times p$ complex matrix. Let $k \in \mathbb{C}$.

(1) $\overline{(\overline{\mathbf{Z}})} = \mathbf{Z}$, and $(\mathbf{Z}^*)^* = \mathbf{Z}$

(2) $(\mathbf{Z} + \mathbf{Y})^* = \mathbf{Z}^* + \mathbf{Y}^*$

(3) $(k\mathbf{Z})^* = \overline{k}(\mathbf{Z}^*)$

(4) $\overline{\mathbf{ZW}} = \overline{\mathbf{Z}}\,\overline{\mathbf{W}}$

(5) $(\mathbf{ZW})^* = \mathbf{W}^*\mathbf{Z}^*$

Note the use of $\overline{k}$ in part (3). The proof of this theorem is easy, and parts of it are left for you to do in Exercise 4.

As before, you can use the Gauss-Jordan row reduction method to solve linear systems involving complex numbers. Also, the determinant of a complex matrix is defined just as it is for real matrices, and it has the same uses and properties.

Special Types of Complex Matrices

Symmetric and skew-symmetric real matrices have complex analogs:

DEFINITION

Let $\mathbf{Z}$ be a square complex matrix. Then $\mathbf{Z}$ is **Hermitian** if and only if $\mathbf{Z}^* = \mathbf{Z}$, and $\mathbf{Z}$ is **skew-Hermitian** if and only if $\mathbf{Z}^* = -\mathbf{Z}$.

Notice that an $n \times n$ complex matrix $\mathbf{Z}$ is Hermitian if and only if $z_{ij} = \overline{z_{ji}}$, for $1 \leq i, j \leq n$. When $i = j$, we have $z_{ii} = \overline{z_{ii}}$, for all i, and so *all main diagonal entries of a Hermitian matrix are real*. Similarly, $\mathbf{Z}$ is skew-Hermitian if and only if $z_{ij} = -\overline{z_{ji}}$, for $1 \leq i, j \leq n$. When $i = j$, we have $z_{ii} = -\overline{z_{ii}}$, for all i, and so *all main diagonal entries of a skew-Hermitian matrix are pure imaginary.*

Example 4

Consider the matrix

$$\mathbf{H} = \begin{bmatrix} 3 & 2-i & 1-2i \\ 2+i & -1 & -3i \\ 1+2i & 3i & 4 \end{bmatrix}.$$

Notice that

$$\overline{\mathbf{H}} = \begin{bmatrix} 3 & 2+i & 1+2i \\ 2-i & -1 & 3i \\ 1-2i & -3i & 4 \end{bmatrix}, \text{ and so } \mathbf{H}^* = (\overline{\mathbf{H}})^T = \begin{bmatrix} 3 & 2-i & 1-2i \\ 2+i & -1 & -3i \\ 1+2i & 3i & 4 \end{bmatrix}.$$

Since $\mathbf{H}^* = \mathbf{H}$, $\mathbf{H}$ is Hermitian. Similarly, it is easy to show that the matrix

$$\mathbf{K} = \begin{bmatrix} -2i & 5+i & -1-3i \\ -5+i & i & 6 \\ 1-3i & -6 & 3i \end{bmatrix}$$

is skew-Hermitian. (Verify.) ■

Some other useful results concerning Hermitian and skew-Hermitian matrices are left for you to prove in Exercises 6 and 7.

Another very important type of complex matrix is the following:

DEFINITION

A square complex matrix $\mathbf{Z}$ is **normal** if and only if $\mathbf{Z}\mathbf{Z}^* = \mathbf{Z}^*\mathbf{Z}$.

The next theorem gives two important classes of normal matrices.

THEOREM 7.3

If $\mathbf{Z}$ is a Hermitian or skew-Hermitian matrix, then $\mathbf{Z}$ is normal.

The proof of this theorem is easy and is left for you to do in Exercise 8. The next example gives a normal matrix that is neither Hermitian nor skew-Hermitian, thus illustrating that the converse to Theorem 7.3 is false.

Example 5

Consider $\mathbf{Z} = \begin{bmatrix} 1-2i & -i \\ 1 & 2-3i \end{bmatrix}$. Now, $\mathbf{Z}^* = \begin{bmatrix} 1+2i & 1 \\ i & 2+3i \end{bmatrix}$, and so

$$\mathbf{Z}\mathbf{Z}^* = \begin{bmatrix} 1-2i & -i \\ 1 & 2-3i \end{bmatrix}\begin{bmatrix} 1+2i & 1 \\ i & 2+3i \end{bmatrix} = \begin{bmatrix} 6 & 4-4i \\ 4+4i & 14 \end{bmatrix}.$$

Also, $\mathbf{Z}^*\mathbf{Z} = \begin{bmatrix} 1+2i & 1 \\ i & 2+3i \end{bmatrix}\begin{bmatrix} 1-2i & -i \\ 1 & 2-3i \end{bmatrix} = \begin{bmatrix} 6 & 4-4i \\ 4+4i & 14 \end{bmatrix}$.

Since $\mathbf{Z}\mathbf{Z}^* = \mathbf{Z}^*\mathbf{Z}$, $\mathbf{Z}$ is normal. ■

In general, a matrix $\mathbf{Z}$ is normal if and only if $\mathbf{Z} = \mathbf{H}_1 + i\mathbf{H}_2$, where $\mathbf{H}_1$ and $\mathbf{H}_2$ are Hermitian matrices such that $\mathbf{H}_1\mathbf{H}_2 = \mathbf{H}_2\mathbf{H}_1$ (see Exercise 30). For example, the normal matrix $\mathbf{Z}$ from Example 5 equals $\begin{bmatrix} 1 & \frac{1}{2}-\frac{1}{2}i \\ \frac{1}{2}+\frac{1}{2}i & 2 \end{bmatrix}$ $+\, i\begin{bmatrix} -2 & -\frac{1}{2}+\frac{1}{2}i \\ -\frac{1}{2}-\frac{1}{2}i & -3 \end{bmatrix}$.

Complex Vector Spaces

We define **complex vector spaces** exactly the same way that we defined real vector spaces in Section 4.1, except that the set of scalars is enlarged to the complex numbers rather than just the real numbers. Clearly, $\mathbb{C}^n$ is an example (in fact, the most important one) of a complex vector space. Also, under regular addition and complex scalar multiplication, both $\mathcal{M}_{mn}^{\mathbb{C}}$ and $\mathcal{P}_n^{\mathbb{C}}$ (polynomials of degree $\leq n$ with complex coefficients) are complex vector spaces (see Exercise 9).

The concepts of subspace, span, linear independence, basis, and dimension for real vector spaces carry over to complex vector spaces in an analogous way. All the results in Chapter 4 have complex counterparts. In particular, if $\mathcal{W}$ is any subspace of a finite n-dimensional complex vector space (for example, $\mathbb{C}^n$), then $\mathcal{W}$ has a finite basis, and $\dim(\mathcal{W}) \leq n$.

We sometimes have to be careful about whether a vector space is being considered as a *real* or a *complex* vector space—that is, whether complex scalars are to be used or just real scalars. For instance, as a *real* vector space, $\mathbb{C}^3$ has $\{[1,0,0], [i,0,0], [0,1,0], [0,i,0], [0,0,1], [0,0,i]\}$ as a basis, and $\dim(\mathbb{C}^3) = 6$. But as a *complex* vector space, $\mathbb{C}^3$ has $\{[1,0,0], [0,1,0], [0,0,1]\}$ as a basis (since i can now be used as a scalar), and so $\dim(\mathbb{C}^3) = 3$. In general, $\dim(\mathbb{C}^n) = 2n$ as a *real* vector space, but $\dim(\mathbb{C}^n) = n$ as a *complex* vector space. As usual, we let $\mathbf{e}_1 = [1,0,0,\ldots,0]$, $\mathbf{e}_2 = [0,1,0,\ldots,0], \ldots, \mathbf{e}_n = [0,0,0,\ldots,1]$ represent the **standard basis vectors** for the complex vector space $\mathbb{C}^n$.

Coordinatization in a complex vector space is done in the usual manner, as the following example indicates.

Example 6

Consider the subspace $\mathcal{W}$ of the complex vector space $\mathbb{C}^4$ spanned by the vectors $\mathbf{x}_1 = [1+i, 3, 0, -2i]$ and $\mathbf{x}_2 = [-i, 1-i, 3i, 1+2i]$. Since these vectors are linearly independent (why?), the set $B = (\mathbf{x}_1, \mathbf{x}_2)$ is an ordered basis for $\mathcal{W}$, and $\dim(\mathcal{W}) = 2$. The linear combination $\mathbf{z} = (1-i)\mathbf{x}_1 + 3\mathbf{x}_2$ of these basis vectors is equal to

$$\begin{aligned}\mathbf{z} &= (1-i)\mathbf{x}_1 + 3\mathbf{x}_2\\ &= [2, 3-3i, 0, -2-2i] + [-3i, 3-3i, 9i, 3+6i]\\ &= [2-3i, 6-6i, 9i, 1+4i].\end{aligned}$$

Of course, the coordinatization of $\mathbf{z}$ with respect to B is $[\mathbf{z}]_B = [1-i, 3]$. ■

Linear transformations between complex vector spaces are defined just as they are between real vector spaces, except that complex scalars are used in the rule $L(k\mathbf{v}) = kL(\mathbf{v})$. The properties of complex linear transformations are completely analogous to those for linear transformations between real vector spaces. As before, we say that a linear transformation L from a complex vector space $\mathcal{V}$ to a complex vector space $\mathcal{W}$ is an **isomorphism** if $L^{-1}: \mathcal{W} \longrightarrow \mathcal{V}$ exists (that is, if L has an inverse). It can be shown that two finite dimensional complex vector spaces are isomorphic if and only if they have the same dimension. Hence, every complex vector space of dimension n is isomorphic to $\mathbb{C}^n$.

Orthogonal Bases and the Gram-Schmidt Process

DEFINITION

A subset $\{\mathbf{v}_1, \mathbf{v}_2, \ldots, \mathbf{v}_n\}$ of vectors of $\mathbb{C}^n$ is **orthogonal** if and only if the *complex* dot product of any two distinct vectors in the set is zero. An orthogonal set of vectors in $\mathbb{C}^n$ is **orthonormal** if and only if each vector in the set is a unit vector.

The Gram-Schmidt Process for finding an orthogonal basis extends to the complex case, as in the next example.

Example 7

We find an orthogonal basis for the complex vector space $\mathbb{C}^3$ containing $\mathbf{w}_1 = [i, 1+i, 1]$. Let $\mathbf{w}_2 = [1, 0, 0]$ and $\mathbf{w}_3 = [0, 1, 0]$. (You can easily check that $\{\mathbf{w}_1, \mathbf{w}_2, \mathbf{w}_3\}$ is linearly independent and hence a basis for $\mathbb{C}^3$.) Let $\mathbf{v}_1 = \mathbf{w}_1$. Following the steps of the Gram-Schmidt Process, we obtain

$$\mathbf{v}_2 = \mathbf{w}_2 - \left(\frac{\mathbf{w}_2 \cdot \mathbf{v}_1}{\mathbf{v}_1 \cdot \mathbf{v}_1}\right)\mathbf{v}_1 = [1, 0, 0] - \left(\frac{-i}{4}\right)[i, 1+i, 1].$$

Multiplying by 4 to avoid fractions, we get

$$\mathbf{v}_2 = [4, 0, 0] + i[i, 1+i, 1] = [3, -1+i, i].$$

Continuing, we get

$$\begin{aligned}\mathbf{v}_3 &= \mathbf{w}_3 - \left(\frac{\mathbf{w}_3 \cdot \mathbf{v}_1}{\mathbf{v}_1 \cdot \mathbf{v}_1}\right)\mathbf{v}_1 - \left(\frac{\mathbf{w}_3 \cdot \mathbf{v}_2}{\mathbf{v}_2 \cdot \mathbf{v}_2}\right)\mathbf{v}_2 \\ &= [0, 1, 0] - \left(\frac{1-i}{4}\right)[i, 1+i, 1] - \left(\frac{-1-i}{12}\right)[3, -1+i, i].\end{aligned}$$

Multiplying by 12 to avoid fractions, we get

$$\begin{aligned}\mathbf{v}_3 &= [0, 12, 0] + 3(-1+i)[i, 1+i, 1] + (1+i)[3, -1+i, i] \\ &= [0, 12, 0] + [-3-3i, -6, -3+3i] + [3+3i, -2, -1+i] \\ &= [0, 4, -4+4i].\end{aligned}$$

We can divide by 4 to avoid multiples, and so we finally get $\mathbf{v}_3 = [0, 1, -1+i]$. Hence, $\{\mathbf{v}_1, \mathbf{v}_2, \mathbf{v}_3\} = \{[i, 1+i, 1], [3, -1+i, i], [0, 1, -1+i]\}$ is an orthogonal basis for $\mathbb{C}^3$ containing $\mathbf{w}_1$. (You should verify that $\mathbf{v}_1$, $\mathbf{v}_2$, and $\mathbf{v}_3$ are mutually orthogonal.) ■

Unitary Matrices

We now examine the complex analog of orthogonal matrices:

DEFINITION

A square (nonsingular) complex matrix $\mathbf{A}$ is **unitary** if and only if $\mathbf{A}^* = \mathbf{A}^{-1}$ (that is, if and only if $(\overline{\mathbf{A}})^T = \mathbf{A}^{-1}$).

It follows immediately from this definition that every unitary matrix is a normal matrix. (Why?)

Example 8

The matrix

$$\mathbf{A} = \begin{bmatrix} \frac{1-i}{\sqrt{3}} & 0 & \frac{i}{\sqrt{3}} \\ \frac{-1+i}{\sqrt{15}} & \frac{3}{\sqrt{15}} & \frac{2i}{\sqrt{15}} \\ \frac{1-i}{\sqrt{10}} & \frac{2}{\sqrt{10}} & \frac{-2i}{\sqrt{10}} \end{bmatrix}$$

is easily seen to be unitary, since

$$\mathbf{A}^* = (\overline{\mathbf{A}})^T = \begin{bmatrix} \frac{1+i}{\sqrt{3}} & \frac{-1-i}{\sqrt{15}} & \frac{1+i}{\sqrt{10}} \\ 0 & \frac{3}{\sqrt{15}} & \frac{2}{\sqrt{10}} \\ -\frac{i}{\sqrt{3}} & -\frac{2i}{\sqrt{15}} & \frac{2i}{\sqrt{10}} \end{bmatrix}$$

and $\mathbf{AA}^* = \mathbf{I}_3$. ■

It is easy to show that if $\mathbf{A}$ is unitary, then so is $\mathbf{A}^{-1}$, and that $|\mathbf{A}| = \pm 1$ (see Exercise 17). The next three results are the analogs of Theorems 5.29, 5.30, and 5.31; they are left for you to prove in Exercises 18, 20, and 21. You should verify that the unitary matrix of Example 8 satisfies Theorem 7.5.

THEOREM 7.4

Let $\mathbf{A}$ and $\mathbf{B}$ be unitary matrices of the same size. Then $\mathbf{AB}$ is unitary.

THEOREM 7.5

Let $\mathbf{A}$ be an $n \times n$ complex matrix. Then $\mathbf{A}$ is unitary
(1) if and only if the rows of $\mathbf{A}$ form an orthonormal basis for $\mathbb{C}^n$ (as a complex vector space).
(2) if and only if the columns of $\mathbf{A}$ form an orthonormal basis for $\mathbb{C}^n$ (as a complex vector space).

THEOREM 7.6

Let B and C be ordered orthonormal bases for $\mathbb{C}^n$ (as a complex vector space). Then the transition matrix from B to C is a unitary matrix.

Complex Eigenvalues and Complex Eigenvectors

We can also extend the concept of eigenvalues and eigenvectors to complex vector spaces. If $\mathbf{A}$ is an $n \times n$ complex matrix, then $\lambda \in \mathbb{C}$ is an **eigenvalue** for $\mathbf{A}$ if and only if there is a nonzero vector $\mathbf{v} \in \mathbb{C}^n$ such that $\mathbf{Av} = \lambda\mathbf{v}$. As before, the nonzero vector $\mathbf{v}$ is called an **eigenvector** for $\mathbf{A}$ associated with λ. The **characteristic polynomial** of $\mathbf{A}$, defined as $p_{\mathbf{A}}(x) = |x\mathbf{I}_n - \mathbf{A}|$, is used to find the eigenvalues of $\mathbf{A}$, just as in Chapter 6.

If $\mathcal{V}$ is a finite dimensional complex vector space, then eigenvalues and eigenvectors for a linear operator $L: \mathcal{V} \longrightarrow \mathcal{V}$ are defined analogously. All of the theorems in Section 6.1 regarding eigenvalues and eigenvectors carry over to the complex vector space setting.

Example 9

For the matrix

$$\mathbf{A} = \begin{bmatrix} -4+7i & 2+i & 7+7i \\ 1-3i & 1-i & -3-i \\ 5+4i & 1-2i & 7-5i \end{bmatrix},$$

we have

$$x\mathbf{I}_3 - \mathbf{A} = \begin{bmatrix} x+4-7i & -2-i & -7-7i \\ -1+3i & x-1+i & 3+i \\ -5-4i & -1+2i & x-7+5i \end{bmatrix}.$$

After some calculation, you can verify that $p_{\mathbf{A}}(x) = |x\mathbf{I}_3 - \mathbf{A}| = x^3 - (4+i)x^2 + (5+5i)x - (6+6i)$. You can also check that $p_{\mathbf{A}}(x)$ factors as $(x-(1-i))(x-2i)(x-3)$. Hence, the eigenvalues of $\mathbf{A}$ are $\lambda_1 = 1-i$, $\lambda_2 = 2i$, and $\lambda_3 = 3$. To find an eigenvector for λ_1, we look for a nontrivial solution $\mathbf{v}$ of the system $((1-i)\mathbf{I}_3 - \mathbf{A})\mathbf{v} = \mathbf{0}$, which is

$$\begin{bmatrix} 5-8i & -2-i & -7-7i \\ -1+3i & 0 & 3+i \\ -5-4i & -1+2i & -6+4i \end{bmatrix}\begin{bmatrix} z_1 \\ z_2 \\ z_3 \end{bmatrix} = \begin{bmatrix} 0 \\ 0 \\ 0 \end{bmatrix}.$$

After row reduction, we obtain the augmented matrix

$$\left[\begin{array}{ccc|c} 1 & 0 & -i & 0 \\ 0 & 1 & 1 & 0 \\ 0 & 0 & 0 & 0 \end{array}\right].$$

It is then easy to see that $\{[-1, 1, i]\}$ is a basis for the solution set and thus is a basis for E_{λ_1}. A similar analysis shows that $\{[3i, -i, 2]\}$ is a basis for E_{λ_2} and $\{[i, 0, 1]\}$ is a basis for E_{λ_3}. ■

Diagonalizable Complex Matrices

We say a complex matrix $\mathbf{A}$ is **diagonalizable** if and only if $\mathbf{A}$ is similar to a diagonal matrix (that is, if and only if there is a nonsingular complex matrix $\mathbf{P}$ such that $\mathbf{P}^{-1}\mathbf{A}\mathbf{P}$ is a diagonal matrix). Also, we say a linear operator L on a finite dimensional complex vector space $\mathcal{V}$ is **diagonalizable** if and only if the matrix for L with respect to some ordered basis for $\mathcal{V}$ is a diagonal matrix. (As in Chapter 6, we can show a linear operator L is diagonalizable if and only if the matrix for L with respect to any ordered basis for $\mathcal{V}$ is diagonalizable.) A proof similar to that of Theorem 6.10 gives:

THEOREM 7.7

Let L be a linear operator on an n-dimensional complex vector space $\mathcal{V}$. Then L is diagonalizable if and only if there is a set of n linearly independent complex eigenvectors for L.

Example 10

The linear operator L on $\mathbb{C}^3$ given by $L(\mathbf{v}) = \mathbf{A}\mathbf{v}$, for $\mathbf{v} \in \mathbb{C}^3$, where $\mathbf{A}$ is the matrix of Example 9, is diagonalizable, since we have already seen that $\mathbf{A}$ has three distinct eigenvalues $\lambda_1 = 1-i$, $\lambda_2 = 2i$, and $\lambda_3 = 3$. In fact, $\mathbf{A}$ is similar to the diagonal matrix

$$\begin{bmatrix} \lambda_1 & 0 & 0 \\ 0 & \lambda_2 & 0 \\ 0 & 0 & \lambda_3 \end{bmatrix} = \begin{bmatrix} 1-i & 0 & 0 \\ 0 & 2i & 0 \\ 0 & 0 & 3 \end{bmatrix}.$$ ■

Recall that linear transformations between real vector spaces (and their associated real matrices) do not necessarily have any real eigenvalues. However, linear transformations between complex vector spaces (and their associated complex matrices) must have at least one complex eigenvalue. In particular, for any $n \times n$ matrix $\mathbf{A}$, the equation $p_{\mathbf{A}}(x) = |x\mathbf{I}_n - \mathbf{A}| = 0$ must have at least one complex root, by the **Fundamental Theorem of Algebra**. This theorem states that every complex polynomial of degree ≥ 1 has at least one complex root. (We do not prove this theorem here.) Hence $\mathbf{A}$ has at least one eigenvalue. In fact, $p_{\mathbf{A}}(x)$ must factor into a product of *linear* factors. Even so, every complex $n \times n$ matrix need not be diagonalizable, because the total of all the geometric multiplicities of the eigenvalues of a given matrix may be less than n (although the algebraic multiplicities must add up to n).

Unitarily Diagonalizable Matrices

We now consider the complex analog of orthogonal diagonalization:

DEFINITION

An $n \times n$ complex matrix $\mathbf{A}$ is **unitarily diagonalizable** if and only if there is a unitary matrix $\mathbf{P}$ such that $\mathbf{P}^{-1}\mathbf{AP}$ is diagonal.

Example 11

Consider the matrix

$$\mathbf{P} = \frac{1}{3}\begin{bmatrix} -2i & 2 & 1 \\ 2i & 1 & 2 \\ 1 & -2i & 2i \end{bmatrix}.$$

Notice that $\mathbf{P}$ is a unitary matrix, since the columns of $\mathbf{P}$ form an orthonormal basis for $\mathbb{C}^3$.

Next, consider the matrix

$$\mathbf{A} = \frac{1}{3}\begin{bmatrix} -1+3i & 2+2i & -2 \\ 2+2i & 2i & -2i \\ 2 & 2i & 1+4i \end{bmatrix}.$$

Now, **A** is unitarily diagonalizable because

$$\mathbf{P}^{-1}\mathbf{A}\mathbf{P} = \mathbf{P}^*\mathbf{A}\mathbf{P} = \begin{bmatrix} -1 & 0 & 0 \\ 0 & 2i & 0 \\ 0 & 0 & 1+i \end{bmatrix},$$

a diagonal matrix. ■

We saw in Section 6.3 that a matrix is orthogonally diagonalizable if and only if it is symmetric. What types of matrices are unitarily diagonalizable? The following theorem characterizes them:

THEOREM 7.8

An $n \times n$ matrix **A** is unitarily diagonalizable if and only if **A** is normal.

It is easy to see that the matrix **A** in Example 11 is normal (see Exercise 24).

Example 12

A direct computation shows that $\mathbf{A} = \begin{bmatrix} -48+18i & -24+36i \\ 24-36i & -27+32i \end{bmatrix}$ is normal. (Verify.) Therefore, **A** is unitarily diagonalizable by Theorem 7.8. After some calculation, you can verify that the eigenvalues of **A** are $\lambda_1 = 50i$ and $\lambda_2 = -75$. Hence, **A** is unitarily diagonalizable to $\mathbf{D} = \begin{bmatrix} 50i & 0 \\ 0 & -75 \end{bmatrix}$.

In fact, λ_1 and λ_2 have associated eigenvectors $[\frac{3}{5}, -\frac{4}{5}i]$ and $[-\frac{4}{5}i, \frac{3}{5}]$. Since these eigenvectors are orthonormal, the matrix $\mathbf{P} = \begin{bmatrix} \frac{3}{5} & -\frac{4}{5}i \\ -\frac{4}{5}i & \frac{3}{5} \end{bmatrix}$, whose columns are these eigenvectors, is a unitary matrix, and $\mathbf{P}^{-1}\mathbf{A}\mathbf{P} = \mathbf{P}^*\mathbf{A}\mathbf{P} = \mathbf{D}$. ■

An immediate corollary of Theorems 7.3 and 7.8 is:

COROLLARY 7.9

If **A** is a Hermitian matrix, then **A** is unitarily diagonalizable.

Even more can be proved about Hermitian matrices. First, consider the following lemma:

LEMMA 7.10

If $\mathbf{A}$ is any $n \times n$ complex matrix and $\mathbf{x}$ and $\mathbf{y}$ are complex (column) n-vectors, then $(\mathbf{Ax}) \cdot \mathbf{y} = \mathbf{x} \cdot (\mathbf{A}^*\mathbf{y})$.

Proof of Lemma 7.10

$$(\mathbf{Ax}) \cdot \mathbf{y} = (\mathbf{Ax})^T\overline{\mathbf{y}} = \mathbf{x}^T\mathbf{A}^T\overline{\mathbf{y}} = \mathbf{x}^T(\overline{\mathbf{A}^*\mathbf{y}}) = \mathbf{x} \cdot (\mathbf{A}^*\mathbf{y}). \quad \blacksquare$$

If operators L and M on $\mathbb{C}^n$ have the property $L(\mathbf{x}) \cdot \mathbf{y} = \mathbf{x} \cdot M(\mathbf{y})$, for all $\mathbf{x}, \mathbf{y} \in \mathbb{C}^n$, then M is called an **adjoint** of L. Now, suppose that $L: \mathbb{C}^n \longrightarrow \mathbb{C}^n$ is the linear operator $L(\mathbf{x}) = \mathbf{Ax}$, where $\mathbf{A}$ is an $n \times n$ matrix, and let $L^*: \mathbb{C}^n \longrightarrow \mathbb{C}^n$ be given by $L^*(\mathbf{x}) = \mathbf{A}^*\mathbf{x}$. By Lemma 7.10, $(L(\mathbf{x})) \cdot \mathbf{y} = \mathbf{x} \cdot (L^*(\mathbf{y}))$ for all $\mathbf{x}, \mathbf{y} \in \mathbb{C}^n$, and so L^* is an adjoint of L.

If $\mathbf{A}$ is a Hermitian matrix, then $L = L^*$, and so $(L(\mathbf{x})) \cdot \mathbf{y} = \mathbf{x} \cdot (L(\mathbf{y}))$ for all $\mathbf{x}, \mathbf{y} \in \mathbb{C}^n$. Such an operator is called **self-adjoint**, since it equals its adjoint. It can be shown that every self-adjoint operator on $\mathbb{C}^n$ has a Hermitian matrix as its matrix representation with respect to any orthonormal basis. Self-adjoint operators are the complex analogs of the symmetric operators you studied in Section 6.3. Corollary 7.9 tells us that all self-adjoint operators are unitarily diagonalizable. The final theorem of this section shows that any diagonal matrix representation for a self-adjoint operator has all real entries:

THEOREM 7.11

All eigenvalues of a Hermitian matrix are real.

Proof of Theorem 7.11

Let λ be an eigenvalue for a Hermitian matrix $\mathbf{A}$, and let $\mathbf{u}$ be a unit eigenvector for λ. Then $\lambda = \lambda\|\mathbf{u}\|^2 = \lambda(\mathbf{u} \cdot \mathbf{u}) = (\lambda\mathbf{u}) \cdot \mathbf{u} = (\mathbf{Au}) \cdot \mathbf{u} = \mathbf{u} \cdot (\mathbf{Au})$ (by Lemma 7.10) $= \mathbf{u} \cdot \lambda\mathbf{u} = \overline{\lambda}(\mathbf{u} \cdot \mathbf{u})$ (by part (5) of Theorem 7.1) $= \overline{\lambda}$. Hence, $\lambda = \overline{\lambda}$, which means that λ is real. $\blacksquare$

Example 13

Let $\mathbf{A}$ and $\mathbf{P}$ be the Hermitian and unitary matrices (respectively) given by

$$\mathbf{A} = \begin{bmatrix} 17 & -24 + 8i & -24 - 32i \\ -24 - 8i & 53 & 4 + 12i \\ -24 + 32i & 4 - 12i & 11 \end{bmatrix}$$

$$\text{and} \quad \mathbf{P} = \frac{1}{9}\begin{bmatrix} 4 & 6-2i & -3-4i \\ 6+2i & 1 & 2+6i \\ -3+4i & 2-6i & 4 \end{bmatrix}.$$

By Theorem 7.11, the matrix **A** has all eigenvalues real. It can be shown that these eigenvalues are $\lambda_1 = 27$, $\lambda_2 = -27$, and $\lambda_3 = 81$. By Corollary 7.9, **A** is unitarily diagonalizable. In fact, the product $\mathbf{P}^{-1}\mathbf{AP}$ yields the diagonal matrix with eigenvalues λ_1, λ_2, and λ_3 on the diagonal. (Verify.) ■

Exercises—Section 7.1

1. Perform the following computations involving complex vectors.

★(a) $[2+i, 3, -i] + [-1+3i, -2+i, 6]$

★(b) $(-8+3i)[4i, 2-3i, -7+i]$

(c) $\overline{[5-i, 2+i, -3i]}$

★(d) $\overline{(-4)[6-3i, 7-2i, -8i]}$

★(e) $[-2+i, 5-2i, 3+4i] \cdot [1+i, 4-3i, -6i]$

(f) $[5+2i, 6i, -2+i] \cdot [3-6i, 8+i, 1-4i]$

2. (a) Prove parts (1) and (2) of Theorem 7.1.

(b) Prove part (5) of Theorem 7.1.

3. Perform the computations below with these matrices:

$$\mathbf{A} = \begin{bmatrix} 2+5i & -4+i \\ -3-6i & 8-3i \end{bmatrix} \qquad \mathbf{B} = \begin{bmatrix} 9-i & -3i \\ 5+2i & 4+3i \end{bmatrix}$$

$$\mathbf{C} = \begin{bmatrix} 1+i & -2i & 6+4i \\ 0 & 3+i & 5 \\ -10i & 0 & 7-3i \end{bmatrix} \qquad \mathbf{D} = \begin{bmatrix} 5-i & -i & -3 \\ 2+3i & 0 & -4+i \end{bmatrix}$$

★(a) $\mathbf{A}+\mathbf{B}$ (b) $\overline{\mathbf{C}}$ ★(c) $\mathbf{C}^*$ ★(d) $(-3i)\mathbf{D}$

(e) $\mathbf{A}-\mathbf{B}^T$ ★(f) $\mathbf{AB}$ (g) $\mathbf{D}(\mathbf{C}^*)$ (h) $\mathbf{B}^2$

★(i) $\mathbf{C}^T\mathbf{D}^*$ (j) $(\mathbf{C}^*)^2$

4. (a) Prove part (3) of Theorem 7.2.

(b) Prove part (5) of Theorem 7.2.

★**5.** Determine which of the following matrices are Hermitian or skew-Hermitian.

$$\text{(a)} \begin{bmatrix} -4i & 6-2i & 8 \\ -6-2i & 0 & -2-i \\ -8 & 2-i & 5i \end{bmatrix} \qquad \text{(b)} \begin{bmatrix} 2+3i & 6i & 1+i \\ -6i & 4 & 8-3i \\ 1-i & -8+3i & 5i \end{bmatrix}$$

$$\text{(c)} \begin{bmatrix} 2 & 0 & 0 \\ 0 & -3 & 0 \\ 0 & 0 & 4 \end{bmatrix} \qquad \text{(d)} \begin{bmatrix} 5i & 0 & 0 \\ 0 & -2i & 0 \\ 0 & 0 & 6i \end{bmatrix} \qquad \text{(e)} \begin{bmatrix} 2 & -2i & 2 \\ 2i & -2 & -2i \\ 2 & 2i & -2 \end{bmatrix}$$

6. (a) Prove that if $\mathbf{Z}$ is any square complex matrix, then $\mathbf{H} = \frac{1}{2}(\mathbf{Z} + \mathbf{Z}^*)$ is a Hermitian matrix and $\mathbf{K} = \frac{1}{2}(\mathbf{Z} - \mathbf{Z}^*)$ is a skew-Hermitian matrix.
(b) Prove that if $\mathbf{Z}$ is any square complex matrix, then $\mathbf{Z}$ can be expressed uniquely as the sum of a Hermitian matrix $\mathbf{H}$ and a skew-Hermitian matrix $\mathbf{K}$. (Hint: Use part (a).)

7. (a) Let $\mathbf{H}$ and $\mathbf{J}$ be Hermitian matrices of the same size. Prove that $\mathbf{HJ}$ is Hermitian if and only if $\mathbf{HJ} = \mathbf{JH}$.
(b) Prove that if $\mathbf{H}$ is Hermitian, then $\mathbf{H}^k$ is Hermitian, for all integers $k \geq 1$. (Hint: Use part (a) and a proof by induction.)
(c) Prove that for any square complex matrix $\mathbf{H}$, $\mathbf{HH}^*$ and $\mathbf{H}^*\mathbf{H}$ are both Hermitian.
(d) Prove that if $\mathbf{H}$ is an $n \times n$ Hermitian matrix, then $\mathbf{P}^*\mathbf{HP}$ is Hermitian for any $n \times n$ complex matrix $\mathbf{P}$.

8. Prove Theorem 7.3.

9. (a) Show that the set $\mathcal{P}_n^{\mathbb{C}}$ of all polynomials of degree $\leq n$ under addition and complex scalar multiplication is a complex vector space.
(b) Show that the set $\mathcal{M}_{mn}^{\mathbb{C}}$ of all $m \times n$ complex matrices under addition and complex scalar multiplication is a complex vector space.

10. Determine which of the following subsets of the complex vector space $\mathbb{C}^3$ are linearly independent. Also, in each case find the dimension of the span of the subset.
(a) $\{[2+i, -i, 3], [-i, 3+i, -1]\}$
★(b) $\{[2+i, -i, 3], [-3+6i, 3, 9i]\}$
(c) $\{[3-i, 1+2i, -i], [1+i, -2, 4+i], [1-3i, 5+2i, -8-3i]\}$
★(d) $\{[3-i, 1+2i, -i], [1+i, -2, 4+i], [3+i, -2+5i, 3-8i]\}$

11. Repeat Exercise 10 considering $\mathbb{C}^3$ as a *real* vector space.

12. (a) Show that $B = ([2i, -1+3i, 4], [3+i, -2, 1-i], [-3+5i, 2i, -5+3i])$ is an ordered basis for the complex vector space $\mathbb{C}^3$.
★(b) Let $\mathbf{z} = [3-i, -5-5i, 7+i]$. For the ordered basis B in part (a), find $[\mathbf{z}]_B$.

13. (a) Show that the mapping $L\colon \mathbb{C}^2 \longrightarrow \mathbb{C}^2$ given by $L([z_1, z_2]) = [\overline{z_1}, \overline{z_2}]$ is a linear transformation when considering $\mathbb{C}^2$ as a real vector space but is not a linear transformation when considering $\mathbb{C}^2$ as a complex vector space.
★(b) With $\mathbb{C}^2$ as a real vector space, give an ordered basis for $\mathbb{C}^2$ and a matrix for the linear transformation L in part (a) with respect to this basis. (Hint: What is the dimension of $\mathbb{C}^2$ as a real vector space?)

★**14.** Give the matrix with respect to the standard bases for the linear transformation $L\colon \mathbb{C}^2 \longrightarrow \mathbb{C}^3$ (considered as complex vector spaces) such that $L([1+i, -1+3i]) = [3-i, 5, -i]$ and $L([1-i, 1+2i]) = [2+i, 1-3i, 3]$.

15. Determine whether the following sets of vectors are orthogonal.

★(a) In $\mathbb{C}^2$: $\{[1+2i, -3-i], [4-2i, 3+i]\}$

(b) In $\mathbb{C}^3$: $\{[1-i, -1+i, 1-i], [i, -2i, 2i]\}$

★(c) In $\mathbb{C}^3$: $\{[2i, -1, i], [1, -i, -1], [0, 1, i]\}$

(d) In $\mathbb{C}^4$: $\{[1, i, -1, 1+i], [4, -i, 1, -1-i], [0, 3, -i, -1+i]\}$

★**16.** (a) Use the Gram-Schmidt Process to find an orthogonal basis for $\mathbb{C}^3$ containing $[1+i, i, 1]$.

(b) Find a 3×3 unitary matrix having a multiple of $[1+i, i, 1]$ as its first row.

17. Prove that if $\mathbf{A}$ is a unitary matrix, then $\mathbf{A}^{-1}$ is unitary and $|\mathbf{A}| = \pm 1$.

18. Prove Theorem 7.4.

19. (a) Prove that a complex matrix $\mathbf{A}$ is unitary if and only if $\overline{\mathbf{A}}$ is unitary.

(b) Let $\mathbf{A}$ be a unitary matrix. Prove that $\mathbf{A}^k$ is unitary for all integers $k \geq 1$.

(c) Let $\mathbf{A}$ be a unitary matrix. Prove that $\mathbf{A}^2 = \mathbf{I}_n$ if and only if $\mathbf{A}$ is Hermitian.

20. (a) Without using Theorem 7.5, prove that $\mathbf{A}$ is a unitary matrix if and only if $\mathbf{A}^T$ is unitary.

(b) Prove Theorem 7.5. (Hint: First prove part (1) of Theorem 7.5, and then use part (a) of this exercise to prove part (2). Modify the proof of Theorem 5.30. For instance, when $i \neq j$, to show that the ith row of $\mathbf{A}$ is orthogonal to the jth column of $\mathbf{A}$, we must show that the *complex* dot product of the ith row of $\mathbf{A}$ with the jth column of $\mathbf{A}$ equals zero.)

21. Prove Theorem 7.6. (Hint: Modify the proof of Theorem 5.31.)

22. Each of the following matrices represents a linear operator $L: \mathbb{C}^n \longrightarrow \mathbb{C}^n$, for some n, with respect to the standard basis. In each case, find all eigenvalues and a basis for each eigenspace.

★(a) $\begin{bmatrix} 4+3i & -1-3i \\ 8-2i & -5-2i \end{bmatrix}$ (b) $\begin{bmatrix} 11 & 2 & -7 \\ 0 & 6 & -5 \\ 10 & 2 & -6 \end{bmatrix}$

★(c) $\begin{bmatrix} 4+3i & -4-2i & 4+7i \\ 2-4i & -2+5i & 7-4i \\ -4-2i & 4+2i & -4-6i \end{bmatrix}$ (d) $\begin{bmatrix} -i & 2i & -1+2i \\ 1 & -1+i & -i \\ -2+i & 2-i & 3+2i \end{bmatrix}$

23. ★(a) Explain why the linear transformation in part (a) of Exercise 22 is diagonalizable. If $\mathbf{A}$ is the matrix for this linear transformation, find a nonsingular $\mathbf{P}$ and diagonal $\mathbf{D}$ such that $\mathbf{P}^{-1}\mathbf{A}\mathbf{P} = \mathbf{D}$.

(b) Show that the linear transformation in part (d) of Exercise 22 is not diagonalizable.

(c) Show that the linear transformation $L: \mathbb{C}^3 \longrightarrow \mathbb{C}^3$ from part (b) of Exercise 22 is diagonalizable but that the linear transformation $L': \mathbb{R}^3 \longrightarrow \mathbb{R}^3$ having the same matrix with respect to the standard basis is not diagonalizable.

24. Show that the following matrix, from Example 11, is normal:

$$\mathbf{A} = \frac{1}{3}\begin{bmatrix} -1+3i & 2+2i & -2 \\ 2+2i & 2i & -2i \\ 2 & 2i & 1+4i \end{bmatrix}.$$

25. (a) Show that the linear transformation $L: \mathbb{C}^2 \longrightarrow \mathbb{C}^2$ given by $L\left(\begin{bmatrix} z_1 \\ z_2 \end{bmatrix}\right) = \begin{bmatrix} 1-6i & -10-2i \\ 2-10i & 5 \end{bmatrix}\begin{bmatrix} z_1 \\ z_2 \end{bmatrix}$ is unitarily diagonalizable.

★(b) If $\mathbf{A}$ is the matrix for L (with respect to the standard basis for $\mathbb{C}^2$), find a unitary matrix $\mathbf{P}$ such that $\mathbf{P}^{-1}\mathbf{AP}$ is diagonal.

26. (a) Show that the following matrix is unitarily diagonalizable:

$$\mathbf{A} = \begin{bmatrix} -4+5i & 2+2i & 4+4i \\ 2+2i & -1+8i & -2-2i \\ 4+4i & -2-2i & -4+5i \end{bmatrix}.$$

(b) Find a unitary matrix $\mathbf{P}$ such that $\mathbf{P}^{-1}\mathbf{AP}$ is diagonal.

27. (a) Let $\mathbf{A}$ be a unitary matrix. Show that $|\lambda| = 1$ for every eigenvalue λ of $\mathbf{A}$. (Hint: Suppose $\mathbf{Az} = \lambda\mathbf{z}$, for some $\mathbf{z}$. Use Lemma 7.10 to calculate $\mathbf{Az} \cdot \mathbf{Az}$ two different ways to show that $\lambda\overline{\lambda} = 1$.)

(b) Prove that a unitary matrix $\mathbf{A}$ is Hermitian if and only if the eigenvalues of $\mathbf{A}$ are 1 and/or -1.

★**28.** Verify directly that all of the eigenvalues of the following Hermitian matrix are real:

$$\begin{bmatrix} 1 & 2+i & 1-2i \\ 2-i & -3 & -i \\ 1+2i & i & 2 \end{bmatrix}.$$

29. (a) Prove that if $\mathbf{A}$ is normal and has real eigenvalues, then $\mathbf{A}$ is Hermitian. (Hint: Use Theorem 7.8 to express $\mathbf{A}$ as $\mathbf{PDP}^*$ for some unitary $\mathbf{P}$ and diagonal $\mathbf{D}$. Calculate $\mathbf{A}^*$.)

(b) Prove that if $\mathbf{A}$ is normal and all eigenvalues have absolute value equal to 1, then $\mathbf{A}$ is unitary. (Hint: With $\mathbf{A} = \mathbf{PDP}^*$ as in part (a), show $\mathbf{DD}^* = \mathbf{I}$ and use this to calculate $\mathbf{AA}^*$.)

30. Let $\mathbf{Z}$ be a square complex matrix. Prove that $\mathbf{Z}$ is normal if and only if there exist two Hermitian matrices $\mathbf{H}_1$ and $\mathbf{H}_2$ such that $\mathbf{Z} = \mathbf{H}_1 + i\mathbf{H}_2$ and $\mathbf{H}_1\mathbf{H}_2 = \mathbf{H}_2\mathbf{H}_1$. (Hint: If $\mathbf{Z}$ is normal, let $\mathbf{H}_1 = (\mathbf{Z} + \mathbf{Z}^*)/2$.)

31. Use Theorem 7.11 to give an alternate proof of Lemma 6.13 in Section 6.3.

7.2 Inner Product Spaces

In general, a vector space has two basic operations: addition and scalar multiplication. However, in $\mathbb{R}^n$ and $\mathbb{C}^n$ we have an additional operation: the dot product. In other vector spaces, we can often create a similar type of product, known as an inner product. Many properties of the usual dot products in $\mathbb{R}^n$ and $\mathbb{C}^n$ hold for these more general inner products as well.

Inner Products

DEFINITION

Let $\mathcal{V}$ be a real [complex] vector space with operations $+$ and $\cdot$. Let $<,>$ be an operation that assigns to each pair of vectors $\mathbf{x}, \mathbf{y} \in \mathcal{V}$ a real [complex] number, denoted $<\mathbf{x}, \mathbf{y}>$. Then $<,>$ is a **real [complex] inner product** for $\mathcal{V}$ if and only if the following properties hold for all $\mathbf{x}, \mathbf{y} \in \mathcal{V}$ and all $k \in \mathbb{R}$ [$k \in \mathbb{C}$]:

(1) $<\mathbf{x}, \mathbf{x}>$ is always real, and $<\mathbf{x}, \mathbf{x}> \geq 0$
(2) $<\mathbf{x}, \mathbf{x}> = 0$ if and only if $\mathbf{x} = \mathbf{0}$
(3) $<\mathbf{x}, \mathbf{y}> = <\mathbf{y}, \mathbf{x}>$ $\quad$ [$<\mathbf{x}, \mathbf{y}> = \overline{<\mathbf{y}, \mathbf{x}>}$]
(4) $<\mathbf{x} + \mathbf{y}, \mathbf{z}> = <\mathbf{x}, \mathbf{z}> + <\mathbf{y}, \mathbf{z}>$
(5) $<k\mathbf{x}, \mathbf{y}> = k <\mathbf{x}, \mathbf{y}>$

A vector space together with a real [complex] inner product operation is known as a **real [complex] inner product space**.

Example 1

Consider the real vector space $\mathbb{R}^n$. Let $\mathbf{x} = [x_1, \ldots, x_n]$ and $\mathbf{y} = [y_1, \ldots, y_n]$ be vectors in $\mathbb{R}^n$. By Theorem 1.4, it is easy to see that the operation $<\mathbf{x}, \mathbf{y}> = \mathbf{x} \cdot \mathbf{y} = x_1y_1 + \cdots + x_ny_n$ (usual real dot product) is a real inner product. (Verify.) Hence, $\mathbb{R}^n$ together with the dot product is a real inner product space.

Similarly, let $\mathbf{x} = [x_1, \ldots, x_n]$ and $\mathbf{y} = [y_1, \ldots, y_n]$ be vectors in the complex vector space $\mathbb{C}^n$. By Theorem 7.1, the operation $<\mathbf{x}, \mathbf{y}> = \mathbf{x} \cdot \mathbf{y} = x_1\overline{y_1} + \cdots + x_n\overline{y_n}$ (usual complex dot product) is an inner product on $\mathbb{C}^n$. Thus, $\mathbb{C}^n$ together with the complex dot product is a complex inner product space. ■

Example 2

Consider the real vector space $\mathbb{R}^2$. For $\mathbf{x} = [x_1, x_2]$ and $\mathbf{y} = [y_1, y_2]$ in $\mathbb{R}^2$, define $<\mathbf{x}, \mathbf{y}> = x_1y_1 - x_1y_2 - x_2y_1 + 2x_2y_2$. Let us verify the five properties in the definition of an inner product space:

Property (1): $< \mathbf{x}, \mathbf{x} > = x_1x_1 - x_1x_2 - x_2x_1 + 2x_2x_2 = x_1^2 - 2x_1x_2 + x_2^2 + x_2^2 = (x_1 - x_2)^2 + x_2^2 \geq 0$.

Property (2): $< \mathbf{x}, \mathbf{x} > = 0$ exactly when $x_1 = x_2 = 0$ (that is, when $\mathbf{x} = \mathbf{0}$).

Property (3): $< \mathbf{y}, \mathbf{x} > = y_1x_1 - y_1x_2 - y_2x_1 + 2y_2x_2 = x_1y_1 - x_1y_2 - x_2y_1 + 2x_2y_2 = < \mathbf{x}, \mathbf{y} >$.

Property (4): Let $\mathbf{z} = [z_1, z_2]$. Then

$$\begin{aligned} < \mathbf{x} + \mathbf{y}, \mathbf{z} > &= (x_1 + y_1)z_1 - (x_1 + y_1)z_2 - (x_2 + y_2)z_1 + 2(x_2 + y_2)z_2 \\ &= x_1z_1 + y_1z_1 - x_1z_2 - y_1z_2 - x_2z_1 - y_2z_1 + 2x_2z_2 + 2y_2z_2 \\ &= (x_1z_1 - x_1z_2 - x_2z_1 + 2x_2z_2) + (y_1z_1 - y_1z_2 - y_2z_1 + 2y_2z_2) \\ &= < \mathbf{x}, \mathbf{z} > + < \mathbf{y}, \mathbf{z} > . \end{aligned}$$

Property (5): $< k\mathbf{x}, \mathbf{y} > = (kx_1)y_1 - (kx_1)y_2 - (kx_2)y_1 + 2(kx_2)y_2 = k[x_1y_1 - x_1y_2 - x_2y_1 + 2x_2y_2] = k < \mathbf{x}, \mathbf{y} >$.

Hence, $<,>$ is a real inner product on $\mathbb{R}^2$, and $\mathbb{R}^2$ together with this operation $<,>$ is a real inner product space. ■

Example 3

Consider the real vector space $\mathbb{R}^n$. Let $\mathbf{A}$ be a nonsingular $n \times n$ real matrix. Let $\mathbf{x}, \mathbf{y} \in \mathbb{R}^n$, and define $< \mathbf{x}, \mathbf{y} > = (\mathbf{A}\mathbf{x}) \cdot (\mathbf{A}\mathbf{y})$ (the usual dot product of $\mathbf{Ax}$ and $\mathbf{Ay}$). It is easy to show (see Exercise 1) that $<,>$ is a real inner product on $\mathbb{R}^n$, and so $\mathbb{R}^n$ together with this operation $<,>$ is a real inner product space. ■

Example 4

Consider the real vector space $\mathcal{P}_n$. Let $\mathbf{p}_1 = a_nx^n + \cdots + a_1x + a_0$ and $\mathbf{p}_2 = b_nx^n + \cdots + b_1x + b_0$ be in $\mathcal{P}_n$. Define $< \mathbf{p}_1, \mathbf{p}_2 > = a_nb_n + \cdots + a_1b_1 + a_0b_0$. It is easy to show (see Exercise 2) that $<,>$ is a real inner product on $\mathcal{P}_n$, and so $\mathcal{P}_n$ together with this operation $<,>$ is a real inner product space. ■

Example 5

Let $a, b \in \mathbb{R}$, with $a < b$, and consider the real vector space $\mathcal{V}$ of all real continuous functions defined on the interval $[a, b]$ (for example, polynomials, $\sin(x)$, e^x, and so on). Let $\mathbf{f}, \mathbf{g} \in \mathcal{V}$. Define $< \mathbf{f}, \mathbf{g} > = \int_a^b f(t)g(t)\,dt$. It can be shown (see Exercise 3) that $<,>$ is a real inner product on $\mathcal{V}$, and so $\mathcal{V}$ together with this operation $<,>$ is a real inner product space.

Analogously, the operation $< \mathbf{f}, \mathbf{g} > = \int_a^b f(t)\overline{g(t)}\,dt$ makes the complex vector space of all complex-valued continuous functions on $[a, b]$ into a complex inner product space. ■

Of course, not every operation is an inner product. For example, for the vectors $\mathbf{x} = [x_1, x_2]$ and $\mathbf{y} = [y_1, y_2]$ in $\mathbb{R}^2$, consider the operation $< \mathbf{x}, \mathbf{y} > = x_1^2 + y_1^2$.

But then with $\mathbf{x} = \mathbf{y} = [1, 0]$, we have $< 2\mathbf{x}, \mathbf{y} > = 2^2 + 1^2 = 5$, but $2 < \mathbf{x}, \mathbf{y} > = 2(1^2 + 1^2) = 4$, so property (5) fails to hold.

The next theorem lists some useful results for inner product spaces.

THEOREM 7.12

Let $\mathcal{V}$ be a real [complex] inner product space with inner product $<,>$. Then, for all $\mathbf{x}, \mathbf{y} \in \mathcal{V}$ and all $k \in \mathbb{R}$ $[k \in \mathbb{C}]$, we have:
(1) $< \mathbf{0}, \mathbf{x} > = < \mathbf{x}, \mathbf{0} > = 0$.
(2) $< \mathbf{x}, \mathbf{y} + \mathbf{z} > = < \mathbf{x}, \mathbf{y} > + < \mathbf{x}, \mathbf{z} >$.
(3) $< \mathbf{x}, k\mathbf{y} > = k < \mathbf{x}, \mathbf{y} > \qquad [< \mathbf{x}, k\mathbf{y} > = \overline{k} < \mathbf{x}, \mathbf{y} >]$.

Note the use of $\overline{k}$ in part (3) for complex vector spaces. The proof of this theorem is easy, and parts are left for you to do in Exercise 5.

Length and Distance in Inner Product Spaces

The next definition extends the concept of the length of a vector to any inner product space.

DEFINITION

If $\mathbf{x}$ is a vector in an inner product space, then the **norm** (**length**) of $\mathbf{x}$ is $\|\mathbf{x}\| = \sqrt{< \mathbf{x}, \mathbf{x} >}$.

This definition yields a nonnegative real number for $\|\mathbf{x}\|$, since by definition $< \mathbf{x}, \mathbf{x} >$ is always real and nonnegative for any vector $\mathbf{x}$. Also note that this definition agrees with the earlier definition of length in $\mathbb{R}^n$ based on the usual dot product in $\mathbb{R}^n$. We also have:

THEOREM 7.13

Let $\mathcal{V}$ be a real [complex] inner product space, with $\mathbf{x} \in \mathcal{V}$. Let $k \in \mathbb{R}$ $[k \in \mathbb{C}]$. Then, $\|k\mathbf{x}\| = |k|\,\|\mathbf{x}\|$.

The proof of this theorem is easy and is left for you to do in Exercise 6.

As before, we say that a vector of length 1 in an inner product space is a **unit vector**. For instance, in the inner product space of Example 4, the polynomial $\mathbf{p} = (\sqrt{2}/2)x + (\sqrt{2}/2)$ is a unit vector, since $\|\mathbf{p}\| = \sqrt{< \mathbf{p}, \mathbf{p} >} = \sqrt{(\sqrt{2}/2)^2 + (\sqrt{2}/2)^2} = 1$.

We define the distance between two vectors in the general inner product space setting as we did for $\mathbb{R}^n$:

DEFINITION

Let $\mathbf{x}, \mathbf{y} \in \mathcal{V}$, an inner product space. Then the **distance between x and y** is $\|\mathbf{x} - \mathbf{y}\|$.

Example 6

Consider the real vector space $\mathcal{V}$ of real continuous functions from Example 5, with $a = 0$ and $b = \pi$. That is, $< \mathbf{f}, \mathbf{g} > = \int_0^\pi f(t)g(t)\,dt$, for all $\mathbf{f}, \mathbf{g} \in \mathcal{V}$. Let $\mathbf{f} = \cos t$ and $\mathbf{g} = \sin t$. Then the distance between $\mathbf{f}$ and $\mathbf{g}$ is $\|\mathbf{f} - \mathbf{g}\| = \sqrt{< \cos t - \sin t, \cos t - \sin t >} = \sqrt{\int_0^\pi (\cos t - \sin t)^2\,dt} = \sqrt{\int_0^\pi (\cos^2 t - 2\cos t \sin t + \sin^2 t)\,dt} = \sqrt{\int_0^\pi (1 - \sin 2t)\,dt} = \sqrt{(t + \frac{1}{2}\cos 2t)\Big|_0^\pi} = \sqrt{\pi}$. Hence, the distance between $\cos t$ and $\sin t$ is $\sqrt{\pi}$ under this inner product. ■

The next theorem shows that some other familiar results from the ordinary dot product carry over to the general inner product.

THEOREM 7.14

Let $\mathbf{x}, \mathbf{y} \in \mathcal{V}$, an inner product space, with inner product $<,>$. Then
(1) $|< \mathbf{x}, \mathbf{y} >| \leq \|\mathbf{x}\|\,\|\mathbf{y}\|$ Cauchy-Schwarz Inequality
(2) $\|\mathbf{x} + \mathbf{y}\| \leq \|\mathbf{x}\| + \|\mathbf{y}\|$ Triangle Inequality

The proofs of these statements are analogous to the proofs for the ordinary dot product and are left for you to do in Exercise 11.

From the Cauchy-Schwarz Inequality, we have $-1 \leq < \mathbf{x}, \mathbf{y} > / (\|\mathbf{x}\|\,\|\mathbf{y}\|) \leq 1$, for any vectors $\mathbf{x}$ and $\mathbf{y}$ in a *real* inner product space. Hence, we can make the following definition:

DEFINITION

Let $\mathbf{x}, \mathbf{y} \in \mathcal{V}$, a *real* inner product space. Then the **angle θ between x and y** is the angle from 0 to π such that $\cos\theta = < \mathbf{x}, \mathbf{y} > / (\|\mathbf{x}\|\,\|\mathbf{y}\|)$.

Example 7

Consider again the inner product space of Example 5, where $< \mathbf{f}, \mathbf{g} > = \int_0^\pi f(t)g(t)\,dt$. Let $\mathbf{f} = t$ and $\mathbf{g} = \sin t$. Then $< \mathbf{f}, \mathbf{g} > = \int_0^\pi t \sin t\,dt$. Using integration by parts, we get $-t\cos t\Big|_0^\pi + \int_0^\pi \cos t\,dt = \pi + \left(\sin t\Big|_0^\pi\right) = \pi$. Also,

$\|\mathbf{f}\|^2 = <\mathbf{f},\mathbf{f}> = \int_0^\pi (f(t))^2\,dt = \int_0^\pi t^2\,dt = t^3/3\big|_0^\pi = \pi^3/3$, and so $\|\mathbf{f}\| = \sqrt{\pi^3/3}$. Similarly, $\|\mathbf{g}\|^2 = <\mathbf{g},\mathbf{g}> = \int_0^\pi (g(t))^2\,dt = \int_0^\pi \sin^2 t\,dt = \int_0^\pi \frac{1}{2}(1-\cos 2t)\,dt = (\frac{1}{2}t - \frac{1}{4}\sin 2t)\big|_0^\pi = \pi/2$, and so $\|\mathbf{g}\| = \sqrt{\pi/2}$. Hence the angle θ between t and $\sin t$ has cosine $= <\mathbf{f},\mathbf{g}>/(\|\mathbf{f}\|\,\|\mathbf{g}\|) = \pi/(\sqrt{\pi^3/3}\sqrt{\pi/2}) = \sqrt{6}/\pi \approx 0.78$. Hence, $\theta \approx 0.676$ radians (38.8°). ■

Orthogonality

We next define orthogonal vectors in a general inner product space setting and show that nonzero orthogonal vectors are linearly independent.

DEFINITION

Let $\mathbf{x},\mathbf{y} \in \mathcal{V}$, an inner product space with inner product $<,>$. Then $\mathbf{x}$ and $\mathbf{y}$ are **orthogonal** if and only if $<\mathbf{x},\mathbf{y}> = 0$. Also, $\mathbf{x}$ and $\mathbf{y}$ are **orthonormal** if and only if they are orthogonal unit vectors.

As before, we say that a set of [unit] vectors is orthogonal [orthonormal] if and only if every distinct pair $\mathbf{x}$ and $\mathbf{y}$ of vectors in the set has the property $<\mathbf{x},\mathbf{y}> = 0$. The next theorem is the analog of Theorem 5.24, and its proof is left for you to do in Exercise 17.

THEOREM 7.15

If $\mathcal{V}$ is an inner product space and T is an orthogonal set of nonzero vectors in $\mathcal{V}$, then T is a linearly independent set.

Example 8

Consider again the inner product space $\mathcal{V}$ of Example 5 of real continuous functions with inner product $<\mathbf{f},\mathbf{g}> = \int_a^b f(t)g(t)\,dt$, with $a = -\pi$ and $b = \pi$. The set $\{1, \cos t, \sin t\}$ is an orthogonal set in $\mathcal{V}$ since each of the following definite integrals equals zero (verify):

$$\int_{-\pi}^{\pi} (1)\cos t\,dt, \qquad \int_{-\pi}^{\pi} (1)\sin t\,dt, \qquad \int_{-\pi}^{\pi} (\cos t)(\sin t)\,dt.$$

Also, note that $\|1\|^2 = <1,1> = \int_{-\pi}^{\pi}(1)(1)\,dt = 2\pi$, $\|\cos t\|^2 = <\cos t,\cos t> = \int_{-\pi}^{\pi}\cos^2 t\,dt = \pi$ (why?) and $\|\sin t\|^2 = <\sin t,\sin t> = \int_{-\pi}^{\pi}\sin^2 t\,dt = \pi$. (Why?) Therefore, the set

$$\left\{\frac{1}{\sqrt{2\pi}}, \frac{\cos t}{\sqrt{\pi}}, \frac{\sin t}{\sqrt{\pi}}\right\}$$

is an orthonormal set in $\mathcal{V}$. ■

Example 8 can be generalized. The set $\{1, \cos t, \sin t, \cos 2t, \sin 2t, \cos 3t, \sin 3t, \ldots\}$ is an orthogonal set (see Exercise 18) and therefore linearly independent by Theorem 7.15. The functions in this set are important in the theory of partial differential equations. It can be shown that every continuously differentiable function on the interval $[-\pi, \pi]$ can be represented as the (infinite) sum of constant multiples of these functions, known as the **Fourier series** of the function.

A basis for an inner product space $\mathcal{V}$ is an **orthogonal [orthonormal] basis** if the vectors in the basis form an orthogonal [orthonormal] set.

Example 9

Consider again the inner product space $\mathcal{P}_n$ with the inner product of Example 4: that is, if $\mathbf{p}_1 = a_n x^n + \cdots + a_1 x + a_0$ and $\mathbf{p}_2 = b_n x^n + \cdots + b_1 x + b_0$ are in $\mathcal{P}_n$, then $< \mathbf{p}_1, \mathbf{p}_2 > = a_n b_n + \cdots + a_1 b_1 + a_0 b_0$. Now, $\{x^n, x^{n-1}, \ldots, x, 1\}$ is an orthogonal basis for $\mathcal{P}_n$ with this inner product, since $< x^k, x^l > = 0$, for $0 \le k, l \le n$, with $k \ne l$. (Why?) Since $\|x^k\| = \sqrt{< x^k, x^k >} = 1$, for all k, $0 \le k \le n$ (why?), the set $\{x^n, x^{n-1}, \ldots, x, 1\}$ is also an orthonormal basis for this inner product space. ■

A proof analogous to that of Theorem 5.26 gives us the next theorem (see Exercise 19).

THEOREM 7.16

> If $B = (\mathbf{v}_1, \mathbf{v}_2, \ldots, \mathbf{v}_k)$ is an orthogonal ordered basis for a subspace $\mathcal{W}$ of an inner product space $\mathcal{V}$ and if $\mathbf{v}$ is any vector in $\mathcal{W}$, then $[\mathbf{v}]_B = [< \mathbf{v}, \mathbf{v}_1 >/\|\mathbf{v}_1\|^2, < \mathbf{v}, \mathbf{v}_2 >/\|\mathbf{v}_2\|^2, \ldots, < \mathbf{v}, \mathbf{v}_k >/\|\mathbf{v}_k\|^2]$. In particular, if B is an orthonormal ordered basis for $\mathcal{W}$, then $[\mathbf{v}]_B = [< \mathbf{v}, \mathbf{v}_1 >, < \mathbf{v}, \mathbf{v}_2 >, \ldots, < \mathbf{v}, \mathbf{v}_k >]$.

Example 10

Recall the inner product space $\mathbb{R}^2$ in Example 2, with inner product given as follows: if $\mathbf{x} = [x_1, x_2]$ and $\mathbf{y} = [y_1, y_2]$, then $< \mathbf{x}, \mathbf{y} > = x_1 y_1 - x_1 y_2 - x_2 y_1 + 2x_2 y_2$. An ordered orthogonal basis for this space is $B = (\mathbf{v}_1, \mathbf{v}_2) = ([2, 1], [0, 1])$. (Verify.) Recall from Example 2 that if $\mathbf{x} = [x_1, x_2]$, then $< \mathbf{x}, \mathbf{x} > = (x_1 - x_2)^2 + x_2^2$, and so $\| < \mathbf{x}, \mathbf{x} > \| = \sqrt{(x_1 - x_2)^2 + x_2^2}$. Thus, $\|\mathbf{v}_1\| = \sqrt{(2-1)^2 + 1^2} = \sqrt{2}$, and $\|\mathbf{v}_2\| = (0-1)^2 + 1^2 = \sqrt{2}$.

Next, suppose that $\mathbf{v} = [a, b]$ is any vector in $\mathbb{R}^2$. Now, $< \mathbf{v}, \mathbf{v}_1 > = < [a, b], [2, 1] > = (a)(2) - (a)(1) - (b)(2) + 2(b)(1) = a$. Also, $< \mathbf{v}, \mathbf{v}_2 > = < [a, b], [0, 1] > = (a)(0) - (a)(1) - (b)(0) + 2(b)(1) = -a + 2b$. Then, $[\mathbf{v}]_B = [< \mathbf{v}, \mathbf{v}_1 >/\|\mathbf{v}_1\|^2, < \mathbf{v}, \mathbf{v}_2 >/\|\mathbf{v}_2\|^2] = [a/2, (-a+2b)/2]$. Notice that $(a/2)[2, 1] + ((-a+2b)/2)[0, 1]$ does equal $[a, b] = \mathbf{v}$. ■

The Generalized Gram-Schmidt Process

We can generalize the Gram-Schmidt method of Section 5.5 to any inner product space. That is, we can replace any linearly independent set of k vectors with an orthogonal set of k vectors that spans the same subspace.

GENERALIZED GRAM-SCHMIDT PROCESS

Let $\{\mathbf{w}_1, \ldots, \mathbf{w}_k\}$ be a linearly independent subset of an inner product space $\mathcal{V}$, with inner product $<,>$. We create a new set $\{\mathbf{v}_1, \ldots, \mathbf{v}_k\}$ of vectors as follows:

Let $\mathbf{v}_1 = \mathbf{w}_1$.

Let $\mathbf{v}_2 = \mathbf{w}_2 - \left(\dfrac{<\mathbf{w}_2, \mathbf{v}_1>}{<\mathbf{v}_1, \mathbf{v}_1>}\right)\mathbf{v}_1$.

Let $\mathbf{v}_3 = \mathbf{w}_3 - \left(\dfrac{<\mathbf{w}_3, \mathbf{v}_1>}{<\mathbf{v}_1, \mathbf{v}_1>}\right)\mathbf{v}_1 - \left(\dfrac{<\mathbf{w}_3, \mathbf{v}_2>}{<\mathbf{v}_2, \mathbf{v}_2>}\right)\mathbf{v}_2$.

$\vdots$

Let $\mathbf{v}_k = \mathbf{w}_k - \left(\dfrac{<\mathbf{w}_k, \mathbf{v}_1>}{<\mathbf{v}_1, \mathbf{v}_1>}\right)\mathbf{v}_1 - \left(\dfrac{<\mathbf{w}_k, \mathbf{v}_2>}{<\mathbf{v}_2, \mathbf{v}_2>}\right)\mathbf{v}_2 - \cdots - \left(\dfrac{<\mathbf{w}_k, \mathbf{v}_{k-1}>}{<\mathbf{v}_{k-1}, \mathbf{v}_{k-1}>}\right)\mathbf{v}_{k-1}$.

A similar proof to that of Theorem 5.27 (see Exercise 23) gives:

THEOREM 7.17

Let $B = \{\mathbf{w}_1, \ldots, \mathbf{w}_k\}$ be a basis for an inner product space $\mathcal{V}$. Then the set $\{\mathbf{v}_1, \ldots, \mathbf{v}_k\}$ obtained by applying the Generalized Gram-Schmidt Process to B is an orthogonal basis for $\mathcal{V}$.

Since the Generalized Gram-Schmidt Process can be applied to any finite dimensional inner product space to obtain an orthogonal basis, we have the following immediate corollary:

COROLLARY 7.18

Every (nonzero) finite dimensional inner product space has a (finite) orthogonal basis.

Similarly, every finite dimensional inner product space has a (finite) orthonormal basis. (Why?)

Example 11

Recall the inner product space $\mathcal{V}$ from Example 5 of real continuous functions with $a = -1$ and $b = 1$—that is, with inner product $<\mathbf{f}, \mathbf{g}> = \int_{-1}^{1} f(t)g(t)\,dt$. Now, $\{1, t, t^2, t^3\}$ is a linearly independent set in $\mathcal{V}$. We use this set to find four orthogonal vectors in $\mathcal{V}$.

Let $\mathbf{w}_1 = 1$, $\mathbf{w}_2 = t$, $\mathbf{w}_3 = t^2$, and $\mathbf{w}_4 = t^3$. Using the Generalized Gram-Schmidt Process, we start with $\mathbf{v}_1 = \mathbf{w}_1 = 1$ and obtain

$$\mathbf{v}_2 = \mathbf{w}_2 - \left(\frac{<\mathbf{w}_2, \mathbf{v}_1>}{<\mathbf{v}_1, \mathbf{v}_1>}\right)\mathbf{v}_1 = t - \left(\frac{<t, 1>}{<1, 1>}\right)1.$$

Now, $<t, 1> = \int_{-1}^{1}(t)(1)\,dt = (t^2/2)\big|_{-1}^{1} = 0$. Hence $\mathbf{v}_2 = t$. Next,

$$\mathbf{v}_3 = \mathbf{w}_3 - \left(\frac{<\mathbf{w}_3, \mathbf{v}_1>}{<\mathbf{v}_1, \mathbf{v}_1>}\right)\mathbf{v}_1 - \left(\frac{<\mathbf{w}_3, \mathbf{v}_2>}{<\mathbf{v}_2, \mathbf{v}_2>}\right)\mathbf{v}_2 = t^2 - \left(\frac{<t^2, 1>}{<1, 1>}\right)1 - \left(\frac{<t^2, t>}{<t, t>}\right)t.$$

After a little calculation, we obtain $<t^2, 1> = \frac{2}{3}$, $<1, 1> = 2$, and $<t^2, t> = 0$. Hence, $\mathbf{v}_3 = t^2 - \left((\frac{2}{3})/2\right)1 = t^2 - \frac{1}{3}$. Finally,

$$\begin{aligned}\mathbf{v}_4 &= \mathbf{w}_4 - \left(\frac{<\mathbf{w}_4, \mathbf{v}_1>}{<\mathbf{v}_1, \mathbf{v}_1>}\right)\mathbf{v}_1 - \left(\frac{<\mathbf{w}_4, \mathbf{v}_2>}{<\mathbf{v}_2, \mathbf{v}_2>}\right)\mathbf{v}_2 - \left(\frac{<\mathbf{w}_4, \mathbf{v}_3>}{<\mathbf{v}_3, \mathbf{v}_3>}\right)\mathbf{v}_3 \\ &= t^3 - \left(\frac{<t^3, 1>}{<1, 1>}\right)1 - \left(\frac{<t^3, t>}{<t, t>}\right)t - \left(\frac{<t^3, t^2>}{<t^2, t^2>}\right)t^2.\end{aligned}$$

Now, $<t^3, 1> = 0$, $<t^3, t> = \frac{2}{5}$, $<t, t> = \frac{2}{3}$, and $<t^3, t^2> = 0$. Hence, $\mathbf{v}_4 = t^3 - ((\frac{2}{5})/(\frac{2}{3}))t = t^3 - \frac{3}{5}t$.

Thus the set $\{\mathbf{v}_1, \mathbf{v}_2, \mathbf{v}_3, \mathbf{v}_4\} = \{1, t, t^2 - \frac{1}{3}, t^3 - \frac{3}{5}t\}$ is an orthogonal set of vectors in this inner product space.[†] ■

You saw in Theorem 5.31 that the transition matrix between orthonormal bases of $\mathbb{R}^n$ is an orthogonal matrix. This result generalizes to inner product spaces as follows:

THEOREM 7.19

Let $\mathcal{V}$ be a finite dimensional real [complex] inner product space, and let B and C be ordered orthonormal bases for $\mathcal{V}$. Then the transition matrix from B to C is an orthogonal [unitary] matrix.

[†]The polynomials 1, t, $t^2 - \frac{1}{3}$, and $t^3 - \frac{3}{5}t$ from Example 11 are multiples of the first four Legendre polynomials: 1, t, $\frac{3}{2}t^2 - \frac{1}{2}$, $\frac{5}{2}t^3 - \frac{3}{2}t$. All Legendre polynomials equal 1 when $t = 1$. To find the complete set of Legendre polynomials, we can continue the Generalized Gram-Schmidt Process with t^4, t^5, t^6, and so on, and take appropriate multiples so that the resulting polynomials equal 1 when $t = 1$. These polynomials form an (infinite) orthogonal set for the inner product space of Example 11.

Orthogonal Complements

We can generalize the notion of an orthogonal complement of a subspace to inner product spaces as follows:

DEFINITION

Let $\mathcal{W}$ be a subspace of a real (or complex) inner product space $\mathcal{V}$. Then the **orthogonal complement** $\mathcal{W}^\perp$ of $\mathcal{W}$ in $\mathcal{V}$ is the set of all vectors $\mathbf{x} \in \mathcal{V}$ with the property that $< \mathbf{x}, \mathbf{w} > = 0$, for all $\mathbf{w} \in \mathcal{W}$.

Example 12

Consider again the real vector space $\mathcal{P}_n$ with the inner product of Example 4: for $\mathbf{p}_1 = a_n x^n + \cdots + a_1 x + a_0$ and $\mathbf{p}_2 = b_n x^n + \cdots + b_1 x + b_0$, $< \mathbf{p}_1, \mathbf{p}_2 > = a_n b_n + \cdots + a_1 b_1 + a_0 b_0$. Example 9 showed that $\{x^n, x^{n-1}, \ldots, x, 1\}$ is an orthogonal basis for $\mathcal{P}_n$ under this inner product. Now, consider the subspace $\mathcal{W}$ spanned by $\{x, 1\}$. It is fairly easy to see that $\mathcal{W}^\perp = \text{span}\{x^n, x^{n-1}, \ldots, x^2\}$ and that $\dim(\mathcal{W}^\perp) = (n+1) - 2 = n - 1$. ■

The following properties of orthogonal complements are the analogs to Theorems 5.34 through 5.37 and are proved in a similar manner (see Exercise 24):

THEOREM 7.20

Let $\mathcal{W}$ be a subspace of a real (or complex) inner product space $\mathcal{V}$. Then $\mathcal{W}^\perp$ is a subspace of $\mathcal{V}$, and $\mathcal{W} \cap \mathcal{W}^\perp = \{\mathbf{0}\}$.

THEOREM 7.21

Let $\mathcal{W}$ be a subspace of a finite dimensional real (or complex) inner product space $\mathcal{V}$. Then $\dim(\mathcal{W}) + \dim(\mathcal{W}^\perp) = \dim(\mathcal{V})$. In particular, if $\{\mathbf{v}_1, \ldots, \mathbf{v}_k\}$ is an orthogonal basis for $\mathcal{W}$ contained in an orthogonal basis $\{\mathbf{v}_1, \ldots, \mathbf{v}_k, \mathbf{v}_{k+1}, \ldots, \mathbf{v}_n\}$ for $\mathcal{V}$, then $\{\mathbf{v}_{k+1}, \ldots, \mathbf{v}_n\}$ is an orthogonal basis for $\mathcal{W}^\perp$.

THEOREM 7.22

Let $\mathcal{W}$ be a subspace of a finite dimensional real (or complex) inner product space $\mathcal{V}$. Then $(\mathcal{W}^\perp)^\perp = \mathcal{W}$.

Note that Theorem 7.22 is stated only for the case where $\mathcal{V}$ is finite dimensional. Otherwise, $(\mathcal{W}^\perp)^\perp$ is not necessarily equal to $\mathcal{W}$, although it is always true that $\mathcal{W} \subseteq (\mathcal{W}^\perp)^\perp$.

The next theorem is the analog of Theorem 5.38. It holds for any inner product space $\mathcal{V}$ where the subspace $\mathcal{W}$ is finite dimensional. The proof is left for you to do in Exercise 27.

THEOREM 7.23: Projection Theorem

Let $\mathcal{W}$ be a finite dimensional subspace of an inner product space $\mathcal{V}$. Then every vector $\mathbf{v} \in \mathcal{V}$ can be expressed in a unique way as $\mathbf{w}_1 + \mathbf{w}_2$, where $\mathbf{w}_1 \in \mathcal{W}$ and $\mathbf{w}_2 \in \mathcal{W}^\perp$.

As before, we define the **projection vector** of a vector $\mathbf{v}$ onto a subspace $\mathcal{W}$ as follows:

DEFINITION

If $\{\mathbf{v}_1, \ldots, \mathbf{v}_k\}$ is an orthonormal basis for $\mathcal{W}$, a subspace of an inner product space $\mathcal{V}$, then the vector $\mathbf{proj}_{\mathcal{W}}\mathbf{v} = <\mathbf{v}, \mathbf{v}_1> \mathbf{v}_1 + \cdots + <\mathbf{v}, \mathbf{v}_k> \mathbf{v}_k$ is called the **projection vector** of $\mathbf{v}$ onto $\mathcal{W}$.

It is easy to show that the formula for $\mathbf{proj}_{\mathcal{W}}\mathbf{v}$ yields the unique vector $\mathbf{w}_1$ in the Projection Theorem. Therefore, the choice of orthonormal basis in the definition does not matter, because any choice leads to the same vector for $\mathbf{proj}_{\mathcal{W}}\mathbf{v}$. Hence, the Projection Theorem can be restated as follows:

If $\mathcal{W}$ is a finite dimensional subspace of an inner product space $\mathcal{V}$, and if $\mathbf{v} \in \mathcal{V}$, then $\mathbf{v}$ can be uniquely expressed as $\mathbf{w}_1 + \mathbf{w}_2$, where $\mathbf{w}_1 = \mathbf{proj}_{\mathcal{W}}\mathbf{v} \in \mathcal{W}$ and $\mathbf{w}_2 = \mathbf{v} - \mathbf{proj}_{\mathcal{W}}\mathbf{v} \in \mathcal{W}^\perp$.

Example 13

Consider again the real vector space $\mathcal{V}$ of real continuous functions in Example 8, where $<\mathbf{f}, \mathbf{g}> = \int_{-\pi}^{\pi} f(t)g(t)\,dt$. Notice from that example that the set $\{(1/\sqrt{2\pi}), ((\sin t)/\sqrt{\pi})\}$ is an orthonormal (and hence, linearly independent) set of vectors in $\mathcal{V}$. Let $\mathcal{W} = \text{span}(\{(1/\sqrt{2\pi}), ((\sin t)/\sqrt{\pi})\})$ in $\mathcal{V}$. Then any continuous function $\mathbf{f}$ in $\mathcal{V}$ can be expressed uniquely as $\mathbf{f}_1 + \mathbf{f}_2$, where $\mathbf{f}_1 \in \mathcal{W}$ and $\mathbf{f}_2 \in \mathcal{W}^\perp$.

We illustrate this decomposition for the function $\mathbf{f} = t + 1$. Now,

$$\mathbf{f}_1 = \mathbf{proj}_{\mathcal{W}}\mathbf{f} = \left\langle (t+1), \frac{1}{\sqrt{2\pi}} \right\rangle \frac{1}{\sqrt{2\pi}} + \left\langle (t+1), \frac{\sin t}{\sqrt{\pi}} \right\rangle \frac{\sin t}{\sqrt{\pi}}.$$

Let $c_1 = < (t+1), 1/\sqrt{2\pi} >$ and $c_2 = < (t+1), (\sin t)/\sqrt{\pi} >$. Then

$$c_1 = \int_{-\pi}^{\pi} (t+1)\left(\frac{1}{\sqrt{2\pi}}\right) dt = \frac{1}{\sqrt{2\pi}} \int_{-\pi}^{\pi} (t+1)\, dt$$
$$= \frac{1}{\sqrt{2\pi}}\left(\frac{t^2}{2} + t\right)\bigg|_{-\pi}^{\pi} = \frac{2\pi}{\sqrt{2\pi}} = \sqrt{2\pi}.$$

Also,

$$c_2 = \int_{-\pi}^{\pi} (t+1)\left(\frac{\sin t}{\sqrt{\pi}}\right) dt = \frac{1}{\sqrt{\pi}} \int_{-\pi}^{\pi} (t+1)\sin t\, dt$$
$$= \frac{1}{\sqrt{\pi}}\left(\int_{-\pi}^{\pi} t \sin t\, dt + \int_{-\pi}^{\pi} \sin t\, dt\right).$$

The very last integral equals zero, and using integration by parts on the next-to-last integral, we obtain

$$c_2 = \frac{1}{\sqrt{\pi}}\left((-t\cos t)\big|_{-\pi}^{\pi} + \int_{-\pi}^{\pi} \cos t\, dt\right) = \frac{1}{\sqrt{\pi}} 2\pi = 2\sqrt{\pi}.$$

Hence,

$$\mathbf{f}_1 = c_1\left(\frac{1}{\sqrt{2\pi}}\right) + c_2\left(\frac{\sin t}{\sqrt{\pi}}\right) = \sqrt{2\pi}\left(\frac{1}{\sqrt{2\pi}}\right) + 2\sqrt{\pi}\left(\frac{\sin t}{\sqrt{\pi}}\right) = 1 + 2\sin t.$$

Then, by the Projection Theorem, $\mathbf{f}_2 = \mathbf{f} - \mathbf{f}_1 = (t+1) - (1 + 2\sin t) = t - 2\sin t$ is orthogonal to $\mathcal{W}$. We check that $\mathbf{f}_2 \in \mathcal{W}^{\perp}$ by showing that $\mathbf{f}_2$ is orthogonal to both $(1/\sqrt{2\pi})$ and $((\sin t)/\sqrt{\pi})$:

$$< \mathbf{f}_2, \frac{1}{\sqrt{2\pi}} > = \int_{-\pi}^{\pi} (t - 2\sin t)\left(\frac{1}{\sqrt{2\pi}}\right) dt = \left(\frac{1}{\sqrt{2\pi}}\right)\left(\frac{t^2}{2} + 2\cos t\right)\bigg|_{-\pi}^{\pi} = 0.$$

Also,

$$< \mathbf{f}_2, \frac{\sin t}{\sqrt{\pi}} > = \int_{-\pi}^{\pi} (t - 2\sin t)\left(\frac{\sin t}{\sqrt{\pi}}\right) dt$$
$$= \frac{1}{\sqrt{\pi}} \int_{-\pi}^{\pi} t \sin t\, dt - \frac{2}{\sqrt{\pi}} \int_{-\pi}^{\pi} \sin^2 t\, dt,$$

which equals $2\sqrt{\pi} - 2\sqrt{\pi} = 0$. ■

Exercises—Section 7.2

1. (a) Let $\mathbf{A}$ be a nonsingular $n \times n$ real matrix. For $\mathbf{x}, \mathbf{y} \in \mathbb{R}^n$, define an operation $< \mathbf{x}, \mathbf{y} > = (\mathbf{Ax}) \cdot (\mathbf{Ay})$ (dot product). Prove that this operation is a real inner product on $\mathbb{R}^n$.

★(b) For the inner product in part (a) with $\mathbf{A} = \begin{bmatrix} 5 & 4 & 2 \\ -2 & 3 & 1 \\ 1 & -1 & 0 \end{bmatrix}$, find $< \mathbf{x}, \mathbf{y} >$ and $\|\mathbf{x}\|$ for $\mathbf{x} = [3, -2, 4]$ and $\mathbf{y} = [-2, 1, -1]$.

2. Define an operation $<,>$ on $\mathcal{P}_n$ as follows: if $\mathbf{p}_1 = a_n x^n + \cdots + a_1 x + a_0$ and $\mathbf{p}_2 = b_n x^n + \cdots + b_1 x + b_0$, let $< \mathbf{p}_1, \mathbf{p}_2 > = a_n b_n + \cdots + a_1 b_1 + a_0 b_0$. Prove that this operation is a real inner product on $\mathcal{P}_n$.

3. (a) Let a and b be fixed real numbers with $a < b$, and let $\mathcal{V}$ be the set of all real continuous functions on $[a, b]$. Define $<,>$ on $\mathcal{V}$ by $< \mathbf{f}, \mathbf{g} > = \int_a^b f(t)g(t)\,dt$. Prove that this operation is a real inner product on $\mathcal{V}$.

★(b) For the inner product of part (a) and $a = 0$ and $b = \pi$, find $< \mathbf{f}, \mathbf{g} >$ and $\|\mathbf{f}\|$ for $\mathbf{f} = e^t$ and $\mathbf{g} = \sin t$.

4. On the real vector space $\mathcal{M}_{mn}$, define $<,>$ by $< \mathbf{A}, \mathbf{B} > = \text{trace}(\mathbf{A}^T\mathbf{B})$. Prove that this operation is a real inner product on $\mathcal{M}_{mn}$. (Hint: Refer to Exercise 14 in Section 1.4 and Exercise 21 in Section 1.5.)

5. (a) Prove part (1) of Theorem 7.12. (Hint: $\mathbf{0} = \mathbf{0} + \mathbf{0}$. Use property (4) in the definition of an inner product space.)

(b) Prove part (3) of Theorem 7.12. (Be sure to prove this part for both real and complex vector spaces.)

6. Prove Theorem 7.13.

7. Let $\mathbf{x}, \mathbf{y} \in \mathcal{V}$, a real inner product space.

(a) Prove that $\|\mathbf{x} + \mathbf{y}\|^2 = \|\mathbf{x}\|^2 + 2 < \mathbf{x}, \mathbf{y} > + \|\mathbf{y}\|^2$.

(b) Show that $\mathbf{x}$ and $\mathbf{y}$ are orthogonal in $\mathcal{V}$ if and only if $\|\mathbf{x} + \mathbf{y}\|^2 = \|\mathbf{x}\|^2 + \|\mathbf{y}\|^2$.

(c) Show that $\frac{1}{2}\left(\|\mathbf{x} + \mathbf{y}\|^2 + \|\mathbf{x} - \mathbf{y}\|^2\right) = \|\mathbf{x}\|^2 + \|\mathbf{y}\|^2$.

8. (a) Let $\mathbf{x}, \mathbf{y} \in \mathcal{V}$, a real inner product space. Prove the following **(real) polarization identity**: $< \mathbf{x}, \mathbf{y} > = \frac{1}{4}\left(\|\mathbf{x} + \mathbf{y}\|^2 - \|\mathbf{x} - \mathbf{y}\|^2\right)$.

(b) Let $\mathbf{x}, \mathbf{y} \in \mathcal{V}$, a complex inner product space. Prove the following **complex polarization identity**:

$$< \mathbf{x}, \mathbf{y} > = \frac{1}{4}\left(\left(\|\mathbf{x} + \mathbf{y}\|^2 - \|\mathbf{x} - \mathbf{y}\|^2\right) + i\left(\|\mathbf{x} + i\mathbf{y}\|^2 - \|\mathbf{x} - i\mathbf{y}\|^2\right)\right).$$

These formulas show how the value of the inner product can be derived from the norm (length).

★**9.** Find the distance between $\mathbf{f} = t$ and $\mathbf{g} = \sin t$ in the inner product space of Example 5, with $a = 0$ and $b = \pi$.

10. Find the distance between $\mathbf{x} = [2, -1, 3]$ and $\mathbf{y} = [5, -2, 2]$ in the inner product space of Example 3, using

$$\mathbf{A} = \begin{bmatrix} -2 & 0 & 1 \\ 1 & -1 & 2 \\ 3 & -1 & -1 \end{bmatrix}.$$

11. Let $\mathcal{V}$ be an inner product space.
(a) Prove part (1) of Theorem 7.14. (Hint: Modify the proof of Theorem 1.5.)
(b) Prove part (2) of Theorem 7.14. (Hint: Modify the proof of Theorem 1.6.)

12. Let $\mathcal{V}$ be the inner product space of Example 5. Show that if $\mathbf{f}, \mathbf{g} \in \mathcal{V}$, then

$$\left(\int_a^b f(t)g(t)\,dt\right)^2 \leq \int_a^b (f(t))^2\,dt \int_a^b (g(t))^2\,dt.$$

(Hint: Use the Cauchy-Schwarz Inequality.)

13. Find the angle between $\mathbf{f} = e^t$ and $\mathbf{g} = \sin t$ in the inner product space of Example 5, with $a = 0$ and $b = \pi$.

★**14.** Find the angle between $\mathbf{f} = [2, -1, 3]$ and $\mathbf{g} = [5, -2, 2]$ in the inner product space of Example 3, using

$$\mathbf{A} = \begin{bmatrix} -2 & 0 & 1 \\ 1 & -1 & 2 \\ 3 & -1 & -1 \end{bmatrix}.$$

15. A **metric space** is a set in which every pair of elements x, y has been assigned a real number distance d with the following properties:
(i) $d(x, y) = d(y, x)$.
(ii) $d(x, y) \geq 0$, with $d(x, y) = 0$ if and only if $x = y$.
(iii) $d(x, y) \leq d(x, z) + d(z, y)$, for all z in the set.

Prove that every inner product space is a metric space with $d(\mathbf{x}, \mathbf{y})$ taken to be $\|\mathbf{x} - \mathbf{y}\|$ for all vectors $\mathbf{x}$ and $\mathbf{y}$ in the space.

16. Determine whether the following sets of vectors are orthogonal.
★(a) $\{t^2, t + 1, t - 1\}$, in $\mathcal{P}_3$, under the inner product of Example 4
(b) $\{[15, 9, 19], [-2, -1, -2], [-12, -9, -14]\}$, in $\mathbb{R}^3$, under the inner product of Example 3, with

$$\mathbf{A} = \begin{bmatrix} -3 & 1 & 2 \\ 0 & -2 & 1 \\ 2 & -1 & -1 \end{bmatrix}$$

★(c) $\{[5, -2], [3, 4]\}$, in $\mathbb{R}^2$, under the inner product of Example 2
(d) $\{3t^2 - 1, 4t, 5t^3 - 3t\}$, in $\mathcal{P}_3$, under the inner product of Example 11

17. Prove Theorem 7.15. (Hint: Modify the proof of Result 7 in Section 1.3.)

18. (a) Show that $\int_{-\pi}^{\pi} \cos mt \, dt = 0$ and $\int_{-\pi}^{\pi} \sin nt \, dt = 0$, for all integers m, $n \geq 1$.

(b) Show that $\int_{-\pi}^{\pi} \cos mt \cos nt dt = 0$ and $\int_{-\pi}^{\pi} \sin mt \sin nt \, dt = 0$, for any *distinct* integers $m, n \geq 1$. (Hint: Use trigonometric identities.)

(c) Show that $\int_{-\pi}^{\pi} \cos mt \sin nt \, dt = 0$, for any integers $m, n \geq 1$.

(d) Conclude from parts (a), (b), and (c) that $\{1, \cos t, \sin t, \cos 2t, \sin 2t, \cos 3t, \sin 3t, \ldots\}$ is an orthogonal subset of real continuous functions on $[-\pi, \pi]$.

19. Prove Theorem 7.16. (Hint: Modify the proof of Theorem 5.26.)

20. Let $\{\mathbf{v}_1, \ldots, \mathbf{v}_k\}$ be an orthonormal basis for a complex inner product space $\mathcal{V}$. Prove that for all $\mathbf{v}, \mathbf{w} \in \mathcal{V}$,

$$< \mathbf{v}, \mathbf{w} > = < \mathbf{v}, \mathbf{v}_1 > \overline{< \mathbf{w}, \mathbf{v}_1 >} + < \mathbf{v}, \mathbf{v}_2 > \overline{< \mathbf{w}, \mathbf{v}_2 >} + \cdots + < \mathbf{v}, \mathbf{v}_k > \overline{< \mathbf{w}, \mathbf{v}_k >}.$$

★21. Use the Generalized Gram-Schmidt Process to find an orthogonal basis for $\mathcal{P}_2$ containing $t^2 - t + 1$ under the inner product of Example 11.

22. Use the Generalized Gram-Schmidt Process to find an orthogonal basis for $\mathbb{R}^3$ containing $[-9, -4, 8]$ under the inner product of Example 3 with the matrix

$$\mathbf{A} = \begin{bmatrix} 2 & 1 & 3 \\ 3 & -1 & 3 \\ 2 & -1 & 2 \end{bmatrix}.$$

23. Prove Theorem 7.17. (Hint: Modify the proof of Theorem 5.27.)

24. (a) Prove Theorem 7.20. (Hint: Modify the proof of Theorem 5.34.)

(b) Prove Theorem 7.21. (Hint: Modify the proofs of Theorem 5.35 and Corollary 5.36.)

(c) Prove Theorem 7.22. (Hint: Note that $\mathcal{W} \subseteq \left(\mathcal{W}^{\perp}\right)^{\perp}$. Use Theorem 7.21 to show $\dim(\mathcal{W}) = \dim\left(\left(\mathcal{W}^{\perp}\right)^{\perp}\right)$. Then apply Theorem 4.16, or its complex analog.)

★25. Find $\mathcal{W}^{\perp}$ if $\mathcal{W} = \text{span}\left(\{t^3 + t^2, t - 1\}\right)$ in $\mathcal{P}_3$ with the inner product of Example 4.

26. Find an orthogonal basis for $\mathcal{W}^{\perp}$ if $\mathcal{W} = \text{span}\left(\{(t-1)^2\}\right)$ in $\mathcal{P}_2$, with the inner product $< \mathbf{f}, \mathbf{g} > = \int_0^1 f(t)g(t)\,dt$, for all $\mathbf{f}, \mathbf{g} \in \mathcal{P}_2$.

27. Prove Theorem 7.23. (Hint: Choose an orthonormal basis $\{\mathbf{v}_1, \ldots, \mathbf{v}_k\}$ for $\mathcal{W}$. Then define $\mathbf{w}_1 = \mathbf{proj}_{\mathcal{W}}\mathbf{v} = < \mathbf{v}, \mathbf{v}_1 > \mathbf{v}_1 + \cdots + < \mathbf{v}, \mathbf{v}_k > \mathbf{v}_k$. Let $\mathbf{w}_2 = \mathbf{v} - \mathbf{w}_1$, and prove $\mathbf{w}_2 \in \mathcal{W}^{\perp}$. Finally, see the proof of Theorem 5.38 for uniqueness.)

★28. In the inner product space of Example 8, decompose $\mathbf{f} = \frac{1}{k}e^t$, where $k = e^{\pi} - e^{-\pi}$, as $\mathbf{w}_1 + \mathbf{w}_2$, where $\mathbf{w}_1 \in \mathcal{W} = \text{span}(\{\cos t, \sin t\})$ and $\mathbf{w}_2 \in \mathcal{W}^{\perp}$. Check that $< \mathbf{w}_1, \mathbf{w}_2 > = 0$. (Hint: First find an orthonormal basis for $\mathcal{W}$.)

29. Decompose $\mathbf{v} = 4t^2 - t + 3$ in $\mathcal{P}_2$ as $\mathbf{w}_1 + \mathbf{w}_2$, where $\mathbf{w}_1 \in \mathcal{W} = \text{span}(\{2t^2 - 1, t + 1\})$ and $\mathbf{w}_2 \in \mathcal{W}^\perp$, under the inner product of Example 11. Check that $< \mathbf{w}_1, \mathbf{w}_2 > = 0$. (Hint: First find an orthonormal basis for $\mathcal{W}$.)

30. Bessel's Inequality: Let $\mathcal{V}$ be a real inner product space, and let $\{\mathbf{v}_1, \ldots, \mathbf{v}_k\}$ be an orthonormal set in $\mathcal{V}$. Prove that for any vector $\mathbf{v} \in \mathcal{V}$, $\sum_{i=1}^{k} < \mathbf{v}, \mathbf{v}_i >^2 \leq \|\mathbf{v}\|^2$. (Hint: Let $\mathcal{W} = \text{span}(\{\mathbf{v}_1, \ldots, \mathbf{v}_k\})$. Now, $\mathbf{v} = \mathbf{w}_1 + \mathbf{w}_2$, where $\mathbf{w}_1 = \mathbf{proj}_{\mathcal{W}}\mathbf{v} \in \mathcal{W}$ and $\mathbf{w}_2 \in \mathcal{W}^\perp$. Expand $< \mathbf{v}, \mathbf{v} > = < \mathbf{w}_1 + \mathbf{w}_2, \mathbf{w}_1 + \mathbf{w}_2 >$. Show that $\|\mathbf{v}\|^2 \geq \|\mathbf{w}_1\|^2$, and use the definition of $\mathbf{proj}_{\mathcal{W}}\mathbf{v}$.)

31. Let $\mathcal{W}$ be a finite dimensional subspace of an inner product space $\mathcal{V}$. Consider the mapping $L: \mathcal{V} \longrightarrow \mathcal{W}$ given by $L(\mathbf{v}) = \mathbf{proj}_{\mathcal{W}}\mathbf{v}$.

(a) Prove that L is a linear transformation.

★(b) What are the kernel and range of L?

(c) Show that $L \circ L = L$.

8

Applications

In this chapter, we present several additional practical applications of linear algebra to show its usefulness in mathematics and the sciences.

8.1 Graph Theory

Prerequisite: Section 1.5, Matrix Multiplication

Multiplication of matrices is widely used in graph theory, a branch of mathematics that has come into prominence for modeling many situations in computer science, business, and the social sciences. We begin by introducing graphs and digraphs and then examine their relationship with matrices. Our main goal is to show how matrices are used to calculate the number of paths of a certain length between vertices of a graph or digraph.

Graphs and Digraphs

DEFINITION

A **graph** is a finite collection of **vertices** (points) together with a finite collection of **edges** (curves), each of which has two (not necessarily distinct) vertices as endpoints.

For example, Figure 8.1 depicts two graphs. Note that a graph may have an edge connecting some vertex to itself. Such edges are called **loops**. A graph with no loops, such as G_1 in Figure 8.1, is said to be **loop-free**.

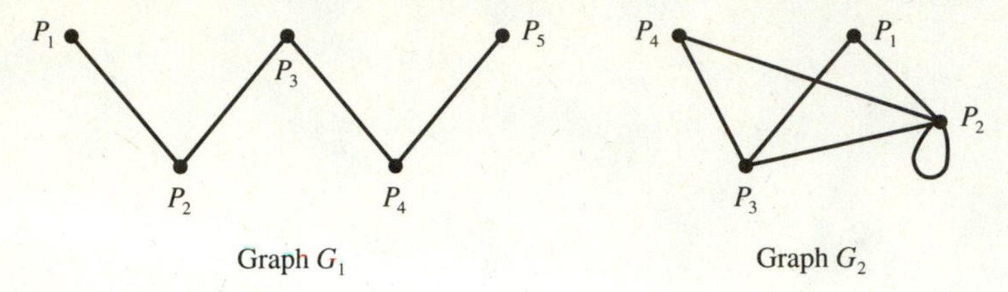

Figure 8.1
Two examples of graphs

A **digraph**, or **directed graph**, is a special type of graph in which each edge is assigned a "direction." Some examples of digraphs appear in Figure 8.2.

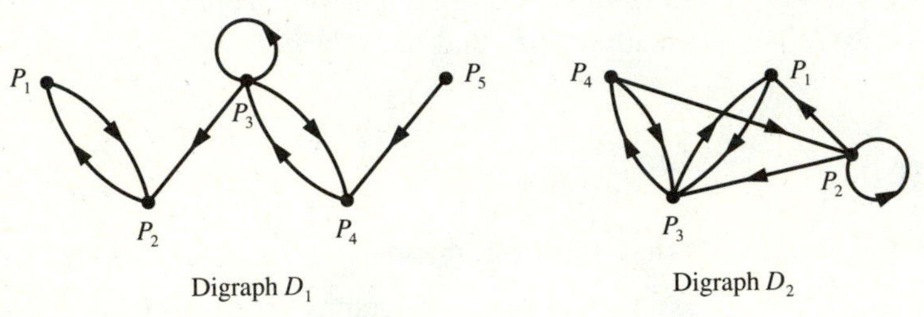

Figure 8.2
Two examples of digraphs

Although the edges in a digraph may resemble vectors, they are not necessarily vectors since there is usually no coordinate system present. One interpretation for graphs and digraphs is to consider the vertices as towns and the edges as roads connecting them. In the case of a digraph, we can think of the roads as one-way streets. Notice that some pairs of towns may not be connected by roads. Another interpretation for graphs and digraphs is to consider the vertices as relay stations and the edges as communication channels (for example, phone lines) between the stations. The stations could be individual people, homes, radio/TV installations, or even computer terminals hooked into a network. You will see additional interpretations for graphs and digraphs in the exercises. The wide applicability of graphs and digraphs makes them very important in computer science.

In this section, we consider only "simple" graphs and digraphs. A **simple** graph is one having at most one edge between each pair of vertices. Similarly, a **simple** digraph is one having at most one edge in each direction between each pair of vertices. All of the graphs and digraphs pictured in the figures are simple.

The Adjacency Matrix

The pattern of edges between the vertices in a graph or digraph can be summarized in an algebraic way using matrices:

DEFINITION

The **adjacency matrix** of a graph having vertices $P_1, P_2, \ldots, P_n$ is the $n \times n$ matrix whose (i, j) entry is 1 if there is an edge between P_i and P_j and is zero otherwise.

The **adjacency matrix** of a digraph having vertices $P_1, P_2, \ldots, P_n$ is the $n \times n$ matrix whose (i, j) entry is 1 if there is an edge directed from P_i to P_j and is zero otherwise.

Example 1

The adjacency matrices for the two graphs in Figure 8.1 and the two digraphs in Figure 8.2 are as follows:

$$\underbrace{\begin{array}{c} \\ P_1 \\ P_2 \\ P_3 \\ P_4 \\ P_5 \end{array}\begin{array}{c} \begin{array}{ccccc} P_1 & P_2 & P_3 & P_4 & P_5 \end{array} \\ \begin{bmatrix} 0 & 1 & 0 & 0 & 0 \\ 1 & 0 & 1 & 0 & 0 \\ 0 & 1 & 0 & 1 & 0 \\ 0 & 0 & 1 & 0 & 1 \\ 0 & 0 & 0 & 1 & 0 \end{bmatrix} \end{array}}_{\text{Adjacency matrix for } G_1} \qquad \underbrace{\begin{array}{c} \\ P_1 \\ P_2 \\ P_3 \\ P_4 \end{array}\begin{array}{c} \begin{array}{cccc} P_1 & P_2 & P_3 & P_4 \end{array} \\ \begin{bmatrix} 0 & 1 & 1 & 0 \\ 1 & 1 & 1 & 1 \\ 1 & 1 & 0 & 1 \\ 0 & 1 & 1 & 0 \end{bmatrix} \end{array}}_{\text{Adjacency matrix for } G_2}$$

$$\underbrace{\begin{array}{c} \\ P_1 \\ P_2 \\ P_3 \\ P_4 \\ P_5 \end{array}\begin{array}{c} \begin{array}{ccccc} P_1 & P_2 & P_3 & P_4 & P_5 \end{array} \\ \begin{bmatrix} 0 & 1 & 0 & 0 & 0 \\ 1 & 0 & 0 & 0 & 0 \\ 0 & 1 & 1 & 1 & 0 \\ 0 & 0 & 1 & 0 & 0 \\ 0 & 0 & 0 & 1 & 0 \end{bmatrix} \end{array}}_{\text{Adjacency matrix for } D_1} \qquad \underbrace{\begin{array}{c} \\ P_1 \\ P_2 \\ P_3 \\ P_4 \end{array}\begin{array}{c} \begin{array}{cccc} P_1 & P_2 & P_3 & P_4 \end{array} \\ \begin{bmatrix} 0 & 0 & 1 & 0 \\ 1 & 1 & 1 & 0 \\ 1 & 0 & 0 & 1 \\ 0 & 1 & 1 & 0 \end{bmatrix} \end{array}}_{\text{Adjacency matrix for } D_2}.$$

■

Notice that the adjacency matrix of any graph is symmetric, for the obvious reason that there is an edge between P_i and P_j if and only if there is an edge (the same one) between P_j and P_i. However, the adjacency matrix for a digraph is usually not symmetric, since the existence of an edge from P_i to P_j does not necessarily imply the existence of an edge in the reverse direction.

Paths in a Graph or Digraph

We often want to know how many different routes exist between two given vertices in a graph or digraph. We need the following definition:

DEFINITION

A **path** (or **chain**) between two vertices P_i and P_j in a graph or digraph is a finite sequence of edges with the following properties:

(1) The first edge "begins" at P_i.
(2) The last edge "ends" at P_j.
(3) Each edge after the first one in the sequence "begins" at the vertex where the previous edge "ended."

The **length** of a path is the number of edges in the path.

Example 2

Consider the digraph pictured in Figure 8.3. There are many different types of paths from P_1 to P_5. For example:

(1) $P_1 \longrightarrow P_2 \longrightarrow P_5$
(2) $P_1 \longrightarrow P_2 \longrightarrow P_3 \longrightarrow P_5$
(3) $P_1 \longrightarrow P_4 \longrightarrow P_3 \longrightarrow P_5$
(4) $P_1 \longrightarrow P_4 \longrightarrow P_4 \longrightarrow P_3 \longrightarrow P_5$
(5) $P_1 \longrightarrow P_2 \longrightarrow P_5 \longrightarrow P_4 \longrightarrow P_3 \longrightarrow P_5$

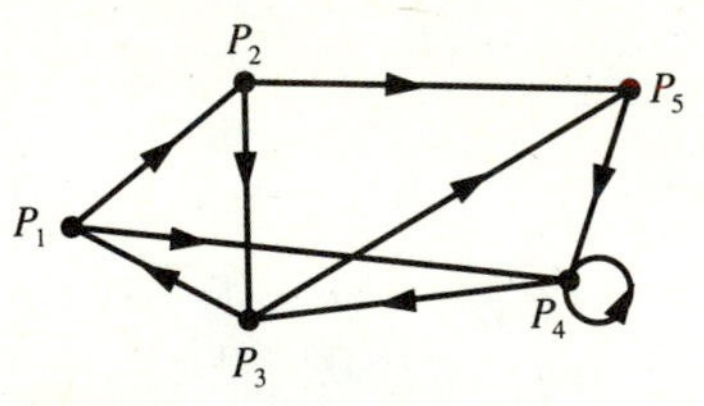

Figure 8.3
Digraph for Examples 2, 3, and 4

(Can you find other paths from P_1 to P_5?) Path (1) is a path of length 2 (or a 2-chain); paths (2), (3), (4), and (5) are paths of lengths 3, 3, 4, and 5, respectively. ■

Counting Paths

Our goal is to calculate exactly how many paths of a given length exist between two vertices in a graph or digraph. For example, suppose we want to know precisely how many paths there are of length 4 from vertex P_2 to vertex P_4 in the digraph of Figure 8.3. We could attempt to list them, but the chance of making a mistake in enumerating them all is great enough to cast doubt on our final total. However, the next theorem, which you are asked to prove in Exercise 11, gives an algebraic way to get the exact count using the adjacency matrix.

THEOREM 8.1

Let $\mathbf{A}$ be the adjacency matrix for a graph or digraph having vertices $P_1, P_2, \ldots, P_n$. Then the total number of paths from P_i to P_j of length k is given by the (i, j) entry in the matrix $\mathbf{A}^k$.

Example 3

Consider again the digraph in Figure 8.3. The adjacency matrix for this digraph is

$$\mathbf{A} = \begin{array}{c} \\ P_1 \\ P_2 \\ P_3 \\ P_4 \\ P_5 \end{array} \begin{array}{c} \begin{array}{ccccc} P_1 & P_2 & P_3 & P_4 & P_5 \end{array} \\ \begin{bmatrix} 0 & 1 & 0 & 1 & 0 \\ 0 & 0 & 1 & 0 & 1 \\ 1 & 0 & 0 & 0 & 1 \\ 0 & 0 & 1 & 1 & 0 \\ 0 & 0 & 0 & 1 & 0 \end{bmatrix} \end{array}.$$

If we want to find the number of paths of length 4 from P_1 to P_4, we need to calculate the first-row, fourth-column entry of $\mathbf{A}^4$. Now,

$$\mathbf{A}^4 = (\mathbf{A}^2)^2 = \left(\begin{bmatrix} 0 & 0 & 2 & 1 & 1 \\ 1 & 0 & 0 & 1 & 1 \\ 0 & 1 & 0 & 2 & 0 \\ 1 & 0 & 1 & 1 & 1 \\ 0 & 0 & 1 & 1 & 0 \end{bmatrix} \right)^2 = \begin{array}{c} \\ P_1 \\ P_2 \\ P_3 \\ P_4 \\ P_5 \end{array} \begin{array}{c} \begin{array}{ccccc} P_1 & P_2 & P_3 & P_4 & P_5 \end{array} \\ \begin{bmatrix} 1 & 2 & 2 & 6 & 1 \\ 1 & 0 & 4 & 3 & 2 \\ 3 & 0 & 2 & 3 & 3 \\ 1 & 1 & 4 & 5 & 2 \\ 1 & 1 & 1 & 3 & 1 \end{bmatrix} \end{array}.$$

Since the first-row, fourth-column entry is 6, there are exactly six paths of length 4 from P_1 to P_4. Looking at the digraph, we can see that these paths are:

$$\begin{array}{l} P_1 \longrightarrow P_2 \longrightarrow P_3 \longrightarrow P_5 \longrightarrow P_4 \\ P_1 \longrightarrow P_2 \longrightarrow P_3 \longrightarrow P_1 \longrightarrow P_4 \\ P_1 \longrightarrow P_4 \longrightarrow P_3 \longrightarrow P_5 \longrightarrow P_4 \\ P_1 \longrightarrow P_4 \longrightarrow P_3 \longrightarrow P_1 \longrightarrow P_4 \\ P_1 \longrightarrow P_2 \longrightarrow P_5 \longrightarrow P_4 \longrightarrow P_4 \\ P_1 \longrightarrow P_4 \longrightarrow P_4 \longrightarrow P_4 \longrightarrow P_4 \end{array}$$

■

Of course, we can generalize the result in Theorem 8.1 easily. A little thought will convince you of the following:

> The total number of paths of length $\leq k$ from a vertex P_i to a vertex P_j in a graph or digraph is the sum of the (i, j) entries of the matrices $\mathbf{A}, \mathbf{A}^2, \mathbf{A}^3, \ldots, \mathbf{A}^k$.

Example 4

For the digraph in Figure 8.3, we will calculate the total number of paths of length ≤ 4 from P_2 to P_3. We listed the adjacency matrix $\mathbf{A}$ for this digraph in Example 3, as well as the products $\mathbf{A}^2$ and $\mathbf{A}^4$. You can easily verify that $\mathbf{A}^3$ is given by

$$\mathbf{A}^3 = \begin{array}{c} \\ P_1 \\ P_2 \\ P_3 \\ P_4 \\ P_5 \end{array} \begin{array}{c} \begin{array}{ccccc} P_1 & P_2 & P_3 & P_4 & P_5 \end{array} \\ \begin{bmatrix} 2 & 0 & 1 & 2 & 2 \\ 0 & 1 & 1 & 3 & 0 \\ 0 & 0 & 3 & 2 & 1 \\ 1 & 1 & 1 & 3 & 1 \\ 1 & 0 & 1 & 1 & 1 \end{bmatrix} \end{array}.$$

Then, a quick calculation gives

$$\mathbf{A} + \mathbf{A}^2 + \mathbf{A}^3 + \mathbf{A}^4 = \begin{array}{c} \\ P_1 \\ P_2 \\ P_3 \\ P_4 \\ P_5 \end{array} \begin{array}{c} \begin{array}{ccccc} P_1 & P_2 & P_3 & P_4 & P_5 \end{array} \\ \begin{bmatrix} 3 & 3 & 5 & 10 & 4 \\ 2 & 1 & 6 & 7 & 4 \\ 4 & 1 & 5 & 7 & 5 \\ 3 & 2 & 7 & 10 & 4 \\ 2 & 1 & 3 & 6 & 2 \end{bmatrix} \end{array}.$$

Hence, the number of paths of length ≤ 4 from P_2 to P_3 is the second-row, third-column entry of this matrix, which is 6. A listing of these paths is as follows:

$$\begin{array}{l} P_2 \longrightarrow P_3 \\ P_2 \longrightarrow P_5 \longrightarrow P_4 \longrightarrow P_3 \\ P_2 \longrightarrow P_5 \longrightarrow P_4 \longrightarrow P_4 \longrightarrow P_3 \\ P_2 \longrightarrow P_3 \longrightarrow P_1 \longrightarrow P_2 \longrightarrow P_3 \\ P_2 \longrightarrow P_3 \longrightarrow P_5 \longrightarrow P_4 \longrightarrow P_3 \\ P_2 \longrightarrow P_3 \longrightarrow P_1 \longrightarrow P_4 \longrightarrow P_3 \end{array}$$

In fact, since we calculated all of the entries of the matrix $\mathbf{A} + \mathbf{A}^2 + \mathbf{A}^3 + \mathbf{A}^4$, we can easily find the total number of paths of length ≤ 4 between any pair of given vertices. For example, the total number of paths of length ≤ 4 between P_3 and P_5 is 5, because that is the third-row, fifth-column entry of the sum. ■

Exercises—Section 8.1

Note: You may want to use a computer with appropriate software to perform the matrix calculations in these exercises.

★**1.** For each of the graphs and digraphs in Figure 8.4, give the corresponding adjacency matrix. Which of these matrices are symmetric?

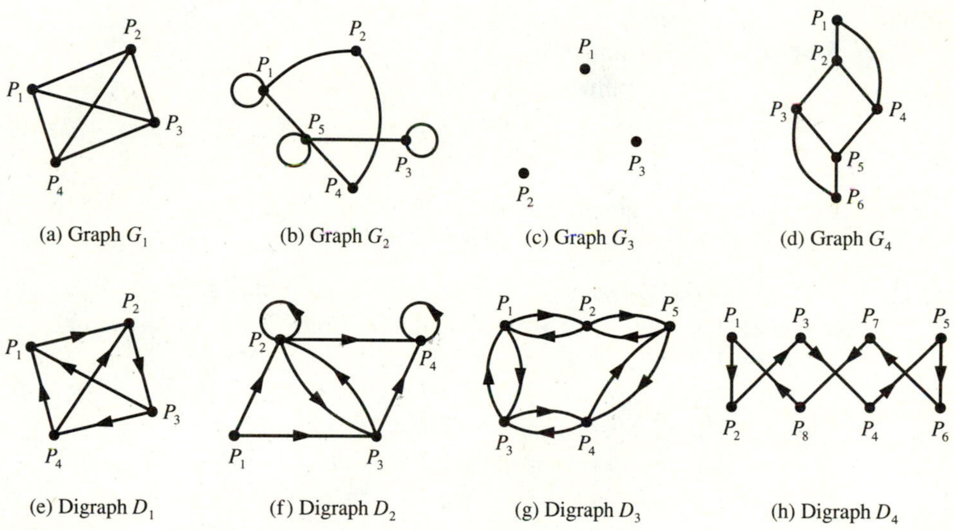

Figure 8.4
Graphs and digraphs for Exercise 1

★**2.** Which of the given matrices could be the adjacency matrix for a simple graph or digraph? Draw the corresponding graph and/or digraph when appropriate.

$$\mathbf{A} = \begin{bmatrix} -1 & 4 \\ 0 & 1 \\ 6 & 0 \end{bmatrix} \quad \mathbf{B} = \begin{bmatrix} 2 & 0 \\ 0 & -1 \end{bmatrix} \quad \mathbf{C} = \begin{bmatrix} 6 & 0 & 0 & 0 \\ 0 & 6 & 0 & 0 \\ 0 & 0 & 6 & 0 \\ 0 & 0 & 0 & 6 \end{bmatrix} \quad \mathbf{D} = \begin{bmatrix} -1 \\ 4 \\ 2 \end{bmatrix}$$

$$\mathbf{E} = \begin{bmatrix} 0 & 0 & 0 & 6 \\ 0 & 0 & 6 & 0 \\ 0 & -6 & 0 & 0 \\ -6 & 0 & 0 & 0 \end{bmatrix} \quad \mathbf{F} = \begin{bmatrix} 1 & 0 & 1 & 0 & 1 \\ 0 & 1 & 0 & 0 & 1 \\ 1 & 0 & 0 & 1 & 1 \\ 0 & 0 & 1 & 0 & 0 \\ 1 & 1 & 1 & 0 & 1 \end{bmatrix} \quad \mathbf{G} = \begin{bmatrix} 1 & 1 & 1 \\ 0 & 1 & 1 \\ 0 & 0 & 1 \end{bmatrix}$$

$$\mathbf{H} = \begin{bmatrix} 0 & 0 & 0 \\ 1 & 0 & 0 \\ 1 & 1 & 0 \end{bmatrix} \quad \mathbf{I} = \begin{bmatrix} 1 & 0 & 0 \\ 0 & 1 & 0 \\ 0 & 0 & 1 \end{bmatrix} \quad \mathbf{J} = \begin{bmatrix} 1 & 2 & 3 & 4 \\ -2 & 1 & 5 & 6 \\ -3 & -5 & 1 & 7 \\ -4 & -6 & -7 & 1 \end{bmatrix}$$

$$\mathbf{K} = \begin{bmatrix} 0 & 1 \\ 1 & 0 \end{bmatrix} \quad \mathbf{L} = \begin{bmatrix} 0 & 1 & 0 & 0 \\ 1 & 0 & 1 & 1 \\ 0 & 1 & 1 & 1 \\ 0 & 1 & 1 & 0 \end{bmatrix} \quad \mathbf{M} = \begin{bmatrix} -2 & 0 & 0 \\ 4 & 0 & 0 \\ -1 & 2 & 3 \end{bmatrix}$$

★**3.** Suppose the writings of six authors—labeled A, B, C, D, E, and F—have been influenced by one another in the following ways:

A has been influenced by D and E.
B has been influenced by C and E.
C has been influenced by A.
D has been influenced by B, E, and F.
E has been influenced by B and C.
F has been influenced by D.

Draw the digraph that represents these relationships. What is its adjacency matrix? What would the transpose of this adjacency matrix represent?

4. Using the adjacency matrix for the digraph in Figure 8.5, find the following:

★(a) The number of paths of length 3 from P_2 to P_4
(b) The number of paths of length 4 from P_1 to P_5
★(c) The number of paths of length ≤ 3 from P_3 to P_2
(d) The number of paths of length ≤ 4 from P_3 to P_1
★(e) The length of the shortest path from P_4 to P_5
(f) The length of the shortest path from P_4 to P_1

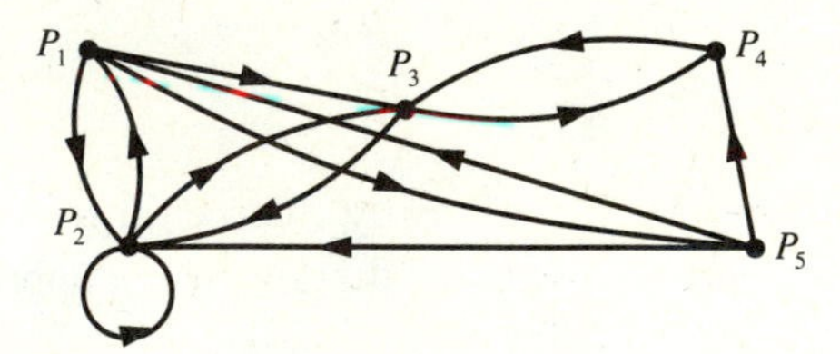

Figure 8.5
Digraph for Exercises 4, 6, and 9

5. For the digraph in Figure 8.6, find the following:

★(a) The number of paths of length 3 from P_2 to P_4

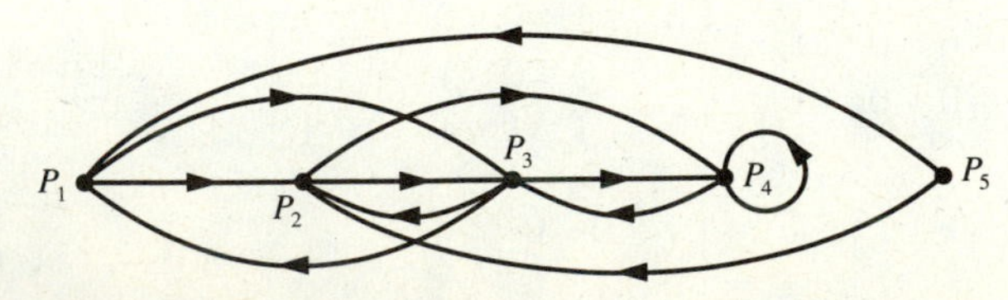

Figure 8.6
Digraph for Exercises 5, 6, and 9

(b) The number of paths of length 4 from P_1 to P_5

★(c) The number of paths of length ≤ 3 from P_3 to P_2

(d) The number of paths of length ≤ 4 from P_3 to P_1

★(e) The length of the shortest path from P_4 to P_5

(f) The length of the shortest path from P_4 to P_1

6. A **cycle** in a graph or digraph is a path connecting a vertex to itself. For the digraphs in each of Figures 8.5 and 8.6, find the following:

★(a) The number of cycles of length 3 connecting P_2 to itself

(b) The number of cycles of length 4 connecting P_1 to itself

★(c) The number of cycles of length ≤ 4 connecting P_4 to itself

★7. (a) Suppose that one vertex is not connected to any other in a graph. How will this situation be reflected in the adjacency matrix for the graph?

(b) Suppose that one vertex is not directed to any other in a digraph. How will this situation be reflected in the adjacency matrix for the digraph?

★8. Recall the definition of the trace of a matrix (Exercise 14 of Section 1.4). What information does the trace of the adjacency matrix of a graph or digraph give?

9.★(a) A **strongly connected digraph** is a digraph in which, given any pair of vertices, there is a directed path (of some length) from each of these two vertices to the other. Determine whether the digraphs in Figures 8.5 and 8.6 are strongly connected.

(b) Prove that a digraph with n vertices having adjacency matrix $\mathbf{A}$ is strongly connected if and only if $\mathbf{A} + \mathbf{A}^2 + \mathbf{A}^3 + \cdots + \mathbf{A}^{n-1}$ has no zero entries.

10. (a) A **dominance digraph** is one with no loops in which, for any two distinct vertices P_i and P_j, there is either an edge from P_i to P_j, or an edge from P_j to P_i—but *not both*. (Dominance digraphs are useful in psychology, sociology, and communications.) Show that the following matrix is the adjacency matrix for a dominance digraph:

$$\begin{array}{c} \\ P_1 \\ P_2 \\ P_3 \\ P_4 \end{array} \begin{array}{c} \begin{array}{cccc} P_1 & P_2 & P_3 & P_4 \end{array} \\ \left[\begin{array}{cccc} 0 & 1 & 0 & 1 \\ 0 & 0 & 1 & 0 \\ 1 & 0 & 0 & 1 \\ 0 & 1 & 0 & 0 \end{array}\right] \end{array}.$$

★(b) Consider a league of six teams in which each team plays every other team once in a tournament (with no tie games possible). Represent this situation with a digraph by drawing an edge from the vertex for Team A to the vertex for Team B if Team A defeats Team B. Is this a dominance digraph? Why or why not?

(c) Suppose that $\mathbf{A}$ is a square matrix with all entries a_{ij} equal to zero or to 1. Show that $\mathbf{A}$ is the adjacency matrix for a dominance digraph if and only if $\mathbf{A} + \mathbf{A}^T$ has all main diagonal entries equal to zero, and all other entries equal to 1.

11. Prove Theorem 8.1. (Hint: Use a proof by induction on the length of the path between vertices P_i and P_j. In the inductive step, assume that the given statement is true for paths of length t, and prove that it holds for paths of length $t + 1$. Use the fact that the total number of paths from P_i to P_j of length $t + 1$ is the sum of n products, where each product is the number of paths of length t from P_i to some vertex P_q $(1 \leq q \leq n)$ times the number of paths of length 1 from P_q to P_j. Use the induction assumption and the definition of matrix multiplication to show that this sum is the (i, j) entry of $\mathbf{A}^{k+1}$.)

8.2 Ohm's Law

Prerequisite: Section 2.1, Solving Systems of Linear Equations

In this section, we examine an important application of systems of linear equations to circuit theory in physics.

Circuit Fundamentals and Ohm's Law

In a simple electrical circuit, such as the one in Figure 8.7, **voltage sources** (for example, batteries) stimulate electric current to flow through the circuit. **Voltage** (V) is measured in **volts**, and **current** (I) is measured in **amperes**. The circuit in Figure 8.7 has two voltage sources: $48V$ and $9V$. Current flows from the positive (+) end of the voltage source to the negative (−) end.

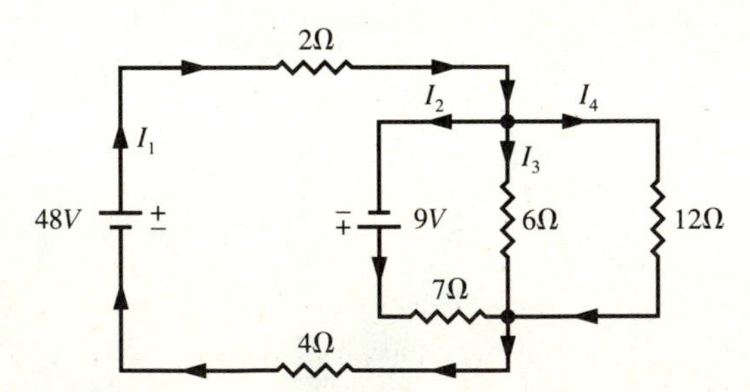

Figure 8.7
Electrical circuit

In contrast to voltage sources, there are **voltage drops**, or **sinks**, when **resistors** are present, because resistors impede the flow of current. In particular, the following principle holds:

OHM'S LAW

At any resistor, the amount of voltage V dropped is proportional to the amount of current I flowing through the resistor. That is,

$$V = IR,$$

where the proportionality constant R is a measure of the resistance to the current.

Resistance (R) is measured in **ohms**, or volts/ampere. The Greek letter Ω is used to denote ohms.

Any point in the circuit where current-carrying branches meet is called a **junction**. Any path along the branches of a circuit that begins and ends at the same junction is called a **loop**. The following two principles involving junctions and loops are very important:

KIRCHOFF'S LAWS

First Law: The sum of the currents flowing into a junction must equal the sum of the currents leaving a junction.

Second Law: The sum of the voltage sources and drops around any loop of a circuit is zero.

Example 1

Consider the electrical circuit in Figure 8.7. We will use Ohm's Law to find the amount of current flowing through each branch of the circuit. Notice that the circuit has two junctions: the first where current I_1 branches into the three currents I_2, I_3, and I_4 and the second where these last three currents merge again into I_1. Notice also that three different loops start and end at the voltage source $48V$:

$$\begin{aligned}
&(1)\quad I_1 \longrightarrow I_2 \longrightarrow I_1\\
&(2)\quad I_1 \longrightarrow I_3 \longrightarrow I_1\\
&(3)\quad I_1 \longrightarrow I_4 \longrightarrow I_1
\end{aligned}$$

By Kirchoff's First Law, we have $I_1 = I_2 + I_3 + I_4$. Kirchoff's Second Law gives an Ohm's Law equation for each possible loop around the circuit:

$$\begin{aligned}
48V + 9V - I_1(2\Omega) - I_2(7\Omega) - I_1(4\Omega) &= 0 \qquad \text{(loop 1)}\\
48V - I_1(2\Omega) - I_3(6\Omega) - I_1(4\Omega) &= 0 \qquad \text{(loop 2)}\\
48V - I_1(2\Omega) - I_4(12\Omega) - I_1(4\Omega) &= 0 \qquad \text{(loop 3)}
\end{aligned}$$

These equations lead to the following system of four equations and four variables:

$$\begin{cases} -I_1 + I_2 + I_3 + I_4 = 0 \\ 6I_1 + 7I_2 = 57 \\ 6I_1 + 6I_3 = 48 \\ 6I_1 + 12I_4 = 48 \end{cases}.$$

After applying the Gauss-Jordan algorithm to the augmented matrix for this system, we obtain

$$\left[\begin{array}{cccc|c} 1 & 0 & 0 & 0 & 6 \\ 0 & 1 & 0 & 0 & 3 \\ 0 & 0 & 1 & 0 & 2 \\ 0 & 0 & 0 & 1 & 1 \end{array}\right].$$

Hence, $I_1 = 6$ amperes, $I_2 = 3$ amperes, $I_3 = 2$ amperes, and $I_4 = 1$ ampere. ■

Exercise—Section 8.2

1. Use Ohm's Law to find the current in each branch of the electrical circuits in Figure 8.8, with the indicated voltage sources and resistances.

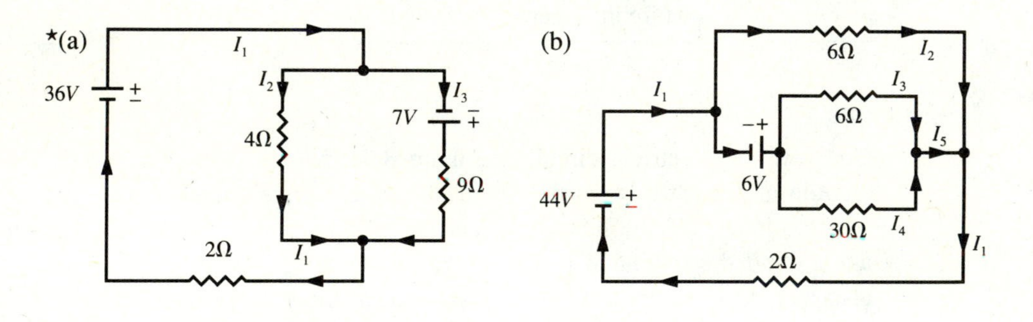

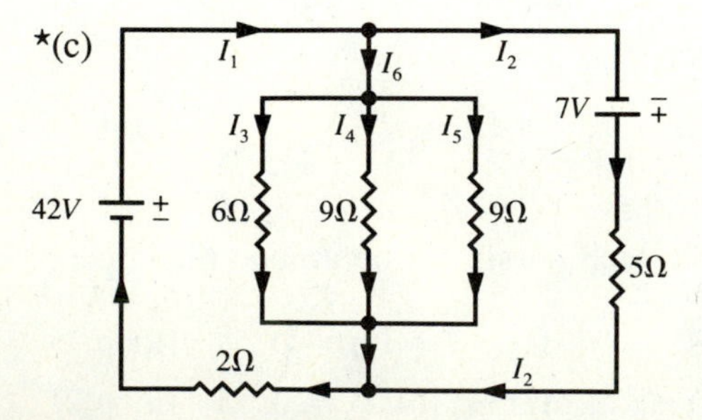

Figure 8.8
Electrical circuits for Exercise 1

8.3 Least-Squares Approximations

Prerequisite: Section 2.1, Solving Systems of Linear Equations

In this section, we present the least-squares method for finding the "closest" polynomial to a given set of data points. You should have a calculator or computer handy as you work through some of the examples and exercises.

Least-Squares Polynomials

In science and business, we often need to predict the relationship between two given variables. In many cases, we begin by performing an appropriate laboratory experiment or statistical analysis to obtain the necessary data. However, even if a simple law governs the behavior of the variables, this law may not be easy to find because of errors introduced in measuring or sampling. In practice, therefore, we are often content with a polynomial equation that provides a close approximation to the data.

Suppose we have a set of points $(a_1, b_1), (a_2, b_2), (a_3, b_3), \ldots, (a_n, b_n)$. We want a method for finding polynomial equations $y = f(x)$ to fit these points as "closely" as possible. One approach would be to minimize the sum of the vertical distances $|f(a_1) - b_1|, |f(a_2) - b_2|, \ldots, |f(a_n) - b_n|$ between the graph of $y = f(x)$ and the data points.† These distances are the lengths of the line segments in Figure 8.9. However, absolute values are hard to work with algebraically. Instead, the

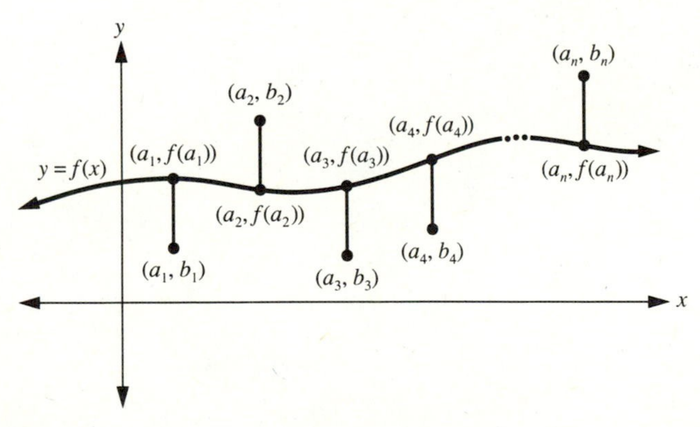

Figure 8.9
Vertical distances from data points (a_k, b_k) to $y = f(x)$, for $1 \leq k \leq n$

†Notice that minimizing the sum of the differences $f(a_k) - b_k$ (without the absolute values) is not enough, since some of the quantities $f(a_k) - b_k$ may be positive and others may be negative. If one of the positive terms, say $f(a_l) - b_l$, had the same magnitude as one of the negative terms, say $f(a_m) - b_m$, then they would cancel each other out if summed directly. Thus, the distances between $y = f(x)$ and the points (a_l, b_l) and (a_m, b_m) would not be represented in the sum.

method most commonly used involves minimizing the sum of the *squares* of the vertical distances.

DEFINITION

A **least-squares polynomial of degree** t for the points $(a_1, b_1), (a_2, b_2), \ldots, (a_n, b_n)$ is a polynomial $y = f(x) = c_t x^t + \cdots + c_2 x^2 + c_1 x + c_0$ for which the sum

$$(f(a_1) - b_1)^2 + (f(a_2) - b_2)^2 + (f(a_3) - b_3)^2 + \cdots + (f(a_n) - b_n)^2$$

of the squares of the vertical distances from each of the given points to the polynomial is a minimum.

A method for calculating least-squares polynomials is stated in Theorem 8.2, which we present without proof. This method is usually used to find a least-squares polynomial whose degree is less than the given number n of data points. (Generally, there is no unique least-squares polynomial when the degree is greater than or equal to n or if some of the given data points have identical x-coordinates.)

THEOREM 8.2

Let $(a_1, b_1), (a_2, b_2), \ldots, (a_n, b_n)$ be n points, and let $\mathbf{A}$ be the $n \times (t+1)$ matrix

$$\begin{bmatrix} 1 & a_1 & a_1^2 & \cdots & a_1^t \\ 1 & a_2 & a_2^2 & \cdots & a_2^t \\ \vdots & \vdots & \vdots & \ddots & \vdots \\ 1 & a_n & a_n^2 & \cdots & a_n^t \end{bmatrix}.$$

A polynomial

$$c_t x^t + \cdots + c_2 x^2 + c_1 x^1 + c_0$$

whose coefficients $c_0, c_1, \ldots, c_t$ satisfy the linear system

$$(\mathbf{A}^T \mathbf{A}) \begin{bmatrix} c_0 \\ c_1 \\ \vdots \\ c_t \end{bmatrix} = \mathbf{A}^T \begin{bmatrix} b_1 \\ b_2 \\ \vdots \\ b_n \end{bmatrix}$$

is a least-squares polynomial for the given points. In addition, if $\mathbf{A}^T\mathbf{A}$ row reduces to $\mathbf{I}_{t+1}$, there is a unique least-squares polynomial of degree t for the given points.

Notice in Theorem 8.2 that $\mathbf{A}^T$ is a $(t+1) \times n$ matrix. Then, $\mathbf{A}^T\mathbf{A}$ is a $(t+1) \times (t+1)$ matrix, and so the matrix products in Theorem 8.2 make sense.

The following two examples show how this theorem is used to calculate least-squares polynomials. The first example calculates a least-squares line for a given set of points. Such a line is often called a **line of best fit**, or a **linear regression**.

Example 1

We will find the least-squares line $y = c_1x + c_0$ through the points $(a_1, b_1) = (-4, 6)$, $(a_2, b_2) = (-2, 4)$, $(a_3, b_3) = (1, 1)$, $(a_4, b_4) = (2, -1)$, and $(a_5, b_5) = (4, -3)$.

Since the degree of the desired polynomial is 1 in this case, the matrix $\mathbf{A}$ has only two columns, as follows:

$$\mathbf{A} = \begin{bmatrix} 1 & a_1 \\ 1 & a_2 \\ 1 & a_3 \\ 1 & a_4 \\ 1 & a_5 \end{bmatrix} = \begin{bmatrix} 1 & -4 \\ 1 & -2 \\ 1 & 1 \\ 1 & 2 \\ 1 & 4 \end{bmatrix}.$$

Then, $\mathbf{A}^T = \begin{bmatrix} 1 & 1 & 1 & 1 & 1 \\ -4 & -2 & 1 & 2 & 4 \end{bmatrix}$ and $\mathbf{A}^T\mathbf{A} = \begin{bmatrix} 5 & 1 \\ 1 & 41 \end{bmatrix}$.

$$\text{Since} \quad \begin{bmatrix} b_1 \\ b_2 \\ b_3 \\ b_4 \\ b_5 \end{bmatrix} = \begin{bmatrix} 6 \\ 4 \\ 1 \\ -1 \\ -3 \end{bmatrix}, \quad \text{we have} \quad \mathbf{A}^T \begin{bmatrix} b_1 \\ b_2 \\ b_3 \\ b_4 \\ b_5 \end{bmatrix} = \begin{bmatrix} 7 \\ -45 \end{bmatrix}.$$

Hence the equation

$$(\mathbf{A}^T\mathbf{A}) \begin{bmatrix} c_0 \\ c_1 \end{bmatrix} = \mathbf{A}^T \begin{bmatrix} b_1 \\ b_2 \\ b_3 \\ b_4 \\ b_5 \end{bmatrix} \quad \text{becomes} \quad \begin{bmatrix} 5 & 1 \\ 1 & 41 \end{bmatrix} \begin{bmatrix} c_0 \\ c_1 \end{bmatrix} = \begin{bmatrix} 7 \\ -45 \end{bmatrix}.$$

Row reducing the augmented matrix

$$\left[\begin{array}{cc|c} 5 & 1 & 7 \\ 1 & 41 & -45 \end{array}\right] \quad \text{gives} \quad \left[\begin{array}{cc|c} 1 & 0 & 1.63 \\ 0 & 1 & -1.14 \end{array}\right],$$

and so the least-squares line is $y = -1.14x + 1.63$ (see Figure 8.10).

Notice that, for each given a_i value, this line produces a value "close" to the given b_i value. For example, when $x = a_1 = -4$, $y = -1.14(-4) + 1.63 = 6.19$, which is close to $b_1 = 6$.

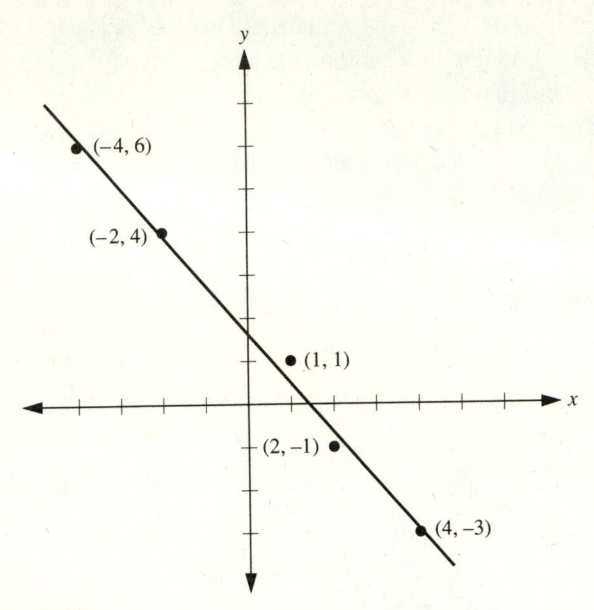

Figure 8.10
Least-squares line for the data points in Example 1 ■

Once we have calculated the least-squares line, we can use it to find the values of other potential data points. This technique is called **extrapolation**. For instance, in Example 1, if x is allowed to have the value 7, the value of y is $-1.14(7) + 1.63 = -6.35$. Thus, we might predict that there is a data point close to $(7, -6.35)$.

In the next example, we use a set of data points that seems to suggest a parabolic rather than a linear shape and find a second-degree least-squares polynomial to fit the data.

Example 2

We will find the quadratic least-squares polynomial for the points $(-3, 7)$, $(-1, 4)$, $(2, 0)$, $(3, 1)$, and $(5, 6)$. As in Example 1, we label these points (a_1, b_1) through (a_5, b_5), respectively. Since we want a quadratic polynomial, the degree $t = 2$. The matrix $\mathbf{A}$ then has three columns:

$$\mathbf{A} = \begin{bmatrix} 1 & -3 & 9 \\ 1 & -1 & 1 \\ 1 & 2 & 4 \\ 1 & 3 & 9 \\ 1 & 5 & 25 \end{bmatrix}.$$

Then,

$$\mathbf{A}^T = \begin{bmatrix} 1 & 1 & 1 & 1 & 1 \\ -3 & -1 & 2 & 3 & 5 \\ 9 & 1 & 4 & 9 & 25 \end{bmatrix}, \quad \text{and so} \quad \mathbf{A}^T\mathbf{A} = \begin{bmatrix} 5 & 6 & 48 \\ 6 & 48 & 132 \\ 48 & 132 & 804 \end{bmatrix}.$$

Also, the matrix

$$\begin{bmatrix} b_1 \\ b_2 \\ b_3 \\ b_4 \\ b_5 \end{bmatrix} = \begin{bmatrix} 7 \\ 4 \\ 0 \\ 1 \\ 6 \end{bmatrix}, \quad \text{and so} \quad \mathbf{A}^T \begin{bmatrix} b_1 \\ b_2 \\ b_3 \\ b_4 \\ b_5 \end{bmatrix} = \begin{bmatrix} 18 \\ 8 \\ 226 \end{bmatrix}.$$

Therefore, the coefficients of the least-squares polynomial $c_2x^2 + c_1x + c_0$ are the solutions of the system

$$(\mathbf{A}^T\mathbf{A}) \begin{bmatrix} c_0 \\ c_1 \\ c_2 \end{bmatrix} = \mathbf{A}^T \begin{bmatrix} b_1 \\ b_2 \\ b_3 \\ b_4 \\ b_5 \end{bmatrix},$$

which is

$$\begin{bmatrix} 5 & 6 & 48 \\ 6 & 48 & 132 \\ 48 & 132 & 804 \end{bmatrix} \begin{bmatrix} c_0 \\ c_1 \\ c_2 \end{bmatrix} = \begin{bmatrix} 18 \\ 8 \\ 226 \end{bmatrix}.$$

Solving, we find $c_0 = 1.21$, $c_1 = -1.02$, and $c_2 = 0.38$. Hence, the least-squares quadratic polynomial is $y = 0.38x^2 - 1.02x + 1.21$ (see Figure 8.11).

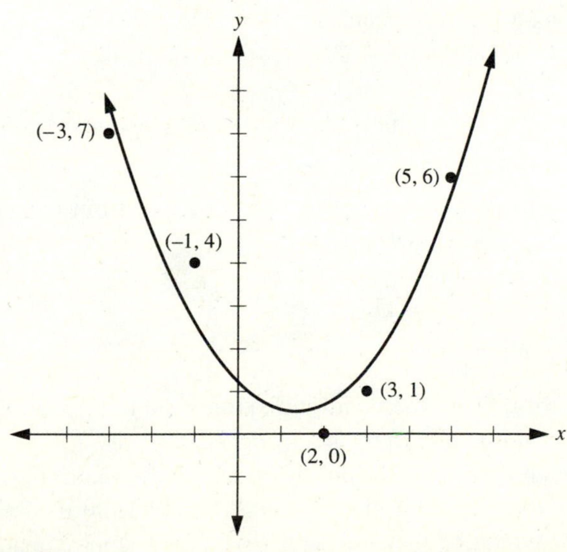

Figure 8.11
Least-squares quadratic polynomial for the data points in Example 2 ■

Exercises—Section 8.3

Note: You should have a calculator or computer handy for the computations in many of these exercises.

1. For each of the following sets of points, find the line of best fit (that is, the least-squares line). In each case, extrapolate to find the approximate y-value when $x = 5$.

★(a) $(3, -8), (1, -5), (0, -4), (2, -1)$

(b) $(-6, -6), (-4, -3), (-1, 0), (1, 2)$

★(c) $(-4, 10), (-3, 8), (-2, 7), (-1, 5), (0, 4)$

2. For each of the following sets of points, find the least-squares quadratic polynomial.

★(a) $(-4, 8), (-2, 5), (0, 3), (2, 6)$

(b) $(-1, -4), (0, -2), (2, -2), (3, -5)$

★(c) $(-4, -3), (-3, -2), (-2, -1), (0, 0), (1, 1)$

3. For each of the following sets of points, find the least-squares cubic (degree 3) polynomial.

★(a) $(-3, -3), (-2, -1), (-1, 0), (0, 1), (1, 4)$

(b) $(-2, 5), (-1, 4), (0, 3), (1, 3), (2, 1)$

4. Use the points given for each function to find the desired approximation.

★(a) Least-squares quadratic polynomial for $y = x^4$, using $x = -2, -1, 0, 1, 2$

(b) Least-squares quadratic polynomial for $y = e^x$, using $x = -2, -1, 0, 1, 2$

★(c) Least-squares quadratic polynomial for $y = \ln x$, using $x = 1, 2, 3, 4$

(d) Least-squares cubic polynomial for $y = \sin x$, using $x = -\frac{\pi}{2}, -\frac{\pi}{4}, 0, \frac{\pi}{4}, \frac{\pi}{2}$

★(e) Least-squares cubic polynomial for $y = \cos x$, using $x = -\frac{\pi}{2}, -\frac{\pi}{4}, 0, \frac{\pi}{4}, \frac{\pi}{2}$

5. An engineer is monitoring a leaning tower whose angle from the vertical over a period of months is given below:

Month	1	2	3	4	5
Angle from vertical	3°	3.3°	3.7°	4.1°	4.6°

★(a) Find the line of best fit for the data, and extrapolate to predict the month in which the angle will be 20° from the vertical.

★(b) Find a least-squares quadratic approximation for the data, and extrapolate to predict the month in which the angle will be 20° from the vertical.

(c) Compare your answers to parts (a) and (b). Which approximation do you think is more accurate? Why?

6. The population of the U.S. (in millions), according to the Census Bureau, is given below:

Year	1940	1950	1960	1970	1980	1990
Population	131.7	150.7	179.3	203.3	226.5	249.6

(a) Find the line of best fit for the data, and extrapolate to predict the population in 2010. (Hint: Renumber the years as 1 through 6 to simplify the computation.)
(b) Find a least-squares quadratic approximation for the data, and extrapolate to predict the population in 2010.
(c) Compare your answers to parts (a) and (b). Which approximation do you think is more accurate? Why?

★**7.** Since three points with distinct x-values in the plane determine a unique quadratic polynomial, show that the method of least-squares gives the *exact* quadratic polynomial that goes through the points $(-2, 6)$, $(0, 2)$, and $(3, 8)$.

8. Show that the following system has the same solutions for c_0 and c_1 as the system in Theorem 8.2 when $t = 1$:

$$\begin{cases} nc_0 + \left(\sum_{i=1}^{n} a_i\right)c_1 = \sum_{i=1}^{n} b_i \\ \left(\sum_{i=1}^{n} a_i\right)c_0 + \left(\sum_{i=1}^{n} a_i^2\right)c_1 = \sum_{i=1}^{n} a_i b_i \end{cases}.$$

9. Although an inconsistent system $\mathbf{AX} = \mathbf{B}$ has no solutions, the least-squares method is sometimes used to find values that come "close" to satisfying all the equations in the system. Solutions to the related system $\mathbf{A}^T\mathbf{AX} = \mathbf{A}^T\mathbf{B}$ (obtained by multiplying on the left by $\mathbf{A}^T$) are called **least-squares "solutions"** for the inconsistent system $\mathbf{AX} = \mathbf{B}$. For each inconsistent system, find a least-squares "solution," and check that it comes close to satisfying each equation in the system.

★(a) $\begin{cases} 4x_1 - 3x_2 = 12 \\ 2x_1 + 5x_2 = 32 \\ 3x_1 + x_2 = 21 \end{cases}$

(b) $\begin{cases} 2x_1 - x_2 + x_3 = 11 \\ -x_1 + 3x_2 - x_3 = -9 \\ x_1 - 2x_2 + 3x_3 = 12 \\ 3x_1 - 4x_2 + 2x_3 = 21 \end{cases}$

8.4 Markov Chains

Prerequisite: Section 2.1, Solving Systems of Linear Equations

In this section, we introduce Markov chains and demonstrate how they are used to predict the future states of an interdependent system. You may want to have a calculator or computer handy as you work through some of the examples and exercises in this section.

An Introductory Example

The following example will introduce many of the ideas associated with Markov chains.

Example 1

Suppose that three banks in a certain town are competing for investors. Currently Bank A has 40% of the investors, Bank B has 10%, and Bank C has the remaining 50%. We can set up the following **probability** (or **state**) vector **p** to represent this distribution:

$$\mathbf{p} = \begin{bmatrix} .4 \\ .1 \\ .5 \end{bmatrix}.$$

Suppose the townsfolk are tempted by various promotional campaigns to switch banks. Records show that each year Bank A keeps half of its investors, with the remainder switching equally to Banks B and C. However, Bank B keeps two-thirds of its investors, with the remainder switching equally to Banks A and C. Finally, Bank C keeps half of its investors, with the remainder switching equally to Banks A and B. The following **transition matrix M** (rounded to three decimal places) keeps track of the changing investment patterns:

$$\mathbf{M} = \text{Next year} \begin{matrix} \\ \text{A} \\ \text{B} \\ \text{C} \end{matrix} \overset{\text{Current year}}{\overset{\begin{matrix} \text{A} & \text{B} & \text{C} \end{matrix}}{\begin{bmatrix} .500 & .167 & .250 \\ .250 & .667 & .250 \\ .250 & .167 & .500 \end{bmatrix}}}.$$

The (i, j) entry of **M** represents the fraction of current investors going *from* Bank j *to* Bank i next year.†

To find the distribution of investors after one year, consider

$$\mathbf{p}_1 = \mathbf{M}\mathbf{p} = \text{Next year} \begin{matrix} \\ \text{A} \\ \text{B} \\ \text{C} \end{matrix} \overset{\text{Current year}}{\overset{\begin{matrix} \text{A} & \text{B} & \text{C} \end{matrix}}{\begin{bmatrix} .500 & .167 & .250 \\ .250 & .667 & .250 \\ .250 & .167 & .500 \end{bmatrix}}} \begin{bmatrix} .4 \\ .1 \\ .5 \end{bmatrix} = \begin{bmatrix} .342 \\ .292 \\ .367 \end{bmatrix}.$$

†It may seem more natural to let the (i, j) entry of **M** represent the fraction going *from* Bank i *to* Bank j. However, the matrix is set up the way it is to make certain matrix multiplications easier later in this section.

The entries of $\mathbf{p}_1$ give the distribution of investors after one year. For example, the first entry of this product, .342, is obtained by taking the dot product of the first row of $\mathbf{M}$ with $\mathbf{p}$ as follows:

$$\underbrace{(.500)}_{\substack{\text{At Bank A,}\\ \text{fraction of}\\ \text{investors}\\ \text{who stay at}\\ \text{Bank A}}} \underbrace{(.4)}_{\substack{\text{Fraction of}\\ \text{investors}\\ \text{currently at}\\ \text{Bank A}}} + \underbrace{(.167)}_{\substack{\text{At Bank B,}\\ \text{fraction of}\\ \text{investors}\\ \text{who switch}\\ \text{to Bank A}}} \underbrace{(.1)}_{\substack{\text{Fraction of}\\ \text{investors}\\ \text{currently at}\\ \text{Bank B}}} + \underbrace{(.250)}_{\substack{\text{At Bank C,}\\ \text{fraction of}\\ \text{investors}\\ \text{who switch}\\ \text{to Bank A}}} \underbrace{(.5)}_{\substack{\text{Fraction of}\\ \text{investors}\\ \text{currently at}\\ \text{Bank C}}}$$

which gives .342, the total fraction of investors at Bank A after one year.

We can continue this process for another year, as follows:

$$\mathbf{p}_2 = \mathbf{M}\mathbf{p}_1 = \begin{array}{c} \\ A \\ B \\ C \end{array}\!\!\begin{array}{c} \begin{array}{ccc} A & \;\;B & \;\;C \end{array} \\ \begin{bmatrix} .500 & .167 & .250 \\ .250 & .667 & .250 \\ .250 & .167 & .500 \end{bmatrix} \end{array} \begin{bmatrix} .342 \\ .292 \\ .367 \end{bmatrix} = \begin{bmatrix} .312 \\ .372 \\ .318 \end{bmatrix}.$$

Since multiplication by $\mathbf{M}$ gives the yearly change and the entries of $\mathbf{p}_1$ represent the distribution of investors at the end of the first year, we see that the entries of $\mathbf{p}_2$ represent the correct distribution of investors at the end of the second year. That is, after two years, 31.2% of the investors are at Bank A, 37.2% are at Bank B, and 31.8% are at Bank C. Notice that

$$\mathbf{p}_2 = \mathbf{M}\mathbf{p}_1 = \mathbf{M}(\mathbf{M}\mathbf{p}) = \mathbf{M}^2\mathbf{p}.$$

In other words, the matrix $\mathbf{M}^2$ takes us directly from $\mathbf{p}$ to $\mathbf{p}_2$. Similarly, if $\mathbf{p}_3$ is the distribution after three years, then

$$\mathbf{p}_3 = \mathbf{M}\mathbf{p}_2 = \mathbf{M}(\mathbf{M}^2\mathbf{p}) = \mathbf{M}^3\mathbf{p}.$$

A simple induction proof shows that, in general, if $\mathbf{p}_n$ represents the distribution after n years, then $\mathbf{p}_n = \mathbf{M}^n\mathbf{p}$.

We can use the last formula to find the distribution of investors after six years. After tedious calculation (rounding to three decimal places at each step), we find

$$\mathbf{M}^6 = \begin{bmatrix} .288 & .285 & .288 \\ .427 & .432 & .427 \\ .288 & .285 & .288 \end{bmatrix}.$$

Then

$$\mathbf{p}_6 = \mathbf{M}^6\mathbf{p} = \begin{bmatrix} .288 & .285 & .288 \\ .427 & .432 & .427 \\ .288 & .285 & .288 \end{bmatrix} \begin{bmatrix} .4 \\ .1 \\ .5 \end{bmatrix} = \begin{bmatrix} .288 \\ .428 \\ .288 \end{bmatrix}.$$

■

Formal Definitions

We now recap many of the ideas presented in Example 1 and give them a more formal treatment.

The notion of probability is important when discussing Markov chains. Probabilities of events are always given as values between $0 = 0\%$ and $1 = 100\%$, where a probability of 0 indicates no possibility and a probability of 1 indicates certainty. For example, if we draw a random card from a standard deck of fifty-two playing cards, the probability that the card is an ace is $\frac{4}{52} = \frac{1}{13}$, because exactly four of the fifty-two cards are aces. The probability that the card is a red card is $\frac{26}{52} = \frac{1}{2}$, since there are twenty-six red cards in the deck. The probability that the card is both red and black (at the same time) is $\frac{0}{52} = 0$, since this event is impossible. Finally, the probability that the card is red or black is $\frac{52}{52} = 1$, since this event is certain.

Now consider a set of events that are completely "distinct" and "exhaustive" (that is, one and only one of them must occur at any time). The sum of all of their probabilities must total $100\% = 1$. For example, if we select a card at random, we have a $\frac{13}{52} = \frac{1}{4}$ chance each of choosing a club, diamond, heart, or spade. These represent the only distinct suit possibilities, and the sum of these four probabilities is 1.

Now recall that each column of the matrix $\mathbf{M}$ in Example 1 represents the probabilities that an investor switches assets to Bank A, B, or C. Since these were the only banks in town, the sum of the probabilities in each column of $\mathbf{M}$ must total 1, or Example 1 would not make sense as stated. Hence, $\mathbf{M}$ is a matrix of the following type:

DEFINITION

A **stochastic matrix** is a square matrix in which all entries are nonnegative and the entries of each column add up to 1.

A single-column stochastic matrix is often called a **stochastic vector**. The next theorem is easy to prove by induction (see Exercise 9).

THEOREM 8.3

The product of any finite number of stochastic matrices is a stochastic matrix.

Now we are ready to formally define a Markov chain.

DEFINITION

A **Markov chain** (or **Markov process**) is a system containing a finite number of distinct states $S_1, S_2, \ldots, S_n$ on which steps are performed such that

(1) At any time, each element of the system resides in exactly one of the states.
(2) At each step in the process, elements in the system can move from one state to another.
(3) The probabilities of moving from state to state are fixed—that is, they are the same at each step in the process.

In Example 1, the distinct states of the Markov chain were the three banks—A, B, and C—and the elements of the system were the investors, who kept their money in only one of the three banks at any given time. Each new year represented another step in the process, during which time investors could switch banks or remain with their current bank. Finally, we assumed that the probabilities of switching banks did not change from year to year.

DEFINITION

A **probability** (or **state**) **vector p** for a Markov chain is a stochastic vector whose i^{th} entry is the probability that an element in the system is currently in state S_i. A **transition matrix M** for a Markov chain is a stochastic matrix whose (i, j) entry is the probability that an element in state S_j will move to state S_i during the next step of the process.

The next theorem is easily proved by induction (see Exercise 10).

THEOREM 8.4

Let $\mathbf{p}$ be the (current) probability vector and $\mathbf{M}$ be the transition matrix for a Markov chain. After n steps in the process, where $n \geq 1$, the (new) probability vector is given by $\mathbf{p}_n = \mathbf{M}^n\mathbf{p}$.

Theorem 8.4 asserts that once the initial probability vector $\mathbf{p}$ and the transition matrix $\mathbf{M}$ for a Markov chain are known, all future steps of the Markov chain are determined.

Steady-State Vectors for Markov Chains

A natural question to ask about a given Markov chain is whether we can discern any long-term trend.

Example 2

Consider the Markov chain from Example 1, with transition matrix

$$\mathbf{M} = \begin{bmatrix} .500 & .167 & .250 \\ .250 & .667 & .250 \\ .250 & .167 & .500 \end{bmatrix}.$$

What happens in the long run? It turns out that, as k gets larger, $\mathbf{M}^k$ approaches[†] the matrix

$$\mathbf{W} = \begin{bmatrix} .286 & .286 & .286 \\ .429 & .429 & .429 \\ .286 & .286 & .286 \end{bmatrix},$$

where we are again rounding to three decimal places.[‡] This matrix $\mathbf{W}$ can be obtained by multiplying out higher powers of $\mathbf{M}$ until successive powers agree to the desired number of decimal places. (A computer is extremely useful here.) The new probability vector, which we denote by $\mathbf{p}_f$, is then found by

$$\mathbf{p}_f = \mathbf{W}\mathbf{p} = \begin{bmatrix} .286 & .286 & .286 \\ .429 & .429 & .429 \\ .286 & .286 & .286 \end{bmatrix} \begin{bmatrix} .4 \\ .1 \\ .5 \end{bmatrix} = \begin{bmatrix} .286 \\ .429 \\ .286 \end{bmatrix}.$$

Hence, ultimately, Banks A and C will each capture 28.6%, or $\frac{2}{7}$, of the investors, and Bank B will capture 42.9%, or $\frac{3}{7}$, of the investors. The vector $\mathbf{p}_f$ is called the **steady-state vector** of the Markov chain. ■

We now give a formal definition for a steady-state vector of a Markov chain.

DEFINITION

Let $\mathbf{M}$ be the transition matrix, and let $\mathbf{p}$ be the current probability vector for a Markov chain. Let $\mathbf{p}_k$ represent the probability vector after k steps of the Markov chain. If the sequence $\mathbf{p}, \mathbf{p}_1, \mathbf{p}_2, \ldots$ of vectors approaches some vector $\mathbf{p}_f$, then $\mathbf{p}_f$ is called a **steady-state vector** for the Markov chain.

[†] The intuitive concept of a sequence of matrices approaching a matrix can be defined precisely using limits. We say that $\lim_{k\to\infty} \mathbf{M}^k = \mathbf{F}$ if and only if the largest of the absolute values of the differences between the corresponding entries of $\mathbf{M}^k$ and $\mathbf{F}$ approaches zero as k grows larger. That is, $\lim_{k\to\infty} \mathbf{M}^k = \mathbf{F}$ if and only if $\lim_{k\to\infty} \max |m_{ij}^{(k)} - f_{ij}| = 0$, where $m_{ij}^{(k)}$ is used to denote the (i, j) entry of $\mathbf{M}^k$.

[‡] When raising matrices, such as $\mathbf{M}$, to high powers, roundoff error can quickly compound. Although we have printed $\mathbf{M}$ rounded to 3 significant digits, we actually performed the computations using $\mathbf{M}$ rounded to 12 digits of accuracy. In general, minimize your roundoff error by using as many digits as your calculator or software will provide.

The computation of large powers of the transition matrix **M** is not always an easy task, even with the use of a computer. We now show a quicker method to obtain the steady-state vector $\mathbf{p}_f$ for the Markov chain of Example 2. Notice that this vector $\mathbf{p}_f$ has the property that

$$\mathbf{M}\mathbf{p}_f = \begin{bmatrix} .500 & .167 & .250 \\ .250 & .667 & .250 \\ .250 & .167 & .500 \end{bmatrix}\begin{bmatrix} .286 \\ .429 \\ .286 \end{bmatrix} = \begin{bmatrix} .286 \\ .429 \\ .286 \end{bmatrix} = \mathbf{p}_f.$$

This remarkable property says that $\mathbf{p}_f$ is a vector that satisfies the equation $\mathbf{Mx} = \mathbf{x}$. Therefore, if we did not know $\mathbf{p}_f$, we could solve the equation

$$\mathbf{M}\begin{bmatrix} x_1 \\ x_2 \\ x_3 \end{bmatrix} = \begin{bmatrix} x_1 \\ x_2 \\ x_3 \end{bmatrix}$$

to find it. We can rewrite this as

$$\mathbf{M}\begin{bmatrix} x_1 \\ x_2 \\ x_3 \end{bmatrix} - \begin{bmatrix} x_1 \\ x_2 \\ x_3 \end{bmatrix} = \begin{bmatrix} 0 \\ 0 \\ 0 \end{bmatrix}, \quad \text{or} \quad (\mathbf{M} - \mathbf{I}_3)\begin{bmatrix} x_1 \\ x_2 \\ x_3 \end{bmatrix} = \begin{bmatrix} 0 \\ 0 \\ 0 \end{bmatrix}.$$

The augmented matrix for this system is

$$\left[\begin{array}{ccc|c} .500-1 & .167 & .250 & 0 \\ .250 & .667-1 & .250 & 0 \\ .250 & .167 & .500-1 & 0 \end{array}\right] = \left[\begin{array}{ccc|c} -.500 & .167 & .250 & 0 \\ .250 & -.333 & .250 & 0 \\ .250 & .167 & -.500 & 0 \end{array}\right].$$

We can also add another condition, since we know that $x_1 + x_2 + x_3 = 1$. Thus, the augmented matrix gets a fourth row:

$$\left[\begin{array}{ccc|c} -.500 & .167 & .250 & 0 \\ .250 & -.333 & .250 & 0 \\ .250 & .167 & -.500 & 0 \\ 1.000 & 1.000 & 1.000 & 1 \end{array}\right].$$

After row reducing, we find that the solution set is $x_1 = .286$, $x_2 = .429$, and $x_3 = .286$, as expected.

This shortcut is especially useful in situations where we know that the initial state converges to a steady-state vector, as we will see later. However, we must be careful, because the current state in a given Markov chain may not converge to a steady state, as the next example shows.

Example 3

Suppose that W, X, Y, and Z represent four train stations linked as shown in Figure 8.12.

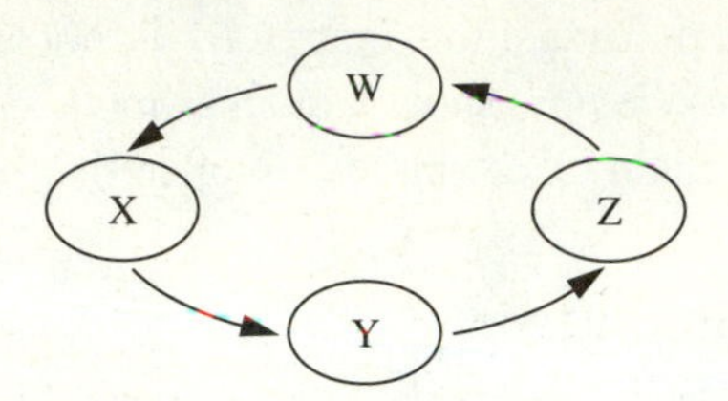

Figure 8.12
Four train stations

Suppose that ten trains shuttle between these stations. Currently, there are four trains at station W, three trains at station X, one train at station Y, and two trains at station Z. The probability that a randomly chosen train is at each station is given by the probability vector

$$\mathbf{p} = \begin{matrix} W \\ X \\ Y \\ Z \end{matrix} \begin{bmatrix} .4 \\ .3 \\ .1 \\ .2 \end{bmatrix}.$$

Suppose that each train moves to the next station in Figure 8.12 each hour. Then we have a Markov chain whose transition matrix is

$$\mathbf{M} = \text{Next state} \begin{matrix} \\ W \\ X \\ Y \\ Z \end{matrix} \begin{matrix} \text{Current state} \\ \begin{matrix} W & X & Y & Z \end{matrix} \\ \begin{bmatrix} 0 & 0 & 0 & 1 \\ 1 & 0 & 0 & 0 \\ 0 & 1 & 0 & 0 \\ 0 & 0 & 1 & 0 \end{bmatrix} \end{matrix}.$$

Intuitively, we can see there is no steady-state vector for this system, since the number of trains in each station never settles down to a fixed number but keeps rising and falling as the trains go around the "loop." This notion is borne out when we consider that the powers of the transition matrix are

$$\mathbf{M}^2 = \begin{bmatrix} 0 & 0 & 1 & 0 \\ 0 & 0 & 0 & 1 \\ 1 & 0 & 0 & 0 \\ 0 & 1 & 0 & 0 \end{bmatrix}, \quad \mathbf{M}^3 = \begin{bmatrix} 0 & 1 & 0 & 0 \\ 0 & 0 & 1 & 0 \\ 0 & 0 & 0 & 1 \\ 1 & 0 & 0 & 0 \end{bmatrix}, \quad \text{and } \mathbf{M}^4 = \mathbf{I}_4 = \begin{bmatrix} 1 & 0 & 0 & 0 \\ 0 & 1 & 0 & 0 \\ 0 & 0 & 1 & 0 \\ 0 & 0 & 0 & 1 \end{bmatrix}.$$

Since $\mathbf{M}^4 = \mathbf{I}_4$, all higher powers of $\mathbf{M}$ are equal to $\mathbf{M}$, $\mathbf{M}^2$, $\mathbf{M}^3$, or $\mathbf{I}_4$. (Why?) Therefore, the only probability vectors produced by this Markov chain are $\mathbf{p}$,

$$\mathbf{p}_1 = \mathbf{M}\mathbf{p} = \begin{bmatrix} .2 \\ .4 \\ .3 \\ .1 \end{bmatrix}, \quad \mathbf{p}_2 = \mathbf{M}^2\mathbf{p} = \begin{bmatrix} .1 \\ .2 \\ .4 \\ .3 \end{bmatrix}, \quad \text{and} \quad \mathbf{p}_3 = \mathbf{M}^3\mathbf{p} = \begin{bmatrix} .3 \\ .1 \\ .2 \\ .4 \end{bmatrix},$$

since $\mathbf{p}_4 = \mathbf{M}^4\mathbf{p} = \mathbf{I}_4\mathbf{p} = \mathbf{p}$ again. Since $\mathbf{p}_k$ keeps changing to one of four distinct vectors, the initial state $\mathbf{p}$ does not converge to a steady state. ■

Regular Transition Matrices

DEFINITION

A square matrix $\mathbf{R}$ is **regular** if and only if $\mathbf{R}$ is a stochastic matrix and some power $\mathbf{R}^k$, for $k \geq 1$, has all entries positive (nonzero).

Example 4

The transition matrix $\mathbf{M}$ in Example 1 is a regular matrix, since $\mathbf{M}^1 = \mathbf{M}$ has all entries positive. However, the transition matrix $\mathbf{M}$ in Example 3 is not regular, because, as you saw in that example, all positive powers of $\mathbf{M}$ are equal to one of four matrices, each containing zero entries. Finally,

$$\mathbf{R} = \begin{bmatrix} 0 & \frac{1}{2} & 0 \\ \frac{1}{2} & 0 & 1 \\ \frac{1}{2} & \frac{1}{2} & 0 \end{bmatrix}$$

is regular since it is stochastic and since

$$\mathbf{R}^4 = (\mathbf{R}^2)^2 = \left(\begin{bmatrix} \frac{1}{4} & 0 & \frac{1}{2} \\ \frac{1}{2} & \frac{3}{4} & 0 \\ \frac{1}{4} & \frac{1}{4} & \frac{1}{2} \end{bmatrix}\right)^2 = \begin{bmatrix} \frac{3}{16} & \frac{1}{8} & \frac{3}{8} \\ \frac{1}{2} & \frac{9}{16} & \frac{1}{4} \\ \frac{5}{16} & \frac{5}{16} & \frac{3}{8} \end{bmatrix},$$

which has all entries positive. ■

The next theorem, stated without proof, shows that Markov chains with regular transition matrices always have a steady-state vector $\mathbf{p}_f$ for every choice of an initial probability vector $\mathbf{p}$.

THEOREM 8.5

If $\mathbf{R}$ is a regular $n \times n$ transition matrix for a Markov chain, then

(1) $\lim_{k\to\infty} \mathbf{R}^k$ exists.
(2) If $\mathbf{W} = \lim_{k\to\infty} \mathbf{R}^k$, then $\mathbf{W}$ has all entries positive, and every column of $\mathbf{W}$ will be identical.
(3) For all initial probability vectors $\mathbf{p}$, the Markov chain has a steady-state vector $\mathbf{p}_f$. Also, the steady-state vector $\mathbf{p}_f$ is the same for all $\mathbf{p}$.

(4) $\mathbf{p}_f$ is equal to any of the identical columns of $\mathbf{W}$.
(5) $\mathbf{p}_f$ is the unique stochastic n-vector such that $\mathbf{R}\mathbf{p}_f = \mathbf{p}_f$.

Example 5

Consider a school of fish hunting for food in three adjoining lakes—L_1, L_2, and L_3. Each day, the fish select a different lake to hunt in than the previous day, with probabilities given in the transition matrix below:

$$\mathbf{M} = \text{Next day} \begin{matrix} \\ \\ L_1 \\ L_2 \\ L_3 \end{matrix} \overset{\text{Current day}}{\begin{matrix} L_1 & L_2 & L_3 \\ \end{matrix}} \begin{bmatrix} 0 & .5 & 0 \\ .5 & 0 & 1 \\ .5 & .5 & 0 \end{bmatrix}.$$

Can we determine what percentage of time the fish will spend in each lake in the long run?

Notice that $\mathbf{M}$ is equal to the matrix $\mathbf{R}$ in Example 4, and so $\mathbf{M}$ is regular. Theorem 8.5 asserts that this Markov chain has a steady-state vector. To find this vector, we use the technique described after Example 2, which involves solving the system

$$(\mathbf{M} - \mathbf{I}_3)\begin{bmatrix} x_1 \\ x_2 \\ x_3 \end{bmatrix} = \begin{bmatrix} -1 & .5 & 0 \\ .5 & -1 & 1 \\ .5 & .5 & -1 \end{bmatrix}\begin{bmatrix} x_1 \\ x_2 \\ x_3 \end{bmatrix} = \begin{bmatrix} 0 \\ 0 \\ 0 \end{bmatrix},$$

with the extra condition that $x_1 + x_2 + x_3 = 1$. The solution is $x_1 = .222$, $x_2 = .444$, and $x_3 = .333$. Therefore, in the long run, the fish will hunt in lake L_1 22.2%, or $\frac{2}{9}$, of the time, in lake L_2 44.4%, or $\frac{4}{9}$, of the time, and in lake L_3 33.3%, or $\frac{1}{3}$, of the time. ■

Notice that Example 5 was solved without even knowing the current probability vector $\mathbf{p}$. Notice also that the steady-state solution could have been found by calculating larger and larger powers of $\mathbf{M}$ to see that they converge to the matrix

$$\mathbf{W} = \begin{bmatrix} .222 & .222 & .222 \\ .444 & .444 & .444 \\ .333 & .333 & .333 \end{bmatrix}.$$

Each of the identical columns of $\mathbf{W}$ is the steady-state vector for this Markov chain.

Exercises—Section 8.4

Note: You should have a calculator or computer handy for the computations in many of these exercises.

★**1.** Which of the following matrices are stochastic? Which are regular? Why?

$$\mathbf{A} = \begin{bmatrix} \frac{1}{4} \\ \frac{1}{2} \\ \frac{1}{4} \end{bmatrix} \quad \mathbf{B} = \begin{bmatrix} .2 & .4 & .5 \\ .5 & .1 & .4 \\ .3 & .4 & .1 \end{bmatrix} \quad \mathbf{C} = \begin{bmatrix} \frac{1}{5} & \frac{2}{3} \\ \frac{4}{5} & \frac{1}{3} \end{bmatrix} \quad \mathbf{D} = \begin{bmatrix} 0 & 1 & 0 \\ 0 & 0 & 1 \\ 1 & 0 & 0 \end{bmatrix}$$

$$\mathbf{E} = \begin{bmatrix} \frac{1}{3} & \frac{2}{3} \\ \frac{1}{4} & \frac{3}{4} \end{bmatrix} \quad \mathbf{F} = \begin{bmatrix} \frac{1}{3} & \frac{1}{3} & 1 \\ 0 & 0 & 0 \\ \frac{2}{3} & \frac{2}{3} & 0 \end{bmatrix} \quad \mathbf{G} = \begin{bmatrix} \frac{1}{3} & 0 \\ 0 & \frac{2}{3} \\ \frac{2}{3} & \frac{1}{3} \end{bmatrix} \quad \mathbf{H} = \begin{bmatrix} \frac{1}{2} & 0 & \frac{1}{2} \\ 0 & \frac{1}{2} & \frac{1}{2} \\ \frac{1}{2} & \frac{1}{2} & 0 \end{bmatrix}$$

2. Suppose that each of the following represents the transition matrix $\mathbf{M}$ and the initial probability vector $\mathbf{p}$ for a Markov chain. Find the probability vectors $\mathbf{p}_1$ (after one step of the process) and $\mathbf{p}_2$ (after two steps).

★(a) $\mathbf{M} = \begin{bmatrix} \frac{1}{4} & \frac{1}{3} \\ \frac{3}{4} & \frac{2}{3} \end{bmatrix}, \quad \mathbf{p} = \begin{bmatrix} \frac{2}{3} \\ \frac{1}{3} \end{bmatrix}$

(b) $\mathbf{M} = \begin{bmatrix} \frac{1}{2} & \frac{1}{3} & 0 \\ 0 & \frac{2}{3} & \frac{1}{2} \\ \frac{1}{2} & 0 & \frac{1}{2} \end{bmatrix}, \quad \mathbf{p} = \begin{bmatrix} \frac{1}{3} \\ \frac{1}{6} \\ \frac{1}{2} \end{bmatrix}$

★(c) $\mathbf{M} = \begin{bmatrix} \frac{1}{4} & \frac{1}{3} & \frac{1}{2} \\ \frac{1}{2} & \frac{1}{3} & \frac{1}{6} \\ \frac{1}{4} & \frac{1}{3} & \frac{1}{3} \end{bmatrix}, \quad \mathbf{p} = \begin{bmatrix} \frac{1}{4} \\ \frac{1}{2} \\ \frac{1}{4} \end{bmatrix}$

3. Suppose that each of the following regular matrices represents the transition matrix $\mathbf{M}$ for a Markov chain. Find the steady-state solution for the Markov chain by solving the appropriate system of linear equations.

★(a) $\begin{bmatrix} \frac{1}{2} & \frac{1}{3} \\ \frac{1}{2} & \frac{2}{3} \end{bmatrix}$ (b) $\begin{bmatrix} \frac{1}{3} & \frac{1}{4} & \frac{1}{3} \\ \frac{1}{6} & \frac{1}{2} & \frac{1}{3} \\ \frac{1}{2} & \frac{1}{4} & \frac{1}{3} \end{bmatrix}$ (c) $\begin{bmatrix} \frac{1}{5} & \frac{1}{2} & 0 & \frac{1}{3} \\ \frac{3}{5} & 0 & \frac{1}{2} & 0 \\ 0 & \frac{1}{2} & \frac{1}{2} & 0 \\ \frac{1}{5} & 0 & 0 & \frac{2}{3} \end{bmatrix}$

4. Find the steady-state solution for each of the following (which are parts (a) and (b) of Exercise 3) by calculating large powers of the transition matrix (using a computer or calculator).

(a) $\begin{bmatrix} \frac{1}{2} & \frac{1}{3} \\ \frac{1}{2} & \frac{2}{3} \end{bmatrix}$ (b) $\begin{bmatrix} \frac{1}{3} & \frac{1}{4} & \frac{1}{3} \\ \frac{1}{6} & \frac{1}{2} & \frac{1}{3} \\ \frac{1}{2} & \frac{1}{4} & \frac{1}{3} \end{bmatrix}$

★**5.** Suppose that the citizens in a certain community tend to switch their votes among political parties, as shown in the following transition matrix:

		Current election			
		Party A	Party B	Party C	Nonvoting
Next election	Party A	.7	.2	.2	.1
	Party B	.1	.6	.1	.1
	Party C	.1	.2	.6	.1
	Nonvoting	.1	0	.1	.7

(a) Suppose that in the last election 30% of the citizens voted for Party A, 15% voted for Party B, and 45% voted for Party C. What is the likely outcome of the next election? the election after that?

(b) If current trends continue, what percentage of the citizens will vote for Party A one century from now? Party C?

★**6.** In a psychology experiment, a rat wanders in the maze in Figure 8.13. During each time interval, the rat is allowed to pass through exactly one doorway. Assume there is a 50% probability that the rat will switch rooms during an interval. If it does switch rooms, assume that it has an equally likely chance of using any doorway out of its current room.

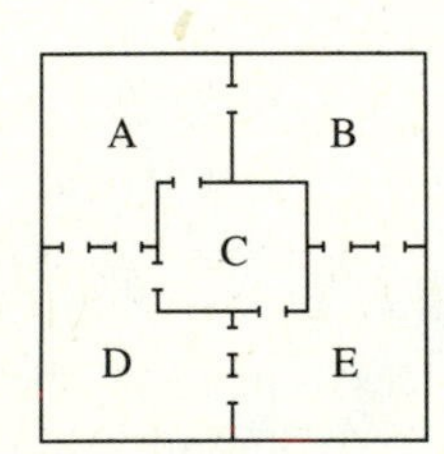

Figure 8.13
Maze with five rooms

(a) What is the transition matrix for this Markov chain?

(b) If the rat is known to be in room C, what is the probability it will be in room D after two time intervals have passed?

(c) What is the steady-state solution for this Markov chain? Over time, which room does the rat frequent the least? the most?

7. Show that the converse to part (3) of Theorem 8.5 is not true by demonstrating that the transition matrix

$$\mathbf{M} = \begin{bmatrix} 1 & \frac{1}{2} & \frac{1}{4} \\ 0 & \frac{1}{2} & \frac{1}{4} \\ 0 & 0 & \frac{1}{2} \end{bmatrix}$$

has the same steady-state solution for any initial input but is not regular.

8. (a) Show that the transition matrix $\begin{bmatrix} 1-a & b \\ a & 1-b \end{bmatrix}$ has $1/(a+b)\begin{bmatrix} b \\ a \end{bmatrix}$ as a steady-state solution if neither a nor b equals zero or 1.

(b) Use the result in part (a) to check that your answer for part (a) of Exercise 3 is correct.

9. Prove Theorem 8.3.

10. Prove Theorem 8.4.

8.5 Hill Substitution: An Introduction to Coding Theory

Prerequisite: Section 2.4, Inverses of Matrices

In this section we show how matrix inverses can be used in a simple manner to encode and decode textual information.

Substitution Ciphers

The coding and decoding of secret messages has long been important in times of warfare, of course, but it is also quite valuable in peacetime for keeping government and business secrets under tight security. Throughout history, many ingenious coding mechanisms have been proposed. One of the simplest is the **substitution cipher**, in which an array of nonrepeating symbols is used to assign each character of a given text (**plaintext**) to a corresponding character in coded text (**ciphertext**). For example, consider the **cipher array** in Figure 8.14. A message can be enciphered by replacing every instance of the kth letter of the alphabet with the kth character in the cipher array, for $1 \le k \le 26$. For example, the message

LINEAR ALGEBRA IS EXCITING

is encoded as

FXUSRI RFTSWIR XG SNEXVXUT.

This type of substitution can be extended to other characters, such as punctuation symbols and blanks.

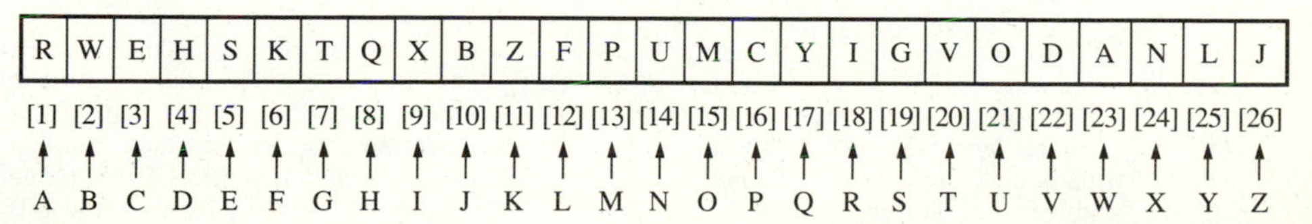

Figure 8.14
A cipher array

Messages can be decoded by reversing the process. In fact, we can create an "inverse" array, or **decipher array**, as in Figure 8.15 to restore the symbols of FXUSRI RFTSWIR XG SNEXVXUT back to LINEAR ALGEBRA IS EXCITING.

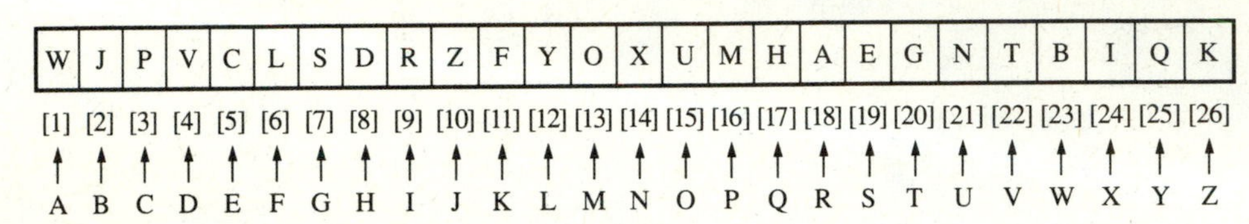

Figure 8.15
A decipher array

Cryptograms, a standard feature in newspapers and puzzle magazines, are substitution ciphers. However, these ciphers are relatively easy to "crack" because the relative frequencies (occurrences per length of text) of the letters of the English alphabet have been studied extensively.†

Hill Substitution

We now illustrate a method that uses matrices to create codes that are harder to break. This technique is known as **Hill substitution** after the mathematician Lester Hill, who developed it between the world wars. The method is relatively simple and can easily be implemented on a computer. To begin, we choose any nonsingular $n \times n$ matrix $\mathbf{A}$. (Usually $\mathbf{A}$ is chosen with integer entries.) We split the message into blocks of n symbols each and replace each symbol with an integer value. To simplify things, we replace each letter by its position in the alphabet. The last block may have to be "padded" with random values to ensure that each block contains exactly n integers. In effect, we are creating a set of n-vectors that we can label as $\mathbf{x}_1, \mathbf{x}_2$, and so on. We then multiply the matrix $\mathbf{A}$ by each one of these vectors in turn to produce a new set of n-vectors: $\mathbf{Ax}_1, \mathbf{Ax}_2$, and so on. When these vectors are concatenated together, they form the coded message. The matrix $\mathbf{A}$ used in the process is often called the **encoding matrix**, or the **key matrix**.

---

†The longer the enciphered text is, the easier it is to decode by comparing the number of times each letter appears. The actual frequency of the letters depends on the type of text, but the letters *E*, *T*, *A*, *O*, *I*, *N*, *S*, *H*, and *R* typically appear most often (about 70% of the time), with *E* usually the most common (about 12–13% of the time). Once a few letters have been deciphered, the rest of the text is usually easy to determine. Sample frequency tables can be found on p. 219 of *Cryptanalysis* by Gaines (published by Dover, 1956) and on p. 16 of *Cryptography: A Primer* by Konheim (published by Wiley, 1981).

Example 1

Suppose we wish to encode the message LINEAR ALGEBRA IS EXCITING using the key matrix

$$\mathbf{A} = \begin{bmatrix} -7 & 5 & 3 \\ 3 & -2 & -2 \\ 3 & -2 & -1 \end{bmatrix}.$$

Since we are using a 3×3 matrix, we break the characters of the message into blocks of length 3 and replace each character by its position in the alphabet. This procedure gives:

$$\underbrace{\begin{matrix} \text{L} \\ \text{I} \\ \text{N} \end{matrix}\begin{bmatrix} 12 \\ 9 \\ 14 \end{bmatrix}}_{\mathbf{x}_1}, \quad \underbrace{\begin{matrix} \text{E} \\ \text{A} \\ \text{R} \end{matrix}\begin{bmatrix} 5 \\ 1 \\ 18 \end{bmatrix}}_{\mathbf{x}_2}, \quad \underbrace{\begin{matrix} \text{A} \\ \text{L} \\ \text{G} \end{matrix}\begin{bmatrix} 1 \\ 12 \\ 7 \end{bmatrix}}_{\mathbf{x}_3}, \quad \underbrace{\begin{matrix} \text{E} \\ \text{B} \\ \text{R} \end{matrix}\begin{bmatrix} 5 \\ 2 \\ 18 \end{bmatrix}}_{\mathbf{x}_4},$$

$$\underbrace{\begin{matrix} \text{A} \\ \text{I} \\ \text{S} \end{matrix}\begin{bmatrix} 1 \\ 9 \\ 19 \end{bmatrix}}_{\mathbf{x}_5}, \quad \underbrace{\begin{matrix} \text{E} \\ \text{X} \\ \text{C} \end{matrix}\begin{bmatrix} 5 \\ 24 \\ 3 \end{bmatrix}}_{\mathbf{x}_6}, \quad \underbrace{\begin{matrix} \text{I} \\ \text{T} \\ \text{I} \end{matrix}\begin{bmatrix} 9 \\ 20 \\ 9 \end{bmatrix}}_{\mathbf{x}_7}, \quad \underbrace{\begin{matrix} \text{N} \\ \text{G} \\ - \end{matrix}\begin{bmatrix} 14 \\ 7 \\ 27 \end{bmatrix}}_{\mathbf{x}_8},$$

where the last entry of the last vector was chosen outside the range from 1 to 26. Now, forming the products with $\mathbf{A}$, we have

$$\mathbf{A}\mathbf{x}_1 = \begin{bmatrix} -7 & 5 & 3 \\ 3 & -2 & -2 \\ 3 & -2 & -1 \end{bmatrix}\begin{bmatrix} 12 \\ 9 \\ 14 \end{bmatrix} = \begin{bmatrix} 3 \\ -10 \\ 4 \end{bmatrix},$$

$$\mathbf{A}\mathbf{x}_2 = \begin{bmatrix} -7 & 5 & 3 \\ 3 & -2 & -2 \\ 3 & -2 & -1 \end{bmatrix}\begin{bmatrix} 5 \\ 1 \\ 18 \end{bmatrix} = \begin{bmatrix} 24 \\ -23 \\ -5 \end{bmatrix}, \quad \text{and so on.}$$

The final encoded text is

$$\begin{matrix} 3 & -10 & 4 & 24 & -23 & -5 & 74 & -35 & -28 & 29 & -25 & -7 \\ 95 & -53 & -34 & 94 & -39 & -36 & 64 & -31 & -22 & 18 & -26 & 1. \end{matrix}$$ ■

The code produced by a Hill substitution is much harder to break than a simple substitution cipher, since the coding of a given letter depends not only on the way the text is broken into blocks but also on the letters adjacent to it. (Nevertheless, there are techniques to decode Hill substitutions using high-speed computers.) However, a Hill substitution is easy to decode if you know the inverse of the key matrix. In Example 6 of Section 2.4, we noted that

$$\mathbf{A}^{-1} = \begin{bmatrix} 2 & 1 & 4 \\ 3 & 2 & 5 \\ 0 & -1 & 1 \end{bmatrix}.$$

Breaking the encoded text back into 3-vectors and multiplying $\mathbf{A}^{-1}$ times each of these vectors in turn restores the original message. For example,

$$\mathbf{A}^{-1}(\mathbf{A}\mathbf{x}_1) = \begin{bmatrix} 2 & 1 & 4 \\ 3 & 2 & 5 \\ 0 & -1 & 1 \end{bmatrix}\begin{bmatrix} 3 \\ -10 \\ 4 \end{bmatrix} = \begin{bmatrix} 12 \\ 9 \\ 14 \end{bmatrix} = \mathbf{x}_1,$$

which represents the first three letters LIN.

Exercises—Section 8.5

1. Encode each message with the given key matrix:

★(a) PROOF BY INDUCTION using the matrix $\begin{bmatrix} 3 & -4 \\ 5 & -7 \end{bmatrix}$

(b) CONTACT HEADQUARTERS using the matrix $\begin{bmatrix} 4 & 1 & 5 \\ 7 & 2 & 9 \\ 6 & 2 & 7 \end{bmatrix}$

2. Each of the following coded messages was produced with the key matrix shown. In each case, find the inverse of the key matrix, and use it to decode the message:

★(a)

−62 116 107 −32 59 67 −142 266 223 −160 301 251
−122 229 188 −122 229 202 −78 148 129 −111 207 183

with key matrix $\begin{bmatrix} -8 & 1 & -1 \\ 15 & -2 & 2 \\ 12 & -1 & 2 \end{bmatrix}$

(b) 162 108 23 303 206 33 276 186 33 170 116 21
281 191 36 576 395 67 430 292 51 340 232 45

with key matrix $\begin{bmatrix} -10 & 19 & 16 \\ -7 & 13 & 11 \\ -1 & 2 & 2 \end{bmatrix}$

(c) 69 44 −28 −43 104 53 −38 −25
71 38 −3 −7 58 32 −11 −14

with key matrix $\begin{bmatrix} 1 & 2 & 5 & 1 \\ 0 & 1 & 3 & 1 \\ -2 & 0 & 0 & -1 \\ 0 & 0 & -1 & -2 \end{bmatrix}$

8.6 Function Spaces

Prerequisite: Section 4.7, Coordinatization

In this section, we apply the techniques of Chapter 4 to vector spaces whose elements are functions. The vector spaces $\mathcal{P}_n$ and $\mathcal{P}$ are familiar examples of such spaces. Other important examples are $C^0(\mathbb{R})$ = {all continuous real-valued functions on $\mathbb{R}$} and $C^1(\mathbb{R})$ = {all continuously differentiable real-valued functions on $\mathbb{R}$}.

Linear Independence in Function Spaces

Proving that a finite subset S of a function space is linearly independent usually requires different strategies from those used in $\mathbb{R}^n$.

Example 1

Consider the subset $S = \{x^3 - x, xe^{-x^2}, \sin(\frac{\pi}{2}x)\}$ of $C^1(\mathbb{R})$. We will show that S is linearly independent using Theorem 4.7. Let a, b, and c be real numbers such that

$$a(x^3 - x) + b(xe^{-x^2}) + c\left(\sin\left(\frac{\pi}{2}x\right)\right) = 0$$

for every value of x. We must show that $a = b = c = 0$.

The above equation must be satisfied for every value of x so that, in particular, it is true for $x = 1$, $x = 2$, and $x = 3$. This yields the following system:

$$\begin{cases} (\text{Letting } x = 1 \longrightarrow) & a(0) + b\left(\frac{1}{e}\right) + c(1) = 0 \\ (\text{Letting } x = 2 \longrightarrow) & a(6) + b\left(\frac{2}{e^4}\right) + c(0) = 0 \\ (\text{Letting } x = 3 \longrightarrow) & a(24) + b\left(\frac{3}{e^9}\right) + c(-1) = 0 \end{cases}$$

Row reducing the matrix

$$\begin{array}{c} \begin{matrix} a & b & c & \end{matrix} \\ \left[\begin{array}{ccc|c} 0 & \frac{1}{e} & 1 & 0 \\ 6 & \frac{2}{e^4} & 0 & 0 \\ 24 & \frac{3}{e^9} & -1 & 0 \end{array}\right] \end{array} \quad \text{to} \quad \begin{array}{c} \begin{matrix} a & b & c & \end{matrix} \\ \left[\begin{array}{ccc|c} 1 & 0 & 0 & 0 \\ 0 & 1 & 0 & 0 \\ 0 & 0 & 1 & 0 \end{array}\right] \end{array}$$

shows that the trivial solution $a = b = c = 0$ is the only solution to this homogeneous system. Hence, the set S is linearly independent by Theorem 4.7. ■

When proving linear independence using the technique of Example 1, we try to choose "nice" values of x to make computations easier. Even so, the use of a computer software package is often desirable when working with function spaces.

Other problems may also occur because of the choice of x-values. For instance, in Example 1, if instead we had plugged in $x = -1$, $x = 0$, and $x = 1$, we would have obtained the system

$$\begin{cases} (x = -1 \longrightarrow) & a(0) + b\left(-\frac{1}{e}\right) + c(-1) = 0 \\ (x = 0 \longrightarrow) & a(0) + b(0) + c(0) = 0 \\ (x = 1 \longrightarrow) & a(0) + b\left(\frac{1}{e}\right) + c(1) = 0 \end{cases}$$

which has infinitely many nontrivial solutions. To prove linear independence, we must examine further values of x, generating more equations for the system, until the new system we obtain has only the trivial solution, as in Example 1.

Suppose, however, that after substituting many values for x and creating a huge homogeneous system we still have nontrivial solutions. We cannot conclude that the set of functions is linearly dependent, although we may suspect that it is. In general, to *prove* that a set of functions $\{\mathbf{f}_1, \ldots, \mathbf{f}_n\}$ is linearly dependent, we must find real numbers $a_1, \ldots, a_n$, not all zero, such that

$$a_1\mathbf{f}_1(x) + a_2\mathbf{f}_2(x) + \cdots + a_n\mathbf{f}_n(x) = 0$$

is a functional identity for every value of x, not just those we have used.

Example 2

Let $S = \{\sin 2x, \cos 2x, \sin^2 x, \cos^2 x\}$, a subset of $C^1(\mathbb{R})$. Suppose we attempt to show that S is linearly independent using Theorem 4.7. Let a, b, c, and d represent real numbers such that

$$a(\sin 2x) + b(\cos 2x) + c(\sin^2 x) + d(\cos^2 x) = 0.$$

Since we have four vectors in S, we substitute four different values for x into this equation to obtain the following system:

$$\begin{cases} (x = 0 \longrightarrow) & a(0) + b(1) + c(0) + d(1) = 0 \\ \left(x = \frac{\pi}{4} \longrightarrow\right) & a(1) + b(0) + c\left(\frac{1}{2}\right) + d\left(\frac{1}{2}\right) = 0 \\ \left(x = \frac{\pi}{2} \longrightarrow\right) & a(0) + b(-1) + c(1) + d(0) = 0 \\ \left(x = \frac{3\pi}{4} \longrightarrow\right) & a(-1) + b(0) + c\left(\frac{1}{2}\right) + d\left(\frac{1}{2}\right) = 0 \end{cases}$$

Since the augmented matrix for this system row reduces to

$$\begin{array}{c} \begin{matrix} a & b & c & d \end{matrix} \\ \left[\begin{array}{cccc|c} 1 & 0 & 0 & 0 & 0 \\ 0 & 1 & 0 & 1 & 0 \\ 0 & 0 & 1 & 1 & 0 \\ 0 & 0 & 0 & 0 & 0 \end{array}\right], \end{array}$$

there are nontrivial solutions to the system, such as $a = 0$, $b = -1$, $c = -1$, $d = 1$.

At this point, we cannot infer that S is linearly independent, because we still have nontrivial solutions. We also cannot conclude that S is linearly dependent, because we have tested only a few values for x. We could try more values, such as $x = \frac{\pi}{6}$ and $x = \pi$, but we would still find that $a = 0$, $b = -1$, $c = -1$, $d = 1$ satisfies each equation we generate. This situation leads us to believe that the set S is linearly dependent. To be certain, we must check that the values $a = 0$, $b = -1$, $c = -1$, and $d = 1$ yield a functional identity when plugged into the original functional equation. Substituting these values yields

$$0(\sin 2x) + (-1)(\cos 2x) + (-1)(\sin^2 x) + (1)(\cos^2 x) = 0,$$

or $\cos 2x = \cos^2 x - \sin^2 x$, a well-known trigonometric identity. Thus, one vector in S can be expressed as a linear combination of the other vectors in S, and S is linearly dependent. ■

Exercises—Section 8.6

1. In each part of this exercise, determine whether the given subset S of $C^1(\mathbb{R})$ is linearly independent. If S is linearly independent, prove that it is. If S is linearly dependent, solve for a functional identity that expresses one function in S as a linear combination of the others.

★(a) $S = \{e^x, e^{2x}, e^{3x}\}$

(b) $S = \{\sin x, \sin 2x, \sin 3x, \sin 4x\}$

★(c) $S = \{(5x - 1)/(1 + x^2), (3x + 1)/(2 + x^2), (7x^3 - 3x^2 + 17x - 5)/(x^4 + 3x^2 + 2)\}$

(d) $S = \{\sin x, \sin(x + 1), \sin(x + 2), \sin(x + 3)\}$

2. Recall that a function $\mathbf{f}(x) \in C^0(\mathbb{R})$ is **even** if $\mathbf{f}(x) = \mathbf{f}(-x)$ for all $x \in \mathbb{R}$ and is **odd** if $\mathbf{f}(x) = -\mathbf{f}(-x)$ for all $x \in \mathbb{R}$. Suppose we want to prove that a finite subset S of $C^0(\mathbb{R})$ is linearly independent by the method of Example 1.

(a) Suppose that every element of S is an odd function of x (as in Example 1). Show why we would not want to substitute both 1 and -1 for x into the appropriate functional equation. Also explain why $x = 0$ would be a poor choice.

(b) Suppose that every element of S is an even function. Would we want to substitute both 1 and -1 for x into the appropriate functional equation? Why or why not? What about $x = 0$?

3. Let S be the subset $\{\cos(x+1), \cos(x+2), \cos(x+3)\}$ of $C^1(\mathbb{R})$.

(a) Show that span(S) has $\{\cos x, \sin x\}$ for a basis. (Hint: The identity $\cos(\alpha+\beta) = \cos\alpha\cos\beta - \sin\alpha\sin\beta$ is useful.)

(b) Use part (a) to prove that S is linearly dependent.

★4. The theory of differential equations shows that the set of solutions in $C^2(\mathbb{R})$ (all real-valued functions on $\mathbb{R}$ whose second derivative is continuous everywhere) for the homogeneous differential equation $y'' + 2y' - 15y = 0$ is a finite dimensional subspace of $C^2(\mathbb{R})$. This subspace $\mathcal{W}$ has a basis B consisting of functions with the form $\mathbf{f}(x) = e^{ax}$. Find the values of a that make $\mathbf{f}$ satisfy the differential equation. Show that the resulting set of functions is linearly independent. Assuming that this set spans $\mathcal{W}$, and hence is a basis for $\mathcal{W}$, find $\dim(\mathcal{W})$.

5. For each given subset S of $C^1(\mathbb{R})$, find a subset B of S that is a basis for $\mathcal{V} = \text{span}(S)$.

★(a) $S = \{\sin 2x, \cos 2x, \sin^2 x, \cos^2 x, \sin x\cos x, 1\}$

(b) $S = \{e^x, 1, e^{-x}\}$

★(c) $S = \{\sin(x+1), \cos(x+1), \sin(x+2), \cos(x+2)\}$

6. In each part of this exercise, let B represent an ordered basis for a subspace $\mathcal{V}$ of $C^1(\mathbb{R})$, and find $[\mathbf{v}]_B$ for the given $\mathbf{v} \in \mathcal{V}$.

★(a) $B = (e^x, e^{2x}, e^{3x})$, $\mathbf{v} = 5e^x - 7e^{3x}$

(b) $B = (\sin 2x, \cos 2x, \sin^2 x)$, $\mathbf{v} = 1$

★(c) $B = (\sin(x+1), \sin(x+2))$, $\mathbf{v} = \cos x$

8.7 Rotation of Axes

Prerequisite: Section 4.7, Coordinatization

Often in mathematics we can replace a complicated equation with a much simpler one by moving to a different coordinate system. In this section, we show how to use rotation of the coordinate axes to simplify the equations of conic sections.

Simplifying the Equation of a Conic Section

The general form for the equation of a conic section in the xy-plane is

$$ax^2 + by^2 + cxy + dx + ey + f = 0.$$

If $c \neq 0$, the term cxy in this equation imposes a rotation on the graph of the conic section, putting its axes of symmetry on a slant. Our goal is to change to a different set of coordinates in $\mathbb{R}^2$ so that the equation of the conic expressed in these coordinates will no longer have an xy term. This will make the axes of symmetry horizontal and/or vertical with respect to the new coordinate axes.[†]

Let θ be the angle between the positive x-axis and the axis of symmetry of the conic section. We will construct a new coordinate grid by rotating the old coordinate grid counterclockwise through the angle θ. This is illustrated in Figure 8.16 for the hyperbola $xy = 1$.

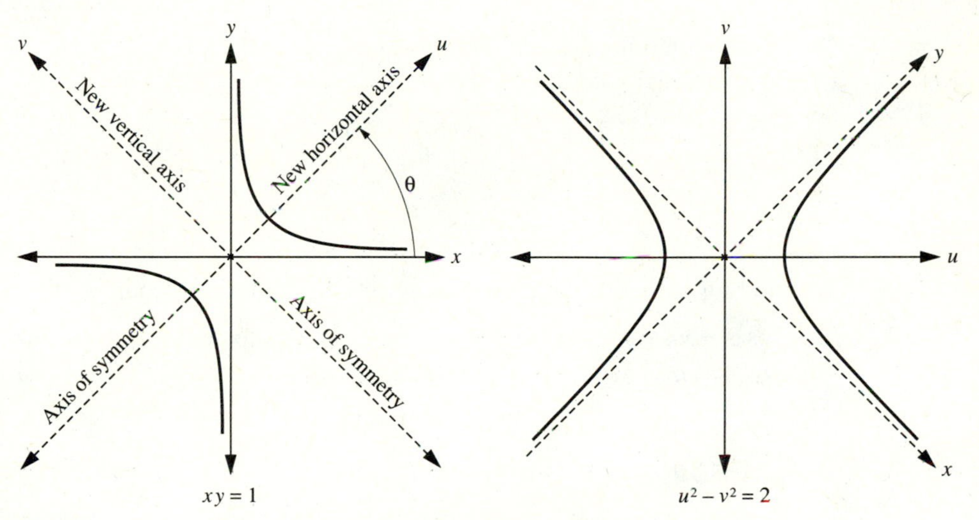

Figure 8.16
Rotation of the hyperbola $xy = 1$

Notice that, in converting from one set of coordinates to another, the standard basis $S = (\mathbf{i}, \mathbf{j})$ is replaced by a new basis $B = ([\cos\theta, \sin\theta], [-\sin\theta, \cos\theta])$ for $\mathbb{R}^2$ (see Figure 8.17). That is, a counterclockwise rotation of the axes through the

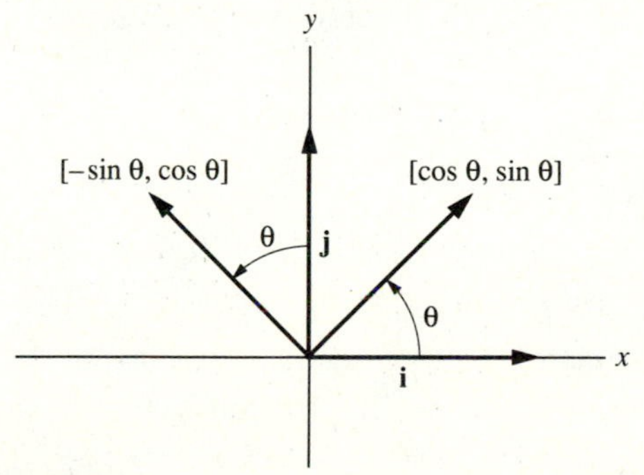

Figure 8.17
Vectors that replace the standard basis vectors in $\mathbb{R}^2$ after a counterclockwise rotation through the angle θ

[†]Actually, if the conic section is a circle, there is an axis of symmetry in every direction. However, the equation of a circle never contains an xy term.

angle θ is equivalent to introducing a new coordinate system with basis B whose vectors are obtained by rotating $\mathbf{i}, \mathbf{j}$ counterclockwise through θ.

We will use the variables u and v to describe the new coordinate system arising from the basis B. Our goal is to find an angle θ for which a given conic section has no uv term. Now, the transition matrix from B to S is

$$\mathbf{P} = \begin{bmatrix} \cos\theta & -\sin\theta \\ \sin\theta & \cos\theta \end{bmatrix}.$$

Thus, we convert points in B-coordinates (u, v coordinates) to points in S-coordinates (x, y coordinates) with the equation

$$\begin{bmatrix} x \\ y \end{bmatrix} = \begin{bmatrix} \cos\theta & -\sin\theta \\ \sin\theta & \cos\theta \end{bmatrix}\begin{bmatrix} u \\ v \end{bmatrix}, \qquad \text{or}$$

$$\begin{cases} x = u\cos\theta - v\sin\theta \\ y = u\sin\theta + v\cos\theta \end{cases}$$

Substituting these equations into the general equation of a conic section, we obtain

$$a(u\cos\theta - v\sin\theta)^2 + b(u\sin\theta + v\cos\theta)^2 + c(u\cos\theta - v\sin\theta)(u\sin\theta + v\cos\theta)$$
$$+ d(u\cos\theta - v\sin\theta) + e(u\sin\theta + v\cos\theta) + f = 0.$$

After expanding, we find that the uv term is

$$(2\sin\theta\cos\theta(b-a) + (\cos^2\theta - \sin^2\theta)c)uv = ((\sin 2\theta)(b-a) + (\cos 2\theta)c)uv.$$

One possible way of setting the coefficient of uv equal to zero in this expression is by choosing the angle of rotation to be

$$\theta = \begin{cases} \dfrac{\pi}{4} & \text{if } a = b \\ \dfrac{1}{2}\arctan(c/(a-b)) & \text{otherwise} \end{cases}.$$

(Adding multiples of $\pi/2$ to this solution yields other solutions.)

Example 1

Consider the conic section whose equation is

$$5x^2 + 7y^2 - 10xy - 3x + 2y - 8 = 0.$$

The preceding formula for the angle of rotation gives us $\theta = \frac{1}{2}\arctan(-10/(-2))$ in this case. Hence, θ is approximately 0.6867 radians,† or about 39°21′. Now, $\cos\theta \approx 0.7733$ and $\sin\theta \approx 0.6340$. Hence, the conversion from uv-coordinates to xy-coordinates is given by

$$\begin{cases} x = 0.7733u - 0.6340v \\ y = 0.6340u + 0.7733v \end{cases}.$$

†All computations in this example were done on a calculator rounding to twelve significant digits. However, we have printed only four significant digits in the text.

Substituting the formulas for x and y in terms of u and v into the equation for the conic section, and simplifying, yields

$$0.9010u^2 + 11.10v^2 - 1.052u + 3.449v - 8 = 0.$$

Completing the squares gives

$$0.9010(u - 0.5838)^2 + 11.10(v + 0.1554)^2 = 8.575,$$

or

$$\frac{(u - 0.5838)^2}{(3.085)^2} + \frac{(v + 0.1554)^2}{(0.8790)^2} = 1.$$

The graph of this equation in the uv-plane, an ellipse centered at $(0.5838, -0.1554)$, is depicted in Figure 8.18.

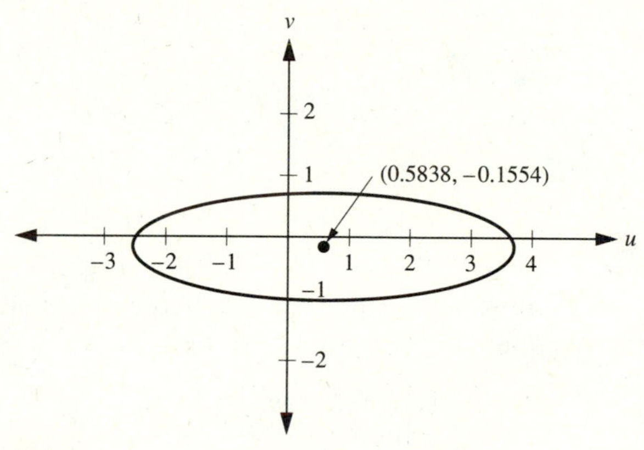

Figure 8.18
The ellipse $\frac{(u - 0.5838)^2}{(3.085)^2} + \frac{(v + 0.1554)^2}{(0.8790)^2} = 1$

The original conic section can be obtained by rotating the coordinate system in Figure 8.18 counterclockwise through the angle $\theta \approx 0.6867$ radians (see Figure 8.19). The center of the original conic section can be found by converting $(0.5838, -0.1554)$, the center of the uv-coordinate system ellipse, into xy-coordinates via the transition matrix

$$\mathbf{P} \approx \begin{bmatrix} \cos\theta & -\sin\theta \\ \sin\theta & \cos\theta \end{bmatrix} \approx \begin{bmatrix} 0.7733 & -0.6340 \\ 0.6340 & 0.7733 \end{bmatrix}.$$

That is, the center of

$$5x^2 + 7y^2 - 10xy - 3x + 2y - 8 = 0$$

is

$$\begin{bmatrix} x_0 \\ y_0 \end{bmatrix} \approx \begin{bmatrix} 0.7733 & -0.6340 \\ 0.6340 & 0.7733 \end{bmatrix} \begin{bmatrix} 0.5838 \\ -0.1554 \end{bmatrix} \approx \begin{bmatrix} 0.5500 \\ 0.2500 \end{bmatrix}.$$

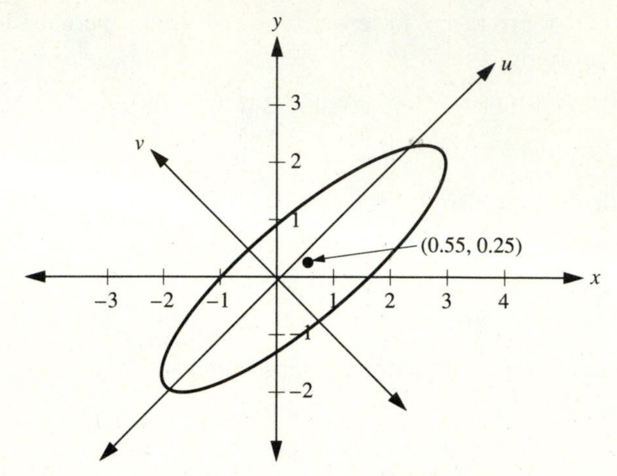

Figure 8.19
The ellipse $5x^2 + 7y^2 - 10xy - 3x + 2y - 8 = 0$ ■

Notice in Example 1 that the transition matrix

$$\mathbf{P} = \begin{bmatrix} \cos\theta & -\sin\theta \\ \sin\theta & \cos\theta \end{bmatrix}$$

allows us to convert directly from uv-coordinates to xy-coordinates. Hence, $\mathbf{P}^{-1}$ provides the means for converting from xy-coordinates to uv-coordinates. Now,

$$\mathbf{P}^{-1} = \begin{bmatrix} \cos\theta & \sin\theta \\ -\sin\theta & \cos\theta \end{bmatrix}.$$

For example, with the angle $\theta = 0.6867$ radian in Example 1, the point $(-1, 0)$ on the ellipse in xy-coordinates corresponds to the point

$$\begin{bmatrix} u_0 \\ v_0 \end{bmatrix} = \begin{bmatrix} \cos\theta & \sin\theta \\ -\sin\theta & \cos\theta \end{bmatrix}\begin{bmatrix} -1 \\ 0 \end{bmatrix} \approx \begin{bmatrix} 0.7733 & 0.6340 \\ -0.6340 & 0.7733 \end{bmatrix}\begin{bmatrix} -1 \\ 0 \end{bmatrix} \approx \begin{bmatrix} -0.7733 \\ 0.6340 \end{bmatrix},$$

in uv-coordinates.

Exercise—Section 8.7

1. For each of the conic sections described below, perform the following steps:

   (i) Find an appropriate angle θ through which to rotate the coordinate system counterclockwise from xy-coordinates into uv-coordinates so that the conic has no uv term.

   (ii) Calculate the transition matrix $\mathbf{P}$ from uv-coordinates to xy-coordinates.

   (iii) Solve for the equation of the conic in uv-coordinates.

   (iv) Determine the center of the conic in uv-coordinates if it is an ellipse or hyperbola or the vertex in uv-coordinates if it is a parabola. Graph the conic in uv-coordinates.

(v) Use the transition matrix $\mathbf{P}$ to solve for the center or vertex of the conic in xy-coordinates. Draw the graph of the conic in xy-coordinates.

(a) $xy - 1 = 0$ (hyperbola)

★(b) $3x^2 + y^2 - 2\sqrt{3}xy - (1 + 12\sqrt{3})x + (12 - \sqrt{3})y + 26 = 0$ (parabola)

(c) $13x^2 + 13y^2 - 10xy - 8\sqrt{2}x - 8\sqrt{2}y - 64 = 0$ (ellipse)

★(d) $8x^2 - 5y^2 + 16xy - 37 = 0$ (hyperbola)

8.8 Differential Equations

Prerequisite: Section 6.2, Diagonalization

In this section, we use the diagonalization process to solve certain first-order linear homogeneous systems of differential equations. We then adjust this technique to solve higher-order homogeneous differential equations as well.

First-Order Linear Homogeneous Systems

DEFINITION

Let

$$\mathbf{F}(t) = \begin{bmatrix} f_1(t) \\ \vdots \\ f_n(t) \end{bmatrix}$$

represent an $n \times 1$ matrix whose entries are real-valued functions, and let $\mathbf{A}$ be an $n \times n$ matrix of real numbers. Then the equation $\mathbf{F}'(t) - \mathbf{AF}(t) = \mathbf{0}$, or $\mathbf{F}'(t) = \mathbf{AF}(t)$, is called a **first-order linear homogeneous system of differential equations**. A **solution** for such a system is a particular function

$$\mathbf{F}(t) = \begin{bmatrix} f_1(t) \\ \vdots \\ f_n(t) \end{bmatrix}$$

that satisfies the equation for all values of t.

For brevity, in the remainder of this section we will refer to an equation of the form $\mathbf{F}'(t) = \mathbf{AF}(t)$ as a **first-order system.** Our goal is to find solutions for systems of this type.

Example 1

Letting $\mathbf{F} = \begin{bmatrix} f_1(t) \\ f_2(t) \end{bmatrix}$ and $\mathbf{A} = \begin{bmatrix} 13 & -45 \\ 6 & -20 \end{bmatrix}$, we get the first-order system $\mathbf{F}'(t) = \mathbf{A}\mathbf{F}(t)$, or

$$\begin{bmatrix} f_1'(t) \\ f_2'(t) \end{bmatrix} = \begin{bmatrix} 13 & -45 \\ 6 & -20 \end{bmatrix} \begin{bmatrix} f_1(t) \\ f_2(t) \end{bmatrix}.$$

Multiplying yields

$$\begin{cases} f_1'(t) = 13f_1(t) - 45f_2(t) \\ f_2'(t) = \ \ 6f_1(t) - 20f_2(t) \end{cases}.$$

A solution for this system consists of a pair of functions, $f_1(t)$ and $f_2(t)$, that satisfy both of these differential equations. One such solution is

$$\mathbf{F}(t) = \begin{bmatrix} f_1(t) \\ f_2(t) \end{bmatrix} = \begin{bmatrix} 5e^{-5t} \\ 2e^{-5t} \end{bmatrix}.$$

(Verify.) You will see how to obtain such solutions later in this section. ■

In what follows, we concern ourselves only with solutions that are *continuously differentiable* (that is, solutions having continuous derivatives). First, we state, without proof, a well-known result from the theory of differential equations about solutions of a single first-order equation:

LEMMA 8.6

A real-valued continuously differentiable function $f(t)$ is a solution to the differential equation $f'(t) = af(t)$ if and only if $f(t) = be^{at}$ for some real number b.

A first-order system of the form $\mathbf{F}'(t) = \mathbf{A}\mathbf{F}(t)$ is more complicated than the differential equation in Lemma 8.6 since it involves a matrix $\mathbf{A}$ instead of a real number a. However, in the special case when $\mathbf{A}$ is a diagonal matrix, the system $\mathbf{F}'(t) = \mathbf{A}\mathbf{F}(t)$ can be written as

$$\begin{cases} f_1'(t) = a_{11}f_1(t) \\ f_2'(t) = a_{22}f_2(t) \\ \quad \vdots \\ f_n'(t) = a_{nn}f_n(t) \end{cases}.$$

Each of the differential equations in this system can be solved separately using Lemma 8.6. Hence, when $\mathbf{A}$ is diagonal, the general solution has the form

$$\mathbf{F}(t) = [b_1 e^{a_{11}t}, b_2 e^{a_{22}t}, \ldots, b_n e^{a_{nn}t}],$$

for some $b_1, \ldots, b_n \in \mathbb{R}$.

Example 2

Consider the first-order system $\mathbf{F}'(t) = \begin{bmatrix} 3 & 0 \\ 0 & -2 \end{bmatrix} \mathbf{F}(t)$, whose matrix is diagonal. This system is equivalent to

$$\begin{cases} f_1'(t) = 3f_1(t) \\ f_2'(t) = -2f_2(t) \end{cases}.$$

Using Lemma 8.6, we see that the solutions are all functions of the form

$$\mathbf{F}(t) = [f_1(t), f_2(t)] = [b_1 e^{3t}, b_2 e^{-2t}].$$

■

Since first-order systems $\mathbf{F}'(t) = \mathbf{AF}(t)$ are easily solved when the matrix $\mathbf{A}$ is diagonal, it is natural to consider the case when $\mathbf{A}$ is diagonalizable. Thus, suppose $\mathbf{A}$ is a diagonalizable $n \times n$ matrix with (not necessarily distinct) eigenvalues $\lambda_1, \ldots, \lambda_n$ corresponding to the eigenvectors in the ordered basis $B = (\mathbf{v}_1, \ldots, \mathbf{v}_n)$ for $\mathbb{R}^n$. The matrix $\mathbf{P}$ having columns $\mathbf{v}_1, \ldots, \mathbf{v}_n$ is the transition matrix from B to standard coordinates, and $\mathbf{P}^{-1}\mathbf{AP} = \mathbf{D}$, the diagonal matrix having eigenvalues $\lambda_1, \lambda_2, \ldots, \lambda_n$ along its main diagonal. Hence,

$$\begin{aligned} \mathbf{F}'(t) = \mathbf{AF}(t) &\Longleftrightarrow \mathbf{F}'(t) = (\mathbf{PP}^{-1}\mathbf{APP}^{-1})\mathbf{F}(t) \\ &\Longleftrightarrow \mathbf{F}'(t) = \mathbf{PDP}^{-1}\mathbf{F}(t) \\ &\Longleftrightarrow \mathbf{P}^{-1}\mathbf{F}'(t) = \mathbf{DP}^{-1}\mathbf{F}(t). \end{aligned}$$

Letting $\mathbf{G}(t) = \mathbf{P}^{-1}\mathbf{F}(t)$, we see that the original system $\mathbf{F}'(t) = \mathbf{AF}(t)$ is equivalent to the system $\mathbf{G}'(t) = \mathbf{DG}(t)$. Since $\mathbf{D}$ is diagonal, with diagonal entries $\lambda_1, \ldots, \lambda_n$, the latter system is easily solved:

$$\mathbf{G}(t) = [b_1 e^{\lambda_1 t}, b_2 e^{\lambda_2 t}, \ldots, b_n e^{\lambda_n t}].$$

But, $\mathbf{F}(t) = \mathbf{PG}(t)$. Since the columns of $\mathbf{P}$ are the eigenvectors $\mathbf{v}_1, \mathbf{v}_2, \ldots, \mathbf{v}_n$, we obtain

$$\mathbf{F}(t) = b_1 e^{\lambda_1 t}\mathbf{v}_1 + b_2 e^{\lambda_2 t}\mathbf{v}_2 + \cdots + b_n e^{\lambda_n t}\mathbf{v}_n.$$

Thus, we have proved the following:

THEOREM 8.7

Let $\mathbf{A}$ be a diagonalizable $n \times n$ matrix and let $(\mathbf{v}_1, \ldots, \mathbf{v}_n)$ be an ordered basis for $\mathbb{R}^n$ consisting of eigenvectors for $\mathbf{A}$ corresponding to the (not necessarily distinct) eigenvalues $\lambda_1, \ldots, \lambda_n$. Then the continuously differentiable solutions for the first-order system $\mathbf{F}'(t) = \mathbf{AF}(t)$ are all functions of the form

$$\mathbf{F}(t) = b_1 e^{\lambda_1 t}\mathbf{v}_1 + b_2 e^{\lambda_2 t}\mathbf{v}_2 + \cdots + b_n e^{\lambda_n t}\mathbf{v}_n,$$

where $b_1, \ldots, b_n \in \mathbb{R}$.

Using this result, we can solve any first-order system whose matrix is diagonalizable.

Example 3

We will solve the first-order system $\mathbf{F}'(t) = \mathbf{AF}(t)$, where

$$\mathbf{A} = \begin{bmatrix} 1 & 0 & -2 & 6 \\ 4 & -1 & -4 & 12 \\ -32 & 9 & 40 & -114 \\ -11 & 3 & 14 & -40 \end{bmatrix}.$$

Following Steps 3 through 5 of the method in Section 6.2 for diagonalizing a linear operator, we find that $\mathbf{A}$ has the following eigenvectors and corresponding eigenvalues:

$$\begin{aligned} \mathbf{v}_1 &= [-2, -4, 5, 2], && \text{corresponding to } \lambda_1 = 0 \\ \mathbf{v}_2 &= [-3, 2, 0, 1], && \text{corresponding to } \lambda_2 = -1 \\ \mathbf{v}_3 &= [1, -1, 1, 0], && \text{corresponding to } \lambda_3 = -1 \\ \mathbf{v}_4 &= [0, 0, 3, 1], && \text{corresponding to } \lambda_4 = 2 \end{aligned}$$

(Notice that $\mathbf{v}_2$ and $\mathbf{v}_3$ were chosen to be linearly independent eigenvectors for the eigenvalue -1, so that $\{\mathbf{v}_2, \mathbf{v}_3\}$ forms a basis for E_{-1}.) Therefore, Theorem 8.7 tells us that the solutions to the first-order system $\mathbf{F}'(t) = \mathbf{AF}(t)$ are all of the functions of the form

$$\begin{aligned} \mathbf{F}(t) &= [f_1(t), f_2(t), f_3(t), f_4(t)] \\ &= b_1[-2, -4, 5, 2] + b_2 e^{-t}[-3, 2, 0, 1] + b_3 e^{-t}[1, -1, 1, 0] \\ &\qquad + b_4 e^{2t}[0, 0, 3, 1] \\ &= [-2b_1 - 3b_2 e^{-t} + b_3 e^{-t}, -4b_1 + 2b_2 e^{-t} - b_3 e^{-t}, 5b_1 + b_3 e^{-t} + 3b_4 e^{2t}, \\ &\qquad 2b_1 + b_2 e^{-t} + b_4 e^{2t}]. \end{aligned}$$

■

Notice that, in order to use Theorem 8.7 to solve a first-order system $\mathbf{F}'(t) = \mathbf{AF}(t)$, $\mathbf{A}$ must be a diagonalizable matrix. If it is not, you can still find some of the solutions to the system using an analogous process. If $\{\mathbf{v}_1, \ldots, \mathbf{v}_k\}$ is a linearly independent set of eigenvectors for $\mathbf{A}$ corresponding to the eigenvalues $\lambda_1, \ldots, \lambda_k$, then functions of the form

$$\mathbf{F}(t) = b_1 e^{\lambda_1 t}\mathbf{v}_1 + b_2 e^{\lambda_2 t}\mathbf{v}_2 + \cdots + b_k e^{\lambda_k t}\mathbf{v}_k$$

are solutions (see Exercise 3). However, these are not all the possible solutions for the system. To find all the solutions, you must use complex eigenvalues and eigenvectors, as well as *generalized eigenvectors*. Complex eigenvalues are studied in Section 7.1; generalized eigenvectors are not covered in this book.

Higher-Order Homogeneous Differential Equations

Our next goal is to solve higher-order homogeneous differential equations of the form

$$y^{(n)} + a_{n-1}y^{(n-1)} + \cdots + a_2y'' + a_1y' + a_0y = 0.$$

Example 4

Consider the differential equation $y''' - 6y'' + 3y' + 10y = 0$. To find solutions for this equation, we define the functions $f_1(t)$, $f_2(t)$, and $f_3(t)$ as follows: $f_1 = y$, $f_2 = y'$, and $f_3 = y''$. We then have the system

$$\begin{cases} f_1' = f_2 \\ f_2' = f_3 \\ f_3' = -10f_1 - 3f_2 + 6f_3 \end{cases}.$$

The first two equations in this system come directly from the definitions of f_1, f_2, and f_3. The third equation is obtained from the original differential equation by moving all terms except y''' to the right side. But this system can be expressed as

$$\begin{bmatrix} f_1'(t) \\ f_2'(t) \\ f_3'(t) \end{bmatrix} = \begin{bmatrix} 0 & 1 & 0 \\ 0 & 0 & 1 \\ -10 & -3 & 6 \end{bmatrix} \begin{bmatrix} f_1(t) \\ f_2(t) \\ f_3(t) \end{bmatrix};$$

that is, as $\mathbf{F}'(t) = \mathbf{AF}(t)$, with

$$\mathbf{F}(t) = \begin{bmatrix} f_1(t) \\ f_2(t) \\ f_3(t) \end{bmatrix} \quad \text{and} \quad \mathbf{A} = \begin{bmatrix} 0 & 1 & 0 \\ 0 & 0 & 1 \\ -10 & -3 & 6 \end{bmatrix}.$$

We now use the method of Theorem 8.7 to solve this first-order system.

A quick calculation yields $p_{\mathbf{A}}(x) = x^3 - 6x^2 + 3x + 10 = (x+1)(x-2)(x-5)$, giving the eigenvalues $\lambda_1 = -1$, $\lambda_2 = 2$, and $\lambda_3 = 5$. Solving for the corresponding eigenvectors for $\mathbf{A}$ produces

$$\begin{array}{lll} \mathbf{v}_1 = [1, -1, 1] & \text{corresponding to} & \lambda_1 = -1 \\ \mathbf{v}_2 = [1, 2, 4] & \text{corresponding to} & \lambda_2 = 2 \\ \mathbf{v}_3 = [1, 5, 25] & \text{corresponding to} & \lambda_3 = 5 \end{array}$$

Hence, Theorem 8.7 gives us the general solution

$$\begin{aligned} \mathbf{F}(t) &= b_1e^{-t}[1, -1, 1] + b_2e^{2t}[1, 2, 4] + b_3e^{5t}[1, 5, 25] \\ &= [b_1e^{-t} + b_2e^{2t} + b_3e^{5t}, -b_1e^{-t} + 2b_2e^{2t} + 5b_3e^{5t}, \\ &\quad b_1e^{-t} + 4b_2e^{2t} + 25b_3e^{5t}]. \end{aligned}$$

Since the first entry of this result equals $f_1(t) = y$, the general solution to the original third-order differential equation is

$$y = b_1 e^{-t} + b_2 e^{2t} + b_3 e^{5t}.$$

■

The method of Example 4 can be generalized to any homogeneous higher-order differential equation $y^{(n)} + a_{n-1}y^{(n-1)} + \cdots + a_1 y' + a_0$. In Exercise 5, you are asked to show that this equation can be represented as a linear system $\mathbf{F}'(t) = \mathbf{AF}(t)$, where $\mathbf{F}(t) = [f_1(t), f_2(t), \ldots, f_n(t)]$, with $f_1(t) = y,\ f_2(t) = y', \ldots,\ f_n(t) = y^{(n-1)}$ and where

$$\mathbf{A} = \begin{bmatrix} 0 & 1 & 0 & 0 & \cdots & 0 \\ 0 & 0 & 1 & 0 & \cdots & 0 \\ \vdots & \vdots & \vdots & \vdots & \ddots & \vdots \\ 0 & 0 & 0 & 0 & \cdots & 1 \\ -a_0 & -a_1 & -a_2 & -a_3 & \cdots & -a_{n-1} \end{bmatrix}.$$

The corresponding linear system can then be solved using the method of Theorem 8.7, as in Example 4.

Several startling patterns were revealed in Example 4. First, notice the similarity between the original differential equation $y''' - 6y'' + 3y' + 10y = 0$ and $p_{\mathbf{A}}(x) = x^3 - 6x^2 + 3x + 10$. This observation leads to the following general principle:

> If $y^{(n)} + a_{n-1}y^{(n-1)} + \cdots + a_1 y' + a_0$ is represented as a linear system $\mathbf{F}'(t) = \mathbf{AF}(t)$, where $\mathbf{F}(t)$ and $\mathbf{A}$ are as described above, then
>
> $$p_{\mathbf{A}}(x) = x^n + a_{n-1}x^{n-1} + \cdots + a_1 x + a_0.$$

You are asked to prove this principle in Exercise 5. Hence, from now on, we can avoid the long calculations necessary to determine $p_{\mathbf{A}}(x)$. When solving differential equations, $p_{\mathbf{A}}(x)$ is always derived using this shortcut. The equation $p_{\mathbf{A}}(x) = 0$ is called the **characteristic equation** of the original differential equation. The roots of this equation, the eigenvalues of $\mathbf{A}$, are frequently called the **characteristic values** of the differential equation.

Also notice in Example 4 that the eigenspace E_λ for each eigenvalue λ is one-dimensional and is spanned by the vector $[1, \lambda, \lambda^2]$. More generally, you are asked to prove the following in Exercise 6:

> If $y^{(n)} + a_{n-1}y^{(n-1)} + \cdots + a_1 y' + a_0$ is represented as a linear system $\mathbf{F}'(t) = \mathbf{AF}(t)$, where $\mathbf{F}(t)$ and $\mathbf{A}$ are as described above, and if λ is any eigenvalue for $\mathbf{A}$, then the eigenspace E_λ is one-dimensional and is spanned by the vector $[1, \lambda, \lambda^2, \ldots, \lambda^{n-1}]$.

Combining the preceding facts, we can state the solution set for many higher-order homogeneous differential equations directly (and avoid linear algebra techniques altogether), as follows:

> Consider the differential equation
>
> $$y^{(n)} + a_{n-1}y^{(n-1)} + \cdots + a_2 y'' + a_1 y' + a_0 = 0.$$
>
> Suppose that $\lambda_1, \ldots, \lambda_n$ are n distinct solutions to the characteristic equation
>
> $$x^n + a_{n-1}x^{n-1} + \cdots + a_2 x^2 + a_1 x + a_0 = 0.$$
>
> Then all continuously differentiable solutions of the differential equation have the form
>
> $$y = b_1 e^{\lambda_1 t} + b_2 e^{\lambda_2 t} + \cdots + b_n e^{\lambda_n t},$$
>
> for $b_1, \ldots, b_n \in \mathbb{R}$.

Example 5

To solve the homogeneous differential equation

$$y'''' + 2y''' - 28y'' - 50y' + 75y = 0,$$

we first find its characteristic values by solving the characteristic equation

$$x^4 + 2x^3 - 28x^2 - 50x + 75 = 0.$$

By factoring, or using an appropriate numerical technique, we obtain the solutions $\lambda_1 = -5$, $\lambda_2 = -3$, $\lambda_3 = 1$, and $\lambda_4 = 5$. Thus, the solution set for the original differential equation is the set of all functions of the form

$$y = b_1 e^{-5t} + b_2 e^{-3t} + b_3 e^t + b_4 e^{5t},$$

for $b_1, b_2, b_3, b_4 \in \mathbb{R}$. ■

Notice that the method in Example 5 cannot be used if the differential equation has fewer than n distinct characteristic values. If you can find only k distinct characteristic values for an nth-order equation with $k < n$, then the method yields only a k-dimensional subspace of the full n-dimensional solution space. As with first-order systems, finding the complete solution set in this case requires the use of complex eigenvalues, complex eigenvectors, and generalized eigenvectors.

Exercises—Section 8.8

1. In each part of this exercise, the given matrix represents $\mathbf{A}$ in a first-order system of the form $\mathbf{F}'(t) = \mathbf{AF}(t)$. Use Theorem 8.7 to find the general form of a solution to each system.

★(a) $\begin{bmatrix} 13 & -28 \\ 6 & -13 \end{bmatrix}$ (b) $\begin{bmatrix} 18 & -15 \\ 20 & -17 \end{bmatrix}$ ★(c) $\begin{bmatrix} 1 & 4 & 4 \\ -1 & 2 & 2 \\ 1 & 1 & 1 \end{bmatrix}$

★(d) $\begin{bmatrix} -5 & -6 & 15 \\ -6 & -5 & 15 \\ -6 & -6 & 16 \end{bmatrix}$ (e) $\begin{bmatrix} -1 & 0 & -2 & 2 \\ -3 & 5 & 1 & -9 \\ 0 & 4 & 5 & -12 \\ -1 & 4 & 3 & -10 \end{bmatrix}$

2. Find the solution set for each given homogeneous differential equation.

★(a) $y'' + y' - 5y = 0$

(b) $y''' - 5y'' - y' + 5y = 0$

★(c) $y'''' - 6y'' + 8y = 0$

3. Let $\mathbf{A}$ be an $n \times n$ matrix with linearly independent eigenvectors $\mathbf{v}_1, \ldots, \mathbf{v}_k$ corresponding, respectively, to the eigenvalues $\lambda_1, \ldots, \lambda_k$. Prove that

$$\mathbf{F}(t) = b_1 e^{\lambda_1 t}\mathbf{v}_1 + b_2 e^{\lambda_2 t}\mathbf{v}_2 + \cdots + b_k e^{\lambda_k t}\mathbf{v}_k$$

is a solution for the first-order system $\mathbf{F}'(t) = \mathbf{AF}(t)$, for every choice of $b_1, \ldots, b_k \in \mathbb{R}$.

4. (a) Let $\mathbf{A}$ be a diagonalizable $n \times n$ matrix, and let $\mathbf{v}$ be a fixed vector in $\mathbb{R}^n$. Show there is a unique function $\mathbf{F}(t)$ that satisfies the first-order system $\mathbf{F}'(t) = \mathbf{AF}(t)$ such that $\mathbf{F}(0) = \mathbf{v}$. (The vector $\mathbf{v}$ is called an **initial condition** for the system.)

★(b) Find the unique solution to $\mathbf{F}'(t) = \mathbf{AF}(t)$ with initial condition $\mathbf{F}(0) = \mathbf{v}$, where

$$\mathbf{A} = \begin{bmatrix} -11 & -6 & 16 \\ -4 & -1 & 4 \\ -12 & -6 & 17 \end{bmatrix} \quad \text{and} \quad \mathbf{v} = [1, -4, 0].$$

5. (a) Verify that the homogeneous differential equation

$$y^{(n)} + a_{n-1}y^{(n-1)} + \cdots + a_1 y' + a_0 = 0$$

can be represented as $\mathbf{F}'(t) = \mathbf{AF}(t)$, where $\mathbf{F}(t) = [f_1(t), f_2(t), \ldots, f_n(t)]$, with $f_1(t) = y$, $f_2(t) = y', \ldots, f_n(t) = y^{(n-1)}$, and where

$$\mathbf{A} = \begin{bmatrix} 0 & 1 & 0 & 0 & \cdots & 0 \\ 0 & 0 & 1 & 0 & \cdots & 0 \\ \vdots & \vdots & \vdots & \vdots & \ddots & \vdots \\ 0 & 0 & 0 & 0 & \cdots & 1 \\ -a_0 & -a_1 & -a_2 & -a_3 & \cdots & -a_{n-1} \end{bmatrix}.$$

(b) If $\mathbf{A}$ is the matrix given in part (a), prove that

$$p_{\mathbf{A}}(x) = x^n + a_{n-1}x^{n-1} + \cdots + a_1 x + a_0.$$

(Hint: Use induction on n and a cofactor expansion on the first column of $(x\mathbf{I}_n - \mathbf{A})$.)

6. Let $\mathbf{A}$ be the $n \times n$ matrix from Exercise 5, for some $a_0, a_1, \ldots, a_{n-1} \in \mathbb{R}$.

(a) Calculate $\mathbf{A}\begin{bmatrix} b_1 \\ b_2 \\ \vdots \\ b_n \end{bmatrix}$, for a general n-vector $\begin{bmatrix} b_1 \\ b_2 \\ \vdots \\ b_n \end{bmatrix}$.

(b) Let λ be an eigenvalue for $\mathbf{A}$. Show that $[1, \lambda, \lambda^2, \ldots, \lambda^{n-1}]$ is an eigenvector corresponding to λ. (Hint: Use part (b) of Exercise 5.)

(c) Show that if $\mathbf{v}$ is a vector with first coordinate c such that $\mathbf{Av} = \lambda\mathbf{v}$, for some $\lambda \in \mathbb{R}$, then $\mathbf{v} = c[1, \lambda, \lambda^2, \ldots, \lambda^{n-1}]$.

(d) Conclude that the eigenspace E_λ for an eigenvalue λ of $\mathbf{A}$ is always one-dimensional.

8.9 Quadratic Forms

Prerequisite: Section 6.3, Orthogonal Diagonalization

In Section 8.7, we used a change of basis to simplify a general second-degree equation (conic section) in two variables x and y. In this section, we generalize this process to any finite number of variables, using orthogonal diagonalization.

Quadratic Forms

DEFINITION

A **quadratic form** on $\mathbb{R}^n$ is a function $Q\colon \mathbb{R}^n \longrightarrow \mathbb{R}$ of the form

$$Q([x_1, \ldots, x_n]) = \sum_{1 \le i \le j \le n} c_{ij} x_i x_j,$$

for some real numbers c_{ij}, $1 \le i \le j \le n$.

Thus, a quadratic form on $\mathbb{R}^n$ is a polynomial in n variables in which each term has degree 2.

Example 1

The function $Q_1([x_1, x_2, x_3]) = 7x_1^2 + 5x_1x_2 - 6x_2^2 + 9x_2x_3 + 14x_3^2$ is a quadratic form on $\mathbb{R}^3$. Q_1 is a polynomial in three variables in which each term has degree 2. Note that the coefficient c_{13} of the x_1x_3 term is zero.

The function $Q_2([x, y]) = 8x^2 - 3y^2 + 12xy$ is a quadratic form on $\mathbb{R}^2$ with coefficients $c_{11} = 8$, $c_{22} = -3$, and $c_{12} = 12$. On $\mathbb{R}^2$, a quadratic form consists of the x^2, y^2, and xy terms from the general form for the equation of a conic section. ■

In general, a quadratic form Q on $\mathbb{R}^n$ can be expressed as $Q(\mathbf{x}) = \mathbf{x}^T\mathbf{C}\mathbf{x}$, where $\mathbf{x}$ is a column matrix and $\mathbf{C}$ is the upper triangular matrix whose entries on and above the main diagonal are given by the coefficients c_{ij} in the definition of a quadratic form above. For example, the quadratic forms Q_1 and Q_2 in Example 1 can be expressed as

$$Q_1\left(\begin{bmatrix} x_1 \\ x_2 \\ x_3 \end{bmatrix}\right) = [x_1, x_2, x_3]\begin{bmatrix} 7 & 5 & 0 \\ 0 & -6 & 9 \\ 0 & 0 & 14 \end{bmatrix}\begin{bmatrix} x_1 \\ x_2 \\ x_3 \end{bmatrix} \quad \text{and}$$

$$Q_2\left(\begin{bmatrix} x, \\ y \end{bmatrix}\right) = [x, y]\begin{bmatrix} 8 & 12 \\ 0 & -3 \end{bmatrix}\begin{bmatrix} x \\ y \end{bmatrix}.$$

However, this upper triangular representation for a quadratic form is not the most useful one for our purposes. Instead, we will replace the matrix $\mathbf{C}$ with a symmetric matrix.

THEOREM 8.8

Let $Q: \mathbb{R}^n \longrightarrow \mathbb{R}$ be a quadratic form. Then there is a unique symmetric $n \times n$ matrix $\mathbf{A}$ such that $Q(\mathbf{x}) = \mathbf{x}^T\mathbf{A}\mathbf{x}$.

Proof of Theorem 8.8 (abridged)

The uniqueness of the matrix $\mathbf{A}$ in the theorem is unimportant in what follows, so we leave its proof for you to provide in Exercise 4.

To prove the existence of $\mathbf{A}$, let $Q([x_1, \ldots, x_n]) = \sum_{1 \le i \le j \le n} c_{ij}x_ix_j$, and let $\mathbf{A}$ be the $n \times n$ matrix whose entries are

$$a_{ij} = \begin{cases} c_{ij} & \text{if } i = j \\ \frac{1}{2}c_{ij} & \text{if } i < j \\ \frac{1}{2}c_{ji} & \text{if } i > j \end{cases}.$$

Then $\mathbf{A}$ is clearly symmetric. A straightforward calculation of $\mathbf{x}^T\mathbf{A}\mathbf{x}$ shows that the coefficient of its x_ix_j term is c_{ij}. (Verify.) Hence, $\mathbf{x}^T\mathbf{A}\mathbf{x} = Q(\mathbf{x})$. ■

Example 2

Let $Q_3\left(\begin{bmatrix} x_1 \\ x_2 \end{bmatrix}\right) = 17x_1^2 + 8x_1x_2 - 9x_2^2$. Then the corresponding symmetric matrix $\mathbf{A}$ for Q_3 is $\begin{bmatrix} c_{11} & \frac{1}{2}c_{12} \\ \frac{1}{2}c_{12} & c_{22} \end{bmatrix} = \begin{bmatrix} 17 & 4 \\ 4 & -9 \end{bmatrix}$. That is, each main diagonal entry of $\mathbf{A}$ is just the coefficient of the corresponding "square" term in Q_3, and each nondiagonal entry of $\mathbf{A}$ is half of the coefficient of the corresponding "mixed-product" term in Q_3. You can easily verify that

$$Q_3\left(\begin{bmatrix} x_1 \\ x_2 \end{bmatrix}\right) = [x_1, x_2]\begin{bmatrix} 17 & 4 \\ 4 & -9 \end{bmatrix}\begin{bmatrix} x_1 \\ x_2 \end{bmatrix}.$$

■

Orthogonal Change of Basis

The next theorem indicates how the symmetric matrix representing a quadratic form changes when we perform an orthogonal change of coordinates.

THEOREM 8.9

Let $Q: \mathbb{R}^n \longrightarrow \mathbb{R}$ be a quadratic form given by $Q(\mathbf{x}) = \mathbf{x}^T\mathbf{A}\mathbf{x}$, for some symmetric matrix $\mathbf{A}$. Let B be an orthonormal basis for $\mathbb{R}^n$. Let $\mathbf{P}$ be the transition matrix from B-coordinates to standard coordinates, and let $\mathbf{K} = \mathbf{P}^{-1}\mathbf{A}\mathbf{P}$. Then $\mathbf{K}$ is symmetric and $Q(\mathbf{x}) = [\mathbf{x}]_B^T\mathbf{K}[\mathbf{x}]_B$.

Proof of Theorem 8.9

Since B is an orthonormal basis, $\mathbf{P}$ is an orthogonal matrix by Theorem 5.31. Hence, $\mathbf{P}^{-1} = \mathbf{P}^T$. Now, $[\mathbf{x}]_B = \mathbf{P}^{-1}\mathbf{x} = \mathbf{P}^T\mathbf{x}$, and thus $[\mathbf{x}]_B^T = (\mathbf{P}^T\mathbf{x})^T = \mathbf{x}^T\mathbf{P}$. Therefore,

$$\begin{aligned} Q(\mathbf{x}) &= \mathbf{x}^T\mathbf{A}\mathbf{x} \\ &= \mathbf{x}^T\mathbf{P}\mathbf{P}^{-1}\mathbf{A}\mathbf{P}\mathbf{P}^{-1}\mathbf{x} \\ &= [\mathbf{x}]_B^T\mathbf{P}^{-1}\mathbf{A}\mathbf{P}[\mathbf{x}]_B. \end{aligned}$$

Letting $\mathbf{K} = \mathbf{P}^{-1}\mathbf{A}\mathbf{P}$, we have $Q(\mathbf{x}) = [\mathbf{x}]_B^T\mathbf{K}[\mathbf{x}]_B$. Finally, notice that $\mathbf{K}$ is symmetric, since

$$\mathbf{K}^T = (\mathbf{P}^{-1}\mathbf{A}\mathbf{P})^T = (\mathbf{P}^T\mathbf{A}\mathbf{P})^T = \mathbf{P}^T\mathbf{A}^T(\mathbf{P}^T)^T = \mathbf{P}^{-1}\mathbf{A}\mathbf{P} = \mathbf{K}.$$

■

Example 3

Consider the quadratic form $Q([x, y, z]) = 2xy + 4xz + 2yz - y^2 + 3z^2$. Then,

$$Q\left(\begin{bmatrix} x \\ y \\ z \end{bmatrix}\right) = [x, y, z]\mathbf{A}\begin{bmatrix} x \\ y \\ z \end{bmatrix}, \quad \text{where} \quad \mathbf{A} = \begin{bmatrix} 0 & 1 & 2 \\ 1 & -1 & 1 \\ 2 & 1 & 3 \end{bmatrix}.$$

Consider the orthonormal basis $B = (\frac{1}{3}[2, 1, 2], \frac{1}{3}[2, -2, -1], \frac{1}{3}[1, 2, -2])$ for $\mathbb{R}^3$. We will find the symmetric matrix for Q with respect to this new basis B.

The transition matrix from B-coordinates to standard coordinates is the orthogonal matrix

$$\mathbf{P} = \frac{1}{3}\begin{bmatrix} 2 & 2 & 1 \\ 1 & -2 & 2 \\ 2 & -1 & -2 \end{bmatrix}.$$

$$\text{Also,} \quad \mathbf{P}^{-1} = \mathbf{P}^T = \frac{1}{3}\begin{bmatrix} 2 & 1 & 2 \\ 2 & -2 & -1 \\ 1 & 2 & -2 \end{bmatrix}.$$

$$\text{Then,} \quad \mathbf{K} = \mathbf{P}^{-1}\mathbf{A}\mathbf{P} = \frac{1}{9}\begin{bmatrix} 35 & -7 & -11 \\ -7 & -13 & 4 \\ -11 & 4 & -4 \end{bmatrix}.$$

Let $[u, v, w]$ be the representation of the vector $[x, y, z]$ in B-coordinates—that is, $[x, y, z]_B = [u, v, w]$. Then, by Theorem 8.9, Q can be expressed as

$$Q\left(\begin{bmatrix} u \\ v \\ w \end{bmatrix}\right) = [u, v, w]\left(\frac{1}{9}\begin{bmatrix} 35 & -7 & -11 \\ -7 & -13 & 4 \\ -11 & 4 & -4 \end{bmatrix}\right)\begin{bmatrix} u \\ v \\ w \end{bmatrix}$$
$$= \frac{35}{9}u^2 - \frac{13}{9}v^2 - \frac{4}{9}w^2 - \frac{14}{9}uv - \frac{22}{9}uw + \frac{8}{9}vw.$$

Let us check this formula for Q in a particular case. If $[x, y, z] = [9, 2, -1]$, then the original formula for Q yields

$$Q([9, 2, -1]) = (2)(9)(2) + (4)(9)(-1) + (2)(2)(-1) - (2)^2 + (3)(-1)^2 = -5.$$

On the other hand,

$$\begin{bmatrix} u \\ v \\ w \end{bmatrix} = \begin{bmatrix} 9 \\ 2 \\ -1 \end{bmatrix}_B = \mathbf{P}^{-1}\begin{bmatrix} 9 \\ 2 \\ -1 \end{bmatrix} = \frac{1}{3}\begin{bmatrix} 2 & 1 & 2 \\ 2 & -2 & -1 \\ 1 & 2 & -2 \end{bmatrix}\begin{bmatrix} 9 \\ 2 \\ -1 \end{bmatrix} = \begin{bmatrix} 6 \\ 5 \\ 5 \end{bmatrix}.$$

Calculating Q using the formula for B-coordinates, we get

$$Q([u, v, w]) = \frac{35}{9}(6)^2 - \frac{13}{9}(5)^2 - \frac{4}{9}(5)^2 - \frac{14}{9}(6)(5)$$
$$- \frac{22}{9}(6)(5) + \frac{8}{9}(5)(5) = -5,$$

which agrees with our previous calculation for Q. ■

The Principal Axes Theorem

We are now ready to prove the main result of this section: given any quadratic form Q on $\mathbb{R}^n$, an orthonormal basis B for $\mathbb{R}^n$ can be chosen so that the expression for Q in B-coordinates contains no "mixed-product" terms (that is, Q contains only "square" terms).

THEOREM 8.10: Principal Axes Theorem

Let Q: $\mathbb{R}^n \longrightarrow \mathbb{R}$ be a quadratic form. Then there is an orthonormal basis B for $\mathbb{R}^n$ such that $Q(\mathbf{x}) = [\mathbf{x}]_B^T \mathbf{D} [\mathbf{x}]_B$ for some diagonal matrix $\mathbf{D}$. That is, if $[\mathbf{x}]_B = \mathbf{y} = [y_1, y_2, \ldots, y_n]$, then

$$Q(\mathbf{x}) = d_{11}y_1^2 + d_{22}y_2^2 + \cdots + d_{nn}y_n^2.$$

Proof of Theorem 8.10

Let Q be a quadratic form on $\mathbb{R}^n$. Then, by Theorem 8.8, there is a symmetric $n \times n$ matrix $\mathbf{A}$ such that $Q(\mathbf{x}) = \mathbf{x}^T\mathbf{A}\mathbf{x}$. Now, by Theorem 6.15, $\mathbf{A}$ can be orthogonally diagonalized; that is, there is an orthogonal matrix $\mathbf{P}$ such that $\mathbf{P}^{-1}\mathbf{A}\mathbf{P} = \mathbf{D}$ is diagonal. Let B be the orthonormal basis for $\mathbb{R}^n$ given by the columns of $\mathbf{P}$. Then Theorem 8.9 implies that $Q(\mathbf{x}) = [\mathbf{x}]_B^T\mathbf{D}[\mathbf{x}]_B$. ■

The process of finding a diagonal matrix for a given quadratic form Q is referred to as **diagonalizing** Q. We now outline the method for diagonalizing a quadratic form, as presented in the proof Theorem 8.10.

METHOD FOR DIAGONALIZING A QUADRATIC FORM

Given a quadratic form Q: $\mathbb{R}^n \longrightarrow \mathbb{R}$,
Step 1: Find a symmetric $n \times n$ matrix $\mathbf{A}$ such that $Q(\mathbf{x}) = \mathbf{x}^T\mathbf{A}\mathbf{x}$.
Step 2: Apply Steps 3 through 8 of the method in Section 6.3 for orthogonally diagonalizing a symmetric operator, using the matrix $\mathbf{A}$. This process yields an orthonormal basis B, an orthogonal matrix $\mathbf{P}$ whose columns are the vectors in B, and a diagonal matrix $\mathbf{D}$ with $\mathbf{D} = \mathbf{P}^{-1}\mathbf{A}\mathbf{P}$.
Step 3: Then $Q(\mathbf{x}) = [\mathbf{x}]_B^T\mathbf{D}[\mathbf{x}]_B$, with $[\mathbf{x}]_B = \mathbf{P}^{-1}\mathbf{x} = \mathbf{P}^T\mathbf{x}$. If $[\mathbf{x}]_B = [y_1, y_2, \ldots, y_n]$, then $Q(\mathbf{x}) = d_{11}y_1^2 + d_{22}y_2^2 + \cdots + d_{nn}y_n^2$.

Example 4

Let $Q([x, y, z]) = \frac{1}{121}(183x^2 + 266y^2 + 35z^2 + 12xy + 408xz + 180yz)$. We will diagonalize Q.

Step 1: Note that $Q(\mathbf{x}) = \mathbf{x}^T\mathbf{A}\mathbf{x}$, where $\mathbf{A}$ is the symmetric matrix

$$\frac{1}{121}\begin{bmatrix} 183 & 6 & 204 \\ 6 & 266 & 90 \\ 204 & 90 & 35 \end{bmatrix}.$$

Step 2: We apply Steps 3 through 8 of the method for orthogonally diagonalizing $\mathbf{A}$. We list the results here but leave the details of the calculations for you to check.

(3) A quick computation gives

$$p_{\mathbf{A}}(x) = x^3 - 4x^2 + x + 6 = (x-3)(x-2)(x+1).$$

Therefore, the eigenvalues of $\mathbf{A}$ are $\lambda_1 = 3$, $\lambda_2 = 2$, and $\lambda_3 = -1$.

(4) Next, we find a basis for each eigenspace for $\mathbf{A}$. To find a basis for E_{λ_1}, we solve the system $(3\mathbf{I}_3 - \mathbf{A})\mathbf{x} = \mathbf{0}$, which yields the basis $\{[7, 6, 6]\}$. Similarly, we solve appropriate systems to find:

Basis for $E_{\lambda_2} = \{[6, -9, 2]\}$
Basis for $E_{\lambda_3} = \{[6, 2, -9]\}$.

(5) Since each eigenspace from (4) is one-dimensional, we need only normalize each basis vector to find orthonormal bases for E_{λ_1}, E_{λ_2}, and E_{λ_3}:

Orthonormal basis for $E_{\lambda_1} = \{\frac{1}{11}[7, 6, 6]\}$
Orthonormal basis for $E_{\lambda_2} = \{\frac{1}{11}[6, -9, 2]\}$
Orthonormal basis for $E_{\lambda_3} = \{\frac{1}{11}[6, 2, -9]\}$

(6) We let B be the ordered orthonormal basis $(\frac{1}{11}[7, 6, 6], \frac{1}{11}[6, -9, 2], \frac{1}{11}[6, 2, -9])$.

(7) The desired diagonal matrix for Q with respect to the basis B is

$$\mathbf{D} = \begin{bmatrix} 3 & 0 & 0 \\ 0 & 2 & 0 \\ 0 & 0 & -1 \end{bmatrix},$$

which has eigenvalues $\lambda_1 = 3$, $\lambda_2 = 2$, and $\lambda_3 = -1$ along the main diagonal.

(8) The transition matrix $\mathbf{P}$ from B-coordinates to standard coordinates is the matrix whose columns are the vectors in B, namely

$$\mathbf{P} = \frac{1}{11}\begin{bmatrix} 7 & 6 & 6 \\ 6 & -9 & 2 \\ 6 & 2 & -9 \end{bmatrix}.$$

Of course, $\mathbf{D} = \mathbf{P}^{-1}\mathbf{A}\mathbf{P}$. In this case, $\mathbf{P}$ is not only orthogonal but is symmetric as well, so $\mathbf{P}^{-1} = \mathbf{P}^T = \mathbf{P}$. (Be careful! $\mathbf{P}$ will not always be symmetric.)

This concludes Step 2.

Step 3: If $[x, y, z]_B = [u, v, w]$, use $\mathbf{D}$ to obtain $Q = 3u^2 + 2v^2 - w^2$. Notice that Q has only "square" terms, since $\mathbf{D}$ is diagonal.

For a particular example, let $[x, y, z] = [2, 6, -1]$. Then

$$\begin{bmatrix} u \\ v \\ w \end{bmatrix} = \mathbf{P}^{-1}\begin{bmatrix} 2 \\ 6 \\ -1 \end{bmatrix} = \frac{1}{11}\begin{bmatrix} 7 & 6 & 6 \\ 6 & -9 & 2 \\ 6 & 2 & -9 \end{bmatrix}\begin{bmatrix} 2 \\ 6 \\ -1 \end{bmatrix} = \begin{bmatrix} 4 \\ -4 \\ 3 \end{bmatrix}.$$

Hence, $Q([2, 6, -1]) = 3(4)^2 + 2(-4)^2 - (3)^2 = 71$. As an independent check, notice that plugging $[2, 6, -1]$ into the original equation for Q produces the same result. ■

Exercises—Section 8.9

1. In each part of this exercise, a quadratic form $Q: \mathbb{R}^n \longrightarrow \mathbb{R}$ is given. Find an upper triangular matrix $\mathbf{C}$ and a symmetric matrix $\mathbf{A}$ such that, for every $\mathbf{x} \in \mathbb{R}^n$, $Q(\mathbf{x}) = \mathbf{x}^T\mathbf{C}\mathbf{x} = \mathbf{x}^T\mathbf{A}\mathbf{x}$.

★(a) $Q([x, y]) = 8x^2 - 9y^2 + 12xy$

(b) $Q([x, y]) = 7x^2 + 11y^2 - 17xy$

★(c) $Q([x_1, x_2, x_3]) = 5x_1^2 - 2x_2^2 + 4x_1x_2 - 3x_1x_3 + 5x_2x_3$

2. In each part of this exercise, diagonalize the given quadratic form $Q: \mathbb{R}^n \longrightarrow \mathbb{R}$ by following the three-step method described in the text. Your answers should include the matrices $\mathbf{A}$, $\mathbf{P}$, and $\mathbf{D}$ defined in that method, as well as the orthonormal basis B. Finally, calculate $Q(\mathbf{x})$ for the given vector $\mathbf{x}$ two different ways: first, using the given formula for Q, and second, calculating $Q = [\mathbf{x}]_B^T\mathbf{D}[\mathbf{x}]_B$, where $[\mathbf{x}]_B = \mathbf{P}^{-1}\mathbf{x}$ and $\mathbf{D} = \mathbf{P}^{-1}\mathbf{A}\mathbf{P}$.

★(a) $Q([x, y]) = 43x^2 + 57y^2 - 48xy$; $\mathbf{x} = [1, -8]$

(b) $Q([x_1, x_2, x_3]) = -5x_1^2 + 37x_2^2 + 49x_3^2 + 32x_1x_2 + 80x_1x_3 + 32x_2x_3$; $\mathbf{x} = [7, -2, 1]$

★(c) $Q([x_1, x_2, x_3]) = 18x_1^2 - 68x_2^2 + x_3^2 + 96x_1x_2 - 60x_1x_3 + 36x_2x_3$; $\mathbf{x} = [4, -3, 6]$

(d) $Q([x_1, x_2, x_3, x_4]) = x_1^2 + 5x_2 + 864x_3^2 + 864x_4^2 - 24x_1x_3 + 24x_1x_4 + 120x_2x_3 + 120x_2x_4 + 1152x_3x_4$; $\mathbf{x} = [5, 9, -3, -2]$

3. Let $\mathbf{A} = \begin{bmatrix} 5 & 4 & 2 \\ 0 & 3 & 8 \\ 0 & 0 & -7 \end{bmatrix}$ and $\mathbf{B} = \begin{bmatrix} 5 & 2 & 1 \\ 2 & 3 & 4 \\ 1 & 4 & -7 \end{bmatrix}$.

(a) Show that if $Q_1(\mathbf{x}) = \mathbf{x}^T\mathbf{A}\mathbf{x}$ and $Q_2(\mathbf{x}) = \mathbf{x}^T\mathbf{B}\mathbf{x}$, then $Q_1(\mathbf{x}) = Q_2(\mathbf{x})$, for all $\mathbf{x} \in \mathbb{R}^3$. That is, show that Q_1 and Q_2 are the same quadratic form.

★(b) Let $L_1(\mathbf{x}) = \mathbf{A}\mathbf{x}$ and $L_2(\mathbf{x}) = \mathbf{B}\mathbf{x}$ be linear transformations on $\mathbb{R}^3$. Does $L_1 = L_2$? Prove your answer.

★(c) Do the matrices $\mathbf{A}$ and $\mathbf{B}$ have the same eigenvalues? Does your answer seem reasonable? Why or why not? (Hint: Find the eigenvalues for $\mathbf{A}$, and then check to see if they are eigenvalues for $\mathbf{B}$.)

4. Let $Q: \mathbb{R}^n \longrightarrow \mathbb{R}$ be a quadratic form, and let $\mathbf{A}$ and $\mathbf{B}$ be symmetric matrices such that $Q(\mathbf{x}) = \mathbf{x}^T\mathbf{A}\mathbf{x} = \mathbf{x}^T\mathbf{B}\mathbf{x}$. Prove that $\mathbf{A} = \mathbf{B}$ (the uniqueness assertion from Theorem 8.8). (Hint: Use $\mathbf{x} = \mathbf{e}_i$ to show that $a_{ii} = b_{ii}$. Then use $\mathbf{x} = \mathbf{e}_i + \mathbf{e}_j$ to prove that $a_{ij} = b_{ij}$ when $i \neq j$.)

★**5.** Let $Q: \mathbb{R}^n \longrightarrow \mathbb{R}$ be a quadratic form. Is the upper triangular representation for Q necessarily unique? That is, if $\mathbf{C}_1$ and $\mathbf{C}_2$ are upper triangular $n \times n$ matrices with $Q(\mathbf{x}) = \mathbf{x}^T\mathbf{C}_1\mathbf{x} = \mathbf{x}^T\mathbf{C}_2\mathbf{x}$, for all $\mathbf{x} \in \mathbb{R}^n$, must $\mathbf{C}_1 = \mathbf{C}_2$? Prove your answer.

6. A quadratic form $Q(\mathbf{x})$ on $\mathbb{R}^n$ is **positive definite** if and only if both of the following conditions hold:

(i) $Q(\mathbf{x}) \geq 0$, for all $\mathbf{x} \in \mathbb{R}^n$.
(ii) $Q(\mathbf{x}) = 0$ if and only if $\mathbf{x} = \mathbf{0}$.

A quadratic form having property (i) is said to be **positive semidefinite.**

Let Q be a quadratic form on $\mathbb{R}^n$, and let $\mathbf{A}$ be the symmetric matrix such that $Q(\mathbf{x}) = \mathbf{x}^T\mathbf{A}\mathbf{x}$.

(a) Prove that Q is positive definite if and only if every eigenvalue of $\mathbf{A}$ is positive.
(b) Prove that Q is positive semidefinite if and only if every eigenvalue of $\mathbf{A}$ is nonnegative.

9

Numerical Methods

Throughout the book, we have urged you to use a computer with appropriate software to perform tedious calculations after you have mastered a computational technique. A computer is especially useful when solving a linear system or when finding eigenvalues and eigenvectors for a linear operator. In this chapter, we discuss additional numerical methods for solving systems and finding eigenvalues that are better suited for the computer. If you have some programming experience, you should find it a straightforward task to write your own programs to implement these algorithms.

9.1 Numerical Methods for Solving Systems

Prerequisite: Section 2.2, Equivalent Systems and Rank

In this section, we discuss some considerations for solving linear systems by computer and investigate some alternate methods for solving systems, including Gaussian elimination, partial pivoting, the Jacobi method, and the Gauss-Seidel method.

Computer Accuracy

One basic problem in using a computer in linear algebra is that real numbers cannot always be represented exactly in computer memory. Computers store information in **binary** (base 2) form as a collection of **bits**: zeroes and ones. Because the physical storage space of a computer is limited, a predetermined number of bits is assigned in the computer memory for the storage of any real number. Thus, only the most significant digits of any real number can be stored by the computer.† Nonterminating decimals, like $\frac{1}{3} = 0.333333\ldots$ or $e = 2.718281828459045\ldots$, can never be rep-

†The first n significant digits of a decimal number are its left-most n digits, beginning with the first nonzero digit. For example, consider the real numbers $r_1 = 47.26835$, $r_2 = 9.00473$, and $r_3 = 0.000456$. Approximating these by stopping after the first three significant digits, and rounding to the nearest digit, we get $r_1 \approx 47.3$, $r_2 \approx 9.00$, and $r_3 \approx 0.000456$ (since the first nonzero digit in r_3 is 4).

resented fully. Using the first few decimal places of such numbers may be enough for most practical purposes, but it is not completely accurate.

As calculations are performed, the computer truncates and rounds all computational results to fit them within the limited storage space allotted. Numerical errors caused by this process are called **roundoff errors**. Unfortunately, if many operations are performed, roundoff errors can compound, thus producing a significant error in the final result. Keep in mind that, although the results of computer calculations are not always exact, they are often adequate for our purposes.

Gaussian Elimination

The first new method we introduce for solving linear systems is known as **Gaussian elimination**. This technique differs from the Gauss-Jordan algorithm in that we "zero out" only those entries of the augmented matrix below the pivots. This process converts the augmented matrix for the system into **row echelon form**. As with reduced row echelon form, a row echelon form matrix has a "staircase pattern," but no zeroes are required above the staircase. For example, the following matrices are in row echelon form:

$$\left[\begin{array}{ccc|c} 1 & 3 & -7 & 3 \\ 0 & 1 & 5 & 2 \\ 0 & 0 & 1 & 8 \end{array}\right] \quad \text{and} \quad \left[\begin{array}{cccc|c} 1 & 2 & 17 & 9 & 6 \\ 0 & 0 & 1 & 3 & 9 \\ 0 & 0 & 0 & 1 & -2 \\ 0 & 0 & 0 & 0 & 0 \end{array}\right].$$

To find the solution set for a system in row echelon form, we translate the matrix back into its associated linear system. Starting with the last equation in the system, we solve for the value of the left-most variable in that equation. We then substitute this value back into the other equations to eliminate this dependent variable from the system. Continuing in this fashion, we work up through the set of equations until we have solved for each dependent variable.

This procedure is called **back substitution**.

Example 1

Consider the following system of equations:

$$\begin{cases} 2w + x + 3y = 16 \\ 3w + 2x + z = 16 \\ 2w + 12y - 5z = 5 \end{cases}.$$

Row reducing the associated augmented matrix for this system produces the following row echelon form matrix:

$$\left[\begin{array}{cccc|c} 1 & \frac{1}{2} & \frac{3}{2} & 0 & 8 \\ 0 & 1 & -9 & 2 & -16 \\ 0 & 0 & 0 & 1 & 9 \end{array}\right].$$

This corresponds to the system

$$\begin{cases} w + \frac{1}{2}x + \frac{3}{2}y \qquad\qquad = \quad 8 \\ \qquad x - 9y + 2z = -16 \\ \qquad\qquad\qquad z = \quad 9 \end{cases}.$$

Solving for the left-most variable in each equation, we obtain

$$\begin{cases} w = \quad 8 - \frac{1}{2}x - \frac{3}{2}y \\ x = -16 + 9y - 2z \\ z = \quad 9 \end{cases}.$$

The third equation gives $z = 9$. The variable y is an independent variable, since we did not pivot in its corresponding column. Thus, we let y take on an arbitrary value, say $y = c$. Substituting the values of y and z into the second equation yields $x = -16 + 9c - 2(9) = 9c - 34$. Plugging all these values into the first equation gives $w = 8 - \frac{1}{2}(9c - 34) - \frac{3}{2}c = 8 - \frac{9}{2}c + 17 - \frac{3}{2}c = 25 - 6c$. Thus, we have the solution set $\{(25 - 6c, 9c - 34, c, 9) \mid c \in \mathbb{R}\}$. As before, particular solutions are found by substituting any chosen value for c. ■

Gaussian elimination is computationally more efficient than the Gauss-Jordan algorithm, since fewer arithmetic operations generally need to be performed. This is especially advantageous when using a computer, because it allows less chance for the compounding of roundoff errors.

Ill-Conditioned Systems

Sometimes the number of significant digits used in computations has a great effect on the answers. For example, consider the similar systems

$$\text{(A)} \quad \begin{cases} 2x_1 + x_2 = 2 \\ 2.005x_1 + x_2 = 7 \end{cases} \quad \text{and} \quad \text{(B)} \quad \begin{cases} 2x_1 + x_2 = 2 \\ 2.01x_1 + x_2 = 7 \end{cases}.$$

The linear equations of these systems are graphed in Figure 9.1.

Even though the coefficients of systems (A) and (B) are almost identical, the solutions to the systems are very different:

$$\text{Solution to (A)} = (1000, -1998) \text{ and solution to (B)} = (500, -998).$$

Systems like these, where a very small change in a coefficient leads to a very large change in the solution set, are called **ill-conditioned systems**. In this case, there is a geometric way to see that these systems are ill-conditioned: the pair of lines in each system are almost parallel. Therefore, a small change in one line can move the point of intersection very far along the other line, as in Figure 9.1.

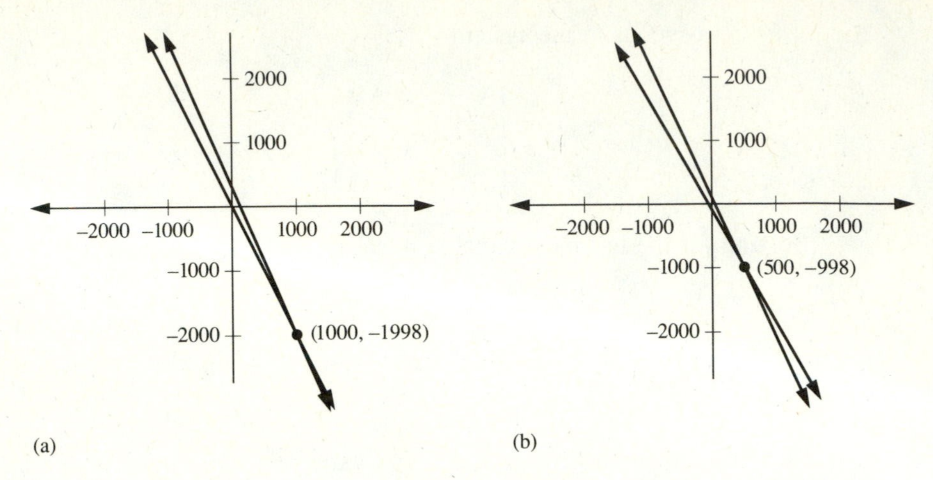

Figure 9.1
(a) Lines of system (A); (b) Lines of system (B)

Suppose the coefficients in system (A) had been obtained after a series of long calculations. A slight difference in the roundoff error of those calculations could lead to a very different final solution set. Thus, we need to be very careful when working with ill-conditioned systems. Special methods have been developed for recognizing ill-conditioned systems, and a technique known as "iterative refinement" is used when the coefficients are known only to a certain degree of accuracy. These methods are beyond the scope of this book, but further details can be found in *Numerical Analysis*, 4th ed., by Burden and Faires (published by PWS, 1989).

Partial Pivoting

A common problem in numerical linear algebra occurs when dividing by real numbers very close to zero—for example, during the row reduction process when a pivot element is extremely small. This small number might itself be inaccurate because of a previous roundoff error. Thus, performing a type (I) row operation with this number might result in additional roundoff error.

Even when dealing with accurate small numbers, we can still have problems. When we divide every entry of a row by a very small pivot value, the rest of those row entries could become much larger (in absolute value) than the other matrix entries. Hence, when these larger row entries are added to the (smaller) entries of another row in a type (II) operation, the most significant digits of the larger row entries may not be affected at all. That is, the data stored in smaller row entries may not be playing their proper role in determining the final solution set. As more computations are performed, these roundoff errors can accumulate, making the final result inaccurate.

Example 2

Consider the linear system

$$\begin{cases} 0.0006x_1 - x_2 + x_3 = 10 \\ 0.03x_1 + 30x_2 - 5x_3 = 15 \\ 0.04x_1 + 40x_2 - 7x_3 = 19 \end{cases}.$$

The unique solution is $(x_1, x_2, x_3) = (5000, -4, 3)$. But if we attempt to solve the system by row reduction and round all computations to four significant figures, we get an inaccurate result. For example, using Gaussian elimination, the augmented matrices are

$$\left[\begin{array}{ccc|c} 0.0006 & -1 & 1 & 10 \\ 0.03 & 30 & -5 & 15 \\ 0.04 & 40 & -7 & 19 \end{array}\right] \quad \text{(I): } <1> \longleftarrow (1/0.0006) <1>$$

$$\left[\begin{array}{ccc|c} 1 & -1667 & 1667 & 16670 \\ 0.03 & 30 & -5 & 15 \\ 0.04 & 40 & -7 & 19 \end{array}\right] \quad \begin{array}{l} \text{(II): } <2> \longleftarrow -0.03 <1> + <2> \\ \text{(II): } <3> \longleftarrow -0.04 <1> + <3> \end{array}$$

$$\left[\begin{array}{ccc|c} 1 & -1667 & 1667 & 16670 \\ 0 & 80.01 & -55.01 & -485.1 \\ 0 & 106.7 & -73.68 & -647.8 \end{array}\right] \quad \text{(I): } <2> \longleftarrow (1/80.01) <2>$$

$$\left[\begin{array}{ccc|c} 1 & -1667 & 1667 & 16670 \\ 0 & 1 & -0.6876 & -6.064 \\ 0 & 106.7 & -73.68 & -647.8 \end{array}\right] \quad \text{(II): } <3> \longleftarrow -106.7 <2> + <3>$$

$$\left[\begin{array}{ccc|c} 1 & -1667 & 1667 & 16670 \\ 0 & 1 & -0.6876 & -6.064 \\ 0 & 0 & -0.3131 & -0.7712 \end{array}\right] \quad \text{(I): } <3> \longleftarrow (-1/0.3131) <3>$$

$$\left[\begin{array}{ccc|c} 1 & -1667 & 1667 & 16670 \\ 0 & 1 & -0.6876 & -6.064 \\ 0 & 0 & 1 & 2.463 \end{array}\right].$$

Back substitution produces the solution $(x_1, x_2, x_3) = (5279, -4.370, 2.463)$. This inaccurate answer is largely the result of dividing row 1 through by 0.0006, a number much smaller than the other entries of the matrix, in the first step of the row reduction. ■

A method known as **partial pivoting** is employed to avoid roundoff errors of the type encountered in Example 2. In this method, when choosing a pivot element, we first determine whether there are any entries below the next pivot candidate that have a greater absolute value. If so, we switch rows to move the entry with the highest absolute value into the pivot position.

Example 3

We use partial pivoting on the system in Example 2. The initial augmented matrix is

$$\left[\begin{array}{ccc|c} 0.0006 & -1 & 1 & 10 \\ 0.03 & 30 & -5 & 15 \\ 0.04 & 40 & -7 & 19 \end{array}\right].$$

The entry in the first column with the largest absolute value is in the third row, so we interchange the first and third rows to obtain

$$\text{(III): } < 1 > \longleftrightarrow < 3 >$$

$$\left[\begin{array}{ccc|c} 0.04 & 40 & -7 & 19 \\ 0.03 & 30 & -5 & 15 \\ 0.0006 & -1 & 1 & 10 \end{array}\right]$$

Continuing the row reduction with the operation (I): $< 1 > \longleftarrow (1/0.04) < 1 >$, we obtain

$$\left[\begin{array}{ccc|c} 1 & 1000 & -175.0 & 475.0 \\ 0.03 & 30 & -5 & 15 \\ 0.0006 & -1 & 1 & 10 \end{array}\right] \quad \begin{array}{l} \text{(II): } < 2 > \longleftarrow -0.03 < 1 > + < 2 > \\ \text{(II): } < 3 > \longleftarrow -0.0006 < 1 > + < 3 > \end{array}$$

$$\left[\begin{array}{ccc|c} 1 & 1000 & -175.0 & 475.0 \\ 0 & 0.000 & 0.2500 & 0.7500 \\ 0 & -1.600 & 1.105 & 9.715 \end{array}\right] \quad \text{(III): } < 2 > \longleftrightarrow < 3 >$$

$$\left[\begin{array}{ccc|c} 1 & 1000 & -175.0 & 475.0 \\ 0 & -1.600 & 1.105 & 9.715 \\ 0 & 0 & 0.2500 & 0.7500 \end{array}\right] \quad \text{(I): } < 2 > \longleftarrow (-1/1.600) < 2 >$$

$$\left[\begin{array}{ccc|c} 1 & 1000 & -175.0 & 475.0 \\ 0 & 1 & -0.6906 & -6.072 \\ 0 & 0 & 0.2500 & 0.7500 \end{array}\right] \quad \text{(I): } < 3 > \longleftarrow (1/0.2500) < 3 >$$

$$\left[\begin{array}{ccc|c} 1 & 1000 & -175.0 & 475.0 \\ 0 & 1 & -0.6906 & -6.072 \\ 0 & 0 & 1 & 3.000 \end{array}\right]$$

Back substitution produces the solution $(x_1, x_2, x_3) = (5000, -4.000, 3.000)$. Therefore, by partial pivoting, we have obtained the correct solution, a big improvement over the answer obtained in Example 2 without partial pivoting. ■

For many systems, the method of partial pivoting is powerful enough to provide reasonably accurate answers. However, in more difficult cases, partial pivoting is not enough. An even more useful technique is **total pivoting** (or **full pivoting** or

complete pivoting), in which columns as well as rows are interchanged. The strategy in total pivoting is to select as the next pivot the entry with the largest absolute value from all the remaining rows and columns.

Iterative Techniques: Jacobi and Gauss-Seidel Methods

When we have a rough approximation of the solution to a certain $n \times n$ linear system, an **iterative method** may be the fastest way to obtain the actual solution. We use the initial approximation to generate a second (preferably better) approximation. We then use the second approximation to generate a third, and so on. The process stops if the approximations "stabilize"—that is, if the difference between successive approximations becomes negligible. In this section, we illustrate two iterative methods: the **Jacobi method** and the **Gauss-Seidel method**.

For these iterative methods, it is convenient to express linear systems in a slightly different form. Suppose we are given the following system of n equations in n unknowns:

$$\begin{cases} a_{11}x_1 + a_{12}x_2 + a_{13}x_3 + \cdots + a_{1n}x_n = b_1 \\ a_{21}x_1 + a_{22}x_2 + a_{23}x_3 + \cdots + a_{2n}x_n = b_2 \\ \quad\vdots \qquad\quad \vdots \qquad\quad \vdots \qquad\quad \vdots \qquad\quad \vdots \qquad\quad \vdots \\ a_{n1}x_1 + a_{n2}x_2 + a_{n3}x_3 + \cdots + a_{nn}x_n = b_n \end{cases}.$$

If the coefficient matrix has rank n, every row and column of the reduced row echelon form of the coefficient matrix contains a (nonzero) pivot. In this case, it is always possible to rearrange the equations so that the coefficient of x_i is nonzero in the ith equation, for $1 \leq i \leq n$. Let us assume that the equations have already been rearranged in this way.† Solving for x_i in the ith equation in terms of the remaining unknowns, we obtain

$$\begin{cases} x_1 = \qquad\qquad c_{12}x_2 + c_{13}x_3 + \cdots + c_{1n}x_n + d_1 \\ x_2 = c_{21}x_1 \qquad\qquad + c_{23}x_3 + \cdots + c_{2n}x_n + d_2 \\ x_3 = c_{31}x_1 + c_{32}x_2 \qquad\qquad + \cdots + c_{3n}x_n + d_3 \\ \;\vdots \qquad \vdots \qquad\quad \vdots \qquad\quad \vdots \qquad\quad \vdots \qquad\quad \vdots \qquad\quad \vdots \end{cases}.$$

For example, suppose we are given the system

$$\begin{cases} 3x_1 - 2x_2 + x_3 = 11 \\ 2x_1 + 7x_2 - 3x_3 = -14 \\ 9x_1 - x_2 - 4x_3 = 17 \end{cases}.$$

Solving for x_1 in the first equation, x_2 in the second equation, and x_3 in the third equation, we obtain

---

†In fact, the Jacobi and Gauss-Seidel methods often require fewer steps if the equations are rearranged so that the coefficient of x_i in the ith row is as large as possible.

$$\begin{cases} x_1 = \frac{2}{3}x_2 - \frac{1}{3}x_3 + \frac{11}{3} \\ x_2 = -\frac{2}{7}x_1 + \frac{3}{7}x_3 - 2 \\ x_3 = \frac{9}{4}x_1 - \frac{1}{4}x_2 - \frac{17}{4} \end{cases}.$$

For the **Jacobi method**, we solve a system in the form

$$\begin{cases} x_1 = \phantom{c_{21}x_1 +} c_{12}x_2 + c_{13}x_3 + \cdots + c_{1n}x_n + d_1 \\ x_2 = c_{21}x_1 \phantom{+ c_{32}x_2} + c_{23}x_3 + \cdots + c_{2n}x_n + d_2 \\ x_3 = c_{31}x_1 + c_{32}x_2 \phantom{+ c_{23}x_3} + \cdots + c_{3n}x_n + d_3 \\ \vdots \quad \vdots \quad \vdots \quad \vdots \quad \vdots \quad \vdots \quad \vdots \end{cases}$$

by substituting an initial approximation for $x_1, x_2, \ldots, x_n$ into the right-hand side to obtain new values for $x_1, x_2, \ldots, x_n$ on the left-hand side. These new values are then substituted into the right-hand side to obtain another set of values for $x_1, x_2, \ldots, x_n$ on the left-hand side. This process is repeated as many times as necessary. If the values on the left-hand side "stabilize," they are a good approximation for a solution.

Example 4

We solve

$$\begin{cases} 8x_1 + x_2 - 2x_3 = -11 \\ 2x_1 + 9x_2 + x_3 = 22 \\ -x_1 - 2x_2 + 11x_3 = -15 \end{cases}$$

with the Jacobi method. The true solution is $(x_1, x_2, x_3) = (-2, 3, -1)$. Let us use $x_1 = -1.5$, $x_2 = 2.5$, and $x_3 = -0.5$ as an initial approximation (or guess) of the solution.

First, we rewrite the system in the form

$$\begin{cases} x_1 = -\frac{1}{8}x_2 + \frac{1}{4}x_3 - \frac{11}{8} \\ x_2 = -\frac{2}{9}x_1 - \frac{1}{9}x_3 + \frac{22}{9} \\ x_3 = \frac{1}{11}x_1 + \frac{2}{11}x_2 - \frac{15}{11} \end{cases}$$

In the following calculations, we round all results to three decimal places. Plugging the initial guess into the right-hand side of each equation, we get

$$x_1 = -\tfrac{1}{8}(2.5) + \tfrac{1}{4}(-0.5) - \tfrac{11}{8},$$

$$x_2 = -\tfrac{2}{9}(-1.5) - \tfrac{1}{9}(-0.5) + \tfrac{22}{9},$$
$$x_3 = \tfrac{1}{11}(-1.5) + \tfrac{2}{11}(2.5) - \tfrac{15}{11},$$

yielding the new values $x_1 = -1.813$, $x_2 = 2.833$, $x_3 = -1.045$. We then plug these values into the right-hand side of each equation to obtain

$$x_1 = -\tfrac{1}{8}(2.833) + \tfrac{1}{4}(-1.045) - \tfrac{11}{8},$$
$$x_2 = -\tfrac{2}{9}(-1.813) - \tfrac{1}{9}(-1.045) + \tfrac{22}{9},$$
$$x_3 = \tfrac{1}{11}(-1.813) + \tfrac{2}{11}(2.833) - \tfrac{15}{11},$$

yielding the values $x_1 = -1.990$, $x_2 = 2.963$, $x_3 = -1.013$. Repeating this process, we get the values in the following chart:

	x_1	x_2	x_3
Initial values	−1.500	2.500	−0.500
After 1 step	−1.813	2.833	−1.045
After 2 steps	−1.990	2.963	−1.013
After 3 steps	−1.999	2.999	−1.006
After 4 steps	−2.001	3.000	−1.000
After 5 steps	−2.000	3.000	−1.000
After 6 steps	−2.000	3.000	−1.000

After six steps, the values for x_1, x_2, and x_3 have stabilized at the true solution. ■

In Example 4, we could have used any starting values for x_1, x_2, and x_3 as the initial approximation. In the absence of any information about the solution, we can begin with $x_1 = x_2 = x_3 = 0$. If we use the Jacobi method on the system in Example 4 with $x_1 = x_2 = x_3 = 0$ as the initial values, we obtain the following chart (again, rounding each result to three decimal places):

	x_1	x_2	x_3
Initial values	0.000	0.000	0.000
After 1 step	−1.375	2.444	−1.364
After 2 steps	−2.022	2.902	−1.044
After 3 steps	−1.999	3.010	−1.020
After 4 steps	−2.006	3.002	−0.998
After 5 steps	−2.000	3.001	−1.000
After 6 steps	−2.000	3.000	−1.000
After 7 steps	−2.000	3.000	−1.000

In this case, the Jacobi method still produces the correct solution, although an extra step is required.

The **Gauss-Seidel method** is similar to the Jacobi method except that, as each new value x_i is obtained, it is used immediately in place of the previous value for x_i when plugging values into the right-hand side of the equations.

Example 5

Consider the system

$$\begin{cases} 8x_1 + x_2 - 2x_3 = -11 \\ 2x_1 + 9x_2 + x_3 = 22 \\ -x_1 - 2x_2 + 11x_3 = -15 \end{cases}$$

of Example 4. We solve this system with the Gauss-Seidel method, using the initial approximation $x_1 = x_2 = x_3 = 0$. Again, we begin by rewriting the system in the form

$$\begin{cases} x_1 = -\frac{1}{8}x_2 + \frac{1}{4}x_3 - \frac{11}{8} \\ x_2 = -\frac{2}{9}x_1 - \frac{1}{9}x_3 + \frac{22}{9} \\ x_3 = \frac{1}{11}x_1 + \frac{2}{11}x_2 - \frac{15}{11} \end{cases}.$$

Plugging the initial approximation into the right-hand side of the first equation, we get

$$x_1 = -\tfrac{1}{8}(0) + \tfrac{1}{4}(0) - \tfrac{11}{8} = -1.375.$$

We now plug this new value for x_1 and the current values for x_2 and x_3 into the right-hand side of the second equation to get

$$x_2 = -\tfrac{2}{9}(-1.375) - \tfrac{1}{9}(0) + \tfrac{22}{9} = 2.750.$$

We then plug the new values for x_1 and x_2 and the current value for x_3 into the right-hand side of the third equation to get

$$x_3 = \tfrac{1}{11}(-1.375) + \tfrac{2}{11}(2.750) - \tfrac{15}{11} = -0.989.$$

The process is then repeated as many times as necessary with the newest values of x_1, x_2, and x_3 used in each case. The results are given in the following chart (rounding all results to three decimal places):

	x_1	x_2	x_3
Initial values	0.000	0.000	0.000
After 1 step	−1.375	2.750	−0.989
After 2 steps	−1.966	2.991	−0.999
After 3 steps	−1.999	3.000	−1.000
After 4 steps	−2.000	3.000	−1.000
After 5 steps	−2.000	3.000	−1.000

After five steps, we see that the values for x_1, x_2, and x_3 have stabilized to the correct solution. ■

The Gauss-Seidel method takes fewer steps than the Jacobi method in most ordinary applications. However, for some systems, the Jacobi method is superior to the Gauss-Seidel method. In addition, for some systems neither method produces the correct answer (see Exercise 9).† On the other hand, for certain classes of linear systems, the Jacobi and Gauss-Seidel methods will always stabilize to the correct solution for any given initial approximation (see Exercise 8).

Comparing Iterative and Row Reduction Methods

When are iterative methods useful? In many applications, the coefficient matrix for a given system contains a large number of zeroes. Such matrices are said to be **sparse**. When a linear system has a sparse matrix, each equation in the system may involve very few variables. If so, each step of the iterative process is relatively easy. However, the Gauss-Jordan and Gaussian elimination methods would not be very attractive in such a case, because the cumulative effect of many row operations would tend to replace the zero coefficients with nonzero numbers. Also, iterative methods can often give more accurate answers when large matrices are involved, because there are fewer overall arithmetic operations performed. Roundoff errors are not given a chance to "accumulate," as they are in the Gauss-Jordan method, because each iteration essentially creates a new approximation to the solution. The only roundoff error that we need to consider with an iterative method is the error involved in the most recent step.

On the other hand, when iterative methods take an extremely large number of steps to stabilize or do not stabilize at all, it is much better to use the Gauss-Jordan or Gaussian elimination method.

---

†In cases where the Jacobi and Gauss-Seidel methods do not stabilize, related iterative methods (known as "relaxation methods") may still work. For further details, see *Numerical Analysis*, 4th ed., by Burden and Faires (published by PWS, 1989).

Exercises—Section 9.1

Note: You should use a calculator or appropriate computer software to solve these problems.

1. In each part of this exercise, find the exact solution sets for the two given systems. Are these systems ill-conditioned? Why or why not?

★(a) $\begin{cases} 5x - 2y = 10 \\ 5x - 1.995y = 17.5 \end{cases}$, $\begin{cases} 5x - 2y = 10 \\ 5x - 1.99y = 17.5 \end{cases}$

(b) $\begin{cases} 6x - z = 400 \\ 3y - z = 400 \\ 25x + 12y - 8z = 3600 \end{cases}$, $\begin{cases} 6x - 1.01z = 400 \\ 3y - z = 400 \\ 25x + 12y - 8z = 3600 \end{cases}$

2. Using Gaussian elimination, solve each of the systems in Section 2.1, Exercise 3, parts (a), (c), (d), and (h).

3. First, use Gaussian elimination *without* partial pivoting to solve each of the following systems. Then solve each system using Gaussian elimination *with* partial pivoting. Which solution is more accurate? In each case, round all numbers in the problem to three significant figures before beginning, and round the results of all computations to three significant figures.

★(a) $\begin{cases} 0.00072x - 4.312y = -0.9846 \\ 2.31x - 9876.0y = -130.8 \end{cases}$

(b) $\begin{cases} 0.0004x_1 - 0.6234x_2 - 2.123x_3 = 5.581 \\ 0.0832x_1 - 26.17x_2 - 1.759x_3 = -3.305 \\ 0.09512x_1 + 0.1458x_2 + 55.13x_3 = 11.168 \end{cases}$

★(c) $\begin{cases} 0.00032x_1 + 0.2314x_2 + 0.127x_3 = -0.03456 \\ -241x_1 - 217x_2 - 8x_3 = -576 \\ 49x_1 + 45x_2 + 2.4x_3 = 283.2 \end{cases}$

4. Repeat Exercise 3, but round all computations to four significant figures.

5. Solve each of the following systems using the Jacobi method. Round all results to three decimal places, and stop when successive values of the variables agree to three decimal places. Let the initial values of all variables be zero. List the values of the variables after each step of the iteration.

★(a) $\begin{cases} 5x_1 + x_2 = 26 \\ 3x_1 + 7x_2 = -42 \end{cases}$

(b) $\begin{cases} 9x_1 - x_2 - x_3 = -7 \\ 2x_1 - 8x_2 - x_3 = 35 \\ x_1 + 2x_2 + 11x_3 = 22 \end{cases}$

★(c) $\begin{cases} 7x_1 + x_2 - 2x_3 = -62 \\ -x_1 + 6x_2 + x_3 = 27 \\ 2x_1 - x_2 - 6x_3 = 26 \end{cases}$

(d) $\begin{cases} 10x_1 + x_2 - 2x_3 + x_4 = 9 \\ -x_1 - 9x_2 + x_3 - 2x_4 = 15 \\ -2x_1 + x_2 + 7x_3 + x_4 = 21 \\ x_1 - x_2 - x_3 + 13x_4 = -27 \end{cases}$

6. Repeat Exercise 5 using Gauss-Seidel iteration instead of Jacobi iteration.

★7. A square matrix is **strictly diagonally dominant** if the absolute value of each diagonal entry is larger than the sum of the absolute values of the remaining entries in its row. That is, if $\mathbf{A}$ is an $n \times n$ matrix, then $\mathbf{A}$ is strictly diagonally dominant if, for $1 \leq i \leq n$, $|a_{ii}| < \sum_{\substack{1 \leq j \leq n \\ j \neq i}} |a_{ij}|$. Which of the following matrices are strictly diagonally dominant?

(a) $\begin{bmatrix} -3 & 1 \\ -2 & 4 \end{bmatrix}$ (b) $\begin{bmatrix} 2 & 2 \\ 4 & 3 \end{bmatrix}$ (c) $\begin{bmatrix} -6 & 2 & 1 \\ 2 & 5 & -2 \\ -1 & 4 & 7 \end{bmatrix}$

(d) $\begin{bmatrix} 15 & 9 & -3 \\ 3 & 6 & 4 \\ 7 & -2 & 11 \end{bmatrix}$ (e) $\begin{bmatrix} 6 & 2 & 3 \\ 4 & 5 & 1 \\ 7 & 1 & 9 \end{bmatrix}$ (f) $\begin{bmatrix} 10 & 5 & 4 \\ -3 & -11 & 6 \\ -4 & 6 & 12 \end{bmatrix}$

(g) $\begin{bmatrix} 20 & -9 & 1 & 9 \\ 7 & 18 & 7 & -3 \\ -3 & 8 & 19 & 7 \\ 3 & 11 & -2 & 17 \end{bmatrix}$

8. The Jacobi and Gauss-Seidel methods stabilize to the correct solution (for any choice of initial values) if the equations can be rearranged to make the coefficient matrix for the system strictly diagonally dominant (see Exercise 7). For the following systems, rearrange the equations accordingly, and then perform the Gauss-Seidel method. Use initial values of zero for all variables. Round all results to three decimal places. List the values of the variables after each step of the iteration, and give the final solution set in each case.

★(a) $\begin{cases} 2x_1 + 13x_2 + x_3 = 0 \\ x_1 - 2x_2 + 15x_3 = 26 \\ 8x_1 - x_2 + 3x_3 = 25 \end{cases}$ (b) $\begin{cases} -3x_1 - x_2 - 7x_3 = -39 \\ 10x_1 + x_2 + x_3 = 37 \\ x_1 + 9x_2 + 2x_3 = -58 \end{cases}$

★(c) $\begin{cases} x_1 + x_2 + 13x_3 + 2x_4 = 120 \\ 9x_1 + 2x_2 - x_3 + x_4 = 49 \\ -2x_1 + 3x_2 - x_3 - 14x_4 = 110 \\ -x_1 - 17x_2 - 3x_3 + 2x_4 = 86 \end{cases}$

★9. Show that neither the Jacobi method nor the Gauss-Seidel method seems to stabilize when applied to the following system by observing what happens during the first six steps of the Jacobi method and the first four steps of the Gauss-Seidel method. Let the initial value of all variables be zero, and round all results to three decimal places. Then, find the solution using Gaussian elimination.

$$\begin{cases} x_1 - 5x_2 - \ \ x_3 = 16 \\ 6x_1 - \ \ x_2 - 2x_3 = 13 \\ 7x_1 + \ \ x_2 + \ \ x_3 = 12 \end{cases}$$

10. (a) For the following system, show that, with initial values of zero for each variable, the Gauss-Seidel method stabilizes to the correct solution. Round all results to three decimal places, and give the values of the variables after each step of the iteration.

$$\begin{cases} 2x_1 + \ \ x_2 + \ \ x_3 = 7 \\ \ \ x_1 + 2x_2 + \ \ x_3 = 8 \\ \ \ x_1 + \ \ x_2 + 2x_3 = 9 \end{cases}$$

(b) Work out the first eight steps of the Jacobi method for the system in part (a) (again using initial values of zero for each variable), and observe that this method does not stabilize: on alternate passes the results oscillate between values near $x_1 = 3$, $x_2 = 4$, $x_3 = 5$ and $x_1 = -1$, $x_2 = 0$, $x_3 = 1$.

9.2 LDU Decomposition

Prerequisite: Section 2.5, Elementary Matrices

In this section, we show that many nonsingular matrices can be written as the product of a lower triangular matrix **L**, a diagonal matrix **D**, and an upper triangular matrix **U**. As you will see, this **LDU** decomposition is useful in solving certain types of linear systems.

LDU decomposition is used here only to solve systems having square coefficient matrices, but the method can be generalized to solve systems with nonsquare coefficient matrices as well.

Calculating the LDU Decomposition

For a given matrix **A**, we can find the matrices **L**, **D**, and **U** by row reduction. However, it is not necessary to bring **A** completely to reduced row echelon form; *row echelon form*, where the entries above the staircase pattern of pivots are not altered (see Section 9.1), is sufficient.

In our discussion, we need to give a name to a row operation of type (II) in which the home row is used to zero out an entry *below* it. Let us call this a **lower type (II)** row operation. Notice that a matrix can be put in row echelon form using only type (I) and lower type (II) operations if you do not need to interchange any rows.

We can now state the **LDU** decomposition theorem:

THEOREM 9.1

Let $\mathbf{A}$ be a nonsingular $n \times n$ matrix. If $\mathbf{A}$ can be row reduced to row echelon form using only type (I) and lower type (II) operations, then $\mathbf{A} = \mathbf{LDU}$, where $\mathbf{L}$ is an $n \times n$ lower triangular matrix, $\mathbf{D}$ is an $n \times n$ diagonal matrix, and $\mathbf{U}$ is an $n \times n$ upper triangular matrix and where all main diagonal entries of $\mathbf{L}$ and $\mathbf{U}$ equal 1.

Furthermore, this decomposition of $\mathbf{A}$ is unique—that is, if $\mathbf{A} = \mathbf{L}'\mathbf{D}'\mathbf{U}'$, where $\mathbf{L}'$ is $n \times n$ lower triangular, $\mathbf{D}'$ is $n \times n$ diagonal, and $\mathbf{U}'$ is $n \times n$ upper triangular, with all main diagonal entries of $\mathbf{L}'$ and $\mathbf{U}'$ equal to 1, then $\mathbf{L}' = \mathbf{L}$, $\mathbf{D}' = \mathbf{D}$, and $\mathbf{U}' = \mathbf{U}$.

We outline the proof of this theorem below, to illustrate how to calculate the **LDU** decomposition for a matrix **A** when it exists. To produce a complete version of the proof, we would first have to prove several results regarding lower triangular matrices. Also, we omit the proof of uniqueness, since that property is not needed for the applications.

Proof of Theorem 9.1 (outline)

Suppose that $\mathbf{A}$ is a nonsingular $n \times n$ matrix and we can reduce $\mathbf{A}$ to row echelon form using only type (I) and lower type (II) row operations. Let $\mathbf{U}$ be the row echelon form matrix obtained from this process. Since $\mathbf{A}$ is nonsingular, all of the main diagonal entries of $\mathbf{U}$ must equal 1. Now, $\mathbf{U} = \mathbf{E}_t \cdots \mathbf{E}_2\mathbf{E}_1\mathbf{A}$, where $\mathbf{E}_1, \ldots, \mathbf{E}_t$ are elementary matrices corresponding to the row operations used. By Theorem 2.17, we have $\mathbf{A} = \mathbf{E}_1^{-1}\mathbf{E}_2^{-1} \cdots \mathbf{E}_t^{-1}\mathbf{U}$. Since only row operations of type (I) and lower type (II) are involved, it can be proved that $\mathbf{E}_i$ is lower triangular for $1 \leq i \leq t$. It can then be shown that each inverse $\mathbf{E}_i^{-1}$ is lower triangular. Proving that the product $\mathbf{K} = \mathbf{E}_1^{-1}\mathbf{E}_2^{-1} \cdots \mathbf{E}_t^{-1}$ is also lower triangular is then straightforward, and so $\mathbf{K}$ has the general form

$$\begin{bmatrix} k_{11} & 0 & 0 & \cdots & 0 \\ k_{21} & k_{22} & 0 & \cdots & 0 \\ k_{31} & k_{32} & k_{33} & \cdots & 0 \\ \vdots & \vdots & \vdots & \ddots & \vdots \\ k_{n1} & k_{n2} & k_{n3} & \cdots & k_{nn} \end{bmatrix}.$$

Since only type (I) and lower type (II) operations were used in the row reduction of $\mathbf{A}$, all main diagonal entries of each $\mathbf{E}_i^{-1}$ are nonzero. It is easy to show that all main diagonal entries of $\mathbf{K}$ are also nonzero. Thus, $\mathbf{K}$ can be expressed as $\mathbf{LD}$, where

$$\mathbf{L} = \begin{bmatrix} 1 & 0 & 0 & \cdots & 0 \\ \frac{k_{21}}{k_{11}} & 1 & 0 & \cdots & 0 \\ \frac{k_{31}}{k_{11}} & \frac{k_{32}}{k_{22}} & 1 & \cdots & 0 \\ \vdots & \vdots & \vdots & \ddots & \vdots \\ \frac{k_{n1}}{k_{11}} & \frac{k_{n2}}{k_{22}} & \frac{k_{n3}}{k_{33}} & \cdots & 1 \end{bmatrix} \quad \text{and} \quad \mathbf{D} = \begin{bmatrix} k_{11} & 0 & 0 & \cdots & 0 \\ 0 & k_{22} & 0 & \cdots & 0 \\ 0 & 0 & k_{33} & \cdots & 0 \\ \vdots & \vdots & \vdots & \ddots & \vdots \\ 0 & 0 & 0 & \cdots & k_{nn} \end{bmatrix},$$

where **L** is lower triangular and **D** is diagonal. Therefore, we have $\mathbf{A} = \mathbf{KU} = \mathbf{LDU}$, with **L** lower triangular, **D** diagonal, **U** upper triangular, and all main diagonal entries of **L** and **U** equal to 1. ■

In the next example, we decompose a nonsingular matrix **A** into **LDU** form. As in the proof of Theorem 9.1, we first decompose **A** into **KU** form, with $\mathbf{K} = \mathbf{LD}$. We then find the matrices **L** and **D** using **K**.

Example 1

Let us express

$$\mathbf{A} = \begin{bmatrix} 2 & 1 & 4 \\ 3 & 2 & 5 \\ 4 & 1 & 9 \end{bmatrix}$$

in **LDU** form. To do this, we convert **A** into row echelon form **U**. For each row operation, we give the corresponding elementary matrix and its inverse. (Notice that only type (I) and lower type (II) row operations are used.)

$$\mathbf{A} = \begin{bmatrix} 2 & 1 & 4 \\ 3 & 2 & 5 \\ 4 & 1 & 9 \end{bmatrix} \quad \left(\begin{array}{l} \text{(I): } <1> \longleftarrow \frac{1}{2} <1>, \\ \mathbf{E}_1 = \begin{bmatrix} \frac{1}{2} & 0 & 0 \\ 0 & 1 & 0 \\ 0 & 0 & 1 \end{bmatrix}, \quad \mathbf{E}_1^{-1} = \begin{bmatrix} 2 & 0 & 0 \\ 0 & 1 & 0 \\ 0 & 0 & 1 \end{bmatrix} \end{array} \right)$$

$$\Longrightarrow \begin{bmatrix} 1 & \frac{1}{2} & 2 \\ 3 & 2 & 5 \\ 4 & 1 & 9 \end{bmatrix} \quad \left(\begin{array}{l} \text{(II): } <2> \longleftarrow -3 <1> + <2>, \\ \mathbf{E}_2 = \begin{bmatrix} 1 & 0 & 0 \\ -3 & 1 & 0 \\ 0 & 0 & 1 \end{bmatrix}, \quad \mathbf{E}_2^{-1} = \begin{bmatrix} 1 & 0 & 0 \\ 3 & 1 & 0 \\ 0 & 0 & 1 \end{bmatrix} \end{array} \right)$$

$$\Longrightarrow \begin{bmatrix} 1 & \frac{1}{2} & 2 \\ 0 & \frac{1}{2} & -1 \\ 4 & 1 & 9 \end{bmatrix} \quad \left(\begin{array}{l} \text{(II): } <3> \longleftarrow -4 <1> + <3>, \\ \mathbf{E}_3 = \begin{bmatrix} 1 & 0 & 0 \\ 0 & 1 & 0 \\ -4 & 0 & 1 \end{bmatrix}, \quad \mathbf{E}_3^{-1} = \begin{bmatrix} 1 & 0 & 0 \\ 0 & 1 & 0 \\ 4 & 0 & 1 \end{bmatrix} \end{array} \right)$$

$$\Longrightarrow \begin{bmatrix} 1 & \frac{1}{2} & 2 \\ 0 & \frac{1}{2} & -1 \\ 0 & -1 & 1 \end{bmatrix} \quad \left(\begin{array}{l} \text{(I): } < 2 > \longleftarrow 2 < 2 >, \\ \mathbf{E}_4 = \begin{bmatrix} 1 & 0 & 0 \\ 0 & 2 & 0 \\ 0 & 0 & 1 \end{bmatrix}, \quad \mathbf{E}_4^{-1} = \begin{bmatrix} 1 & 0 & 0 \\ 0 & \frac{1}{2} & 0 \\ 0 & 0 & 1 \end{bmatrix} \end{array} \right)$$

$$\Longrightarrow \begin{bmatrix} 1 & \frac{1}{2} & 2 \\ 0 & 1 & -2 \\ 0 & -1 & 1 \end{bmatrix} \quad \left(\begin{array}{l} \text{(II): } < 3 > \longleftarrow 1 < 2 > + < 3 >, \\ \mathbf{E}_5 = \begin{bmatrix} 1 & 0 & 0 \\ 0 & 1 & 0 \\ 0 & 1 & 1 \end{bmatrix}, \quad \mathbf{E}_5^{-1} = \begin{bmatrix} 1 & 0 & 0 \\ 0 & 1 & 0 \\ 0 & -1 & 1 \end{bmatrix} \end{array} \right)$$

$$\Longrightarrow \begin{bmatrix} 1 & \frac{1}{2} & 2 \\ 0 & 1 & -2 \\ 0 & 0 & -1 \end{bmatrix} \quad \left(\begin{array}{l} \text{(I): } < 3 > \longleftarrow -1 < 3 >, \\ \mathbf{E}_6 = \begin{bmatrix} 1 & 0 & 0 \\ 0 & 1 & 0 \\ 0 & 0 & -1 \end{bmatrix}, \quad \mathbf{E}_6^{-1} = \begin{bmatrix} 1 & 0 & 0 \\ 0 & 1 & 0 \\ 0 & 0 & -1 \end{bmatrix} \end{array} \right)$$

$$\Longrightarrow \begin{bmatrix} 1 & \frac{1}{2} & 2 \\ 0 & 1 & -2 \\ 0 & 0 & 1 \end{bmatrix} = \mathbf{U}.$$

Thus, we have $\mathbf{U} = \mathbf{E}_6\mathbf{E}_5 \cdots \mathbf{E}_1\mathbf{A}$, so $\mathbf{A} = \mathbf{E}_1^{-1}\mathbf{E}_2^{-1} \cdots \mathbf{E}_6^{-1}\mathbf{U}$. The product $\mathbf{K} = \mathbf{E}_1^{-1}\mathbf{E}_2^{-1} \cdots \mathbf{E}_6^{-1}$ is lower triangular, and after some calculation we find that

$$\mathbf{K} = \begin{bmatrix} 2 & 0 & 0 \\ 3 & \frac{1}{2} & 0 \\ 4 & -1 & -1 \end{bmatrix}.$$

Finally, $\mathbf{K}$ can be broken into a product $\mathbf{LD}$ as follows: take the main diagonal entries of $\mathbf{D}$ to be those of $\mathbf{K}$, and create $\mathbf{L}$ by dividing each column of $\mathbf{K}$ by the main diagonal entry in that column. Performing these steps yields

$$\mathbf{L} = \begin{bmatrix} 1 & 0 & 0 \\ \frac{3}{2} & 1 & 0 \\ 2 & -2 & 1 \end{bmatrix} \quad \text{and} \quad \mathbf{D} = \begin{bmatrix} 2 & 0 & 0 \\ 0 & \frac{1}{2} & 0 \\ 0 & 0 & -1 \end{bmatrix}.$$

You should verify that $\mathbf{A} = \mathbf{LDU}$. ■

We can shorten the tedious process of **LDU** decomposition by skipping over the elementary matrices involved and creating the product $\mathbf{K} = \mathbf{LD}$ directly. Notice in Example 1 that

> The main diagonal entries of **K** are the reciprocals $\frac{1}{c}$ of the values of c used in the type (I) row operations: $< i > \longleftarrow c < i >$. Also, the values of **K** below the main diagonal are the additive inverses $-c$ of the constants c used in the type (II) operations: $< j > \longleftarrow c < i > + < j >$.

Once the matrix **K** is known, **L** and **D** are easily computed, as in Example 1.

Solving a System Using LDU Decomposition

Sometimes it is useful to leave the **LDU** decomposition of **A** in **KU** form when **A** is the coefficient matrix of a system of linear equations. We can then find the solution using substitution techniques, as in the next example.

Example 2

We solve

$$\begin{cases} -4x_1 + 5x_2 - 2x_3 = 5 \\ -3x_1 + 2x_2 - x_3 = 4 \\ x_1 + x_2 = -1 \end{cases}$$

by decomposing the coefficient matrix into **KU** form. Let **A** be the coefficient matrix. First, putting **A** into row echelon form **U**, we have

$$\mathbf{A} = \begin{bmatrix} -4 & 5 & -2 \\ -3 & 2 & -1 \\ 1 & 1 & 0 \end{bmatrix} \qquad (\text{(I): } \langle 1 \rangle \longleftarrow -\tfrac{1}{4}\langle 1 \rangle)$$

$$\Longrightarrow \begin{bmatrix} 1 & -\frac{5}{4} & \frac{1}{2} \\ -3 & 2 & -1 \\ 1 & 1 & 0 \end{bmatrix} \qquad \begin{array}{l} (\text{(II): } \langle 2 \rangle \longleftarrow 3\langle 1 \rangle + \langle 2 \rangle \\ \text{and } \langle 3 \rangle \longleftarrow -1\langle 1 \rangle + \langle 3 \rangle) \end{array}$$

$$\Longrightarrow \begin{bmatrix} 1 & -\frac{5}{4} & \frac{1}{2} \\ 0 & -\frac{7}{4} & \frac{1}{2} \\ 0 & \frac{9}{4} & -\frac{1}{2} \end{bmatrix} \qquad (\text{(I): } \langle 2 \rangle \longleftarrow -\tfrac{4}{7}\langle 2 \rangle)$$

$$\Longrightarrow \begin{bmatrix} 1 & -\frac{5}{4} & \frac{1}{2} \\ 0 & 1 & -\frac{2}{7} \\ 0 & \frac{9}{4} & -\frac{1}{2} \end{bmatrix} \qquad (\text{(II): } \langle 3 \rangle \longleftarrow -\tfrac{9}{4}\langle 2 \rangle + \langle 3 \rangle)$$

$$\Longrightarrow \begin{bmatrix} 1 & -\frac{5}{4} & \frac{1}{2} \\ 0 & 1 & -\frac{2}{7} \\ 0 & 0 & \frac{1}{7} \end{bmatrix} \qquad (\text{(I): } \langle 3 \rangle \longleftarrow 7\langle 3 \rangle)$$

$$\Longrightarrow \begin{bmatrix} 1 & -\frac{5}{4} & \frac{1}{2} \\ 0 & 1 & -\frac{2}{7} \\ 0 & 0 & 1 \end{bmatrix} = \mathbf{U}.$$

Then

$$\mathbf{K} = \begin{bmatrix} -4 & 0 & 0 \\ -3 & -\frac{7}{4} & 0 \\ 1 & \frac{9}{4} & \frac{1}{7} \end{bmatrix},$$

because the main diagonal entries of **K** are the reciprocals of the constants used in the type (I) operations, and the entries of **K** below the main diagonal are the additive inverses of the constants used in the type (II) operations.

Now the original system can be written as

$$\mathbf{A}\begin{bmatrix} x_1 \\ x_2 \\ x_3 \end{bmatrix} = \begin{bmatrix} 5 \\ 4 \\ -1 \end{bmatrix}, \quad \text{or} \quad \mathbf{KU}\begin{bmatrix} x_1 \\ x_2 \\ x_3 \end{bmatrix} = \begin{bmatrix} 5 \\ 4 \\ -1 \end{bmatrix}.$$

If we let

$$\begin{bmatrix} y_1 \\ y_2 \\ y_3 \end{bmatrix} = \mathbf{U}\begin{bmatrix} x_1 \\ x_2 \\ x_3 \end{bmatrix}, \quad \text{then we can write} \quad \mathbf{K}\begin{bmatrix} y_1 \\ y_2 \\ y_3 \end{bmatrix} = \begin{bmatrix} 5 \\ 4 \\ -1 \end{bmatrix}.$$

Each of these last two systems can be solved easily using substitution. We solve the second system for the y-values, and once they are known, we solve the first system for the x-values.

The second system,

$$\mathbf{K}\begin{bmatrix} y_1 \\ y_2 \\ y_3 \end{bmatrix} = \begin{bmatrix} 5 \\ 4 \\ -1 \end{bmatrix},$$

is equivalent to

$$\begin{cases} -4y_1 & = \;\; 5 \\ -3y_1 - \frac{7}{4}y_2 & = \;\; 4 \\ \;\;\; y_1 + \frac{9}{4}y_2 + \frac{1}{7}y_3 & = -1 \end{cases}.$$

The first equation gives $y_1 = -\frac{5}{4}$. Substituting this solution into the second equation and solving for y_2, we get $-3(-\frac{5}{4}) - \frac{7}{4}y_2 = 4$, or $y_2 = -\frac{1}{7}$. Finally, substituting for y_1 and y_2 in the third equation, we get $-\frac{5}{4} + \frac{9}{4}(-\frac{1}{7}) + \frac{1}{7}y_3 = -1$, or $y_3 = 4$. But then the first system,

$$\mathbf{U}\begin{bmatrix} x_1 \\ x_2 \\ x_3 \end{bmatrix} = \begin{bmatrix} y_1 \\ y_2 \\ y_3 \end{bmatrix},$$

is equivalent to

$$\begin{cases} x_1 - \frac{5}{4}x_2 + \frac{1}{2}x_3 & = -\frac{5}{4} \\ \qquad\; x_2 - \frac{2}{7}x_3 & = -\frac{1}{7} \\ \qquad\qquad\quad x_3 & = \;\; 4 \end{cases}.$$

This time, we solve the equations in *reverse* order. The last equation gives $x_3 = 4$. Then $x_2 - \frac{2}{7}(4) = -\frac{1}{7}$, or $x_2 = 1$. Finally, $x_1 - \frac{5}{4}(1) + \frac{1}{2}(4) = -\frac{5}{4}$, or $x_1 = -2$. Therefore $(x_1, x_2, x_3) = (-2, 1, 4)$. ■

Solving a system of linear equations using (**KU** =) **LDU** decomposition has an advantage over the Gauss-Jordan method when there are many systems to be solved with the same coefficient matrix **A**. In that case, **K** and **U** need to be calculated just once, and the solutions to each system can be obtained relatively efficiently using substitution. You saw a similar philosophy in Section 2.4 when we discussed the practicality of solving several systems that had the same coefficient matrix by using the inverse of that matrix.

In our discussion of **LDU** decomposition, we have not encountered type (III) row operations. If we use type (III) row operations in reducing a nonsingular matrix **A** to row echelon form, it turns out that **A** = **PLDU**, for some product **P** of $n \times n$ elementary matrices of type (III), and with **L**, **D**, and **U** as before. However, this **PLDU** decomposition is not necessarily unique.

Exercises—Section 9.2

1. Find the **LDU** decomposition for each of the following matrices.

★(a) $\begin{bmatrix} 2 & -4 \\ -6 & 17 \end{bmatrix}$ (b) $\begin{bmatrix} 3 & 1 \\ \frac{3}{2} & -\frac{3}{2} \end{bmatrix}$ ★(c) $\begin{bmatrix} -1 & 4 & -2 \\ 2 & -6 & -4 \\ 2 & 0 & -25 \end{bmatrix}$

(d) $\begin{bmatrix} 2 & 6 & -4 \\ 5 & 11 & 10 \\ 1 & 9 & -29 \end{bmatrix}$ ★(e) $\begin{bmatrix} -3 & 1 & 1 & -1 \\ 4 & -2 & -3 & 5 \\ 6 & -1 & 1 & -2 \\ -2 & 2 & 4 & -7 \end{bmatrix}$

(f) $\begin{bmatrix} -3 & -12 & 6 & 9 \\ -6 & -26 & 12 & 20 \\ 9 & 42 & -17 & -28 \\ 3 & 8 & -8 & -18 \end{bmatrix}$

2. (a) Show that the matrix $\begin{bmatrix} 0 & 1 \\ 1 & 0 \end{bmatrix}$ has no **LDU** decomposition by showing that there are no values w, x, y, and z such that

$$\begin{bmatrix} 0 & 1 \\ 1 & 0 \end{bmatrix} = \underbrace{\begin{bmatrix} 1 & 0 \\ w & 1 \end{bmatrix}}_{\mathbf{L}} \underbrace{\begin{bmatrix} x & 0 \\ 0 & y \end{bmatrix}}_{\mathbf{D}} \underbrace{\begin{bmatrix} 1 & z \\ 0 & 1 \end{bmatrix}}_{\mathbf{U}}.$$

(b) The result of part (a) does not contradict Theorem 9.1. Why not?

3. For each system, find the **KU** decomposition (where **K** = **LD**) for the coefficient matrix, and use it to solve the system by substitution, as in Example 2.

★(a) $\begin{cases} -x_1 + 5x_2 = -9 \\ 2x_1 - 13x_2 = 21 \end{cases}$ (b) $\begin{cases} 2x_1 - 4x_2 + 10x_3 = 34 \\ 2x_1 - 5x_2 + 7x_3 = 29 \\ x_1 - 5x_2 - x_3 = 8 \end{cases}$

★(c) $\begin{cases} -x_1 + 3x_2 - 2x_3 = -13 \\ 4x_1 - 9x_2 - 7x_3 = 28 \\ -2x_1 + 11x_2 - 31x_3 = -68 \end{cases}$

(d) $\begin{cases} 3x_1 - 15x_2 + 6x_3 + 6x_4 = 60 \\ x_1 - 7x_2 + 8x_3 + 2x_4 = 30 \\ -5x_1 + 24x_2 - 3x_3 - 18x_4 = -115 \\ x_1 - 2x_2 - 7x_3 - x_4 = -4 \end{cases}$

9.3 The Power Method for Finding Eigenvalues

Prerequisite: Section 6.1, Introduction to Eigenvalues

The only method given in Section 6.1 for finding the eigenvalues of an $n \times n$ matrix $\mathbf{A}$ is to calculate the characteristic polynomial of $\mathbf{A}$ and find its roots. However, if n is large, $p_{\mathbf{A}}(x)$ is often difficult to calculate. Also, numerical techniques may be required to find its roots. Finally, if an eigenvalue λ is not known to a high enough degree of accuracy, we may have difficulty finding a corresponding eigenvector $\mathbf{v}$, because the matrix $\lambda\mathbf{I} - \mathbf{A}$ in the equation $(\lambda\mathbf{I} - \mathbf{A})\mathbf{v} = \mathbf{0}$ may not be singular for the given value of λ.

Therefore, in this section we present a numerical technique known as the "power method" for finding the largest eigenvalue (in absolute value) of a matrix and a corresponding eigenvector. Such an eigenvalue is called a **dominant eigenvalue**.

All calculations for the examples and exercises in this section were performed on a calculator that stores numbers with twelve-digit accuracy, but only the first four significant digits are printed here. Your own computations may differ slightly if you are using a different number of significant digits. If you do not have a calculator with the ability to perform matrix calculations, you should use an appropriate linear algebra software package. You might also consider writing your own power method program, since the algorithm involved is not difficult.

The Power Method

Suppose $\mathbf{A}$ is a diagonalizable $n \times n$ matrix having (not necessarily distinct) eigenvalues $\lambda_1, \lambda_2, \ldots, \lambda_n$, with λ_1 being the dominant eigenvalue. The **power method** can be used to find λ_1 and an associated eigenvector. In fact, it often works in cases where $\mathbf{A}$ is not diagonalizable, but there is no guarantee that it will work in such a case.

The idea behind the power method is as follows: choose any nonzero n-vector $\mathbf{v}$ and calculate $\mathbf{A}^k\mathbf{v}/\|\mathbf{A}^k\mathbf{v}\|$ for some large positive integer k. The result should be a good approximation for a unit eigenvector corresponding to λ_1.

To see why, first express the vector $\mathbf{v}$ as $\mathbf{v} = a_1\mathbf{v}_1 + \cdots + a_n\mathbf{v}_n$, where $\{\mathbf{v}_1, \ldots, \mathbf{v}_n\}$ is a basis of eigenvectors for $\mathbf{A}$ corresponding to the eigenvalues $\lambda_1, \ldots, \lambda_n$. Then

$$\begin{aligned}\mathbf{A}^k\mathbf{v} &= a_1\mathbf{A}^k\mathbf{v}_1 + a_2\mathbf{A}^k\mathbf{v}_2 + \cdots + a_n\mathbf{A}^k\mathbf{v}_n \\ &= a_1\lambda_1^k\mathbf{v}_1 + a_2\lambda_2^k\mathbf{v}_2 + \cdots + a_n\lambda_n^k\mathbf{v}_n.\end{aligned}$$

Because $|\lambda_1| > |\lambda_i|$, for $2 \leq i \leq n$, we see that for large k, $|\lambda_1^k|$ is significantly larger than $|\lambda_i^k|$, since the ratio $|\lambda_i|^k/|\lambda_1|^k$ approaches 0 as $k \to \infty$. Thus, the term $a_1\lambda_1^k\mathbf{v}_1$ dominates the expression for $\mathbf{A}^k\mathbf{v}$ for large enough values of k.† If we normalize $\mathbf{A}^k\mathbf{v}$, we have $(\mathbf{A}^k\mathbf{v})/\|\mathbf{A}^k\mathbf{v}\| \approx (a_1\lambda_1^k\mathbf{v}_1)/\|a_1\lambda_1^k\mathbf{v}_1\|$, which is a scalar multiple of $\mathbf{v}_1$ and thus a unit eigenvector corresponding to λ_1.

Finally, if we let $\mathbf{u} = \mathbf{A}^k\mathbf{v}/\|\mathbf{A}^k\mathbf{v}\|$, then $\|\mathbf{A}\mathbf{u}\|$ approximates $|\lambda_1|$. The sign of λ_1 is determined by checking if $\mathbf{A}\mathbf{u}$ is in the same direction as $\mathbf{u}$ or in the opposite direction.

We now outline the power method in detail.

POWER METHOD FOR FINDING THE DOMINANT EIGENVALUE OF A MATRIX

Let $\mathbf{A}$ be an $n \times n$ matrix.

Step 1: Choose an arbitrary unit vector $\mathbf{u}_0$.

Step 2: Create a sequence of unit vectors $\mathbf{u}_1, \ldots, \mathbf{u}_n$ by repeating steps 2(a) to 2(d) until one of the terminal conditions in steps 2(c) or 2(d) is reached or until it becomes clear that the method is not converging to an answer:

(a) Given $\mathbf{u}_{i-1}$, calculate $\mathbf{w}_i = \mathbf{A}\mathbf{u}_{i-1}$.

(b) Calculate $\mathbf{u}_i = \mathbf{w}_i/\|\mathbf{w}_i\|$.

(c) If $\mathbf{u}_{i-1}$ equals $\mathbf{u}_i$ to the desired degree of accuracy, let $\lambda = \|\mathbf{w}_i\|$ and go to step 3.

(d) If $\mathbf{u}_{i-1}$ equals $-\mathbf{u}_i$ to the desired degree of accuracy, let $\lambda = -\|\mathbf{w}_i\|$ and go to step 3.

Step 3: The last vector, $\mathbf{u}_n$, calculated in step 2 is an approximate eigenvector of $\mathbf{A}$ corresponding to the (approximate) eigenvalue λ.

It is possible but unlikely to get $\mathbf{w}_i = \mathbf{0}$ in step 2(a), which makes step 2(b) impossible to perform. In this case, $\mathbf{u}_{i-1}$ is an eigenvector for $\mathbf{A}$ corresponding to $\lambda = 0$. You can then return to step 1, choosing a different $\mathbf{u}_0$, in hopes of finding another eigenvalue for $\mathbf{A}$.

---

†Theoretically, a problem may arise if $a_1 = 0$. However, in most practical situations this will not happen. If the method does not work and you suspect it is because $a_1 = 0$, try using instead some $\mathbf{v}$ that is linearly independent from those you have already tried.

Example 1

Let

$$\mathbf{A} = \begin{bmatrix} -16 & 6 & 30 \\ 4 & 1 & -8 \\ -9 & 3 & 17 \end{bmatrix}.$$

We use the power method to find the dominant eigenvalue for $\mathbf{A}$ and a corresponding eigenvector.

Step 1: We choose $\mathbf{u}_0 = [1, 0, 0]$.

Step 2: A first pass through this step gives the following:

(a) $\mathbf{w}_1 = \mathbf{A}\mathbf{u}_0 \approx [-16, 4, -9]$.

(b) $\|\mathbf{w}_1\| = \sqrt{(-16)^2 + 4^2 + (-9)^2} \approx 18.79$. So $\mathbf{u}_1 = \mathbf{w}_1/\|\mathbf{w}_1\| \approx [-0.8516, 0.2129, -0.4790]$.

Because $\mathbf{u}_0$ and $\pm\mathbf{u}_1$ do not agree to four decimal places, we return to step 2(a). Subsequent iterations of step 2 lead to the results in the following table:

i	$\mathbf{w}_i = \mathbf{A}\mathbf{u}_{i-1}$	$\|\mathbf{w}_i\|$	$\mathbf{u}_i = \frac{\mathbf{w}_i}{\|\mathbf{w}_i\|}$
1	$[-16, 4, -9]$	18.79	$[-0.8516, 0.2129, -0.4790]$
2	$[0.5322, 0.6387, 0.1597]$	0.8466	$[0.6287, 0.7544, 0.1886]$
3	$[0.1257, 1.760, -0.1886]$	1.775	$[0.0708, 0.9918, -0.1063]$
4	$[1.629, 2.125, 0.5313]$	2.730	$[0.5968, 0.7784, 0.1946]$
5	$[0.9601, 1.609, 0.2725]$	1.893	$[0.5071, 0.8498, 0.1439]$
6	$[1.302, 1.727, 0.4317]$	2.205	$[0.5904, 0.7830, 0.1958]$
7	$[1.125, 1.578.0.3635]$	1.972	$[0.5704, 0.8004, 0.1843]$
8	$[1.207, 1.607, 0.4018]$	2.050	$[0.5889, 0.7841, 0.1960]$
9	$[1.164, 1.571, 0.3851]$	1.993	$[0.5840, 0.7884, 0.1932]$
10	$[1.184, 1.578, 0.3946]$	2.012	$[0.5885, 0.7844, 0.1961]$
11	$[1.173, 1.570, 0.3905]$	1.998	$[0.5873, 0.7855, 0.1954]$
12	$[1.179, 1.571, 0.3928]$	2.003	$[0.5884, 0.7844, 0.1961]$
13	$[1.176, 1.569, 0.3918]$	2.000	$[0.5881, 0.7847, 0.1959]$
14	$[1.177, 1.570, 0.3924]$	2.001	$[0.5884, 0.7845, 0.1961]$
15	$[1.176, 1.569, 0.3921]$	2.000	$[0.5883, 0.7845, 0.1961]$
16	$[1.177, 1.569, 0.3923]$	2.000	$[0.5884, 0.7845, 0.1961]$
17	$[1.177, 1.569, 0.3922]$	2.000	$[0.5883, 0.7845, 0.1961]$
18	$[1.177, 1.569, 0.3922]$	2.000	$[0.5883, 0.7845, 0.1961]$

After eighteen iterations, we find that $\mathbf{u}_{17}$ and $\mathbf{u}_{18}$ agree to four decimal places. Therefore, step 2 terminates with $\lambda = 2.000$.

Step 3: Thus, $\lambda = 2.000$ is the dominant eigenvalue for $\mathbf{A}$, and $\mathbf{u}_{18} = [0.5883, 0.7845, 0.1961]$ is a corresponding unit eigenvector. ∎

We can check that the power method gives the correct result in this particular case. A quick calculation shows that, for the given matrix $\mathbf{A}$, $p_{\mathbf{A}}(x) = x^3 - 2x^2 - x + 2 = (x-2)(x-1)(x+1)$. Clearly, $\lambda_1 = 2$ is the dominant eigenvalue for $\mathbf{A}$.

Solving the system $(2\mathbf{I}_3 - \mathbf{A})\mathbf{v} = \mathbf{0}$ produces an eigenvector $\mathbf{v} = [3, 4, 1]$ corresponding to $\lambda_1 = 2$. Normalizing $\mathbf{v}$ yields a unit eigenvector $\mathbf{v}/\|\mathbf{v}\| \approx [0.5883, 0.7845, 0.1961]$.

Unfortunately, the power method does not always work. Note that it depends on the fact that multiplying by $\mathbf{A}$ magnifies the size of an eigenvector for the dominant eigenvalue more than for any other vector in $\mathbb{R}^n$. For example, if $\mathbf{A}$ is a diagonalizable matrix, the power method fails if both $\pm\lambda$ are eigenvalues of $\mathbf{A}$ with the largest absolute value. In particular, suppose $\mathbf{A}$ is a 3×3 matrix with eigenvalues $\lambda_1 = 2$, $\lambda_2 = -2$, and $\lambda_3 = 1$ and corresponding eigenvectors $\mathbf{v}_1$, $\mathbf{v}_2$, and $\mathbf{v}_3$. Multiplying $\mathbf{A}$ by any vector $\mathbf{v} = a_1\mathbf{v}_1 + a_2\mathbf{v}_2 + a_3\mathbf{v}_3$ produces $\mathbf{A}\mathbf{v} = 2a_1\mathbf{v}_1 - 2a_2\mathbf{v}_2 + a_3\mathbf{v}_3$. The contribution of neither eigenvector $\mathbf{v}_1$ nor $\mathbf{v}_2$ dominates over the other, since both terms are doubled simultaneously.

The next example illustrates that the power method is not guaranteed to work for a nondiagonalizable matrix.

Example 2

Consider the matrix

$$\mathbf{A} = \begin{bmatrix} 7 & -15 & -24 \\ -12 & 25 & 42 \\ 6 & -15 & -23 \end{bmatrix}.$$

This matrix has only one eigenvalue, $\lambda = 1$, with a corresponding one-dimensional eigenspace spanned by $\mathbf{v}_1 = [3, -2, 2]$. The power method cannot be used to find this eigenvalue and eigenvector, since some vectors in $\mathbb{R}^3$ have their magnitudes continually increased when successively multiplied by $\mathbf{A}$ while the magnitude of $\mathbf{v}_1$ is fixed by $\mathbf{A}$. If we attempt the power method anyway, starting with $\mathbf{u}_0 = [1, 0, 0]$, the following results are produced:

i	$\mathbf{w}_i = \mathbf{A}\mathbf{u}_{i-1}$	$\|\mathbf{w}_i\|$	$\mathbf{u}_i = \dfrac{\mathbf{w}_i}{\|\mathbf{w}_i\|}$
1	$[7, -12, 6]$	15.13	$[0.4626, -0.7930, 0.3965]$
2	$[5.617, -8.723, 5.551]$	11.77	$[0.4774, -0.7413, 0.4718]$
3	$[3.139, -4.448, 3.134]$	6.282	$[0.4998, -0.7081, 0.4989]$
⋮	⋮	⋮	⋮
25	$[0.3434, 0.3341, 0.3434]$	0.5894	$[0.5825, 0.5668, 0.5825]$
26	$[-18.41, 31.65, -18.41]$	40.98	$[-0.4492, 0.7723, -0.4492]$
27	$[-3.949, 5.833, -3.949]$	8.075	$[-0.4890, 0.7223, -0.4890]$
⋮	⋮	⋮	⋮
50	$[2.589, -5.325, 2.589]$	6.462	$[0.4006, -0.8240, 0.4006]$
51	$[5.551, -8.583, 5.551]$	11.63	$[0.4772, -0.7379, 0.4772]$
52	$[2.957, -4.132, 2.957]$	5.879	$[0.5029, -0.7029, 0.5029]$
⋮	⋮	⋮	⋮

As you can see, there is no evidence of any convergence at all in either the $\|\mathbf{w}_i\|$ or $\mathbf{u}_i$ columns. If the power method were successful, these would be converging to, respectively, the absolute value of the dominant eigenvalue and its corresponding eigenvector. ■

One disadvantage of the power method is that it can only be used to find the dominant eigenvalue for a matrix. There are additional numerical techniques for calculating other eigenvalues. One such technique is the **inverse power method**, which finds the *smallest* eigenvalue of a matrix essentially by using the power method on the inverse of the matrix. If you are interested in learning more about this technique and other more sophisticated methods for finding eigenvalues, check the numerical analysis books in your library. One classic reference is *Numerical Analysis*, 4th ed., by Burden and Faires (published by PWS, 1989).

Exercises—Section 9.3

1. Use the power method on each of the given matrices, starting with the given vector, to find the dominant eigenvalue and a corresponding eigenvector for each matrix. Perform as many iterations as needed until two successive vectors agree in every entry in the first k digits after the decimal point for the given value of k. Carry out all calculations using as many significant digits as are feasible with your computing resources.

★(a) $\begin{bmatrix} 2 & 36 \\ 36 & 23 \end{bmatrix}, \begin{bmatrix} 1 \\ 0 \end{bmatrix}, k = 2$

(b) $\begin{bmatrix} 3 & 5 \\ 2 & 1 \end{bmatrix}, \begin{bmatrix} 0 \\ 1 \end{bmatrix}, k = 2$

★(c) $\begin{bmatrix} 2 & 3 & -1 \\ 1 & 0 & 1 \\ 3 & 1 & 1 \end{bmatrix}, \begin{bmatrix} 0 \\ 1 \\ 0 \end{bmatrix}, k = 2$

(d) $\begin{bmatrix} 3 & 1 & 1 & 2 \\ 1 & 1 & 0 & 4 \\ 0 & 1 & 0 & -1 \\ 2 & 3 & 2 & 1 \end{bmatrix}, \begin{bmatrix} 1 \\ 0 \\ 0 \\ 0 \end{bmatrix}, k = 2$

★(e) $\begin{bmatrix} -10 & 2 & -1 & 11 \\ 4 & 2 & -3 & 6 \\ -44 & 7 & 3 & 28 \\ -17 & 4 & 1 & 12 \end{bmatrix}, \begin{bmatrix} 3 \\ 8 \\ 2 \\ 3 \end{bmatrix}, k = 3$

(f) $\begin{bmatrix} 5 & 3 & -4 & 6 \\ -2 & -1 & 6 & -10 \\ -6 & -6 & 8 & -7 \\ -2 & -2 & 1 & 2 \end{bmatrix}, \begin{bmatrix} 4 \\ -5 \\ -6 \\ -1 \end{bmatrix}, k = 4$

2. In each part of this exercise, show that the power method does not work on the given matrix using $[1, 0, 0]$ as a starting vector. Explain why the method fails in each case.

(a) $\begin{bmatrix} -21 & 10 & -74 \\ 25 & -9 & 80 \\ 10 & -4 & 33 \end{bmatrix}$ (b) $\begin{bmatrix} 13 & -10 & 8 \\ -8 & 11 & -4 \\ -40 & 40 & -23 \end{bmatrix}$

3. (a) Suppose that $\mathbf{A}$ is a diagonalizable 2×2 matrix with eigenvalues λ_1 and λ_2 such that $|\lambda_1| > |\lambda_2| \neq 0$. Let $\{\mathbf{v}_1, \mathbf{v}_2\}$ be a basis of unit eigenvectors for $\mathbb{R}^2$ corresponding to λ_1 and λ_2, respectively. Clearly, each vector $\mathbf{x} \in \mathbb{R}^n$ can be expressed uniquely in the form $\mathbf{x} = a\mathbf{v}_1 + b\mathbf{v}_2$. Finally, suppose $\mathbf{u}_0$ is the initial vector used in the power method for finding the dominant eigenvalue of $\mathbf{A}$. Expressing $\mathbf{u}_i$ in that method as $a_i\mathbf{v}_1 + b_i\mathbf{v}_2$, prove that, for all $i \geq 0$,

$$\frac{|a_i|}{|b_i|} = \left|\frac{\lambda_1}{\lambda_2}\right|^i \cdot \frac{|a_0|}{|b_0|},$$

assuming that $b_i \neq 0$. Explain what this result implies about the rate of convergence of the power method in this case.

★(b) State and prove a similar formula for an $n \times n$ diagonalizable matrix $\mathbf{A}$.

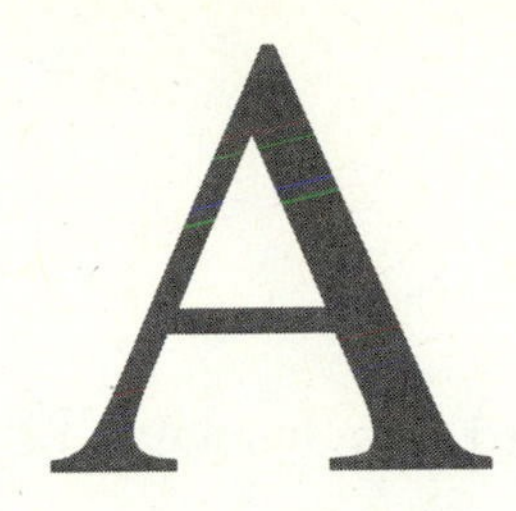

Miscellaneous Proofs

In this appendix, we present some proofs of theorems that were omitted from the text.

Proof of Theorem 1.13, Part (1)

Part (1) of Theorem 1.13 can be restated as follows:

THEOREM 1.13, Part (1)

If **A** is an $m \times n$ matrix, **B** is an $n \times p$ matrix, and **C** is a $p \times r$ matrix, then $\mathbf{A}(\mathbf{BC}) = (\mathbf{AB})\mathbf{C}$.

Proof of Theorem 1.13, Part (1)

We must show that the (i, j) entry of $\mathbf{A}(\mathbf{BC})$ is the same as the (i, j) entry of $(\mathbf{AB})\mathbf{C}$. Now,

(i, j) entry of $\mathbf{A}(\mathbf{BC})$

$$\begin{aligned}
&= [i^{th} \text{ row of } \mathbf{A}] \cdot [j^{th} \text{ column of } \mathbf{BC}] \\
&= [i^{th} \text{ row of } \mathbf{A}] \cdot \left[\sum_{k=1}^{p} b_{1k}c_{kj}, \sum_{k=1}^{p} b_{2k}c_{kj}, \ldots, \sum_{k=1}^{p} b_{nk}c_{kj}\right] \\
&= a_{i1}\left(\sum_{k=1}^{p} b_{1k}c_{kj}\right) + a_{i2}\left(\sum_{k=1}^{p} b_{2k}c_{kj}\right) + \cdots + a_{in}\left(\sum_{k=1}^{p} b_{nk}c_{kj}\right) \\
&= \sum_{k=1}^{p}\left(a_{i1}b_{1k}c_{kj} + a_{i2}b_{2k}c_{kj} + \cdots + a_{in}b_{nk}c_{kj}\right).
\end{aligned}$$

Similarly, we have

(i, j) entry of **(AB)C**

$$
\begin{aligned}
&= [i^{th} \text{ row of } \mathbf{AB}] \cdot [j^{th} \text{ column of } \mathbf{C}] \\
&= \left[\sum_{k=1}^{n} a_{ik}b_{k1}, \sum_{k=1}^{n} a_{ik}b_{k2}, \ldots, \sum_{k=1}^{n} a_{ik}b_{kp}\right] \cdot [j^{th} \text{ column of } \mathbf{C}] \\
&= \left(\sum_{k=1}^{n} a_{ik}b_{k1}\right)c_{1j} + \left(\sum_{k=1}^{n} a_{ik}b_{k2}\right)c_{2j} + \cdots + \left(\sum_{k=1}^{n} a_{ik}b_{kp}\right)c_{pj} \\
&= \sum_{k=1}^{n} \left(a_{ik}b_{k1}c_{1j} + a_{ik}b_{k2}c_{2j} + \cdots + a_{ik}b_{kp}c_{pj}\right).
\end{aligned}
$$

It is easy to see that the final sums for the (i, j) entries of **A(BC)** and **(AB)C** are equal, because both are equal to the giant sum of terms

$$
\left\{
\begin{array}{ccccccccc}
a_{i1}b_{11}c_{1j} & + & a_{i1}b_{12}c_{2j} & + & a_{i1}b_{13}c_{3j} & + & \cdots & + & a_{i1}b_{1p}c_{pj} \\
a_{i2}b_{21}c_{1j} & + & a_{i2}b_{22}c_{2j} & + & a_{i2}b_{23}c_{3j} & + & \cdots & + & a_{i2}b_{2p}c_{pj} \\
 & & & & \vdots & & & & \\
a_{in}b_{n1}c_{1j} & + & a_{in}b_{n2}c_{2j} & + & a_{in}b_{n3}c_{3j} & + & \cdots & + & a_{in}b_{np}c_{pj}
\end{array}
\right\}.
$$

Notice that the ith term in the sum for **A(BC)** represents the ith column of terms in the giant sum, whereas the ith term in the sum for **(AB)C** represents the ith row of terms in the giant sum. Hence, the (i, j) entries of **A(BC)** and **(AB)C** agree. ■

Proof of Theorem 2.8

THEOREM 2.8

> Let **A** and **B** be $n \times n$ matrices. If either product **AB** or **BA** equals $\mathbf{I}_n$, then the other product also equals $\mathbf{I}_n$, and **A** and **B** are inverses of each other.

We say that **B** is a **left inverse** of **A** and **A** is a **right inverse** of **B** whenever $\mathbf{BA} = \mathbf{I}_n$.

Proof of Theorem 2.8

We need to show that any left inverse of a matrix is also a right inverse, and vice versa.

First, suppose that $\mathbf{B}$ is a left inverse of $\mathbf{A}$—that is, $\mathbf{BA} = \mathbf{I}_n$. We will show that $\mathbf{AB} = \mathbf{I}_n$. The remarks in Section 1.5 on calculating a particular column of a matrix product state that it is enough to show, for $1 \leq i \leq n$,

$$\mathbf{A} \bullet [i^{th}\text{column of } \mathbf{B}] = [i^{th}\text{column of } \mathbf{I}_n].$$

Consider the homogeneous system $\mathbf{AX} = \mathbf{O}$ of n equations and n unknowns. This system has only the trivial solution, because multiplying both sides of $\mathbf{AX} = \mathbf{O}$ by $\mathbf{B}$ on the left, we obtain

$$\mathbf{B(AX)} = \mathbf{BO} \Longrightarrow \mathbf{(BA)X} = \mathbf{O} \Longrightarrow \mathbf{I}_n\mathbf{X} = \mathbf{O} \Longrightarrow \mathbf{X} = \mathbf{O}.$$

By Theorem 2.4, every column of $\mathbf{A}$ becomes a pivot column during the Gauss-Jordan row reduction process. Therefore, each of the augmented matrices

$$\left[\begin{array}{c|c} \mathbf{A} & \begin{array}{c} 1^{st} \\ \text{column} \\ \text{of } \mathbf{I}_n \end{array} \end{array}\right], \left[\begin{array}{c|c} \mathbf{A} & \begin{array}{c} 2^{nd} \\ \text{column} \\ \text{of } \mathbf{I}_n \end{array} \end{array}\right], \ldots, \left[\begin{array}{c|c} \mathbf{A} & \begin{array}{c} n^{th} \\ \text{column} \\ \text{of } \mathbf{I}_n \end{array} \end{array}\right]$$

represents a system with a unique solution. Consider the matrix $\mathbf{C}$, whose i^{th} column is the solution to the ith system above. Then $\mathbf{C}$ is a right inverse for $\mathbf{A}$, because the product $\mathbf{AC} = \mathbf{I}_n$. But then

$$\mathbf{B} = \mathbf{B(I}_n) = \mathbf{B(AC)} = \mathbf{(BA)C} = \mathbf{I}_n\mathbf{C} = \mathbf{C}.$$

Hence, $\mathbf{B}$ is also a right inverse for $\mathbf{A}$.

Conversely, suppose that $\mathbf{B}$ is a right inverse for $\mathbf{A}$—that is, $\mathbf{AB} = \mathbf{I}_n$. We must show that $\mathbf{B}$ is also a left inverse for $\mathbf{A}$. By assumption, $\mathbf{A}$ is a left inverse for $\mathbf{B}$. However, we have already shown that any left inverse is also a right inverse. Therefore, $\mathbf{A}$ must be a (full) inverse for $\mathbf{B}$, and $\mathbf{AB} = \mathbf{BA} = \mathbf{I}_n$. Hence, $\mathbf{B}$ is a left (and a full) inverse for $\mathbf{A}$. ■

Proof of Theorem 3.4, Part (2)

THEOREM 3.4, Part (2)

> Let $\mathbf{A}$ be an $n \times n$ matrix. If $\mathbf{A}$ has two equal rows, then $|\mathbf{A}| = 0$.

Proof of Theorem 3.4, Part (2)

Suppose that the ith and j^{th} rows of $\mathbf{A}$ have identical entries, where $i < j$. Consider a typical term T in the formula for $|\mathbf{A}|$:

$$\begin{aligned} T &= \pm a_{1\alpha}a_{2\beta}\cdots a_{i\sigma}\cdots a_{j\tau}\cdots a_{n\omega} \\ &= \pm a_{1\alpha}a_{2\beta}\cdots a_{i\sigma}\cdots a_{i\tau}\cdots a_{n\omega} \qquad \text{since} \quad a_{j\tau} = a_{i\tau} \end{aligned}$$

By switching the column subscripts on the ith and j^{th} factors of T, we obtain the factors of a "corresponding" term U in the formula for $|\mathbf{A}|$:

$$\begin{aligned} U &= \pm a_{1\alpha}a_{2\beta}\cdots a_{i\tau}\cdots a_{j\sigma}\cdots a_{n\omega} \\ &= \pm a_{1\alpha}a_{2\beta}\cdots a_{i\tau}\cdots a_{i\sigma}\cdots a_{n\omega} \end{aligned}$$

since $a_{j\sigma} = a_{i\sigma}$.

Of course, a switch of the column subscripts on the ith and j^{th} factors of U returns us to the factors of the term T, so that T is the "corresponding" term for U as well. (Note that T and U have a different sequence of column subscripts, so they do not represent the same term of $|\mathbf{A}|$.)

In this way, we see that switching the ith and j^{th} column subscripts allows us to pair each term T in the formula for $|\mathbf{A}|$ with *precisely one other term* U in this formula. If we can show that $T + U = 0$, then $|\mathbf{A}| = 0$, since all the terms in the formula for $|\mathbf{A}|$ can be put in pairs, like T and U, whose sum is zero.

Notice that the factors in the product T are identical to the factors in U but appear in a different order. Therefore, to establish that $T = -U$, we need only show that the sign in front of T is positive whenever the sign in front of U is negative, and vice versa. That is, we need to show that the permutation (the arrangement of the column subscripts) for the term T is even if and only if the corresponding permutation for U is odd, and vice versa.

The permutation for T is $t = \alpha\beta\ldots\sigma\ldots\tau\ldots\omega$, and the permutation for U is $u = \alpha\beta\ldots\tau\ldots\sigma\ldots\omega$. Notice that these permutations are the same except that the positions of σ and τ are switched. Hence, if t can be converted to the special permutation $1\ 2\ldots n$ using k interchanges, the permutation u can be converted to $1\ 2\ldots n$ in $k+1$ interchanges: first an interchange of σ and τ to obtain t from u, then k more interchanges to obtain $1\ 2\ldots n$ from t. Obviously, if k is even, $k+1$ is odd, and vice versa. Therefore, t odd implies u even, and t even implies u odd. ■

Proof of Theorem 6.8

THEOREM 6.8: Cayley-Hamilton Theorem

Let $\mathbf{A}$ be an $n \times n$ matrix, and let $p_{\mathbf{A}}(x)$ be its characteristic polynomial. Then $p_{\mathbf{A}}(\mathbf{A}) = \mathbf{O}_n$.

Proof of Theorem 6.8

Let $\mathbf{A}$ be an $n \times n$ matrix with characteristic polynomial $p_{\mathbf{A}}(x) = |x\mathbf{I}_n - \mathbf{A}| = x^n + a_{n-1}x^{n-1} + a_{n-2}x^{n-2} + \cdots + a_1x + a_0$, for some real numbers $a_0, \ldots, a_{n-1}$. Consider the (classical) adjoint $\mathbf{B}(x)$ of $x\mathbf{I}_n - \mathbf{A}$ (see Section 3.3). By Theorem 3.13,

$$(x\mathbf{I}_n - \mathbf{A})\mathbf{B}(x) = p_{\mathbf{A}}(x)\mathbf{I}_n,$$

for every $x \in \mathbb{R}$. We will find an expanded form for $\mathbf{B}(x)$ and then use the preceding equation to expand $p_{\mathbf{A}}(\mathbf{A})$ into a form that obviously results in $\mathbf{O}_n$.

Now, each entry of $\mathbf{B}(x)$ is defined as the determinant of an $(n-1) \times (n-1)$ minor of $x\mathbf{I}_n - \mathbf{A}$ and hence is a polynomial in x of degree $\leq n-1$. So, for each k, $0 \leq k \leq n-1$, we create the matrix $\mathbf{B}_k$ whose (i, j) entry is the coefficient of x^k in the (i, j) entry of $\mathbf{B}(x)$. Thus,

$$\mathbf{B}(x) = x^{n-1}\mathbf{B}_{n-1} + x^{n-2}\mathbf{B}_{n-2} + \cdots + x\mathbf{B}_1 + \mathbf{B}_0.$$

Therefore,

$$\begin{aligned}(x\mathbf{I}_n - \mathbf{A})\mathbf{B}(x) &= (x^n\mathbf{B}_{n-1} - x^{n-1}\mathbf{A}\mathbf{B}_{n-1}) + (x^{n-1}\mathbf{B}_{n-2} - x^{n-2}\mathbf{A}\mathbf{B}_{n-2}) \\ &\quad + \cdots + (x^2\mathbf{B}_1 - x\mathbf{A}\mathbf{B}_1) + (x\mathbf{B}_0 - \mathbf{A}\mathbf{B}_0) \\ &= x^n\mathbf{B}_{n-1} + x^{n-1}(-\mathbf{A}\mathbf{B}_{n-1} + \mathbf{B}_{n-2}) + x^{n-2}(-\mathbf{A}\mathbf{B}_{n-2} + \mathbf{B}_{n-3}) \\ &\quad + \cdots + x(-\mathbf{A}\mathbf{B}_1 + \mathbf{B}_0) + (-\mathbf{A}\mathbf{B}_0).\end{aligned}$$

Setting the coefficient of x^k in this expression equal to the coefficient of x^k in $p_{\mathbf{A}}(x)\mathbf{I}_n$ yields

$$\begin{cases} \mathbf{B}_{n-1} = \mathbf{I}_n \\ -\mathbf{A}\mathbf{B}_k + \mathbf{B}_{k-1} = a_k\mathbf{I}_n, \quad \text{for } 1 \leq k \leq n-1. \\ -\mathbf{A}\mathbf{B}_0 = a_0\mathbf{I}_n \end{cases}$$

Hence,

$$\begin{aligned}p_{\mathbf{A}}(\mathbf{A}) &= \mathbf{A}^n + a_{n-1}\mathbf{A}^{n-1} + a_{n-2}\mathbf{A}^{n-2} + \cdots + a_1\mathbf{A} + a_0\mathbf{I}_n \\ &= \mathbf{A}^n\mathbf{I}_n + \mathbf{A}^{n-1}(a_{n-1}\mathbf{I}_n) + \mathbf{A}^{n-2}(a_{n-2}\mathbf{I}_n) + \cdots + \mathbf{A}(a_1\mathbf{I}_n) + a_0\mathbf{I}_n \\ &= \mathbf{A}^n(\mathbf{B}_{n-1}) + \mathbf{A}^{n-1}(-\mathbf{A}\mathbf{B}_{n-1} + \mathbf{B}_{n-2}) + \mathbf{A}^{n-2}(-\mathbf{A}\mathbf{B}_{n-2} + \mathbf{B}_{n-3}) \\ &\quad + \cdots + \mathbf{A}(-\mathbf{A}\mathbf{B}_1 + \mathbf{B}_0) + (-\mathbf{A}\mathbf{B}_0) \\ &= \mathbf{A}^n\mathbf{B}_{n-1} + (-\mathbf{A}^n\mathbf{B}_{n-1} + \mathbf{A}^{n-1}\mathbf{B}_{n-2}) + (-\mathbf{A}^{n-1}\mathbf{B}_{n-2} + \mathbf{A}^{n-2}\mathbf{B}_{n-3}) \\ &\quad + \cdots + (-\mathbf{A}^2\mathbf{B}_1 + \mathbf{A}\mathbf{B}_0) + (-\mathbf{A}\mathbf{B}_0) \\ &= \mathbf{A}^n(\mathbf{B}_{n-1} - \mathbf{B}_{n-1}) + \mathbf{A}^{n-1}(\mathbf{B}_{n-2} - \mathbf{B}_{n-2}) + \mathbf{A}^{n-2}(\mathbf{B}_{n-3} - \mathbf{B}_{n-3}) \\ &\quad + \cdots + \mathbf{A}^2(\mathbf{B}_1 - \mathbf{B}_1) + \mathbf{A}(\mathbf{B}_0 - \mathbf{B}_0) \\ &= \mathbf{O}_n.\end{aligned}$$

■

Proof of Theorem 6.12

THEOREM 6.12

Let $\mathcal{V}$ be a nontrivial subspace of $\mathbb{R}^n$, and let L be a linear operator on $\mathcal{V}$. Let B be an ordered orthonormal basis for $\mathcal{V}$, and let $\mathbf{A}$ be the matrix for L with respect to B. Then L is a symmetric operator if and only if $\mathbf{A}$ is a symmetric matrix.

Proof of Theorem 6.12

Let $\mathcal{V}$, L, B, and $\mathbf{A}$ be given as in the statement of the theorem, and let $k = \dim(\mathcal{V})$. Also, suppose that $B = (\mathbf{v}_1, \ldots, \mathbf{v}_k)$.

First we claim that, for all $\mathbf{w}_1, \mathbf{w}_2 \in \mathcal{V}$, $[\mathbf{w}_1]_B \cdot [\mathbf{w}_2]_B = \mathbf{w}_1 \cdot \mathbf{w}_2$, where the first dot product is in $\mathbb{R}^k$ and the second is in $\mathbb{R}^n$. To prove this statement, suppose that $[\mathbf{w}_1]_B = [a_1, \ldots, a_k]$ and $[\mathbf{w}_2]_B = [b_1, \ldots, b_k]$. Then, $\mathbf{w}_1 \cdot \mathbf{w}_2 = (a_1\mathbf{v}_1 + \cdots + a_k\mathbf{v}_k) \cdot (b_1\mathbf{v}_1 + \cdots + b_k\mathbf{v}_k) = \sum_{i=1}^{k} \sum_{j=1}^{k} (a_i b_j)\mathbf{v}_i \cdot \mathbf{v}_j = \sum_{i=1}^{k} (a_i b_i)\mathbf{v}_i \cdot \mathbf{v}_i$ (since $\mathbf{v}_i \cdot \mathbf{v}_j = 0$ if $i \neq j$) $= \sum_{i=1}^{k} a_i b_i$ (since $\mathbf{v}_i \cdot \mathbf{v}_i = 1$) $= [\mathbf{w}_1]_B \cdot [\mathbf{w}_2]_B$.

Now suppose that L is a symmetric operator on $\mathcal{V}$. We will prove that $\mathbf{A}$ is symmetric by showing that its (i, j) entry equals its (j, i) entry. We have

$$
\begin{aligned}
(i, j) \text{ entry of } \mathbf{A} &= \mathbf{e}_i \cdot (\mathbf{A}\mathbf{e}_j) = [\mathbf{v}_i]_B \cdot (\mathbf{A}[\mathbf{v}_j]_B) \\
&= [\mathbf{v}_i]_B \cdot [L(\mathbf{v}_j)]_B \\
&= \mathbf{v}_i \cdot L(\mathbf{v}_j) && \text{by the claim verified earlier in this proof} \\
&= L(\mathbf{v}_i) \cdot \mathbf{v}_j && \text{since } L \text{ is symmetric} \\
&= [L(\mathbf{v}_i)]_B \cdot [\mathbf{v}_j]_B && \text{by the claim} \\
&= (\mathbf{A}[\mathbf{v}_i]_B) \cdot [\mathbf{v}_j]_B \\
&= (\mathbf{A}\mathbf{e}_i) \cdot \mathbf{e}_j = (j, i) \text{ entry of } \mathbf{A}.
\end{aligned}
$$

Conversely, if $\mathbf{A}$ is a symmetric matrix and $\mathbf{w}_1, \mathbf{w}_2 \in \mathcal{V}$, we have

$$
\begin{aligned}
L(\mathbf{w}_1) \cdot \mathbf{w}_2 &= [L(\mathbf{w}_1)]_B \cdot [\mathbf{w}_2]_B && \text{by the claim} \\
&= (\mathbf{A}[\mathbf{w}_1]_B) \cdot [\mathbf{w}_2]_B \\
&= (\mathbf{A}[\mathbf{w}_1]_B)^T [\mathbf{w}_2]_B && \text{changing vector dot product to matrix multiplication} \\
&= [\mathbf{w}_1]_B^T \mathbf{A}^T [\mathbf{w}_2]_B \\
&= [\mathbf{w}_1]_B^T \mathbf{A} [\mathbf{w}_2]_B && \text{since } \mathbf{A} \text{ is symmetric} \\
&= [\mathbf{w}_1]_B \cdot (\mathbf{A}[\mathbf{w}_2]_B) && \text{changing matrix multiplication to vector dot product} \\
&= [\mathbf{w}_1]_B \cdot [L(\mathbf{w}_2)]_B \\
&= \mathbf{w}_1 \cdot L(\mathbf{w}_2) && \text{by the claim}
\end{aligned}
$$

The proof is now complete. ■

B

Functions

In this appendix, we include some basic terminology associated with functions: domain, codomain, range, image, pre-image, one-to-one, onto, composition, and inverses. It is a good idea to review this material thoroughly before beginning Chapter 5.

Functions: Domain, Codomain, and Range

A function from a set X to a set Y is a mapping (assignment) of elements of X (called the **domain**) to elements of Y (called the **codomain**) in such a way that each element of X is assigned to some (single) chosen element of Y. That is, every element of X must be assigned to *some* element of Y and *only one* element of Y. For example, $f: \mathbb{Z} \longrightarrow \mathbb{R}$ (where $\mathbb{Z}$ represents the set $\{\ldots, -3, -2, -1, 0, 1, 2, 3, \ldots\}$ of all integers) given by $f(x) = x^2$ is a function, since each integer in $\mathbb{Z}$ is assigned to one and only one element of $\mathbb{R}$.

Notice that the definition of a function allows two different elements of X to **map** (be assigned) to the same element of Y, as in the function $f: \mathbb{Z} \longrightarrow \mathbb{R}$ given by $f(x) = x^2$, where $f(3) = f(-3) = 9$. However, no function allows any member of the domain to map to more than one element of the codomain. Hence, the rule $x \longrightarrow \pm\sqrt{x}$, for $x \in \mathbb{R}^+$ (positive real numbers) is not a function, since, for example, 4 would have to map to both 2 and -2.

The **image** of a domain element is the codomain element to which it is mapped, and the **pre-images** of a codomain element are the domain elements that map to it. With the function $f: \mathbb{Z} \longrightarrow \mathbb{R}$ given by $f(x) = x^2$ above, 4 is the image of 2, and both 2 and -2 are pre-images of 4, since $2^2 = (-2)^2 = 4$.

If $f: X \longrightarrow Y$ is a function, not every element of Y needs to have a pre-image. For the function $f(x) = x^2$ given above, the element 5 in the codomain has no pre-image, because no integer when squared equals 5.

The **image of a subset** S of the domain under a function f is the set of all values in the codomain that are mapped to by elements of S. The **pre-image of a subset** T

of the codomain under f is the set of *all* values in the domain that map to elements of T under S. For example, for the function $f\colon \mathbb{Z} \longrightarrow \mathbb{R}$ given by $f(x) = x^2$, the image of the subset $\{-5, -3, 3, 5\}$ of the domain is $\{9, 25\}$, and the pre-image of $\{15, 16, 17\}$ is $\{4, -4\}$.

The image of the entire domain is called the **range** of the function. For the same function $f\colon \mathbb{Z} \longrightarrow \mathbb{R}$ given by $f(x) = x^2$, the range is the set of all squares of integers. In this case, the range is a proper subset of the codomain. This situation is depicted in Figure B.1. For some functions, however, the range is the whole codomain, as with $g\colon \mathbb{R} \longrightarrow \mathbb{R}$ given by $g(x) = 2x$.

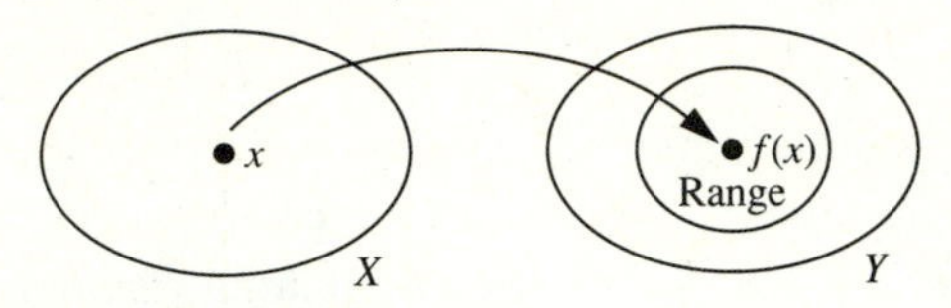

Figure B.1
The domain X, codomain Y, and range of a function $f\colon X \longrightarrow Y$

One-to-One and Onto Functions

We now discuss two very important types of functions: one-to-one and onto functions. We say that a function $f\colon X \longrightarrow Y$ is **one-to-one** if and only if distinct elements of X map to distinct elements of Y. That is, f is one-to-one if and only if no two distinct elements of X map to the same element of Y. For example, $f\colon \mathbb{R} \longrightarrow \mathbb{R}$ given by $f(x) = x^3$ is one-to-one, since no two distinct real numbers have the same cube.

A standard method of proving that a function f is one-to-one is as follows:

> **To show that $f\colon X \longrightarrow Y$ is one-to-one:** Prove that whenever $f(x_1) = f(x_2)$, for arbitrary elements $x_1, x_2 \in X$, we must have $x_1 = x_2$.

In other words, the only way x_1 and x_2 can have the same image is if they are not really distinct. Let us use this technique to show that $f\colon \mathbb{R} \longrightarrow \mathbb{R}$ given by $f(x) = 3x - 7$ is one-to-one. Suppose that $f(x_1) = f(x_2)$, for some $x_1, x_2 \in \mathbb{R}$. Then $3x_1 - 7 = 3x_2 - 7$. Hence, $3x_1 = 3x_2$, which implies $x_1 = x_2$. Thus, f is one-to-one.

On the other hand, we sometimes need to show that a function is *not* one-to-one. The usual method for doing this is as follows:

To show that f: $X \longrightarrow Y$ is not one-to-one: Find two different elements x_1 and x_2 in the domain X such that $f(x_1) = f(x_2)$.

For example, g: $\mathbb{R} \longrightarrow \mathbb{R}$ given by $g(x) = x^2$ is not one-to-one, because $g(3) = g(-3) = 9$. That is, both elements 3 and -3 in the domain $\mathbb{R}$ of g have the same image 9, so g is not one-to-one.

We say that a function f: $X \longrightarrow Y$ is **onto** if and only if every element of Y is an image of some element in X. That is, f is onto if and only if the range of f equals the codomain of f. For example, the function f: $\mathbb{R} \longrightarrow \mathbb{R}$ given by $f(x) = 2x$ is onto, since every real number y in the codomain $\mathbb{R}$ is the image of the real number $x = \frac{1}{2}y$; that is, $f(x) = f(\frac{1}{2}y) = y$. Here we are using the standard method of proving that a given function is onto:

To show that f: $X \longrightarrow Y$ is onto: Choose an arbitrary element $y \in Y$, and show that there is some $x \in X$ such that $y = f(x)$.

On the other hand, we sometimes need to show that a function is *not* onto. The usual method for doing this is as follows:

To show that f: $X \longrightarrow Y$ is not onto: Find an element y in the codomain Y that is not the image of any element x in the domain X.

For example, f: $\mathbb{R} \longrightarrow \mathbb{R}$ given by $f(x) = x^2$ is not onto, since the real number -4 in the codomain $\mathbb{R}$ is never the image of any real number in the domain; that is, for all $x \in \mathbb{R}$, $f(x) \neq -4$.

Composition and Inverses of Functions

If f: $X \longrightarrow Y$ and g: $Y \longrightarrow Z$ are functions, we define the **composition** of f and g to be the function $g \circ f$: $X \longrightarrow Z$ given by $(g \circ f)(x) = g(f(x))$. This composition mapping is pictured in Figure B.2. For example, if f: $\mathbb{R} \longrightarrow \mathbb{R}$ is given by

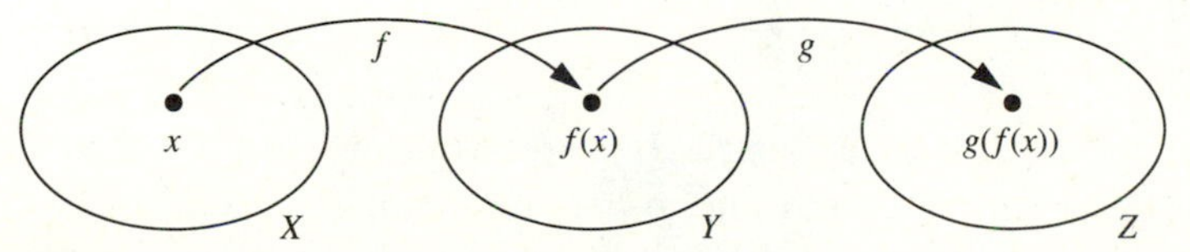

Figure B.2
Composition $g \circ f$ of f: $X \longrightarrow Y$ and g: $Y \longrightarrow Z$

$f(x) = 1 - x^2$ and $g\colon \mathbb{R} \longrightarrow \mathbb{R}$ is given by $g(x) = 5\cos x$, then $(g \circ f)(x) = g(f(x)) = g(1 - x^2) = 5\cos(1 - x^2)$.

THEOREM B.1

(1) If $f\colon X \longrightarrow Y$ and $g\colon Y \longrightarrow Z$ are both one-to-one, then $g \circ f\colon X \longrightarrow Z$ is one-to-one.
(2) If $f\colon X \longrightarrow Y$ and $g\colon Y \longrightarrow Z$ are both onto, then $g \circ f\colon X \longrightarrow Z$ is onto.

Proof of Theorem B.1

Part (1): Assume that f and g are both one-to-one. To prove $g \circ f$ is one-to-one, we assume that $(g \circ f)(x_1) = (g \circ f)(x_2)$, for two elements $x_1, x_2 \in X$, and try to prove that $x_1 = x_2$. However, $(g \circ f)(x_1) = (g \circ f)(x_2)$ implies that $g(f(x_1)) = g(f(x_2))$. Hence, $f(x_1)$ and $f(x_2)$ have the same image under g. Since g is one-to-one, we must have $f(x_1) = f(x_2)$. Then x_1 and x_2 have the same image under f. Since f is one-to-one, $x_1 = x_2$. Hence, $g \circ f$ is one-to-one.

Part (2): Assume that f and g are both onto. To prove that $g \circ f\colon X \longrightarrow Z$ is onto, we choose an arbitrary element $z_1 \in Z$ and try to find some element in X that $g \circ f$ maps to z_1. Now, since g is onto, there is some $y_1 \in Y$ for which $g(y_1) = z_1$. Also, since f is onto, there is some $x_1 \in X$ for which $f(x_1) = y_1$. Therefore, $(g \circ f)(x_1) = g(f(x_1)) = g(y_1) = z_1$, and so $g \circ f$ maps x_1 to z_1. Hence, $g \circ f$ is onto. ■

Two functions $f\colon X \longrightarrow Y$ and $g\colon Y \longrightarrow X$ are **inverses** of each other if $(g \circ f)(x) = x$ and $(f \circ g)(y) = y$, for every $x \in X$ and $y \in Y$. For example, $f\colon \mathbb{R} \longrightarrow \mathbb{R}$ given by $f(x) = x^3$ and $g\colon \mathbb{R} \longrightarrow \mathbb{R}$ given by $g(x) = \sqrt[3]{x}$ are inverses of each other because $(g \circ f)(x) = g(f(x)) = g(x^3) = \sqrt[3]{x^3} = x$, and $(f \circ g)(x) = f(g(x)) = f(\sqrt[3]{x}) = (\sqrt[3]{x})^3 = x$.

The next theorem categorizes those functions that have an inverse.

THEOREM B.2

The function $f\colon X \longrightarrow Y$ has an inverse $g\colon Y \longrightarrow X$ if and only if f is both one-to-one and onto.

Notice that the inverse functions $f\colon \mathbb{R} \longrightarrow \mathbb{R}$ given by $f(x) = x^3$ and $g\colon \mathbb{R} \longrightarrow \mathbb{R}$ given by $g(x) = \sqrt[3]{x}$ are both one-to-one and onto. However, a function such as $f\colon \mathbb{R} \longrightarrow \mathbb{R}$ given by $f(x) = x^2$ has no inverse, since it is not one-to-one. In this case, we could also have shown that f has no inverse since it is not onto.

Proof of Theorem B.2

First, suppose that $f: X \longrightarrow Y$ has an inverse $g: Y \longrightarrow X$. We show that f is one-to-one and onto. To prove f is one-to-one, we assume that $f(x_1) = f(x_2)$, for some $x_1, x_2 \in X$, and try to prove $x_1 = x_2$. Since $f(x_1) = f(x_2)$, we have $g(f(x_1)) = g(f(x_2))$. However, since g is an inverse for f, $x_1 = (g \circ f)(x_1) = g(f(x_1)) = g(f(x_2)) = x_2$, and so $x_1 = x_2$. Hence, f is one-to-one. To prove f is onto, we choose an arbitrary $y_1 \in Y$. We must show that y_1 is the image of some $x_1 \in X$. Now, g maps y_1 to an element x_1 of X; that is, $g(y_1) = x_1$. However, $f(x_1) = f(g(y_1)) = (f \circ g)(y_1) = y_1$, since f and g are inverses. Hence, f maps x_1 to y_1, and f is onto.

Conversely, we assume that $f: X \longrightarrow Y$ is one-to-one and onto and show that f has an inverse $g: Y \longrightarrow X$. Let y_1 be an arbitrary element of Y. Since f is onto, the element y_1 in Y is the image of some element in X. Since f is one-to-one, y_1 is the image of precisely one element, say x_1, in X. Now consider the mapping $g: Y \longrightarrow X$, which maps each element y in Y to its unique pre-image x in X under f. Then $(f \circ g)(y) = f(g(y)) = f(x) = y$. Also since every x in X is a pre-image of some y in Y, $(g \circ f)(x) = g(f(x)) = g(y) = x$. Thus, g and f are inverses. ■

The next result assures us that when inverses exist, they are unique.

THEOREM B.3

If $f: X \longrightarrow Y$ has an inverse $g: Y \longrightarrow X$, then g is the only inverse of f.

Proof of Theorem B.3

Suppose that $g_1: Y \longrightarrow X$ and $g_2: Y \longrightarrow X$ are both inverse functions for f. Our goal is to show that $g_1(y) = g_2(y)$, for all $y \in Y$, for then g_1 and g_2 are identical functions, and hence the inverse of f is unique.

Now, $(g_2 \circ f)(x) = x$, for every $x \in X$, since f and g_2 are inverses. Thus, since $g_1(y) \in X$, $g_1(y) = (g_2 \circ f)(g_1(y)) = g_2(f(g_1(y))) = g_2((f \circ g_1)(y)) = g_2(y)$, since f and g_1 are inverses. ■

Whenever a function $f: X \longrightarrow Y$ has an inverse, we denote this unique inverse by $f^{-1}: Y \longrightarrow X$.

THEOREM B.4

If $f: X \longrightarrow Y$ and $g: Y \longrightarrow Z$ both have inverses, then $g \circ f: X \longrightarrow Z$ has an inverse, and $(g \circ f)^{-1} = f^{-1} \circ g^{-1}$.

Proof of Theorem B.4

Because $g^{-1}\colon Z \longrightarrow Y$ and $f^{-1}\colon Y \longrightarrow X$, it follows that $f^{-1} \circ g^{-1}$ is a well-defined function from Z to X. We need to show that the inverse of $g \circ f$ is $f^{-1} \circ g^{-1}$. If we can show that both

$$((g \circ f) \circ (f^{-1} \circ g^{-1}))(z) = z, \quad \text{for all } z \in Z,$$

and

$$((f^{-1} \circ g^{-1}) \circ (g \circ f))(x) = x, \quad \text{for all } x \in X,$$

then by the definition of inverse functions, $g \circ f$ and $f^{-1} \circ g^{-1}$ are inverses. However,

$$\begin{aligned} ((g \circ f) \circ (f^{-1} \circ g^{-1}))(z) &= g(f(f^{-1}(g^{-1}(z)))) \\ &= g(g^{-1}(z)) && \text{since } f \text{ and } f^{-1} \text{ are inverses} \\ &= z. && \text{since } g \text{ and } g^{-1} \text{ are inverses} \end{aligned}$$

A similar argument establishes the other statement. ■

As an example of Theorem B.4, consider $f\colon \mathbb{R} \longrightarrow \mathbb{R}$ given by $f(x) = x^3$ and $g\colon \mathbb{R} \longrightarrow \mathbb{R}^+$ given by $g(x) = e^x$. Then, $g \circ f\colon \mathbb{R} \longrightarrow \mathbb{R}^+$ is $(g \circ f)(x) = e^{x^3}$. However, since $f^{-1}(x) = \sqrt[3]{x}$ and $g^{-1}(x) = \ln x$, $(g \circ f)^{-1}\colon \mathbb{R}^+ \longrightarrow \mathbb{R}$ is given by

$$(g \circ f)^{-1}(x) = (f^{-1} \circ g^{-1})(x) = f^{-1}(g^{-1}(x)) = \sqrt[3]{\ln x}.$$

Exercises—Appendix B

1. Which of the following are functions? For those that are functions, determine the range and all images and pre-images of the value 2. For those that are not functions, explain why with a precise reason. (Note: $\mathbb{N}$ represents the set $\{0, 1, 2, 3, \ldots\}$ of natural numbers, and $\mathbb{Z}$ represents the set $\{\ldots, -2, -1, 0, 1, 2, \ldots\}$ of integers.)

★(a) $f\colon \mathbb{R} \longrightarrow \mathbb{R}$, given by $f(x) = \sqrt{x-1}$

(b) $g\colon \mathbb{R} \longrightarrow \mathbb{R}$, given by $g(x) = \sqrt{|x-1|}$

★(c) $h\colon \mathbb{R} \longrightarrow \mathbb{R}$, given by $h(x) = \pm\sqrt{|x-1|}$

(d) $j\colon \mathbb{N} \longrightarrow \mathbb{Z}$, given by $j(a) = \begin{cases} a-5, & \text{if } a \text{ is odd} \\ a-4, & \text{if } a \text{ is even} \end{cases}$

★(e) $k\colon \mathbb{R} \longrightarrow \mathbb{R}$, given by $k(\theta) = \tan\theta$ (where θ is in radians)

★(f) $l\colon \mathbb{N} \longrightarrow \mathbb{N}$, where $l(t)$ is the smallest prime number $\geq t$

(g) $m\colon \mathbb{R} \longrightarrow \mathbb{R}$, given by $m(x) = \begin{cases} x-3, & \text{if } x \leq 2 \\ x+4, & \text{if } x \geq 2 \end{cases}$

2. Let $f\colon \mathbb{Z} \longrightarrow \mathbb{N}$ (with $\mathbb{Z}$ and $\mathbb{N}$ as in Exercise 1) be given by $f(x) = 2|x|$.

★(a) Find the pre-image of the set $\{10, 20, 30\}$.

(b) Find the pre-image of the set $\{10, 11, 12, \ldots, 19\}$.

★(c) Find the pre-image of the multiples of 4 in $\mathbb{N}$.

★**3.** Let $f, g: \mathbb{R} \longrightarrow \mathbb{R}$ be given by $f(x) = (5x - 1)/4$ and $g(x) = \sqrt{3x^2 + 2}$. Find $g \circ f$ and $f \circ g$.

★**4.** Let $f: \mathbb{R}^2 \longrightarrow \mathbb{R}^2$ be given by $f\left(\begin{bmatrix} x \\ y \end{bmatrix}\right) = \begin{bmatrix} 3 & -2 \\ 1 & 4 \end{bmatrix}\begin{bmatrix} x \\ y \end{bmatrix}$. Let $g: \mathbb{R}^2 \longrightarrow \mathbb{R}^2$ be given by $g\left(\begin{bmatrix} x \\ y \end{bmatrix}\right) = \begin{bmatrix} -4 & 4 \\ 0 & 2 \end{bmatrix}\begin{bmatrix} x \\ y \end{bmatrix}$. Describe $g \circ f$ and $f \circ g$.

5. Let $A = \{1, 2, 3\}$, $B = \{4, 5, 6, 7\}$, and $C = \{8, 9, 10\}$.
(a) Give an example of functions $f: A \longrightarrow B$ and $g: B \longrightarrow C$ such that $g \circ f$ is onto but f is not onto.
(b) Give an example of functions $f: A \longrightarrow B$ and $g: B \longrightarrow C$ such that $g \circ f$ is one-to-one but g is not one-to-one.

6. For $n \geq 2$, show that $f: \mathcal{M}_{nn} \longrightarrow \mathbb{R}$ given by $f(\mathbf{A}) = |\mathbf{A}|$ is onto but not one-to-one.

7. Show that $f: \mathcal{M}_{33} \longrightarrow \mathcal{M}_{33}$ given by $f(\mathbf{A}) = \mathbf{A} + \mathbf{A}^T$ is neither one-to-one nor onto.

★**8.** Show that the function $f: \mathcal{P}_n \longrightarrow \mathcal{P}_n$ given by $f(\mathbf{p}) = \mathbf{p}'$ is neither one-to-one nor onto. What is the pre-image of the subset $\mathcal{P}_2$ of the codomain?

9. Prove that $f: \mathbb{R} \longrightarrow \mathbb{R}$ given by $f(x) = 3x^3 - 5$ has an inverse by showing that it is both one-to-one and onto. Give a formula for $f^{-1}: \mathbb{R} \longrightarrow \mathbb{R}$.

★**10.** Let $\mathbf{B}$ be a fixed nonsingular matrix in $\mathcal{M}_{nn}$. Show that the map $f: \mathcal{M}_{nn} \longrightarrow \mathcal{M}_{nn}$ given by $f(\mathbf{A}) = \mathbf{B}^{-1}\mathbf{A}\mathbf{B}$ is both one-to-one and onto. What is the inverse of f?

11. Let $f: A \longrightarrow B$ and $g: B \longrightarrow C$ be functions.
(a) Prove that if $g \circ f$ is onto, then g is onto. (Compare this exercise with Exercise 5(a).)
(b) Prove that if $g \circ f$ is one-to-one, then f is one-to-one. (Compare this exercise with Exercise 5(b).)

C

Complex Numbers

In this appendix, we define complex numbers and for reference list their most important operations and properties. In what follows, we introduce a nonreal number i with the property $i^2 = -1$.

DEFINITION

The set $\mathbb{C}$ of **complex numbers** is the set of all numbers of the form $a + bi$, where $i^2 = -1$ and where a and b are real numbers. The **real part** of $a + bi$ is a, and the **imaginary part** of $a + bi$ is b.

Some examples of complex numbers are $2 + 3i$, $-\frac{1}{2} + \frac{1}{4}i$, and $\sqrt{3} - i$. Any real number a can be expressed as $a + 0i$, so the real numbers are a subset of the complex numbers; that is, $\mathbb{R} \subset \mathbb{C}$. A complex number of the form $0 + bi = bi$ is called a **pure imaginary** complex number.

The **magnitude** of $a + bi$ is defined to be $|a + bi| = \sqrt{a^2 + b^2}$. For example, the magnitude of $3 - 2i$ is $|3 - 2i| = \sqrt{3^2 + (-2)^2} = \sqrt{13}$.

We define **addition** of complex numbers by

$$(a + bi) + (c + di) = (a + c) + (b + d)i,$$

where $a, b, c, d \in \mathbb{R}$. Complex number **multiplication** is defined by

$$(a + bi)(c + di) = (ac - bd) + (ad + bc)i.$$

For example,

$$\begin{aligned}(3-2i)[(2-i)+(-3+5i)] &= (3-2i)(-1+4i)\\ &= [(3)(-1)-(-2)(4)]+[(3)(4)+(-2)(-1)]i\\ &= 5+14i.\end{aligned}$$

If $z = a + bi$, we let $-z$ denote the special scalar product $-1z = -a - bi$. The **complex conjugate** of a complex number $a + bi$ is defined as

$$\overline{a+bi} = a - bi.$$

For example, $\overline{-4-3i} = -4+3i$. Notice that if $z = a+bi$, then $\overline{z} = a - bi$, and so $z\overline{z} = (a+bi)(a-bi) = a^2+b^2 = |a+bi|^2 = |z|^2$, a real number. We can use this property to calculate the **reciprocal** of a complex number, as follows:

If $z = a + bi \neq 0$, then

$$\frac{1}{z} = \frac{1}{a+bi} = \frac{1}{a+bi}\cdot\frac{a-bi}{a-bi} = \frac{a-bi}{a^2+b^2} = \frac{\overline{z}}{|z|^2}.$$

It is easy to show that the operations of complex addition and multiplication satisfy the commutative, associative, and distributive laws. Some other useful properties are listed in the next theorem, which is easy to prove.

THEOREM C.1

Let $z_1, z_2, z_3 \in \mathbb{C}$. Then

(1) $\overline{z_1+z_2} = \overline{z_1}+\overline{z_2}$	Additive Conjugate Law
(2) $\overline{(z_1z_2)} = \overline{z_1}\,\overline{z_2}$	Multiplicative Conjugate Law
(3) If $z_1z_2 = 0$, then either $z_1 = 0$ or $z_2 = 0$	Zero Product Property
(4) $z_1 = \overline{z_1}$ if and only if z_1 is real	Condition for Complex Number to Be Real
(5) $z_1 = -\overline{z_1}$ if and only if z_1 is pure imaginary	Condition for Complex Number to Be Pure Imaginary

Exercises—Appendix C

1. Perform the following computations involving complex numbers.

★(a) $(6 - 3i) + (5 + 2i)$ (b) $8(3 - 4i)$

★(c) $4((8 - 2i) - (3 + i))$ (d) $-3((-2 + i) - (4 - 2i))$

★(e) $(5 + 3i)(3 + 2i)$ (f) $(-6 + 4i)(3 - 5i)$

★(g) $(7 - i)(-2 - 3i)$ (h) $\overline{5 + 4i}$ ★(i) $\overline{9 - 2i}$ (j) $\overline{-6}$

★(k) $\overline{(6 + i)(2 - 4i)}$ (l) $|8 - 3i|$ ★(m) $|-2 + 7i|$ (n) $|\overline{3 + 4i}|$

2. Find the multiplicative inverse (reciprocal) of each of the following.

★(a) $6 - 2i$ (b) $3 + 4i$ ★(c) $-4 + i$ (d) $-5 - 3i$

3. (a) Prove parts (1) and (2) of Theorem C.1.

(b) Prove part (3) of Theorem C.1.

(c) Prove parts (4) and (5) of Theorem C.1.

Answers to Selected Exercises

Section 1.1 (pp. 12–15)

1. (a) $[9, -4]$, distance $= \sqrt{97}$
(c) $[-1, -1, 2, -3, -4]$, distance $= \sqrt{31}$

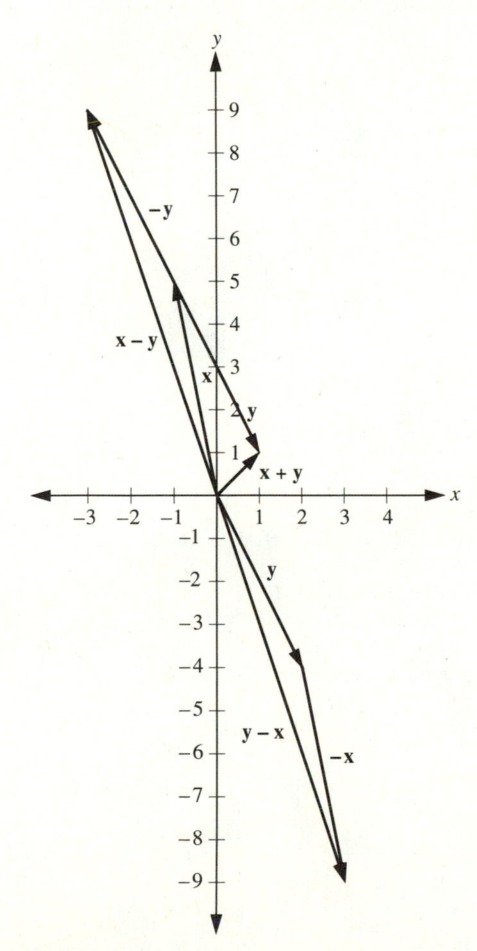

Figure 1
Answer to Exercise 8(a), Section 1.1

2. (a) $(3, 4, 2)$ (c) $(1, -2, 0)$

3. (a) $(7, -13)$ (c) $(-1, 3, -1, 4, 6)$

4. (a) $\left[\frac{16}{3}, -\frac{13}{3}, 8\right]$

5. (a) $\left[3/\sqrt{70}, -5/\sqrt{70}, 6/\sqrt{70}\right]$; shorter, since length of original vector is > 1 (c) $[0.6, -0.8]$; neither, since given vector is a unit vector

6. (a) Parallel (c) Not parallel

7. (a) $[-6, 12, 15]$ (c) $[-3, 4, 8]$ (e) $[6, -20, -13]$

8. (a) $\mathbf{x}+\mathbf{y} = [1, 1]$, $\mathbf{x}-\mathbf{y} = [-3, 9]$ (see Figure 1)
(c) $\mathbf{x}+\mathbf{y} = [1, 8, -5]$, $\mathbf{x}-\mathbf{y} = [3, 2, -1]$ (see Figure 2)

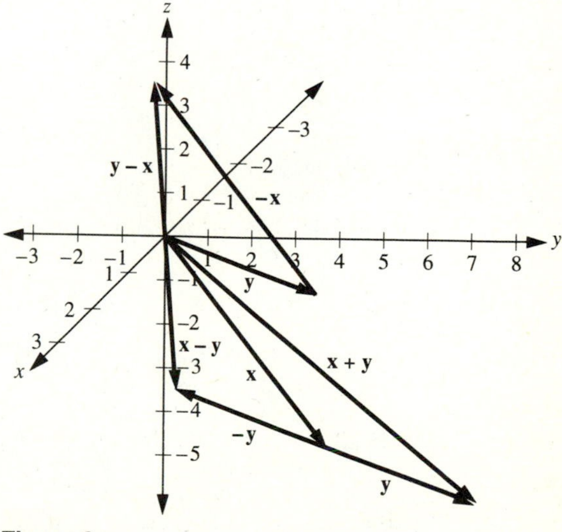

Figure 2
Answer to Exercise 8(c), Section 1.1

10. (a) $[10, -10]$ (b) $\left[-5\sqrt{3}, -15\right]$

13. $\left[0.5 - 0.6\sqrt{2}, -0.4\sqrt{2}\right]$

15. Net velocity $= \left[-2\sqrt{2}, -3 + 2\sqrt{2}\right]$, resultant speed $= 2.83$ km/hr

17. $\left[-8 - \sqrt{2}, -\sqrt{2}\right]$

18. Acceleration $= \frac{1}{20}\left[\frac{12}{13}, -\frac{344}{65}, \frac{392}{65}\right]$

20. $\mathbf{a} = \left[-mg/(1+\sqrt{3}), mg/(1+\sqrt{3})\right]$,

$\mathbf{b} = \left[mg/(1+\sqrt{3}), mg\sqrt{3}/(1+\sqrt{3})\right]$

Section 1.2 (pp. 24–27)

1. (a) $\arccos(-27/(5\sqrt{37}))$, or about 152.6°, or 2.66 radians

(c) $\arccos(0)$, which is 90°, or $\frac{\pi}{2}$ radians

4. $(1040\sqrt{5})/9$, or 258.4 joules

7. No; consider $\mathbf{x} = [1, 0]$, $\mathbf{y} = [0, 1]$, and $\mathbf{z} = [1, 1]$.

13. $\cos\theta_1 = a/\sqrt{a^2+b^2+c^2}$,

$\cos\theta_2 = b/\sqrt{a^2+b^2+c^2}$, and

$\cos\theta_3 = c/\sqrt{a^2+b^2+c^2}$

14. Length of diagonal $= \sqrt{3}s$; angle $= \arccos(\sqrt{3}/3)$, which is about 54.7°, or 0.955 radians

15. (a) $\left[-\frac{3}{5}, -\frac{3}{10}, -\frac{3}{2}\right]$ (c) $\left[\frac{1}{6}, 0, -\frac{1}{6}, \frac{1}{3}\right]$

17. $a\mathbf{i}$, $b\mathbf{j}$, $c\mathbf{k}$

18. (a) Parallel: $\left[\frac{20}{29}, -\frac{30}{29}, \frac{40}{29}\right]$,

orthogonal: $\left[-\frac{194}{29}, \frac{88}{29}, \frac{163}{29}\right]$

(c) Parallel: $\left[\frac{60}{49}, -\frac{40}{49}, \frac{120}{49}\right]$,

orthogonal: $\left[-\frac{354}{49}, \frac{138}{49}, \frac{223}{49}\right]$

Section 1.3 (pp. 41–44)

1. (b) Let $m = \max\{|c|, |d|\}$. Then $\|c\mathbf{x} \pm d\mathbf{y}\| \le m(\|\mathbf{x}\| + \|\mathbf{y}\|)$.

2. (b) Consider the number 4.

5. (a) Consider $\mathbf{x} = [1, 0, 0]$ and $\mathbf{y} = [1, 1, 0]$.

(b) If $\mathbf{x} \ne \mathbf{y}$, then $\mathbf{x} \cdot \mathbf{y} \ne \|\mathbf{x}\|^2$.

(c) Yes

8. (a) Contrapositive: If $\mathbf{x} = \mathbf{0}$, then $\mathbf{x}$ is not a unit vector.

Converse: If $\mathbf{x}$ is nonzero, then $\mathbf{x}$ is a unit vector.

Inverse: If $\mathbf{x}$ is not a unit vector, then $\mathbf{x} = \mathbf{0}$.

(c) (Let $\mathbf{x}$, $\mathbf{y}$ be nonzero vectors.)

Contrapositive: If $\mathbf{proj}_{\mathbf{y}}\mathbf{x} \ne \mathbf{0}$, then $\mathbf{proj}_{\mathbf{x}}\mathbf{y} \ne \mathbf{0}$.

Converse: If $\mathbf{proj}_{\mathbf{y}}\mathbf{x} = \mathbf{0}$, then $\mathbf{proj}_{\mathbf{x}}\mathbf{y} = \mathbf{0}$.

Inverse: If $\mathbf{proj}_{\mathbf{x}}\mathbf{y} \ne \mathbf{0}$, then $\mathbf{proj}_{\mathbf{y}}\mathbf{x} \ne \mathbf{0}$.

10. (b) Converse: Let $\mathbf{x}$ and $\mathbf{y}$ be vectors in $\mathbb{R}^n$. If $\|\mathbf{x}+\mathbf{y}\| \ge \|\mathbf{y}\|$, then $\mathbf{x} \cdot \mathbf{y} = 0$. The original statement is true, but the converse is false in general. Proof of the original statement follows from

$$\begin{aligned}\|\mathbf{x}+\mathbf{y}\|^2 &= (\mathbf{x}+\mathbf{y}) \cdot (\mathbf{x}+\mathbf{y}) \\ &= \|\mathbf{x}\|^2 + 2(\mathbf{x} \cdot \mathbf{y}) + \|\mathbf{y}\|^2 \\ &= \|\mathbf{x}\|^2 + \|\mathbf{y}\|^2 \ge \|\mathbf{y}\|^2.\end{aligned}$$

Counterexample to converse: Let $\mathbf{x} = [1, 0]$, $\mathbf{y} = [1, 1]$.

17. Step 1 cannot be reversed, because y could equal $\pm(x^2+2)$.

Step 2 cannot be reversed, because y^2 could equal $x^4 + 4x^2 + c$.

Step 4 cannot be reversed, because in general y does not have to equal $x^2 + 2$.

Step 6 cannot be reversed, since dy/dx could equal $2x + c$.

All other steps remain true when reversed.

18. (a) For every unit vector $\mathbf{x}$ in $\mathbb{R}^3$, $\mathbf{x} \cdot [1, -2, 3] \ne 0$.

(c) $\mathbf{x} = \mathbf{0}$ or $\|\mathbf{x}+\mathbf{y}\| \ne \|\mathbf{y}\|$, where $\mathbf{x}$ and $\mathbf{y}$ are vectors in $\mathbb{R}^n$.

(e) There is an $\mathbf{x} \in \mathbb{R}^3$ such that, for every nonzero $\mathbf{y} \in \mathbb{R}^3$, $\mathbf{x} \cdot \mathbf{y} \ne 0$.

19. (a) Contrapositive: If $\mathbf{x} \ne \mathbf{0}$ and $\|\mathbf{x}-\mathbf{y}\| \le \|\mathbf{y}\|$, then $\mathbf{x} \cdot \mathbf{y} \ne 0$.

Converse: If $\mathbf{x} = \mathbf{0}$ or $\|\mathbf{x}-\mathbf{y}\| > \|\mathbf{y}\|$, then $\mathbf{x} \cdot \mathbf{y} = 0$.

Inverse: If $\mathbf{x} \cdot \mathbf{y} \ne 0$, then $\mathbf{x} \ne \mathbf{0}$ and $\|\mathbf{x}-\mathbf{y}\| \le \|\mathbf{y}\|$.

Section 1.4 (pp. 52–54)

1. (a) $\begin{bmatrix} 2 & 1 & 3 \\ 2 & 7 & -5 \\ 9 & 0 & -1 \end{bmatrix}$ (c) $\begin{bmatrix} -16 & 8 & 12 \\ 0 & 20 & -4 \\ 24 & 4 & -8 \end{bmatrix}$

(e) Impossible (g) $\begin{bmatrix} -23 & 14 & -9 \\ -5 & 8 & 8 \\ -9 & -18 & 1 \end{bmatrix}$

(i) $\begin{bmatrix} -1 & 1 & 12 \\ -1 & 5 & 8 \\ 8 & -3 & -4 \end{bmatrix}$ (l) Impossible

(n) $\begin{bmatrix} 13 & -6 & 2 \\ 3 & -3 & -5 \\ 3 & 5 & 1 \end{bmatrix}$

2. Square: **B**, **C**, **E**, **F**, **G**, **H**, **J**, **K**, **L**, **M**, **N**, **P**, **Q**

Diagonal: **B**, **G**, **N**

Upper triangular: **B**, **G**, **L**, **N**

Lower triangular: **B**, **G**, **M**, **N**, **Q**

Symmetric: **B**, **F**, **G**, **J**, **N**, **P**

Skew-symmetric: **H** (but not **E**, **C**, **K**)

Transposes: $\mathbf{A}^T = \begin{bmatrix} -1 & 0 & 6 \\ 4 & 1 & 0 \end{bmatrix}$, $\mathbf{B}^T = \mathbf{B}$,

$\mathbf{C}^T = \begin{bmatrix} -1 & -1 \\ 1 & 1 \end{bmatrix}$, and so on

3. (a) $\begin{bmatrix} 3 & -\frac{1}{2} & \frac{5}{2} \\ -\frac{1}{2} & 2 & 1 \\ \frac{5}{2} & 1 & 2 \end{bmatrix} + \begin{bmatrix} 0 & -\frac{1}{2} & \frac{3}{2} \\ \frac{1}{2} & 0 & 4 \\ -\frac{3}{2} & -4 & 0 \end{bmatrix}$

5. (c) The matrix must be a square zero matrix.

14. (a) trace(**B**) = 1, trace(**C**) = 0, trace(**E**) = −6, trace(**F**) = 2, trace(**G**) = 18, trace(**H**) = 0, trace(**J**) = 1, trace(**K**) = 4, trace(**L**) = 3, trace(**M**) = 0, trace(**N**) = 3, trace(**P**) = 0, trace(**Q**) = 1

(c) No; consider matrices **L** and **N** in Exercise 2.

Section 1.5 (pp. 62–66)

1. (b) $\begin{bmatrix} 34 & -24 \\ 42 & 49 \\ 8 & -22 \end{bmatrix}$ (c) Impossible

(e) $[-38]$ (f) $\begin{bmatrix} -24 & 48 & -16 \\ 3 & -6 & 2 \\ -12 & 24 & -8 \end{bmatrix}$

(g) Impossible (j) Impossible

(l) $\begin{bmatrix} 5 & 3 & 2 & 5 \\ 4 & 1 & 3 & 1 \\ 1 & 1 & 0 & 2 \\ 4 & 1 & 3 & 1 \end{bmatrix}$ (n) $\begin{bmatrix} 146 & 5 & -603 \\ 154 & 27 & -560 \\ 38 & -9 & -193 \end{bmatrix}$

2. (a) No (c) No (d) Yes

3. (a) $[15, -13, -8]$ (c) $[4]$

4. (a) Holds, by Theorem 1.13, part (1) (b) Fails
(c) Holds, by Theorem 1.13, part (1)
(d) Holds, by Theorem 1.13, part (2)
(e) Holds, by Theorem 1.15 (f) Fails
(g) Holds, by Theorem 1.13, part (3) (h) Holds
(i) Holds, by Theorem 1.13, part (2) (j) Fails
(k) Holds, by Theorem 1.13, part (3), and Theorem 1.15

5.

	Salary	Fringe benefits
Outlet 1	\$367500	\$78000
Outlet 2	\$225000	\$48000
Outlet 3	\$765000	\$162000
Outlet 4	\$360000	\$76500

6.

	Field 1	Field 2	Field 3
Nitrogen	1.00	0.45	0.65
Phosphate	0.90	0.35	0.75
Potash	0.95	0.35	0.85

(in tons)

7. (a) One example: $\begin{bmatrix} 1 & 1 \\ 0 & -1 \end{bmatrix}$

(b) One example: $\begin{bmatrix} 1 & 1 & 0 \\ 0 & -1 & 0 \\ 0 & 0 & 1 \end{bmatrix}$

(c) Consider $\begin{bmatrix} 0 & 0 & 1 \\ 1 & 0 & 0 \\ 0 & 1 & 0 \end{bmatrix}$.

8. (a) Third-row, fourth-column entry of **AB**
(c) Third-row, second-column entry of **BA**

9. (a) $\sum_{k=1}^{n} a_{3k} b_{k2}$

22. (a) Consider any matrix of the form $\begin{bmatrix} 1 & 0 \\ x & 0 \end{bmatrix}$.

23. (b) Consider $\mathbf{A} = \begin{bmatrix} 1 & 2 & -1 \\ 2 & 4 & -2 \end{bmatrix}$ and $\mathbf{B} = \begin{bmatrix} 1 & -2 \\ 0 & 1 \\ 1 & 0 \end{bmatrix}$.

24. See Exercise 25(c).

Section 2.1 (pp. 86–88)

1. Matrices in (a), (b), (c), (d), and (f) are not in reduced row echelon form.

Matrix in (a) fails condition 2 of the definition.
Matrix in (b) fails condition 4 of the definition.
Matrix in (c) fails condition 1 of the definition.
Matrix in (d) fails conditions 1, 2, and 3 of the definition.
Matrix in (f) fails condition 3 of the definition.

2. (a) $\left[\begin{array}{ccc|c} 1 & 4 & 0 & -13 \\ 0 & 0 & 1 & -3 \\ 0 & 0 & 0 & 0 \end{array}\right]$ (b) $\mathbf{I}_4$

(c) $\left[\begin{array}{cccc|c} 1 & -2 & 0 & 11 & -23 \\ 0 & 0 & 1 & -2 & 5 \\ 0 & 0 & 0 & 0 & 0 \\ 0 & 0 & 0 & 0 & 0 \end{array}\right]$

(e) $\left[\begin{array}{ccccc|c} 1 & -2 & 0 & 2 & -1 & 1 \\ 0 & 0 & 1 & -1 & 3 & 2 \end{array}\right]$

3. (a) Solution set = $\{(-2, 3, 5)\}$ (c) Solution set = $\varnothing$
(e) Solution set = $\{(2b - d - 4, b, 2d + 5, d, 2) \mid b, d \in \mathbb{R}\}$
(g) Solution set = $\{(6, -1, 3)\}$

4. 51 nickels, 62 dimes, 31 quarters

5. $y = 2x^2 - x + 3$

7. $x^2 + y^2 - 6x - 8y = 0$, or $(x-3)^2 + (y-4)^2 = 25$

8. (a) A = 3, B = 4, C = −2

9. (a) $a = 2$, $b = 15$, $c = 12$, $d = 6$
(c) $a = 4$, $b = 2$, $c = 4$, $d = 1$, $e = 4$

Section 2.2 (pp. 95–99)

1. (a) A row operation of type (I) converts **A** to **B**: $\langle 2 \rangle \leftarrow -5 \langle 2 \rangle$.
(c) A row operation of type (II) converts **A** to **B**: $\langle 2 \rangle \leftarrow \langle 3 \rangle + \langle 2 \rangle$.

2. (b) The sequence of row operations converting **B** to **A**:
(II): $\langle 1 \rangle \leftarrow -5 \langle 3 \rangle + \langle 1 \rangle$
(III): $\langle 2 \rangle \leftrightarrow \langle 3 \rangle$
(II): $\langle 3 \rangle \leftarrow 3 \langle 1 \rangle + \langle 3 \rangle$
(II): $\langle 2 \rangle \leftarrow -2 \langle 1 \rangle + \langle 2 \rangle$
(I): $\langle 1 \rangle \leftarrow 4 \langle 1 \rangle$

3. (a) Common reduced row echelon form is $\mathbf{I}_3$.

(b) The sequence of row operations:
(II): $<3> \leftarrow 2<2>+<3>$
(I): $<3> \leftarrow -1<3>$
(II): $<1> \leftarrow -9<3>+<1>$
(II): $<2> \leftarrow 3<3>+<2>$
(II): $<3> \leftarrow -\frac{9}{5}<2>+<3>$
(II): $<1> \leftarrow -\frac{3}{5}<2>+<1>$
(I): $<2> \leftarrow -\frac{1}{5}<2>$
(II): $<3> \leftarrow -3<1>+<3>$
(II): $<2> \leftarrow -2<1>+<2>$
(I): $<1> \leftarrow -5<1>$

5. (a) 2 (c) 2 (e) 3

6. (a) Solution set $= \{(c-2d, -3d, c, d) | c, d \in \mathbb{R}\}$; one particular solution $= (-3, -6, 1, 2)$
(c) Solution set $= \{(-4b+2d-f, b, -3d+2f, d, -2f, f) | b, d, f \in \mathbb{R}\}$; one particular solution $= (-3, 1, 0, 2, -6, 3)$

7. (a) Solution set $= \{(2c, -4c, c) | c \in \mathbb{R}\}$
(c) Solution set $= \{(0, 0, 0, 0)\}$

8. (a) Solution set $= \{(0, 0, 0)\}$

9. In the following answers, the asterisk represents any nonzero real entry:

(a) Smallest rank = 1: $\left[\begin{array}{ccc|c} 1 & * & * & * \\ 0 & 0 & 0 & 0 \\ 0 & 0 & 0 & 0 \\ 0 & 0 & 0 & 0 \end{array}\right]$;

largest rank = 4: $\left[\begin{array}{ccc|c} 1 & 0 & 0 & 0 \\ 0 & 1 & 0 & 0 \\ 0 & 0 & 1 & 0 \\ 0 & 0 & 0 & 1 \end{array}\right]$

(c) Smallest rank = 2: $\left[\begin{array}{cccc|c} 1 & * & * & * & 0 \\ 0 & 0 & 0 & 0 & 1 \\ 0 & 0 & 0 & 0 & 0 \end{array}\right]$;

largest rank = 3: $\left[\begin{array}{cccc|c} 1 & 0 & * & * & 0 \\ 0 & 1 & * & * & 0 \\ 0 & 0 & 0 & 0 & 1 \end{array}\right]$

10. Solution for system $\mathbf{AX} = \mathbf{B}_1$: $(6, -51, 21)$
Solution for system $\mathbf{AX} = \mathbf{B}_2$: $(\frac{35}{3}, -98, \frac{79}{2})$

14. The zero vector is a solution to $\mathbf{AX} = \mathbf{O}$, but it is not a solution for $\mathbf{AX} = \mathbf{B}$.

17. (b) Any nonhomogeneous system with two equations and two unknowns that has a unique solution will serve as a counterexample. For instance, consider

$$\begin{cases} x + y = 1 \\ x - y = 1 \end{cases}.$$

This system has a unique solution: $(1, 0)$. Let (s_1, s_2) and (t_1, t_2) both equal $(1, 0)$. Then the sum of solutions is not a solution in this case. Also, let c be any real number other than 1. The scalar multiple of a solution by c is not necessarily a solution in this case.

19. Consider the systems

$$\begin{cases} x + y = 1 \\ x + y = 0 \end{cases} \quad \text{and} \quad \begin{cases} x - y = 1 \\ x - y = 2 \end{cases}.$$

The reduced row echelon matrices for these inconsistent systems are, respectively,

$$\left[\begin{array}{cc|c} 1 & 1 & 0 \\ 0 & 0 & 1 \end{array}\right] \quad \text{and} \quad \left[\begin{array}{cc|c} 1 & -1 & 0 \\ 0 & 0 & 1 \end{array}\right].$$

Thus, the original augmented matrices are not row equivalent, since their reduced row echelon forms are different.

Section 2.3 (pp. 104–106)

1. (a) $\mathbf{x} = -\frac{21}{11}\mathbf{a}_1 + \frac{6}{11}\mathbf{a}_2$
(c) Not possible
(e) The answer is not unique; one possible answer is $\mathbf{x} = -3\mathbf{a}_1 + 2\mathbf{a}_2 + 0\mathbf{a}_3$.
(g) $\mathbf{x} = 2\mathbf{a}_1 - \mathbf{a}_2 - \mathbf{a}_3$

2. (a) Yes; 5(row 1) − 3(row 2) − 1(row 3)
(c) Not in row space
(e) Yes, but the linear combination of the rows is not unique; one possible expression for the given vector is −3(row 1) + 1(row 2) + 0(row 3).

3. (a) $[13, -23, 60] = -2\mathbf{q}_1 + \mathbf{q}_2 + 3\mathbf{q}_3$
(b) $\mathbf{q}_1 = 3\mathbf{r}_1 - \mathbf{r}_2 - 2\mathbf{r}_3$
$\mathbf{q}_2 = 2\mathbf{r}_1 + 2\mathbf{r}_2 - 5\mathbf{r}_3$
$\mathbf{q}_3 = \mathbf{r}_1 - 6\mathbf{r}_2 + 4\mathbf{r}_3$
(c) $[13, -23, 60] = -\mathbf{r}_1 - 14\mathbf{r}_2 + 11\mathbf{r}_3$

Section 2.4 (pp. 115–118)

2. (a) Rank = 2; nonsingular
(c) Rank = 3; nonsingular
(e) Rank = 3; singular

3. (a) $\begin{bmatrix} \frac{1}{10} & \frac{1}{15} \\ \frac{3}{10} & -\frac{2}{15} \end{bmatrix}$ (c) $\begin{bmatrix} -\frac{2}{21} & -\frac{5}{84} \\ \frac{1}{7} & -\frac{1}{28} \end{bmatrix}$
(e) No inverse exists.

4. (a) $\begin{bmatrix} 1 & 3 & 2 \\ -1 & 0 & 2 \\ 2 & 2 & -1 \end{bmatrix}$ (c) $\begin{bmatrix} \frac{3}{2} & 0 & \frac{1}{2} \\ -3 & \frac{1}{2} & -\frac{1}{2} \\ -\frac{8}{3} & \frac{1}{3} & -\frac{2}{3} \end{bmatrix}$
(e) No inverse exists.

5. (c) $\begin{bmatrix} \frac{1}{a_{11}} & 0 & \cdots & 0 \\ 0 & \frac{1}{a_{22}} & \cdots & 0 \\ \vdots & \vdots & \ddots & \vdots \\ 0 & 0 & \cdots & \frac{1}{a_{nn}} \end{bmatrix}$

6. (a) The general inverse is $\begin{bmatrix} \cos\theta & \sin\theta \\ -\sin\theta & \cos\theta \end{bmatrix}$.

When $\theta = \frac{\pi}{6}$, the inverse is $\begin{bmatrix} \frac{\sqrt{3}}{2} & \frac{1}{2} \\ -\frac{1}{2} & \frac{\sqrt{3}}{2} \end{bmatrix}$.

When $\theta = \frac{\pi}{4}$, the inverse is $\begin{bmatrix} \frac{\sqrt{2}}{2} & \frac{\sqrt{2}}{2} \\ -\frac{\sqrt{2}}{2} & \frac{\sqrt{2}}{2} \end{bmatrix}$.

When $\theta = \frac{\pi}{2}$, the inverse is $\begin{bmatrix} 0 & 1 \\ -1 & 0 \end{bmatrix}$.

(b) The general inverse is $\begin{bmatrix} \cos\theta & \sin\theta & 0 \\ -\sin\theta & \cos\theta & 0 \\ 0 & 0 & 1 \end{bmatrix}$.

When $\theta = \frac{\pi}{6}$, the inverse is $\begin{bmatrix} \frac{\sqrt{3}}{2} & \frac{1}{2} & 0 \\ -\frac{1}{2} & \frac{\sqrt{3}}{2} & 0 \\ 0 & 0 & 1 \end{bmatrix}$.

When $\theta = \frac{\pi}{4}$, the inverse is $\begin{bmatrix} \frac{\sqrt{2}}{2} & \frac{\sqrt{2}}{2} & 0 \\ -\frac{\sqrt{2}}{2} & \frac{\sqrt{2}}{2} & 0 \\ 0 & 0 & 1 \end{bmatrix}$.

When $\theta = \frac{\pi}{2}$, the inverse is $\begin{bmatrix} 0 & 1 & 0 \\ -1 & 0 & 0 \\ 0 & 0 & 1 \end{bmatrix}$.

7. (a) Inverse $= \begin{bmatrix} \frac{2}{3} & \frac{1}{3} \\ \frac{7}{3} & \frac{5}{3} \end{bmatrix}$; solution set $= \{(3, -5)\}$

(c) Inverse $= \begin{bmatrix} 1 & -13 & -15 & 5 \\ -3 & 3 & 0 & -7 \\ -1 & 2 & 1 & -3 \\ 0 & -4 & -5 & 1 \end{bmatrix}$;

solution set $= \{(5, -8, 2, -1)\}$

8. (a) Consider $\begin{bmatrix} 0 & 1 \\ 1 & 0 \end{bmatrix}$.

(b) Consider $\begin{bmatrix} 0 & 1 & 0 \\ 1 & 0 & 0 \\ 0 & 0 & 1 \end{bmatrix}$.

(c) $\mathbf{A} = \mathbf{A}^{-1}$ if $\mathbf{A}$ is involutory.

10. (a) $\mathbf{B}$ must be the zero matrix.

(b) No, since $\mathbf{A}^{-1}$ exists

11. $\ldots, \mathbf{A}^{-11}, \mathbf{A}^{-6}, \mathbf{A}^{-1}, \mathbf{A}^{4}, \mathbf{A}^{9}, \mathbf{A}^{14}, \ldots$

12. $\mathbf{B}^{-1}\mathbf{A}$ is the inverse of $\mathbf{A}^{-1}\mathbf{B}$.

14. (a) All steps in the row reduction process will not alter the column of zeroes—and so the matrix cannot be reduced to $\mathbf{I}_n$.

Section 2.5 (pp. 123–125)

1. (a) (III): $< 2 > \longleftrightarrow < 3 >$; inverse operation is (III): $< 2 > \longleftrightarrow < 3 >$. The matrix is its own inverse.

(b) (I): $< 2 > \longleftarrow -2 < 2 >$; inverse operation is (I): $< 2 > \longleftarrow -\frac{1}{2} < 2 >$. The inverse matrix is $\begin{bmatrix} 1 & 0 & 0 \\ 0 & -\frac{1}{2} & 0 \\ 0 & 0 & 1 \end{bmatrix}$.

(e) (II): $< 3 > \longleftarrow -2 < 4 > + < 3 >$; inverse operation is (II): $< 3 > \longleftarrow 2 < 4 > + < 3 >$. The inverse matrix is $\begin{bmatrix} 1 & 0 & 0 & 0 \\ 0 & 1 & 0 & 0 \\ 0 & 0 & 1 & 2 \\ 0 & 0 & 0 & 1 \end{bmatrix}$.

2. (a) $\begin{bmatrix} 4 & 9 \\ 3 & 7 \end{bmatrix}$

$$= \begin{bmatrix} 4 & 0 \\ 0 & 1 \end{bmatrix} \begin{bmatrix} 1 & 0 \\ 3 & 1 \end{bmatrix} \begin{bmatrix} 1 & 0 \\ 0 & \frac{1}{4} \end{bmatrix} \begin{bmatrix} 1 & \frac{9}{4} \\ 0 & 1 \end{bmatrix} \begin{bmatrix} 1 & 0 \\ 0 & 1 \end{bmatrix}$$

(b) Not possible, since matrix is singular

Section 3.1 (pp. 134–137)

1. (a) -17 (c) 0 (e) -108 (g) -40 (i) 0 (j) -3

2. (a) 0 (two equal rows)

(c) -30 (upper triangular)

(e) 0 (row of zeroes)

(g) -120 (upper triangular)

3. (a) Odd (c) Even (d) Odd

4. (a) Negative sign (since 3 2 1 is an odd permutation); $\mathbf{A}$ is 3×3.

(c) Positive sign; $\mathbf{A}$ is 4×4.

(e) Negative sign; $\mathbf{A}$ is 5×5.

(g) Positive sign; $\mathbf{A}$ is 3×3.

9. (a) 7 (c) 12

11. (a) 18 (c) 63

13. (a) $x = -5$ or $x = 2$ (c) $x = 3$, $x = 1$, or $x = 2$

14. $-a_{16}a_{25}a_{34}a_{43}a_{52}a_{61}$

15. (b) 20

Section 3.2 (pp. 146–150)

1. (a) (II): $< 1 > \longleftarrow -3 < 2 > + < 1 >$; determinant $= 1$

(c) (I): $< 3 > \longleftarrow -4 < 3 >$; determinant $= -4$

(f) (III): $< 1 > \longleftrightarrow < 2 >$; determinant $= -1$

2. (a) 30 (c) -4 (e) 35

3. (a) Determinant $= -2$; determinant of inverse $= -\frac{1}{2}$

(c) Determinant $= -79$; determinant of inverse $= -\frac{1}{79}$

4. (a) Determinant $= -1$; the system has only the trivial solution.

7. Let $\mathbf{A} = \begin{bmatrix} 1 & 1 \\ 1 & 1 \end{bmatrix}$, and let $\mathbf{B} = \begin{bmatrix} 1 & 0 \\ 0 & 1 \end{bmatrix}$.

14. (b) Consider $\mathbf{A} = \begin{bmatrix} 0 & -1 \\ 1 & 0 \end{bmatrix}$.

15. (b) Consider $\begin{bmatrix} 0 & 1 & 0 \\ 1 & 0 & 0 \\ 0 & 0 & 1 \end{bmatrix}$.

18. (b) For example, consider

$$\mathbf{B} = \begin{bmatrix} 1 & -1 \\ -1 & 2 \end{bmatrix} \mathbf{A} \begin{bmatrix} 1 & -1 \\ -1 & 2 \end{bmatrix}^{-1} = \begin{bmatrix} -6 & -4 \\ 16 & 11 \end{bmatrix},$$

or

$$\mathbf{B} = \begin{bmatrix} 3 & 5 \\ 1 & 2 \end{bmatrix} \mathbf{A} \begin{bmatrix} 3 & 5 \\ 1 & 2 \end{bmatrix}^{-1} = \begin{bmatrix} 10 & -12 \\ 4 & -5 \end{bmatrix}.$$

Section 3.3 (pp. 158–161)

1. (a) $\begin{vmatrix} 4 & 3 \\ -2 & 4 \end{vmatrix} = 22$ (c) $\begin{vmatrix} -3 & 0 & 5 \\ 2 & -1 & 4 \\ 6 & 4 & 0 \end{vmatrix} = 118$

2. (a) $(-1)^{2+2} \begin{vmatrix} 4 & -3 \\ 9 & -7 \end{vmatrix} = -1$

(c) $(-1)^{4+3} \begin{vmatrix} -5 & 2 & 13 \\ -8 & 2 & 22 \\ -6 & -3 & -16 \end{vmatrix} = 222$

(d) $(-1)^{1+2} \begin{vmatrix} x-4 & x-3 \\ x-1 & x+2 \end{vmatrix} = -2x + 11$

3. (a) Adjoint $= \begin{bmatrix} -6 & 9 & 3 \\ 6 & -42 & 0 \\ -4 & 8 & 2 \end{bmatrix}$;

determinant $= -6$;

inverse $= \begin{bmatrix} 1 & -\frac{3}{2} & -\frac{1}{2} \\ -1 & 7 & 0 \\ \frac{2}{3} & -\frac{4}{3} & -\frac{1}{3} \end{bmatrix}$

(c) Adjoint $= \begin{bmatrix} -3 & 0 & 3 & -3 \\ 0 & 0 & 0 & 0 \\ -3 & 0 & 3 & -3 \\ 6 & 0 & -6 & 6 \end{bmatrix}$;

determinant $= 0$; no inverse

(e) Adjoint $= \begin{bmatrix} 3 & -1 & -2 \\ 0 & -3 & -6 \\ 0 & 0 & -9 \end{bmatrix}$; determinant $= 9$;

inverse $= \begin{bmatrix} \frac{1}{3} & -\frac{1}{9} & -\frac{2}{9} \\ 0 & -\frac{1}{3} & -\frac{2}{3} \\ 0 & 0 & -1 \end{bmatrix}$

4. (a) $a_{31}(-1)^{3+1}|\mathbf{A}_{31}| + a_{32}(-1)^{3+2}|\mathbf{A}_{32}| + a_{33}(-1)^{3+3}|\mathbf{A}_{33}| + a_{34}(-1)^{3+4}|\mathbf{A}_{34}|$

(c) $a_{14}(-1)^{1+4}|\mathbf{A}_{14}| + a_{24}(-1)^{2+4}|\mathbf{A}_{24}| + a_{34}(-1)^{3+4}|\mathbf{A}_{34}| + a_{44}(-1)^{4+4}|\mathbf{A}_{44}|$

5. (a) -76 (c) 102

6. (a) $\{(-4, 3, -7)\}$ (d) $\{(4, -1, -3, 6)\}$

7. $\frac{\mathcal{BA}}{|\mathbf{AB}|}$

11. (b) Consider $\begin{bmatrix} 0 & 1 & 1 \\ -1 & 0 & 1 \\ -1 & -1 & 0 \end{bmatrix}$.

Section 4.2 (pp. 182–185)

1. (a) Not a subspace; no zero vector
(c) Subspace
(e) Not a subspace; no zero vector
(g) Not a subspace; not closed under addition
(j) Not a subspace; not closed under addition
(l) Not a subspace; not closed under addition

2. (a) Subspace (c) Subspace
(e) Subspace (g) Subspace
(i) Not a subspace; not closed under addition

3. (a) Subspace (c) Subspace
(e) Not a subspace; not closed under addition
(g) Subspace

5. (b) $[a + b, a + c, b + c, c] = a[1, 1, 0, 0] + b[1, 0, 1, 0] + c[0, 1, 1, 1]$, which equals the row space of $\mathbf{A} = \begin{bmatrix} 1 & 1 & 0 & 0 \\ 1 & 0 & 1 & 0 \\ 0 & 1 & 1 & 1 \end{bmatrix}$

(c) $\mathbf{B} = \begin{bmatrix} 1 & 0 & 0 & -\frac{1}{2} \\ 0 & 1 & 0 & \frac{1}{2} \\ 0 & 0 & 1 & \frac{1}{2} \end{bmatrix}$

(d) Row space of $\mathbf{B}$
$= a\left[1, 0, 0, -\frac{1}{2}\right] + b\left[0, 1, 0, \frac{1}{2}\right] + c\left[0, 0, 1, \frac{1}{2}\right]$
$= \{(a, b, c, -\frac{1}{2}a + \frac{1}{2}b + \frac{1}{2}c) \mid a, b, c \in \mathbb{R}\}$

7. (a) $\{(2a + b + 3c, 3a + 2b + c, a + 3b - 16c) \mid a, b, c \in \mathbb{R}\}$

(b) $\mathbf{B} = \begin{bmatrix} 1 & 0 & -7 \\ 0 & 1 & 5 \\ 0 & 0 & 0 \end{bmatrix}$

(c) $\{(a, b, -7a + 5b)\}$ (d) Plane

14. (e) No; if $|\mathbf{A}| \neq 0$ and $c = 0$, then $|c\mathbf{A}| = 0$.

Section 4.3 (pp. 193–195)

1. (a) $\{[a, b, -a + b] \mid a, b \in \mathbb{R}\}$
(c) $\{[a, b, -b] \mid a, b \in \mathbb{R}\}$
(e) $\{[a, b, c, -2a + b + c] \mid a, b, c \in \mathbb{R}\}$

2. (a) $\{ax^3 + bx^2 + cx - (a + b + c) \mid a, b, c \in \mathbb{R}\}$
(c) $\{ax^3 - ax + b \mid a, b \in \mathbb{R}\}$

3. (a) $\left\{ \begin{bmatrix} a & b \\ c & -a - b - c \end{bmatrix} \middle| \, a, b, c \in \mathbb{R} \right\}$

(c) $\left\{ \begin{bmatrix} a & b \\ c & d \end{bmatrix} \middle| \, a, b, c, d \in \mathbb{R} \right\} = \mathcal{M}_{22}$

9. One answer is $-1(x^3 - 2x^2 + x - 3) + 2(2x^3 - 3x^2 + 2x + 5) - 1(4x^2 + x - 3) + 0(4x^3 - 7x^2 + 4x - 1)$.

12. (a) Hint: Use Theorem 1.12.

21. (b) $S_1 = \{[1,0,0],[0,1,0]\}$, $S_2 = \{[0,1,0],[0,0,1]\}$
(c) $S_1 = \{[1,0,0],[0,1,0]\}$, $S_2 = \{[1,0,0],[1,1,0]\}$

22. (c) $S_1 = \{x^5\}$, $S_2 = \{x^4\}$

Section 4.4 (pp. 203–207)

1. (a) Linearly independent (c) Linearly dependent (e) Linearly dependent (g) Linearly dependent

2. (a) Linearly independent (c) Linearly dependent

3. (a) $[0,0,0,0]$

4. (a) Linearly independent (c) Linearly dependent (d) Linearly independent (f) Linearly independent

7. (b) $[0,1,0]$ (c) No; $[0,0,1]$ will also work.
(d) Any linear combination of $[1,1,0]$ and $[-2,0,1]$ will work.

11. (a) One answer is $\{[1,0,0,0],[0,1,0,0],[0,0,1,0],[0,0,0,1]\}$.
(c) One answer is $\{1, x, x^2, x^3\}$.
(e) One answer is

$$\left\{\begin{bmatrix}1&0&0\\0&0&0\\0&0&0\end{bmatrix},\begin{bmatrix}0&1&0\\1&0&0\\0&0&0\end{bmatrix},\begin{bmatrix}0&0&1\\0&0&0\\1&0&0\end{bmatrix},\begin{bmatrix}0&0&0\\0&1&0\\0&0&0\end{bmatrix}\right\}.$$

(Notice that each matrix is symmetric.)

18. (b) Let $\mathbf{A}$ be the zero matrix.

Section 4.5 (pp. 216–219)

4. (a) Not a basis (linearly independent but does not span)
(c) Basis
(e) Not a basis (linearly dependent but spans)

5. (a) $\{[1,0,0,2,-2],[0,1,0,0,1],[0,0,1,-1,0]\}$
(d) $\{[1,0,0,-2,-\frac{13}{4}],[0,1,0,3,\frac{9}{2}],[0,0,1,0,-\frac{1}{4}]\}$

6. $\{x^3-3x, x^2-x, 1\}$

7. $$\left\{\begin{bmatrix}1&0\\\frac{4}{3}&\frac{1}{3}\\2&0\end{bmatrix},\begin{bmatrix}0&1\\-\frac{1}{3}&-\frac{1}{3}\\0&0\end{bmatrix},\begin{bmatrix}0&0\\0&0\\0&1\end{bmatrix}\right\}$$

8. (b) 2 (c) No; $\dim(\text{span}(S)) = 2 \neq 4 = \dim(\mathbb{R}^4)$

14. (b) 5
(c) $\{(x-2)(x-3), x(x-2)(x-3), x^2(x-2)(x-3), x^3(x-2)(x-3)\}$
(d) 4

Section 4.6 (pp. 227–229)

1. (a) One answer is $\{[1,3,-2],[2,1,4],[0,1,-1]\}$.
(c) One answer is $\{[3,-2,2],[1,2,-1],[3,-2,7]\}$.
(f) One answer is $\{[1,-3,0],[0,1,1]\}$.

2. (a) One answer is $\{x^3-8x^2+1, 3x^3-2x^2+x, 4x^3+2x-10, x^3-20x^2-x+12\}$.
(c) One answer is $\{x^3, x^2, x\}$.
(e) One answer is $\{x^3+x^2, x, 1\}$.

3. (a) One answer is

$$\left\{\begin{bmatrix}1&0&0\\0&0&0\\0&0&0\end{bmatrix},\begin{bmatrix}0&1&0\\0&0&0\\0&0&0\end{bmatrix},\begin{bmatrix}0&0&1\\0&0&0\\0&0&0\end{bmatrix},\begin{bmatrix}0&0&0\\1&0&0\\0&0&0\end{bmatrix},\begin{bmatrix}0&0&0\\0&1&0\\0&0&0\end{bmatrix},\begin{bmatrix}0&0&0\\0&0&1\\0&0&0\end{bmatrix},\begin{bmatrix}0&0&0\\0&0&0\\1&0&0\end{bmatrix},\begin{bmatrix}0&0&0\\0&0&0\\0&1&0\end{bmatrix},\begin{bmatrix}0&0&0\\0&0&0\\0&0&1\end{bmatrix}\right\}.$$

(c) One possibility is

$$\left\{\begin{bmatrix}1&0&0\\0&0&0\\0&0&0\end{bmatrix},\begin{bmatrix}0&1&0\\1&0&0\\0&0&0\end{bmatrix},\begin{bmatrix}0&0&1\\0&0&0\\1&0&0\end{bmatrix},\begin{bmatrix}0&0&0\\0&1&0\\0&0&0\end{bmatrix},\begin{bmatrix}0&0&0\\0&0&1\\0&1&0\end{bmatrix},\begin{bmatrix}0&0&0\\0&0&0\\0&0&1\end{bmatrix}\right\}.$$

4. (a) One answer is $\{[1,-3,0,1,4],[2,2,1,-3,1],[1,0,0,0,0],[0,1,0,0,0],[0,0,1,0,0]\}$.
(c) One possibility is $\{[1,0,-1,0,0],[0,1,-1,1,0],[2,3,-8,-1,0],[1,0,0,0,0],[0,0,0,0,1]\}$.

5. (a) One answer is $\{x^3-x^2, x^4-3x^3+5x^2-x, x^4, x^3, 1\}$.
(c) One answer is $\{x^4-x^3+x^2-x+1, x^3-x^2+x-1, x^2-x+1, x^2, x\}$.

6. (a) One possibility is

$$\left\{\begin{bmatrix}1&-1\\-1&1\\0&0\end{bmatrix},\begin{bmatrix}0&0\\1&-1\\-1&1\end{bmatrix},\begin{bmatrix}1&0\\0&0\\0&0\end{bmatrix},\begin{bmatrix}0&1\\0&0\\0&0\end{bmatrix},\begin{bmatrix}0&0\\1&0\\0&0\end{bmatrix},\begin{bmatrix}0&0\\0&0\\1&0\end{bmatrix}\right\}.$$

(c) One answer is

$$\left\{\begin{bmatrix}3&0\\-1&7\\0&1\end{bmatrix},\begin{bmatrix}-1&0\\1&3\\0&-2\end{bmatrix},\begin{bmatrix}2&0\\3&1\\0&-1\end{bmatrix},\begin{bmatrix}6&0\\0&1\\0&-1\end{bmatrix},\begin{bmatrix}0&1\\0&0\\0&0\end{bmatrix},\begin{bmatrix}0&0\\0&0\\1&0\end{bmatrix}\right\}.$$

7. $$\left\{\begin{bmatrix}1&0&0&0\\0&0&0&0\\0&0&0&0\\0&0&0&0\end{bmatrix},\begin{bmatrix}0&1&0&0\\0&0&0&0\\0&0&0&0\\0&0&0&0\end{bmatrix},\begin{bmatrix}0&0&1&0\\0&0&0&0\\0&0&0&0\\0&0&0&0\end{bmatrix},\begin{bmatrix}0&0&0&1\\0&0&0&0\\0&0&0&0\\0&0&0&0\end{bmatrix},\begin{bmatrix}0&0&0&0\\0&1&0&0\\0&0&0&0\\0&0&0&0\end{bmatrix},\begin{bmatrix}0&0&0&0\\0&0&1&0\\0&0&0&0\\0&0&0&0\end{bmatrix},\begin{bmatrix}0&0&0&0\\0&0&0&1\\0&0&0&0\\0&0&0&0\end{bmatrix},\begin{bmatrix}0&0&0&0\\0&0&0&0\\0&0&1&0\\0&0&0&0\end{bmatrix},\begin{bmatrix}0&0&0&0\\0&0&0&0\\0&0&0&1\\0&0&0&0\end{bmatrix},\right.$$

$$\left.\begin{bmatrix} 0 & 0 & 0 & 0 \\ 0 & 0 & 0 & 0 \\ 0 & 0 & 0 & 0 \\ 0 & 0 & 0 & 1 \end{bmatrix}\right\}.$$

8. (b) 8 (d) 3

9. (b) $(n^2 - n)/2$

10. (a) Let $\mathcal{V} = \mathbb{R}^3$, and let $S = \{[1,0,0], [2,0,0], [3,0,0]\}$.

(b) Let $\mathcal{V} = \mathbb{R}^3$, and let $T = \{[1,0,0], [2,0,0], [3,0,0]\}$.

14. (b) No; consider the subspace $\mathcal{W}$ of $\mathbb{R}^3$ given by $\mathcal{W} = \{[a,0,0] \mid a \in \mathbb{R}\}$. No subset of $B = \{[1,1,0], [1,-1,0], [0,0,1]\}$ (a basis for $\mathbb{R}^3$) is a basis for $\mathcal{W}$.

(c) Yes; consider $\mathcal{Y} = \text{span}(B')$.

15. (b) In $\mathbb{R}^3$, consider $\mathcal{W} = \{[a,b,0] \mid a, b \in \mathbb{R}\}$. We could let $\mathcal{W}' = \{[0,0,c] \mid c \in \mathbb{R}\}$ or $\mathcal{W}' = \{[0,c,c] \mid c \in \mathbb{R}\}$.

Section 4.7 (pp. 240–243)

1. (a) $[\mathbf{v}]_B = [7,-1,-5]$ (c) $[\mathbf{v}]_B = [-2,4,-5]$ (e) $[\mathbf{v}]_B = [4,-5,3]$ (g) $[\mathbf{v}]_B = [-1,4,-2]$ (h) $[\mathbf{v}]_B = [2,-3,1]$ (j) $[\mathbf{v}]_B = [5,-2]$

2. (a) $\begin{bmatrix} -102 & 20 & 3 \\ 67 & -13 & -2 \\ 36 & -7 & -1 \end{bmatrix}$ (c) $\begin{bmatrix} 20 & -30 & -69 \\ 24 & -24 & -80 \\ -9 & 11 & 31 \end{bmatrix}$

(d) $\begin{bmatrix} -1 & -4 & 2 & -9 \\ 4 & 5 & 1 & 3 \\ 0 & 2 & -3 & 1 \\ -4 & -13 & 13 & -15 \end{bmatrix}$ (f) $\begin{bmatrix} 6 & 1 & 2 \\ 1 & 1 & 2 \\ -1 & -1 & -3 \end{bmatrix}$

4. (a) $\mathbf{P} = \begin{bmatrix} 13 & 31 \\ -18 & -43 \end{bmatrix}$, $\mathbf{Q} = \begin{bmatrix} -11 & -8 \\ 29 & 21 \end{bmatrix}$,

$\mathbf{T} = \begin{bmatrix} 1 & 3 \\ -1 & -4 \end{bmatrix}$

(c) $\mathbf{P} = \begin{bmatrix} 2 & 8 & 13 \\ -6 & -25 & -43 \\ 11 & 45 & 76 \end{bmatrix}$,

$\mathbf{Q} = \begin{bmatrix} -24 & -2 & 1 \\ 30 & 3 & -1 \\ 139 & 13 & -5 \end{bmatrix}$,

$\mathbf{T} = \begin{bmatrix} -25 & -97 & -150 \\ 31 & 120 & 185 \\ 145 & 562 & 868 \end{bmatrix}$

5. (a) $C = ([1,-4,0,-2,0], [0,0,1,4,0], [0,0,0,0,1])$,

$\mathbf{P} = \begin{bmatrix} 1 & 6 & 3 \\ 1 & 5 & 3 \\ 1 & 3 & 2 \end{bmatrix}$,

$\mathbf{Q} = \mathbf{P}^{-1} = \begin{bmatrix} 1 & -3 & 3 \\ 1 & -1 & 0 \\ -2 & 3 & -1 \end{bmatrix}$,

$[\mathbf{v}]_B = [17,4,-13]$,
$[\mathbf{v}]_C = [2,-2,3]$

(c) $C = ([1,0,0,0], [0,1,0,0], [0,0,1,0], [0,0,0,1])$,

$\mathbf{P} = \begin{bmatrix} 3 & 6 & -4 & -2 \\ -1 & 7 & -3 & 0 \\ 4 & -3 & 3 & 1 \\ 6 & -2 & 4 & 2 \end{bmatrix}$,

$\mathbf{Q} = \mathbf{P}^{-1} = \begin{bmatrix} 1 & -4 & -12 & 7 \\ -2 & 9 & 27 & -\frac{31}{2} \\ -5 & 22 & 67 & -\frac{77}{2} \\ 5 & -23 & -71 & 41 \end{bmatrix}$,

$[\mathbf{v}]_B = [2,1,-3,7]$,
$[\mathbf{v}]_C = [10,14,3,12]$

6. $C = ([-142,64,167], [-53,24,63], [-246,111,290])$

7. (a) Transition matrix to $C_1 = \begin{bmatrix} 0 & 1 & 0 \\ 0 & 0 & 1 \\ 1 & 0 & 0 \end{bmatrix}$

Transition matrix to $C_2 = \begin{bmatrix} 0 & 0 & 1 \\ 1 & 0 & 0 \\ 0 & 1 & 0 \end{bmatrix}$

Transition matrix to $C_3 = \begin{bmatrix} 1 & 0 & 0 \\ 0 & 0 & 1 \\ 0 & 1 & 0 \end{bmatrix}$

Transition matrix to $C_4 = \begin{bmatrix} 0 & 1 & 0 \\ 1 & 0 & 0 \\ 0 & 0 & 1 \end{bmatrix}$

Transition matrix to $C_5 = \begin{bmatrix} 0 & 0 & 1 \\ 0 & 1 & 0 \\ 1 & 0 & 0 \end{bmatrix}$

Section 5.1 (pp. 254–258)

1. Only the starred parts of the exercise are listed:
Linear transformations: (a), (d), (h)
Linear operators: (a), (d)

9. (c) $\begin{bmatrix} \cos\phi & 0 & -\sin\phi \\ 0 & 1 & 0 \\ \sin\phi & 0 & \cos\phi \end{bmatrix}$

25. $L(\mathbf{i}) = \frac{7}{5}\mathbf{i} - \frac{11}{5}\mathbf{j}$, $L(\mathbf{j}) = -\frac{2}{5}\mathbf{i} - \frac{4}{5}\mathbf{j}$

29. (b) Consider the zero linear transformation.

Section 5.2 (pp. 268–272)

2. (a) $\begin{bmatrix} -6 & 4 & -1 \\ -2 & 3 & -5 \\ 3 & -1 & 7 \end{bmatrix}$ (c) $\begin{bmatrix} 4 & -1 & 3 & 3 \\ 1 & 3 & -1 & 5 \\ -2 & -7 & 5 & -1 \end{bmatrix}$

3. (a) $\begin{bmatrix} -47 & 128 & -288 \\ -18 & 51 & -104 \end{bmatrix}$ (c) $\begin{bmatrix} 22 & 14 \\ 62 & 39 \\ 68 & 43 \end{bmatrix}$

(e) $\begin{bmatrix} 5 & 6 & 0 \\ -11 & -26 & -6 \\ -14 & -19 & -1 \\ 6 & 3 & -2 \\ -1 & 1 & 1 \\ 11 & 13 & 0 \end{bmatrix}$

4. (a) $\begin{bmatrix} -202 & -32 & -43 \\ -146 & -23 & -31 \\ 83 & 14 & 18 \end{bmatrix}$

(b) $\begin{bmatrix} 21 & 7 & 21 & 16 \\ -51 & -13 & -51 & -38 \end{bmatrix}$

6. (a) $\begin{bmatrix} 67 & -123 \\ 37 & -68 \end{bmatrix}$ (b) $\begin{bmatrix} -7 & 2 & 10 \\ 5 & -2 & -9 \\ -6 & 1 & 8 \end{bmatrix}$

7. (a) $\begin{bmatrix} 3 & 0 & 0 & 0 \\ 0 & 2 & 0 & 0 \\ 0 & 0 & 1 & 0 \end{bmatrix}$, $12x^2 - 10x + 6$

8. (a) $\begin{bmatrix} \frac{\sqrt{3}}{2} & -\frac{1}{2} \\ \frac{1}{2} & \frac{\sqrt{3}}{2} \end{bmatrix}$

9. (b) $\begin{bmatrix} \frac{1}{2} & 0 & 0 & -\frac{1}{2} & 0 & 0 \\ \frac{1}{2} & 0 & 0 & \frac{1}{2} & 0 & 0 \\ 0 & \frac{1}{2} & \frac{1}{2} & 0 & -\frac{1}{2} & 0 \\ 0 & \frac{1}{2} & \frac{1}{2} & 0 & \frac{1}{2} & 0 \\ 0 & -\frac{1}{2} & 0 & 0 & -\frac{1}{2} & \frac{1}{2} \\ 0 & -\frac{1}{2} & 0 & 0 & \frac{1}{2} & -\frac{1}{2} \end{bmatrix}$

10. $\begin{bmatrix} -12 & 12 & -2 \\ -4 & 6 & -2 \\ -10 & -3 & 7 \end{bmatrix}$

13. (a) $\mathbf{I}_n$ (c) $c\mathbf{I}_n$ (e) The $n \times n$ matrix whose columns are $\mathbf{e}_n, \mathbf{e}_1, \mathbf{e}_2, \ldots, \mathbf{e}_{n-1}$, respectively

Section 5.3 (pp. 279–281)

1. (a) Yes, because $L([1, -2, 3]) = [0, 0, 0]$
(c) No, because the system

$$\begin{cases} 5x_1 + x_2 - x_3 = 2 \\ -3x_1 \qquad + x_3 = -1 \\ x_1 - x_2 - x_3 = 4 \end{cases}$$

has no solutions.

2. (a) No, since $L(x^3 - 5x^2 + 3x - 6) \neq 0$
(c) Yes, because, for example, $L(x^3 + 4x + 3) = 8x^3 - x - 1$

3. (a) $\dim(\ker(L)) = 2$, basis for $\ker(L) = \{[1, 0, 0], [0, 0, 1]\}$, $\dim(\text{range}(L)) = 1$, basis for $\text{range}(L) = \{[0, 1]\}$
(d) $\dim(\ker(L)) = 2$, basis for $\ker(L) = \{x^4, x^3\}$, $\dim(\text{range}(L)) = 3$, basis for $\text{range}(L) = \{x^2, x, 1\}$
(f) $\dim(\ker(L)) = 1$, basis for $\ker(L) = \{[0, 1, 1]\}$, $\dim(\text{range}(L)) = 2$, basis for $\text{range}(L) = \{[1, 0, 1], [0, 0, -1]\}$ (A simpler basis for $\text{range}(L) = \{[1, 0, 0], [0, 0, 1]\}$.)
(g) $\dim(\ker(L)) = 0$, basis for $\ker(L) = \{\ \}$ (empty set), $\dim(\text{range}(L)) = 4$, basis for $\text{range}(L) =$ standard basis for $\mathcal{M}_{22}$
(i) $\dim(\ker(L)) = 1$, basis for $\ker(L) = \{x^2 - 2x + 1\}$; $\dim(\text{range}(L)) = 2$; basis for $\text{range}(L) = \{[1, 2], [1, 1]\}$ (A simpler basis for $\text{range}(L) =$ standard basis for $\mathbb{R}^2$.)

4. (a) $\dim(\ker(L)) = 1$, basis for $\ker(L) = \{[-2, 3, 1]\}$, $\dim(\text{range}(L)) = 2$, basis for $\text{range}(L) = \{[1, -2, 3], [-1, 3, -3]\}$
(d) $\dim(\ker(L)) = 2$, basis for $\ker(L) = \{[1, -3, 1, 0], [-1, 2, 0, 1]\}$, $\dim(\text{range}(L)) = 2$, basis for $\text{range}(L) = \{[-14, -4, -6, 3, 4], [-8, -1, 2, -7, 2]\}$

6. $\ker(L) =$

$$\left\{ \begin{bmatrix} a & b & c \\ d & e & f \\ g & h & -a-e \end{bmatrix} \middle| \ a, b, c, d, e, f, g, h \in \mathbb{R} \right\},$$

$\dim(\ker(L)) = 8$, $\text{range}(L) = \mathbb{R}$, $\dim(\text{range}(L)) = 1$

8. $\ker(L) = \{\mathbf{0}\}$; $\text{range}(L) = \{ax^4 + bx^3 + cx^2\}$; $\dim(\ker(L)) = 0$; $\dim(\text{range}(L)) = 3$

10. When $k \le n$, $\ker(L) =$ all polynomials of degree less than k, $\dim(\ker(L)) = k$, $\text{range}(L) = \mathcal{P}_{n-k}$, and $\dim(\text{range}(L)) = n - k + 1$. When $k > n$, $\ker(L) = \mathcal{P}_n$, $\dim(\ker(L)) = n + 1$, $\text{range}(L) = \{\mathbf{0}\}$, and $\dim(\text{range}(L)) = 0$.

12. $\ker(L) = \{[0, 0, \ldots, 0]\}$, $\text{range}(L) = \mathbb{R}^n$ (Note: Every vector $\mathbf{X}$ is in the range, since $L(\mathbf{A}^{-1}\mathbf{X}) = \mathbf{A}(\mathbf{A}^{-1}\mathbf{X}) = \mathbf{X}$.)

17. Consider $L\left(\begin{bmatrix} x \\ y \end{bmatrix}\right) = \begin{bmatrix} 1 & -1 \\ 1 & -1 \end{bmatrix}\begin{bmatrix} x \\ y \end{bmatrix}$. Then, $\ker(L) = \text{range}(L) = \{[a, a] \mid a \in \mathbb{R}\}$.

Section 5.4 (pp. 294–298)

1. (a) Not one-to-one, because $L([1, 0, 0]) = L([0, 0, 0]) = [0, 0, 0, 0]$; not onto, because $[0, 0, 0, 1]$ is not in $\text{range}(L)$
(c) One-to-one, because $L([x, y, z]) = [0, 0, 0]$ implies that $[2x, x + y + z, -y] = [0, 0, 0]$, which gives $x = y = z = 0$; onto, because every vector $[a, b, c]$ can be expressed as $[2x, x + y + z, -y]$, where $x = \frac{a}{2}$, $y = -c$, and $z = b - \frac{a}{2} + c$
(e) One-to-one, because $L(ax^2 + bx + c) = 0$ implies that $a + b = b + c = a + c = 0$, which gives $b = c$ and hence $a = b = c = 0$; onto, because every polynomial $Ax^2 + Bx + C$ can be expressed as $(a+b)x^2 + (b+c)x + (a+c)$, where

$a = (A - B + C)/2$, $b = (A + B - C)/2$, and $c = (-A + B + C)/2$

(g) Not one-to-one, because $L\left(\begin{bmatrix} 0 & 1 & 0 \\ 1 & 0 & -1 \end{bmatrix}\right) = L\left(\begin{bmatrix} 0 & 0 & 0 \\ 0 & 0 & 0 \end{bmatrix}\right) = \begin{bmatrix} 0 & 0 \\ 0 & 0 \end{bmatrix}$; onto because every 2×2 matrix $\begin{bmatrix} A & B \\ C & D \end{bmatrix}$ can be expressed as $\begin{bmatrix} a & -c \\ 2e & d+f \end{bmatrix}$, where $a = A$, $c = -B$, $e = C/2$, $d = D$, and $f = 0$

(h) One-to-one, because $L(ax^2+bx+c) = \begin{bmatrix} 0 & 0 \\ 0 & 0 \end{bmatrix}$ implies that $a + c = b - c = -3a = 0$, which gives $a = b = c = 0$; not onto, because $\begin{bmatrix} 0 & 1 \\ 0 & 0 \end{bmatrix}$ is not in range(L)

2. (a) One-to-one; onto; isomorphism; the matrix row reduces to $\mathbf{I}_2$, which means that $\dim(\ker(L)) = 0$ and $\dim(\text{range}(L)) = 2$.

(b) One-to-one; not onto; not an isomorphism; the matrix row reduces to

$$\begin{bmatrix} 1 & 0 \\ 0 & 1 \\ 0 & 0 \end{bmatrix},$$

which means that $\dim(\ker(L)) = 0$ and $\dim(\text{range}(L)) = 2$.

(c) Not one-to-one; not onto; not an isomorphism; the matrix row reduces to

$$\begin{bmatrix} 1 & 0 & -\frac{2}{5} \\ 0 & 1 & -\frac{6}{5} \\ 0 & 0 & 0 \end{bmatrix},$$

which means that $\dim(\ker(L)) = 1$ and $\dim(\text{range}(L)) = 2$.

3. (a) One-to-one; onto; isomorphism; the matrix row reduces to

$$\begin{bmatrix} 1 & 0 & 0 \\ 0 & 1 & 0 \\ 0 & 0 & 1 \end{bmatrix},$$

which means that $\dim(\ker(L)) = 0$ and $\dim(\text{range}(L)) = 3$.

(c) Not one-to-one; not onto; not an isomorphism; the matrix row reduces to

$$\begin{bmatrix} 1 & 0 & -\frac{10}{11} & \frac{19}{11} \\ 0 & 1 & \frac{3}{11} & -\frac{9}{11} \\ 0 & 0 & 0 & 0 \\ 0 & 0 & 0 & 0 \end{bmatrix},$$

which means that $\dim(\ker(L)) = 2$ and $\dim(\text{range}(L)) = 2$.

4. (a) No, because $\dim(\mathbb{R}^6) = \dim(\mathcal{P}_5)$

(b) No, because $\dim(\mathcal{M}_{22}) = \dim(\mathcal{P}_3)$

7. (a) $$L_1^{-1}\left(\begin{bmatrix} x_1 \\ x_2 \\ x_3 \end{bmatrix}\right) = \begin{bmatrix} 0 & 0 & 1 \\ 0 & -1 & 0 \\ 1 & -2 & 0 \end{bmatrix}\begin{bmatrix} x_1 \\ x_2 \\ x_3 \end{bmatrix},$$

$$L_2^{-1}\left(\begin{bmatrix} x_1 \\ x_2 \\ x_3 \end{bmatrix}\right) = \begin{bmatrix} 1 & 0 & 0 \\ 0 & 0 & -\frac{1}{3} \\ 2 & 1 & 0 \end{bmatrix}\begin{bmatrix} x_1 \\ x_2 \\ x_3 \end{bmatrix},$$

$$(L_2 \circ L_1)\left(\begin{bmatrix} x_1 \\ x_2 \\ x_3 \end{bmatrix}\right) = \begin{bmatrix} 0 & -2 & 1 \\ 1 & 4 & -2 \\ 0 & 3 & 0 \end{bmatrix}\begin{bmatrix} x_1 \\ x_2 \\ x_3 \end{bmatrix},$$

$$(L_2 \circ L_1)^{-1}\left(\begin{bmatrix} x_1 \\ x_2 \\ x_3 \end{bmatrix}\right) = (L_1^{-1} \circ L_2^{-1})\left(\begin{bmatrix} x_1 \\ x_2 \\ x_3 \end{bmatrix}\right) = \begin{bmatrix} 2 & 1 & 0 \\ 0 & 0 & \frac{1}{3} \\ 1 & 0 & \frac{2}{3} \end{bmatrix}\begin{bmatrix} x_1 \\ x_2 \\ x_3 \end{bmatrix}$$

(c) $$L_1^{-1}\left(\begin{bmatrix} x_1 \\ x_2 \\ x_3 \end{bmatrix}\right) = \begin{bmatrix} 2 & -4 & -1 \\ 7 & -13 & -3 \\ 5 & -10 & -3 \end{bmatrix}\begin{bmatrix} x_1 \\ x_2 \\ x_3 \end{bmatrix}$$

$$L_2^{-1}\left(\begin{bmatrix} x_1 \\ x_2 \\ x_3 \end{bmatrix}\right) = \begin{bmatrix} 1 & 0 & -1 \\ 3 & 1 & -3 \\ -1 & -2 & 2 \end{bmatrix}\begin{bmatrix} x_1 \\ x_2 \\ x_3 \end{bmatrix},$$

$$(L_2 \circ L_1)\left(\begin{bmatrix} x_1 \\ x_2 \\ x_3 \end{bmatrix}\right) = \begin{bmatrix} 29 & -6 & -4 \\ 21 & -5 & -2 \\ 38 & -8 & -5 \end{bmatrix}\begin{bmatrix} x_1 \\ x_2 \\ x_3 \end{bmatrix},$$

$$(L_2 \circ L_1)^{-1}\left(\begin{bmatrix} x_1 \\ x_2 \\ x_3 \end{bmatrix}\right) = (L_1^{-1} \circ L_2^{-1})\left(\begin{bmatrix} x_1 \\ x_2 \\ x_3 \end{bmatrix}\right) = \begin{bmatrix} -9 & -2 & 8 \\ -29 & -7 & 26 \\ -22 & -4 & 19 \end{bmatrix}\begin{bmatrix} x_1 \\ x_2 \\ x_3 \end{bmatrix}$$

Section 5.5 (pp. 307–309)

1. (a) Orthogonal, not orthonormal (c) Neither

(f) Orthogonal, not orthonormal

2. (a) Orthogonal

(c) Not orthogonal; columns not normalized

(e) Orthogonal

3. (a) $[\mathbf{v}]_B = \left[\frac{(2\sqrt{3}+3)}{2}, \frac{(3\sqrt{3}-2)}{2}\right]$

(c) $[\mathbf{v}]_B = \left[3, \frac{13\sqrt{3}}{3}, \frac{5\sqrt{6}}{3}, 4\sqrt{2}\right]$

4. (a) {[5, −1, 2], [5, −3, −14]}

(c) {[2, 1, 0, −1], [−1, 1, 3, −1], [5, −7, 5, 3]}

5. (a) {[2, 2, −3], [13, −4, 6], [0, 3, 2]}

(c) {[1, −3, 1], [2, 5, 13], [4, 1, −1]}

(e) {[2, 1, −2, 1], [3, −1, 2, −1], [0, 5, 2, −1], [0, 0, 1, 2]}

6. (b) No

14. (b) $$\begin{bmatrix} \frac{\sqrt{6}}{6} & \frac{\sqrt{6}}{3} & \frac{\sqrt{6}}{6} \\ \frac{\sqrt{30}}{6} & -\frac{\sqrt{30}}{15} & -\frac{\sqrt{30}}{30} \\ 0 & \frac{\sqrt{5}}{5} & -\frac{2\sqrt{5}}{5} \end{bmatrix}$$

Section 5.6 (pp. 318–320)

1. (a) $\mathcal{W}^\perp = \text{span}(\{[2, 3]\})$
(c) $\mathcal{W}^\perp = \text{span}(\{[2, 3, 7]\})$
(e) $\mathcal{W}^\perp = \text{span}(\{[-2, 5, -1]\})$
(f) $\mathcal{W}^\perp = \text{span}(\{[7, 1, -2, -3], [0, 4, -1, 2]\})$

2. (a) $\mathbf{w}_1 = \mathbf{proj}_{\mathcal{W}}\mathbf{v} = \left[-\frac{33}{35}, \frac{111}{35}, \frac{60}{35}\right]$,
$\mathbf{w}_2 = \mathbf{proj}_{\mathcal{W}}\mathbf{v}\left[-\frac{2}{35}, -\frac{6}{35}, \frac{10}{35}\right]$
(b) $\mathbf{w}_1 = \mathbf{proj}_{\mathcal{W}}\mathbf{v} = \left[-\frac{17}{9}, -\frac{10}{9}, \frac{14}{9}\right]$,
$\mathbf{w}_2 = \left[\frac{26}{9}, -\frac{26}{9}, \frac{13}{9}\right]$

4. (a) $(3\sqrt{129})/43$ (d) $(8\sqrt{17})/17$

5. (a) $\begin{bmatrix} \frac{50}{59} & -\frac{21}{59} & -\frac{3}{59} \\ -\frac{21}{59} & \frac{10}{59} & -\frac{7}{59} \\ -\frac{3}{59} & -\frac{7}{59} & \frac{58}{59} \end{bmatrix}$ (c) $\begin{bmatrix} \frac{2}{9} & \frac{2}{9} & \frac{1}{3} & -\frac{1}{9} \\ \frac{2}{9} & \frac{8}{9} & 0 & \frac{2}{9} \\ \frac{1}{3} & 0 & \frac{2}{3} & -\frac{1}{3} \\ -\frac{1}{9} & \frac{2}{9} & -\frac{1}{3} & \frac{2}{9} \end{bmatrix}$

Section 6.1 (pp. 335–338)

1. (a) $x^2 - 7x + 14$
(c) $x^3 - 8x^2 + 21x - 18$
(e) $x^4 - 3x^3 - 4x^2 + 12x$

2. (a) $\{[1, 1]\}$ (c) $\{[1, 2, 0], [0, 0, 1]\}$

3. (c) Algebraic multiplicity = 2, geometric multiplicity = 1

5. (a) $\lambda = 1$, basis for $E_1 = \{[1, 0]\}$, geometric multiplicity of $\lambda = 1$, algebraic multiplicity of $\lambda = 2$
(c) $\lambda_1 = 1$, basis for $E_1 = \{[1, 0, 0]\}$, geometric multiplicity of $\lambda_1 = 1$, algebraic multiplicity of $\lambda_1 = 1$; $\lambda_2 = 2$, basis for $E_2 = \{[0, 1, 0]\}$, geometric multiplicity of $\lambda_2 = 1$, algebraic multiplicity of $\lambda_2 = 1$; $\lambda_3 = -5$, basis for $E_{-5} = \{[-\frac{1}{6}, \frac{3}{7}, 1]\}$, geometric multiplicity of $\lambda_3 = 1$, algebraic multiplicity of $\lambda_3 = 1$
(e) $\lambda_1 = 0$, basis for $E_0 = \{[1, 3, 2]\}$, geometric multiplicity of $\lambda_1 = 1$, algebraic multiplicity of $\lambda_1 = 1$; $\lambda_2 = 2$, basis for $E_2 = \{[1, 0, 1], [0, 1, 0]\}$, geometric multiplicity = 2, algebraic multiplicity = 2
(h) $\lambda_1 = 0$, basis for $E_0 = \{[-1, 1, 1, 0], [0, -1, 0, 1]\}$, geometric multiplicity of $\lambda_1 = 2$, algebraic multiplicity of $\lambda_1 = 2$; $\lambda_2 = -3$, basis for $E_{-3} = \{[-1, 0, 2, 2]\}$, geometric multiplicity of $\lambda_2 = 1$, algebraic multiplicity of $\lambda_2 = 2$

6. (a) $[-1, 3, 3]$ (c) $[5, 1, 1]$

7. $\frac{1}{9}\begin{bmatrix} 5 & 2 & -4 \\ 2 & 8 & 2 \\ -4 & 2 & 5 \end{bmatrix}$

9. (a) Characteristic polynomial $= x^3 + 10x^2 + 29x + 20$; eigenvalues $\lambda_1 = -1$, $\lambda_2 = -4$, $\lambda_3 = -5$; basis for $E_{-1} = \{2x^2 - 8x + 9\}$, basis for $E_{-4} = \{x\}$, basis for $E_{-5} = \{1\}$
(c) Characteristic polynomial $= x^2 - x + 1$; no eigenvalues
(e) Characteristic polynomial $= x^3 - 2x^2 + x$; eigenvalues $\lambda_1 = 0$, $\lambda_2 = 1$; basis for $E_0 = \{[4, -3, 2]\}$, basis for $E_1 = \{[1, 0, -2], [0, 2, 3]\}$
(g) Characteristic polynomial $= x^3 - x^2 - x + 1$; eigenvalues $\lambda_1 = 1$, $\lambda_2 = -1$; basis for eigenspace $E_1 = \{[1, 0, 3], [0, 1, 5]\}$, basis for eigenspace $E_{-1} = \{[3, 5, -1]\}$

10. (a) The only eigenvalue is $\lambda = 1$; $E_1 = \{1\}$.

12. (b) Consider the matrix $\mathbf{A} = \begin{bmatrix} 0 & -1 \\ 1 & 0 \end{bmatrix}$, which represents a rotation about the origin in $\mathbb{R}^2$ through an angle of $\frac{\pi}{2}$ radians, or 90°. Although $\mathbf{A}$ has no eigenvalues, $\mathbf{A}^4 = \mathbf{I}_2$ has 1 as an eigenvalue.

15. (c) $\begin{bmatrix} 1 & 1 & -1 \\ 0 & 1 & 0 \\ 0 & 0 & 1 \end{bmatrix}$, eigenvalue $\lambda = 1$, basis for $E_1 = \{[0, 1, 1], [1, 0, 0]\}$, λ has algebraic multiplicity 3 and geometric multiplicity 2
(d) $\begin{bmatrix} 1 & 0 & 0 \\ 0 & 1 & 0 \\ 0 & 0 & 0 \end{bmatrix}$, eigenvalues $\lambda_1 = 1$ and $\lambda_2 = 0$, basis for $E_1 = \{[1, 0, 0], [0, 1, 0]\}$, λ_1 has algebraic and geometric multiplicity 2

Section 6.2 (pp. 348–351)

1. (a) Basis of eigenvectors = $\{[3, 1], [2, 1]\}$,
$\mathbf{P} = \begin{bmatrix} 3 & 2 \\ 1 & 1 \end{bmatrix}$, $\mathbf{D} = \begin{bmatrix} 3 & 0 \\ 0 & -5 \end{bmatrix}$
(c) Not diagonalizable
(d) Basis of eigenvectors = $\{[6, 2, 5], [1, 2, 1], [1, 1, 1]\}$,
$\mathbf{P} = \begin{bmatrix} 6 & 1 & 1 \\ 2 & 2 & 1 \\ 5 & 1 & 1 \end{bmatrix}$, $\mathbf{D} = \begin{bmatrix} 1 & 0 & 0 \\ 0 & -1 & 0 \\ 0 & 0 & 2 \end{bmatrix}$
(f) Not diagonalizable
(g) Basis of eigenvectors = $\{[2, 3, 0], [1, 0, 3], [0, 1, -1]\}$,
$\mathbf{P} = \begin{bmatrix} 2 & 1 & 0 \\ 3 & 0 & 1 \\ 0 & 3 & -1 \end{bmatrix}$, $\mathbf{D} = \begin{bmatrix} 2 & 0 & 0 \\ 0 & 2 & 0 \\ 0 & 0 & 3 \end{bmatrix}$
(i) Basis of eigenvectors = $\{[2, 2, 1, 0], [1, 0, 0, 1], [1, 2, 1, 0], [1, -1, 0, 1]\}$,
$\mathbf{P} = \begin{bmatrix} 2 & 1 & 1 & 1 \\ 2 & 0 & 2 & -1 \\ 1 & 0 & 1 & 0 \\ 0 & 1 & 0 & 1 \end{bmatrix}$, $\mathbf{D} = \begin{bmatrix} 1 & 0 & 0 & 0 \\ 0 & 1 & 0 & 0 \\ 0 & 0 & -1 & 0 \\ 0 & 0 & 0 & 0 \end{bmatrix}$

2. (a) $C = (x^2, x, 1)$, $\mathbf{A} = \begin{bmatrix} 2 & 0 & 0 \\ -2 & 1 & 0 \\ 0 & -1 & 0 \end{bmatrix}$,
$B = (x^2 - 2x + 1, x - 1, 1)$,
$\mathbf{P} = \begin{bmatrix} 1 & 0 & 0 \\ -2 & 1 & 0 \\ 1 & -1 & 1 \end{bmatrix}$,

$$\mathbf{D} = \begin{bmatrix} 2 & 0 & 0 \\ 0 & 1 & 0 \\ 0 & 0 & 0 \end{bmatrix}$$

(d) $C = \left(\begin{bmatrix} 1 & 0 \\ 0 & 0 \end{bmatrix}, \begin{bmatrix} 0 & 1 \\ 0 & 0 \end{bmatrix}, \begin{bmatrix} 0 & 0 \\ 1 & 0 \end{bmatrix}, \begin{bmatrix} 0 & 0 \\ 0 & 1 \end{bmatrix} \right)$,

$$\mathbf{A} = \begin{bmatrix} -4 & 0 & 3 & 0 \\ 0 & -4 & 0 & 3 \\ -10 & 0 & 7 & 0 \\ 0 & -10 & 0 & 7 \end{bmatrix},$$

$$B = \left(\begin{bmatrix} 3 & 0 \\ 5 & 0 \end{bmatrix}, \begin{bmatrix} 0 & 3 \\ 0 & 5 \end{bmatrix}, \begin{bmatrix} 1 & 0 \\ 2 & 0 \end{bmatrix}, \begin{bmatrix} 0 & 1 \\ 0 & 2 \end{bmatrix} \right),$$

$$\mathbf{P} = \begin{bmatrix} 3 & 0 & 1 & 0 \\ 0 & 3 & 0 & 1 \\ 5 & 0 & 2 & 0 \\ 0 & 5 & 0 & 2 \end{bmatrix}, \mathbf{D} = \begin{bmatrix} 1 & 0 & 0 & 0 \\ 0 & 1 & 0 & 0 \\ 0 & 0 & 2 & 0 \\ 0 & 0 & 0 & 2 \end{bmatrix}$$

3. (a) $\begin{bmatrix} 32770 & -65538 \\ 32769 & -65537 \end{bmatrix}$ (c) $\mathbf{A}^{49} = \mathbf{A}$

(e) $\begin{bmatrix} 4188163 & 6282243 & -9421830 \\ 4192254 & 6288382 & -9432060 \\ 4190208 & 6285312 & -9426944 \end{bmatrix}$

4. (a) Orthogonal projection onto $3x + y + z = 0$
(d) Neither

6. (b) If $\mathbf{A}$ has all eigenvalues nonnegative, then $\mathbf{A}$ has a square root.

11. (b) Consider the matrix $\mathbf{A} = \begin{bmatrix} 0 & -1 & 0 \\ 1 & 0 & 0 \\ 0 & 0 & 1 \end{bmatrix}$, which represents a rotation of $\frac{\pi}{2}$ radians, or 90°, about the z-axis in the xy-plane. The characteristic polynomial of this matrix is $x^3 - x^2 + x - 1 = (x-1)(x^2+1)$. Hence, the only eigenvalue is $\lambda = 1$, and its eigenspace E_1 is one-dimensional (spanned by $[0, 0, 1]$). Thus, the algebraic and geometric multiplicities of $\lambda = 1$ are both equal to 1. Since λ is the only eigenvalue, $\mathbf{A}$ is not diagonalizable.

Section 6.3 (pp. 361–363)

1. (a) Symmetric, because the matrix for L is symmetric (d) Symmetric (e) Not symmetric (g) Symmetric

2. (a) $\frac{1}{25}\begin{bmatrix} -7 & 24 \\ 24 & 7 \end{bmatrix}$

(d) $\frac{1}{169}\begin{bmatrix} -119 & -72 & -96 & 0 \\ -72 & 119 & 0 & 96 \\ -96 & 0 & 119 & -72 \\ 0 & 96 & -72 & -119 \end{bmatrix}$

3. (a) $B = \left(\frac{1}{13}[5, 12], \frac{1}{13}[12, -5]\right)$,

$$\mathbf{P} = \frac{1}{13}\begin{bmatrix} 5 & 12 \\ 12 & -5 \end{bmatrix},$$

$$\mathbf{D} = \begin{bmatrix} 0 & 0 \\ 0 & 169 \end{bmatrix}$$

(c) $B = \left(\frac{1}{3}[-1, 2, 2], \frac{1}{3}[2, -1, 2], \frac{1}{3}[2, 2, -1]\right)$ (other bases are possible, since E_1 is two-dimensional), $\mathbf{P} = \frac{1}{3}\begin{bmatrix} -1 & 2 & 2 \\ 2 & -1 & 2 \\ 2 & 2 & -1 \end{bmatrix}$,

$$\mathbf{D} = \begin{bmatrix} 1 & 0 & 0 \\ 0 & 1 & 0 \\ 0 & 0 & 3 \end{bmatrix}$$

(e) $B = \left(\frac{1}{\sqrt{14}}[3, 2, 1, 0], \frac{1}{\sqrt{14}}[-2, 3, 0, 1], \frac{1}{\sqrt{14}}[1, 0, -3, 2], \frac{1}{\sqrt{14}}[0, -1, 2, 3]\right)$,

$$\mathbf{P} = \frac{1}{\sqrt{14}}\begin{bmatrix} 3 & -2 & 1 & 0 \\ 2 & 3 & 0 & -1 \\ 1 & 0 & -3 & 2 \\ 0 & 1 & 2 & 3 \end{bmatrix},$$

$$\mathbf{D} = \begin{bmatrix} 2 & 0 & 0 & 0 \\ 0 & 2 & 0 & 0 \\ 0 & 0 & -3 & 0 \\ 0 & 0 & 0 & 5 \end{bmatrix}$$

(g) $B = \left(\frac{1}{\sqrt{5}}[1, 2, 0], \frac{1}{\sqrt{6}}[-2, 1, 1], \frac{1}{\sqrt{30}}[2, -1, 5]\right)$, (other bases are possible since E_{15} is two-dimensional)

$$\mathbf{P} = \begin{bmatrix} \frac{1}{\sqrt{5}} & -\frac{2}{\sqrt{6}} & \frac{2}{\sqrt{30}} \\ \frac{2}{\sqrt{5}} & \frac{1}{\sqrt{6}} & -\frac{1}{\sqrt{30}} \\ 0 & \frac{1}{\sqrt{6}} & \frac{5}{\sqrt{30}} \end{bmatrix},$$

$$\mathbf{D} = \begin{bmatrix} 15 & 0 & 0 \\ 0 & 15 & 0 \\ 0 & 0 & -15 \end{bmatrix}$$

4. (a) $C = \frac{1}{19}[-10, 15, 6], \frac{1}{19}[15, 6, 10]$,

$$\mathbf{A} = \begin{bmatrix} -2 & 2 \\ 2 & 1 \end{bmatrix},$$

$B = \left(\frac{1}{(19\sqrt{5})}[20, 27, 26], \frac{1}{(19\sqrt{5})}[35, -24, -2]\right)$,

$$\mathbf{P} = \frac{1}{\sqrt{5}}\begin{bmatrix} 1 & -2 \\ 2 & 1 \end{bmatrix}, \mathbf{D} = \begin{bmatrix} 2 & 0 \\ 0 & -3 \end{bmatrix}$$

5. (a) $\frac{1}{25}\begin{bmatrix} 23 & -36 \\ -36 & 2 \end{bmatrix}$ (c) $\frac{1}{3}\begin{bmatrix} 11 & 4 & -4 \\ 4 & 17 & -8 \\ -4 & -8 & 17 \end{bmatrix}$

6. For example, the matrix $\mathbf{A}$ in Example 4 of Section 6.2 is diagonalizable but not symmetric and hence not orthogonally diagonalizable.

7.

$$\frac{1}{2}\begin{bmatrix} a + c + \sqrt{(a-c)^2 + 4b^2} & 0 \\ 0 & a + c - \sqrt{(a-c)^2 + 4b^2} \end{bmatrix}$$

8. (b) L must be the zero linear operator. Since L is diagonalizable, the eigenspace for 0 must be all of $\mathcal{V}$.

Section 7.1 (pp. 378–381)

1. (a) $[1+4i, 1+i, 6-i]$
(b) $[-12-32i, -7+30i, 53-29i]$
(d) $[-24-12i, -28-8i, -32i]$ (e) $1+28i$

3. (a) $\begin{bmatrix} 11+4i & -4-2i \\ 2-4i & 12 \end{bmatrix}$

(c) $\begin{bmatrix} 1-i & 0 & 10i \\ 2i & 3-i & 0 \\ 6-4i & 5 & 7+3i \end{bmatrix}$

(d) $\begin{bmatrix} -3-15i & -3 & 9i \\ 9-6i & 0 & 3+12i \end{bmatrix}$

(f) $\begin{bmatrix} 1+40i & -4-14i \\ 13-50i & 23+21i \end{bmatrix}$

(i) $\begin{bmatrix} 4+36i & -5+39i \\ 1-7i & -6-4i \\ 5+40i & -7-5i \end{bmatrix}$

5. (a) Skew-Hermitian (b) Neither (c) Hermitian
(d) Skew-Hermitian (e) Hermitian

10. (b) Not linearly independent, dim = 1
(d) Not linearly independent, dim = 2

11. (b) Linearly independent, dim = 2
(d) Linearly independent, dim = 3

12. (b) $[i, 1+i, -1]$

13. (b) Ordered basis = $([1,0], [i,0], [0,1], [0,i])$,

matrix = $\begin{bmatrix} 1 & 0 & 0 & 0 \\ 0 & -1 & 0 & 0 \\ 0 & 0 & 1 & 0 \\ 0 & 0 & 0 & -1 \end{bmatrix}$

14. $\begin{bmatrix} -3+i & -\frac{2}{5}-\frac{11}{5}i \\ \frac{1}{2}-\frac{3}{2}i & -i \\ -\frac{1}{2}+\frac{7}{2}i & -\frac{8}{5}-\frac{4}{5}i \end{bmatrix}$

15. (a) Not orthogonal (c) Orthogonal

16. (a) $\{[1+i, i, 1], [2, -1-i, -1+i], [0, 1, i]\}$

(b) $\begin{bmatrix} \frac{1+i}{2} & \frac{i}{2} & \frac{1}{2} \\ \frac{2}{\sqrt{8}} & \frac{-1-i}{\sqrt{8}} & \frac{-1+i}{\sqrt{8}} \\ 0 & \frac{1}{\sqrt{2}} & \frac{i}{\sqrt{2}} \end{bmatrix}$

22. (a) Eigenvalues i and -1, with respective eigenvectors $[i, 1+i]$, $[2+i, 4-i]$
(c) Eigenvalues i and -2, with eigenvectors $[3i, 2, -2i]$ and $[1, 1, 0]$ for i and eigenvector $[i, 1, -i]$ for -2. Other bases are possible since E_i is two-dimensional.

23. (a) $\mathbf{P} = \begin{bmatrix} i & 2+i \\ 1+i & 4-i \end{bmatrix}$, $\mathbf{D} = \begin{bmatrix} i & 0 \\ 0 & -1 \end{bmatrix}$

25. (b) $\mathbf{P} = \frac{1}{\sqrt{6}}\begin{bmatrix} -1+i & 2i \\ 2 & -1+i \end{bmatrix}$; the corresponding diagonal matrix is $\begin{bmatrix} 9+6i & 0 \\ 0 & -3-12i \end{bmatrix}$.

28. The eigenvalues are -4, $2+\sqrt{6}$, and $2-\sqrt{6}$.

Section 7.2 (pp. 393–396)

1. (b) $<\mathbf{x}, \mathbf{y}> = -183$, $\|\mathbf{x}\| = \sqrt{314}$

3. (b) $<\mathbf{f}, \mathbf{g}> = \frac{1}{2}(e^{\pi}+1)$; $\|\mathbf{f}\| = \sqrt{\frac{1}{2}(e^{2\pi}-1)}$

9. $\sqrt{(\pi^3/3)-(3\pi/2)}$

14. 0.586 radians, or 33.6°

16. (a) Orthogonal (c) Not orthogonal

21. Using $\mathbf{w}_1 = t^2-t+1$, $\mathbf{w}_2 = 1$, and $\mathbf{w}_3 = t$ yields the orthogonal basis $\{\mathbf{v}_1, \mathbf{v}_2, \mathbf{v}_3\}$, with $\mathbf{v}_1 = t^2-t+1$, $\mathbf{v}_2 = -20t^2+20t+13$, and $\mathbf{v}_3 = 25t^2+8t-8$.

25. $\mathcal{W}^{\perp} = \text{span}(\{t^3-t^2, t+1\})$

28. $\mathbf{w}_1 = \frac{1}{2\pi}(\sin t - \cos t)$, $\mathbf{w}_2 = \frac{1}{k}e^t - \frac{1}{2\pi}\sin t + \frac{1}{2\pi}\cos t$

31. (b) $\ker(L) = \mathcal{W}^{\perp}$, $\text{range}(L) = \mathcal{W}$

Section 8.1 (pp. 403–406)

1. Symmetric: (a), (b), (c), (d), (g)

(a) Matrix for G_1 = $\begin{bmatrix} 0 & 1 & 1 & 1 \\ 1 & 0 & 1 & 1 \\ 1 & 1 & 0 & 1 \\ 1 & 1 & 1 & 0 \end{bmatrix}$

(b) Matrix for G_2 = $\begin{bmatrix} 1 & 1 & 0 & 0 & 1 \\ 1 & 0 & 0 & 1 & 0 \\ 0 & 0 & 1 & 0 & 1 \\ 0 & 1 & 0 & 0 & 1 \\ 1 & 0 & 1 & 1 & 1 \end{bmatrix}$

(c) Matrix for G_3 = $\begin{bmatrix} 0 & 0 & 0 \\ 0 & 0 & 0 \\ 0 & 0 & 0 \end{bmatrix}$

(d) Matrix for G_4 = $\begin{bmatrix} 0 & 1 & 0 & 1 & 0 & 0 \\ 1 & 0 & 1 & 1 & 0 & 0 \\ 0 & 1 & 0 & 0 & 1 & 1 \\ 1 & 1 & 0 & 0 & 1 & 0 \\ 0 & 0 & 1 & 1 & 0 & 1 \\ 0 & 0 & 1 & 0 & 1 & 0 \end{bmatrix}$

(e) Matrix for D_1 = $\begin{bmatrix} 0 & 1 & 0 & 0 \\ 0 & 0 & 1 & 0 \\ 1 & 0 & 0 & 1 \\ 1 & 1 & 0 & 0 \end{bmatrix}$

(f) Matrix for D_2 = $\begin{bmatrix} 0 & 1 & 1 & 0 \\ 0 & 1 & 1 & 1 \\ 0 & 1 & 0 & 1 \\ 0 & 0 & 0 & 1 \end{bmatrix}$

(g) Matrix for D_3 = $\begin{bmatrix} 0 & 1 & 1 & 0 & 0 \\ 1 & 0 & 0 & 0 & 1 \\ 1 & 0 & 0 & 1 & 0 \\ 0 & 0 & 1 & 0 & 1 \\ 0 & 1 & 0 & 1 & 0 \end{bmatrix}$.

(h) Matrix for D_4 = $\begin{bmatrix} 0 & 1 & 0 & 0 & 0 & 0 & 0 & 0 \\ 0 & 0 & 1 & 0 & 0 & 0 & 0 & 0 \\ 0 & 0 & 0 & 1 & 0 & 0 & 0 & 0 \\ 0 & 0 & 0 & 0 & 1 & 0 & 0 & 0 \\ 0 & 0 & 0 & 0 & 0 & 1 & 0 & 0 \\ 0 & 0 & 0 & 0 & 0 & 0 & 1 & 0 \\ 0 & 0 & 0 & 0 & 0 & 0 & 0 & 1 \\ 1 & 0 & 0 & 0 & 0 & 0 & 0 & 0 \end{bmatrix}$

2. See Figure 3 for graphs and digraphs of the following:

F can be the adjacency matrix for either a graph or digraph.

Graph for **F** =

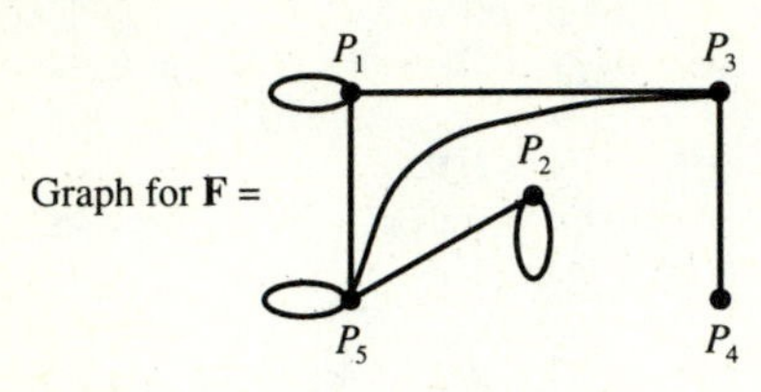

G can be the adjacency matrix for a digraph (only).

Digraph for **G** =

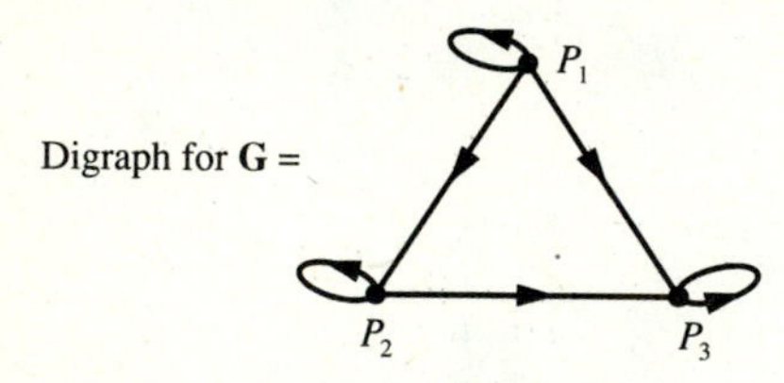

H can be the adjacency matrix for a digraph (only).

Digraph for **H** =

P_1

P_2 P_3

I can be the adjacency matrix for a graph or digraph.

Graph for **I** =

P_1 P_2 P_3

K can be the adjacency matrix for a graph or digraph.

Graph for **K** =

P_1 P_2

L can be the adjacency matrix for a graph or digraph.

Graph for **L** =

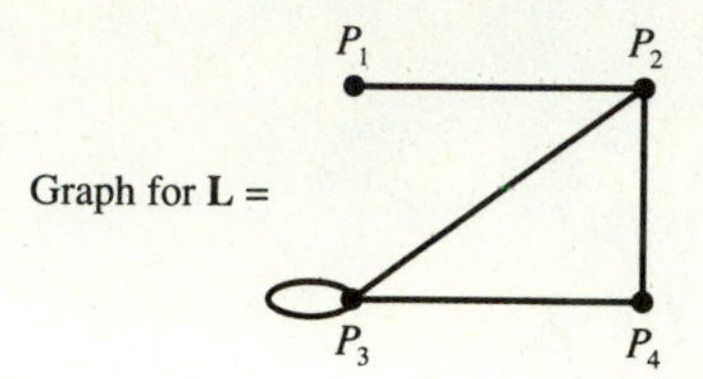

3. The digraph is shown in Figure 4, and the adjacency matrix is

	A	B	C	D	E	F
A	0	0	1	0	0	0
B	0	0	0	1	1	0
C	0	1	0	0	1	0
D	1	0	0	0	0	1
E	1	1	0	1	0	0
F	0	0	0	1	0	0

The transpose gives no new information. But it does suggest a different interpretation of the results: namely, the (i, j) entry of the transpose equals 1 if author j influences author i.

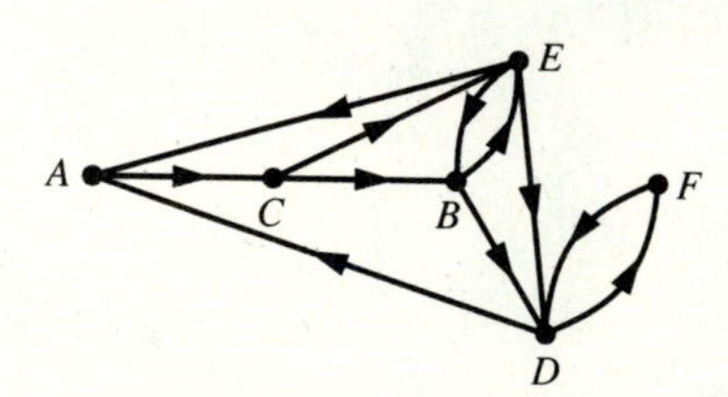

Figure 4
Answer to Exercise 3, Section 8.1

4. (a) 3 (c) $6 = 1 + 1 + 4$ (e) Length 4

5. (a) 4 (c) $5 = 1 + 1 + 3$ (e) No such path exists.

6. (a) Figure 8.5: 7; Figure 8.6: 2

(c) Figure 8.5: $3 = 0 + 1 + 0 + 2$; Figure 8.6: $17 = 1 + 2 + 4 + 10$

7. (a) If the vertex is the ith vertex, then the ith row and i^{th} column entries of the adjacency matrix will all equal zero, except possibly for the (i, i) entry.

(b) If the vertex is the ith vertex, then the ith row entries of the adjacency matrix will all equal zero, except possibly for the (i, i) entry. (Note: The ith column entries may be nonzero.)

8. The trace equals the number of loops in the graph or digraph.

9. (b) The digraph in Figure 8.5 is strongly connected; the digraph in Figure 8.6 is not strongly connected (since there is no path directed to P_5 from any other vertex).

10. (b) Yes, it is a dominance digraph, because no tie games are possible and because each team plays every other team. Thus if P_i and P_j are two given teams, either P_i defeats P_j or vice versa.

Section 8.2 (p. 408)

1. (a) $I_1 = 8,\ I_2 = 5,\ I_3 = 3$
(c) $I_1 = 12,\ I_2 = 5,\ I_3 = 3,\ I_4 = 2,\ I_5 = 2,\ I_6 = 7$

Section 8.3 (pp. 414–415)

1. (a) $y = -0.8x - 3.3$, $y = -7.3$ when $x = 5$
(c) $y = -1.5x + 3.8$, $y = -3.7$ when $x = 5$

2. (a) $y = 0.375x^2 + 0.35x + 3.60$
(c) $y = -0.042x^2 + 0.633x + 0.266$

3. (a) $y = \frac{1}{4}x^3 + \frac{25}{28}x^2 + \frac{25}{14}x + \frac{37}{35}$

4. (a) $y = 4.4286x^2 - 2.0571$
(c) $y = -0.1014x^2 + 0.9633x - 0.8534$
(e) $y = 0x^3 - 0.3954x^2 + 0.9706$

5. (a) $y = 0.4x + 2.54$; the angle reaches 20° in the 44th month.
(b) $y = 0.02857x^2 + 0.2286x + 2.74$; the angle reaches 20° in the 21st month.

7. The least-squares polynomial is $y = \frac{4}{5}x^2 - \frac{2}{5}x + 2$, which is the *exact* quadratic through the given three points.

9. (a) $x_1 = \frac{230}{39}$, $x_2 = \frac{155}{39}$

$$\begin{cases} 4x_1 - 3x_2 = 11\frac{2}{3} & \text{which is almost 12} \\ 2x_1 + 5x_2 = 31\frac{2}{3} \\ 3x_1 + x_2 = 21\frac{2}{3} \end{cases}$$

Section 8.4 (pp. 424–427)

1. **A** is not stochastic, since **A** is not square; **A** is not regular, since **A** is not stochastic.

B is not stochastic, since the entries of column 2 do not sum to 1; **B** is not regular, since **B** is not stochastic.

C is stochastic; **C** is regular, since **C** is stochastic and has all nonzero entries.

D is stochastic; **D** is not regular, since every positive power of **D** is a matrix whose rows are the rows of **D** permuted in some order, and hence every such power contains zero entries.

E is not stochastic, since the entries of column 1 do not sum to 1; **E** is not regular, since **E** is not stochastic.

F is stochastic; **F** is not regular, since every positive power of **F** has all second-row entries zero.

G is not stochastic, since **G** is not square; **G** is not regular, since **G** is not stochastic.

H is stochastic; **H** is regular, since

$$\mathbf{H}^2 = \begin{bmatrix} \frac{1}{2} & \frac{1}{4} & \frac{1}{4} \\ \frac{1}{4} & \frac{1}{2} & \frac{1}{4} \\ \frac{1}{4} & \frac{1}{4} & \frac{1}{2} \end{bmatrix},$$

which has no zero entries.

2. (a) $\mathbf{p}_1 = \left[\frac{5}{18}, \frac{13}{18}\right]$, $\mathbf{p}_2 = \left[\frac{67}{216}, \frac{149}{216}\right]$
(c) $\mathbf{p}_1 = \left[\frac{17}{48}, \frac{1}{3}, \frac{5}{16}\right]$, $\mathbf{p}_2 = \left[\frac{205}{576}, \frac{49}{144}, \frac{175}{576}\right]$

3. (a) $\left[\frac{2}{5}, \frac{3}{5}\right]$

5. (a) $[0.34, 0.175, 0.34, 0.145]$ in the next election, $[0.3555, 0.1875, 0.2875, 0.1695]$ in the election after that
(b) The steady-state vector is $[0.36, 0.20, 0.24, 0.20]$; in a century, the votes would be 36% Party A and 24% Party C.

6. (a) $$\begin{bmatrix} \frac{1}{2} & \frac{1}{6} & \frac{1}{6} & \frac{1}{5} & 0 \\ \frac{1}{8} & \frac{1}{2} & 0 & 0 & \frac{1}{5} \\ \frac{1}{8} & 0 & \frac{1}{2} & \frac{1}{10} & \frac{1}{10} \\ \frac{1}{4} & 0 & \frac{1}{6} & \frac{1}{2} & \frac{1}{5} \\ 0 & \frac{1}{3} & \frac{1}{6} & \frac{1}{5} & \frac{1}{2} \end{bmatrix}$$
(b) $\frac{29}{120}$, since the probability vector after two time intervals is $\left[\frac{1}{5}, \frac{13}{240}, \frac{73}{240}, \frac{29}{120}, \frac{1}{5}\right]$
(c) $\left[\frac{1}{5}, \frac{3}{20}, \frac{3}{20}, \frac{1}{4}, \frac{1}{4}\right]$; over time, the rat frequents rooms B and C the least and rooms D and E the most.

Section 8.5 (p. 430)

1. (a) −24 −46 −15 −30 10 16 39 62 26 42 51 84 24 37 −11 −23

2. (a) HOMEWORK IS GOOD FOR THE SOUL

Section 8.6 (pp. 433–434)

1. (a) Linearly independent; to prove that it is, substitute the values $x = 0$, $x = 1$, and $x = 2$, and follow the method of Example 1.
(c) Linearly dependent ($a = 2$, $b = -1$, $c = -1$)

4. $B = \{e^{-5x}, e^{3x}\}$, $\dim(\mathcal{W}) = 2$; to prove that B is linearly independent, substitute $x = 0$ and $x = 1$, and follow the method in Example 1.

5. (a) $B = \{\sin(2x), \cos(2x), \sin^2 x\}$
(c) $B = \{\sin(x + 1), \cos(x + 1)\}$

6. (a) $[\mathbf{v}]_B = [5, 0, -7]$
(c) $[\mathbf{v}]_B = [-(\cos 2/\sin 1), (\cos 1/\sin 1)]$, or approximately $[0.4945, 0.6421]$. If your answer seems more complicated, compare numerical approximations.

Section 8.7 (pp. 438–439)

1. (b) $\theta = \frac{1}{2}\arctan(-\sqrt{3}) = -\frac{\pi}{6}$; $\mathbf{P} = \begin{bmatrix} \frac{\sqrt{3}}{2} & \frac{1}{2} \\ -\frac{1}{2} & \frac{\sqrt{3}}{2} \end{bmatrix}$; equation in uv-coordinates: $v = 2u^2 - 12u + 13$, or $(v+5) = 2(u-3)^2$; vertex in uv-coordinates: $(3, -5)$; vertex in xy-coordinates: $(0.0981, -5.830)$; see figure below:

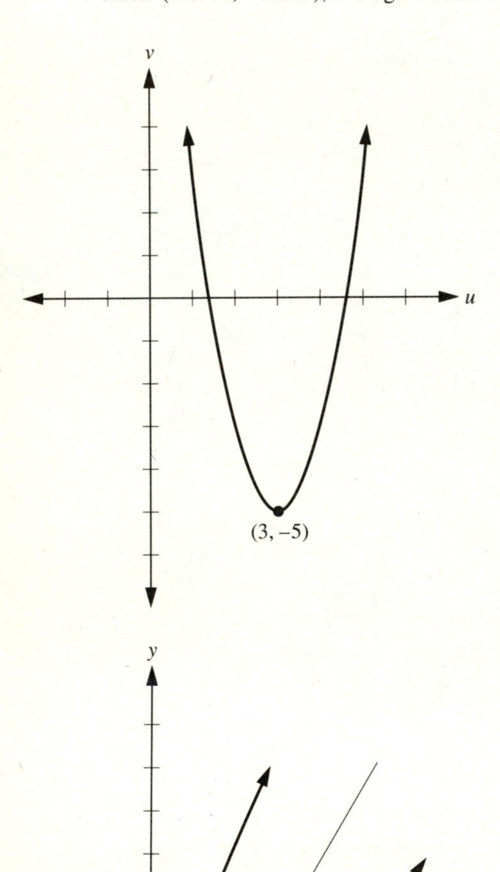

(d) All answers are rounded to four significant digits: $\theta = 0.4442$ radians (about $25°27'$); $\mathbf{P} = \begin{bmatrix} 0.9029 & -0.4298 \\ 0.4298 & 0.9029 \end{bmatrix}$; equation in uv-coordinates: $u^2/(1.770)^2 - v^2/(2.050)^2 = 1$; center in uv-coordinates: $(0, 0)$; center in xy-coordinates $= (0, 0)$; see figure below:

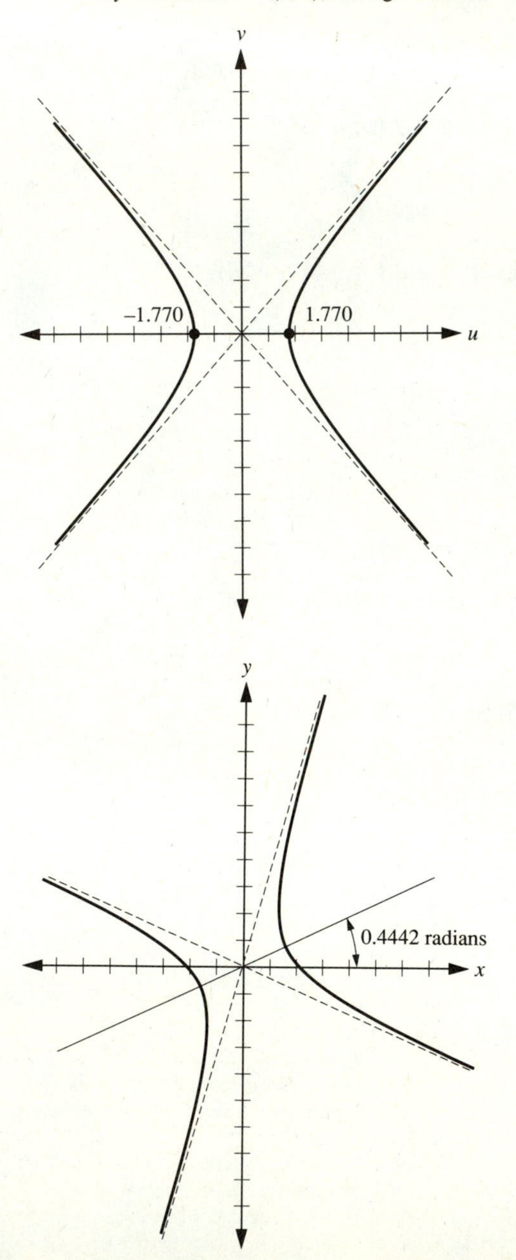

Section 8.8 (pp. 445–447)

1. (a) $b_1e^t\begin{bmatrix}7\\3\end{bmatrix} + b_2e^{-t}\begin{bmatrix}2\\1\end{bmatrix}$

(c) $b_1\begin{bmatrix}0\\-1\\1\end{bmatrix} + b_2e^t\begin{bmatrix}1\\-1\\1\end{bmatrix} + b_3e^{3t}\begin{bmatrix}2\\0\\1\end{bmatrix}$

(d) $b_1e^t\begin{bmatrix}6\\-1\\2\end{bmatrix} + b_2e^t\begin{bmatrix}1\\-1\\0\end{bmatrix} + b_3e^{4t}\begin{bmatrix}1\\1\\1\end{bmatrix}$ (There are other possible answers. For example, the first two vectors in the sum could be any basis for the two-dimensional eigenspace corresponding to the eigenvalue 1.)

2. (a) $y = b_1e^{2t} + b_2e^{-3t}$

(c) $y = b_1e^{2t} + b_2e^{-2t} + b_3e^{\sqrt{2}t} + b_4e^{-\sqrt{2}t}$

4. (b) $\mathbf{F}(t) = 2e^{5t}\begin{bmatrix}1\\0\\1\end{bmatrix} + e^t\begin{bmatrix}1\\-2\\0\end{bmatrix} - 2e^{-t}\begin{bmatrix}1\\1\\1\end{bmatrix}$

Section 8.9 (pp. 453–454)

1. (a) $\mathbf{C} = \begin{bmatrix}8 & 12\\0 & -9\end{bmatrix}$, $\mathbf{A} = \begin{bmatrix}8 & 6\\6 & -9\end{bmatrix}$

(c) $\mathbf{C} = \begin{bmatrix}5 & 4 & -3\\0 & -2 & 5\\0 & 0 & 0\end{bmatrix}$, $\mathbf{A} = \begin{bmatrix}5 & 2 & -\frac{3}{2}\\2 & -2 & \frac{5}{2}\\-\frac{3}{2} & \frac{5}{2} & 0\end{bmatrix}$

2. (a) $\mathbf{A} = \begin{bmatrix}43 & -24\\-24 & 57\end{bmatrix}$, $\mathbf{P} = \frac{1}{5}\begin{bmatrix}3 & 4\\-4 & 3\end{bmatrix}$,

$\mathbf{D} = \begin{bmatrix}75 & 0\\0 & 25\end{bmatrix}$, $B = \left(\frac{1}{5}[3, -4], \frac{1}{5}[4, 3]\right)$,

$[\mathbf{x}]_B = [7, -4]$, $Q(x) = 4075$

(c) $\mathbf{A} = \begin{bmatrix}18 & 48 & -30\\48 & -68 & 18\\-30 & 18 & 1\end{bmatrix}$,

$\mathbf{P} = \frac{1}{7}\begin{bmatrix}2 & 6 & -3\\3 & 2 & 6\\6 & -3 & -2\end{bmatrix}$,

$\mathbf{D} = \begin{bmatrix}0 & 0 & 0\\0 & 49 & 0\\0 & 0 & -98\end{bmatrix}$,

$B = \left(\frac{1}{7}[2, 3, 6], \frac{1}{7}[6, 2, -3], \frac{1}{7}[-3, 6, -2]\right)$,
$[\mathbf{x}]_B = [5, 0, -6]$, $Q(x) = -3528$

3. (b) No, since $\mathbf{A} \neq \mathbf{B}$ (c) No, since the eigenvalues for $\mathbf{A}$ are $5, 3, -7$, but none of these are eigenvalues for $\mathbf{B}$.

5. Yes; if $Q(x) = \sum a_{ij}x_ix_j$, $1 \le i \le j \le n$, then $\mathbf{x}^T\mathbf{C}_1\mathbf{x}$ and $\mathbf{C}_1$ upper triangular imply that the (i, j) entry for $\mathbf{C}_1$ is zero if $i > j$ and a_{ij} if $i \le j$. A similar argument describes $\mathbf{C}_2$. Thus, $\mathbf{C}_1 = \mathbf{C}_2$.

Section 9.1 (pp. 466–468)

1. (a) Solution to first system: (602, 1500); solution to second system: (302, 750). The system is ill-conditioned, because a very small change in the coefficient of y leads to a very large change in the solution.

3. Your answers to this problem may differ significantly from the following, depending on where you do your rounding in the algorithm:

(a) Without partial pivoting: (3210, 0.765); with partial pivoting: (3230, 0.767). Actual solution is (3214, 0.765).

(c) Without partial pivoting: $(0.45, 1.01, -2.11)$; with partial pivoting: $(277, -327, 595)$. Actual solution is $(267, -315, 573)$.

4. Your answers to this problem may differ significantly from the following, depending on where you do your rounding in the algorithm:

(a) Without partial pivoting: (3214, 0.7651); with partial pivoting: (3213, 0.7648). Actual solution is (3214, 0.765).

(c) Without partial pivoting: $(-2.533, 8.801, -16.30)$; with partial pivoting: $(267.7, -315.9, 574.6)$. Actual solution is $(267, -315, 573)$.

5. (a)

	x_1	x_2
Initial values	0.000	0.000
After 1 step	5.200	−6.000
After 2 steps	6.400	−8.229
After 3 steps	6.846	−8.743
After 4 steps	6.949	−8.934
After 5 steps	6.987	−8.978
After 6 steps	6.996	−8.994
After 7 steps	6.999	−8.998
After 8 steps	7.000	−9.000
After 9 steps	7.000	−9.000

(c)

	x_1	x_2	x_3
Initial values	0.000	0.000	0.000
After 1 step	−8.857	4.500	−4.333
After 2 steps	−10.738	3.746	−8.036
After 3 steps	−11.688	4.050	−8.537
After 4 steps	−11.875	3.975	−8.904
After 5 steps	−11.969	4.005	−8.954
After 6 steps	−11.988	3.998	−8.991
After 7 steps	−11.997	4.001	−8.996
After 8 steps	−11.999	4.000	−8.999
After 9 steps	−12.000	4.000	−9.000
After 10 steps	−12.000	4.000	−9.000

6. (a)

	x_1	x_2
Initial values	0.000	0.000
After 1 step	5.200	−8.229
After 2 steps	6.846	−8.934
After 3 steps	6.987	−8.994
After 4 steps	6.999	−9.000
After 5 steps	7.000	−9.000
After 6 steps	7.000	−9.000

(c)

	x_1	x_2	x_3
Initial values	0.000	0.000	0.000
After 1 step	−8.857	3.024	−7.790
After 2 steps	−11.515	3.879	−8.818
After 3 steps	−11.931	3.981	−8.974
After 4 steps	−11.990	3.997	−8.996
After 5 steps	−11.998	4.000	−8.999
After 6 steps	−12.000	4.000	−9.000
After 7 steps	−12.000	4.000	−9.000

7. Strictly diagonally dominant: (a), (c), (f), (g)

8. (a) Put the third equation first, and move the other two down to get the following:

	x_1	x_2	x_3
Initial values	0.000	0.000	0.000
After 1 step	3.125	−0.481	1.461
After 2 steps	2.517	−0.500	1.499
After 3 steps	2.500	−0.500	1.500
After 4 steps	2.500	−0.500	1.500

(c) Put the second equation first, the fourth equation second, the first equation third, and the third equation fourth to get the following:

	x_1	x_2	x_3	x_4
Initial values	0.000	0.000	0.000	0.000
After 1 step	5.444	−5.379	9.226	−10.447
After 2 steps	8.826	−8.435	10.808	−11.698
After 3 steps	9.820	−8.920	10.961	−11.954
After 4 steps	9.973	−8.986	10.994	−11.993
After 5 steps	9.995	−8.998	10.999	−11.999
After 6 steps	9.999	−9.000	11.000	−12.000
After 7 steps	10.000	−9.000	11.000	−12.000
After 8 steps	10.000	−9.000	11.000	−12.000

9. The Jacobi method yields the following:

	x_1	x_2	x_3
Initial values	0.0	0.0	0.0
After 1 step	16.0	−13.0	12.0
After 2 steps	−37.0	59.0	−87.0
After 3 steps	224.0	−61.0	212.0
After 4 steps	−77.0	907.0	−1495.0
After 5 steps	3056.0	2515.0	−356.0
After 6 steps	12235.0	19035.0	−23895.0

The Gauss-Seidel method yields the following:

	x_1	x_2	x_3
Initial values	0.0	0.0	0.0
After 1 step	16.0	83.0	−183.0
After 2 steps	248.0	1841.0	−3565.0
After 3 steps	5656.0	41053.0	−80633.0
After 4 steps	124648.0	909141.0	−1781665.0

The actual solution is $(2, -3, 1)$.

Section 9.2 (pp. 474–475)

1. (a) $\mathbf{LDU} = \begin{bmatrix} 1 & 0 \\ -3 & 1 \end{bmatrix}\begin{bmatrix} 2 & 0 \\ 0 & 5 \end{bmatrix}\begin{bmatrix} 1 & -2 \\ 0 & 1 \end{bmatrix}$

(c) $\mathbf{L} = \begin{bmatrix} 1 & 0 & 0 \\ -2 & 1 & 0 \\ -2 & 4 & 1 \end{bmatrix}$, $\mathbf{D} = \begin{bmatrix} -1 & 0 & 0 \\ 0 & 2 & 0 \\ 0 & 0 & 3 \end{bmatrix}$,

$\mathbf{U} = \begin{bmatrix} 1 & -4 & 2 \\ 0 & 1 & -4 \\ 0 & 0 & 1 \end{bmatrix}$

(e) $\mathbf{L} = \begin{bmatrix} 1 & 0 & 0 & 0 \\ -\frac{4}{3} & 1 & 0 & 0 \\ -2 & -\frac{3}{2} & 1 & 0 \\ \frac{2}{3} & -2 & 0 & 1 \end{bmatrix}$, $\mathbf{D} = \begin{bmatrix} -3 & 0 & 0 & 0 \\ 0 & -\frac{2}{3} & 0 & 0 \\ 0 & 0 & \frac{1}{2} & 0 \\ 0 & 0 & 0 & 1 \end{bmatrix}$,

$\mathbf{U} = \begin{bmatrix} 1 & -\frac{1}{3} & -\frac{1}{3} & \frac{1}{3} \\ 0 & 1 & \frac{5}{2} & -\frac{11}{2} \\ 0 & 0 & 1 & 3 \\ 0 & 0 & 0 & 1 \end{bmatrix}$

3. (a) $\mathbf{KU} = \begin{bmatrix} -1 & 0 \\ 2 & -3 \end{bmatrix}\begin{bmatrix} 1 & -5 \\ 0 & 1 \end{bmatrix}$; the solution is $\{(4, -1)\}$.

(c) $\mathbf{KU} = \begin{bmatrix} -1 & 0 & 0 \\ 4 & 3 & 0 \\ -2 & 5 & -2 \end{bmatrix}\begin{bmatrix} 1 & -3 & 2 \\ 0 & 1 & -5 \\ 0 & 0 & 1 \end{bmatrix}$; the solution is $\{(2, -3, 1)\}$.

Section 9.3 (pp. 479–480)

1. (a) After nine iterations, eigenvector $= [0.60, 0.80]$ and eigenvalue $= 50$.
(c) After seven iterations, eigenvector $= [0.41, 0.41, 0.82]$ and eigenvalue $= 3.0$.
(e) After fifteen iterations, eigenvector $= [0.346, 0.852, 0.185, 0.346]$ and eigenvalue $= 5.405$.

3. (b) Let $\lambda_1, \ldots, \lambda_n$ be the nonzero eigenvalues of $\mathbf{A}$ with $|\lambda_1| > |\lambda_j|$, for $2 \le j \le n$. Let $\{\mathbf{v}_1, \ldots, \mathbf{v}_n\}$ be a basis of unit eigenvectors for $\mathbb{R}^n$ corresponding to $\lambda_1, \ldots, \lambda_n$, respectively. Suppose the initial vector in the power method is $\mathbf{u}_0 = a_{01}\mathbf{v}_1 + \cdots + a_{0n}\mathbf{v}_n$ and after the ith iteration we have $\mathbf{u}_i = a_{i1}\mathbf{v}_1 + \cdots + a_{in}\mathbf{v}_n$. Then, for j with $2 \le j \le n$ and $\lambda_j \ne 0$, we have $|a_{i1}|/|a_{ij}| = |\lambda_1/\lambda_j|^i\, |a_{01}|/|a_{0j}|$.

Appendix B (pp. 492–493)

1. (a) Not a function; undefined for $x < 1$
(c) Not a function; two values assigned to each $x \ne 1$
(e) Not a function (k undefined at $\theta = \frac{\pi}{2}$)
(f) Function; range = all prime numbers; image of $2 = 2$; pre-image of $2 = \{0, 1, 2\}$

2. (a) $\{-15, -10, -5, 5, 10, 15\}$
(c) $\{\ldots, -8, -6, -4, -2, 0, 2, 4, 6, 8, \ldots\}$

3. $(g \circ f)(x) = \frac{1}{4}\sqrt{75x^2 - 30x + 35}$, $(f \circ g)(x) = \frac{1}{4}(5\sqrt{3x^2+2} - 1)$

4. $(g \circ f)\left(\begin{bmatrix} x \\ y \end{bmatrix}\right) = \begin{bmatrix} -8 & 24 \\ 2 & 8 \end{bmatrix}\begin{bmatrix} x \\ y \end{bmatrix}$,

$(f \circ g)\left(\begin{bmatrix} x \\ y \end{bmatrix}\right) = \begin{bmatrix} -12 & 8 \\ -4 & 12 \end{bmatrix}\begin{bmatrix} x \\ y \end{bmatrix}$

8. f is not one-to-one, because $f(x^2+1) = f(x^2+2) = 2x$; f is not onto, because there is no pre-image for x^n. The pre-image of $\mathcal{P}_2$ is $\mathcal{P}_3$.

10. f is one-to-one, because if $f(\mathbf{A}_1) = f(\mathbf{A}_2)$, then $\mathbf{B}^{-1}\mathbf{A}_1\mathbf{B} = \mathbf{B}^{-1}\mathbf{A}_2\mathbf{B}$, which implies that $\mathbf{A}_1 = \mathbf{A}_2$. f is onto, because for any $\mathbf{C} \in \mathcal{M}_{nn}$, $f(\mathbf{BCB}^{-1}) = \mathbf{B}^{-1}(\mathbf{BCB}^{-1})\mathbf{B} = \mathbf{C}$. $f^{-1}(\mathbf{A}) = \mathbf{BAB}^{-1}$.

Appendix C (p. 496)

1. (a) $11 - i$ (c) $20 - 12i$ (e) $9 + 19i$
(g) $-17 - 19i$ (i) $9 + 2i$ (k) $16 + 22i$
(m) $\sqrt{53}$

2. (a) $\frac{3}{20} + \frac{1}{20}i$ (c) $-\frac{4}{17} - \frac{1}{17}i$

Subject Index

A

B

C

J

K

L

M

Symbol Index

Equivalent Conditions on Square Matrices

Let **A** be an $n \times n$ matrix. Any pair of statements in the same column are equivalent.

A is singular ($\mathbf{A}^{-1}$ does not exist).	**A** is nonsingular ($\mathbf{A}^{-1}$ exists).
$\text{Rank}(\mathbf{A}) \neq n$.	$\text{Rank}(\mathbf{A}) = n$.
$\lvert\mathbf{A}\rvert = 0$.	$\lvert\mathbf{A}\rvert \neq 0$.
A is not row equivalent to $\mathbf{I}_n$.	**A** is row equivalent to $\mathbf{I}_n$.
$\mathbf{AX} = \mathbf{O}$ has a nontrivial solution for **X**.	$\mathbf{AX} = \mathbf{O}$ has only the trivial solution for **X**.
$\mathbf{AX} = \mathbf{B}$ does not have a unique solution (no solutions or infinitely many solutions).	$\mathbf{AX} = \mathbf{B}$ has a unique solution for **X** ($\mathbf{X} = \mathbf{A}^{-1}\mathbf{B}$).
A is not equal to a (finite) product of elementary matrices.	$\mathbf{A} = \mathbf{E}_1\mathbf{E}_2 \cdots \mathbf{E}_k$, for some elementary matrices $\mathbf{E}_1, \mathbf{E}_2, \ldots, \mathbf{E}_k$.

Tests for Linear Independence

Linear Independence of S	**Linear Dependence of S**
When $S = \{\mathbf{v}_1, \ldots, \mathbf{v}_n\}$, for each k, $\mathbf{v}_k \notin \text{span}(\{\mathbf{v}_1, \ldots, \mathbf{v}_{k-1}\})$. (Each $\mathbf{v}_k$ is not a linear combination of the *previous* vectors in S.)	When $S = \{\mathbf{v}_1, \ldots, \mathbf{v}_n\}$, some $\mathbf{v}_k$ can be expressed as $\mathbf{v}_k = a_1\mathbf{v}_1 + \cdots + a_{k-1}\mathbf{v}_{k-1}$. (Some $\mathbf{v}_k$ can be expressed as a linear combination of *previous* vectors.)
$\mathbf{v}_1, \ldots, \mathbf{v}_n \in S$ and $a_1\mathbf{v}_1 + \cdots + a_n\mathbf{v}_n = \mathbf{0}$ implies that $a_1 = a_2 = \cdots = a_n = 0$. (The zero vector requires zero coefficients.)	There are vectors $\mathbf{v}_1, \ldots, \mathbf{v}_n$ in S such that $a_1\mathbf{v}_1 + \cdots + a_n\mathbf{v}_n = \mathbf{0}$, for scalars $a_1, a_2, \ldots, a_n$, with some $a_i \neq 0$. (The zero vector does not require all coefficients to be zero.)